INDUSTRIAL ORGANIC CHEMICALS

Starting Materials
and Intermediates

VOLUME **5**

Weinheim · New York · Chichester · Brisbane · Singapore · Toronto

INDUSTRIAL ORGANIC CHEMICALS

VOLUME 1

Acetaldehyde to Aniline

VOLUME 2

Anthracene
to Cellulose Ethers

VOLUME 3

Chlorinated Hydrocarbons
to Dicarboxylic Acids, Aliphatic

VOLUME 4

Dimethyl Ether
to Fatty Acids

VOLUME 5

Fatty Alcohols
to Melamine and Guanamines

VOLUME 6

Mercaptoacetic Acid and Derivatives
to Phosphorus Compounds, Organic

VOLUME 7

Phthalic Acid and Derivatives
to Sulfones and Sulfoxides

VOLUME 8

Sulfonic Acids, Aliphatic
to Xylidines

Index

INDUSTRIAL ORGANIC CHEMICALS

Starting Materials and Intermediates

VOLUME 5
Fatty Alcohols
to **Melamine and Guanamines**

Weinheim · New York · Chichester · Brisbane · Singapore · Toronto

> This book was carefully produced. Nevertheless, authors and publisher do not warrant the information contained therein to be free of errors. Readers are advised to keep in mind that statements, data, illustrations, procedural details or other items may inadvertently be inaccurate.

Library of Congress Card No.: Applied for.
British Library Cataloguing-in-Publication Data: A catalogue record for this book is available from the British Library.

Die Deutsche Bibliothek – CIP-Einheitsaufnahme
Industrial organic chemicals : starting materials and intermediates ;
an Ullmann's encyclopedia. – Weinheim ; New York ;
Chichester ; Brisbane ; Singapore ; Toronto : Wiley-VCH
   ISBN 3-527-29645-X
Vol. 5. Fatty Alcohols to Melamine and Guanamines. – 1. Aufl. – 1999.

© WILEY-VCH Verlag GmbH, D-69469 Weinheim (Federal Republic of Germany), 1999
Printed on acid-free and chlorine-free paper.
All rights reserved (including those of translation in other languages). No part of this book may be reproduced in any form – by photoprinting, microfilm, or any other means – nor transmitted or translated into machine language without written permission from the publishers. Registered names, trademarks, etc. used in this book, even when not specifically marked as such, are not to be considered unprotected by law.

Composition and Printing: Rombach GmbH, Druck- und Verlagshaus, D-79115 Freiburg
Bookbinding: Wilhelm Osswald & Co., D-67433 Neustadt (Weinstraße)
Cover design: mmad, Michel Meyer, D-69469 Weinheim
Printed in the Federal Republic of Germany

Contents

1 Fatty Alcohols

1. Introduction 2533
2. Saturated Fatty Alcohols 2534
3. Unsaturated Fatty Alcohols 2548
4. Guerbet Alcohols 2550
5. Bifunctional Fatty Alcohols 2552
6. Quality Specifications 2552
7. Storage and Transportation 2555
8. Economic Aspects.............. 2556
9. Toxicology.................... 2558
10. References.................... 2560

2 Fluorine Compounds, Organic

1. Introduction 2566
2. Production Processes 2568
3. Fluorinated Alkanes 2572
4. Fluorinated Olefins............. 2583
5. Fluorinated Alcohols............ 2590
6. Fluorinated Ethers 2591
7. Fluorinated Ketones and Aldehydes. 2596
8. Fluorinated Carboxylic Acids and Fluorinated Alkanesulfonic Acids .. 2600
9. Fluorinated Tertiary Amines...... 2607
10. Aromatic Compounds with Fluorinated Side-Chains............. 2609
11. Ring-Fluorinated Aromatic, Heterocyclic, and Polycyclic Compounds .. 2612
12. Economic Aspects.............. 2621
13. Toxicology and Occupational Health 2622
14. References.................... 2626

3 Formaldehyde

1. Introduction 2640
2. Physical Properties............. 2641
3. Chemical Properties 2646
4. Production 2647
5. Environmental Protection 2660
6. Quality Specifications and Analysis . 2664
7. Storage and Transportation 2666
8. Uses......................... 2667
9. Economic Aspects.............. 2668
10. Toxicology and Occupational Health 2670
11. Low Molecular Mass Polymers.... 2671
12. Formaldehyde Cyanohydrin 2681
13. References.................... 2683

4 Formamides

1. Introduction 2691
2. Formamide 2692
3. N,N-Dimethylformamide 2698
4. Other Formamides 2704
5. Toxicology and Occupational Health 2705
6. References.................... 2708

V

5 Formic Acid

1. Introduction 2711
2. Physical Properties 2712
3. Chemical Properties 2716
4. Production 2718
5. Environmental Protection 2732
6. Quality Specifications 2732
7. Chemical Analysis 2732
8. Storage and Transportation 2733
9. Legal Aspects 2734
10. Uses 2734
11. Economic Aspects 2735
12. Derivatives 2735
13. Toxicology and Occupational Health 2740
14. References 2741

6 Furan and Derivatives

1. Introduction 2745
2. Furan 2746
3. Furfural 2750
4. Furfuryl Alcohol 2756
5. 2-Methylfuran 2760
6. 2-Methyltetrahydrofuran 2761
7. Tetrahydrofurfuryl Alcohol 2762
8. 2,3-Dihydropyran 2763
9. Benzofuran 2764
10. Dibenzofuran 2765
11. Toxicology and Occupational Health 2767
12. References 2768

7 Gluconic Acid

1. Introduction 2773
2. Physical Properties 2774
3. Chemical Properties 2775
4. Production 2776
5. Uses 2780
6. Economic Aspects 2782
7. Physiology, Product Specifications . . 2782
8. References 2783

8 Glycerol

1. Introduction 2788
2. Physical Properties 2788
3. Chemical Properties 2792
4. Production 2793
5. Environmental Protection 2801
6. Quality Specifications and Analysis . 2802
7. Storage and Transportation 2803
8. Uses 2804
9. Derivatives 2804
10. Economic Aspects 2806
11. Toxicology and Occupational Health 2806
12. References 2807

9 Glyoxal

1. Introduction 2809
2. Physical Properties 2809
3. Chemical Properties 2809
4. Production 2812
5. Uses 2813
6. Toxicology and Occupational Health 2814
7. References 2815

10 Glyoxylic Acid

1. Introduction 2817
2. Physical Properties 2817
3. Chemical Properties 2818
4. Production 2821
5. Toxicology................... 2821
6. References................... 2822

11 Guanidine and Derivatives

1. Guanidine and Guanidine Salts ... 2823
2. Derivatives 2832
3. Toxicology and Occupational Health 2841
4. References................... 2842

12 Hexamethylenediamine

1. Introduction 2845
2. Physical Properties 2846
3. Chemical Properties 2846
4. Production 2847
5. Quality Specifications and Analysis . 2848
6. Storage and Transportation 2849
7. Economic Aspects............. 2850
8. Toxicology and Occupational Health 2850
9. References................... 2851

13 Hydrocarbons

1. Saturated Hydrocarbons 2854
2. Olefins 2870
3. Alkylbenzenes 2893
4. Biphenyls and Polyphenyls....... 2906
5. Hydrocarbons from Coal Tar 2913
6. Toxicology and Occupational Health 2924
7. References................... 2928

14 Hydroquinone

1. Introduction 2941
2. Physical Properties 2942
3. Chemical Properties 2942
4. Production 2945
5. Environmental Protection 2948
6. Quality Specifications and Analysis . 2948
7. Uses....................... 2949
8. Toxicology and Occupational Health 2950
9. References................... 2951

15 Hydroxycarboxylic Acids, Aliphatic

1. Introduction 2955
2. General Characteristics 2956
3. Preparation.................. 2960
4. Analysis 2964
5. Specific Aliphatic Hydroxycarboxylic Acids of Commercial Significance .. 2964
6. Toxicology................... 2970
7. References................... 2971

16 Hydroxycarboxylic Acids, Aromatic

1. Introduction 2973
2. General Properties 2974
3. Production 2975
4. Individual Aromatic Hydroxycarboxylic Acids................ 2977
5. Toxicology.................... 2983
6. References.................... 2984

17 Imidazole and Derivatives

1. Introduction 2987
2. Physical Properties............ 2987
3. Chemical Properties 2988
4. Production 2989
5. Quality Specifications and Analysis . 2991
6. Uses.......................... 2991
7. Toxicology and Occupational Health 2992
8. References.................... 2998

18 Indole

1. Introduction 3001
2. Properties.................... 3001
3. Production 3002
4. Derivatives 3002
5. Uses.......................... 3003
6. Toxicology.................... 3004
7. References.................... 3005

19 Iodine Compounds, Organic

1. The Individual Compounds 3007
2. References.................... 3008

20 Isocyanates, Organic

1. Introduction 3009
2. Physical Properties............ 3010
3. Chemical Properties 3010
4. Production 3015
5. Environmental Protection 3023
6. Quality Specifications and Analysis . 3024
7. Storage and Transportation 3024
8. Uses.......................... 3025
9. Economic Aspects............. 3026
10. Toxicology and Occupational Health 3027
11. References.................... 3030

21 Isoprene

1. Introduction 3033
2. Properties.................... 3034
3. Production 3036
4. Quality Specifications 3046
5. Storage and Transportation 3047
6. Uses.......................... 3048
7. Economic Aspects............. 3050
8. Toxicity and Occupational Health .. 3052
9. References.................... 3055

22 Ketenes

1. Introduction 3061
2. Ketene 3062
3. Diketene..................... 3068
4. Higher Ketenes 3076
5. Toxicology and Occupational Health 3078
6. References.................... 3079

23 Ketones

1. Introduction 3084
2. Methyl Alkyl Ketones 3085
3. Dialkyl Ketones 3095
4. Cyclic Ketones 3099
5. Unsaturated Ketones........... 3101
6. Diketones 3106
7. Aromatic Ketones.............. 3110
8. Toxicology and Occupational Health 3113
9. References.................... 3115

24 Lactic Acid

1. Introduction 3119
2. Properties.................... 3120
3. Production 3123
4. Environmental Protection 3127
5. Quality Specifications 3127
6. Analysis 3127
7. Storage and Transportation 3128
8. Legal Aspects.................. 3128
9. Uses........................ 3128
10. Derivatives 3129
11. Economic Aspects.............. 3131
12. Toxicology and Occupational Health 3131
13. References.................... 3131

25 Lactose and Derivatives

1. Lactose..................... 3135
2. Derivatives 3142
3. References.................... 3146

26 Lecithin

1. Introduction 3149
2. Production 3151
3. Commercial Grades of Lecithin.... 3152
4. Physical Properties............. 3154
5. Chemical Properties 3155
6. Uses........................ 3156
7. Quality Specifications and Analysis . 3157
8. References.................... 3157

27 Maleic and Fumaric Acids

1. Maleic Acid................... 3159
2. Maleic Anhydride (MA) 3161
3. Citraconic and Mesaconic Acids ... 3169
4. Fumaric Acid.................. 3170
5. Toxicology................... 3172
6. References.................... 3173

28 Malonic Acid and Derivatives

1. Introduction 3177
2. Malonic Acid 3178
3. Malonates 3179
4. Cyanoacetic Acid 3185
5. Cyanoacetates 3187
6. Malononitrile 3188
7. Quality Specifications and Analysis . 3192
8. Economic Aspects 3194
9. Toxicology and Occupational Health 3194
10. References 3195

29 Melamine and Guanamines

1. Introduction 3197
2. Physical Properties 3198
3. Chemical Properties 3199
4. Production 3202
5. Quality Specifications 3210
6. Chemical Analysis 3210
7. Storage and Transportation 3211
8. Uses 3211
9. Economic Aspects 3211
10. Toxicology 3212
11. Guanamines 3214
12. References 3218

Fatty Alcohols

KLAUS NOWECK, Condea Chemie GmbH, Brunsbüttel, Federal Republic of Germany

HEINZ RIDDER, Condea Chemie GmbH, Brunsbüttel, Federal Republic of Germany

1.	Introduction 2533	2.4.2.	Oxo Process 2544
2.	Saturated Fatty Alcohols 2534	2.4.3.	Hydrogenation of Fatty Acids Produced by Oxidation of Paraffinic Hydrocarbons 2545
2.1.	Physical Properties 2534		
2.2.	Chemical Properties 2536	2.4.4.	Bashkirov Oxidation 2546
2.3.	Production from Natural Sources 2536	2.4.5.	Other Processes 2546
		2.5.	Uses 2547
2.3.1.	Hydrolysis of Wax Esters...... 2537	3.	Unsaturated Fatty Alcohols... 2548
2.3.2.	Reduction of Wax Esters with Sodium 2537	4.	Guerbet Alcohols.......... 2550
2.3.3.	Hydrogenation of Natural Raw Materials 2538	5.	Bifunctional Fatty Alcohols .. 2552
2.3.3.1.	Raw Materials and Pretreatment 2538	6.	Quality Specifications....... 2552
2.3.3.2.	Hydrogenation Processes...... 2539	7.	Storage and Transportation .. 2555
2.4.	Fatty alcohols, saturated, synthesis from Petrochemical Feedstocks 2542	8.	Economic Aspects 2556
		9.	Toxicology............... 2558
2.4.1.	Ziegler Alcohol Processes 2542	10.	References............... 2560

1. Introduction

Fatty alcohols are aliphatic alcohols with chain lengths between C_6 and C_{22}:

$CH_3(CH_2)_nCH_2OH$ ($n = 4-20$)

They are predominantly straight-chain and monohydric, and can be saturated or have one or more double bonds. Alcohols with a carbon chain length above C_{22} are referred to as wax alcohols. Diols whose chain length exceeds C_8 are regarded as substituted fatty alcohols. The character of the fatty alcohols (primary or secondary, linear or branched-chain, saturated or unsaturated) is determined by the manufacturing process and the raw materials used. Natural products, such as fats, oils, and waxes, and the Ziegler alcohol process give straight-chain, primary, and even-numbered alcohols; those obtained from natural sources may be unsaturated. In contrast, the oxo process yields 20 – 60 % branched-chain fatty alcohols, and also some odd-numbered ones. Guerbet dimerization gives α -branched, primary alcohols, whereas Bashkirov oxidation yields secondary alcohols.

Depending on the raw materials used, fatty alcohols are classified as natural or synthetic. Natural fatty alcohols are based on renewable resources such as fats, oils, and waxes of plant or animal origin, whereas synthetic fatty alcohols are produced from petrochemicals such as olefins and paraffins. Up to 1930, when catalytic high-pressure hydrogenation was developed [7] – [10], the manufacture of fatty alcohols was based almost exclusively on the splitting of sperm oil. By 1962, the world production capacity from natural raw materials had grown to ca. 200 000 t/a. New processes based on petrochemical raw materials, e.g., the Ziegler alcohol process, allowed a further increase. In 1985, the world nameplate production capacity of fatty alcohols was estimated to be 1.3×10^6 t/a, of which ca. 60 % was based on petrochemicals. However, production and consumption were estimated to be only 750 000 t/a. Fatty alcohols and their derivatives are used in synthetics, surfactants, oil additives, and cosmetics and have many specialty uses.

2. Saturated Fatty Alcohols

2.1. Physical Properties

Saturated fatty alcohols up to dodecanol are clear, colorless liquids. The next higher homologues are soft materials; octadecanol and higher alcohols have a waxy consistency. The saturated alcohols crystallize in a nearly orthorhombic lattice [11] and all have a lower specific density than water. The lower members of the series have characteristic odors; the higher fatty alcohols are odorless, except for traces of impurities such as carbonyl compounds and hydrocarbons, which are usually present.

Physical properties of straight-chain, primary fatty alcohols are summarized in Table 1 (for additional data on pure alcohols and commercially important mixtures, see [12] – [15]).

Boiling points and melting points increase uniformly with chain length. Both are significantly higher than those of the hydrocarbons with the same number of carbon atoms. The influence of the polarizing hydroxyl function diminishes with increasing chain length. Thus hexanol and even octanol show some water solubility, but decanol and the higher fatty alcohols can be considered as immiscible with water. However, a slight hygroscopicity is observed even with octadecanol and higher fatty alcohols, which can absorb water vapor from air during storage. Common organic solvents such as petroleum ether, lower alcohols, and diethyl ether are suitable solvents for fatty alcohols.

Table 1. Physical properties of fatty alcohols

IUPAC name	Common name	CAS registry number	Molecular formula	M_r	Hydroxyl number	mp, °C	bp, °C (p, kPa)	Density, g/cm³ (t, °C)	Refractive index (t, °C)
1-Hexanol	caproic alcohol	[111-27-3]	$C_6H_{14}O$	102.2	548	−52	157	0.819 (20)	1.4181 (20)
1-Heptanol	enanthic alcohol	[111-70-6]	$C_7H_{16}O$	116.2	482	−30	176	0.822 (20)	1.4242 (20)
1-Octanol	caprylic alcohol	[111-87-5]	$C_8H_{18}O$	130.2	430	−16	195	0.825 (20)	1.4296 (20)
1-Nonanol	pelargonic alcohol	[143-08-8]	$C_9H_{20}O$	144.3	388	−4	213	0.828 (20)	1.4338 (20)
1-Decanol	capric alcohol	[112-30-1]	$C_{10}H_{22}O$	158.3	354	7	230	0.829 (20)	1.4371 (20)
1-Undecanol		[112-42-5]	$C_{11}H_{24}O$	172.3	326	16	245	0.830 (20)	1.4402 (20)
1-Dodecanol	lauryl alcohol	[112-53-8]	$C_{12}H_{26}O$	186.3	300	23	260	0.822 (40)	1.4428 (20)
1-Tridecanol		[112-70-9]	$C_{13}H_{28}O$	200.4	280	30	276		
1-Tetradecanol	myristyl alcohol	[112-72-1]	$C_{14}H_{30}O$	214.4	261	38	172 (2.67)	0.823 (40)	1.4358 (50)
1-Pentadecanol		[629-76-5]	$C_{15}H_{32}O$	228.4	245	44			1.4408 (50)
1-Hexadecanol	cetyl alcohol	[36653-82-4]	$C_{16}H_{34}O$	242.5	230	49	194 (2.67)	0.812 (60)	1.4392 (60)
1-Heptadecanol	margaryl alcohol	[1454-85-9]	$C_{17}H_{36}O$	256.5	218	54			
1-Octadecanol	stearyl alcohol	[112-92-5]	$C_{18}H_{38}O$	270.5	207	58	214 (2.67)	0.815 (60)	1.4388 (60)
1-Nonadecanol		[1454-84-8]	$C_{19}H_{40}O$	284.5	196	62			1.4328 (70)
1-Eicosanol	arachidyl alcohol	[629-96-9]	$C_{20}H_{42}O$	298.6	187	64	215 (1.33)	0.806 (70)	
1-Heneicosanol		[15594-90-8]	$C_{21}H_{44}O$	312.6	179	68			
1-Docosanol	behenyl alcohol	[661-19-8]	$C_{22}H_{46}O$	326.6	171	71	241 (1.33)	0.807 (80)	
1-Tricosanol		[3133-01-5]	$C_{23}H_{48}O$	340.6	164	74			
1-Tetracosanol	lignoceryl alcohol	[506-51-4]	$C_{24}H_{50}O$	354.7	158	77			
1-Pentacosanol		[26040-98-2]	$C_{25}H_{52}O$	368.7	152	78			
1-Hexacosanol	ceryl alcohol	[506-52-5]	$C_{26}H_{54}O$	382.7	146	81			
1-Heptacosanol		[2004-39-9]	$C_{27}H_{56}O$	396.8	141	82			
1-Octacosanol	montanyl alcohol	[557-61-9]	$C_{28}H_{58}O$	410.8	136	84			
1-Nonacosanol		[6624-76-6]	$C_{29}H_{60}O$	424.8	132	85			
1-Triacontanol	myricyl alcohol	[593-50-0]	$C_{30}H_{62}O$	438.8	128	87			
1-Hentriacontanol	melissyl alcohol	[544-86-5]	$C_{31}H_{64}O$	452.9	124	87			
1-Dotriacontanol	lacceryl alcohol	[6624-79-9]	$C_{32}H_{66}O$	466.9	120	89			
1-Tritriacontanol		[71353-61-2]	$C_{33}H_{68}O$	480.9	116				
1-Tetratriacontanol	geddyl alcohol	[28484-70-0]	$C_{34}H_{70}O$	494.9	113	92			

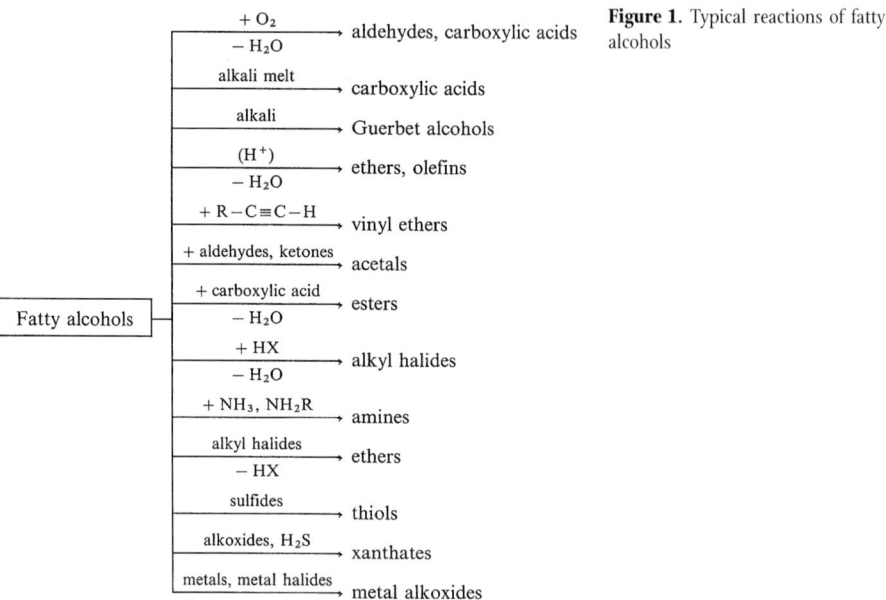

Figure 1. Typical reactions of fatty alcohols

2.2. Chemical Properties

The industrial importance of the fatty alcohols is due to the large number of reactions that the hydroxyl group may undergo. Figure 1 lists some typical examples (see also → Alcohols, Aliphatic); many of the resulting derivatives are intermediates of commercial importance (see Section 2.5). Ethoxylation with ethylene oxide yields fatty alcohol polyglycol ethers:

$$\mathrm{RCH_2OH} + n\,\mathrm{H_2C\!\!-\!\!CH_2} \longrightarrow \mathrm{RCH_2\!\!-\!\!(O\!-\!CH_2CH_2)_{\mathit{n}}OH}$$

Under normal conditions, fatty alcohols are resistant to oxidation. However, they can be converted into aldehydes or carboxylic acids using strong oxidants or by catalytic oxidation with air or oxygen [16]–[20]. This reaction is important only for C_6-C_{10} fatty acids, which are not readily available from natural sources [21] and are therefore produced from synthetic alcohols.

2.3. Production from Natural Sources

Two groups of natural raw materials are used for the production of fatty alcohols: (1) fats and oils of plant or animal origin, which contain fatty acids in the form of triglycerides that can be hydrogenated after suitable pretreatment (→ Fats and Fatty Oils) to yield fatty alcohols; and (2) wax esters from whale oil (sperm oil), from which the fatty alcohols are obtained by simple hydrolysis or reduction with sodium. The commercial exploitation of sperm oil has led to depletion of whale populations and is banned in some countries. Attention has therefore turned to the jojoba plant whose oil also consists of wax esters. The first successful attempts to cultivate this desert shrub and develop it as a source of raw material are currently being made.

Most fatty chemicals obtained from natural sources have chain lengths of C_{16}–C_{18}. The limited availability of compounds with 12–14 carbon atoms, which are important in surfactants, was one of the driving forces behind the development of petrochemical processes for the production of fatty alcohols (see Section 2.4). Higher alcohols, such as C_{20}–C_{22} alcohols, can be produced from rapeseed oils rich in erucic acid and fish oils. Unsaturated fatty alcohols may be manufactured in the presence of selective catalysts. For detailed reviews of large-scale industrial processes for the production of fatty alcohols, see [1], [22], [23].

2.3.1. Hydrolysis of Wax Esters

The hydrolysis of wax esters is of only limited importance today. It is carried out by heating sperm oil with concentrated sodium hydroxide at ca. 300 °C and distilling the alcohol from the sodium soap.

$$R^1-\overset{O}{\underset{\|}{C}}-OR^2 + NaOH \longrightarrow R^1-\overset{O}{\underset{\|}{C}}-ONa + R^2OH$$

The distillate consists of partially unsaturated C_{16}–C_{20} alcohols, which are hardened by catalytic hydrogenation to prevent autoxidation. Since sperm oil contains only ca. 70 % wax esters, the alcohol yield is about 35 %.

2.3.2. Reduction of Wax Esters with Sodium

The reduction of esters with sodium was first described in 1902 by BOUVEAULT and BLANC (for a review, see [24]). Large-scale application of this process was achieved in 1928 (Dehydag).

$$R^1-\overset{O}{\underset{\|}{C}}-OR^2 + 4\,Na + 2\,R^3OH \longrightarrow$$
$$R^1CH_2ONa + R^2ONa + 2\,R^3ONa \xrightarrow[-4\,NaOH]{+4\,H_2O}$$
$$R^1CH_2OH + R^2OH + 2\,R^3OH$$

Molten sodium is dispersed in an inert solvent and the carefully dried ester and alcohol are added. When the reaction is complete, the alkoxides are split by stirring in water, and the alcohols are washed and distilled [25].

The added alcohol R^3OH, preferably a secondary alcohol, acts as a hydrogen donor. Because of side reactions, the consumption of sodium can be as much as 20% above the stoichiometric requirement.

The reduction proceeds selectively without production of hydrocarbons and isomerization or hydrogenation of double bonds. Extensive safety measures are required due to the large quantity of metallic sodium used. The process was used until the 1950s to produce unsaturated fatty alcohols, especially oleyl alcohol from sperm oil. These alcohols can now be produced by selective catalytic hydrogenation processes using cheap raw materials, and the sodium reduction process is of interest only in special cases. A fully continuous plant with a production capacity of 3600 t/a was built in Japan in 1973 for the reduction of sperm oil [26].

2.3.3. Hydrogenation of Natural Raw Materials

2.3.3.1. Raw Materials and Pretreatment

For the production of C_{12}–C_{14} alcohols, only coconut oil and palm kernel oil can be used. Palm oil, soybean oil, and tallow are the main sources for C_{16}–C_{18} alcohols. Rapeseed oil rich in erucic acid yields fatty alcohols with 20 or 22 carbon atoms. Bifunctional fatty alcohols can be obtained from castor oil and other special oils (see Chap. 5).

Before hydrogenation, contaminants such as phosphatides, sterols, or oxidation products and impurities such as seed particles, dirt, and water are removed in a cleaning stage, which includes refining with an adsorption agent (→ Fats and Fatty Oils). The refined triglycerides are then hydrolyzed to yield fatty acids (→ Fatty Acids) or transesterified with lower alcohols to yield fatty acid esters.

Both refined free fatty acids and fatty acid esters (mostly methyl esters and, more rarely, butyl esters) are used for hydrogenation. Direct hydrogenation of triglycerides is also possible; however, under the reaction conditions, glycerol is reduced to propylene glycol and propanol and therefore makes no commercial contribution as a byproduct. More hydrogen is needed and catalyst costs increase. Therefore, triglyceride hydrogenation is not used industrially.

Fatty acid esters are produced either by esterification of free fatty acids or by transesterification of triglycerides (see also → Esters, Organic).

Esterification of Fatty Acids. Esterification is an equilibrium reaction:

$$R^1-CH_2-\overset{O}{\underset{\|}{C}}-OH + R^2OH \rightleftharpoons R^1-CH_2-\overset{O}{\underset{\|}{C}}-OR^2 + H_2O$$

Excess alcohol or removal of water shifts the equilibrium toward ester formation. Industrial esterification is carried out in a column at 200–250 °C under slight pressure and with excess methanol. Distilled fatty acids, which no longer have the composition of the original natural product, are predominantly used. Methanol reacts with the fatty acid in a countercurrent. The acid number (milligrams of KOH needed to neutralize one gram of substance) is used as a quality characteristic and for process control. Batch and continuous processes give esters of similar quality, but the continuous process uses less methanol and the residence time is reduced. The methyl ester is subsequently distilled for purification.

Transesterification of Triglycerides. This reaction is carried out continuously with alkaline catalysts. Like esterification, transesterification is an equilibrium reaction and is shifted toward the desired ester by excess methanol or removal of glycerol:

$$\begin{array}{c} CH_2OCOR \\ | \\ CHOCOR \\ | \\ CH_2OCOR \end{array} + 3\,CH_3OH \rightleftharpoons \begin{array}{c} CH_2OH \\ | \\ CHOH \\ | \\ CH_2OH \end{array} + 3\,RCOOCH_3$$

If the reaction is carried out under mild conditions (50–70 °C, atmospheric pressure, excess methanol), free fatty acids present in the oils must first be removed or pre-esterified, e.g., with glycerol. Under more severe conditions (i.e., at 9–10 MPa and 220–250 °C), pre-esterification is unnecessary, and less pure, cheaper raw materials can be used. Disadvantages of this process are the need for high-pressure equipment, a greater excess of methanol, and the energy-intensive further processing of the aqueous methanol. The methyl esters are purified by distillation.

2.3.3.2. Hydrogenation Processes

Three large-scale hydrogenation processes are used commercially: (1) *suspension hydrogenation*, (2) *gas-phase hydrogenation*, and (3) *trickle-bed hydrogenation*. In all variants, hydrogenation is carried out with copper-containing, mixed-oxide catalysts at 200–300 °C and 20–30 MPa.

Suspension Hydrogenation. This process is applicable to fatty acid methyl esters as well as to fatty acids.

Hydrogen (ca. 50 mol per mole of ester) and the heated methyl ester are fed separately into the bottom of a narrow reactor. The reaction is carried out at ca. 25 MPa and 250–300 °C in the presence of a fine powdery copper catalyst; the LHSV (liquid hourly space velocity) is approximately 1. The excess hydrogen serves to circulate the reaction mixture. The product mixture is separated into a gas phase, which is recycled to the reactor, and a liquid phase, from which the methanol is stripped. The crude product, which still has a saponification number of 6–10 (milligrams of KOH required to saponify one gram of substance), is distilled off after removal of the catalyst. If alkali is added during the distillation, the alcohol moiety of the ester can also be obtained.

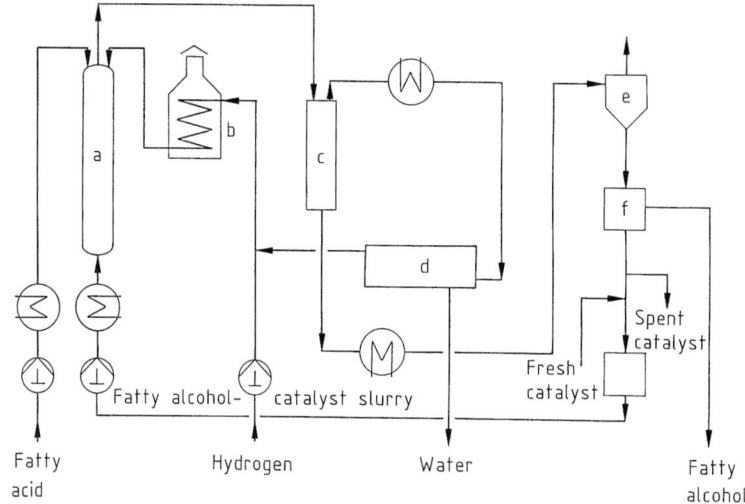

Figure 2. Suspension hydrogenation of fatty acids (Lurgi process)
a) Reactor; b) Heater; c) Hot separator; d) Cold separator; e) Flash drum; f) Catalyst separation

If a stainless steel reactor is used, this process can be applied to the direct hydrogenation of fatty acids. In this case, an acid-resistant catalyst is required, and catalyst consumption is increased.

A variant that is particularly suitable for the hydrogenation of fatty acids has been developed by Lurgi (Fig. 2) [27], [28] and is used by several manufacturers, including Condea Chemie [29]. This process employs a large excess of fatty alcohol in the hydrogenation reactor. Hydrogen, fatty alcohol–catalyst slurry, and fatty acids are fed separately into the reactor; the ester forms almost instantaneously [30] and is then hydrogenolyzed in the same reactor in a slower, second reaction step. Hydrogenolysis is carried out at ca. 30 MPa and 260–300 °C. Catalyst consumption is 0.5–0.7%. The catalyst is separated by centrifugation, and the crude fatty alcohol, which has an acid number of <0.1 and a saponification number of 2–5, is purified by distillation. The hydrocarbon content is less than 1%.

By continuously replacing part of the spent catalyst, the activity of the copper chromite contact can be held constant.

Gas-Phase Hydrogenation. This process requires a vaporized substrate and is therefore particularly suitable for methyl esters (Fig. 3). Characteristics are an extremely large excess of recycle gas (ca. 600 mol of H_2 per mole of ester), high gas speeds, and the addition of methanol to aid evaporation [31], [32]. Decomposition of methanol creates significant quantities of carbon monoxide, water, and dimethyl ether. Admixture of an inert gas to the hydrogen is claimed to make the addition of methanol unnecessary and to reduce the excess of recycle gas [33]. Catalysts (e.g., mixed oxides of copper and zinc) are used in a fixed bed. The conditions required are 20–25 MPa and 230–250 °C, with

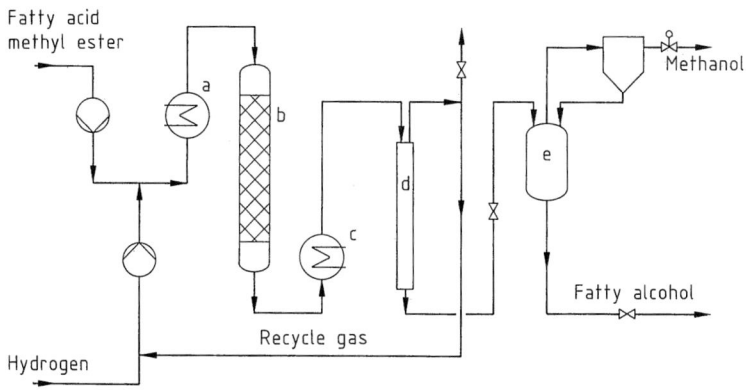

Figure 3. Gas-phase and trickle-bed hydrogenation of fatty acid methyl esters
a) Heater; b) Reactor; c) Cooler; d) Separator; e) Flash drum

an LHSV of about 0.3. Alcohol yields of 99% are achieved in a single passage or in a series of reactors; byproducts account for less than 1%. Catalyst consumption is about 0.3%. The product mixture is separated into gas and liquid phases; the hydrogen is recycled and the methanol stripped from the fatty alcohol.

Trickle-Bed Hydrogenation. In this process, the products to be reduced are used in their liquid form. It is therefore also suitable for non-vaporizable substrates such as wax esters and fatty acids. Corrosive effects of acids can be neutralized by hydrogenation in the presence of amines [34]. If a considerably lower excess of recycle gas (ca. 100 mol of H_2 per mole of ester) is used, a different plant design results. The reaction is carried out at 20–30 MPa and about 250 °C, with an LHSV of ca. 0.2. Supported catalysts, e.g., copper chromite on silicon dioxide, are often used; catalyst consumption is about 0.3%. Further treatment of the product is identical to that described for gas-phase hydrogenation (see Fig. 3).

Comparison of Hydrogenation Processes. With the fixed-bed processes (gas-phase and trickle-bed hydrogenation), the catalyst need not be separated from the crude fatty alcohol. However, a gradual decrease in the hydrogenating activity due to catalyst poisons such as sulfur, phosphorus, or chlorine is observed, whereas continuous replacement of the catalyst in the suspension process ensures constant activity. If methyl esters are employed, separation and further processing of the methanol is necessary.

Catalysts that contain noble metals, especially rhenium, may allow hydrogenation at lower pressures, which would reduce capital and operating costs [35]–[37].

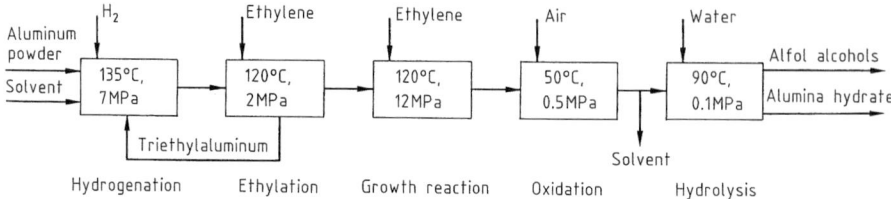

Figure 4. Alfol alcohol process

2.4. Fatty alcohols, saturated, synthesis from Petrochemical Feedstocks

2.4.1. Ziegler Alcohol Processes

Two processes for the production of synthetic fatty alcohols are based on the work of ZIEGLER on organic aluminum compounds: the Alfol process, developed by Conoco, and Ethyl Corporation's Epal process [38], [39]. Fatty alcohols synthesized by these processes are structurally similar to natural fatty alcohols and are thus ideal substitutes for natural products.

Conoco started the first Alfol plant in the United States in 1962. This plant is now operated by Vista Chemical. In 1964 Condea Chemie installed a similar plant in Brunsbüttel, Federal Republic of Germany. Ethyl Corporation developed its own process (Epal process) and began operations in 1964. Another Alfol alcohol plant has been in operation in the Soviet Union for some years.

Alfol Process. Figure 4 shows a simplified diagram of the Alfol process [40]. A hydrocarbon is used as solvent. The process involves five steps: hydrogenation, ethylation, growth reaction, oxidation, and hydrolysis.

Hydrogenation

$$2\,Al(CH_2CH_3)_3 + Al + 1.5\,H_2 \longrightarrow 3\,HAl(CH_2CH_3)_2 \quad (a)$$

Ethylation

$$3\,HAl(CH_2CH_3)_2 + 3\,CH_2=CH_2 \longrightarrow 3\,Al(CH_2CH_3)_3 \quad (b)$$

Of the triethylaluminum produced in the ethylation reaction (b), two-thirds are recycled to the hydrogenation stage (a), and one-third enters the growth reaction (c):

Growth Reaction

$$Al(CH_2CH_3)_3 + (x+y+z)\,CH_2=CH_2$$
$$\longrightarrow Al\begin{array}{l}\diagup (CH_2CH_2)_{x+1}H \\ -(CH_2CH_2)_{y+1}H \\ \diagdown (CH_2CH_2)_{z+1}H\end{array} \quad (c)$$

Table 2. Typical alcohol distributions in Alfol and Epal processes

Alcohol	Distribution, %	
	Alfol process	Epal process
Ethanol	0.5	traces
Butanol	3.4	0.1
Hexanol	9.5	1.5
Octanol	16.1	3.5
Decanol	19.5	8.0
Dodecanol	18.4	34.0
Tetradecanol	14.1	26.0
Hexadecanol	9.1	16.0
Octadecanol	5.1	8.8
Eicosanol	2.5	1.9
Docosanol	1.1	0.2

Insertion of the ethylene molecule into the aluminum–carbon bonds occurs as a statistical process and leads to a broad distribution (Poisson distribution [41], [42]) of chain lengths, ranging from C_2 to beyond C_{26} [40]. An optimal yield of the C_{12}–C_{14} alcohols, which are important in the surfactant sector, requires addition of about four molecules of ethylene per aluminum–carbon bond (see also Table 2). A small percentage of olefins are formed as byproducts.

Oxidation

$$\text{Al}\begin{array}{c}R^1\\-R^2\\R^3\end{array} + 1.5\,O_2 \longrightarrow \text{Al}\begin{array}{c}OR^1\\-OR^2\\OR^3\end{array} \quad (d)$$

Because of the varying reactivity of partially oxidized trialkylaluminum compounds, oxidation is carried out stepwise by passing through carefully dried air. Cooling is necessary, especially at the start of the reaction. Alkanes and oxygen-containing compounds are formed as byproducts [38].

Hydrolysis

$$\text{Al}\begin{array}{c}OR^1\\-OR^2\\OR^3\end{array} + 2\,H_2O \longrightarrow \text{AlO(OH)} + R^1OH + R^2OH + R^3OH \quad (e)$$

Before hydrolysis, the solvent is removed by distillation. Hydrolysis with water gives high-purity hydrated alumina (Pural by Condea, Catapal by Vista Chemical) as a coproduct, which has many industrial applications, e.g., in catalytic processes and in ceramics. In the 1960s, hydrolysis was carried out with hot sulfuric acid at Conoco's and Ethyl Corporation's plants. Conoco changed to neutral hydrolysis, but the sulfuric acid method is still used in the Epal process and leads to high-purity aluminum sulfate as a coproduct. The crude alcohols are finally fractionated into marketable cuts.

Epal Process. Many attempts have been made to achieve a narrower distribution of chain lengths in the growth reaction [39]. The only process that has been used on an industrial scale is the Epal process developed by the Ethyl Corporation. The reaction steps resemble those of the Alfol process, but the growth reaction is not carried as far. The product of the growth reaction is subjected to transalkylation (290 °C, 3.5 MPa) with C_4-C_{10} olefins. The chain lengths of the resulting trialkylaluminum compounds are predominantly C_4-C_{10}. Excess olefins are removed in a stripping column and then fractionated. The trialkylaluminum compound is subjected to a second growth reaction and then transalkylated (200 °C, 35 kPa) with $C_{12}-C_{18}$ olefins. Again the olefins are separated in a stripper and fractionated. At this stage, the trialkylaluminum compound consists largely of alkyl chains with 12–18 carbon atoms.

The alcohol distributions achieved in the Epal and Alfol processes are shown in Table 2 [43]. The Epal process offers greater flexibility than the Alfol process because both the alcohols and the α-olefins that are formed as intermediates can be marketed [44]. Disadvantages of the Epal process are the higher capital and operating costs, a considerably more complicated process control, and an increased proportion of branched-chain olefins and alcohols.

2.4.2. Oxo Process

The oxo process (hydroformylation) consists of the reaction of olefins with an H_2–CO gas mixture in the presence of suitable catalysts. This reaction was discovered in 1938 by O. ROELEN at Ruhrchemie during work on the Fisher–Tropsch synthesis. Although the first production plant was constructed before 1945 [45], the oxo process achieved industrial importance only in the 1950s, with the increasing demand for plasticizers and detergents. The process is based on the following reaction (for details on the reaction mechanism, see [46]):

$$2R-CH=CH_2 + 2CO + 2H_2 \xrightarrow{HCo(CO)_4} R-CH_2CH_2-CHO + R-\underset{\underset{CH_3}{|}}{C}H-CHO$$

α-Olefins yield approximately equal amounts of straight-chain and branched aldehydes (see also Fig. 5). Internal and branched alkenes can also be used in this reaction. Internal olefins give a product containing some primary aldehyde because the catalyst effects double-bond isomerization.

For a long time paraffin-based processes were predominant as a source of olefins, especially for detergents [44], [53]. With the development of the SHOP process (Shell's Higher Olefin Process), ethylene has become the preferred raw material [54]. The principal steps in the SHOP process are ethylene oligomerization, isomerization, and metathesis. The products are $C_{12}-C_{18}$ α-olefins and $C_{11}-C_{14}$ internal olefins, which are all important in the area of surfactants [55], [56].

Heterogeneous hydrogenation of the oxo aldehydes at 5–20 MPa and 150–250 °C in the presence of catalysts based on nickel, molybdenum, copper, or cobalt yields the

Table 3. Typical oxo process parameters

Parameter	Oxo process		
	Classical	Shell	Union Carbide
Catalyst	cobalt carbonyl	cobalt carbonyl–phosphine complex	rhodium carbonyl–phosphine complex
Catalyst concentration, %	0.1–1.0	ca. 0.5	0.01–0.1
$CO:H_2$ ratio	1:1–1.2	1:2–2.5	excess hydrogen
Temperature, °C	150–180	170–210	100–120
Pressure, MPa	20–30	5–10	2–4
LHSV*	0.5–1.0	0.1–0.2	0.1–0.25
Primary products	aldehydes	alcohols	aldehydes
Linearity, %	40–50	80–85	ca. 90 (n-butanol)

* Liquid hourly space velocity

corresponding alcohols. Alternatively, the aldehyde can be subjected to an aldol reaction as in the production of 2-ethylhexanol (→ 2-Ethylhexanol). Worldwide there are three variants of the oxo process [48]: (1) the classical process using $HCo(CO)_4$ as catalyst; (2) the Shell process based on a cobalt carbonyl–phosphine complex [47]; and (3) a process using a rhodium catalyst. The key parameters of these processes are compared in Table 3.

The classical, cobalt-catalyzed oxo process involves the following steps: oxo reaction, catalyst separation and regeneration, aldehyde hydrogenation, and alcohol distillation. Variants of this process differ mainly in the catalyst separation and regeneration steps [49], [50].

In the Shell process, alcohols are obtained directly because of the greater hydrogenating activity of the catalyst; the aldehyde hydrogenation step is unnecessary. Linearity is improved and the 2-methyl isomer is the main byproduct. A disadvantage is the loss of olefins due to hydrogenation to alkanes.

Hydroformylation based on a rhodium catalyst has been used by Union Carbide since the 1970s for the production of n-butanol and 2-ethylhexanol [51]. The higher activity of the catalyst enables operation at lower temperature and pressure. For butanol, the linearity is over 90%. The disadvantage is the high price of rhodium.

For a recent review of the oxo synthesis, see [52].

2.4.3. Hydrogenation of Fatty Acids Produced by Oxidation of Paraffinic Hydrocarbons

The process for the oxidation of paraffinic hydrocarbons, developed in Germany before 1940, is used on an industrial scale in the Comecon countries, particularly in the Soviet Union, for the manufacture of fatty acids. About 5–10% of these synthetic fatty acids are converted to fatty alcohols. The products are mainly linear, primary alcohols, with 5–15% branched alcohols. An overview is given in [57].

A mixture of paraffins is oxidized above 100 °C in the presence of manganese catalysts. The complex product mixture consists of aldehydes, ketones, esters, carboxylic acids, and other compounds. Since the byproducts cannot be completely removed during further processing and distillation of the carboxylic acids, the uses of fatty alcohol produced by this method are limited. As with the natural fatty acids, hydrogenation is carried out after esterification with methanol or butanol. The suspension hydrogenation process is used (see Section 2.3.3.2). The distillation residue contains esters of $C_{10}-C_{20}$ alcohols, 25 % of which are secondary. These alcohols can be obtained by hydrolysis.

2.4.4. Bashkirov Oxidation

A variant of paraffinic hydrocarbon oxidation was developed in the 1950s in the Soviet Union by BASHKIROV [1], [3]. The paraffins are oxidized in the presence of boric acid, which scavenges the hydroperoxides that are formed as intermediates. This results in the formation of boric acid esters of secondary alcohols. These esters are relatively stable to heat and oxidation. Hydrolysis leads to a statistical distribution of secondary alcohols in which the hydroxyl function may occupy any position on the carbon chain.

Industrial oxidation is carried out at about 160 °C with a nitrogen–air mixture containing about 3.5 % oxygen [58], [59] (→ Alcohols, Aliphatic). The conversion of paraffins is limited to a maximum of 20 % in order to minimize side reactions.

The process is used in the Soviet Union and in Japan (Nippon Shokubai). A plant operated by Union Carbide in the United States since 1964 is said to have been closed.

2.4.5. Other Processes

Fatty alcohols can also be obtained by reaction of α-olefins with hydroperoxides in the presence of transition-metal catalysts, especially molybdenum [61]:

$$R-CH=CH_2 + R-OOH \longrightarrow R-\underset{\underset{O}{\diagdown\diagup}}{CH-CH_2} + ROH$$

If *tert*-butyl hydroperoxide is used, the coproduct isobutanol can be readily separated from the epoxide.

After separation of the low molecular mass alcohol and purification, the epoxide is hydrogenolyzed in the presence of a nickel catalyst to form the primary alcohol. About 10–15 % of the secondary alcohol and 2 % paraffin are obtained as byproducts [62].

Mixtures of straight-chain, primary alcohols with average molecular masses between 400 and 700 (corresponding to a chain length of 30–50 carbon atoms) are marketed by Petrolite under the trade name Unilin [63]. The hydrocarbon content of these mixtures is about 20 %. Petrolite oxidizes ethylene oligomers to produce oxygen-containing products on an industrial scale [64].

1-Triacontanol, which does not occur in the common natural raw materials and is produced only in minute quantities in petrochemical processes such as the Alfol process, is of interest because of its ability to stimulate plant growth [65]. Several research groups are investigating synthetic pathways based on cheap raw materials [66]–[71].

2.5. Uses

Fatty alcohols are mainly employed as intermediates. In Western Europe, only 5% are used directly and ca. 95% in the form of derivatives [72].

The amphiphilic character of fatty alcohols, which results from the combination of a nonpolar, lipophilic carbon chain with a polar, hydrophilic hydroxyl group, confers surface activity upon these compounds. Surfactants account for 70–75% of fatty alcohol production [73].

Fatty alcohols orient themselves at phase interfaces and can therefore be used in emulsions and microemulsions. They impart body to cosmetic creams and lotions and solvency to industrial emulsions. They also serve as lubricants in polymer processing. If larger hydrophilic groups are substituted for the hydroxyl group of the fatty alcohols, the polar character is increased [4]. The most important surfactants derived from fatty alcohols are described in the following.

Alkyl polyglycol ethers, fatty alcohol polyglycol ethers, fatty alcohol ethoxylates, $RCH_2(OCH_2CH_2)_nOH$, were the first nonionic surfactants produced on an industrial scale. They are synthesized in a base-catalyzed reaction between fatty alcohols and ethylene oxide. The products obtained from fatty alcohols, ethylene oxide, and propylene oxide are weakly-foaming surfactants. A collection of data of interest in industrial applications of ethoxylates of secondary alcohols is given in [60].

Alkyl sulfates, fatty alcohol sulfates, RCH_2OSO_3Na, belong to the group of anionic surfactants and have been used as detergents since the 1930s. They are synthesized by the reaction of fatty alcohols with sulfur trioxide, chlorosulfuric acid, oleum, or sulfuric acid and subsequent neutralization with alkali, usually sodium hydroxide.

Alkyl polyglycol ether sulfates, fatty alcohol ether sulfates, $RCH_2(OCH_2CH_2)_nOSO_3Na$, are obtained by the reaction of fatty alcohol ethoxylates with sulfur trioxide or chlorosulfuric acid and subsequent neutralization with caustic soda, ammonia, or ethanolamine.

Other surfactants based on fatty alcohols include alkyl sulfosuccinates, alkyl glyceryl ether sulfonates, alkyl polyglycol ether carboxylates, alkyl phosphates, and alkyl ether phosphates.

Many fatty alcohol derivatives are used as intermediates (see Fig. 1).

Alkylamines (fatty amines) can be converted into cationic surfactants (→ Amines, Aliphatic). Alkyl chlorides serve as alkylating agents. Alkyltrimethylammonium halides and alkylpyridinium halides are used as bactericides and extreme-pressure additives of lubricants. Aldehydes are components of perfumes. Alkylthiols are used predominantly as additives in the production of synthetic rubber.

Esters of fatty alcohols with fatty acids, the so-called wax esters, are used as lubricants in polymer processing and as raw materials for waxes and polishes. Liquid esters based on unsaturated fatty alcohols and unsaturated fatty acids are preferred in cosmetics. Acrylic and methacrylic acid esters of fatty alcohols are precursors of the polymethacrylates used as flow improvers and viscosity index improvers in oils. Fumaric acid esters have similar applications. Esters of adipic acid and sebacic acid are used as solvents and, together with phthalic acid esters, as plasticizers for poly(vinyl chloride).

3. Unsaturated Fatty Alcohols

Unsaturated fatty alcohols can only be obtained from natural sources; petrochemical processes for their manufacture do not exist. The physical properties of the most important unsaturated fatty alcohols are listed in Table 4. The melting points are below those of the corresponding saturated alcohols and are influenced by the configuration of the double bond.

Production. The first large-scale hydrogenation plant (Henkel) went into operation in the late 1950s. Previously, unsaturated fatty alcohols could be obtained only by hydrolysis of whale oil (Section 2.3.1) or by the Bouveault–Blanc reduction (Section 2.3.2).

Today, many raw materials are available; both market factors and the degree of unsaturation (iodine number) required in the final product influence the selection.

For products with iodine numbers around 50, cheap beef tallow is available. Polyunsaturated fatty alcohols that tend to autoxidize are undesirable in these products. For iodine numbers of 95–150, soybean oil and sunflower seed oil are frequently used. Linseed oil gives fatty alcohols with iodine numbers over 150.

The hydrogenation processes described in Section 2.3.3.2 are suitable for the large-scale production of unsaturated fatty alcohols. The fixed-bed processes are preferred because of the mild reaction conditions. In suspension hydrogenation, the prolonged contact between fatty alcohol and catalyst results in side reactions such as saturation of the double bond and formation of trans isomers, which leads to a higher solidification point and, hence, loss of quality. With polyunsaturated fatty acids, the formation of conjugated double bonds cannot be completely prevented.

Hydrogenation is generally carried out at 250–280 °C and a pressure of 20–25 MPa. Catalysts include zinc oxide in conjunction with aluminum oxide, chromium oxide, or iron oxide, and possibly other promoters [80]–[85]; copper chromite whose activity has been reduced by the addition of cadmium compounds; and cadmium oxide on an alumina carrier [86]. Selective hydrogenation can also be carried out in a homogeneous phase with metallic soaps as catalysts. An overview of early catalyst

Table 4. Physical properties of primary unsaturated fatty alcohols

IUPAC name	Common name	CAS registry number	Molecular formula	M_r	Hydroxyl number	Iodine number	mp, °C	bp, °C (p, kPa)	Density, g/cm³ (t, °C)	Refractive index (t, °C)
10-Undecen-1-ol		[112-43-6]	$C_{11}H_{22}O$	170.3	329	149	−2	133 (2.1)	0.8495 (15)	1.4509 (20)
(Z)-9-Octadecen-1-ol	oleyl alcohol	[143-28-2]	$C_{18}H_{36}O$	268.4	209	95	−7.5	208–210 (2.0)	0.8489 (20)	1.4606 (20)
(E)-9-Octadecen-1-ol	elaidyl alcohol	[506-42-3]	$C_{18}H_{36}O$	268.4	209	95	36–37	216 (2.4)	0.8388 (20)	1.4552 (40)
(Z,Z)-9,12-Octadecadien-1-ol	linoleyl alcohol	[506-43-4]	$C_{18}H_{34}O$	266.5	211	191	−5 to −2	153–154 (0.4)	0.8612 (20)	1.4782 (20)
(Z,Z,Z)-9,12,15-Octadecatrien-1-ol	linolenyl alcohol	[506-44-5]	$C_{18}H_{32}O$	264.5	212	288		133 (0.27)	0.8708 (25)	1.4775 (25)
(Z)-13-Docosen-1-ol	erucyl alcohol	[629-98-1]	$C_{22}H_{44}O$	324.6	173	78	34–35	240–242 (1.3)	0.8416 (33)	
(E)-13-Docosen-1-ol	brassidyl alcohol	[5634-26-4]	$C_{22}H_{44}O$	324.6	173	78	53–54	238–243 (1.05)		

developments is given in [74]; further references can be found in [75]–[79]. Recent patent applications indicate great interest in selective catalysts [87]–[92].

Unsaturated fatty alcohols are produced in the Federal Republic of Germany (Henkel), Japan (New Japan Chemical), the United States (Sherex), the German Democratic Republic, and Poland.

Uses. Unsaturated fatty alcohols are used in detergents, in cosmetic ointments and creams, as plasticizers and antifoaming agents, and in textile and leather processing [23], [93]–[95]. Oleyl alcohol is also used as an additive in petroleum and lubricating oils.

4. Guerbet Alcohols

Dimerization of primary alcohols at 180–300 °C in the presence of alkaline condensation agents such as potassium hydroxide yields primary, α-branched alcohols [3], [5]:

$$2\,RCH_2CH_2OH \longrightarrow RCH_2CH_2\overset{R}{\underset{|}{C}}HCH_2OH$$

This reaction was first observed by M. GUERBET [96]; it is carried out commercially on a small scale (ca. 2000–3000 t/a). The alcohol is heated under reflux with the alkaline condensation agent and a hydrogenation–dehydrogenation catalyst. The water formed is distilled off continuously with the monomer alcohols. The reaction of lower monomer alcohols (C_6–C_{10}) is carried out under pressure. However, alcohols such as 2-ethyl-1-hexanol can be produced more cheaply by other methods (→ 2-Ethylhexanol).

Table 5 lists the most important characteristics of the common 2-alkyl (Guerbet) alcohols.

The physicochemical properties of Guerbet alcohols can be summarized as follows [97]:

– Melting point or freezing point and viscosity are lower than those of the straight-chain isomers.
– Volatility and vapor pressure are considerably less than those of linear alcohols with comparable viscosity (and correspondingly lower molecular mass).

Guerbet alcohols are less readily oxidized than unsaturated alcohols, which are also liquid. They can therefore be used as components of cosmetic and pharmaceutical oils which must be highly stable to oxidation (rancidity).

Other uses for C_{16}–C_{24} Guerbet alcohols are as plasticizers for synthetic resins, e.g., nitrocellulose (for the manufacture of stencils); solvents for printing inks and specialty inks; lubricants, e.g., for metal working fluids and textile fibers; and intermediates, e.g., in the production of branched carboxylic acids. The C_{32}–C_{36} Guerbet alcohols,

Table 5. Physical properties of Guerbet alcohols

IUPAC name	CAS registry number	Molecular formula	M_r	Hydroxyl number	mp, °C	bp, °C (p, kPa)	Density, g/cm^3 (t, °C)	Refractive index (t, °C)
2-Methyl-1-pentanol	[105-30-6]	$C_6H_{14}O$	102.18	549		147.9 (101.3)	0.8263 (20)	1.4182 (20)
2-Ethyl-1-hexanol	[104-76-7]	$C_8H_{18}O$	130.23	431	<−76	118 (10.7)	0.8328 (20)	1.4328 (20)
2-Propyl-1-heptanol	[10042-59-8]	$C_{10}H_{22}O$	158.29	354		117 (2.7)		
2-Butyl-1-octanol	[3913-02-8]	$C_{12}H_{26}O$	186.34	301		126–128 (1.5)		1.4457 (20)
2-Pentyl-1-nonanol	[5333-48-2]	$C_{14}H_{30}O$	214.39	262		154 (1.7)	0.8352 (24)	1.4460 (24)
2-Hexyl-1-decanol	[2425-77-6]	$C_{16}H_{34}O$	242.45	231	−30 to −26	175 (1.5)	0.8380 (20)	1.4476 (25)
2-Heptyl-1-undecanol	[5333-44-8]	$C_{18}H_{38}O$	270.50	207	−26	198 (2.0)	0.8446 (15)	1.4550 (15)
2-Octyl-1-dodecanol	[5333-42-6]	$C_{20}H_{42}O$	298.56	188	−20	135–137 (0.007)	0.8329 (21)	1.4545 (19)
2-Nonyl-1-tridecanol	[54439-52-0]	$C_{22}H_{46}O$	326.61	172		164–167 (0.013)	0.8476 (17.5)	1.4582 (17)
2-Decyl-1-tetradecanol	[58670-89-6]	$C_{24}H_{50}O$	354.67	158		173–175 (0.007)	0.8413 (17)	1.4606 (17)
2-Undecyl-1-pentadecanol	[79864-02-1]	$C_{26}H_{54}O$	382.72	147				
2-Dodecyl-1-hexadecanol	[22388-18-2]	$C_{28}H_{58}O$	410.77	137		203–207 (0.007)		
2-Tridecyl-1-heptadecanol		$C_{30}H_{62}O$	438.83	128				
2-Tetradecyl-1-octadecanol	[32582-32-4]	$C_{32}H_{66}O$	466.88	120	38–39	308–310 (2.0)		
2-Pentadecyl-1-nonadecanol		$C_{34}H_{70}O$	494.94	113				
2-Hexadecyl-1-eicosanol	[17658-63-8]	$C_{36}H_{74}O$	522.99	107	43–45	270–280 (0.013)		
2-Heptadecyl-1-heneicosanol		$C_{38}H_{78}O$	551.05	102				
2-Octadecyl-1-docosanol		$C_{40}H_{82}O$	579.11	98				
2-Nonadecyl-1-tricosanol		$C_{42}H_{86}O$	607.16	95				
2-Eicosyl-1-tetracosanol	[73761-81-6]	$C_{44}H_{90}O$	635.22	93				

which are obtained from tallow alcohol, can be used to produce waxes and cosmetic sticks.

5. Bifunctional Fatty Alcohols

Some natural oils which contain double bonds, hydroxyl groups, or other functional groups can be converted into long-chain diols. These are usually α,ω-diols or diols whose hydroxyl groups lie far apart. Table 6 lists physical properties of bifunctional fatty alcohols.

1,2-Diols can be produced by epoxidation of internal or α-olefins with subsequent hydrolytic cleavage of the epoxide ring ($\rightarrow$ Alcohols, Polyhydric).

Dimerization of unsaturated fatty acids such as soybean, linseed, and tallow fatty acid, followed by esterification and catalytic hydrogenation, gives saturated diols [98].

Thermal or catalytic dimerization of unsaturated alcohols yields viscous dimers with complex structure and an average of two hydroxyl groups per molecule [99], [100].

1,12-Octadecanediol, 1,10-decanediol, and 9-octadecene-1,12-diol are obtained from castor oil by transesterification and hydrogenation, or by alkali splitting, esterification, and hydrogenation. Oiticica oil and the seed oil of *Malotus philippinensis* (Euphorbiaceae) are also suitable raw materials.

Bifunctional fatty alcohols are mainly used in the production of polyesters and polyamides and as intermediates.

6. Quality Specifications

Analytical methods defined by DIN [101] and ASTM [102] standards and by the Deutsche Gesellschaft für Fettwissenschaft (DGF) [103] are used for the quality control of fatty alcohols (see Table 7). Composition of fatty alcohol mixtures is determined by gas chromatography, which can be combined with mass spectrometry. Alcohols from different sources can thus be identified; typical examples are given in Figure 5.

Linearity L is defined as the percentage of normal alcohols present in the mixture:

$$L = \frac{n-\text{alcohol}}{n-\text{alcohol} + \text{isoalcohol}} \times 100$$

Coconut alcohol gives very few impurity peaks. It contains <0.1% *n*-tridecanol and varying amounts of *n*-alkanes, depending on the hydrogenation process.

Tallow alcohol contains up to 2% isomeric pentadecanols (isopentadecanol, anteisopentadecanol, and *n*-pentadecanol) and *n*-hexadecane, and up to 4% isomeric hepta-

Table 6. Physical properties of bifunctional fatty alcohols

IUPAC name	CAS registry number	Molecular formula	M_r	Hydroxyl number	mp, °C	bp, °C (p, kPa)	Density, g/cm^3 (t, °C)
1,6-Hexanediol	[629-11-8]	$C_6H_{14}O_2$	118.17	949	42	134 (1.3)	0.9580 (45)
1,7-Heptanediol	[629-30-1]	$C_7H_{16}O_2$	132.20	849	18	151 (1.9)	0.9570 (25)
1,8-Octanediol	[629-41-4]	$C_8H_{18}O_2$	146.23	767	61	167–168 (2.4)	0.9200 (60)
1,9-Nonanediol	[3937-56-2]	$C_9H_{20}O_2$	160.26	700	45	173.2 (1.9)	0.9160 (50)
1,10-Decanediol	[112-47-0]	$C_{10}H_{22}O_2$	174.29	644	73	175–176 (1.9)	0.8960 (75)
1,11-Undecanediol	[765-04-8]	$C_{11}H_{24}O_2$	188.31	596	63	178 (1.6)	
1,12-Dodecanediol	[5675-51-4]	$C_{12}H_{26}O_2$	202.34	555	81	183–184 (1.25)	
1,13-Tridecanediol	[13362-52-2]	$C_{13}H_{28}O_2$	216.37	519	75–76	195–197 (1.3)	
1,14-Tetradecanediol	[19812-64-7]	$C_{14}H_{30}O_2$	230.39	487	85	200 (1.2)	
1,15-Pentadecanediol	[14722-40-8]	$C_{15}H_{32}O_2$	244.42	459	70.6–71.6	205–207 (1.3)	
1,16-Hexadecanediol	[7735-42-4]	$C_{16}H_{34}O_2$	258.45	434	91.4	195–200 (0.53)	
1,17-Heptadecanediol	[66577-59-1]	$C_{17}H_{36}O_2$	272.48	412	96–96.5	204–205 (0.27)	
1,18-Octadecanediol	[3155-43-9]	$C_{18}H_{38}O_2$	286.50	392	97–98	210–211 (0.27)	
1,19-Nonadecanediol	[2268-65-7]	$C_{19}H_{40}O_2$	300.53	373	101	212–214 (0.2)	
1,20-Eicosanediol	[7735-43-5]	$C_{20}H_{42}O_2$	314.56	357	102.4–102.6	215–217 (0.2)	
1,21-Heneicosanediol	[95008-70-1]	$C_{21}H_{44}O_2$	328.58	342	105–105.5	223–224 (0.2)	
1,22-Docosanediol	[22513-81-1]	$C_{22}H_{46}O_2$	342.61	328	105.7–106.2		
1,23-Tricosanediol	[95491-58-0]	$C_{23}H_{48}O_2$	356.64	314			
1,24-Tetracosanediol	[22513-82-2]	$C_{24}H_{50}O_2$	370.67	302			
1,25-Pentacosanediol	[92238-33-0]	$C_{25}H_{52}O_2$	384.69	292	109		
(Z)-9-Octadecene-1,12-diol	[540-11-4]	$C_{18}H_{36}O_2$	284.49	394	89	182/0.06	

Quality Specifications

Table 7. Analytical methods for the characterization of fatty alcohols

	DGF*	DIN	ASTM
Composition		by gas chromatography	
Hydrocarbon content		by column chromatography on silica	
Color (Hazen, APHA)		53409	D 1686-81
Color (Lovibond)	C-IV 4b		D 1209-84
Refractive Index	C-IV 5	53491	D 1218-82
Density	C-IV 2b	51757	D 1298-80
			D 891-59 (1976)
Viscosity	C-IV 7	53015	D 445-83
Solidification point	M-III 4a	51570	D 87-77 (1982)
Boiling range		51751	D 1078-83
			E 133-78 (1984)
Flash point	C-IV 8	51758	D 56-82
			D 93-80
Ignition temperature		51794	D 2155-66 (1976)
Hydroxyl number	C-V 17a/b		D 1957 (1984)
Carbonyl number	C-V 18		E 411
Peroxide number	C-VI 6a		D 1022-76
Iodine number (Kaufmann)	C-V 11b		
Iodine number (Wijs)	C-V 11d		D 1959 (1984)
Acid number	C-V 2	53402	D 1613-81
Saponification number	C-V 3	53401	D 94-80
Water content	C-III 13a	51777	D 1744-83

* DGF = Deutsche Gesellschaft für Fettwissenschaft.

decanols (isoheptadecanol, anteisoheptadecanol, and n-heptadecanol) and n-octadecane.

Ziegler alcohols (Alfol and Epal alcohols) are primary, straight-chain alcohols with an even carbon number. Gas chromatography shows up to 1% impurities, consisting of numerous even-numbered, isomeric fatty alcohols.

Oxo alcohols from statistical, cracked olefins contain a large number of isomers that cannot be completely separated, even in the most efficient capillary columns. Both odd- and even-numbered isoalcohols occur. Oxo alcohol mixtures can be identified by spectroscopic determination of the degree of branching, i.e., the ratio of n-alcohols to isoalcohols, which depends on the production process and the raw materials used.

Fatty acids obtained by *oxidation of paraffinic hydrocarbons* yield even- and odd-numbered primary alcohols with a low degree of branching. Branched alkanes present in the original paraffins are lost by oxidative degradation.

The boric acid-catalyzed process (Bashkirov oxidation) yields secondary alcohols with a statistical distribution of isomers.

Guerbet alcohols are primary, α-branched fatty alcohols with two straight-chain alkyl groups of approximately equal length.

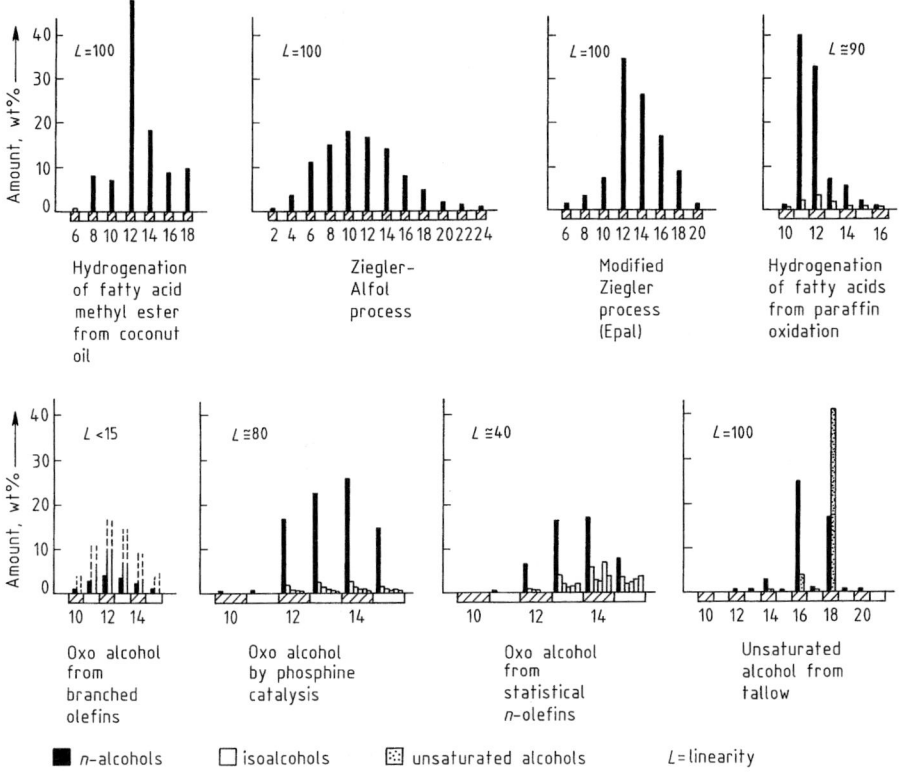

Figure 5. Typical fatty alcohol compositions obtained by different processes

7. Storage and Transportation

During production, processing, and storage, fatty alcohols are handled as liquids. Products with a melting point above 40 °C can also be handled as flakes.

Fatty alcohols are sensitive to oxidation and should be stored under an inert gas; the temperature should not be more than 20 °C above the melting point. Low-pressure steam or warm water are recommended for heating. Water increases stability to autoxidation; as little as 0.1 % has a stabilizing effect. The use of antioxidants depends on further processing and quality requirements.

Stainless steel or an Al–Mg–Mn alloy (DIN 1725/1745) is used as container material; ordinary steel should have a zinc silicate coating, e.g., Dimetcote. Pumps, valves, and pipes exposed to the product should be made of stainless steel.

Fatty alcohols are considered as flammable materials and are classified according to properties such as flash point and boiling point. Their transport is governed by national and international regulations dealing with volatile and combustible materials.

Table 8. Estimated 1985 nameplate production capacity of fatty alcohols, 10^3 t

Country	Natural alcohols, $>C_7$	Synthetic alcohols, $>C_{12}$	Total fatty alcohols
United States	150	420	570
Federal Republic of Germany	175	45	
United Kingdom	25	65	
France	40	10	
Italy		50	
Denmark	5		
Western Europe, total	*245*	*170*	*415*
Eastern bloc countries	25	120	145
Japan	40	65	
Philippines	52		
India	8	5	
Far East, total	*100*	*70*	*170*
Total, worldwide	520	780	1300

8. Economic Aspects

The worldwide production capacity of fatty alcohols is ca. 1.3×10^6 t/a (Table 8). About 60% of this is based on petrochemical feedstocks. World production and consumption in 1985 were estimated at 700 000 – 800 000 t. In Western Europe, only ca. 5% is used directly, and 95% in the form of derivatives. About 70 – 75% is used in surfactants [72], [73], [104] – [108].

In the United States, 70% of production capacity is based on petrochemical feedstocks. The largest producers of synthetic alcohols are Shell (SHOP/oxo process), Ethyl Corporation (Ziegler process), and Vista Chemical (Ziegler/Alfol process). However, 95% of natural alcohols is produced by Procter & Gamble.

In Western Europe, ca. 60% of the production capacity is based on natural raw materials, with the Henkel group being the largest producer. The second-largest producer, Condea Chemie, employs natural (Lurgi process [107]) as well as petrochemical (Ziegler/Alfol process) raw materials, whereas the other European producers predominantly use the oxo process (BASF, Chimica Augusta, ICI, PCUK, and Shell).

Approximately 60% of production capacity in the Far East is also accounted for by natural alcohols. With a planned additional capacity of 100 000 t/a, this will reach 75% by the beginning of the 1990s. The largest manufacturers of natural fatty alcohols are Kao Soap, United Coconut, and New Japan Chemical. The oxo process is used by Mitsubishi Chemical, Mitsubishi Petrochemical, Nissan Chemical, and Nippon Shokubai.

Eastern Europe favors oxidation of paraffins for the manufacture of fatty alcohols. However, there is a Ziegler plant in Ufa (Soviet Union); Poland, the German Democratic Republic, and Romania have plants based on natural raw materials [24], [109] – [112].

Table 9. Commercial detergent-range fatty alcohols

Manufacturer	Country	Trade name	Feedstock
Aarhus Oliefabrik	Denmark	Hyfatol	natural oils
BASF	FRG		olefins
Chimica Augusta	Italy	Lial	olefins
Condea	FRG	Alfol, Nacol	ethylene and natural oils
Ethyl Corp.	USA	Epal	ethylene
Henkel	FRG	Lorol, Stenol, Ocenol	natural oils
Hüls	FRG		olefins, natural oils
ICI	UK	Synprol	olefins
Kao Soap	Japan	Kalcolol	natural oils
Marchon	UK	Laurex	natural oils
Mitsubishi Chemical	Japan	Diadol	olefins
Mitsubishi Gas	Japan		ethylene
New Japan Chemical	Japan	Conol	natural oils
Nippon Shokubai	Japan	Softanol	paraffins
Nissan	Japan	Oxocol	olefins
Procter & Gamble	USA	TA, Co	natural oils
VEB Rodleben	GDR	Rofanol	paraffins, natural oils
Shell	USA, UK	Dobanol, Neodol	ethylene, olefins
Sherex	USA	Adol	natural oils
Sidobre	France	Sipol	natural oils
Ugine Kuhlmann (CdF)	France	Acropol	olefins
Unicom	Philippines		natural oils
Ufa	USSR		ethylene
Vista Chemical	USA	Alfol	ethylene

Economic Aspects

Table 9 gives a global overview of manufacturers, trade names, and raw materials of detergent-range alcohol mixtures.

Because existing overcapacity is some 40%, only 75% of capacity will be used by 1990, even if an annual growth rate of 3% is assumed [72], [73], [108], [111], [113]. In spite of this, new facilities based on fats and oils are planned in Malaysia, Indonesia, Belgium, and Brazil. The capacity utilization of the different plants depends on the general economic development, the product mix, and the prices of the raw materials used. Figure 6 shows the price fluctuations in natural and petrochemical raw materials.

Before 1973 synthetic fatty alcohol plants that used crude oil enjoyed a clear advantage. However, prices of crude oil and petrochemicals such as naphtha, paraffins, and ethylene increased markedly in 1973–1974 and 1979–1980. The long-term price increase is expected to be lower for renewable fats and oils than for crude oil [72], [73], [110], [111], [114]–[121]. Increased yields, new plant breeds, and a wider geographic distribution of the areas under cultivation secure a more regular supply compared with past years. Therefore, the ratio of synthetic to natural fatty alcohols has been gradually shifting from 2:1 to 1:1. However, despite the increased production of coconut, palm kernel, and babassu oils and breeding of C_{12}-rich *Cuphea* varieties, 90% of the natural sources have carbon chain lengths in the C_{16}–C_{18} range [122]. On the other hand, synthetic processes yield predominantly C_{12}–C_{15} alcohols, which play an increasingly important role in surfactants. Consequently, natural raw materials are used mainly for the C_{16}–C_{18} chain range and petrochemical raw materials for the C_{12}–C_{15} range [72], [110], [111].

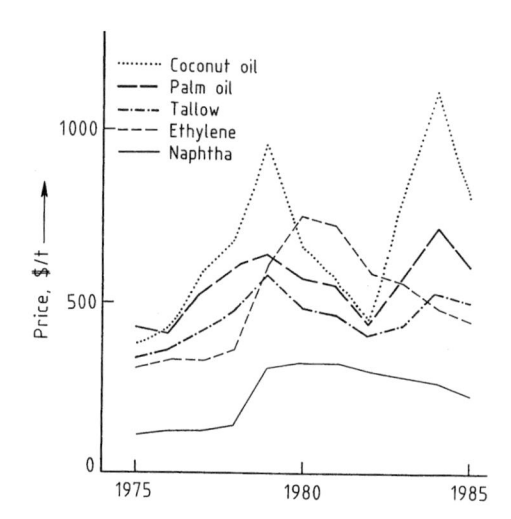

Figure 6. Prices of raw materials used in fatty alcohol production

9. Toxicology

Most of the straight-chain fatty alcohols are considered nontoxic according to EEC regulations [123] (oral LD_{50} in rats >2000 mg/kg). An exception is the C_6 alcohol [124].

The systemic toxicity generally decreases with increasing chain length. Inconsistencies in the literature may be due to differences in test methods, test animals, and product composition, especially with technical-grade mixtures.

Straight-chain C_8-C_{10} alcohols produce slight skin irritation in humans, whereas rabbits are far more sensitive. This difference is even greater with unsaturated C_{18} fatty alcohols [125]. Hypersensitivity to fatty alcohols used in cosmetics and pharmaceuticals is rare [126]–[128]. No sensitization has been observed with C_{16} and C_{18} alcohols [130].

Straight-chain C_6-C_8 fatty alcohols have a more or less irritating effect on mucous membranes [124], [129].

Certain branched oxo alcohols may cause toxicological problems [133]–[141].

1-Hexanol. 1-Hexanol is classified as moderately toxic [123], [131]. For the oral LD_{50} in rats, 0.7 g/kg [124], 3.2 g/kg [131], and 4.87 g/kg [142] have been measured. The value for mice is 1.9 g/kg [124] or 4.0 g/kg [143]. An LD_{50} (dermal, rabbits) of 3.1 g/kg [124] has been found. 1-Hexanol causes strong eye irritation [124], [144] and slight skin irritation in rabbits [124], [145]. In the Epstein test, 1-hexanol causes neither irritation [146] nor sensitization [147] of human skin.

1-Octanol. The LD_{50} is 18.2 g/kg for rats [131] and 1.7 g/kg for mice [124]. In rabbits, slight skin irritation is observed [124].

Nonanol Mixtures. The LD_{50} for rats ranges from 2.98 g/kg [133] to 6.4 g/kg [148]. The percutaneous LD_{50} falls within the same range [149]. Inhalation of 21.7 mg/L over 4 h in rats is lethal [133]. Strong irritation is produced in the rabbit eye. Inhalation of 33, 99, and 136 ppm by rabbits leads to degeneration [150].

1-Decanol. The oral LD_{50} for rats ranges from 4.7 g/kg [124] to 26.4 g/kg [131]; the oral LD_{50} for mice is 6.4 g/kg [124]. A dermal LD_{50} for rabbits of 3.5 g/kg has been found [124]. Strong irritation is produced in the rabbit eye, whereas primary skin irritation is mild [124].

Mixtures of 1-Octanol and 1-Decanol. The oral LD_{50} in rats is >5 g/kg for products from natural raw materials [129] and >26 g/kg for Alfol alcohols [131]; primary skin irritation is slight [129].

Decyl Alcohol Mixtures. The oral LD_{50} for rats ranges from 4.72 g/kg [133] to 25.6 g/kg [148]. The dermal LD_{50} for rabbits lies between 3.56 mL/kg and 18.8 mL/kg [151]. The LD_{50} by inhalation in mice is 525 ppm [150].

1-Dodecyl Alcohol. The oral LD_{50} is >36 g/kg [140] or >26.5 g/kg [131] for rats; >36 g/kg for rabbits, and >10 g/kg for guinea pigs. With guinea pigs, there was no primary skin irritation [140].

Mixtures of 1-Dodecanol and 1-Tetradecanol. The oral LD_{50} for rats is >7.9 g/kg [131] or >10 g/kg [129]. In rabbits slight eye irritation is produced [129]. No primary skin irritation has been observed in guinea pigs [124].

1-Tetradecanol. The oral LD_{50} for rats is >5 g/kg [129], [131]. A 50% solution reddens rabbit skin; a 25% solution does not affect human skin [129].

1-Hexadecanol. The oral LD_{50} for rats is between >6.4 g/kg and 20 g/kg [129]–[131], [140]. For mice, the oral LD_{50} is 3.2–6.4 g/kg [129]; the intraperitoneal LD_{50} is 1.6–3.2 g/kg [140]. 1-Hexadecanol is slightly irritating to the eye and skin of rabbits [124] and to the skin of guinea pigs [140], but a 25% solution does not irritate human skin [129]. Inhalation of 0.41 mg/L over 6 h has no effect on rats [140].

Mixtures of 1-Hexadecanol and 1-Octadecanol. The oral LD_{50} for rats is >10 g/kg. With rabbits, only slight eye irritation and no skin irritation is observed. Inhalation of a saturated vapor atmosphere for 6 h does not produce any effects in rats [131].

1-Octadecanol. The oral LD_{50} for rats is between >5 g/kg and 20 g/kg [129]–[131]; the dermal LD_{50} for rabbits is >1 g/kg [129]. Slight eye irritation is produced in rabbits [130], whereas a 5% solution does not irritate human skin [129].

Mixtures of 1-Eicosanol and 1-Docosanol. The oral LD_{50} for rats is >12.6 g/kg [132].

Oleyl Alcohol. The oral LD_{50} for rats is >25 g/kg. Repeated skin contact does not produce irritation in humans [129].

Mixtures of Oleyl and 1-Hexadecyl Alcohol. Slight eye irritation is observed in rabbits, but skin contact in humans does not cause irritation [129].

2-Hexyldecanol. The oral LD_{50} for rats ranges from >10 g/kg to 40 g/kg [129], [132]. Skin contact does not produce irritation in humans, rabbits, or mice [129], [132]. A single application to the rabbit eye leads to a barely perceptible temporary redness of the conjunctiva [129], [132].

2-Octyldodecanol. The oral LD_{50} for rats ranges from 30 g/kg to >50 g/kg [129], [132]. A 50% solution is tolerated by rabbit skin and eye and by the skin of hairless mice and rats [129], [132].

10. References

General References

[1] *Ullmann*, 4th ed., **11**, 427–445.
[2] *ACS Symp. Ser.* **159** (1981).
[3] F. Korte (ed.): *Methodicum Chimicum*, vol. **5**, Stuttgart 1975.
[4] H. Stache, *Tensid Taschenbuch*, Carl Hanser Verlag, München 1981.
[5] *Houben-Weyl*, **VI/1a/b**, 1 ff.
[6] *Fettalkohole*, 2nd ed., Henkel, Düsseldorf 1982.

Specific References

[7] H. Adkins, K. Folkers, *J. Am. Chem. Soc.* **53** (1931) 1095–1097.
[8] W. Normann, *Angew. Chem.* **44** (1931) 714–717. DE 617 542, 1930 (W. Normann, H. Prückner); Böhme Fettchemie GmbH, DE 639 527, 1930 (W. Normann).
[9] W. Schrauth, O. Schenk, K. Stickdorn, *Ber. Dtsch. Chem. Ges.* **64** (1931) 1314–1318.
[10] O. Schmidt, *Ber. Dtsch. Chem. Ges.* **64** (1931) 2051–2053.
[11] D. Precht, *Fette, Seifen, Anstrichm.* **78** (1976) no. 4, 145.
[12] *Alfol-Alkohole, Typische Analysendaten*, Condea Chemie, Brunsbüttel.
[13] *VDI-Wärmeatlas*, Düsseldorf 1984.
[14] *D'Ans Lax – Taschenbuch für Chemiker und Physiker*, vol. **II**, Springer Verlag, Berlin 1964.
[15] *CRC Handbook*, 55th ed., Cleveland 1974/75.
[16] *Houben-Weyl*, **E3**, 265–300.
[17] *Houben-Weyl*, **E5/1**, 202–212.
[18] K. Heyns, L. Blasejewicz, *Tetrahedron* **9** (1960) 67.
[19] E. S. Gore, *Platinum Met. Rev.* **27** (1983) no. 3, 111.
[20] Atlantic Richfield Co., US 3 997 578, 1976 (Ming Nan Sheng).
[21] *Kirk-Othmer*, **9**, 795 ff.
[22] U. R. Kreutzer, *J. Am. Oil Chem. Soc.* **61** (1984) no. 2, 343.
[23] H.-D. Komp, M. P. Kubersky in: *Fettalkohole*, 2nd ed., Henkel, Düsseldorf 1982, p. 51 ff.
[24] E. F. Hill, G. R. Wilson, E. C. Steinle, Jr., *Ind. Eng. Chem.* **46** (1954) 1917.
[25] M. L. Karstens, H. Peddicord, *Ind. Eng. Chem.* **41** (1949) 438.
[26] H. Igo, *CEER Chem. Econ. Eng. Rev.* **8** (1976) no. 3, 31.
[27] H. Buchold, *Chem. Eng. (N.Y.)* **90** (1983) no. 4, 42.
[28] Th. Voeste, H. Buchold, *J. Am. Oil Chem. Soc.* **61** (1984) no. 2, 350.
[29] *Chem. Ind. (Düsseldorf)* **32** (1980) 739.
[30] Metallgesellschaft AG, DE 2 853 990, 1978 (Th. Voeste, H.-J. Schmidt, F. Marschner).
[31] Dehydag GmbH, DE 1 005 497, 1954 (W. Rittmeister).
[32] Dehydag GmbH, US 3 193 586, 1965 (W. Rittmeister).
[33] Henkel KGaA, DE-AS 2 613 226, 1976 (G. Demmering)
[34] VEB Deutsches Hydrierwerk Rodleben, DD 213 430, 1983 (H. Aring, K. Busch, P. Franke, G. Konetzke et al.)
[35] Institut Français du Pétrole, DE-OS 3 217 429, 1982 (R. Snappe, J.-P. Bournonville).
[36] B. C. Triverdi, D. Grote, Th. O. Mason, *J. Am. Oil Chem. Soc.* **58** (1981) no. 1, 17.
[37] Ashland Oil, Inc. US 4 104 478, 1978 (B. C. Triverdi).
[38] *Houben-Weyl*, **13/4**, 1 ff.

[39] T. Mole, E. A. Jeffery: *Organoaluminium Compounds*, Elsevier, Amsterdam 1972.
[40] A. Lundeen, R. Poe: "Alpha-Alcohols," in J. J. Mc Ketta, W. A. Cunningham (eds.): *Encyclopedia of Chemical Processing and Design*, vol. **2**, Marcel Dekker, New York 1977, p. 465.
[41] K. Ziegler, H.-G. Gellert, K. Zosel, E. Holzkamp et al., *Justus Liebigs Ann. Chem.* **629** (1960) 121.
[42] H. Wesslau, *Justus Liebigs Ann. Chem.* **629** (1960) 198.
[43] P. H. Washecheck, *ACS Symp. Ser.* **159** (1981) 87.
[44] K. L. Lindsay: "Alpha-Olefins," in J. J. Mc Ketta, W. A. Cunningham (eds.): *Encyclopedia of Chemical Processing and Design*, vol. **2**, Marcel Dekker, New York 1977, p. 482.
[45] *Produkte aus der Oxosynthese*, Ruhrchemie AG, Frankfurt 1969.
[46] J. Falbe: *Synthesen mit Kohlenmonoxid*, Berlin 1967.
[47] R. E. Vincent, *ACS Symp. Ser.* **159** (1981) 159.
[48] B. Cornils, *Compend. Dtsch. Ges. Mineralölwiss. Kohlechem.* **78/79** (1978) no. 1, 463.
[49] B. Cornils, "New Syntheses with Carbon Monoxide," *React. Struct. Concepts Org. Chem.* **11** (1980).
[50] H. Lemke, *Hydrocarbon Process.* **45** (1966) no. 2, 148.
[51] C. E. O'Rourke, P. R. Kavasmaneck, R. E. Uhl, *ACS Symp. Ser.* **159** (1981) 71.
[52] B. A. Murrer, J. H. Russel, *Catalysis* **6** (1983) 169.
[53] A. H. Turner, *J. Am. Oil Chem. Soc.* **60** (1983) no. 3, 623.
[54] R. L. Banks, *Appl. Ind. Catal.* 1984, no. 3, 215.
[55] E. R. Freitas, C. R. Gum, *Chem. Eng. Prog.* **75** (1979) 73.
[56] E. L. T. M. Spitzer, *Seifen, Oele, Fette, Wachse* **107** (1981) no. 6, 141.
[57] H. Stage, *Seifen, Oele, Fette, Wachse* **99** (1973) no. 6/7, 143; no. 8, 185; no. 9, 217; no. 11, 299.
[58] N. Kurata, K. Koshida, *Hydrocarbon Process.* **57** (1978) 145.
[59] N. J. Stevens, J. R. Livingstone, Jr., *Chem. Eng. Prog.* **64** (1968) no. 7, 61.
[60] N. Kurata, K. Koshida, H. Yokoyama, T. Goto, *ACS Symp. Ser.* **159** (1981) 113.
[61] L. Marko, *J. Organomet. Chem.* **283** (1985) 221.
[62] B. N. Bobylev, M. J. Farberov, S. A. Kesarev, *Khim. Promst. (Moscow)* 1979 no. 3, 138.
[63] *Unilin Alcohols*, Petrolite Corp., St. Louis 1985.
[64] Petrolite Corp., US 4 426 329, 1984 (J. H. Woods, C. E. Laughlin, T. R. Graves).
[65] S. K. Ries, V. Wert, C. S. Sweely, R. A. Leavitt, *Science* **195** (1977) 1339.
[66] J. Penninger, M. Biermann, H.-J. Krause, *Fette, Seifen, Anstrichm.* **85** (1983) no. 6, 239.
[67] T. Gibson, L. Tulich, *J. Org. Chem.* **46** (1981) no. 9, 1821.
[68] A. V. Rama Rao, J. S. Yadav, G. S. Annapurna, *Synth. Commun.* **13** (1983) no. 4, 331.
[69] S. M. Kulkarni, V. R. Mamdapur, M. S. Chadha, *Indian J. Chem., Sect. B* **22** (1983) no. 7, 683.
[70] U. T. Bhalerao, S. J. Rao, B. D. Tilak, *Tetrahedron Lett.* **25** (1984) no. 47, 5439.
[71] G. Kirchner, H. Weidmann, *Liebigs Ann. Chem.* 1985, no. 1, 214.
[72] H. J. Richtler, J. Knaut, *J. Am. Oil Chem. Soc.* **61** (1984) no. 2, 160–175.
[73] J. Knaut, H. J. Richtler, *J. Am. Oil Chem. Soc.* **62** (1985) no. 2, 317.
[74] H. Bertsch, H. Reinheckel, K. Haage, *Fette, Seifen, Anstrichm.* **66** (1964) 763.
[75] H. Bertsch, H. Reinheckel, K. Haage, *Fette, Seifen, Anstrichm.* **71** (1969) 357, 785.
[76] H. Bertsch, K. Haage, H. Reinheckel, *Fette, Seifen, Anstrichm.* **71** (1969) 851.
[77] K. Lindner: *Tenside, Textilhilfsmittel, Waschrohstoffe*, vol. **I**, 2nd ed., Wissenschaftl. Verlags GmbH, Stuttgart 1964, p. 144.
[78] J. Richter: Dissertation, TH Delft, 1968.
[79] J. W. E. Coenen, *Fette, Seifen, Anstrichm.* **77** (1975) 431.
[80] Dehydag, US 3 193 586, 1965 (W. Rittmeister).
[81] Henkel, US 3 729 520, 1973 (H. Rutzen, W. Rittmeister).

[82] Henkel, DE-AS 2 513 377, 1975 (G. Demmering, H. Schütt, H. Rutzen).
[83] J. Cieslar, W. Bulanda, *Przem. Chem.* **63** (1984) 375; *Chem. Abstr.* **101** (1984) 153 810 t.
[84] Institute Blachownia and Nitrogen Works Kedzierzyn, PL 118 880, 1978 (E. Fabisz, K. Chmielewski, A. Jakubowicz, A. Mankowski et al.).
[85] Institute Blachownia and Nitrogen Works Kedzierzyn, PL 118 979, 1978 (E. Fabisz, K. Chmielewski, A. Jakubowicz, A. Mankowski et al.).
[86] New Japan Chemical Ltd., GB 1 335 173, 1972.
[87] VEB Deutsches Hydrierwerk Rodleben, DD 213 429, 1983 (H. Aring, K. Busch, P. Franke, G. Konetzke et al.).
[88] UOP Inc., US 4 340 546, 1982 (G. Qualeatti, D. Germanas).
[89] UOP Inc., US 4 446 073, 1984 (G. Qualeatti, D. Germanas).
[90] Kao Soap, JP 58 210 035, 1983 (K. Kokubo, K. Tsukada, Y. Miyabata, Y. Kazama).
[91] Kao Soap, JP 5 995 227, 1984 (K. Tsukada, Y. Miyabata).
[92] Kao Soap, JP 59 106 431, 1984 (K. Tsukada, Y. Miyabata, K. Fukuoka).
[93] H. Schütt in W. Foerst, H. Buchholz-Meisenheimer (eds.): *Neue Verfahren, Neue Produkte, Wirtschaftliche Entwicklung*, Urban und Schwarzenberg, München 1970.
[94] U. Ploog, *Seifen, Oele, Fette, Wachse* **109** (1983) no. 8, 225.
[95] R. R. Egan, G. W. Earl, J. Ackermann, *J. Am. Oil Chem. Soc* **61** (1984) no. 2, 324.
[96] M. Guerbet, *C. R. Hebd. Séances Acad. Sci.* **128** (1898) 511; M. Guerbet, *Bull. Soc. Chim. Fr.* **21** (1899) no. 3, 487.
[97] J. Glasl in: *Fettalkohole*, 2nd ed., Henkel, Düsseldorf 1982, p. 169.
[98] Henkel & Cie GmbH, DE-OS 1 768 313, 1971 (H. Rutzen).
[99] Henkel & Cie GmbH, DE 1 198 348, 1961 (W. Stein, M. Walther).
[100] Henkel & Cie GmbH, DE 1 207 371, 1963 (H. Hennig, W. Stein, M. Walther).
[101] "Mineral- und Brennstoffnormen," *DIN-Taschenbuch*, Beuth-Vertrieb, Berlin-Köln-Frankfurt 1984.
[102] *Annual Book of ASTM Standards*, Sections 5, 6, 15, 1985.
[103] *Deutsche Einheitsmethoden zur Untersuchung von Fetten, Fettprodukten und verwandten Stoffen*, Wissenschaftl. Verlags GmbH, Stuttgart 1950–1984.
[104] H. J. Richtler, J. Knaut, *Chem. Ind. (Düsseldorf)* **36** (1984) 131–134.
[105] H. J. Richtler, J. Knaut, *Chem. Ind. (Düsseldorf)* **36** (1984) 199–201.
[106] L. Marcou, *Rev. Fr. Corps. Gras.* **30** (1983) no. 1, 3–6.
[107] *J. Am. Oil Chem. Soc.* **58** (1981) no. 11, 873–874A.
[108] A. L. de Jong, *J. Am. Oil Chem. Soc.* **60** (1983) no. 3, 634–639.
[109] *Kirk-Othmer*, **1**, 740–754.
[110] E. C. Leonard, *J. Am. Oil Chem. Soc.* **60** (1983) no. 6, 1160–1161.
[111] C. A. Houston, *J. Am. Oil Chem. Soc.* **61** (1984) no. 2, 179–184.
[112] *J. Am. Oil Chem. Soc.* **62** (1985) no. 4, 668.
[113] J. C. Dean, *Chem. Week* 1985, May 15, 3–34.
[114] J. W. E. Coenen, *J. Am. Oil Chem. Soc.* **53** (1976) no. 6, 382–389.
[115] H. Klimmek, *Chem. Ind. (Düsseldorf)* **33** (1981) 136–141.
[116] H. Fochem, *Fette, Seifen, Anstrichm.* **87** (1985) no. 2, 47–52.
[117] *Chem. Week* 1982, April 28, 40–44.
[118] *Chem. Ind. (Düsseldorf)* **35** (1983) 251–255.
[119] *J. Am. Oil Chem. Soc.* **61** (1984) no. 8, 1298–1314.
[120] H. Lindemann, *Chem. Ind. (Düsseldorf)* **34** (1982) 713–716.
[121] S. Field, *Hydrocarbon Process.* 1984, Oct., 34 G–N.

[122] W. Stein, *Fette, Seifen, Anstrichm.* **84** (1982) no. 2, 45–54.
[123] *Richtlinie des Rates 79/831, EWG,* Anhang 6; Directive 79/831, EEC, Appendix 6.
[124] NIOSH, *Registry of Toxic Effects of Chemical Substances,* Publ. US 79–100, Government Printing Office, Washington, D.C., 1978.
[125] W. Kästner, *J. Soc. Cosmet. Chem.* **28** (1977) 741–754.
[126] I. Pevny, M. Uhlich, *Hautarzt* **26** (1975) 252–254.
[127] B. Swoboda, M. Ludvan, *Z. Hautkr.* **53** (1978) 485–488.
[128] D. Lämmer, *Z. Hautkr.* **54** (1979) 571–579.
[129] M. Potokar, C. Gloxhuber in: *Fettalkohole, Rohstoffe, Verfahren, Verwendung,* 2nd ed., Henkel, Düsseldorf 1982, pp. 173–178.
[130] *Kirk-Othmer,* **1,** 727 ff.
[131] *Study on Alfol Alcohols,* Scientific Associates Inc., S. Lindbergh, St. Louis, 1965
[132] F. Leuschner: *Studie zu Alfol-Alkoholen,* Laboratorium für Pharmakologie und Toxikologie, Hamburg 1975.
[133] R. A. Scala, E. G. Burtis, *Am. Ind. Hyg. Assoc. J.* **34** (1973) 493–499.
[134] W. F. von Oettingen, *US Public Health Bull.* **281** (1943).
[135] H. P. Rang, *Br. J. Pharmacol.* **15** (1960) 185.
[136] H. W. Gerarde, D. B. Ahlstrom, *Arch. Environ. Health* **13** (1966) 457–461.
[137] H. F. Smyth, C. P. Carpenter, C. S. Weil, U. C. Pozzani, *Arch. Ind. Hyg. Occup. Med.* **10** (1954) 61 ff.
[138] J. C. Gage, *Br. J. Ind. Med.* **27** (1970) 1–18.
[139] N. Nelson, *Med. Bull.* **11** (1951) 1.
[140] F. A. Patty: *Industrial Hygiene and Toxicology,* vol. **II,** Interscience, New York 1965.
[141] L. B. Aust, H. I. Maibach: *Contact Dermatitis* **6** (1980) 269–271.
[142] E. Bar, F. Griepentrog: *Med. Ernähr.* **8** (1967) 244.
[143] G. N. Zaeva, V. J. Fedorova, *Toksikol. Nov. Prom. Khim. Veshchestv* **5** (1963) 51.
[144] D. L. Opdyke, *Food Cosmet. Toxicol.* **13** (1975) 695.
[145] H. J. Smyth, C. P. Carpenter, C. S. Weil, *AMA, Arch. Ind. Hyg. Occup. Med.* **4** (1951) 119.
[146] W. L. Epstein, *Report to RIFM,* August 1974 (cited in [144]).
[147] A. M. Kligman, *J. Invest. Dermatol.* **47** (1966) 393.
[148] G. D. Clayton, F. E. Clayton (eds.): *Patty's Industrial Hygiene and Toxicology,* 3rd ed., vol. **2A,** Wiley-Interscience, New York 1981, p. 4622.
[149] H. J. Smyth, C. P. Carpenter, C. S. Weil, *J. Ind. Hyg. Toxicol.* **31** (1949) 60.
[150] Y. L. Egorov, L. A. Andrianov, *Uch. Zap. Mosk. Nauchno Issled. Inst. Gig. im. F. F. Erismana* **9** (1967) 47.
[151] Y. L. Egorov, *Toksikol. Gig. Prod. Neftekhim. Neftekhim. Proizvod. Vses. Konf.* [Dokl.] 2nd (1972) 98.

Fluorine Compounds, Organic

GÜNTER SIEGEMUND, Hoechst Aktiengesellschaft, Frankfurt, Federal Republic of Germany (Chaps. 1, 2, 10, 11, and 12)

WERNER SCHWERTFEGER, Hoechst Aktiengesellschaft, Frankfurt, Federal Republic of Germany (Chaps. 3 and 4)

ANDREW FEIRING, E. I. Du Pont de Nemours & Co., Wilmington, Delaware, United States (Chaps. 5–7)

BRUCE SMART, E. I. Du Pont de Nemours & Co., Wilmington, Delaware, United States (Chaps. 5–7)

FRED BEHR, Minnesota Mining and Manufacturing Company, St. Paul, Minnesota, United States (Chap. 8)

HERWARD VOGEL, Minnesota Mining and Manufacturing Company, St. Paul, Minnesota, United States (Chap. 9)

BLAINE MCKUSICK, E. I. Du Pont de Nemours & Co., Wilmington, Delaware, United States (Chap. 13)

1.	Introduction	2566	4.6.	3,3,3-Trifluoro-2-(trifluoromethyl)-prop-1-ene	2589
2.	Production Processes	2568	4.7.	Chlorofluoroolefins	2589
2.1.	Substitution of Hydrogen	2569	5.	Fluorinated Alcohols	2590
2.2.	Halogen–Fluorine Exchange	2570	6.	Fluorinated Ethers	2591
2.3.	Synthesis from Fluorinated Synthons	2571	6.1.	Perfluoroethers	2591
2.4.	Addition of Hydrogen Fluoride to Unsaturated Bonds	2571	6.1.1.	Low Molecular Mass Perfluoroethers	2592
2.5.	Miscellaneous Methods	2572	6.1.2.	Perfluorinated Epoxides	2592
2.6.	Purification and Analysis	2572	6.1.3.	High Molecular Mass Perfluoroethers	2593
3.	Fluorinated Alkanes	2572	6.2.	Perfluorovinyl Ethers	2594
3.1.	Fluoroalkanes and Perfluoroalkanes	2573	6.3.	Partially Fluorinated Ethers	2595
3.2.	Chlorofluoroalkanes	2575	7.	Fluorinated Ketones and Aldehydes	2596
3.3.	Bromofluoroalkanes	2580	7.1.	Fluoro- and Chlorofluoroacetones	2596
3.4.	Iodofluoroalkanes	2581	7.2.	Perhaloacetaldehydes	2598
4.	Fluorinated Olefins	2583	7.3.	Fluorinated 1,3-Diketones	2599
4.1.	Tetrafluoroethylene	2584	8.	Fluorinated Carboxylic Acids and Fluorinated Alkanesulfonic Acids	2600
4.2.	Hexafluoropropene	2586			
4.3.	1,1-Difluoroethylene	2587			
4.4.	Monofluoroethylene	2587	8.1.	Fluorinated Carboxylic Acids	2600
4.5.	3,3,3-Trifluoropropene	2588	8.1.1.	Fluorinated Acetic Acids	2600

8.1.2.	Long-Chain Perfluorocarboxylic Acids 2601	11.1.1.	Properties 2613
8.1.3.	Fluorinated Dicarboxylic Acids . . 2604	11.1.2.	Production 2614
8.1.4.	Tetrafluoroethylene – Perfluorovinyl Ether Copolymers with Carboxylic Acid Groups 2604	11.1.3.	Uses 2616
		11.2.	**Highly Fluorinated Aromatic Compounds** 2617
8.2.	**Fluorinated Alkanesulfonic Acids.** 2605	11.3.	**Perhaloaromatic Compounds** . 2618
8.2.1.	Perfluoroalkanesulfonic Acids . . . 2605	11.4.	**Fluorinated Heterocyclic and Polycyclic Compounds** 2619
8.2.2.	Fluorinated Alkanedisulfonic Acids 2606	11.4.1.	Ring-Fluorinated Pyridines 2619
		11.4.2.	Trifluoromethylpyridines 2620
8.2.3.	Tetrafluoroethylene – Perfluorovinyl Ether Copolymers with Sulfonic Acid Groups 2607	11.4.3.	Fluoropyrimidines 2620
		11.4.4.	Fluorotriazines 2621
		11.4.5.	Polycyclic Fluoroaromatic Compounds 2621
9.	**Fluorinated Tertiary Amines** . . 2607	12.	**Economic Aspects** 2621
10.	**Aromatic Compounds with Fluorinated Side-Chains** 2609	13.	**Toxicology and Occupational Health** 2622
10.1.	**Properties** 2609	13.1.	**Fluorinated Alkanes** 2623
10.2.	**Production** 2610	13.2.	**Fluorinated Olefins** 2624
10.3.	**Uses** 2611	13.3.	**Fluorinated Alcohols** 2624
11.	**Ring-Fluorinated Aromatic, Heterocyclic, and Polycyclic Compounds** 2612	13.4.	**Fluorinated Ketones** 2625
		13.5.	**Fluorinated Carboxylic Acids** . 2625
11.1.	**Mono- and Difluoroaromatic Compounds** 2613	13.6.	**Other Classes** 2625
		14.	**References** 2626

1. Introduction

Organic fluorine compounds are characterized by their carbon–fluorine bond. Fluorine can replace any hydrogen atom in linear or cyclic organic molecules because the difference between the van der Waals radii for hydrogen (0.12 nm) and fluorine (0.14 nm) is small compared to that of other elements (e.g., chlorine 0.18 nm). Thus, as in hydrocarbon chemistry, organic fluorine chemistry deals with a great variety of species. When all valences of a carbon chain are satisfied by fluorine, the zig-zag-shaped carbon skeleton is twisted out of its plane in the form of a helix. This situation allows the electronegative fluorine substituents to envelop the carbon skeleton completely and shield it from chemical (especially nucleophilic) attack. Several other properties of the carbon–fluorine bond contribute to the fact that highly fluorinated alkanes are the most stable organic compounds. These include low polarizability and high bond energies, which increase with increasing substitution by fluorine (bond energies: C–F bond in CH_3F, 448 kJ/mol; C–H bond in CH_4, 417 kJ/mol; C–Cl bond in CH_3Cl, 326 kJ/mol; and C–F bond in CF_4, 486 kJ/mol).

The cumulative negative inductive effect of the fluorine in perfluoroalkyl groups may reverse the polarity of adjacent single bonds (e.g., in the pair $H_3C \triangleleft I$ and $F_3C \triangleright I$) or double bonds (e.g., $CH_3C^{\delta+}H=C^{\delta-}H_2$ and $CF_3-C^{\delta-}H=C^{\delta+}H_2$). Fluorine substitution changes the reactivity of olefins and carbonyl compounds. Polyfluorinated olefins possess an electron-deficient double bond, which reacts preferentially with nucleophiles. Carboxy groups are affected by the presence of an adjacent perfluoroalkyl radical. In carboxylic acids, the acidity is markedly increased. In other carbonyl compounds, the reactivity is increased without any fundamental change in the chemistry of the compound. Correspondingly, the basicity of amines is reduced by the introduction of fluorine. Fluorine attached to the ring of aromatic compounds acts mainly as a para-directing substituent, whereas perfluoroalkyl groups behave as meta-directing substituents.

Naturally, the influence of fluorine is greatest in highly fluorinated and perfluorinated compounds. The fact that these compounds have a high thermal stability and chemical resistance and are physiologically inert makes them suitable for many applications for which hydrocarbons are not. Properties that are exploited commercially include high thermal and chemical stability, low surface tension, and good dielectric properties, for example, in fluoropolymers, perfluorinated oils and inert fluids.

Individual fluorine atoms or perfluoroalkyl groups do not change the technical properties of a hydrocarbon fundamentally. However, this is not the case with physiological properties. A fluorine atom in a bioactive material may simulate a hydrogen atom, and although this does not prevent metabolic processes from occurring, the end products may be ineffective or toxic. Accordingly, such fluorine compounds are important in, for example, pesticides and pharmaceuticals.

A bibliography of the scientific literature of organofluorine chemistry was published in 1986 [16]; commercial applications of fluorine products are reviewed in [7] and [17].

Nomenclature. Any organic fluorine compound can be named according to the rules of the International Union for Pure and Applied Chemistry (IUPAC) [18]. However, for highly fluorinated molecules with several carbon atoms, this nomenclature can be confusing. Therefore, the term "perfluoro" may be used when all hydrogen atoms bonded to the carbon skeleton have been replaced by fluorine. The designation of hydrogen atoms belonging to functional groups (e.g., CHO or COOH), of the functional groups themselves, and of other substituents is not affected [18]. Examples are given in Table 1.

In the case of highly fluorinated compounds with few hydrogen atoms (1–4), the perfluoro compound can be taken as the parent compound. The hydrogen atoms are

Table 1. Nomenclature of organic fluorine compounds

Formula	CAS registry no.	IUPAC designation	Perfluoro designation
CF_3CF_3	[76-16-4]	Hexafluoroethane	Perfluoroethane
$CF_3CF_2CF_2CHO$	[375-02-0]	Heptafluoro-n-butyraldehyde	Perfluoro-n-butyraldehyde
$CF_3(CF_2)_6COOH$	[335-67-1]	Pentadecafluoro-n-octanoic acid	Perfluoro-n-octanoic acid
$CF_3(CF_2)_2CHF_2$	[375-17-7]	1,1,1,2,2,3,3,4,4-Nonafluoro-n-butane	1H-Perfluoro-n-butane
$CF_3(CF_2)_4CH_2OH$	[423-46-1]	2,2,3,3,4,4,5,5,6,6,6-Undecafluoro-n-hexanol	1H,1H-Perfluoro-n-hexanol, 1,1-dihydroperfluoro-n-hexanol

named according to their number and position; the letter H or the prefix hydro are used for hydrogen.

Historical Development. The pioneering work in organofluorine chemistry dates from 1835 to 1940 [19]. Controlled production of organic fluorine compounds was started in 1892 by exchanging halogen for fluorine in hydrocarbons, using antimony(III) fluoride. The industrial phase began in 1929 in the United States with the discovery of the nonflammable, nontoxic refrigerants CCl_3F and CCl_2F_2 [20]. In Germany, commercial production of aromatic fluorine compounds started in 1930.

The first fluoropolymer, polychlorotrifluoroethylene, was synthesized in 1934 in Germany, followed by the discovery of polytetrafluoroethylene in 1938 in the United States. During World War II, thermally and chemically stable working materials for the separation of uranium isotopes were investigated by the United States Manhattan Project [21]. After World War II, numerous novel applications were discovered. The development of new organic fluorine compounds with novel applications continues undiminished.

2. Production Processes

The four principal methods for the preparation of organic fluorine compounds are as follows [1], [2], [22], [23]:

1) substitution of hydrogen in hydrocarbons using fluorine, high-valency metal or nonmetal fluorides, or electrochemical fluorination
2) halogen – fluorine exchange with hydrogen fluoride or metal fluorides
3) synthesis of higher molecular mass fluorine compounds from reactive fluorinated synthons
4) addition of fluorine, hydrogen fluoride, or reactive nonmetal fluorides to unsaturated bonds

Only a few of the many possibilities in each group have been developed commercially, with varying degrees of success.

2.1. Substitution of Hydrogen

Fluorination with Elemental Fluorine [24], [25]. The action of elemental fluorine on organic compounds normally leads to violent, mainly explosive, reactions. The substrate fragments into units with a varying degree of fluorination because the heats of formation of the C–F bond (ca. 460 kJ/mol) and the H–F bond (566 kJ/mol) are greater than the heat of formation of the C–C bond (ca. 348 kJ/mol).

Therefore, direct fluorinations must take place with strict control of the reaction and removal of the heat generated. This may be achieved by dilution of the fluorine with inert gases (e.g., N_2 or CO_2), dilution of the organic substrates with inert solvents [26], intensive mixing, and reduction of the temperature to as low as −150 °C.

Direct fluorination can also be carried out in the gas phase in a tubular reactor packed with silver- or gold-plated copper turnings [27]. Specialized methods are based on LaMar fluorination [24], aerosol fluorination [28], porous-tube fluorination [29], and jet fluorination [30]; high product selectivities are achieved at a laboratory scale. Commercial operation remains to be developed.

Fluorination with Metal Fluorides [31]. Metal fluorides that can transfer fluorine to organic substrates by changing the oxidative state of the metal, such as cobalt(III) fluoride (CoF_3) and silver(II) fluoride (AgF_2), serve as fluorinating agents in an oxidizing fluorination. The spent metal fluoride is regenerated with elemental fluorine.

$$-\overset{|}{\underset{|}{C}}-H + 2\,CoF_3 \longrightarrow -\overset{|}{\underset{|}{C}}-F + HF + 2\,CoF_2$$

$$2\,CoF_2 + F_2 \longrightarrow 2\,CoF_3$$

Fluorination and regeneration can be cyclical, permitting a commercial operation.

Electrochemical Fluorination. The *Simons process* [32] is used commercially for the production of perfluorinated compounds. Solutions of organic compounds (mainly carboxylic acids, sulfonic acids, and tertiary amines) are electrolyzed in anhydrous hydrogen fluoride in a single cell without intermediate formation of free fluorine. Fluorination takes place at a nickel anode by a free-radical mechanism at current densities of $10-20$ mA/cm^2 [33]. Selectivity decreases sharply as the number of carbons increases.

Volatile, hydrogen-containing compounds (hydrocarbons and chlorohydrocarbons) can be electrofluorinated on porous graphite anodes in KF-containing hydrogen fluoride, using a process recently developed by Phillips Petroleum [34], [35]. The organic compound is introduced into the cell through the anode. In the pores of the anode, i.e., at the phase boundary, partial or complete exchange of the hydrogen, but not the chlorine, takes place. To date this process has been used only on a small scale.

2.2. Halogen–Fluorine Exchange

Exchange of chlorine with hydrogen fluoride is used in many commercial processes both for the production of chlorofluoroalkanes and for the side-chain fluorination of aromatic and N-heterocyclic compounds [36]. The method involves exchange of chlorine for fluorine in polyhalogenated compounds using hydrogen fluoride [37]:

$$-\overset{|}{\underset{|}{C}}-Cl + HF \longrightarrow -\overset{|}{\underset{|}{C}}-F + HCl$$

Whether the process takes place with or without a catalyst depends on the reactivity of the chlorine atoms to be exchanged. With compounds containing several chlorine atoms of differing reactivity, selective fluorination can be achieved by selecting suitable process conditions.

Fluorinations without a catalyst such as

$$CCl_3CH_3 \longrightarrow CF_3CH_3 \text{ or } C_6H_5-CCl_3 \longrightarrow C_6H_5-CF_3$$

are carried out in liquid, anhydrous hydrogen fluoride at 100–150 °C in pressure vessels made of steel, alloy steels, or nickel. The process can be carried out by batch (e.g., autoclave) or continuous methods (autoclaves in series or a tubular reactor). In either case, the hydrogen chloride that is generated during fluorination is removed from the reactor to maintain the desired pressure.

Most liquid-phase fluorinations, e.g., of CCl_4, $CHCl_3$, CCl_3CCl_3, CCl_3CHCl_2, or CCl_3CH_2Cl, are carried out in the presence of a catalyst [38] to promote the exchange, which becomes increasingly difficult as fluorination progresses. The main catalysts used are antimony(III) and antimony(V) halides with low volatility. Addition of chlorine oxidizes the antimony to the pentavalent state.

In addition to the liquid-phase processes, many commercially important gas-phase fluorinations employ hydrogen fluoride [38]. The components in the gas phase are passed through a tubular reactor containing the catalyst. The composition of the product can be controlled within wide limits by varying temperature, pressure, residence time, catalyst, and the proportions of the reactants. Various metal fluorides are suitable catalysts, e.g., aluminum fluoride [39] or basic chromium fluoride [40].

For further processing of the mixture produced by gas- or liquid-phase fluorination, the following criteria should be satisfied [41]:

1) The hydrogen chloride generated should be separated in a pure form to permit further use.
2) Unreacted hydrogen fluoride should be recovered.
3) Acid residues, water, and other impurities must be removed from the product.

Usually, the hydrogen chloride is separated from the crude fluorination mixture by fractional distillation. The bulk of the hydrogen fluoride may then be separated from the residue. Further treatment includes washing to remove traces of acid, drying, and fractional distillation.

Exchange of Chlorine with Nonoxidizing Metal Fluorides [36]. Alkali fluorides, especially potassium fluoride, are often used to exchange chlorine in carboxylic acid chlorides, sulfonic acid chlorides, α-chlorocarboxylic acid derivatives (esters, amides, and nitriles), aliphatic monochloro compounds, or activated aromatic chloro compounds (Halex process) [42].

The dry, finely powdered metal fluoride is employed in a solvent-free process at 400–600 °C, e.g., with polychlorinated aromatic compounds, or, in most other cases, in the presence of a solvent. For slow reactions, polar, aprotic solvents are used.

2.3. Synthesis from Fluorinated Synthons

The variety of organic fluorine compounds can be greatly increased by the use of easily accessible, low molecular mass fluoroalkanes and olefins to synthesize higher molecular mass products. Halofluoromethanes add to halogenated ethylenes to form halogenated fluoropropanes [43]. An industrially applied reaction is the addition of iodopentafluoroethane to tetrafluoroethylene yielding a homologous series of long-chain 1-iodoperfluoroalkanes (see Section 3.4). The pyrolysis of chlorodifluoromethane is the industrial source of tetrafluoroethylene, hexafluoropropene, and the corresponding oligomers and polymers (see Sections 4.1 and 4.2). These examples illustrate the importance of this synthetic method, especially for the production of organic fluorine compounds containing more than two carbon atoms where the above-mentioned fluorination methods fail to give high yields of the desired products.

2.4. Addition of Hydrogen Fluoride to Unsaturated Bonds

Addition of hydrogen fluoride to alkenes and alkynes takes place below 0 °C with formation of mono- or difluoroalkanes; ethylene and acetylene are exceptions [22]. Ethyl fluoride is produced from ethylene and hydrogen fluoride at 90 °C; catalytic processes have been developed for the addition of hydrogen fluoride to acetylene to produce vinyl fluoride or 1,1-difluoroethane. Unsymmetrical olefins obey Markovnikov's rule. Chloroolefins can undergo chlorine–fluorine exchange after addition of hydrogen fluoride.

2.5. Miscellaneous Methods

Substitution of Amino Groups in Aromatic Compounds [44]. Introduction of one or two fluorine atoms into aromatic rings is carried out commercially by diazotization of aromatic amines in anhydrous hydrogen fluoride with solid sodium nitrite and decomposition of the dissolved diazonium salt (see Section 11.1.2).

Fluorination with Nonmetal Fluorides. Reactions of nonmetal fluorides with certain substrates are predominantly restricted to laboratory operations. Sulfur tetrafluoride and the following compounds can be used for the controlled introduction of fluorine into organic compounds: dialkylaminosulfur(IV) fluorides (R_2NSF_3) [45]–[47], fluoroalkylamines (e.g., 2-chloro-1,1,2-trifluoroethyldiethylamine or 1,1,2,3,3,3-hexafluoropropyldiethylamine), tetra-n-butylammonium fluoride, nitrosyl fluoride, perchloryl fluoride, fluoroxyfluoroalkanes (e.g., CF_3OF) [48], xenon difluoride [49], or CH_3COOF [50]. They are of commercial value for the fluorination of complex organic compounds such as pharmaceuticals.

2.6. Purification and Analysis

Impurities are usually removed from organic fluorine compounds by fractional distillation, fractional crystallization, or chromatographic methods. This does not apply to fluoropolymers and high-boiling perfluorinated oils, which require special measures, i.e., the use of extremely pure starting materials.

Quantitative determination of fluorine is possible in most cases by combustion and subsequent analysis of the hydrogen fluoride generated. Wet chemical methods are used to determine fluoride ions [51].

Because of the high volatility of organic fluorine compounds, purity can be readily determined by gas chromatography. ^{19}F-Nuclear magnetic resonance spectroscopy is a valuable tool for determining the structure of organic fluorine compounds. Structure determinations, even of mixtures, are often easier using this method than with ^{1}H-NMR spectra due to the larger chemical shifts of ^{19}F-NMR spectra. The ^{19}F signals can be integrated and used for quantitative analysis [52].

3. Fluorinated Alkanes

The hydrogen atoms of alkanes may be partially or totally replaced by fluorine. Partially fluorinated alkanes are fluorohydrocarbons; fully fluorinated alkanes (perfluoroalkanes) are fluorocarbons. In chlorofluorocarbons (CFCs) and chlorofluorohydrocarbons, the alkane hydrogens are replaced by both chlorine and fluorine.

A special nomenclature has been introduced to identify smaller chain length fluoroalkanes (up to four carbon atoms) used in refrigerants. It consists of a three-digit number combined with various letters. The first figure of the three-digit number indicates the number of carbon atoms minus one (for methane derivatives, the figure 0 is omitted); the second figure indicates the number of the hydrogen atoms plus one; and the third figure indicates the number of fluorine atoms. All other bonds are saturated with chlorine. The letter R before the code number is an abbreviation for refrigerant; the letter C indicates a cyclic compound. The complete number is called the refrigerant number. The American Society of Heating, Refrigerating, and Air Conditioning Engineers (Atlanta, Georgia) ASHRAE Standard 34–78 describes the method of coding. The abbreviation F is sometimes used and stands for fluorohydrocarbon.

This system does not allow isomerisms to be expressed for ethane derivatives and higher homologues; for such cases, a letter (a, b, ... etc.) is added to isomers as their asymmetry increases, e.g.,

$CF_2Cl–CCl_2F$ = R 113
$CF_3–CCl_3$ = R 113 a

The compound with the highest degree of symmetry is not given a letter.

3.1. Fluoroalkanes and Perfluoroalkanes

Properties. *Monofluoroalkanes* are attacked by bases and sometimes by heat; however, chemical resistance increases with increasing fluorine substitution, especially multiple substitution at the same carbon atom.

Perfluoroalkanes have distinct properties [53], [54]. Their physical properties differ from those of the corresponding hydrocarbons: densities and viscosities are higher, whereas surface tensions, refractive indices, and dielectric constants are lower. At room temperature perfluoroalkanes are attacked only by sodium in liquid ammonia. At 400–500 °C, they are degraded by alkali metals and silicon dioxide; the former produce a metal fluoride and carbon, while the latter produces silicon tetrafluoride and carbon dioxide. Thermal decomposition starts above 800 °C (compounds with tertiary carbon atoms above 600 °C) with the formation of saturated and unsaturated decomposition products and some carbon. In addition to their chemical and physical stability, fluoroalkanes and perfluoroalkanes are characterized by low flammability (or non-flammability) and physiological inertness.

Whereas partially fluorinated alkanes dissolve in many common solvents, perfluoroalkanes have low solubility, which decreases with their chain length. Only ethers, ketones, esters, chlorohydrocarbons, and chlorofluorocarbons have the power to dissolve perfluoroalkanes [53]. For other physical properties, see Table 2.

Table 2. Boiling points, melting points, and densities of fluoroalkanes and perfluoroalkanes

Compound	CAS registry no.	Formula	M_r	Refrigerant no.	bp, °C	mp, °C	d_4^θ, g/cm³ (θ, °C)
Fluoromethane	[593-53-3]	CH_3F	34.03	R 41	−78.5	−141.8	0.8428 (−60)
Difluoromethane	[75-10-5]	CH_2F_2	52.03	R 32	−51.7	−136	1.100 (20)
Trifluoromethane	[75-46-7]	CHF_3	70.02	R 23	−82.1	−160	1.246 (−34)
Tetrafluoromethane	[75-73-0]	CF_4	88.01	R 14	−128	−83.6	1.33 (−80)
Fluoroethane	[353-36-6]	CH_3CH_2F	48.06	R 161	−37.1	−143.2	0.8176 (−37)
1,1-Difluoroethane	[75-37-6]	CH_3CHF_2	66.05	R 152	−24.7	−117	0.966 (19)
1,1,1-Trifluoroethane	[420-46-2]	CH_3CF_3	84.04	R 143	−47.6	−111	0.942 (30)
Pentafluoroethane	[354-33-6]	CF_3CHF_2	120.03	R 125	−48.5	−103	
Hexafluoroethane	[76-16-4]	CF_3CF_3	138.02	R 116	−78.1	−100.6	1.607 (−78)
Octafluoropropane	[76-19-7]	$CF_3CF_2CF_3$	188.03	R 218	−36.7	−183	1.350 (20)
Octafluorocyclobutane	[115-25-3]	F_2C-CF_2 $\|\ \ \ \ \|$ F_2C-CF_2	200.04	RC 318	−6.06	−40.7	1.5241 (20)

Production. Mono- and difluoroalkanes can be produced by addition of hydrogen fluoride to olefins or alkynes. Another synthetic pathway is the exchange of chlorine (or bromine) for fluorine using hydrogen fluoride or metal fluorides such as antimony fluoride. Monohydroperfluoroalkanes can be obtained by adding hydrogen fluoride to perfluoroalkenes or by decarboxylation of perfluorocarboxylates in the presence of proton donors. The commercial production of trifluoromethane (CHF_3) is carried out by reaction of chloroform with hydrogen fluoride over a chromium catalyst [40].

Perfluoroalkanes can be produced by a variety of routes. Indirect fluorination of hydrocarbons with cobalt(III) fluoride or silver(II) fluoride is carried out in a steel or nickel tube with stirring. The hydrocarbon vapors are passed at 150–450 °C over the fluorinating agent, which is regenerated in a fluorine stream [55], [56]. This process is suitable for the production of perfluoroalkanes containing up to 20 carbon atoms. It affords better process control and yields than the direct gas-phase fluorination using dilute fluorine and a metal catalyst, especially for longer-chain compounds [57].

Fluoroalkanes and perfluoroalkanes can also be produced electrochemically by the Phillips Petroleum process or the electrochemical fluorination of alcohols, amines, carboxylic acids, and nitriles by the Simons process (see Section 2.1).

Tetrafluoromethane (carbon tetrafluoride, CF_4) can be produced by reaction of CCl_2F_2 or CCl_3F and hydrogen fluoride in the gas phase [58]. Octafluorocyclobutane is obtained by dimerization of tetrafluoroethylene [59] or by passing 1,2-dichloro-1,1,2,2-tetrafluoroethane, $CClF_2CClF_2$, over a nickel catalyst at 590 °C [60].

Uses. Trifluoromethane (R 23) and tetrafluoromethane (R 14) are used commercially as low-temperature refrigerants; the less important hexafluoroethane (R 116) and octafluoropropane (R 218) serve as gaseous dielectrics. Octafluorocyclobutane (RC 318) exhibits unusually high stability and low toxicity and is used as an aerosol propellant and as a refrigerant. 1,1,1-Trifluoroethane (R 143a) can be used for the production of vinylidene fluoride. Perfluoroalkanes are used as refrigerants for cooling by evaporation and as heat-transfer media, lubricants, and hydraulic fluids. Per- and polyfluoroalkanes are also used in the electronics industry as etching agents (plasma etching) in the production of electronic components [61], as insulators, and as heat-transfer media in vapor-phase soldering [62].

Perfluoroalkanes, e.g., perfluorodecalin [*306-94-5*], are used in the production of blood substitutes [63].

Trade Names. Freon C-51-12; Perfluorokerosene FCX-329, FCX-330; Perfluorolube oil FCX-512, FCX-412 (Du Pont); Flutec PP-1, PP-2, PP-3, PP-9 (Imperial Smelting).

3.2. Chlorofluoroalkanes

Chlorofluoroalkanes are commercially the most important organic fluorine compounds. They are mostly used directly, but are also important as intermediates in the production of fluorine chemicals, especially fluoroplastics. The following five products are by far the most important commercial organic fluorine compounds:

trichlorofluoromethane (R 11)
dichlorodifluoromethane (R 12)
chlorodifluoromethane (R 22)
1,2,2-trichloro-1,1,2-trifluoroethane (R 113)
1,2-dichloro-1,1,2,2-tetrafluoroethane (R 114)

The atmospheric changes caused by fluorochloroalkanes with respect to the ozone balance and the greenhouse effect are under investigation [64], [65]. As a result of this possibility, voluntary reductions in their applications have been made; some countries (Canada, the United States, and Sweden) prohibit the use of certain chlorofluoroalkanes as aerosol propellants, e.g., R 11 and R 12. Further restrictions may, in the near future, lead to major changes in the production of chlorofluoroalkanes.

Properties. Chlorofluoroalkanes are characterized by high chemical and thermal stabilities, which increase with their fluorine content. Low flammability (or nonflammability) and low toxicity are additional commercial advantages. Most of these compounds have a pleasant, weak odor; some are mild anesthetics [66].

Boiling points, freezing points, and densities of commercially important fluoroalkanes are shown in Table 3; other physical properties are listed in Table 4.

Production. Commercial production of chlorofluoroalkanes employs halogen exchange, with hydrogen fluoride in the liquid phase in the presence of a catalyst. The production scheme for dichlorodifluoromethane shown in Figure 1 is typical [68]:

$$CCl_4 + 2\ HF \longrightarrow CCl_2F_2 + 2\ HCl$$

A steam-heated steel autoclave (a) lined with stainless steel (V2A) serves as the reactor. The seals are made from aluminum or copper. The autoclave (capacity 2–5 m^3) is filled with 500 kg of hydrogen fluoride, 1540 kg of carbon tetrachloride, 220 kg of antimony(III) chloride, and 20 kg of chlorine, and the mixture is heated to 100 °C. After ca. 2 h and an increase in pressure to ca. 3 MPa, the fluorination

Table 3. Boiling points, melting points, and densities of chlorofluoroalkanes

Compound	CAS registry no.	Formula	M_r	Refrigerant number	bp, °C	mp, °C	d_4^θ, g/cm^3 (θ, °C)
Trichlorofluoromethane	[75-69-4]	CCl$_3$F	137.38	R 11	23.7	−111	1.490 (20)
Dichlorodifluoromethane	[75-71-8]	CCl$_2$F$_2$	120.93	R 12	−29.8	−155	1.328 (20)
Chlorotrifluoromethane	[75-72-9]	CClF$_3$	104.47	R 13	−81.1	−181	0.924 (20)
Dichlorofluoromethane	[75-43-4]	CHCl$_2$F	102.93	R 21	8.9	−135	1.366 (20)
Chlorodifluoromethane	[75-45-6]	CHClF$_2$	86.48	R 22	−40.8	−160	1.213 (20)
Tetrachloro-1,2-difluoroethane	[76-12-0]	CCl$_2$FCCl$_2$F	203.85	R 112	92	27.4	1.634 (30)
Tetrachloro-1,1-difluoroethane	[76-11-9]	CClF$_2$CCl$_3$	203.85	R 112 a	91.5	40.8	1.649 (20)
1,1,2-Trichlorotrifluoroethane	[76-13-1]	CCl$_2$FCClF$_2$	187.39	R 113	47.7	−33	1.582 (20)
1,1,1-Trichlorotrifluoroethane	[354-58-5]	CF$_3$CCl$_3$	187.39	R 113 a	45.9	14	1.579 (20)
1,2-Dichlorotetrafluoroethane	[76-14-2]	CClF$_2$CClF$_2$	170.94	R 114	3.8	−94	1.473 (20)
1,1-Dichlorotetrafluoroethane	[374-07-2]	CF$_3$CCl$_2$F	170.94	R 114 a	−2	−56.6	1.478 (21)
Chloropentafluoroethane	[76-15-3]	CF$_3$CClF$_2$	154.48	R 115	−38	−106	1.291 (25)
1,1,2-Trichloro-2,2-difluoroethane	[354-21-2]	CClF$_2$CHCl$_2$	169.39	R 122	71.9	−140	1.544 (25)
1,1-Dichlor-2,2,2-trifluoroethane	[306-83-2]	CF$_3$CHCl$_2$	152.94	R 123	28.7	−107	1.475 (15)
1-Chloro-1,2,2,2-tetrafluoroethane	[2837-89-0]	CF$_3$CHClF	136.48	R 124	−12		
1,2-Dichloro-1,1-difluoroethane	[1649-08-7]	CClF$_2$CH$_2$Cl	134.94	R 132 b	46.8	−101.2	1.4163 (20)
1-Chloro-2,2,2-trifluoroethane	[75-88-7]	CF$_3$CH$_2$Cl	118.49	R 133 a	6.9	−101	1.389 (0)
1,1-Dichloro-1-fluoroethane	[1717-00-6]	CCl$_2$FCH$_3$	116.95	R 141 b	32	−103.5	1.250 (10)
1-Chloro-1,1-difluoroethane	[75-68-3]	CClF$_2$CH$_3$	100.49	R 142 b	−9.2	−130.8	1.120 (25)

Table 4. Physical properties of chlorofluoroalkanes

Property	CCl_3F	CCl_2F_2	$CClF_3$	$CHClF_2$	CCl_2FCClF_2	$CClF_2CClF_2$
Critical temperature, °C	198.0	112.0	28.8	96.0	214.1	145.7
Critical pressure, MPa	4.40	4.21	3.86	4.94	3.41	3.27
Critical density, g/cm³	0.548	0.558	0.581	0.525	0.576	0.578
Heat of evaporation at bp, J/kg	18 216	16 688	14 850	23 412	14 570	13 942
Specific heat at 101.3 kPa, J kg⁻¹ K⁻¹	871	854	850	1088	946	971
Refractive index, $n_D^{26.5}$	1.384	1.285	1.263	1.252	1.355	1.290
Surface tension, N/cm	19×10^{-5}	9×10^{-5}		9×10^{-5}	19×10^{-5}	13×10^{-5}
Solubility in water, g/100 g						
at 0 °C	0.0036	0.0025	0.0019	0.060	0.0036	0.0026
at 30 °C	0.013	0.0125	0.0065	0.15	0.013	0.011
Dielectric strength at 101.3 kPa, 23 °C, nitrogen = 1	3.1	2.4	1.4	1.3	2.6 (39.2 kPa)	2.8
Dielectric constant						
liquid at 25 °C*	2.5	2.1	2.3 (−30 °C)	6.6	2.6	2.2
vapor (t, °C)	1.0019 (26)	1.0016 (29)	1.0013 (29)	1.0035 (25.4)	1.0024 (27.5)	1.0021 (26.8)
Vapor pressure (kPa) at						
−120 °C			6.96			
−100 °C		1.18	33.14			
−80 °C		6.12	109.8			
−60 °C		22.7	282.5			3.6
−40 °C	5.1	64.3	607			13.0
−20 °C	15.7	151.0	1146	246		37.0
0 °C	40.2	308.7	1969	500	5.08	87.9
20 °C	89.0	566.9	3177	917	14.79	182
40 °C	176	958.5		1549	36.4	340
60 °C	316	1518.7		2459	78.3	583
80 °C	528	2284.7			151.3	935
100 °C	830	3297.5			268.0	1423
120 °C	1242				442.2	2083
140 °C	1785					2964

* Except where otherwise stated.

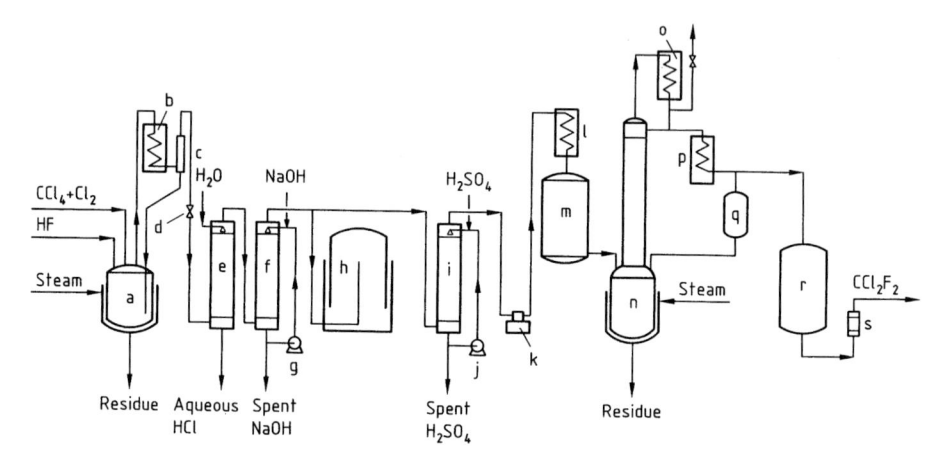

Figure 1. Manufacture of dichlorodifluoromethane (R 12) by fluorination of carbon tetrachloride in the liquid phase
a) Autoclave; b) Reflux condenser; c) Separator; d) Pressure valve; e) Wash column (water); f) Wash column (NaOH); g) Pump for NaOH; h) Gasometer; i) Wash column (H_2SO_4); j) Pump for H_2SO_4; k) Compressor; l) Condenser; m) Receiver for crude product; n) Distillation pot; o) Dephlegmator; p) Condenser; q) Forerun receiver; r) Tank for pure product; s) Caustic filter

products with lower boiling points are removed together with the hydrogen chloride that is generated and some hydrogen fluoride; higher-boiling products in the exit gases are condensed and recycled. The low-boiling fraction is first washed with water in a tower (e) lined with poly(vinyl chloride) and packed with graphite; it is then washed with caustic in a tower (f) filled with porcelain packing. After being washed to neutrality, the product is dried in a tower (i) containing concentrated sulfuric acid, compressed to a liquid, and fed into an intermediate storage tank (m). Each batch takes ca. 24 h to process. The antimony catalyst remains in the reactor and is regenerated before each subsequent batch by adding a small amount of chlorine to convert it to the catalytically active Sb(V) form.

The crude product is fractionally distilled under pressure (0.6–0.8 MPa). The lower-boiling fraction contains some chlorotrifluoromethane and most of the dichlorodifluoromethane (yield 90% based on carbon tetrachloride, 80% based on hydrogen fluoride). The higher-boiling fraction consists of trichlorofluoromethane (5–10% based on carbon tetrachloride), which can be recycled. The distilled product is passed through a caustic filter (s). Steel bottles, pressure vessels, tank cars, and tank trucks are used for transport.

More recently developed exchange processes are carried out continuously in the gas phase at 100–400 °C, using catalysts based on chromium [40], aluminum [39], or iron [69]. Starting materials, which include carbon tetrachloride, chloroform, tetrachloroethylene, and trichloroethylene, are passed over the catalyst with excess hydrogen fluoride and, where necessary, chlorine. Further processing follows the same principles as in the liquid-phase process.

In the Montedison chlorofluorination process, reaction of C_1- and C_2-hydrocarbons with chlorine and hydrogen fluoride takes place in a single step in a fluidized-bed reactor. A suitable catalyst is a combination of aluminum chloride and other metals [70]–[72]:

$$CH_4 + 4\,Cl_2 + 2\,HF \xrightarrow{Cat.} CCl_2F_2 + 6\,HCl$$

Commercial production of chlorofluoroalkanes is also possible by the electrochemical fluorination process developed by Phillips Petroleum (see Section 2.1).

High molecular mass chlorofluoroalkanes are produced by fluorination with chlorine trifluoride [73], [74].

Other processes are based on the oligomerization of chlorotrifluoroethylene to low molecular mass polymers followed by treatment with cobalt(III) fluoride [75], [76].

Specifications. Chlorofluoroalkanes produced on an industrial scale are subject to stringent standards. Impurities must not exceed the following limits (vol %):

acids	0
moisture	< 0.001
higher-boiling fractions	< 0.05
other gases	2

Uses. Currently, chlorofluoroalkanes are used mainly as aerosol propellants and as spraying and foam blowing agents (R 11, R 12, R 114). Further applications are in the area of refrigerants, where R 11, R 12, R 13, R 22, R 113, R 114, R 115, and the chlorine-free compounds R 23 and RC 318 are preferred.

Chlorofluoroalkanes, especially R 11 and R 113, are also employed as solvents and degreasing and cleaning agents for textiles; R 22, R 113, R 132, and R 142 b are important intermediates for the production of fluoroolefins.

Higher molecular mass perchlorofluoroalkanes are used as oils, greases, and waxes, as lubricants, hydraulic fluids, damping oils, heat-transfer media, impregnating agents, and plasticizers. Oligomers of chlorotrifluoroethylene have achieved special importance in this area [77].

Trade Names. Chlorofluoroalkanes are sold worldwide under protected trade names; the refrigerant numbers describing the chemical composition (see Table 3) are included to specify individual compounds. Some of the trade names that apply worldwide are:

Australia
    Australian Fluorine Chemicals   Isceon
Czechoslovakia
    Slovek Pro Chemickov   Ledon
Federal Republic of Germany
    Hoechst   Frigen
    Kali-Chemie   Kaltron
France
    Rhône Poulenc   Flugene
    Ugine Kuhlmann   Forane
German Democratic Republic
    Volkseigener Betrieb Chemiewerk Nunchritz   Fridohna
    Volkseigener Betrieb Fluorwerke Dohna   Frigedohn

Italy	
Montedison	Algofrene
Japan	
Asahi Glass	Asahiflon
Daikin Kogyo	Daiflon
Mitsui Fluorochemicals	Flon
North America	
Allied Chemical	Genetron
Du Pont	Freon
Kaiser Chemicals	Kaiser
Pennwalt	Isotron
Racon	Racon
Union Carbide	Ucon
Soviet Union	
	Khladon
	Eskimon
The Netherlands	
AKZO	FCC
United Kingdom	
ICI	Arcton
Imperial Smelting Corporation	Isceon

Trade names of higher molecular mass chlorofluoroalkanes include Florubes (ICI), Fluorolube oils, Fluorolube greases (Hooker Industrial Chemicals Division), Halocarbon oils, greases, and waxes (Halocarbon Products), Kel-F oils, greases, and waxes (3M), and Voltalef (Ugine Kuhlmann).

3.3. Bromofluoroalkanes

Bromofluoro compounds of practical importance are found in the methane and ethane series.

In the refrigerant (R) numbering system for bromofluoroalkanes, the corresponding chlorofluoroalkanes are taken as the basic structures (see also Section 3); the substitution of chlorine by bromine is expressed by the addition of B1, B2, etc. For example, bromotrifluoromethane is denoted as R 13B1 and 1,2-dibromotetrafluoroethane as R 114B2.

Properties. Fully halogenated compounds with a high fluorine content have excellent thermal stability; they are nonflammable and some (e.g., CF_3Br) are physiologically inert [67]. At high temperature, thermal cleavage of the C—Br bond into radicals occurs, which is responsible for the utility of some of these compounds in extinguishing fires [78]. Their chemical stability is slightly lower than that of the corresponding chlorofluoroalkanes. However, as with the chlorofluoroalkanes, stability increases with the fluorine:bromine ratio. Some compounds have a marked anesthetic effect [66]. Physical properties are listed in Table 5.

Table 5. Boiling points, melting points, and densities of bromofluoroalkanes

Compound	CAS registry no.	Formula	M_r	bp, °C	mp, °C	d_4^θ, g/m^3 (θ,°C)
Tribromofluoromethane	[353-54-8]	CBr$_3$F	270.76	106	−74.5	2.7648 (20)
Dibromodifluoromethane	[75-61-6]	CBr$_2$F$_2$	209.84	24.5	−110	2.3063 (15)
Dibromochlorofluoromethane	[353-55-9]	CBr$_2$ClF	226.30	80		
Bromotrifluoromethane	[75-63-8]	CBrF$_3$	148.93	−57.8	−168	1.58 (21)
Bromochlorodifluoromethane	[353-59-3]	CBrClF$_2$	165.38	−4	−160.5	1.850 (15)
1,2-Dibromotetrafluoroethane	[124-73-2]	CBrF$_2$CBrF$_2$	259.85	47.5	−110.4	2.18 (20)
1-Bromo-2-chloro-1,1,2-trifluoroethane	[354-06-3]	CHClFCBrF$_2$	197.40	51.7		1.864 (20)
2-Bromo-2-chloro-1,1,1-trifluoroethane	[151-67-7]	CF$_3$CHBrCl	197.40	50.2		1.861 (25)

Production. Bromofluoromethanes are obtained by bromination of a stream of the appropriate fluoromethane [79] or chlorofluoromethane [80] at 300–600 °C. Ethane derivatives can also be obtained by thermal bromination [81] or by addition of bromine or hydrogen bromide [82] to fluoroolefins. In some cases hydrogen bromide can be used to exchange a chlorine atom in a chlorofluoroalkane for a bromine atom [83]. Iodine–bromine exchange in a fluoroiodoalkane can be effected with bromine [84].

Uses. The lower-boiling compounds CBrF$_3$ (R 13B1; Halon 1301) and CBrClF$_2$ (R 12B1; Halon 1211) are used as fire extinguishing agents.

Perfluoro-1-bromo-*n*-octane [423-55-2] is physiologically inert and is useful as an X-ray contrast agent, especially for lung examinations [85]. With its low surface tension it penetrates small spaces and evenly wets healthy lung tissue.

2-Bromo-2-chloro-1,1,1-trifluoroethane [151-67-7], also known as halothane, has been used worldwide since 1956 as an effective, nonflammable inhalation anesthetic. It is commonly produced by the ICI process:

$$CF_3CH_2Cl \xrightarrow[-HBr]{Br_2} CF_3CHBrCl$$

or the Hoechst process [38, pp. 208–210]:

$$CF_2=CClF \xrightarrow[h\nu]{+HBr} CBrF_2CHClF \xrightarrow{Cat.} CF_3CHBrCl$$

Trade Names. Fluothane (ICI), Halothane "HOECHST" (Hoechst), Halan (GDR).

3.4. Iodofluoroalkanes

Recently iodofluoroalkanes have become important intermediates in the commercial production of compounds containing a perfluorinated moiety.

Table 6. Boiling points and densities of iodoperfluoroalkanes

Compound	CAS registry no.	Formula	M_r	bp, °C	$d_4 \theta$, g/cm^3 (θ, °C)
Trifluoroiodomethane	[2314-97-8]	CF$_3$I	195.9	−22.5	2.3608 (−32)
Pentafluoroiodoethane	[354-64-3]	CF$_3$CF$_2$I	245.9	12.5	2.0850 (20)
Perfluoro-1-iodopropane	[27636-85-7]	CF$_3$(CF$_2$)$_2$I	295.9	41.2	2.0626 (20)
Perfluoro-2-iodopropane	[677-69-0]	CF$_3$CFICF$_3$	295.9	38	
Perfluoro-1-iodo-n-butane	[423-39-2]	CF$_3$(CF$_2$)$_3$I	345.9	67	
Perfluoro-1-iodo-n-pentane	[638-79-9]	CF$_3$(CF$_2$)$_4$I	395.9	94.4	2.0349 (27.8)
Perfluoro-1-iodo-n-hexane	[355-43-1]	CF$_3$(CF$_2$)$_5$I	445.9	117	
Perfluoro-1-iodo-n-heptane	[335-58-0]	CF$_3$(CF$_2$)$_6$I	495.9	137−138	
Perfluoro-1-iodo-n-decane	[423-62-1]	CF$_3$(CF$_2$)$_9$I	645.9	195−200	
1,2-Diiodotetrafluoroethane	[354-65-4]	CF$_2$ICF$_2$I	353.8	112	2.629 (25)

Properties. In contrast to chloro- and bromofluoroalkanes, iodofluoroalkanes readily undergo chemical reactions [22, Chap. 6], reacting preferentially by homolytic cleavage of the C−I bond.

The radical intermediates $C_nF_{2n+1}\cdot$ and $I\cdot$ can add to double bonds; thus, reaction with ethylene yields 1H,1H,2H,2H-1-iodoperfluoroalkanes, which are commercially important intermediates [86]:

$$C_nF_{2n+1}I + CH_2=CH_2 \longrightarrow C_nF_{2n+1}CH_2CH_2I$$

Control of the reaction between iodofluoroalkanes and fluoroolefins, especially tetrafluoroethylene, can result in oligomerization (telomerization) of the olefin [87]:

$$CF_3I + n\,CF_2=CF_2 \longrightarrow CF_3(CF_2CF_2)_nI$$

This reaction is employed commercially and is initiated by free radicals, UV irradiation, or heat [88].

Iodofluoroalkanes also form organometallic compounds, some of which are useful intermediates, e.g., for Grignard reactions [89].

Iodoperfluoroalkanes cannot be used as alkylating agents and their applications are therefore limited. However, derivatives of the FITS-type reagent are alkylating agents [90]:

$$C_nF_{2n+1}\!-\!\underset{\underset{C_6H_5}{|}}{I}\!-\!OSO_2\!-\!CF_3 \quad \text{FITS reagent (perfluoroalkyl)phenyl-iodonium triflate}$$

Physical constants of some iodofluoroalkanes are shown in Table 6.

Production. Iodofluoroalkanes can be produced by heating the silver salts of the perfluorocarboxylic acids with iodine [91] or the corresponding sodium salts with iodine in dimethylformamide [92]. Of commercial importance is the production of pentafluoroiodoethane (CF$_3$CF$_2$I) by reaction of tetrafluoroethylene with a mixture of iodine pentafluoride and iodine [93]:

$$5\,CF_2=CF_2 + IF_5 + 2\,I_2 \longrightarrow 5\,CF_3CF_2I$$

Heptafluoro-2-iodopropane (CF_3CFICF_3) is obtained similarly from hexafluoropropene.

The higher homologues $C_nF_{2n+1}I$ ($n = 4-12$) are produced commercially by reaction of the lower members with tetrafluoroethylene (telomerization).

α,ω-Diiodoperfluoroalkanes are obtained from tetrafluoroethylene and iodine [94]:

$$I_2 + CF_2=CF_2 \longrightarrow ICF_2CF_2I \xrightarrow{CF_2=CF_2} I(CF_2CF_2)_nI$$

Uses. 1-Iodoperfluoroalkanes and 1H,1H, 2H,2H-1-iodoperfluoroalkanes are intermediates in the production of surfactants and textile finishes. Perfluorocarboxylic acids, especially perfluorooctanoic acid, are obtained from perfluoroiodoalkanes [95], and perfluorinated dicarboxylic acids are obtained from α,ω-diiodoperfluoroalkanes [96].

4. Fluorinated Olefins

The commercial importance of fluoro- and chlorofluoroolefins lies in the production of fluorinated plastics and inert fluids.

Properties. The chemical behavior of fluoroolefins [97] is governed by the number of vinylic fluorine atoms. In contrast to their hydrocarbon analogues, fluoroolefins are attacked by electrophiles only with difficulty [98], which increases with the degree of fluorination. However, fluoroolefins react readily with nucleophiles [99], [100], because as the number of vinylic fluorine atoms increases, the π-electron system of the double bond is destabilized. Thermodynamic calculations have shown that the strength of the $C-C$ π-bond in tetrafluoroethylene

$$CF_2=CF_2 \rightleftharpoons \cdot CF_2 - CF_2 \cdot$$

is only ca. 160 kJ/mol as opposed to 241 kJ/mol in ethylene [101]. In unsymmetrically substituted fluoroolefins, the nucleophile attacks the carbon atom that is made strongly positive by the neighboring fluorine atoms and is shielded only weakly (sp^2 hybridization). The reactivity of fluoroolefins toward nucleophiles increases as follows:

$$CF_2=CF_2 < CF_2=CF-CF_3 < CF_2=C(CF_3)_2$$

Fluoroolefins are slightly to highly toxic and must be handled with care. The toxicity of fluorinated olefins is apparently proportional to their reactivity toward nucleophiles [102]. Perfluoroisobutene, $CF_2=C(CF_3)_2$ [382-21-8], for example, is far more toxic than its lower homologues.

Physical properties of commercial fluoro- and chlorofluoroolefins are given in Table 7.

Table 7. Physical properties of fluoroolefins and chlorofluoroolefins

Property	Tetrafluoro-ethylene [116-14-3]	Hexafluoro-propene [116-15-4]	1,1-Difluoro-ethylene [75-38-7]	Fluoro-ethylene [75-02-5]	Chlorotri-fluoroethylene [359-29-5]	3,3,3-Trifluo-roprop-1-ene [677-21-4]
M_r	100.02	150.02	64.03	46.04	116.47	96.05
bp, °C	−75.6	−29.4	−82	−72.2	−28.36	−27
mp, °C	−142.5	−156.2	−144	−160.5	−158.2	
d_4^θ, g/cm^3 (θ, °C)	1.519 (−76.3)	1.292 (−29.4)	0.617 (23.6)	0.775 (−30)	1.51 (−40)	
Critical temperature, °C	33.3	86.2	30.1	54.7	105.8	107
Critical pressure, MPa	3.82	2.75	4.29	5.43	3.93	4.14
Critical density, g/cm^3	0.58		0.417	0.320	0.55	
Heat of evaporation, J/mol (t, °C)	16 821 (−75.6)	20 100 (−29)	13 189 (−40)	13 494 (−20)	20 893 (−28.4)	

Fluoroolefins also differ from hydrogen-containing olefins in their marked tendency to undergo cycloaddition [103].

Production. Fluoroolefins are produced by dehalogenation of chlorofluoro-, bromofluoro-, or iodofluoroalkanes with zinc and alcohol, by dehydrohalogenation of hydrogen-containing haloalkanes with alcoholic alkali, or by heating. Other common methods include addition of hydrogen halides to alkynes, decarboxylation of fluorocarboxylic acid salts, and pyrolysis of fluorohydrocarbons [104]–[106].

4.1. Tetrafluoroethylene

Properties. Tetrafluoroethylene (TFE), perfluoroethylene, $CF_2=CF_2$, a colorless, odorless gas, is flammable in oxygen, producing tetrafluoromethane and carbon dioxide. For physical properties, see Table 7. At low temperature in the presence of oxygen, explosive peroxides are formed [107], [108]. Tetrafluoroethylene must be handled with great care since, even in the absence of oxygen, it can decompose explosively into carbon and tetrafluoromethane under pressure above −20 °C ($\Delta H = -276$ kJ/mol at 298 K). If the polymerization to polytetrafluoroethylene (PTFE) ($\Delta H = -172$ kJ/mol at 298 K) is uncontrolled, a more strongly exothermic decomposition reaction can occur. Polymerization inhibitors include dipentene [138-86-3], α-pinene [80-56-8], which are added to liquid tetrafluoroethylene during purification and storage (at −30 °C) [109]. In the United States, the transportation of liquid TFE containing stabilizers is permitted.

In the gas phase at ca. 300–500 °C, tetrafluoroethylene dimerizes to perfluorocyclobutane [110]. Above 600 °C, hexafluoropropylene (see Section 4.2) and the highly toxic perfluoroisobutene are formed [110].

Production. Many commercial processes for the production of tetrafluoroethylene are known, e.g., reaction of tetrafluoromethane in an electric arc [111], dechlorination

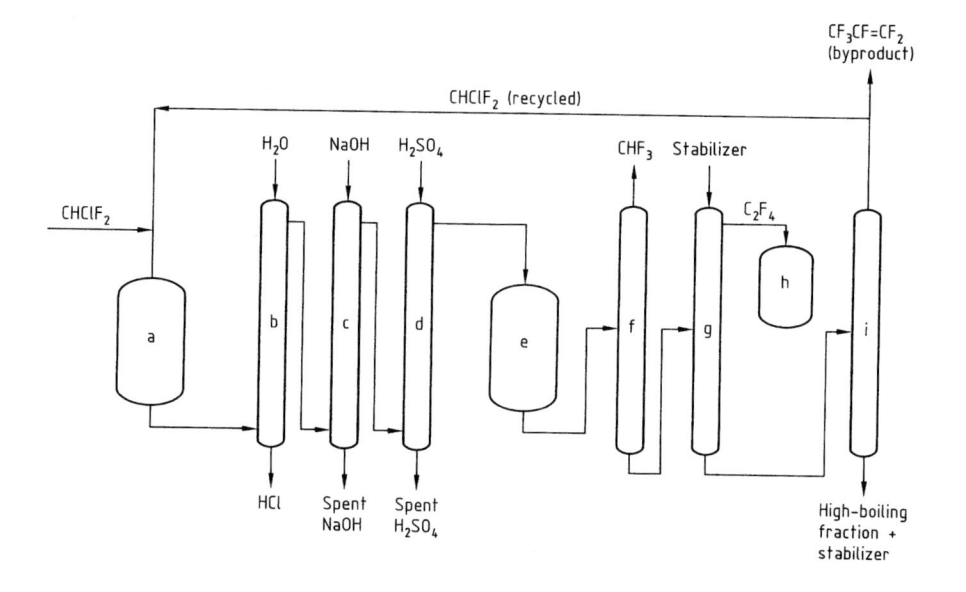

Figure 2. Flow sheet for tetrafluoroethylene production
a) Pyrolysis reactor; b) Quench column (water); c) Wash column (NaOH); d) Drier (conc. H_2SO_4); e) Intermediate storage tank for crude tetrafluoroethylene; f) Fractionation column for low-boiling constituents; g) Product distillation column; h) Tetrafluoroethylene storage tank; i) Fractionation column

of CF_2Cl-CF_2Cl with a metal [112], and thermal decomposition of trifluoroacetic acid [113]:

$$2\,CF_3COOH \longrightarrow CF_2{=}CF_2 + 2\,HF + 2\,CO_2$$

The two principal commercial methods are pyrolysis of trifluoromethane [114]:

$$2\,CHF_3 \xrightarrow{700-800\,°C} CF_2{=}CF_2 + 2\,HF$$

and pyrolysis of chlorodifluoromethane [115]:

$$2\,CHClF_2 \xrightarrow{600-800\,°C} CF_2{=}CF_2 + 2\,HCl$$

In the second method (Fig. 2), the chlorodifluoromethane gas is passed at atmospheric or reduced pressure through a heated platinum, silver, or carbon tubular reactor (a). The 28% conversion obtained under these conditions is low (yield, 83%), but can be increased to ca. 65% with the same yield by adding steam [116]. In modification of this method, chlorodifluoromethane is treated with superheated steam at ca. 700 °C, which results in a conversion rate of 60 – 80% and a selectivity of 84 – 93% [117]. The pyrolysis gas is washed with water (b) to cool it and to remove HCl. After being washed with caustic soda (c) and dried with concentrated sulfuric acid (d), the crude product can be stored in liquid or gaseous form (e). It is a complex mixture from which tetrafluoroethylene is separated by distillation in the presence of dipentene or a similar stabilizer to prevent polymerization.

An overhead fraction containing inert gases and trifluoromethane is obtained in a low-boiling fractionation column (f) before isolating pure TFE (column g).

The higher-boiling fractions are processed (i) to recover unreacted $CHCl_2F$ and to isolate hexafluoropropene, a byproduct. Extractants such as methanol are added because of the formation of azeotropes during distillation [118]. Methanol also reacts with the toxic perfluoroisobutene, $(CF_3)_2C=CF_2$, to give an addition product.

Uses. Currently, tetrafluoroethylene is the most important fluoroolefin; it is used mainly for the production of fluoropolymers. It reacts with perfluoronitrosoalkanes to produce so-called nitroso rubbers [119]. Tetrafluoroethylene is also used in the production of low molecular mass compounds and intermediates, e.g., for the manufacture of iodoperfluoroalkanes (see Section 3.4).

Polytetrafluoroethylene [9002-84-0], (PTFE) is a homopolymer sold under the trade names Algoflon (Montedison), Fluon (ICI), Halon (Allied Chemical), Hostaflon (Hoechst) and Teflon (Du Pont). Copolymers with fluorinated or fluorine-free olefins or vinyl ethers are also commercially available.

4.2. Hexafluoropropene

Properties. Hexafluoropropene (HFP), $CF_3CF=CF_2$, a colorless, odorless gas, is nonflammable in air at room temperature. For physical properties, see Table 7. It exhibits no special tendency toward radical homopolymerization [120], but like other fluoroolefins, it reacts readily with nucleophiles [99], [100].

Liquid hexafluoropropene has unlimited storage life under pressure at room temperature in steel containers, even without stabilizers. In many countries the liquid can be transported in steel cylinders or tank cars. Hexafluoropropene is toxic (LC_{50} 3000 ppm) and can decompose thermally to form highly toxic perfluoroisobutene.

Production. Hexafluoropropene is produced commercially by temperature-controlled pyrolysis of chlorodifluoromethane (cf. production of tetrafluoroethylene, p. 2585) [121]. Hexafluoropropene can also be obtained from tetrafluoroethylene by heating at normal or reduced pressure, preferably in the presence of an inert gas (e.g., CO_2) or water vapor [122]:,

$$3\ CF_2=CF_2 \xrightarrow{700-900\ °C} 2\ CF_3CF=CF_2$$

Uses. An important application of hexafluoropropene is the production of copolymers, e.g., with tetrafluoroethylene or 1,1-difluoroethylene. The versatile epoxide, hexafluoropropylene oxide [428-59-1] (see Section 6.1.2), can be obtained from hexafluoropropene by oxidation.

4.3. 1,1-Difluoroethylene

1,1-Difluoroethylene, vinylidene fluoride (VDF) $CH_2=CF_2$, is a commercially important, partially fluorinated olefin. It is a colorless, flammable gas that undergoes homopolymerization and copolymerization. For physical properties, see Table 7.

Production. Currently, three basic methods are used for the commercial production of vinylidene fluoride:

1) Dechlorination of 1,2-dichloro-1,1-difluoroethane [1649-08-7], R 132b in the gas phase and on a metal catalyst [113], [123]:

$$CH_2ClCClF_2 \xrightarrow[Ni]{500\,°C} CH_2=CF_2 + Cl_2$$

2) Dehydrochlorination of 1-chloro-1,1-difluoroethane [75-68-3], R 142b [115], [124]:

$$CH_3CClF_2 \xrightarrow{700-900\,°C} CH_2=CF_2 + HCl$$

In the presence of steam the temperature can be reduced to 500–650 °C [125].

3) Dehydrofluorination of 1,1,1-trifluoroethane [420-46-2], R 143a [126]:

$$CH_3CF_3 \xrightarrow{1100-1300\,°C} CH_2=CF_2 + HF$$

Vinylidene fluoride is transported as a liquid in steel cylinders without stabilizers.

Uses. 1,1-Difluoroethylene is the starting material for the commercially important homopolymer poly(vinylidene fluoride) [24937-79-9], (PVDF).

Trade Names (PVDF). Kynar (Pennsalt Chemical), Vidar (Süddeutsche Kalkstickstoffwerke). The copolymer with hexafluoropropene [116-15-4] is marketed as Viton (Du Pont) and Fluorel (3M).

4.4. Monofluoroethylene

Monofluoroethylene, vinyl fluoride (VF), $CH_2=CHF$, is a colorless, highly flammable gas; up to 0.2 % of polymerization inhibitors are added for stabilization during transport and storage. For physical properties, see Table 7.

Production. In the past, vinyl fluoride was produced by dehydrofluorination of 1,1-difluoroethane [75-37-6], obtained in two steps by addition of hydrogen fluoride to acetylene [22, p. 59–66], [127]:

$$\text{CH} \equiv \text{CH} + 2\,\text{HF} \longrightarrow \text{CH}_3\text{CHF}_2 \xrightarrow{\text{Cat.}} \text{CH}_2=\text{CHF} + \text{HF}$$

However, with mercury catalysts, vinyl fluoride can be produced directly from acetylene and hydrogen fluoride [22, p. 59–66], [128]:

$$\text{CH} \equiv \text{CH} + \text{HF} \xrightarrow{\text{Hg Cat.}} \text{CH}_2=\text{CHF}$$

The dehydrochlorination of 1-chloro-1-fluoroethane [*1615-75-4*], CHClFCH$_3$, and 1-chloro-2-fluoroethane [*762-50-5*], CH$_2$FCH$_2$Cl, are also utilized commercially [129].

Uses. The main use of monofluoroethylene is in the production of poly(vinyl fluoride) [*24981-14-4*] (PVF).

Trade Names (PVF). Tedlar (Du Pont), Dalvor (Diamond Shamrock, USA).

4.5. 3,3,3-Trifluoropropene

3,3,3-Trifluoropropene, CF$_3$CH=CH$_2$ (TFP), is produced almost exclusively by fluorination and dehalogenation of 1,1,1,3-tetrachloropropane [*1070-78-6*] (TCP), CCl$_3$CH$_2$CH$_2$Cl. With sodium fluoride at 400–475 °C, trifluoropropene is obtained in a single step, involving chlorine–fluorine exchange and dehydrochlorination [130]. Using hydrogen fluoride and oxygen, the reaction can be carried out at 300 °C over a chromium fluoride catalyst [131]. Liquid-phase fluorination of 1,1,1,3-tetrachloropropane with hydrogen fluoride in the presence of an antimony catalyst yields 1,1,1-trifluoro-3-chloropropane, CF$_3$CH$_2$CH$_2$Cl, which gives 3,3,3-trifluoropropene when treated with a base [132]. The conversion is effected in a single operation with a mixture of hydrogen fluoride and a tertiary amine [133].

$$\text{CCl}_3\text{CH}_2\text{CH}_2\text{Cl} \longrightarrow \text{CF}_3\text{CH}_2\text{CH}_2\text{Cl} \longrightarrow \text{CF}_3\text{CH}=\text{CH}_2$$

A multistep synthesis starts from vinylidene fluoride (CF$_2$=CH$_2$) [134].

Uses. The main use of trifluoropropene is in the production of fluorine-containing silicones used in hydraulic fluids [130], [135].

4.6. 3,3,3-Trifluoro-2-(trifluoromethyl)-prop-1-ene

3,3,3-Trifluoro-1-(trifluoromethyl)prop-1-ene [382-10-5], hexafluoroisobutene, (HFIB), $(CF_3)_2C=CH_2$, is a colorless, toxic gas (4-h LC_{50} 1700 ppm) with a bp of 14.1 °C at 101.3 kPa. A number of processes for its production have been described; the most important use hexafluoroacetone [136] or perfluoroisobutene [137] as the starting material.,

$$(CF_3)_2=O + CH_2=C=O \longrightarrow \underset{\underset{H_2C-C=O}{||}}{(CF_3)_2C-O}$$

$$\longrightarrow (CF_3)_2C=CH_2 + CO_2$$

$$(CF_3)_2C=CF_2 + CH_3OH \longrightarrow (CF_3)_2CHCF_2OCH_3$$

$$\xrightarrow[\text{steps}]{\text{Several}} (CF_3)_2C=CH_2$$

Acetic anhydride can be used with hexafluoroacetone instead of ketene. The multistep reaction takes place in a single operation in a copper reactor above 300 °C [138]. Presumably, processes based on the highly toxic perfluoroisobutene are designed to remove it as a harmful byproduct of tetrafluoroethylene or hexafluoropropene production.

Hexafluoroisobutene is used for the production of fluoropolymers. The trade name of the copolymer [34149-71-8] with vinylidene fluoride is CM-X (Ausimont) [139].

4.7. Chlorofluoroolefins

Among the numerous known chlorofluoroolefins [104], [105], 1,1-dichloro-2,2-difluoroethylene [79-35-6] has some importance as a starting material in the production of methoxyflurane [76-38-0], an inhalation anesthetic. However, chlorotrifluoroethylene is the most important member of this class.

Chlorotrifluoroethylene [359-29-5] (CTFE), $CF_2=CFCl$, a colorless, flammable gas, is less reactive than tetrafluoroethylene. For physical properties, see Table 7. Although chlorotrifluoroethylene is more stable than tetrafluoroethylene, stabilizers such as tributylamine are used during transportation and storage in steel cylinders [140]. Chlorotrifluoroethylene is toxic.

Production. Chlorotrifluoroethylene is produced commercially by dechlorination of 1,1,2-trichloro-1,2,2-trifluoroethane [76-13-1], (R 113) with zinc in methanol [141]:

$$CClF_2CCl_2F + Zn \xrightarrow{CH_3OH} CF_2=CClF + ZnCl_2$$

Table 8. pvalues of alcohols

Alcohol	R = H	R = F
CR$_3$CH$_2$OH	15.9	12.8
(CR$_3$)$_2$CHOH	17.1	9.3
(CR$_3$)$_3$COH	19.2	5.4

An alternative route is dechlorination in the gas phase, e.g., on an aluminum fluoride–nickel phosphate catalyst; this catalyst is highly stable [142].

Uses. Chlorotrifluoroethylene is a starting material for homopolymers and copolymers the latter are commercially available under the trade name Halar (Allied). In addition, chlorotrifluoroethylene is an intermediate in the production of the inhalation anesthetic halothane. (2-Chloro-1,1,2-trifluoroethyl)-diethylamine [*357-83-5*], ClFHCCF$_2$N(C$_2$H$_5$)$_2$ (an addition product of chlorotrifluoroethylene with diethylamine), is used as a fluorinating agent to replace hydroxyl groups in steroids and carbohydrates with fluorine [48], [143]. Another use of chlorotrifluoroethylene is in telomerization with carbon tetrachloride or chloroform. The products are stabilized with fluorine or CoF$_3$ and are used as inert fluids, hydraulic fluids, or lubricants [77].

5. Fluorinated Alcohols [144]

Primary and secondary alcohols that have fluorine and hydroxyl groups on the same carbon atom are unstable, and readily lose hydrogen fluoride to form carbonyl compounds. Primary and secondary perfluoroalkoxides, however, can be prepared in polar solvents under aprotic conditions from carbonyl compounds and a source of ionic fluoride [145]; typical counterions are alkali-metal, tetraalkylammonium, or tris(dialkylamino)sulfonium cations. The perfluoroalkoxides are moderately nucleophilic and are used in situ to prepare compounds containing perfluoroalkoxy groups (see Section 6.3). Tris(dialkylamino)sulfonium perfluoroalkoxides are unusually stable and, in some cases, have been isolated as crystalline solids [146]. Tertiary perfluoro alcohols, e.g., perfluoro-*tert*-butanol (1,1,1,3,3,3-hexafluoro-2-trifluoromethyl-2-propanol) [*2378-02-1*] [147] and alcohols containing CH$_2$ groups between the fluorinated segment and the OH group are also stable. Fluorinated alcohols are more acidic than their nonfluorinated analogues because fluorine is highly electronegative (see Table 8) [148].

Short-Chain Fluoroalcohols. Of the short-chain fluorinated alcohols (C$_2$–C$_4$), only 2,2,2-trifluoroethanol [*75-89-8*], bp 73.6 °C, and 1,1,1,3,3,3-hexafluoro-2-propanol [*920-66-1*], bp 58.2 °C, have achieved commercial importance. The former is prepared by catalytic hydrogenation of trifluoroacetamide [149], trifluoroacetyl chloride [150], or trifluoroacetic acid esters [151], or by reduction of trifluoroacetic acid with metal hydrides [152]. Similar processes are used to prepare the latter from hexafluoroacetone [153]. Because of their strong tendency to form hydrogen bonds, the fluorinated alcohols form stable complexes with hydrogen bond acceptors and are excellent solvents for polar polymers [154]. Ethers derived from these alcohols are used as inhalation anesthetics (see Section 6.3). 2-Aryl-1,1,1,3,3,3-hexafluoro-2-propanols are prepared by alkylation of aromatic compounds with hexafluoroacetone [155]. Tertiary diols, e.g., 1,3-bis(2-hydroxyhexafluoro-2-propyl) benzene [*802-93-7*] are used to prepare fluoroepoxy resins [156].

$$HO(CF_3)_2C\text{—}C_6H_4\text{—}C(CF_3)_2OH$$

Long-Chain Fluoroalcohols. The unique hydro- and oleophobic properties of long perfluoroalkyl chains lead to many applications of long-chain fluoroalcohols and their derivatives as surfactants, antisoilants, and surface-treatment agents [157]. Structures can be represented by,

$$X-(CF_2)_n(CH_2)_mOH$$

where X is typically F, H, or CH_3, n is 6–12, and m is 1 or 2. The most important compounds are the $1H,1H$-perfluoroalcohols (X = F, m = 1), prepared by reduction of the corresponding perfluorinated carboxylic acids [158], and the $1H,1H,2H,2H$-perfluoroalcohols (X = F, m = 2); the latter are prepared by treatment of the corresponding 1-iodo-$1H,1H,2H,2H$-perfluoroalkanes with oleum [159]. Alcohols of the formula $H(CF_2)_nCH_2OH$, where n is an even number, are prepared by telomerization of tetrafluoroethylene with methanol [160].

For use as soil-repellent finishes, the long-chain fluoroalcohols are converted into derivatives to adjust the hydrophobicity, solubility in aqueous and organic solutions, surface retention, and stability for each specific application. These derivatives include esters, phosphates, carboxylates, and polyoxyethylenes. Polymeric alcohols are prepared by esterification with (meth)acryloyl chloride, followed by polymerization.

Trade Names. Teflon, Zepel, and Zonyl (Du Pont), Scotchgard (3M), and Oleophobal (Chemische Fabrik Pfersee, Ciba-Geigy).

6. Fluorinated Ethers [161]

6.1. Perfluoroethers

Perfluorinated ethers are less basic than their hydrogen-containing analogues [162]. The saturated aliphatic and cycloaliphatic perfluoroethers are noncombustible, and, with the exception of the perfluorinated epoxides, display high chemical and thermal stability. Other properties of the stable ethers, such as a large difference between the melting and boiling points at ambient pressure, and a low pour point, surface tension, and dielectric constant, are the basis of applications such as dielectric and heat-exchange fluids in, for example, high power transformers. Higher molecular mass perfluoroethers are used as lubricants and hydraulic fluids in extreme service conditions, i.e., at high temperature and/or in a corrosive environment.

6.1.1. Low Molecular Mass Perfluoroethers

The low molecular mass perfluoroethers are usually prepared by electrochemical fluorination of aliphatic ethers, alcohols, or carboxylic acids [163]. Typical empirical formulas are $C_6F_{12}O$, $C_7F_{14}O$, and $C_8F_{16}O$; these ethers often consist of isomer mixtures. Compound FC-75 [*11072-16-5*] (3M), mostly perfluorobutyltetrahydrofurans, is a useful solvent that has been explored as an oxygen-transport agent in artificial blood [164]. Perfluorinated ethers can also be prepared from partially fluorinated or nonfluorinated ethers by fluorination with cobalt fluoride or elemental fluorine under carefully controlled conditions [165].

Trade Names. Fluorinert liquids (3M) and Galden fluorinated fluids (Montedison).

6.1.2. Perfluorinated Epoxides

In contrast to other perfluorinated cyclic ethers, ring strain in perfluorinated epoxides results in high reactivity, making them versatile precursors to other fluorinated compounds [166]. The most important is hexafluoropropylene oxide [*428-59-1*] (HFPO), trifluoro(trifluoromethyl)oxirane [167]. It is prepared from hexafluoropropene by reaction with elemental oxygen [168], by electrochemical oxidation [169], or by reaction with hypochlorites [170] or hydrogen peroxide [171] in alkaline media.

Hexafluoropropylene oxide (bp – 27.4 °C) is stable at room temperature, but decomposes above 150 °C to form trifluoroacetyl fluoride and difluorocarbene [172]. In the presence of strong Brønsted or Lewis acids, such as alumina or aluminum chloride, HFPO undergoes catalytic rearrangement to hexafluoroacetone, constituting a convenient synthesis of this compound [173]. Most significantly, HFPO reacts readily with nucleophiles. Attack usually occurs at the central carbon atom [174], resulting in formation of an acid fluoride by loss of a fluoride ion from the intermediate perfluoroalkoxide, which can react further with HFPO to form higher oligomers. Acid fluorides are precursors to the commercially important perfluorovinyl ethers and higher molecular mass perfluoroethers.

$$\text{CF}_3\overset{O}{\overset{\diagup\diagdown}{\text{CF}-\text{CF}_2}} + \text{X:} \longrightarrow \underset{\text{XCFCF}_2\text{O}^-}{\overset{\text{CF}_3}{|}} \xrightarrow{-\text{F}^-} \underset{\text{XCFCOF}}{\overset{\text{CF}_3}{|}}$$

$$\downarrow n \text{ HFPO}$$

$$\text{X} - (\underset{\overset{|}{\text{CF}_3}}{\text{CFCF}_2\text{O}})_n - \underset{\overset{|}{\text{CF}_3}}{\text{CFCOF}}$$

where X = nucleophile, e.g., fluoride ion.

Both tetrafluoroethylene oxide [*694-17-7*] (TFEO), also called tetrafluorooxirane [175], and epoxides of longer chain-length perfluoroolefins [176] are known; however, the former is unstable at room temperature and rearranges to form trifluoroacetyl fluoride, whereas the latter are prepared from inaccessible perfluoroolefins. Only HFPO has achieved commercial significance because it is used for the synthesis of hexafluoroacetone (see p. 2596), high molecular mass perfluoroethers, Section 6.1.3, and fluorinated vinyl ethers (Section 6.2).

6.1.3. High Molecular Mass Perfluoroethers

High molecular mass perfluorinated ethers are prepared by fluoride-catalyzed oligomerization of HFPO [177]. The resulting terminal acid fluoride group is removed by hydrolysis and decarboxylative fluorination with elemental fluorine. Chemically inert ethers are produced which have the formula,

$$\text{CF}_3\text{CF}_2\text{CF}_2\text{O}(\underset{\overset{|}{\text{CF}_3}}{\text{CFCF}_2\text{O}})_n\text{CF}_2\text{CF}_3$$

These ethers are obtained in various molecular mass and viscosity ranges by controlling oligomerization conditions or by partial distillation of the oligomeric mixture. In an alternative method [178], perfluorinated olefins (e.g., tetrafluoroethylene or hexafluoropropene) react photochemically with oxygen to form oligomeric perfluoroethers with terminal acid fluoride groups and peroxide bonds [179]. These end groups and the unstable peroxide linkages are eliminated by fluorination. Perfluoroethers with molecular masses of ca. 500–6000 are used as inert fluids, lubricants, and hydraulic fluids in applications that require resistance to high temperature or strongly corrosive environments.

Trade Names. Krytox (Du Pont), Aflunox (SCM Specialty Chemicals), and Fomblin (Montedison).

6.2. Perfluorovinyl Ethers

Perfluorovinyl ethers are comonomers used in the preparation of melt-processable perfluoroplastics, fluorinated elastomers, and perfluoropolymers containing functional groups [180]. The ethers are synthesized by reaction of a fluorinated alkoxide generated in situ, or of other nucleophiles with HFPO. The resulting acid fluorides are converted to acid salts, which lose carbon dioxide and metal fluoride when heated in an aprotic environment. For example, HFPO is treated with cesium trifluoromethoxide (from cesium fluoride and carbonyl fluoride). Hydrolysis and decarboxylation of the resulting acid fluoride gives perfluoro(methyl vinyl ether) [1187-93-5].

$$CF_2O + CsF \longrightarrow CF_3OCs \xrightarrow{HFPO}$$

$$CF_3OCFCOF \underset{CF_3}{|} \xrightarrow[\text{2) heat}]{\text{1) }H_2O} CF_3OCF=CF_2$$

This compound is a comonomer with tetrafluoroethylene in a perfluorinated elastomer [181].

Perfluoro(propyl vinyl ether) [1623-05-8], $CF_3CF_2CF_2OCF=CF_2$, prepared from the dimer of HFPO [182], is copolymerized with tetrafluoroethylene to a melt-processable perfluoroplastic.

Perfluorovinyl ethers containing specific functional groups and usually two or more moles of hexafluoropropylene oxide are prepared in a similar fashion. These compounds have the structure $RO(C_3F_6O)CF=CF_2$; they are used as functional groups in perfluoroelastomers which can undergo a crosslinking reaction, i.e., cure-site monomers ($R = CF_2CF_2CN$ or pentafluorophenyl) [183] and as monomers that provide the ionic groups in perfluorinated ion-exchange resins ($R = CF_2CF_2SO_2F$, $CF_2CF_2CF_2CO_2CH_3$, or $CF_2CF_2CO_2CH_3$) [184]. Synthesis of the sulfonyl fluoride-substituted vinyl ether is illustrative. Reaction of tetrafluoroethylene with sulfur trioxide gives a sultone, which rearranges to fluorosulfonyldifluoroacetyl fluoride. The anion formed after addition of fluoride ion to this acid fluoride gives a 2:1 adduct with HFPO [185]. Pyrolysis of the sodium salt over sodium carbonate gives the functional monomer [186].

$$CF_2=CF_2 + SO_3 \longrightarrow \underset{O \text{---} SO_2}{\overset{CF_2 \text{---} CF_2}{|\quad\quad|}} \longrightarrow FO_2SCF_2COF$$

$$FO_2SCF_2CFO + 2\ CH_3\overset{O}{\overset{/\ \backslash}{CF}}-CF_2 \xrightarrow{F^-}$$

$$FO_2SCF_2CF_2OCFCF_2OCFCFO \underset{CF_3\quad\ CF_3}{|\quad\quad\ |} \xrightarrow[\text{heat}]{Na_2CO_3}$$

$$FO_2SCF_2CF_2OCFCF_2OCF=CF_2 \underset{CF_3}{|}$$

Perfluorovinyl ethers can also be prepared by deiodofluorination of iodine-containing ethers, $XR_fCF_2OCF_2CF_2I$, where X is hydrogen, halogen, CO_2R, $CONR_2$, SO_2F, or $PO(OR)_2$ (R = alkyl) and R_f is a perfluoroalkyl group. Initially, an organometallic derivative of the iodide is formed by reaction with metals, such as Mg, Cu, Zn, Sn, or Sb. Heating in the absence of a proton source affords the vinyl ethers [187].

6.3. Partially Fluorinated Ethers

Partially fluorinated ethers are synthesized by several methods. Fluoroalkyl alkyl or fluoroalkyl aryl ethers are prepared from alkoxides or phenoxides and fluoroolefins [188]; for example, 1,1,2,2-tetrafluoro-1-methoxyethane [425-88-7] is prepared from tetrafluoroethylene and sodium methoxide – methanol. Higher perfluoroolefins can also be employed [189]. The methyl ethers are valuable intermediates, which can be converted to acid fluorides with sulfur trioxide [190] or other Lewis acids [191]. Primary and secondary perfluorinated alkoxides (see Chap. 5) are not sufficiently nucleophilic to react with fluorinated or nonfluorinated olefins. However, they react with a variety of olefins in the presence of halogen to give ethers [192], e.g.,

$$(CF_3)_2CFO^- + CH_2=CF_2 + ICl \longrightarrow$$
$$(CF_3)_2CFOCF_2CH_2I + Cl^-$$

Perfluoroalkoxides react with benzyl halides to give perfluoroalkyl benzyl ethers [193].

In contrast to the physiological inertness of perfluoroethers, some partially fluorinated ethers act as inhalation anesthetics. Many cyclic and acyclic ethers containing varying amounts of fluorine, hydrogen, and sometimes other halogens have been evaluated as anesthetics in search of the ideal combination of potency, volatility, lack of short- and long-term toxicity, and nonflammability [194]. Some ethers have achieved clinical significance.

The first fluorinated ether anesthetic to be introduced clinically was 2,2,2-trifluoro-1-vinyloxyethane [406-90-6] [trade name: Fluromar (Anaquest)] [195]. It is prepared by addition of 2 mol of trifluoroethanol to acetylene, followed by thermolysis:

$$2\, CF_3CH_2OH + HC \equiv CH \longrightarrow CH_3CH(OCH_2CF_3)_2$$
$$\longrightarrow CH_2=CHOCH_2CF_3$$

The flammability behavior of the compound is not significantly better than that of nonfluorinated anesthetics. Recent results indicate that it behaves as a mutagen in the Ames test [196]; therefore, the compound is not currently produced commercially.

2-Chloro-1-(difluoromethoxy)-1,1,2-trifluoroethane [13838-16-9] [trade name: Ethrane (Anaquest)], which is nonflammable, is prepared by successive chlorination and fluorination of the hydrocarbon ether [197]. 1-Chloro-1-(difluoromethoxy)-2,2,2-trifluoroethane [26675-46-7] [trade name: Forane (Anaquest)] is prepared in a similar fashion [198]. Other possible anesthetics include 2,2-dichloro-1,1-difluoro-1-methoxyethane [*76-38-0*] [199] and 1,1,1,3,3,3-hexafluoro-2-(fluoromethoxy)propane [*28523-86-6*] [200].

7. Fluorinated Ketones and Aldehydes

The carbonyl groups in partially or perfluorinated ketones and aldehydes are electron deficient owing to the inductive effect of the highly electronegative fluorine atom. Compared with their hydrocarbon counterparts, fluorinated ketones and aldehydes are, consequently, much more reactive toward nucleophilic reagents. Stable addition compounds with water, alcohols, and amines are commonly formed. By contrast, fluorinated ketones and aldehydes are relatively unreactive toward electrophilic reagents. An extreme case is hexafluoroacetone, which is not protonated by the FSO_3H-SbF_5 superacid [201].

A wide variety of fluoroalkyl ketones and aldehydes has been synthesized, often by special methods that are unique to polyfluorinated compounds [202]. Of the many known examples, a few fluorinated acetones, aldehydes, and 1,3-diketones are of practical importance.

7.1. Fluoro- and Chlorofluoroacetones

Several fluoro- and chlorofluoroacetones (propanones) are listed in Table 9. Hexafluoroacetone and, to a lesser extent, chloropentafluoroacetone and *sym*-dichlorotetrafluoroacetone are commercially important. Hexafluoroacetone is manufactured by Du Pont in the United States and by Hoechst in the Federal Republic of Germany. The two chlorofluoroacetones were made in pilot-plant quantities by Allied in the 1960s [203], but their production has been discontinued.

Hexafluoroacetone, 1,1,1,3,3,3-hexafluoro-2-propanone, is made industrially by the vapor-phase reaction of hexachloroacetone with hydrogen fluoride in the presence of a chromium catalyst [204], [205]. The rearrangement of hexafluoropropylene oxide induced by Lewis acids [173] is an attractive new route that avoids the highly toxic *sym*-dichlorotetrafluoro- and chloropentafluoroacetones. A convenient laboratory synthesis uses hexafluoropropylene as a starting material [206]:

$$2\ CF_3CF=CF_2 + \tfrac{1}{4}S_8 \xrightarrow{KF, \text{dimethylformamide}}$$

$$(CF_3)_2C\overset{S}{\underset{S}{\diagup\!\!\!\diagdown}}C(CF_3)_2 \xrightarrow{KF, KIO_3} 2\ CF_3COCF_3$$

Hexafluoroacetone is used mainly for the manufacture of the solvent hexafluoro-2-

Table 9. Molecular masses and boiling points of fluoro- and chlorofluoropropanones

Name	CAS registry no.	Formula	M_r	bp, °C (101.3 kPa)
1,1,1-Trifluoro-2-propanone	[421-50-1]	CF_3COCH_3	112.05	21.5 – 22.5
1,1,3,3-Tetrafluoro-2-propanone	[360-52-1]	CHF_2COCHF_2	130.05	58
1,1,1,3,3,3-Hexafluoro-2-propanone	[684-16-2]	CF_3COCF_3	166.03	−27.4
1-Chloro-1,1,3,3,3-pentafluoro-2-propanone	[79-53-8]	$CClF_2COCF_3$	182.48	7.8
1,3-Dichloro-1,1,3,3-tetrafluoro-2-propanone	[127-21-9]	$CClF_2COCClF_2$	198.93	45.2
1,1,3-Trichloro-1,3,3-trifluoro-2-propanone	[79-52-7]	$CCl_2FCOCClF_2$	215.38	84.5
1,1,3,3-Tetrachloro-1,3-difluoro-2-propanone	[79-51-6]	$CCl_2FCOCCl_2F$	231.83	123.9
1,1,1,3,3-Pentachloro-3-fluoro-2-propanone	[2378-08-7]	CCl_3COCCl_2F	248.28	163.7
1,1,1,3,3,3-Hexachloro-2-propanone	[116-16-5]	CCl_3COCCl_3	264.73	203.6

propanol and high-performance fluoropolymers. For the properties, chemistry, and uses of hexafluoroacetone, see → Acetone.

Chloropentafluoroacetone, 1-chloro-1,1,3,3,3-pentafluoro-2-propanone [trade name: 5FK (Allied)] and *sym*-dichlorotetrafluoroacetone, 1,3-dichloro-1,1,3,3-tetrafluoro-2-propanone (4FK, Allied) can be made by the incomplete exchange of chlorine in hexachloroacetone, using hydrogen fluoride and a Cr^{3+} or Cr^{5+} catalyst [204], [205], [207]. Their properties and chemical reactivities are similar to those of hexafluoroacetone [208]. Like hexafluoroacetone, chloropentafluoroacetone and *sym*-dichlorotetrafluoroacetone form stable, acidic hydrates and hemiacetals, e.g., $CClF_2COCF_3 \cdot 3\,H_2O$ [34202-28-3], *bp* 105 °C, and $CClF_2COCClF_2 \cdot 2\,H_2O$ [34202-29-4], *bp* 106 °C. The hydrates are powerful solvents for acetal resins and various polar polymers [209], [210]. These chlorofluoroacetones and some of their derivatives possess herbicidal or fungicidal activity [211] – [213] and are useful intermediates for synthesizing repellants for textile fibers [214], [215], specialty polycarbonates [216], [217], and inhalation anesthetics [218].

1,1,1-Trifluoro-2-propanone can be prepared in quantitative yield by the acid hydrolysis of ethyl trifluoroacetoacetate, the ethyl ester of 4,4,4-trifluoro-3-oxobutanoic acid [372-31-6], which is made by the alkali-promoted condensation of ethyl trifluoroacetate with ethyl acetate [219]. 1,1,1-Trifluoro-2-propanone is easily made in the laboratory by the reaction of trifluoroacetic acid with methylmagnesium iodide [220]. Its aryl hydrazone derivatives show nematocidal and acaricidal activity [221].

1,1,3,3-Tetrafluoro-2-propanone is made by the acid hydrolysis of the ethyl ester of 2,2,4,4-tetrafluoro-3-oxobutanoic acid, $CHF_2COCF_2\,CO_2C_2H_5$ [219]. The ketone is used as an intermediate in the synthesis of inhalation anesthetics [222], [223].

Table 10. Molecular mass and boiling points of perchlorofluoroacetaldehydes

Name	Formula	M_r	CAS registry no.	bp, °C (101.3 kPa)
Trichloroacetaldehyde	CCl_3CHO	147.38	[75-87-6]	97.8
Dichlorofluoroacetaldehyde	CCl_2FCHO	130.93	[63034-44-6]	56
Chlorodifluoroacetaldehyde	$CClF_2CHO$	114.48	[811-96-1]	17.8
Trifluoroacetaldehyde	CF_3CHO	98.03	[75-90-1]	−18 to −19*

* At 99.7 kPa

7.2. Perhaloacetaldehydes

Perchlorfluoroacetaldehydes. Some physical properties of the *perchlorofluoroacetaldehydes* are shown in Table 10. Their chemical reactivities are similar to those of the perchlorofluoroacetones.

Chlorodifluoroacetaldehyde and dichlorofluoroacetaldehyde can be prepared by the lithium aluminum hydride reduction of the corresponding methyl chlorofluoroacetates [224], [225].

Trifluoroacetaldehyde (fluoral) and some of its derivatives have found practical importance as monomers and intermediates for biologically active compounds.

Properties. Trifluoroacetaldehyde is a colorless gas at ambient temperature and pressure. Like hexafluoroacetone, it reacts with water to form a stable, solid hydrate—1,1-dihydroxy-2,2,2-trifluoroethane [421-53-4], $CF_3CH(OH)_2$, mp 69–70 °C. It reacts in a manner similar to hexafluoroacetone with alcohols to give stable hemiacetals: 2,2,2-trifluoro-1-methoxyethanol [431-46-9], $CF_3CH(OH)OCH_3$, bp 96–96.5 °C; and 2,2,2-trifluoro-1-ethoxyethanol [433-27-2], $CF_3CH(OH)OC_2H_5$, bp 104–105 °C (99.3 kPa). Unlike hexafluoroacetone, it readily homopolymerizes upon cationic, anionic, or free-radical initiation [224], [225].

Production. Several laboratory syntheses of trifluoroacetaldehyde have been developed, including reduction of trifluoroacetic acid and its alkyl esters, trifluoroacetic anhydride, and trifluoroacetyl chloride [226]. Trifluoroacetaldehyde can be manufactured on a large scale by the reaction of trichloroacetaldehyde (chloral) with hydrogen fluoride in the presence of chromium catalysts [205]. The product is a $CF_3CHO \cdot HF$ complex, bp 38 °C (133.3 kPa), which requires treatment with a hydrogen fluoride acceptor (e.g., sodium fluoride) to give free trifluoroacetaldehyde. Hydrolysis of the inhalation anesthetic Halothane, $CF_3CHBrCl$, by a mixture of 65 % oleum, mercuric oxide, and silver oxide also produces trifluoroacetaldehyde in high yield [227].

Trifluoroacetaldehyde is commercially available as its hydrate or methyl and ethyl hemiacetals, which liberate the pure aldehyde in polyphosphoric acid at 150–180 °C. A Hoechst process for the manufacture of its hemiacetals involves treating the product from the gas-phase fluorination of chloral with tetraalkoxyl silanes, or with alcohols and silicon tetrachloride. This process avoids the need to isolate or handle free trifluoroacetaldehyde [228], [229].

$$CCl_3CHO \xrightarrow[CrO_2F_2]{HF} CF_3CHO \cdot HF \xrightarrow{(RO)_4Si} CF_3CH(OH)OR$$

R = alkyl

Uses of Perhaloacetaldehydes. Certain oximes [230] and hydrazones [221] of trifluoroacetaldehyde have insecticidal or acaricidal activity. Its hemiacetal, 2,2,2-trifluoro-1-methoxyethanol, has been used as a starting material for the preparation of fluoroether inhalation anesthetics [231] (see Section 6.3), including isoflurane, $CF_3CHClOCHF_2$ [232], and its isomer $CF_3CHFOCHClF$ [233], [234].

The stereoregularity of perchlorofluoroacetaldehyde polymerizations has been an area of active research, although no commercial uses of the polymers have yet appeared.

Trifluoroacetaldehyde can be homopolymerized to give insoluble crystalline, or soluble, amorphous, polyoxymethylene polymers depending upon the polymerization conditions [224], [225]. This is in contrast with trichloroacetaldehyde which can only be polymerized to a crystalline, apparently isotactic polymer. Copolymers of perhaloacetaldehydes have been prepared [224], [225].

7.3. Fluorinated 1,3-Diketones

Fluorinated 1,3-diketones in which the two carbonyl groups are separated by a methylene or methine group form complexes with a wide variety of metal ions. This property is the basis of their utility in chromatographic analysis of metals, laser technology, NMR spectroscopy, and hydrometallurgical separations. The properties, preparation, and uses of fluorinated 1,3-diketones and their metal complexes have been extensively reviewed [235]–[237].

Properties and Production. Some physical properties of the industrially important fluorinated 1,3-diketones are shown in Table 11. These compounds are considerably more acidic than their nonfluorinated analogues, viz., 1,3-pentanedione [123-54-6], $CH_3COCH_2COCH_3$ ($pK_a = 8.9$), $CF_3COCH_2COCH_3$ ($pK_a = 6.7$), and $CF_3COCH_2COCF_3$ ($pK_a = 4.6$) [237]. Fluorinated 1,3-diketones have a high enolic content, typically 92–100%, in comparison with ca. 80% for 1,3-pentanedione. Their enolic protons can be readily replaced by metals or metal salts to form 1,3-diketonates of the type

Fluorinated 1,3-diketones form hydrates with water and hemiketals with shorter alcohols. They are usually obtained by a Claisen condensation of a fluorinated carboxylic acid ester with a ketone, using a strong base such as a sodium alkoxide or sodium hydride.

Uses. The 1,3-diketones RCOCH$_2$COR' (R = CF$_3$, n-C$_3$F$_7$ and R' = t-C$_4$H$_9$, CF$_3$) that form volatile, thermally and hydrolytically stable complexes are useful for liquid or gas chromatographic analysis of metals. These diketones, especially 1,1,1,2,2,3,3-heptafluoro-7,7-dimethyl-4,6-octanedione are widely used for the extraction of metal ions from aqueous solutions at variable pH [238].

4,4,4-Trifluoromethyl-1-(2-thienyl)-1,3-butanedione is especially suited for the analysis of uranium and other radioactive elements. The *tetrakis*-chelates of 4,4,4-trifluoro-1-phenyl-1,3-butanedione with rare-earth elements are potential laser materials.

The paramagnetic lanthanide complexes of heptafluoro-7,7-dimethyl-4,6-octanedione have gained considerable importance as NMR shift reagents for simplifying the interpretation of complex NMR spectra [239]; the europium complex is the most widely used.

M = Ag; n = 1
M = Dy, Eu, Gd, Ho, Pr, Yb; n = 3

The chiral-shift reagents derived from 3-heptafluorobutyryl-(+)- or -(−)-camphor and 3-trifluoroacetyl-(+)- or -(−)-camphor are very useful for the NMR assay of enantiomeric purity in solution [239].

R = CF$_3$, n-C$_3$F$_7$
M = Eu, Pr, Yb

8. Fluorinated Carboxylic Acids and Fluorinated Alkanesulfonic Acids

8.1. Fluorinated Carboxylic Acids

8.1.1. Fluorinated Acetic Acids

Fluorination increases the strength of acetic acid as seen in the pK_a values of monofluoroacetic acid (2.66), difluoroacetic acid (1.24), and trifluoroacetic acid (0.23) compared to the pK_a of 4.74 for acetic acid [240].

Production. Trifluoroacetic acid has been prepared by the electrochemical fluorination of acetyl chloride or acetic anhydride in anhydrous hydrogen fluoride using the Simons process (Section 2.1) followed by hydrolysis of the resulting trifluoroacetyl fluoride. The yield is excellent (> 90%) [241], [242]. The Phillips Petroleum electrochemical process (Section 2.1) employs acetyl fluoride as feed to produce trifluoroacetyl fluoride along with mono and difluoroacetyl fluoride [243]. Sulfur trioxide treatment of CF_3CCl_3, obtained by isomerization of Freon 113, $CF_2ClCFCl_2$, yields CF_3COCl [244]. The acid is purified by hydrolysis of the acyl halide with alkali, followed by acidification and distillation.

Uses. Most uses of fluorinated acetic acids are confined to trifluoroacetic acid, its anhydride, and its derivative, 1,1,1-trifluoroethanol. Monofluoroacetic acid derivatives (salts, esters, amides, and alcohols) are toxic because they are metabolized to fluorocitric acid which inhibits respiration (see Section 13.5) [245]. ω-Fluoro acids of the formula $F(CH_2)_nCOOH$, where n is an odd number, are extremely toxic, because of degradation in vivo to monofluoroacetic acid and finally to fluorocitric acid [245]. Sodium monofluoroacetate has been used as a rodenticide, but is now banned.

8.1.2. Long-Chain Perfluorocarboxylic Acids

Properties. The boiling points and densities of straight-chain perfluorocarboxylic acids are shown in Table 12.

Production. Long-chain perfluorocarboxylic acids are prepared by the Simons electrochemical fluorination (see Section 2.1.) of the corresponding acyl halide:,

$$C_nH_{2n+1}COF + (2n+1)HF \longrightarrow C_nF_{2n+1}COF + (2n+1)H_2$$

The acids are obtained by hydrolysis of the perfluoroacyl fluoride, followed by distillation. Some carbon–carbon bond scission occurs to form lower homologous acids along with inert fluorocarbons and cyclic ethers. The acid yield decreases with increasing chain length; for example, perfluorobutyric acid yields are ca. 36% compared to ca. 20% for the industrially important perfluorooctanoic acid [241], [242]. Trifluoroacetic acid is also prepared by the Phillips electrochemical method, employing a $KF \cdot HF$ molten salt electrolyte [243]. Acids with higher boiling points are not easily or efficiently prepared by the Phillips process.

Perfluorocarboxylic acids are also prepared by nonelectrochemical methods. Treatment of perfluoroalkyl iodides (R_fI) with sulfur trioxide [246] or chlorosulfonic acid [247] gives the carboxylic acid in good yield. A recently reported method employs fuming sulfuric acid (oleum) [248]. Another process involves the preparation of perfluorotetrahydroalkyl iodides, $R_fCH_2CH_2I$, obtained by the free-radical addition of ethylene to perfluoroalkyl iodides, followed by dehydroiodination and oxidation by dichromate [249] or ozonolysis [250]:

Table 11. Molecular masses and boiling points of fluorinated 1,3-diketones

Name	CAS registry no.	Formula	M_r	bp, °C (kPa)
1,1,1-Trifluoro-2,4-pentanedione	[367-57-7]	$CF_3COCH_2COCH_3$	154.09	105–107 (101.3)
1,1,1,5,5,5-Hexafluoro-2,4-pentanedione	[1522-22-1]	$CF_3COCH_2COCF_3$	208.06	70–71 (101.3)
1,1,1-Trifluoro-5,5-dimethyl-2,4-hexanedione	[22767-90-4]	$CF_3COCH_2COC(CH_3)_3$	196.17	138–141 (101.3)
1,1,1,2,2,3,3-Heptafluoro-7,7-dimethyl-4,6-octanedione	[17587-22-3]	$CF_3CF_2CF_2COCH_2COC(CH_3)_3$	296.19	46–47 (0.67)
4,4,4-Trifluoro-1-phenyl-1,3-butanedione	[326-06-7]	Ph-$COCH_2COCF_3$	216.16	224 (101.3)[a]
4,4,4-Trifluoro-1-(2-thienyl)-1,3-butanedione	[326-91-0]	(2-thienyl)-$COCH_2COCF_3$	222.18	96–98 (1.07)[b]
3-(Trifluoroacetyl)camphor, 1,7,7-trimethyl-3-(trifluoroacetyl)-bicyclo[2.2.1]heptan-2-one	[51800-98-7]	camphor-$COCF_3$	248.25	100–101 (2.13)
3-(Heptafluorobutyryl)camphor, 3-(2,2,3,3,4,4,4-heptafluoro-1-oxo-butyl)-bicyclo[2.2.1]heptan-2-one	[51800-99-8]	camphor-$COCF_2CF_2CF_3$	348.26	60–70 (0.03)

[a] mp 38–40 °C. [b] mp 42–43 °C.

Table 12. Boiling points and densities of perfluorocarboxylic acids

Acid	CAS registry no.	Formula	bp, °C (kPa)	d_4^{20}, g/cm^3
Perfluoroacetic	[76-05-1]	CF_3CO_2H	72.4	1.489
Perfluoropropionic	[422-64-0]	$C_2F_5CO_2H$	96	1.561
Perfluorobutyric	[375-22-4]	$C_3F_7CO_2H$	120	1.651
Perfluorovaleric	[2706-90-3]	$C_4F_9CO_2H$	130	
Perfluorocaproic	[307-24-4]	$C_5F_{11}CO_2H$	157	1.762
Perfluoroheptanoic	[375-85-9]	$C_6F_{13}CO_2H$	175	1.792
Perfluorocaprylic	[335-67-1]	$C_7F_{15}CO_2H$	189	
Perfluorononanoic	[375-95-1]	$C_8F_{17}CO_2H$	110 (2.1)	
Perfluorocapric	[335-76-2]	$C_9F_{19}CO_2H$	121 (2.0)	
Perfluoroundecanoic	[2058-94-8]	$C_{10}F_{21}CO_2H$	245	
Perfluorododecanoic	[307-55-1]	$C_{11}F_{23}CO_2H$	270	

$$C_nF_{2n+1}I + CH_2=CH_2 \longrightarrow C_nF_{2n+1}CH_2CH_2I \xrightarrow{-HI}$$

$$C_nF_{2n+1}CH=CH_2 \xrightarrow[\text{or } O_3]{Cr_2O_7^{2-}} C_nF_{2n+1}COOH$$

This process produces carboxylic acids having one more carbon atom than the starting perfluoroalkyl iodide, in contrast to the electrochemical and oleum routes, which produce acids containing the same number of carbon atoms as the original acyl halide or telomeric iodide. In all processes utilizing fluorinated iodine-containing telomers, it is important to find uses for the telomers and to recover the expensive iodine.

Polyfluoroalkoxyacyl fluorides of the type

$$R_fCF_2O(\overset{CF_3}{\underset{|}{C}}FCF_2O)_n\overset{CF_3}{\underset{|}{C}}FCFO$$

are prepared by the addition of perfluoroacyl fluorides and hexafluoropropylene oxide catalyzed by alkali-metal fluorides [251]. Acids, salts, or esters are obtained by hydrolysis, neutralization, or esterification, respectively.

Addition reactions of perfluorodiacyl fluorides with hexafluoropropylene oxide [252] give ether-containing diacyl fluorides, such as

$$FCOCFO(CF_2)_nO(\overset{CF_3}{\underset{|}{C}}FCF_2O)_m\overset{CF_3}{\underset{|}{C}}FCOF$$

where n is usually 2–4, and at least two units are derived from hexafluoropropylene oxide (i.e., $m \geq 2$) [252]. The addition reaction may occur at one or both acyl groups of the starting diacyl fluoride. Selectivity can be maintained by esterifying one of the acyl fluoride groups. Thus, $CH_3OCO(CF_2)_2COF$, prepared by addition of methanol to perfluoro-γ-butyrolactone or perfluorosuccinyl fluoride, reacts with hexafluoropropylene oxide to give ester acyl fluorides of the formula

$$CH_3OCO(CF_2)_3O(\overset{CF_3}{\underset{|}{C}}FCF_2O)_m\overset{CF_3}{\underset{|}{C}}FCOF$$

[253], [254]. The acyl halides are converted to acids or salts by hydrolysis or neutralization, respectively.

Uses. Long-chain perfluoroalkanecarboxylic acids and their salts are surface-active chemicals (surfactants), which greatly reduce the surface tension (surface energy) of water, aqueous solutions, and organic liquids even at low concentrations. These acids (C_6–C_{12}) and derivatives are used as wetting, dispersing, emulsifying, and foaming agents.

Ammonium perfluorooctanoate (FC-143, 3M) is used as an emulsifier in the polymerization of fluorinated monomers, especially tetrafluoroethylene. It has exceptional chemical stability and lowers the surface tension of water to ca. 3 $\mu N \cdot m$ at 0.5 wt%. In contrast with hydrocarbon emulsifiers, ammonium perfluorooctanoate does not interfere with the emulsion polymerization of tetrafluoroethylene.

Trade Names. Fluorad FC-26, 126, 143 (3M), Fluorowet CP (Hoechst), RM 350, 370 (Rimar), and Surflon S-111P (Asahi Glass).

Table 13. Physical properties of perfluorodicarboxylic acids

Acid	CAS registry no.	Formula	mp, °C	bp, °C (kPa)
Perfluoromalonic	[1514-85-8]	$CF_2(CO_2H)_2$	117–118	
Perfluorosuccinic	[377-38-8]	$(CF_2)_2(CO_2H)_2$	86–87	150 (2.0)
Perfluoroglutaric	[376-73-8]	$(CF_2)_3(CO_2H)_2$	88	134–138 (0.4)
Perfluoroadipic	[336-08-3]	$(CF_2)_4(CO_2H)_2$	132–134	138 (0.5)
Perfluorosuberic	[678-45-5]	$(CF_2)_6(CO_2H)_2$	154–158	

8.1.3. Fluorinated Dicarboxylic Acids

Properties. The boiling points and densities of perfluorodicarboxylic acids are shown in Table 13.

Production. α,ω-Perfluoroalkanedicarboxylic acids are prepared by the electrochemical method, followed by hydrolysis, acidification, and extraction. Other methods include the oxidation of the appropriate chlorofluoroolefin or perfluoroolefin. Perfluoroglutaric acid is prepared from hexachlorocyclopentadiene by halogen exchange, followed by oxidation and acidification [255]. Perfluoroadipic acid is prepared by an analogous method from hexachlorobenzene [256]. The cyclic anhydrides from perfluorosuccinic and perfluoroglutaric acid are prepared by dehydration of the acids with phosphorus pentoxide. They are used to prepare the corresponding alcohols by reduction.

8.1.4. Tetrafluoroethylene – Perfluorovinyl Ether Copolymers with Carboxylic Acid Groups

Ion-exchange membranes, used in fuel cells and chloralkali production, are copolymers of tetrafluoroethylene and perfluorovinyl ethers that contain esters or other acid precursor groups [257]. These membranes have excellent thermal and chemical resistance to hot concentrated alkali (up to 40%).

Production. The vinyl ethers are prepared by the reaction of hexafluoropropylene oxide and methyl-3-fluorocarbonyl perfluoropropionate [255], followed by pyrolysis to give:

$$CF_2=CFO(CF_2\overset{\underset{\displaystyle CF_3}{|}}{C}FO)_n(CF_2)_3COOCH_3$$

Copolymerization with tetrafluoroethylene followed by saponification produces a polymer with terminal carboxylic acid groups.

Trade Names. Flemin (Asahi Glass), Nafion (Du Pont), Neosepta (Tokuyama Soda), and Aciplex (Asahi Chemical).

Table 14. Boiling points of perfluoroalkane sulfonic acids

Acid Formula	CAS registry no.	bp, °C (kPa)
CF_3SO_3H	[1493-13-6]	60 (0.4)
$C_2F_5SO_3H$	[354-88-1]	81 (0.29)
$n\text{-}C_4F_9SO_3H$	[59933-66-3]	76–84 (0.13)
$n\text{-}C_5F_{11}SO_3H$	[3872-25-1]	110 (0.67)*
$n\text{-}C_6F_{13}SO_3H$	[355-46-4]	95 (0.46)
$n\text{-}C_8F_{17}SO_3H$	[1763-23-1]	133 (0.8)

* $n\text{-}C_5F_{11}SO_3H \cdot H_2O$

8.2. Fluorinated Alkanesulfonic Acids

8.2.1. Perfluoroalkanesulfonic Acids

Properties. The first member of the series, trifluoromethanesulfonic acid, was reported in 1954 [258]. Perfluoroalkanesulfonic acids are among the strongest acids known. Conductivity measurements in acetic acid show that the acid strength of trifluoromethanesulfonic acid is comparable to that of fluorosulfonic and perchloric acids [259]. Boiling points are listed in Table 14. Because of their ability to lower surface energy, the longer-chain perfluoroalkanesulfonic acids and sulfonyl fluoride derivatives have found utility as surface-active agents and water repellents and for antisoiling treatment of textiles and fabrics.

Production. Perfluoroalkanesulfonyl fluorides are usually made by the Simons electrochemical fluorination process (see Section 2.1), in which a hydrocarbon sulfonyl fluoride is electrolyzed in anhydrous hydrogen fluoride at nickel electrodes:

$$C_nH_{2n+1}SO_2F + (2n+1)HF \longrightarrow C_nF_{2n+1}SO_2F + (2n+1)H_2$$

The electrochemical yield is excellent for the first member of the series and decreases progressively with the increasing length of the carbon chain; the yield for octanesulfonyl fluoride is ca. 40 % [241], [260]. Alkaline hydrolysis of perfluoroalkanesulfonyl fluorides gives the corresponding salts, which when acidified and distilled from concentrated sulfuric acid yield the anhydrous sulfonic acids [258].

A nonelectrochemical method for the preparation of trifluoromethanesulfonic acid derivatives is shown below:

$$CF_3SSCF_3 \xrightarrow{Cl_2} 2\ CF_3SCl \xrightarrow[H_2O]{Cl_2} 2\ CF_3SO_2Cl$$

Alkaline hydrolysis, followed by acidification of the sulfonate salt, gives the acid [261].

Uses. Short-chain perfluoroalkanesulfonyl fluorides are used to prepare sulfonamides, which are employed as plant growth regulators and herbicides. Trifluoromethanesulfonic acid is used as an esterification catalyst (FC-24, 3M), and the lithium salt has been investigated as a fuel cell and battery electrolyte (FC-124, 3M). Perfluorobutanesulfonate salts are used as antistatic agents [262], [263]. The higher homologues exhibit good surfactant properties; the derivatives of perfluorohexanesulfonyl fluoride are employed in fire extinguishing formulations. Useful derivatives containing the sulfonamido group can be prepared by reaction of the sulfonyl fluoride with a diamine, followed by quaternization with alkylating agents such as methyl iodide to give a cationic surfactant. As an example, reaction of perfluorohexanesulfonyl fluoride with 3-dimethylaminopropylamine gives

$$C_6F_{13}SO_2NH(CH_2)_3N(CH_3)_3{}^+I^-.$$

The properties related to the low surface energy of the perfluorooctanesulfonyl fluoride, $C_8F_{17}SO_2F$, are utilized in many derivatives. These derivatives include alcohols and their acrylate and methacrylate esters; they are used as comonomers in polymers that impart oil-, water-, and soil-repellent properties to porous substrates such as paper and textiles [264], [265]. A typical reaction sequence for the synthesis of a perfluoroalkanesulfonamide acrylate monomer is shown below:

$$C_8F_{17}SO_2F + RCH_2NH_2 \longrightarrow C_8F_{17}SO_2NHCH_2R$$

$$\xrightarrow{ClCH_2CH_2OH,\ OH^-} C_8F_{17}SO_2N(CH_2R)CH_2CH_2OH$$

$$\xrightarrow{CH_2=CHCO_2H} C_8F_{17}SO_2N(CH_2R)CH_2CH_2OCOCH=CH_2$$

Perfluoroalkanesulfonamido alcohols are also used as mold-release agents. Esterification with phosphoric acid gives phosphate ester salts of the type

$$[R_fSO_2N(C_nH_{2n+1})CH_2CH_2O]_mPO(ONH_4)_l$$
$$R_f = \text{perfluoroalkyl};\ l = 1, 2;\ m = 1, 2;\ \text{and}\ n = 1, 2$$

which are useful oil repellents for paper products [266].

Trade Names. Paper Treatment FC-807, 808 (3M), Textile Treatment Dic-Guard (Dainippon Ink), and Scotchgard (3M).

8.2.2. Fluorinated Alkanedisulfonic Acids

α,ω-Perfluoroalkanedisulfonyl fluorides are prepared by the electrochemical fluorination of hydrocarbon disulfonyl fluorides. The corresponding acids are prepared by alkaline hydrolysis of the fluoride and acidification, or by aqueous permanganate oxidation of the disulfone, $CH_3SO_2(CF_2CF_2)_nSO_2CH_3$ [267]. Distillation of the acids from phosphorus pentoxide yields the cyclic anhydrides [268].

8.2.3. Tetrafluoroethylene – Perfluorovinyl Ether Copolymers with Sulfonic Acid Groups

Ion-exchange membranes are prepared by the copolymerization of tetrafluoroethylene with perfluorovinyl ethers containing sulfonyl halide groups, followed by hydrolysis to yield sulfonic acids. These membranes have excellent chemical and thermal properties similar to those of ion-exchange membranes with terminal carboxylic acid groups (see Section 8.1.4) [269] – [271].

Production. These ethers are prepared by the condensation of hexafluoropropylene oxide with fluorosulfonyldifluoroacetyl fluoride, FSO_2CF_2COF, which is prepared from C_2F_4 and SO_3, followed by isomerization [272], [273]. Condensation of hexafluoropropylene oxide with γ-fluorosulfonylperfluoroalkyl carbonyl fluorides, $FSO_2(CF_2)_nCOF$ (prepared by electrochemical fluorination of the respective aliphatic sulfone [274], [275]) gives perfluoroether fluorosulfonyl acyl fluorides, e.g., [276]

$$\text{FCOCFO(CF}_2\text{CFO)}_m\text{(CF}_2\text{)}_{n+1}\text{SO}_2\text{F}$$
$$\overset{|}{\text{CF}_3} \quad \overset{|}{\text{CF}_3}$$

Subsequent conversion to the vinyl ether

$$\text{CF}_2\text{=CFO(CF}_2\text{CFO)}_m\text{(CF}_2\text{)}_{n+1}\text{SO}_2\text{F}$$
$$\overset{|}{\text{CF}_3}$$

followed by copolymerization with tetrafluoroethylene gives polymers with fluorosulfonyl side chains, which are hydrolyzed to sulfonate group side-chains [277].

9. Fluorinated Tertiary Amines

Physical Properties. Relative to their molecular mass, perfluoroalkyl-*tert*-amines, like the perfluoroethers, have low boiling points and low freezing or pour points, as shown in Table 15. Low polarity and weak intermolecular forces are responsible for other unusually low values for properties such as viscosity, solubility, heat of vaporization, refractive index, dielectric constant, and surface tension [278]. Some perfluorobis(dialkylaminoalkyl) ethers have even lower pour points and liquid ranges as broad as 250 °C. These ethers exhibit increased internal flexibility through the combined effect of the nitrogen and oxygen atoms [281].

Chemical Properties. Perfluorinated *tert*-amines are chemically inert and thermally stable [278], [282]. The electron-withdrawing nature of the perfluoroalkyl substituents deprives the nitrogen atom of its basic character and reactivity. Fluorinated *tert*-amines do not form salts or complexes with strong acids and are not attacked by most oxidizing or reducing agents. With aluminum chloride they form chlorinated imines

Table 15. Physical properties of fluorinated amines

Compound	CAS registry no.	Formula	M_r	bp, °C	Pour point, °C	d_4^{25}
Perfluorotrimethylamine	[432-03-1]	$N(CF_3)_3$	221	−11		
Perfluorotriethylamine	[359-70-6]	$N(C_2F_5)_3$	371	69		1.74
Perfluorotripropylamine	[338-83-0]	$N(C_3F_7)_3$	521	130	−52	1.82
Perfluorotributylamine	[311-89-7]	$N(C_4F_9)_3$	671	178	−50	1.88
Perfluorotriamylamine	[338-84-1]	$N(C_5F_{11})_3$	821	215	−25	1.93
Perfluorotrihexylamine	[432-08-6]	$N(C_6F_{13})_3$	971	256	33*	1.90**

* Freezing point. ** d_4^{35}

[283]. Because of their nonpolar nature, fluorinated *tert*-amines are poor solvents and are immiscible with water and alcohols [278]. Gases such as oxygen, nitrogen, and carbon dioxide have unusually high solubility in perfluorinated *tert*-amines. For example, perfluorotributylamine dissolves 40 vol% of oxygen at ambient conditions and has been used in artificial blood as an effective oxygen-transport medium [283].

Production. Electrochemical fluorination via the Simons process (see Section 2.1.) is the preferred route to fluorinated tertiary alkylamines. The hydrogen atoms are completely replaced by fluorine atoms. Perfluorotributylamine, for example, is synthesized as follows:

$$(C_4H_9)_3N + 27\,HF \xrightarrow{e^-} (C_4F_9)_3N + 27\,H_2$$

The crude product contains a significant amount of perfluorinated isomers and cleavage products because of molecular rearrangement during electrolysis; it is purified by fractional distillation and treatment with base.

Uses. The combination of unusual physical and chemical properties, excellent dielectric properties, nonflammability, and lack of toxicity make the perfluorinated *tert*-amines useful for many fluid applications that involve direct contact with sensitive materials [278], [283]. The electronic industry relies heavily on these fluids in reliability testing of electronic components, as direct-contact coolants for integrated circuits, and as heating media in vapor-phase reflow soldering. Perfluorinated tributylamine, triamylamine, and trihexylamine are the main constituents of Fluorinert electronic liquids FC-43, FC-70, and FC-71 (3M).

Information on the preparation and utility of other nitrogen-containing fluoroaliphatic compounds can be found in the general references [1]–[15].

Table 16. Physical properties of benzotrifluorides

Compound	CAS registry number	Empirical formula	M_r	bp, °C	d_4^θ (θ, °C)	n_D^θ (θ, °C)
Benzotrifluoride	[98-08-8]	$C_7H_5F_3$	146.11	102.03	1.188 (20)	1.4114 (25)
2-Chlorobenzotrifluoride	[88-16-4]	$C_7H_4ClF_3$	180.56	152.5	1.367 (20)	1.4550 (25)
4-Chlorobenzotrifluoride	[98-56-6]	$C_7H_4ClF_3$	180.56	140	1.35 (15)	1.4444 (25)
2,4-Dichlorobenzotrifluoride	[320-60-5]	$C_7H_3Cl_2F_3$	215.0	117–118	1.377 (20)	1.4802 (20)
1,3-Bis(trifluoromethyl)benzene	[402-31-3]	$C_8H_4F_6$	214.11	116.1	1.379 (25)	1.3791 (20)
1,4-Bis(trifluoromethyl)benzene	[433-19-2]	$C_8H_4F_6$	214.11	117.1	1.381 (25)	1.3792 (20)
3-Trifluoromethylbenzoyl fluoride	[328-99-4]	$C_8H_4F_4O$	192.11	159–163		1.4350 (20)

10. Aromatic Compounds with Fluorinated Side-Chains

The first aromatic compound with a fluorinated side-chain, benzotrifluoride [98-08-8], trifluoromethylbenzene, was synthesized in 1898 [284]. The perfluoroalkyl substituents of aromatic compounds are meta-directing; this influence can, of cause, be overcome by a stronger ortho- or para-directing substituent. These compounds are usually prepared from iodoperfluoroalkanes and halogenated aromatic compounds in the presence of a copper catalyst [285]–[287]. Except for the benzotrifluorides, aromatic compounds with fluorinated side-chains have only scientific interest. Araliphatic compounds with one or two trifluoromethyl substituents on the benzene ring gained commercial importance in the early 1930s for two reasons: 1) the recognition of the advantageous properties of aromatic dyes with CF_3 substituents, and 2) the development of an economical production process [4]. Since then, their importance in the production of dyes, pharmaceuticals, and pesticides has increased.

10.1. Properties

The physical properties of benzotrifluorides are shown in Table 16.

If no other substituents are present in the benzene ring, the trifluoromethyl group is thermally stable up to 350 °C and resistant to bases up to 130 °C. It is inert toward reducing agents [288], [289] and inhibits oxidation of the benzene ring. However, in the presence of aluminum chloride the trifluoromethyl group undergoes chlorolysis to produce a trichloromethyl group [290], and acid hydrolysis forms a carboxy group [291]. Substituents such as amino or hydroxyl groups destabilize the trifluoromethyl group [292].

The characteristic reaction of benzotrifluorides is electrophilic substitution of the benzene ring; chlorination and nitration are commercially important. *Chlorination* of benzotrifluoride at 65 °C with $FeCl_3$ as a catalyst yields 83 % 3-chlorobenzotrifluoride

[*328-99-4*] [293]; chlorination of 4-chlorobenzotrifluoride gives 3,4-dichlorobenzotrifluoride [*328-84-7*]. *Nitration* of benzotrifluoride results in a 6:3:91 mixture of 2-nitro- [*384-22-5*], 4-nitro- [*402-54-0*], and 3-nitrobenzotrifluoride [*98-46-4*] [294]. Other commercially important nitrations are the conversion of 2-chlorobenzotrifluoride [*88-16-4*] to 2-chloro-5-nitrobenzotrifluoride [*777-37-7*]and the conversion of 4-chlorobenzotrifluoride to yield 99% 4-chloro-3-nitrobenzotrifluoride [*121-17-5*] or 4-chloro-3,5-dinitrobenzotrifluoride [*393-75-9*] [295]. These derivatives are reduced to amines or are further processed.

10.2. Production

Benzotrifluorides are prepared on a laboratory scale by the reaction of aromatic compounds with iodotrifluoromethane [*2314-97-8*] [285]; by the reaction of aromatic carboxylic acids and their derivatives with sulfur tetrafluoride [45]–[47]; or by chlorine–fluorine exchange in trichloromethyl aromatic compounds with metal fluorides [296].

In the commercial process, known since 1931, chlorine is exchanged for fluorine with use of anhydrous hydrogen fluoride in the presence or absence of catalyst [297]:

$$\text{PhCCl}_3 + 3\,\text{HF} \longrightarrow \text{PhCF}_3 + 3\,\text{HCl}$$

This reaction can be carried out as a batch process in autoclaves or continuously in a series of autoclaves [298] (Fig. 3) or tubular reactors [299]. Typical conditions for the production of benzotrifluoride are a temperature of 80–110 °C, pressure of 1.2–1.4 MPa, and a molar ratio of HF:benzotrichloride of 4:1. A yield of 70% is obtained within 3–4 h [300]. In continuous processing, a yield of over 90% is obtained in a nickel flow tube with a residence time of 1 h at 90–130 °C and at 3–5 MPa [301]. Yields are increased by using additives such as hexamethylenetetraamine [302] or by employing chlorine–fluorine exchange in the gas phase in the presence of a transition metal–aluminum oxide catalyst [303]. Corrosion is reduced by lowering the reaction temperature ($< 60\,°C$) and adding iron or iron compounds [304], [305].

This method is used to produce benzotrifluoride, 2-chlorobenzotrifluoride, 4-chlorobenzotrifluoride, 2,4-dichlorobenzotrifluoride, 1,3-bis(trifluoromethyl)benzene, 1,4-bis(trifluoromethyl)benzene, 3-(trifluoromethyl)benzoyl fluoride, and 4-(trifluoromethyl)benzoyl fluoride [*368-94-5*] from the corresponding trichloromethyl compounds.

The trifluoromethyl group can also be introduced into aromatic compounds by the Friedel–Crafts reaction with carbon tetrachloride in the presence of hydrogen fluoride [306].

$$\text{Ph-R} + \text{CCl}_4 + 3\,\text{HF} \longrightarrow \text{R-C}_6\text{H}_4\text{-CF}_3 + 4\,\text{HCl}$$

R = alkyl, aryl, halogen

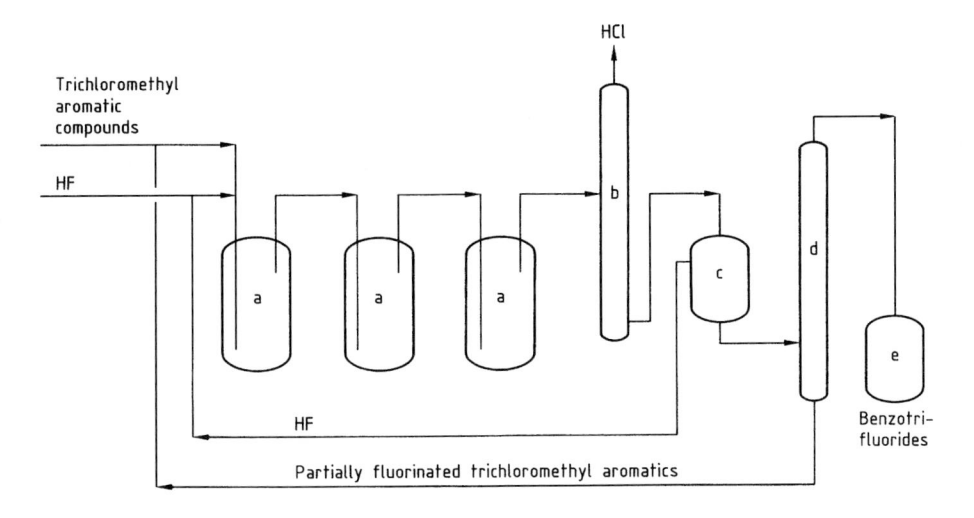

Figure 3. Production of benzotrifluorides in a series of autoclaves [298]
a) Reactor; b) Pressure distillation column; c) Separator; d) Product distillation column; e) Product storage tank

10.3. Uses

Benzotrifluoride and 4-chlorobenzotrifluoride are key intermediates for the synthesis of dyes, pharmaceuticals, and pesticides.

Dyes. Trifluoromethyl aromatic compounds were first used in dyes [4] and are still important intermediates for azo, anthraquinone, and tri-phenylmethane dyes. The strongly electronegative trifluoromethyl group improves color clarity and fastness to light and washing; it also shifts light absorption to the visible and ultraviolet ranges.

Some of these dye intermediates [307] for the production of anthraquinone and azo dyes (Naphtol AS, Hoechst AG) and azo pigments are still used, especially for polyesters and polyamides. These compounds include 2-aminobenzotrifluoride [88-17-5] for C.I. Pigment Yellow 154 [63661-02-9]; 3-amino-4-chlorobenzotrifluoride [121-50-6] for C.I. Pigment Orange 60 [68399-99-5]; 3-(trifluoromethyl)benzoyl fluoride for Indanthren Blue CLB [6492-78-0] (BASF) as well as 2-amino-5-chlorobenzotrifluoride [445-03-4], 3,5-bis(trifluoromethyl)aniline [328-74-5], and 3-amino-4-ethylsulfonylbenzotrifluoride [382-85-4].

Pharmaceuticals. The trifluoromethyl substituent is highly lipophilic; it increases the lipid solubility of pharmaceuticals and thus accelerates their absorption and transport in a living organism [308], [309]. In some cases, introduction of the CF_3 group also increases drug effectiveness and reduces undesirable side effects; therefore, benzotrifluorides are used in the synthesis of many pharmaceuticals [310]. These include the analgesics flufenamic acid [530-78-9] and niflumic acid [4394-00-7], the antidepressant fluoxetine [54910-89-3], the muscle relaxant flumetramide [7125-73-7], the appetite depressants fenfluramine [458-24-2] and fludorex [15221-81-5], and the tranquilizers triflupromazine [146-54-3] and fluphenazine [69-23-8]. Bendroflumethiazide [73-48-3] is an effective diuretic and antihypertensive agent.

Pesticides. Benzotrifluorides are also important in the production of pesticides [311]. The key intermediate, 4-chloro-3,5-dinitrobenzotrifluoride [393-75-9], is obtained by dinitration of 4-chlorobenzotrifluoride. Reaction with secondary amines gives trifluralin [1582-09-8], profluralin [26399-36-0], and benfluralin [1861-40-1]. 4-Chlorobenzotrifluoride and 3,4-dichlorobenzotrifluoride are intermediates for herbicides with a diphenyl ether structure—fluorodifen [15457-05-3] and acifluorfen [50594-66-6] and the insecticide fluvalinate [69409-94-5]. 3-Aminobenzotrifluoride [98-16-8] is obtained from benzotrifluoride by nitration and hydrogenation. It is used to make the selective herbicide fluometuron [2164-17-2] or is used as a component of the herbicide norflurazon [27314-13-2].

11. Ring-Fluorinated Aromatic, Heterocyclic, and Polycyclic Compounds

The compounds discussed in this chapter contain one or more fluorine atoms that are directly attached to aromatic, heterocyclic, or polycyclic rings. Unlike chlorination and bromination, fluorination with elemental fluorine is rarely employed for their production because its violence produces many side reactions (ring-opening, coupling, polymerization, and charring). Therefore, indirect methods are used, distributing the reaction enthalpy over several controllable steps.

Diazonium salts are suitable for introducing up to three fluorine atoms per ring. Chlorine–fluorine exchange (halogen exchange, Halex process) with alkali–metal fluorides is suitable in cases where activating substituents are present.

Ring-fluorinated aromatic, heterocyclic, and polycyclic compounds are predominantly used as intermediates for pharmaceuticals, pesticides, dyes and other products. The exceptional properties of fluorinated cyclic compounds (bioactivity spectrum, effectiveness, and solublity) justify the higher production costs compared to those of fluorine-free compounds.

Table 17. Physical properties of ring-fluorinated aromatic compounds

Compound	CAS registry number	Empirical formula	M_r	bp, °C (101.3 kPa)	d_4^θ (θ, °C)	n_D^θ (θ, °C)
Fluorobenzene	[462-06-6]	C_6H_5F	96.1	84.7	1.083 (25)	1.4629 (25)
2-Fluorotoluene	[95-52-3]	C_7H_7F	110.13	113–114	1.003 (21)	1.4727 (20)
3-Fluorotoluene	[352-70-5]	C_7H_7F	110.13	115	0.991 (25)	1.4691 (20)
4-Fluorotoluene	[352-32-9]	C_7H_7F	110.13	116	0.991 (25)	1.4688 (20)
4,4'-Difluorodiphenyl-methane	[457-68-1]	$C_{13}H_{10}F_2$	204.22	263.5*	1.145 (20)	1.5362 (20)
1,3-Difluorobenzene	[372-18-9]	$C_6H_4F_2$	114.09	82–83	1.1572 (20)	1.4410 (20)
1,4-Difluorobenzene	[540-36-3]	$C_6H_4F_2$	114.09	88–89	1.176 (20)	1.4421 (20)

* At 100.5 kPa

11.1. Mono- and Difluoroaromatic Compounds

Research on ring-substituted aromatic fluorine compounds began in 1870. The fluorine atom attached to the benzene ring is strongly electronegative; it preferentially directs new substituents into the para position and rarely, if ever, into the ortho position [312]. Important aspects, especially for biological applications, are (1) the influence of the strongly electronegative fluorine substituent on adjacent groups, (2) its participation in the resonance system of the aromatic compound by returning electron density, and (3) the simulation of hydrogen or hydroxy substituents that is due to the similarity in space requirements and has led to the synthesis of a series of monofluoro aromatic enzyme inhibitors. The varying lability of the carbon–fluorine bond in mono- and difluoro aromatic compounds has been exploited to prepare indicators and analytical reagents, especially for biogenic and metabolic studies.

11.1.1. Properties

Replacement of aromatic hydrogen by fluorine has only a minor effect on boiling points. Density increases, whereas refractive index and surface tension decrease. Physical constants are shown in Table 17.

The reactivity of fluorine in benzene derivatives depends on the nature of the other ring substituents. Nucleophilic substitution occurs only where activating groups (e.g. nitro) are present in the ortho or para position. In other cases, e.g., with 1-bromo-4-fluorobenzene [460-00-4], the carbon–fluorine bond is stronger; hydrolysis gives 4-fluorophenol [371-41-5] [313], [314].

11.1.2. Production

Among the published processes for the production of mono- or difluoro aromatic compounds, ring fluorination with dilute fluorine or fluorinating agents such as xenon difluoride [315]–[318] does not follow a clear course (i.e., without side reactions) and has no commercial value. Fluorobenzene itself can be prepared by pyrolysis of chlorodifluoromethane or chlorotrifluoroethylene in the presence of cyclodipentadiene at 300–800 °C [319] and by anodic fluorination of benzene with tetraethylammonium fluoride in acetonitrile [320]. Promising methods are the decarbonylation of benzoyl fluorides in the presence of tris(triphenylphosphine)rhodium(I) chloride in boiling xylene [321] and the thermal decarboxylation of aryl fluoroformates [322].

Diazotization. Aromatic compounds containing one or two fluorine substituents are produced commercially by diazotization of aromatic amines and decomposition of the resulting diazonium fluorides. In one process, aromatic amines are diazotized with dry sodium nitrite in anhydrous hydrogen fluoride at 0–20 °C [323].,

$$\text{Ar-NH}_2 + \text{NaNO}_2 + 2\,\text{HF} \xrightarrow[-\text{NaF},\, -2\,\text{H}_2\text{O}]{0-20\,°\text{C}} [\text{Ar-N}{\equiv}\text{N}]^+ \text{F}^- \longrightarrow \text{Ar-F} + \text{N}_2$$

The temperature is increased to 30–120 °C, and the formed diazonium fluorides decompose to form fluoroaromatic compounds. A yield of over 90 % of mono- and difluorinated compounds is obtainable. Yields of ca. 81 % are reported for batch operations on a 1-t scale.

In a continuous process the three exothermic steps are controlled by separation; i.e., hydrofluorination of the aromatic amine, diazotization, and thermal decomposition [324]. This permits safe operation on a large scale. Diazotization with nitrosyl chloride [325] and nitrosyl fluoride–HF complexes [326] is also possible. Problems associated with these processes are hydrogen fluoride recovery and waste gas treatment.

In another method (Balz–Schiemann reaction), water-insoluble diazonium fluoroborates are prepared by diazotization of aromatic amines with sodium nitrite in the presence of 40 % fluoroboric acid or sodium or ammonium tetrafluoroborate in HCl [296], [327]. After filtration the diazonium salts are dried and thermolyzed [328]:

$$\underset{R}{\text{C}_6\text{H}_4\text{NH}_2} \xrightarrow[40\% \text{ HBF}_4]{\text{NaNO}_2/\text{HCl}} \left[\underset{R}{\text{C}_6\text{H}_4\text{N}\equiv\text{N}}\right]^+ \text{BF}_4^-$$

$$\xrightarrow{\text{Heat}} \underset{R}{\text{C}_6\text{H}_4\text{F}} + \text{N}_2 + \text{BF}_3$$

This thermal decomposition must be strictly controlled, especially when nitro substituents are present, to avoid an explosion. Diazonium tetrafluoroborates usually decompose at a higher temperature than the corresponding diazonium fluorides. The Balz–Schiemann process has rarely been used on a large scale because of the difficulties in handling diazonium tetrafluoroborates. However, it is convenient as a laboratory process because it does not require specialized apparatus. A plant of several 100 t/a has been reported [329].

A large number of fluoroaromatic compounds have been produced by using the two diazotization routes described [296], [330]. Those of commercial interest include fluorobenzene from aniline; 2-, 3-, and 4-fluorotoluene from the appropriate toluidines; 4,4′-difluorodiphenylmethane from 4,4′-diaminodiphenylmethane; 1,3-difluorobenzene from *m*-phenylenediamine; and 1,4-difluorobenzene from *p*-phenylenediamine.

The reaction sequence of nitration, reduction, diazotization, and thermal decomposition can be repeated to introduce up to four fluorine atoms into the benzene ring.

Chlorine–Fluorine Exchange. Of similar commercial importance to diazotization is the replacement of activated chlorine atoms with the aid of alkali-metal fluorides. The usual activating groups are ortho and para nitro, cyano, and trifluoromethyl groups [331]. Aprotic–polar solvents are preferred, such as dimethylformamide, dimethylacetamide, dimethyl sulfoxide, *N*-methyl-2-pyrrolidone, and tetrahydrothiophene-1,1-dioxide (sulfolane).,

$$\underset{R}{\text{C}_6\text{H}_4\text{Cl}} + \text{MF} \xrightarrow[\text{Solvent}]{150-250\,°\text{C}} \underset{R}{\text{C}_6\text{H}_4\text{F}} + \text{MCl}$$

R = NO$_2$, CN, CF$_3$; M = K, Cs

Phase-transfer catalysts are sometimes used [332]. The effectiveness of the fluoride source decreases in the order CsF > KF > NaF > LiF.

In commercial batch or semicontinuous operations, the *Halex process* (Fig. 4), potassium fluoride and the activated chloroaromatic compound are thoroughly mixed (1:1) with a large volume of an aprotic solvent (dimethyl sulfoxide or sulfolane) and are heated to 150–250 °C [331], [333]. The reaction is 90% complete within 48 h. Removal of KCl and distillation of the product present no difficulty, but efficient solvent recovery is important to reduce costs.

Cost is also reduced by regenerating the potassium fluoride by treating the KCl with HF [334]. The Halex process has the clear advantage of readily accessible raw materials,

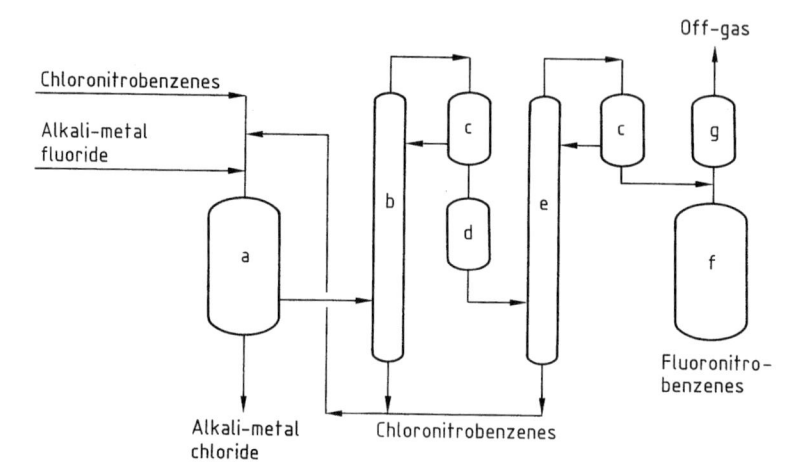

Figure 4. Flow sheet of the Halex process for the manufacture of fluoronitrobenzene [331]
a) Reactor; b) Distillation column; c) Condenser; d) Drier; e) Product distillation column; f) Product storage tank; g) Condenser for off-gas

simplicity of a single-step procedure, and structural specificity. Certain products are accessible in no other way.

The following commercially important intermediates are obtained by the Halex process: 1-fluoro-2-nitrobenzene [1493-27-2] and 1-fluoro-4-nitrobenzene [350-46-9] from the corresponding chloronitrobenzenes; 1-chloro-2-fluoro-5-nitrobenzene [350-30-1] from 1,2-dichloro-4-nitrobenzene; 1-chloro-4-fluoro-3-nitrobenzene [345-18-6] from 1,4-dichloro-2-nitrobenzene; 1-fluoro-2,4-dinitrobenzene [70-34-8] from 1-chloro-2,4-dinitrobenzene; 5-fluoro-2-nitrobenzotrifluoride [393-09-9] from 5-chloro-2-nitrobenzotrifluoride [118-83-2]; 1,3-difluoro-4-nitrobenzene [446-35-5] from 1,3-dichloro-4-nitrobenzene; and 2,6-difluorobenzonitrile [1897-52-5] from 2,6-dichlorobenzonitrile. These intermediates are mostly used for substituted anilines and compounds with carbonyl and carboxy functions.

11.1.3. Uses

Fluorobenzene, difluorobenzenes, and their derivatives are used widely in the synthesis of pharmaceuticals and pesticides, and as fine chemicals. Fluoroaromatics have played a special role in the development of drugs that act on the central nervous system [335]. The increase in lipid solubility due to fluorine atoms facilitates the absorption and transport of drugs through the blood–brain barrier into the central nervous system.

Fluorobenzene derivatives are used in neuroleptics such as haloperidol [52-86-8], tranquilizers such as fluspirilene [1841-19-6], sedatives such as flurazepam [17617-23-1], and antidepressants such as fluroxamin [54739-18-3].

Pesticidal [311] and fungicidal fluorobenzene derivatives [336] include the insecticide diflubenzuron [35367-38-5] (made from 2,6-difluorobenzonitrile), the herbicides flamprop [58677-63-3] and fluoronitrofen [13738-63-1], and the fungicides nuarimol [63284-71-9] and flurimid [41205-21-4].

4,4'-Difluorobenzophenone [345-92-6], a starting material for aromatic polycondensates, is produced by oxidation of 4,4'-difluorodiphenylmethane, which is obtained by diazotization of the corresponding diamine. 4,4'-Difluorobenzophenone undergoes polycondensation with hydroquinone, yielding a polyetheretherketone (PEEK) resin, a thermoplastic. 4-Fluoroaniline [371-40-4], 4-fluorobenzaldehyde [459-57-4], and 4-fluorobenzoic acid [456-22-4] are intermediates for liquid crystal polymers [337].

Aromatic compounds with reactive fluorine substituents are used for the characterization of amino acids (Sanger's reagent, 1-fluoro-2,4-dinitrobenzene [70-34-8]), the immobilization of enzymes (4-fluoro-3-nitrophenylazide [28166-06-5]), and peptide cross-linkage (1,5-difluoro-2,4-dinitrobenzene [327-92-4]).

Aromatic fluorine compounds have recently been developed for medical applications, such as ^{19}F-magnetic resonance imaging (MRI) and in ^{18}F-positron emission tomography (PET) [338].

11.2. Highly Fluorinated Aromatic Compounds

Compounds in which an aromatic ring is substituted by three to five fluorine atoms have little commercial importance. 1,2,4-Trifluorobenzene [367-23-7], bp 88 °C, and 1,3,5-trifluorobenzene [372-38-3], bp 75.5 °C, are produced by the Balz–Schiemann reaction (see Section 11.1.2) from 2,4-difluoroaniline [367-25-9] and 3,5-difluoroaniline [372-39-4], respectively. 1,3,5-Tricyano-2,4,6-trifluorobenzene [3638-97-9], mp 148–150 °C, is used as an intermediate for pesticides; it is obtained by chlorine–fluorine exchange from 1,3,5-tricyano-2,4,6-trichlorobenzene [339].

1,2,3,5-Tetrafluorobenzene [2367-82-0], bp 83 °C, is obtained by the Balz–Schiemann reaction from 2,3,5-trifluoroaniline [363-80-4], and 1,2,4,5-tetrafluorobenzene [327-54-8], bp 90 °C, from 2,4,5-trifluoroaniline [57491-45-9]. 1,2,3,4-Tetrafluorobenzene [551-62-2], bp 95 °C, pentafluorobenzene [363-72-4], bp 85 °C, and hexafluorobenzene are produced by the CoF$_3$ method (see Section 11.3).

The Halex process (Section 11.1.2) is used commercially to produce 1,4-dicyano-2,3,5,6-tetrafluorobenzene [1835-49-0], mp 197–199 °C, from the corresponding tetrachloro derivative [340]. This compound is used as a monomer for thermostable polymers.

11.3. Perhaloaromatic Compounds

Hexafluorobenzene [392-56-3], bp 80.3 °C, has been thoroughly investigated [341]. Nucleophilic substitution produces pentafluoroaromatic compounds such as

bromopentafluorobenzene [344-04-7], bp 135.6 °C
pentafluorophenol [771-56-2], bp 117 – 118 °C
pentafluoroaniline [771-60-8], bp 153 °C
pentafluorobenzene [771-61-9], bp 143 °C
pentafluorotoluene [771-56-2], bp 117 °C
pentafluorobenzaldehyde [653-37-2], bp 164 – 166 °C
pentafluorobenzoic acid [602-94-8], bp 220 °C

Production. Hexafluorobenzene was first produced in 1955 by pyrolysis of tribromofluoromethane [353-54-8] in a platinum tube at 640 °C under atmospheric pressure [342]:,

$$6\, CBr_3F \longrightarrow C_6F_6 + 9\, Br_2$$

Of greater commercial interest is the pyrolysis of an equimolar mixture of dichlorofluoromethane [75-43-40] and chlorofluoromethane [593-70-4] at 600 – 800 °C [343]:

$$3\, CHCl_2F + 3\, CH_2ClF \longrightarrow C_6F_6 + 9\, HCl$$

For many years a commercial multistep process employed CoF_3 as the fluorinating agent [344]. For example, at 150 °C benzene gives a mixture of cyclohexanes containing 8 – 11 fluorine atoms.

$$C_6H_6 \xrightarrow[CoF_2]{CoF_3} C_6H_{12-n}F_n \xrightarrow{MOH} C_6H_{11-n}F_{n-1} + C_6H_{10-n}F_{n-2}$$
$$\downarrow Fe\ or\ Fe_3O_4$$
$$C_6F_6,\ C_6HF_5,\ C_6H_2F_4$$

$n = 8$–11,
M = alkali metal

The CoF_2 formed in this reaction is fluorinated to CoF_3 with elemental fluorine, and reused. The organic products are heated with an alkali-metal hydroxide to form a mixture of polyfluorocyclohexenes and polyfluorocyclohexadienes. These products are aromatized by passage over iron or iron compounds at 400 – 600 °C to give hexafluorobenzene, pentafluorobenzene, and tetrafluorobenzenes. Octafluorotoluene [434-64-0] and the three perfluoroxylenes are also obtained by this method.

The disadvantages of this process are the technological difficulties and the low utilization of fluorine, of which a large part is converted to hydrogen fluoride and alkali-metal fluorides. As a consequence, aromatic fluorine compounds containing other halogens are currently produced by the Halex method (Section 11.1.2). Reaction of hexachlorobenzene [333] with potassium fluoride at 450 °C and 1.03 MPa gives a yield of 21% hexafluorobenzene together with chloropentafluorobenzene [*344-07-0*] (20%), 1,3-dichloro-2,4,5,6-tetrafluorobenzene [*1198-61-4*] (14%), and 1,3,5-trichloro-2,4,6-trifluorobenzene [*319-88-0*](12%).

$$C_6Cl_6 + KF \longrightarrow C_6F_6, C_6ClF_5, C_6Cl_2F_4, C_6Cl_3F_3$$

The yield of hexafluorobenzene can be increased by recycling the other products for further reaction with potassium fluoride. A yield of 42% hexafluorobenzene is obtained from chloropentafluorobenzene with the more reactive, but more expensive, cesium fluoride [345].

Hexachlorobenzene reacts with KF in aprotic solvents such as dimethylformamide, dimethyl sulfoxide, N-methyl-2-pyrrolidone, and sulfolane to give not hexafluorobenzene, but the above-mentioned mixed products. The reaction temperature is 150–250 °C and the residence time 5–36 h. Detailed discussions of the process are given in [331], [333].

Uses. Hexafluorobenzene has been investigated as an inhalation anesthetic in veterinary medicine [346] and as a working fluid in Rankine-cycle engines for temperatures above 350 °C [347]. Derivatives such as pentafluorobenzaldehyde or pentafluorophenyl dimethylsilyl ether are used in the chromatographic analysis of steroids [348] and catecholamines [349], and as intermediates for the production of liquid crystal polymers [350]. Pentafluorophenoxy perfluorovinyl ethers are cross-linking comonomers for perfluorinated elastomers [351].

11.4. Fluorinated Heterocyclic and Polycyclic Compounds

Fluorinated pyridines, pyrimidines, and triazines are heterocycles with commercial importance. Only some representative polycyclic fluorine compounds are discussed as examples.

11.4.1. Ring-Fluorinated Pyridines

Introduction of fluorine into the pyridine ring reduces the basicity of the latter [352]. 2-Fluoropyridine [*372-48-5*], bp 126 °C (100.4 kPa), has a labile fluorine substituent. It is produced in 74% yield by reaction of 2-chloropyridine with potassium bifluoride at 315 °C in 4 h [353]. 3-Fluoropyridine [*372-47-4*], bp 105–107 °C (100.3 kPa), is produced in 50% yield from 3-aminopyridine using the Balz–Schiemann process (see p. 2614). 4-Fluoropyridine [*694-52-0*], bp 108 °C (100 kPa) is produced in 54% yield by diazotization of 4-aminopyridine in anhydrous hydrogen fluoride [354].

2,4-Difluoropyridine [*34941-90-7*], *bp* 104–105 °C, is obtained by the Halex process (see p. 2615) from 2,4-dichloropyridine and potassium fluoride in sulfolane [355]. 2,6-Difluoropyridine [*1513-65-1*], *bp* 124.5 °C (99.1 kPa), is obtained by the reaction of 2,6-dichloropyridine with potassium fluoride in the absence of solvent. The yield is 80% after 18 h at 400 °C [356].

For pentafluoropyridine [*700-16-3*], *bp* 83 °C, the preferred method is the reaction of pentachloropyridine with KF at 480–500 °C; the yield is 83% [357]. Nucleophilic substitution reactions have been thoroughly investigated [358].

Uses. Fluoropyridines are intermediates for pesticides. For example, 2-fluoro-4-hydroxypyridine [*22282-69-5*] is a precursor of 2-fluoro-3,5-dihalo-4-hydroxypyridine herbicides [359]; 2-fluoro-6-hydroxypyridine [*55758-32-2*] is used for insecticides and nematocides [360] and various derivatives of fluorinated pentahalogenpyridine herbicides [361]. Pentafluoropyridine in mixtures with hexafluorobenzene is used as the working fluid in Rankine-cycle engines up to 382 °C [347].

11.4.2. Trifluoromethylpyridines

2-Trifluoromethyl- [*368-48-9*], 3-trifluoromethyl- [*3796-23-4*], and 4-trifluoromethylpyridine [*3796-24-5*] can be obtained by the reaction of picolinic, nicotinic, or isonicotinic acid, respectively, with sulfur tetrafluoride [362]. Commercial processes use side-chain chlorination of the methylpyridines followed by chlorine–fluorine exchange with antimony chlorofluorides or anhydrous hydrogen fluoride [363]. 2-Chloro-5-trifluoromethylpyridine, *bp* 190 °C, is produced commercially from 2-chloro-5-trichloromethylpyridine and hydrogen fluoride at 180–200 °C and at a pressure of 3–4.5 MPa. It is an intermediate for the synthesis of the selective herbicide fluazifop [*69335-91-7*], [364].

11.4.3. Fluoropyrimidines

Certain fluoropyrimidines have gained commercial importance [365]. 5-Fluoropyrimidines are employed in cancer chemotherapy; their biochemistry and pharmacology have been intensively studied [366]. 5-Fluorouracil [*51-21-8*] is obtained in a 80–92% yield by the fluorination of 2,4-dihydroxypyrimidine with fluorine or trifluoromethyl hypofluorite [367], [368].

5-Chloro-2,4,6-trifluoropyrimidine [*697-83-6*], *bp* 114.5 °C (100 kPa), is produced commercially from 2,4,5,6-tetrachloropyrimidine by chlorine–fluorine exchange, using sodium fluoride at 300 °C [369] or anhydrous hydrogen fluoride in the liquid phase [370] or gas phase [371].

The 5-chloro-2,4-difluoropyrimidinyl radical acts as the reactive group in reactive dyes [365] for cellulose and cotton fibers such as Levafix EA (Bayer) and Drimarene K (Sandoz) and for wool, e.g., Verofix (Bayer) and Drimalene (Sandoz).

11.4.4. Fluorotriazines

2,4,6-Trifluoro-1,3,5-triazine [*675-14-9*], *bp* 72.4 °C (101.7 kPa), is produced commercially from 2,4,6-trichloro-1,3,5-triazine (cyanuric chloride) with either anhydrous hydrogen fluoride [372] or sodium fluoride in sulfolane [373]. It is used to manufacture reactive dyes by reaction between one or two fluorine substituents with amino groups of chromophores; the remaining fluorine binds to the fiber [374]. The reactive intermediates can be synthesized e.g., from the corresponding 6-substituted 2,4-dichlorotriazines and an alkali-metal fluoride or by reaction of cyanuric fluoride with anilines or phenols in organic solvents [374], [375]. Like the pyrimidine reactive dyes, the fluorotriazines are used on cellulose, polyesters, polyamides, and wool [365].

11.4.5. Polycyclic Fluoroaromatic Compounds

Polycyclic fluoroaromatic compounds are used as intermediates for the production of pharmaceuticals.

4-Fluoro- [*324-74-3*] and 4,4′-difluorobiphenyl [*398-23-2*] are obtained by diazotization from the corresponding amines in yields of up to 80 % [376]. Other biphenyl derivatives are prepared from fluorobenzenes. The analgesic diflunisal [*22494-42-4*] is produced from 2,4-difluoroaniline [*367-25-9*] by diazotization and coupling with salicylic acid. The anti-inflammatory drugs flurbiprofen [*5104-49-4*]and fluprofen [*17692-38-5*] are prepared from 2-fluoroaniline [*348-54-9*].

The anti-inflammatory drug sulindac [*38194-50-2*] is a monofluorinated indole-3-acetic acid.

12. Economic Aspects

The compounds discussed in Chapters 10 and 11 are not produced in large quantities and often command high prices. 4-Chlorobenzotrifluoride, 3,4-dichlorobenzotrifluoride, 3-trifluoromethylphenyl isocyanate, and benzotrifluoride are produced in the United States by Occidental Chemical, in Western Europe by Hoechst, Rhône Poulenc, Finchimica and Rimar, and in Japan by Daikin. The worldwide production capacity is ca. 35 000 – 40 000 t/a.

Heterocyclic fluorine compounds have economic importance in the production of reactive dyes. The main producers in Western Europe are Bayer AG, Ciba-Geigy, ICI, and Sandoz.

Fluorobenzene and its derivatives are produced in Western Europe by Riedel-de-Haën, Rhône-Poulenc, and Wendstone Chemicals. Fluoronitrobenzenes and fluoroanilines are produced by the Halex process by Hoechst, ISC Chemicals, Shell, and others. The total capacity for fluoroaromatic intermediates is estimated to be several thousand tons per year.

13. Toxicology and Occupational Health

With few exceptions, organic fluorine compounds are physiologically inert and display insignificant toxicity. This is a consequence of the chemical stability of the carbon–fluorine bond and the increased stability of hydrogen and halogen bonds attached to a fluorinated carbon atom. Low toxicity is an important factor in many applications of these compounds.

The difference in toxicity between a chloro or bromo compound and the corresponding fluoro compound is often striking. Thus, carbon tetrachloride [56-23-5] is a powerful liver and kidney toxin and a weak carcinogen, as reflected in its TLV of 5 ppm. However, the product obtained by replacing one chlorine atom with fluorine, trichlorofluoromethane [75-69-4], has no adverse effects on the liver, kidney, or other organs, and no carcinogenic effects on animals exposed to high concentrations for a lifetime; the TLV is 1000 ppm. Another example is the chemical warfare agent mustard gas, $S(CH_2CH_2Cl)_2$ [505-60-2], which is a strong alkylating agent and notorious vesicant. The fluoro analog, bis(2-fluoroethyl) sulfide [373-25-1], is chemically and physiologically inert, with no vesicant properties [377].

The few highly toxic organofluorine compounds usually have easily replaceable fluorine atoms. Examples are diisopropyl fluorophosphate [55-91-4], a potent cholinesterase inhibitor, and perfluoroisobutylene [382-21-8], which causes pulmonary edema at low concentrations. Sodium fluoroacetate [62-74-8], a potent rodenticide, is an exception; it does not liberate fluorine, but interferes with metabolism by mimicking acetic acid.

13.1. Fluorinated Alkanes

Fluoroalkanes [378], [379]. Perfluoroalkanes have very low toxicity. Thus, rats exposed to an 80:20 mixture of perfluorocyclobutane [115-25-3] and oxygen for 4 h survived with no ill effects. Likewise, no ill effects were seen in four species of animals exposed to a 10% concentration in air 6 h/d for 90 d. Partially fluorinated alkanes have similarly low toxicity, as shown by similar experiments with difluoromethane [75-10-5] and 1,1-difluoroethane [75-37-6].

Chlorofluoroalkanes are more toxic than the corresponding fluoroalkanes; nevertheless, most chlorofluoroalkanes have low toxicity [378] – [380]. High concentrations (10 – 50% in air) of many of them, like lower concentrations of many hydrocarbon and chlorohydrocarbon solvents, can cause cardiac sensitization, i.e., sensitization of the heart to the body's adrenalin. This can lead to cardiac arrhythmia (heartbeat irregularity) and sometimes cardiac arrest. Deaths have been caused by "aerosol sniffing".

The toxicity of dichlorodifluoromethane [75-71-8] has been thoroughly investigated. Rats survived a 6-h exposure to an 80% mixture with oxygen. Five species of animals continuously exposed to 810 ppm for 90 d showed no effects except for slight liver damage in guinea pigs. Rats and dogs showed no significant health effects when fed a diet containing 0.3% for 2 years. Teratogenic and reproductive tests in rats were also negative. Repeated exposure caused little or no irritation to rat skin or the rabbit eye. In a screening test, dogs injected with adrenalin showed cardiac sensitization on exposure to 50 000 ppm in air, but not to 25 000. On the basis of animal data and human experience, a TLV of 1000 ppm has been selected to provide an ample margin of safety against cardiac sensitization and other injury [381]. The MAK is also 1000 ppm [382].

Compared to dichlorodifluoromethane, trichlorofluoromethane [75-69-4] and 1,1,2-trichloro-1,2,2-trifluoroethane [76-13-1] are slightly more toxic, whereas 1,2-dichloro-1,1,2,2-tetrafluoroethane [76-14-2] is slightly less toxic. However, for all three a TLV of 1000 ppm is judged to provide an adequate margin of safety. On the other hand, 1,1,1,2-tetrachloro-2,2-difluoroethane [76-11-9] and 1,1,2,2-tetrachloro-1,2-difluoroethane [76-12-0] cause liver and lung damage to rats subjected to repeated exposure at 1000 ppm; therefore, a TLV of 500 ppm is recommended for these compounds. These data indicate that the toxicity of chlorofluorocarbons tends to increase with the chlorine:fluorine ratio and the number of carbon atoms.

Although the toxicity of most chlorofluoroalkanes bearing hydrogen (chlorofluorohydrocarbons) is also low, it tends to be higher than that of the closely related chlorofluorocarbons. The difference is usually slight, as in the case of 2-chloro-1,1,1,2-tetrafluoroethane [2837-89-0], which is a slightly stronger cardiac sensitizer than 1,2-dichloro-1,1,2,2-tetrafluoroethane, but is otherwise similar to it in toxic properties. Chlorodifluoromethane [75-45-6] is similar in toxicity to dichlorodifluoromethane in most respects, and has the same TLV of 1000 ppm. At 50 000 ppm it

Table 18. Toxicities of fluorinated olefins*

Compound	CAS registry no.	Lethal conc., ppm
Vinyl fluoride	[75-02-5]	> 800 000 (ALC)**
Vinylidene fluoride	[75-38-7]	> 800 000 (ALC)**
Tetrafluoroethylene	[116-14-3]	40 000 (LC$_{50}$)
Hexafluoropropene	[116-15-4]	3 000 (LC$_{50}$)
Chlorotrifluoroethylene	[79-38-9]	1 000 (LC$_{50}$)

* Inhalation by rats, 4-h exposure.
** Approximate lethal concentration.

has a weak carcinogenic effect in male rats, but not at lower concentrations or in mice or female rats; therefore, this is considered of no practical significance [383].

The toxicity of dichlorofluoromethane [75-43-4] is more like that of chloroform than of dichlorodifluoromethane or chlorotrifluoromethane [75-72-9], especially with respect to injury on repeated exposure; its TLV, 10 ppm, is low for a chlorofluoroalkane.

Bromofluoroalkanes,/title>some of which are fire extinguishing agents and anesthetics, are more toxic than the corresponding chlorofluoroalkanes, but generally are low in toxicity compared to other fire extinguishing agents and anesthetics [377], [381]. Trifluorobromomethane [75-63-8] produced no adverse effects on dogs and rats exposed to 23 000 ppm 6 h/d, 5 d/week, for 18 weeks; its TLV is 1000 ppm [381].

13.2. Fluorinated Olefins

Most fluorinated olefins have halogen atoms of low reactivity and a correspondingly low-to-moderate toxicity [377], [380]. The toxicities of the five most common members of the class, typical in these respects, are shown in Table 18.

In perfluoroisobutylene and 2,3-dichloro-1,1,1,4,4,4-hexafluoro-2-butene [303-04-8], halogens are readily displaced by nucleophilic reactants; thus, these two compounds exhibit high acute toxicity. Perfluoroisobutene, with a LC$_{50}$ of 0.5 ppm, acts much like phosgene in causing death by pulmonary edema. However, perfluoroisobutene is ca. 10 times as toxic as phosgene, so exposure to it must be carefully avoided. Pyrolysis of tetrafluoroethylene or its polymers above 400 °C is one of its sources.

13.3. Fluorinated Alcohols

2-Fluoroethanol [371-62-0] has a high acute toxicity (LD$_{50}$ 10 mg/kg), a consequence of its ready biological oxidation to fluoroacetic acid (see Section 13.6). Thus, contact with the skin and inhalation of vapors must be avoided. The di- and trifluoroethanols ([359-13-7] and [75-89-8], respectively) have relatively low acute toxicity, similar to the corresponding acetic acids [380]. The acute toxicity of 1,1,1,3,3,3-hexafluoro-2-propanol [*920-66-1*] is also low, but the substance is a strong skin and eye irritant.

13.4. Fluorinated Ketones

In 90-d inhalation studies in animals, hexafluoroacetone [*684-16-2*] caused severe damage to kidneys and other organs at 12 ppm, moderate damage at 1 ppm, and no damage at 0.1 ppm. Repeated skin exposure led to testicular damage in rats. These data and plant experience led to selection of a TLV of 0.1 ppm, with a warning against skin exposure [381]. Studies of hexafluoroacetone and three fully halogenated chlorofluoroketones indicated moderate acute toxicity [384].

13.5. Fluorinated Carboxylic Acids

Fluoroacetic acid [*144-49-0*] is highly toxic to mammals; its sodium salt is an effective, but indiscriminate, rodenticide; in rats the LD_{50} of the salt is only 1.7 mg/kg [381]. In contrast, difluoroacetic acid [*381-73-7*] and perfluoroalkanoic acids have low acute toxicity.

The unusually high toxicity of fluoroacetic acid compared to the more highly fluorinated acids is due to its unique ability to interfere with the citric acid cycle, the oxidation pathway used for energy production from amino acids, fatty acids, and carbohydrates. Fluoroacetic acid enters the cycle at the same site as acetic acid and is converted to fluorocitric acid [*387-89-1*] analogously to the conversion of acetic acid to citric acid. The fluorocitric acid inhibits aconitase, a key enzyme for the breakdown of citric acid, with the result that the citric acid concentration soon rises to lethal levels [385].

Substances that yield fluoroacetic acid on biochemical oxidation, such as straight-chain, even-numbered, ω-fluoro alcohols or alkanoic acids, are also very toxic.

13.6. Other Classes

Simple perfluoroethers, and the oily oligomers of hexafluoropropylene oxide with modified end groups, have the low toxicity expected from their chemical inertness. Hexafluoropropylene oxide itself, a reactive substance, is moderately toxic to rats (4-h LC_{50}, 3700 ppm) [386]. Several partially fluorinated ethers are used as anesthetics, e.g., Enflurane [*13838-16-9*], F_2CHOCF_2CHFCl, and Methoxyflurane [*76-38-0*], $CH_3OCF_2CHCl_2$. Their toxicity is low compared to that of most anesthetics.

Perfluorinated tertiary aliphatic amines are inert both chemically and biologically. This is illustrated by perfluorotripropylamine [*338-83-0*]; in emulsion with perfluorodecalin [*306-94-5*], it has shown promise as a blood substitute in clinical trials [387].

Fluorine substituents usually have little effect on the toxicity of aromatic compounds, whether monocyclic, polycyclic or heterocyclic [384].

14. References

General References

[1] *Houben-Weyl*, **5/3**, pp. 1–502.
[2] *Kirk-Othmer*, **10**, 829–962.
[3] R. C. Downing, F. W. Mader, T. W. Tomkowik: "Fluorocarbons" in J. J. McKetta, *Encyclopedia of Chemical Processing and Design*, vol. **23**, Marcel Dekker, New York 1985, pp. 216–269.
[4] O. Scherer: "Technische Organische Fluorverbindungen" in *Topics in Current Chem.*, vol. **14**, Springer Verlag, Berlin—Heidelberg—New York 1970, pp. 127–234.
[5] R. E. Banks, D. W. A. Sharp, J. C. Tatlow: *Fluorine, The First Hundred Years (1886–1986)*, Elsevier Sequoia, Lausanne—New York 1986; *J. Fluorine Chem.* **33** (1986) 1–399.
[6] R. E. Banks (ed.): *Organofluorine Chemicals and their Industrial Applications*, Ellis Horwood Ltd., Chichester 1979.
[7] R. E. Banks (ed.): *Preparation, Properties and Industrial Applications of Organofluorine Compounds*, Ellis Horwood Ltd. & Wiley Halsted Press, New York—Brisbane—Chichester—Toronto 1982.
[8] M. Sittig: "Fluorinated Hydrocarbons and Polymers," *Chemical Process Monograph* No. 22, Noyes Development Corp., Park Ridge N.J., 1966.
[9] Process Economics Program (PEP) Report No. 166: *Fluorinated Polymers*, SRI International, Menlo Park, Ca. 1983.
[10] J. H. Simons (ed.): *Fluorine Chemistry*, vols. **1–5**, Academic Press, New York 1950–1964.
[11] M. Stacey, J. C. Tatlow, A. G. Sharpe (eds.): *Advances in Fluorine Chemistry*, vols. **1–7**, Butterworths, London 1960–1973.
[12] P. Tarrant (ed.): *Fluorine Chemistry Reviews*, vols. **1–8**, Dekker, New York 1967–1977.
[13] R. E. Banks, M. G. Barlow (eds.): *Fluorocarbons and Related Chemistry*, vols. **1–3**, The Chemical Society, London 1971–1976.
[14] H. J. Emeleus, J. C. Tatlow (eds.): *J. Fluorine Chem.* Elsevier Sequoia S.A., Lausanne.
[15] Chemical Abstracts Selects (CAS[R]): *Organofluorine Chemistry*, American Chemical Society.

Specific References

[16] M. Hudlický, *J. Fluorine Chem.* **33** (1986) 302.
[17] R. E. Banks (ed.): *Organofluorine Chemicals and their Industrial Applications*, Ellis Horwood Ltd., Chichester 1979.
[18] R. E. Banks, J. C. Tatlow, *J. Fluorine Chem.* **33** (1986) 285–286.
[19] R. E. Banks, J. C. Tatlow, *J. Fluorine Chem.* **33** (1986) 71–108.
[20] T. Midgley, A. L. Henne, *Ind. Eng. Chem.* **22** (1930) 542; T. Midgley, *Ind. Eng. Chem.* **29** (1937) 239.
[21] H. Goldwhite, *J. Fluorine Chem.* **33** (1986) 109.

[22] W. A. Sheppard, C. M. Sharts: *Organic Fluorine Chemistry*, W. A. Benjamin Inc., New York 1969.
[23] A. Haas, M. R. C. Gerstenberger, *Angew. Chem.* **93** (1981) 659; *Angew. Chem. Int. Ed. Engl.* **20** (1981) 647.
[24] R. J. Lagow, J. L. Margrave, *Prog. Inorg. Chem.* **26** (1979) 161.
[25] R. J. Lagow, *J. Fluorine Chem.* **33** (1986) 321.
[26] W. Bockemüller, *Justus Liebigs Ann. Chem.* **506** (1933) 20; W. T. Miller, *J. Am. Chem. Soc.* **62** (1940) 341.
[27] G. H. Cady, et al., *Ind. Eng. Chem.* **39** (1947), 290. R. N. Haszeldine, F. Smith, *J. Chem. Soc.* 1950, 2689–94, 2787–89.
[28] J. L. Adcock, K. Horita, E. B. Renk, *J. Am. Chem. Soc.* **103** (1981) 371. J. L. Adcock, *J. Fluorine Chem.* **33** (1986) 327.
[29] Allied, EP 0 032 210, 1984 (H. R. Nykka, J. B. Hino, R. E. Eibeck, J. I. Braumen); EP-A 0 031 519, 1980 (H. R. Nykko, J. B. Hino, R. E. Eibeck, M. A. Robinson).
[30] E. A. Tyczkowski, L. A. Bigelow, *J. Am. Chem. Soc.* **77** (1955) 3007. A. F. Maxwell, F. E. Detoro, L. A. Bigelow, *J. Am. Chem. Soc.* **82** (1960) 5287.
[31] M. Stacey, J. C. Tatlow in [11] vol. **1**, pp. 166–188.
[32] T. Abe, S. Nagase in [7] pp. 19–43.
[33] G. P. Gambaretto, et al., *J. Fluorine Chem.* **27** (1985) 149.
[34] H. M. Fox, F. N. Ruehlen, W. V. Childs, *J. Electrochem. Soc.* **118** (1971) 1246.
[35] Phillips Petroleum Co., US 3 511 760, 1967 (H. M. Fox, F. N. Ruehlen); US 3 686 082, 1970 (F. N. Ruehlen); US 3 709 800, 1971 (H. M. Fox); US 3 882 001, 1973 (K. L. Mills).
[36] A. K. Barbour, L. J. Belf, M. W. Buxton in [11] vol. **3**, pp. 181–270.
[37] Kinetic Chemicals Inc., DE 552 919, 1930.
[38] O. Scherer in [4] vol. **14**, p. 136.
[39] Hoechst, DE 1 000 798, 1952 (O. Scherer, H. Kühn, E. Forche).
[40] Hoechst, GB 1 025 759, 1963 (O. Scherer J. Korinth, P. Frisch).
[41] J. M. Hamilton, Jr. in [11] vol. **3**, pp. 117–180.
[42] L. Dolby-Glover, *Chem. Ind. (London)* 1986, 518–523.
[43] O. Paleta in [11] vol. **8**, pp. 39–71.
[44] H. Suschitzky in [11] vol. **4**, pp. 1–30.
[45] G. A. Boswell, Jr., W. C. Ripka, R. M. Scribner, C. W. Tullock in Wm. G. Dauben et al. (eds.): *Organic Reactions*, vol. **21**, J. Wiley & Sons, New York 1974, pp. 1–124.
[46] Ch.-L. J. Wang in A. S. Kende et al. (eds.): *Organic Reactions*, vol. **34**, J. Wiley & Sons, New York 1985, pp. 319–400.
[47] Y. A. Fialkov, L. M. Yagupolski in I. L. Knunyants, G. G. Yakobson (eds.): *Syntheses of Fluoroorganic Compounds*, Springer Verlag, Berlin—Heidelberg—New York—Tokyo 1985, pp. 233–289.
[48] C. M. Sharts, W. A. Sheppard in [45] pp. 125–406.
[49] R. Filler, *Isr. J. Chem.* **17** (1978) 71.
[50] D. Hebel, O. Lerman, S. Rozen, *J. Fluorine Chem.* **30** (1985) 141.
[51] R. Perry, *J. Fluorine Chem.* **33** (1986) 293.
[52] R. Fields, *J. Fluorine Chem.* **33** (1986) 287.
[53] T. J. Brice in [10] vol. **1**, p. 423.
[54] T. M. Reed III in [10] vol. **5**, p. 133.
[55] R. D. Fowler et al., *Ind. Eng. Chem.* **39** (1947) 292, 319.
[56] A. K. Barbour et al., *J. Appl. Chem.* **2** (1952) 127.
[57] J. M. Tedder in [11] vol. **2**, p. 104.

[58] Du Pont, US 2 458 551, 1947 (A. F. Benning, J. D. Park, S. E. Krahler).
[59] Du Pont, US 2 404 374, 1943 (J. Harmon).
[60] The National Smelting Comp., DE-OS 1 205 533, 1962 (J. C. Tatlow, P. L. Coe).
[61] A. J. Woytek, *J. Fluorine Chem.* **33** (1986) 331.
[62] D. S. L. Slinn, S. W. Green in [7] pp. 45–82.
[63] M. LeBlanc, J. G. Riess in [7] pp. 83–138.
[64] R. T. Watson, M. A. Geller, R. S. Stolarski, R. F. Hampson: *Processes That Control Ozone and Other Climatically Important Trace Gases,* NASA Reference Publication 1162, NASA, Washington 1986.
[65] J. K. Hammitt, et al.: *Product Uses and Market Trends for Potential Ozone Depleting Substances 1985–2000, (R-3386-EPA),* Rand Corp., Santa Monica 1986.
[66] E. R. Larsen in [12] vol. **3**, p. 1.
[67] Hoechst AG, *Frigen, Handbuch für Kälte- und Klimatechnik,* 1969.
[68] BIOS Final Report Nr. 112, I.G. Werke Hoechst (1945).
[69] VEB-Werk Nunchritz, DD 215014, 1983 (I. Lehms, R. Kaden, D. Mross, D. Hass).
[70] M. Vecchio, *Hydrocarbon Process.* **49** (1970) 116.
[71] M. Vecchio, *Hydrocarbon Process.* **52** (1973) 97.
[72] Montecatini Edison S.p.A., US 3 442 962, 1966 (M. Vecchio, L. Lodi).
[73] A. J. Rudge, *Chem. Ind. (London)* 1955, 452–461.
[74] ICI, GB 633 678, 1947 (H. R. Leech, R. L. Burnett).
[75] W. T. Miller, Jr., et al., *Ind. Eng. Chem.* **39** (1947) 333–337.
[76] E. A. Belmore et al., *Ind. Eng. Chem.* **39** (1947) 338–342.
[77] Occidental Chem. Corp., EP-A 93579, 93580, 1982 (M. S. Saran).
[78] L. Scheichl, *VFDB* Z. 3 (1954) 37.
[79] Hoechst, DE-OS 1 155 104, 1960 (O. Scherer, H. Kühn, G. Hörlein, K. H. Grafen); DE-OS 3 046 330, 1980 (H. Bock, J. Mintzer, J. Wittmann, J. Russow).
[80] Hoechst, DE-OS 1 168 404, 1962 (O. Scherer, H. Kühn, G. Hörlein).
[81] ICI, GB 767 779, 1954 (C. W. Suckling, J. Raventos).
[82] *Houben-Weyl,* **5/4,** 51–55, 116–117.
[83] Du Pont, US 2 729 687, 1951 (J. D. Sterling).
[84] Daikin Kogyo KK, JP-Kokai 85 184 033, 1984 (Y. Furutaka, T. Nakada).
[85] *Chem. Eng. News* **53** (1975) no. 36, 18.
[86] Du Pont, US 3 145 222, 1961 (N. O. Brace).
[87] R. N. Haszeldine, *J. Chem. Soc.* 1949, 2856; *J. Chem. Soc.* 1953, 3761.
[88] Du Pont, US 3 234 294, 1962 (R. E. Parsons); DE-OS 1 229 506, 1963 (R. E. Parsons).
[89] R. E. Banks: *Fluorocarbons and their Derivatives,* McDonald Technical & Scientific, London 1970, pp. 102–202.
[90] T. Umemoto et al., *J. Fluorine Chem.* **31** (1986) 37.
[91] R. N. Haszeldine, *J. Chem. Soc.* 1951, 584.
[92] D. Paskovich, P. Gaspar, G. S. Hammond, *J. Org. Chem.* **32** (1967) 833.
[93] Du Pont, US 3 132 185, 1962 (R. E. Parsons).
[94] C. D. Bedford, K. Baum, *J. Org. Chem.* **45** (1980) 347.
[95] Hoechst, EP-A 52285, 1980 (K. v. Werner, A. Gisser).
[96] Daikin Kogyo KK, JP-Kokai 84 048436, 1982 (S. Misaki, T. Kamifukikoshi, M. Suefuji).
[97] *Beilstein* **I (4),** 694 ff.
[98] B. L. Dyatkin, E. P. Mochalina, I. L. Knunyants in [12] vol. 3, p. 45.
[99] R. D. Chambers, R. H. Mobbs in [11] vol. **4,** p. 50.

[100] A. Young in P. Tarrant (ed.): *Fluorine Chemistry Reviews*, vols. **1** Dekker, New York 1967 p. 359.J. D. Park R. J. McMurtry, J. H. Adams in P. Tarrant (ed.): *Fluorine Chemistry Reviews*, vols. **2** Dekker, New York p. 55.

[101] H. E. O'Neal, S. W. Benson, *J. Phys. Chem.* **72** (1968) 1866.

[102] E. W. Cook, J. S. Pierce, *Nature* **242** (1973) 337.

[103] W. H. Sharkey in *Flourine Chemistry Reviews*, vol. 2, pp. 1–53. Du Pont, US 3 316 312, 1959 (D. I. McCane, I. M. Robinson).

[104] A. M. Lovelace, D. A. Rausch, W. Postelnek in: *Aliphatic Fluorine Compounds, A.C.S. Monograph* **138,** Reinhold Publ. Corp., New York 1958, Chap. III.

[105] D. Osteroth: *Chemie und Technologie aliphatischer fluororganischer Verbindungen*, Chap. IB, Ferdinand Enke Verlag, Stuttgart 1964.

[106] R. D. Chambers: *Fluorine in Organic Chemistry*, J. Wiley & Sons, New York—London—Sidney—Toronto 1973, pp. 144–148.

[107] A. Pajaczkowski, J. W. Spoors, *Chem. Ind. (London)* 1964, no. 16, 659.

[108] R. Müller, H. Fischer, *Z. Chem.* **7** (1967) 314.

[109] Du Pont, US 2 407 396, 1943 (M. M. Brubaker); US 2 407 405, 1943 (M. A. Deitrich, R. M. Joyce, Jr.).

[110] B. Atkinson, V. A. Atkinson, *J. Chem. Soc.* 1957, 2086.

[111] Du Pont, US 3 081 245, 1960 (M. W. Farlow); Dow Chemical, US 3 133 871, 1963 (W. R. von Trees).

[112] Kinetic Chemicals, US 2 401 897, 1946 (A. F. Benning, F. B. Downing, R. J. Plunkett).

[113] Du Pont, US 2 894 996, 1955 (M. W. Farlow, L. Muetterties).

[114] Pennsalt Chem. Corp., US 3 009 966, 1960 (M. Hauptschein, A. H. Fainberg).

[115] Du Pont, US 2 551 573, 1945 (F. B. Downing, A. F. Benning, R. C. McHarness).

[116] Hoechst, DE-OS 1 073 475, 1958 (O. Scherer, A. Steinmetz, H. Kühn, W. Wetzel, K. Grafen).

[117] ICI, GB 960 309, 1962 (J. W. Edwards, S. Sherratt, P. A. Small).

[118] Daikin Kogyo KK, GB 1 027 435, 1963.

[119] M. C. Henry, C. B. Griffis, E. C. Stump in [12] vol. **1,** pp. 1–75.

[120] Du Pont, US 2 958 685, 1957 (H. S. Eleuterio).

[121] Du Pont, DE-OS 1 170 935, 1961 (R. H. Halliwell); Asahi Glass Co., US 3 459 818, 1966 (H. Ukihashi, M. Hisasue).

[122] Du Pont, US 2 758 138, 1954 (D. A. Nelson); US 2 970 176, 1957 (E. H. Ten Eyck, G. P. Larson). NL 7 308 149, 1972 (N. E. West); Daikin, Kogyo KK, GB 1 016 016, 1963 (H. Niimiya).

[123] Allied, US 2 734 090, 1950 (J. D. Calfee, C. B. Miller).

[124] Kellogg Co., US 2 774 799, 1954 (M. M. Russel, W. S. Bernhart). Allied, US 2 628 989, 1951 (C. B. Miller). Hoechst, US 3 183 277, 1961 (O. Scherer, A. Steinmetz, H. Kühn, W. Wetzel, K. Grafen).

[125] Bayer, DE-OS 2044370, 1970 (J. N. Meussdoerfer, H. Niederprüm).

[126] Pennsalt Chem. Corp., US 3 188 356, 1961 (M. Hauptschein, A. H. Fainberg).

[127] Du Pont, US 2 599 631, 1945 (J. Harmon).

[128] Du Pont, US 2 419 010, 1943 (D. D. Coffman, T. A. Ford); Dynamit Nobel AG, DE-OS 1 259 329, 1965 (A. Bluemcke, P. Fischer, W. Krings), DE-OS 1 768 543, 1968 (A. Bluemcke, P. Fischer, M. Augenstein, H. J. Vahlensieck).

[129] Montecatini, US 3 200 160, 1962 (D. Sianesi, G. Nelli). Atlantic Richfield Co., US 3 621 067, 1969 (J. W. Hamersma), US 3 642 917, 1969 (J. W. Hamersma).

[130] Dow Corning Corp., US 3 739 036, 1971 (J. A. Valicenti, R. L. Halm, F. O. Stark).

[131] Central Glass Co., JP-Kokai 84 108726, 1982 (T. Nakasora, H. Nakahara).

[132] Halocarbon Products Corp., DE-OS 2 644 595, 1975 (L. L. Ferstanding).
[133] Du Pont, US 4 220 608, 1979 (A. E. Feiring).
[134] A. E. Feiring, *J. Fluorine Chem.* **12** (1978) 471.
[135] Dow Corning Corp., US 3 660 453, 1966 (E. D. Groenhof, H. M. Schiefer).
[136] I. L. Knunyants, Y. A. Cheburkov, *Otdel. Khim. Nauk* 1960, 678.
[137] Daikin Kogyo KK, JP-Kokai 81 090026, 1979 (S. Misaki, S. Takamatsu); JP-Kokai 81 138127, 1980 (S. Takamatsu, S. Misaki).
[138] Allied, US 3 894 097, 1974 (N. Vanderkooi, H. J. Huthwaite).
[139] A. B. Robertson, S. Chandrasekaran, C. S. Chen, *Int. SAMPE Tech. Conf. 1986,* 18 (Mater. Space-Gathering Momentum), 490–9.
[140] *Matheson Gas Data Book,* 4th, 5th ed., Supplements, The Matheson Company, East Rutherford, N.J. 1966–1971.
[141] Kellogg Co., US 2 590 433, 1949 (O. A. Blum); 3M, US 3 014 015, 1953 (J. W. Jewell).
[142] Phillips Petroleum Co., US 3 789 016, 1966 (L. E. Gardner).
[143] F. Liska, *Chem. Listy* **66** (1972) 189. E. D. Bergmann, A. M. Cohen, *Isr. J. Chem.* **8** (1971) 925.
[144] S. K. De, S. R. Palit in J. C. Tatlow, R. D. Peacock, H. H. Hyman (eds.): *Advances in Fluorine Chemistry,* vol. **6,** Butterworths, London 1970, pp. 69–81.
[145] J. A. Young, *Fluorine Chem. Rev.* **1** (1967) 359. M. E. Redwood, C. J. Willis, *Can. J. Chem.* **45** (1967) 389.
[146] W. B. Farnham et al., *J. Amer. Chem. Soc.* **107** (1985) 4565.
[147] Du Pont, US 3 317 616 (V. Weinmayr). R. Filler, R. M. Shure, *J. Org. Chem.* **32** (1967) 1217. B. L. Dyatkin, E. P. Mochalina, I. L. Knunyants, *Tetrahedron* **21** (1965) 2991.
[148] R. D. Chambers: *Fluorine in Organic Chemistry,* Wiley-Interscience, New York 1973, pp. 65–66.
[149] H. Gillman, R. G. Jones, *J. Amer. Chem. Soc.* **70** (1948) 1281.
[150] Abbott Laboratories, US 3 970 710, 1976 (S. Wolownik).
[151] Allied, US 4 072 726, 1978 (H. R. Nychka, R. E. Eibeck, M. A. Robinson, E. S. Jones).
[152] D. R. Husted, A. H. Albrecht, *J. Amer. Chem. Soc.* **74** (1952) 5422.
[153] I. L. Knunyants, SU 138 604, 1961. Allied, BE 634 368, 1963 (J. Hollander, C. Woolf).
[154] Hoechst, DE 1 017 782, 1957 (P. Schlack).M. Goodman, I. G. Rosen, M. Safdy, *Biopolymers* **2**(1964) 503, 519, 537. Du Pont, US 3 418 337, 1968 (W. J. Middleton).
[155] Du Pont, US 3 148 220, 1964 (D. C. England). B. S. Farah, E. E. Gilbert, J. Sibilia, *J. Org. Chem.* **30** (1965) 998.
[156] J. R. Griffith, *Chemtech.* **12** (1982) 290.
[157] H. C. Fielding in [6] Chap. 11.
[158] Allied, US 4 273 947, 1981 (M. Novotny). Phillips Petroleum Co., US 4 396 784, 1983 (M. M. Johnson, G. P. Nowack, P. S. Hudson, B. H. Ashe, Jr.). Hoechst, DE-OS 1 944 381, 1971 (H. Fischer).
[159] Du Pont, US 3 283 012, 1966 (R. Day). F. Mares, B. C. Oxenrider, *J. Fluorine Chem.* **8** (1976) 373. Hoechst, DE-OS 2 318 677, 1974 (H. Millauer). N. O. Brace, *J. Fluorine Chem.* **31** (1986) 151.
[160] T. Nguyen, M. Rubinstein, C. Wakselman, *Synth. Commun.* **13** (1983) 81. Daikin Kogyo Co., US 4 346 250, A 1982 (T. Satokawa, T. Fujii, A. Ohmori, Y. Fujita).
[161] A. M. Lovelace, D. A. Rausch, W. Postelnek: *Aliphatic Fluorine Compounds,* Reinhold Publ. Co., New York 1958, Chap. 5.
[162] A. L. Henne, S. B. Richter, *J. Amer. Chem. Soc.* **74** (1952) 5420.
[163] T. Abe, S. Nagase in [7] Chap. 1.

[164] L. C. Clark, F. Becattini, S. Kaplan, *Ala. J. Med. Sci.* **9** (1972) 16. E. P. Wesseler, R. Iltis, L. C. Clark, *J. Fluorine Chem.* **9** (1977) 137.

[165] G. G. Gerhardt, R. J. Lagow, *J. Chem. Soc., Chem. Commun.* 1976, 259. R. D. Chambers, B. Grievson, F. G. Drakesmith, R. L. Powell, *J. Fluorine Chem.* **29** (1985) 323.

[166] P. Tarrant, C. G. Allison, K. P. Barthold, E. C. Stump, Jr., *Fluorine Chem. Rev.* **5** (1971) 77.

[167] H. Millauer, W. Schwertfeger, G. Siegemund, *Angew. Chem. Int. Ed. Engl.* **24** (1985) 161.

[168] Du Pont, FR 1 322 597, 1963 (H. H. Biggs, J. L. Warnell). Du Pont, US 3 536 733, 1970 (D. P. Carlson). Nippon Mektron, JP 57 134 473, 1982.

[169] Hoechst, DE-OS 2 658 328, 1978 (M. Millauer, W. Lindner). M. Millauer, *Chem. Ing. Tech.* **52** (1980) 53.

[170] Asahi Chemical, EP-OS 64 293, 1982 (M. Ikeda, M. Miura, A. Aoshima). Daikin Kogyo, JP 58 39 472, 1983; JP 58 49 372, 1983. Asahi Chemical, JP 59 78 176, 1984.

[171] Du Pont, US 3 358 003, 1967 (H. S. Eleuterio, R. W. Meschke). Hoechst, DE-OS 2 557 655, 1977 (R. Sulzbach, F. Heller).

[172] R. C. Kennedy, J. B. Levy, *J. Fluorine Chem.* 7 (1974) 101. P. B. Sargeant, *J. Amer. Chem. Soc.* **91** (1969) 3061.

[173] Asahi Glass, JP 58 62 131 A2, 1983; JP 58 62 130 A2, 1983. Hoechst, EP-A 54 227, 1982 (P. P. Rammelt, G. Siegemund). Du Pont, US 4 302 608, 1981 (E. N. Squire). Du Pont, GB 1 019 788, 1966 (E. P. Moore, A. S. Milian). 3M, US 3 213 134, 1965 (D. E. Morin).

[174] R. Watts, P. Tarrant, *J. Fluorine Chem.* **6** (1975) 481.

[175] Du Pont, US 3 125 599, 1964 (J. L. Warnell).

[176] P. L. Coe, A. Sellers, J. C. Tatlow, *J. Fluorine Chem.* **23** (1983) 102. Du Pont, GB 904 877, 1962. 3M, US 3 213 134, 1965 (D. E. Morin). E. M. Rokhlin et al., *Dokl. Akad. Nauk SSSR* **161** (1965) 1356.

[177] H. S. Eleuterio, *J. Macromol. Sci. Chem.* **A-6** (1972) 1027. J. T. Hill, *J. Macromol. Sci. [A] Chem.* **8** (1974) 499.

[178] D. Sianesi, A. Pasetti, R. Fontanelli, G. C. Bernardi, G. Caporiccio, *Chim. Ind. (Milan)* **55** (1973) 221. D. Sianesi, A. Pasetti, C. Corti, *Macromol. Chem.* **86** (1965) 308. D. Sianesi, R. Tontanelli, *Macromol. Chem.* **102** (1967) 115.

[179] A. Faucitano, A. Buttafava, G. Caporiccio, C. T. Viola, *J. Amer. Chem. Soc.* **106** (1984) 4172.

[180] B. C. Anderson, L. R. Barton, J. W. Collette, *Science* **208** (1980) 807.

[181] R. E. Uschold, *Polym. J.* **17** (1985) 253. A. L. Barney, W. J. Keller, N. M. von Gulick, *J. Polym. Sci. Part A* **1** (1970) 1091.

[182] Du Pont, US 3 250 808, 1966 (E. P. Moore, A. S. Milian, Jr., H. S. Eleuterio). Du Pont, US 3 291 843, 1966 (C. G. Fritz, S. Selman). Hoechst, US 4 118 421, 1978 (T. Martini).

[183] Du Pont, US 4 281 092, 1981 (A. F. Breazeale). Du Pont, US 3 467 638, 1969 (D. B. Pattison). G. H. Kalb, A. A. Khan, R. W. Quarles, A. L. Barney: *Advances in Chemistry Series, 129*, Amer. Chem. Soc., Washington, D.C., 1973.

[184] A. Eisenberg, H. Yeager (eds.): "Perfluorinated Ionomer Membranes," *ACS Symposium Series, 180*, Amer. Chem. Soc., Washington, D.C., 1982.

[185] Du Pont, US 3 301 893, 1967 (R. E. Putnam, W. D. Nicoll).

[186] Du Pont US 3 282 875, 1966 (D. J. Connally, W. F. Gresham). W. Grot, *Chem.-Ing.-Tech.* **47** (1975) 617.

[187] Asahi Glass, GB 2 029 827 A, 1979 (M. Yamabe, S. Kumai, S. Munekata).

[188] D. C. England, L. R. Melby, M. A. Dietrich, R. V. Lindsey, Jr., *J. Amer. Chem. Soc.* **82** (1960) 5116.

[189] I. L. Knunyants, A. I. Shchekotiklin, A. V. Fokin, *Izv. Akad. Nauk SSSR, Otd. Khim. Nauk* 1953, 282.
[190] D. C. England, L. Solomon, C. G. Krespan, *J. Fluorine Chem.* **3** (1973/1974) 4.
[191] D. C. England, *J. Org. Chem.* **49** (1984) 4007.
[192] F. W. Evans, M. H. Litt, A.-M. Weidler-Kubanek, F. P. Avonda, *J. Org. Chem.* **33** (1968) 1839.
[193] Dow Chemical, US 3 549 711, 1970 (C. I. Merrill, N. L. Madison).
[194] W. G. M. Jones in [7] Chap. 5.R. Filler in R. E. Banks (ed.): *Organofluorine Chemicals and their Industrial Applications*, Ellis Horwood Ltd., Chichester 1979, Chap. 6; W. G. M. Jones in [6] Chap. 7. E. R. Larsen, *Fluorine Chem. Rev.* **3** (1969) 1.
[195] Air Reduction Co., US 2 870 218, 1959 (P. W. Townsend).
[196] J. M. Baden, M. Kelley, R. S. Wharton, B. A. Hitt, V. F. Simon, R. I. Mazze, *Anesthesiology* **46** (1977) 346.
[197] Air Reduction Co., GB 1 138 406, 1969 (R. C. Terrell).
[198] Air Reduction Co., US 3 535 425, 1970 (R. C. Terrell).A. P. Adams, *Br. J. Anaesth.* **53** (1981) Suppl. 1, 27.
[199] Dow Chemical, US 3 264 356, 1966 (E. R. Larsen).
[200] M. J. Halsey, *Br. J. Anaesth.* **53** (1981) Suppl. 1, 43.
[201] G. A. Olah, C. V. Pittman, *J. Am. Chem. Soc.* **88** (1966) 3310.
[202] *Houben-Weyl* **7 (2C)**, 2147.
[203] *Chem. Eng.* **75** (1968) 22.
[204] Allied, US 3 164 637, 1965 (H. R. Nychka, C. Woolf).
[205] Du Pont, FR 1 372 549, 1964 (F. W. Swamer). *Chem. Abstr.* **62** (1965) 6397 a.
[206] M. Van Der Puy, L. G. Anello, *Org. Synth.* **63** (1985) 154.
[207] Allied, FR 1 369 784, 1964 (L. G. Anello, H. R. Nychka, C. Woolf). *Chem. Abstr.* **62** (1965) 1570 f.
[208] I. Mellan: *Ketones*, Chemical Publ. Co., New York 1968, pp. 64–68.
[209] W. J. Middleton, R. V. Lindsey, Jr., *J. Am. Chem. Soc.* **86** (1964) 4948.
[210] Allied, US 3 620 666, 1971 (R. W. Lenz, L. Barish, V. L. Lyons).
[211] USDA, US 3 998 620, 1976 (A. G. Pittman, W. L. Wasley).
[212] Stauffer Chemical Co., US 3 594 362, 1971 (K. Szabo).
[213] Stauffer Chemical Co., NE-A 6 410 536, 1965 (K. Szabo, G. P. Willsy, Jr.). *Chem. Abstr.* **63** (1965) 11577 g.
[214] Allied, US 3 177 187, 1965 (J. Hollander, C. Woolf).
[215] Allied, US 3 177 185, 1965 (J. Hollander, C. Woolf).
[216] Allied, NL-A 6 407 623, 1965 (E. E. Gilbert, J. A. Otto).
[217] Allied, NL-A 6 407 548, 1965 (E. E. Gilbert, B. C. Oxenrider, G. J. Schmitt).
[218] L. Speers, A. J. Szur, R. C. Terrell, *J. Med. Chem.* **15** (1972) 606.
[219] M. Hudlicky: *Chemistry of Organic Fluorine Compounds*, 2nd ed., Ellis Horwood, Chichester 1976, pp. 353, 712.
[220] [219] p. 714.
[221] Commonwealth Scientific and Industrial Research Organization, JP 54 64 728, 1979 (G. Holan). *Chem. Abstr.* **91** (1979) 56635 g.
[222] Baxter, US 3 659 023, 1972 (B. M. Regan).
[223] Allied, US 3 450 773, 1969 (R. E. A. Dear, E. E. Gilbert).
[224] D. W. Lipp, O. Vogl in T. Saegusa, E. Goethals (eds.): *ACS Symposium Series 59*, Amer. Chem. Soc., Washington, D.C., 1977, Chapter 8, pp. 111–128.
[225] B. Yamada, R. W. Campbell, O. Vogl, *Polym. J.* **9** (1977) 23.

[226] *Beilstein* **1 (4)**, 3133.
[227] CS 183 467, 1980 (A. Posta et al.). *Chem. Abstr.* **94** (1981) 156316j.
[228] Hoechst, DE-OS 2 139 211, 1973 (G. Siegemund).
[229] G. Siegemund, *Chem. Ber.* **106** (1973) 2960.
[230] Sumitomo, JP-Kokai 60 64 949, 1985.
[231] Hoechst, DE-OS 2 656 548, 1978 (G. Siegemund).
[232] Hoechst, DE-OS 2 344 442, 1975 (H. Kuehn, F. Kluge, G. Siegemund).
[233] Hoechst, DE-OS 2 361 058, 1975 (G. Siegemund, R. Muschaweck).
[234] Hoechst, DE-OS 2 340 560, 1975 (G. Siegemund, R. Muschaweck).
[235] K. C. Joshi, *J. Indian Chem. Soc.* **59** (1982) 1375.
[236] K. C. Joshi, V. N. Pathak, *Coord. Chem. Rev.* **22** (1977) 37.
[237] P. Mushak, M. T. Glenn, J. Savory, *Fluorine Chem. Rev.* **6** (1973) 43.
[238] K. G. Seeley, W. J. McDowell, L. K. Felker, *J. Chem. Eng. Data* **31** (1986) 136.
[239] K. A. Kline, R. E. Sievers, *Aldrichimica Acta* **10** (1977) 54.
[240] A. L. Henne, C. J. Fox, *J. Am. Chem. Soc.* **73** (1955) 2323.
[241] S. Nagase in [12] vol. **1,** p. 77.
[242] J. Burdon, J. C. Tatlow in [11] vol. **1,** p. 129.
[243] V. Childs in N. Weinberg, B. Tilak (eds.): *Technique in Electroorganic Synthesis,* vol. **5,** Wiley-Interscience, New York 1982, Part 3, p. 341.
[244] Kali-Chem, US 3 725 475, 1973 (H. Paucksch, J. Massonne, H. Bohm, G. Fernschild).
[245] F. L. M. Pattison: *Toxic Aliphatic Fluorine Compounds,* Elsevier, Amsterdam 1959, p. 387.
[246] M. Hauptschein, *J. Am. Chem. Soc.* **83** (1961) 2500.
[247] Pennwalt, US 3 328 240, 1966 (M. Hauptschein).
[248] Hoechst, US 4 400 325, 1983 (K. von Werner, A. Gisser).
[249] Daikin Kogyo, US 3 525 758, 1970 (A. Katsushima).
[250] Asahi Glass, US 4 138 417, 1979 (H. Ukihashi, T. Hayashi, Y. Takashi et al.).
[251] Du Pont, US 3 250 808, 1966 (E. Moore, A. Milian, E. Eleuterio); US 3 322 826, 1967 (E. Moore).
[252] Du Pont, GB 1 108 128, 1968.
[253] S. Kumai, S. Samejima, M. Yamabe, *Rep. Res. Lab. Asahi Glass Co. Ltd.* **29** (1979) 105.
[254] Asahi Glass, US 4 151 200, 1979 (M. Yamabe, S. Kumai).
[255] A. L. Henne, W. J. Zimmersheid, *J. Am. Chem. Soc.* **69** (1947) 281.
[256] E. T. McBee, P. A. Wiseman, G. B. Bachman, *Ind. Eng. Chem.* **39** (1947) 415.
[257] S. Stinson, *Chem. Eng. News* 1982 (March 15) 22.
[258] R. N. Hazeldine, J. M. Kidd, *J. Chem. Soc.* 1954, 4228.
[259] T. Gramstad, *Tidsskr. Kjemi Bergves. Metall.* **19** (1959) 62.
[260] Bayer, DE-OS 2 201 649, 1973 (P. Voss, H. Niederpruem, M. Wechsberg).
[261] R. N. Hazeldine, J. M. Kidd, *J. Chem. Soc.* 1955, 2901.
[262] Hercules, US 4 505 990, 1985 (S. Dasgupta et al.).
[263] Eastman Kodak, US 4 582 781, 1986 (J. Chen, J. Kelly, J. Plakunov et al.).
[264] 3M, US 2 803 656, 1957 (A. Ahlbrecht, H. Brown et al.).
[265] 3M, US 2 803 615, 1957 (A. Ahlbrecht, H. Brown, S. Smith et al.).
[266] 3M, US 3 094 547, 1963 (R. Heine et al.).
[267] R. B. Ward, *J. Org. Chem.* **30** (1965) 3009.
[268] 3M, US 4 329 478, 1982 (F. Behr).
[269] Du Pont, US 3 301 893, 1967 (R. Putnam, W. Nicoll et al.).

[270] S. Torrey (ed.): *Membrane and Ultrafiltration Technology*, Noyes Data Corp., Park Ridge, N.J., 1980, p. 132.
[271] A. Eisenberg, H. Yeager (eds.): *Perfluorinated Ionene Membranes*, Amer. Chem. Soc., Washington, D.C., 1982.
[272] Du Pont, US 3 282 875, 1964 (D. Connolly, W. Gresham).
[273] Du Pont, FR 1 406 778, 1963 (R. Putnam, W. Nicoll).
[274] F. E. Behr, R. J. Koshar: *Abstracts Fifth Winter Fluorine Conference*, Daytona Beach, Fla., 1981, p. 5.
[275] 3M, EP 0 058 466, 1982 (F. Behr, R. Koshar et al.).
[276] Asahi Chemical Industry, GB 2 051 831, 1980 (K. Kimoto, H. Miyauchi, J. Ohmura, M. Ebisawa et al.).
[277] C. Jackson (ed.): "Modern Chloro-alkali Technology," *3rd Int. Chlorine Symp. 1982*, Horwood, United Kingdom 1983, p. 79.
[278] "Fluorinert Electronic Liquids," Product Manual, 3M Co., St. Paul, Minn., 1985.
[279] R. N. Haszeldine, *Research (London)* **3** (1950) 430. R. N. Hazeldine, *Chem. Soc.* 1951, 102.
[280] 3M Co., US 2 616 927, 1952 (E. A. Kauck, J. H. Simons).
[281] G. G. I. Moore et al., *J. Fluorine Chem.* **32** (1986) no. 1, 41.
[282] H. G. Bryce: "Industrial and Utilitarian Aspects of Fluorine Chemistry," in [10] vol. **5**, pp. 302–346.
[283] T. Abe, S. Nagase, R. E. Banks (ed.): *Preparation, Properties and Industrial Applications of Organofluorine Compounds*, Chap. 1, Ellis Horwood Ltd. & Wiley Halsted Press, New York—Brisbane—Chichester—Toronto 1982, p. 35; D. S. L. Slinn, S. W. Green,
R. E. Banks (ed.): *Preparation, Properties and Industrial Applications of Organofluorine Compounds*, Chap. 2, Ellis Horwood Ltd. & Wiley Halsted Press, New York—Brisbane—Chichester—Toronto 1982, pp. 45–80; M. Le Blanc, J. E. Riess,
R. E. Banks (ed.): *Preparation, Properties and Industrial Applications of Organofluorine Compounds*, Chap. 3, Ellis Horwood Ltd. & Wiley Halsted Press, New York—Brisbane—Chichester—Toronto 1982, pp. 83–116.
[284] F. Swarts, *Bull. Cl. Sci. Acad. R. Belg.* **35** (1898) 375.
[285] Y. Kobayashi, I. Kumadaki, *Tetrahedron Lett.* 1969, 4095.
[286] Y. Kobayashi, I. Kumadaki, *Chem. Pharm. Bull.* **18** (1970) 2334.
[287] V. C. R. McLaughlin, J. Thrower, *Tetrahedron* **25** (1969) 5921.
[288] F. Swarts, *Bull. Cl. Sci. Acad. R. Belg.* 1920, 399.
[289] Y. Kobayashi, I. Kumadaki, *Acc. Chem. Res.* **11** (1978) 197.
[290] A. L. Henne, M. S. Newman, *J. Am. Chem. Soc.* **60** (1938) 1697.
[291] R. Filler, H. Novar, *Chem. Ind. (London)* 1960, 1273.
[292] R. Grinter et al., *Tetrahedron Lett.* 1968, 3845.
[293] Hooker Chemical, US 3 234 292, 1966 (St. Robota, E. A. Belmore).
[294] R. J. Albers, E. C. Kooyman, *Rec. Trav. Chim. Pays Bas* **83** (1964) 930.
[295] Montedison, DE-OS 2 746 787, 1976 (M. Bornengo).
[296] G. Schiemann, B. Cornils: *Chemie und Technologie cyclischer Fluorverbindungen*, Ferdinand Enke Verlag, Stuttgart 1969, pp. 188–193.
[297] Hoechst (IG Farbenindustrie), DE 575 593, 1931 (P. Osswald, F. Müller, F. Steinhäuser).
[298] Hoechst, DE 2 356 257, 1973 (R. Lademann, H. Lindner, Th. Martini, P. Rammelt).
[299] A. K. Barbour et al. in [11] vol. **3**, p. 181.
[300] Kinetic Chem. Inc., US 1 967 244, 1932 (L. C. Holt, E. Lorenzo). J. H. Brown et al., *J. Chem. Soc. Suppl. Issue I*, 1949, 95.

[301] Hoechst, DE 1 618 390, 1967 (O. Scherer, G. Schneider, R. Hähnle).
[302] Hoechst, DE 1 965 782, 1969 (H. Hermann, H. Lindner).
[303] Hooker Chem. Corp., DE-AS 1 543 015, 1965 (St. Robota).
[304] Kali-Chemie, DE-OS 2 161 995, 1971 (H. Böhm, J. Massonne).
[305] Olin Mathieson Chem. Corp., US 3 136 822, 1961 (L. J. Frainier).
[306] Bayer, DE-OS 2 837 499, 1978 (A. Marhold, E. Klauke).
[307] *Ullmann*, 3rd ed., **7** 636–639. G. Wolfrum in [6] pp. 208–213.
[308] T. Fujita, J. Iwasa, C. Hansch, *J. Am. Chem. Soc.* **86** (1964) 5175.
[309] C. Hansch, R. M. Muir, T. Fujita et al., *J. Am. Chem. Soc.* **85** (1963) 2817. C. Hansch, J. Fukunaga, *CHEMTECH* **7** (1977) 120.
[310] R. Filler in [7] pp. 123–153.
[311] G. T. Newbold in [6] pp. 169–187.
[312] W. A. Sheppard, C. M. Sharts: *Organic Fluorine Chemistry*, W. A. Benjamin, New York 1969, pp. 18–49.
[313] G. G. Yakobson, T. D. Rubina, N. N. Vorozhtsov, Jr., *Dokl. Akad. Nauk SSSR* **141** (1961) 1395.
[314] M. M. Boudakian et al., *J. Org. Chem.* **26** (1961) 4641.
[315] US Atomic Energy Com., US 3 833 581, 1971 (D. R. McKenzie, J. Fajer, R. Snol).
[316] V. Grakauskas, *J. Org. Chem.* **35** (1970) 723; **34** (1969) 2835.
[317] M. J. Shaw et al., *Chem. Eng. News* **47** (1969) no. 40, 45; *J. Am. Chem. Soc.* **92** (1970) 6498.
[318] K. Bottenberg, *Chem. Ztg.* **96** (1972) 84.
[319] Institut Organicheskoi Khimii, Imeni Zelinskogo, DE-OS 1 618 564, 1967 (O. M. Nefedov, A. A. Ivashenko).
[320] I. N. Rozhkov, A. V. Bukhtiarov, I. L. Knunyants, *Iszv. Akad. Nauk SSSR Ser. Khim.* 1972, 1130.
[321] G. A. Olah, P. Kreienbühl, *J. Org. Chem.* **32** (1967) 1614.
[322] K. O. Christe, A. E. Pavlath, *J. Org. Chem.* **31** (1966) 559.
[323] Hoechst, DE 600 706, 1933 (P. Osswald, O. Scherer). FIAT Final Rep. Nr. 998, I.G. Werke Hoechst (1947). R. L. Ferm, C. A. v. d. Werf, *J. Am. Chem. Soc.* **72** (1950) 4809.
[324] Riedel-de-Haen, DE-OS 3 520 316, 1985 (E. Begemann, H. Schmand).
[325] Harshaw Chemical Company, US 2 563 796, 1951 (W. J. Shenk, G. R. Pellon).
[326] Allied, FR 611 545, 1960 (L. G. Anello, C. Woolf).
[327] H. Suschitzky in [11] vol. **4**, p. 1.
[328] D. T. Flood in A. H. Blatt (ed.): *Organic Syntheses*, Collective Volume II, J. Wiley & Sons, New York 1943, p. 295.
[329] J. P. Regan: *Meeting on Fluoro-organic Compounds in Industry*, Society of Chemical Industry, London 23rd January 1986.
[330] A. E. Pavlath, A. J. Leffler: "Aromatic Fluorine Compounds," *ACS Monogr.* **155** (1962).
[331] W. Prescott, *Chem. & Ind. (London)* 1986, 56.
[332] Boots Co., DE-OS 2 938 939, 1978 (R. A. North).
[333] L. Dolby-Glover, *Chem. & Ind. (London)* 1986, 518.
[334] ISC Chemicals Ltd., GB 1 143 088, 1966 (D. W. Cottrell).
[335] A. J. Elliott in R. Filler, Y. Kobayashi (eds.): *Biomedicinal Aspects of Fluorine Chemistry*, Kodansha Ltd., Tokyo & Elsevier Biomedical, Amsterdam—New York—Oxford 1982, pp. 55–74.
[336] G. C. Finger, F. H. Reed, L. R. Tehon, *Illinois State Geol. Survey*, Circular No. 199, Urbana (1955). R. S. Shiley et al., *J. Fluorine Chem.* **5** (1975) 371.

[337] Thomson-CSF, US 3 979 321, 1976 (A. Couttet, J. C. Dubois, A. Zaum).American Mikro-System, US 4 018 507, 1977 (N. V. V. Raaghavan).J. P. Van Meter, *Eastman Org. Chem. Bull.* **45** (1973) 1.
[338] R. Filler, *J. Fluorine Chem.* **33** (1986) 361.
[339] K. Wallenfels, F. Witzler, K. Friedrich, *Tetrahedron* **23** (1967) 1845.
[340] Keishin Matsumoto, JP-Kokai 51 6940, 1974 (S. Fujii, K. Inoukai).
[341] *Kirk-Othmer,* **10,** 915. L. S. Kobrina [12] vol. **7,** pp. 1–114.
[342] Y. Desirant, *Bull Cl. Sci. Acad. R. Belg.* **41** (1955) 759.
[343] Sperry Rand. Corp., US 3 158 657, 1962 (F. R. Callihan, Ch. L. Quatela).
[344] M. Stacey, J. C. Tatlow in [11] vol. **1,** p. 166.
[345] G. Fuller, *J. Chem. Soc.* 1965, 6264.
[346] L. W. Hall, S. R. K. Jackson, G. M. Massey, *Int. Congr. Ser. Exerpta Med.* 1974,347.
[347] D. R. Miller et al., *Intersoc. Energy Convers. Engl. Conf. Conf. Proc. 7th,* (1972) 315.
[348] E. D. Morgan, C. F. Poole, *J. Chromatogr.* **104** (1975) 351.
[349] J. C. Lhuguenot, B. F. Maume, *J. Chromatogr. Sci.* **12** (1974) 411.
[350] M. M. Murza et al., *Zh. Org. Khim.* **13** (1977) 1046.
[351] Du Pont, FR 1 555 360, 1969 (D. B. Pattison).
[352] H. C. Brown, D. McDaniel, *J. Am. Chem. Soc.* **77** (1955) 3752.
[353] Olin, US 3 296 269, 1967 (M. M. Boudakian).
[354] Olin, US 3 703 521, 1972 (M. M. Boudakian).
[355] ICI, DE-OS 2 128 540, 1971 (A. Nicolson).
[356] M. M. Boudakian, *J. Heterocycl. Chem.* **5** (1968) 683.
[357] J. A. Burdon et al., *Nature (London)* **186** (1960) 231.
[358] G. G. Yakobson, T. D. Petrova, L. S. Kobrina in [7] vol. **7,** pp. 115–223.
[359] ICI, GB 1 479 721, 1973 (D. W. R. Headford, J. W. Slater, R. L. Sunley, R. D. Bowden, M. B. Green).
[360] Dow, DE 2 263 429, 1975 (R. H. Rigterink); US 3 810 902, 1974 (R. H. Rigterink).
[361] ICI, US 4 063 926, 1977 (Sp. Tomlin, J. W. Slater, D. Hartley); US 3 850 943, 1974 (R. D. Bowdon, J. W. Slater, B. G. White).
[362] M. S. Raasch, *J. Org. Chem.* **27** (1962) 1406.
[363] E. T. McBee, H. B. Hass, E. M. Hodnett, *Ind. Eng. Chem.* 2nd ed. 39 (1947) 389.
[364] Ishihara, DE-OS 2 812 571, 1979 (R. Nishiyama, T. Haga, N. Sakashita).
[365] W. Harms in [6] pp. 188–207.
[366] Acad. Sci. USSR, DE 2 014 216, 1971 (J. L. Knuny-ants, L. S. German, N. B. Kazmina).
[367] PCR, US 3 954 758, 1976 (P. D. Schuman, P. Tarrant, D. A. Warner, G. Westmoreland); US 4 113 949, 1978 (P. D. Schumann, P. Tarrant, D. A. Warner, G. Westmoreland).
[368] D. V. Santi, A. L. Pogolotti, Jr., E. M. Newman, Y. Wataya in [335] pp. 123–142.
[369] Bayer, FR 1 546 305, 1968 (E. Klauke, H. S. Bien); LU 58 622, 1969 (E. Klauke, H. S. Bien).
[370] Bayer, GB 1 158 300, 1966 (E. Klauke, H. S. Bien).
[371] Bayer, GB 1 273 914, 1969 (H. U. Alles, E. Klauke, H. S. Bien).
[372] Bayer, DE 1 044 091, 1957 (A. Dorlars).
[373] C. W. Tullock, A. D. Coffman, *J. Org. Chem.* **25** (1960) 2016.
[374] Bayer, BE 713 937, 1967 (H. S. Bien, E. Klauke, K. Wunderlich); BE 716 013, 1967 (H. S. Bien, E. Klauke, K. Wunderlich); BE 716 014, 1967 (H. S. Bien, E. Klauke, K. Wunderlich).
[375] Bayer, GB 872 313, 1957 (A. Dorlars).
[376] G. Schiemann, E. Bolsted, *Chem. Ber.* **61** (1928) 1403.

[377] J. W. Clayton: "Fluorocarbon Toxicity and Biological Action," in P. Tarrant (ed.): *Fluorine Chemistry Reviews,* vol. **1,** Dekker, New York 1967, pp. 197–252.

[378] R. R. Montgomery, C. F. Reinhardt: "Toxicity of Fluorocarbon Propellants," in P. A. Sanders (ed.): *Handbook of Aerosol Technology,* 2nd ed., Van Nostrand Reinhold, New York 1979, pp. 466–511.

[379] D. M. Aviado: "Fluorine-Containing Organic Compounds," in G. D. Clayton, F. E. Clayton (eds.): *Patty's Industrial Hygiene and Toxicology,* 3rd ed., vol. **2B,** Wiley-Interscience, New York 1981, pp. 3071–3115.

[380] J. W. Clayton: "The Mammalian Toxicology of Organic Compounds Containing Fluorine," in F. A. Smith (ed.): *Handbook of Experimental Pharmacology,* vol. **20/1,** Springer Verlag, New York 1966, pp. 459–500.

[381] ACGIH (ed.): *Documentation of the Threshold Limit Values (TLV),* ACGIH, Cincinnati, Ohio 1986.

[382] DFG (ed.): *Maximum Concentrations at the Workplace and Biological Tolerance Values for Working Materials,* Verlag Chemie, Weinheim 1984, Report 20.

[383] M. H. Litchfield, E. Longstaff, *Food Chem. Toxicol.* **22** (1984) 465.

[384] J. F. Borzelleca, D. Lister, *Toxicol. Appl. Pharmacol.* **7** (1965) 592.

[385] L. Stryer: *Biochemistry,* Freeman, San Francisco 1981, p. 298.

[386] H. Millauer, W. Schwertfeger, G. Siegemund, *Angew. Chem. Int. Ed. Engl.* **24** (1985) 161.

[387] R. Frey, H. Beisbarth, K. Stosseck: *5th International Symposium on Perfluorochemical Blood Substitutes,* Mainz, Mar. 1981, Zuckschwerdt Verlag, Munich 1981, pp. 86–90.

Formaldehyde

GÜNTHER REUSS, BASF Aktiengesellschaft, Ludwigshafen, Federal Republic of Germany (Chaps. 1, 2, 4, and 5)

WALTER DISTELDORF, BASF Aktiengesellschaft, Ludwigshafen, Federal Republic of Germany (Chaps. 3, 6–9)

ARMIN OTTO GAMER, BASF Aktiengesellschaft, Ludwigshafen, Federal Republic of Germany (Chap. 10)

ALBRECHT HILT, Ultraform GmbH, Ludwigshafen, Federal Republic of Germany (Chaps. 11, 12)

1.	Introduction	2640
2.	Physical Properties	2641
2.1.	Monomeric Formaldehyde . . .	2641
2.2.	Aqueous Solutions	2642
3.	Chemical Properties.	2646
4.	Production	2647
4.1.	Silver Catalyst Processes	2648
4.1.1.	Complete Conversion of Methanol (BASF Process)	2650
4.1.2.	Incomplete Conversion and Distillative Recovery of Methanol	2652
4.2.	Formox Process	2654
4.3.	Comparison of Process Economics	2656
4.4.	Distillation of Aqueous Formaldehyde Solutions	2658
4.5.	Preparation of Liquid Monomeric Formaldehyde . . .	2659
5.	Environmental Protection . . .	2660
6.	Quality Specifications and Analysis	2664
6.1.	Quality Specifications.	2664
6.2.	Analysis	2665
7.	Storage and Transportation . .	2666
8.	Uses	2667
9.	Economic Aspects	2668
10.	Toxicology and Occupational Health.	2670
11.	Low Molecular Mass Polymers	2671
11.1.	Linear Polyoxymethylenes . . .	2672
11.2.	Cyclic Polyoxymethylenes	2677
11.2.1.	Trioxane	2677
11.2.2.	Tetraoxane	2680
11.2.3.	Higher Cyclic Polyoxymethylenes	2681
12.	Formaldehyde Cyanohydrin . .	2681
13.	References.	2683

1. Introduction

Formaldehyde occurs in nature and is formed from organic material by photochemical processes in the atmosphere as long as life continues on earth. Formaldehyde is an important metabolic product in plants and animals (including humans), where it occurs in low but measurable concentrations. It has a pungent odor and is an irritant to the eye, nose, and throat even at a low concentration; the threshold concentration for odor detection is 0.05 – 1 ppm. However, formaldehyde does not cause any chronic damage to human health. Formaldehyde is also formed when organic material is incompletely combusted; therefore, formaldehyde is found in combustion gases from, for example, automotive vehicles, heating plants, gas-fired boilers, and even in cigarette smoke. Formaldehyde is an important industrial chemical and is employed in the manufacture of many industrial products and consumer articles. More than 50 branches of industry now use formaldehyde, mainly in the form of aqueous solutions and formaldehyde-containing resins. In 1995, the demand for formaldehyde in the three major markets — Northern America, Western Europe, Japan — was 4.1×10^6 t/a [1].

History. Formaldehyde was first synthesized in 1859, when BUTLEROV hydrolyzed methylene acetate and noted the characteristic odor of the resulting solution. In 1867, HOFMANN conclusively identified formaldehyde, which he prepared by passing methanol vapor and air over a heated platinum spiral. This method, but with other catalysts, still constitutes the principal method of manufacture. The preparation of pure formaldehyde was described later by KEKULÉ in 1882.

Industrial production of formaldehyde became possible in 1882, when TOLLENS discovered a method of regulating the methanol vapor : air ratio and affecting the yield of the reaction. In 1886 LOEW replaced the platinum spiral catalyst by a more efficient copper gauze. The German firm, Mercklin und Lösekann, started to manufacture and market formaldehyde on a commercial scale in 1889. Another German firm, Hugo Blank, patented the first use of a silver catalyst in 1910.

Industrial development continued from 1900 to 1905, when plant sizes, flow rates, yields, and efficiency were increased. In 1905, Badische Anilin & Soda-Fabrik started to manufacture formaldehyde by a continuous process employing a crystalline silver catalyst. Formaldehyde output was 30 kg/d in the form of an aqueous 30 wt% solution.

The methanol required for the production of formaldehyde was initially obtained from the timber industry by carbonizing wood. The development of the high-pressure synthesis of methanol by Badische Anilin & Soda-Fabrik in 1925 allowed the production of formaldehyde on a true industrial scale.

2. Physical Properties

2.1. Monomeric Formaldehyde

Formaldehyde [50-00-0], CH$_2$O, M_r 30.03, is a colorless gas at ambient temperature that has a pungent, suffocating odor and an irritant action on the eyes and skin.

Formaldehyde liquefies at −19.2 °C, the density of the liquid being 0.8153 g/cm^3 at −20 °C and 0.9172 g/cm^3 at −80 °C. It solidifies at −118 °C to give a white paste. The liquid and gas polymerize readily at low and ordinary temperatures up to 80 °C. Pure formaldehyde gas does not polymerize between 80 and 100 °C and behaves as an ideal gas. For the UV absorption spectra of formaldehyde, see [2]. Structural information about the formaldehyde molecule is provided by its fluorescence [3], IR [4], RAMAN [5], and microwave spectra [6]. Following are some of the thermodynamic properties of gaseous formaldehyde:

Heat of formation at 25 °C	−115.9 ± 6.3 kJ/mol
Gibbs energy at 25 °C	−109.9 kJ/mol
Entropy at 25 °C	218.8 ± 0.4 kJ/mol^{-1}K^{-1}
Heat of combustion at 25 °C	−561.5 kJ/mol
Heat of vaporization at −19.2 °C	23.32 kJ/mol
Specific heat capacity at 25 °C, c_p	35.425 J/mol^{-1} K^{-1}
Heat of solution at 23 °C	
in water	−62 kJ/mol
in methanol	−62.8 kJ/mol
in 1-propanol	−59.5 kJ/mol
in 1-butanol	−62.4 kJ/mol
Cubic expansion coefficient	2.83 × 10^{-3} K^{-1}
Specific magnetic susceptibility	−0.62 × 10^6
Vapor density relative to air	1.04

The vapor pressure p of liquid formaldehyde has been measured from −109.4 to −22.3 °C [7] and can be calculated for a given temperature T (K) from the following equation:

$$p(\text{kPa}) = 10^{[5.0233 - (1429/T) + 1.75 \log T - 0.0063\,T]}$$

Polymerization in either the gaseous or the liquid state is influenced by wall effects, pressure, traces of humidity, and small quantities of formic acid. Formaldehyde gas obtained by vaporization of paraformaldehyde or more highly polymerized α-polyoxymethylenes, which is ca. 90–100 % pure, must be stored at 100–150 °C to prevent polymerization. Chemical decomposition is insignificant below 400 °C.

Formaldehyde gas is flammable, its ignition temperature is 430 °C [8]; mixtures with air are explosive. At ca. 20 °C the lower and upper explosive limits of formaldehyde are ca. 7 and 72 vol % (87 and 910 g/m^3), respectively [9]. Flammability is particularly high at a formaldehyde concentration of 65–70 vol %.

At a low temperature, liquid formaldehyde is miscible in all proportions with nonpolar solvents such as toluene, ether, chloroform, or ethyl acetate. However, solubility decreases with increasing temperature and at room temperature polymerization and volatilization occur, leaving only a small amount of dissolved gas. Solutions of liquid formaldehyde in acetaldehyde behave as ideal solutions [10]. Liquid formaldehyde is slightly miscible with petroleum ether and p-cymene [11].

Polar solvents, such as alcohols, amines or acids, either catalyze the polymerization of formaldehyde or react with it to form methylol compounds or methylene derivatives.

2.2. Aqueous Solutions

At room temperature, pure aqueous solutions contain formaldehyde in the form of methylene glycol $HOCH_2OH$ [463-57-0] and its oligomers, namely the low molecular mass poly(oxymethylene) glycols with the following structure

$HO(CH_2O)_nH$ ($n = 1-8$)

Monomeric, physically dissolved formaldehyde is only present in low concentrations of up to 0.1 wt%. The polymerization equilibrium

$HOCH_2OH + n\ CH_2O \rightleftharpoons HO(CH_2O)_{n+1}-H$

is catalyzed by acids and is shifted toward the right at lower temperature and/or higher formaldehyde concentrations, and toward the left if the system is heated and/or diluted [12], [13] (see also Section 11.1).

Dissolution of formaldehyde in water is exothermic, the heat of solution (−62 kJ/mol) being virtually independent of the solution concentration [14]. Clear, colorless solutions of formaldehyde in water can exist at a formaldehyde concentration of up to 95 wt%, but the temperature must be raised to 120 °C to obtain the highest concentrations. Concentrated aqueous solutions containing more than 30 wt% formaldehyde become cloudy on storage at room temperature, because larger poly(oxymethylene) glycols ($n \geq 8$) are formed which then precipitate out (the higher the molecular mass of the polymers, the lower is their solubility).

Equilibrium constants have been determined for the physical dissolution of formaldehyde in water and for the reaction of formaldehyde to give methylene glycol and its oligomers [12]. These parameters can be combined with other data to calculate the approximate equilibria at any temperature from 0 to 150 °C and at a formaldehyde concentration of up to 60 wt% [13]. Table 1 gives the calculated oligomer distribution in an aqueous 40 wt% solution of formaldehyde.

A kinetic study of the formation of methylene glycol from dissolved formaldehyde and water shows that the reverse reaction is 5×10^3 to 6×10^3 times slower than the

Table 1. Calculated distribution of oligomers of methylene glycol, HO(CH$_2$O)H, in an aqueous 40 wt% formaldehyde solution at 35 °C

n	Proportion, %	n	Proportion, %
1	26.80	7	3.89
2	19.36	8	2.50
3	16.38	9	1.59
4	12.33	10	0.99
5	8.70	> 10	1.58
6	5.89		

forward reaction [15], and that it increases greatly with the acidity of the solution. This means that the distribution of the higher mass oligomers ($n > 3$) does not change rapidly when the temperature is increased or the solution is diluted; the methylene glycol content then rises at the expense of the smaller oligomers ($n = 2$ or 3). In aqueous solutions containing ≤ 2 wt% formaldehyde, formaldehyde is entirely monomeric.

Methylene glycol can be determined by the bisulfite method [16] or by measuring the partial pressure of formaldehyde [17]. Molecular masses and monomer contents can be determined by NMR spectroscopy [13], [18].

The approximate amount of monomeric formaldehyde present as formaldehyde hemiformal and methylene glycol in aqueous solutions containing formaldehyde and methanol, can be calculated from data at 25–80 °C [19] by using the following equation:

$$\text{Monomer (mol\%)} = 100 - 12.3 \sqrt{F} + (1.44 - 0.0164\,F)\,M$$

where F is the formaldehyde concentration (7–55 wt%) and M is the methanol concentration (0–14 wt%).

The partial pressure p_F of formaldehyde above aqueous solutions has been measured by LEDBURY and BLAIR and computed by WALKER and LACY [20]. The parameter p_F for solutions in which F is in the range 0–40 wt% can be calculated with a relative error of 5–10% in the temperature range $T = 273 - 353$ K by using the following equation:

$$p_F(\text{kPa}) = 0.1333\,F\,e^{-F^A\left(a_0 + a_1/T + a_2/T^2\right)}$$

$A = 0.08760 \pm 0.00950$

$a_0 = -12.0127 \pm 0.0550$

$a_1 = 3451.72 \pm 17.14$

$a_2 = 248\,257.3 \pm 5296.8$

Results of such calculations are given in Table 2 and agree well with the measured values.

Table 3 gives the partial pressures and concentrations of formaldehyde in the liquid and gaseous phases of aqueous formaldehyde solutions. The partial pressures and

Table 2. Partial pressure p_F of formaldehyde (kPa) above aqueous formaldehyde solutions

t, °C	Formaldehyde concentration, wt%								
	1	5	10	15	20	25	30	35	40
5	0.003	0.011	0.016	0.021	0.025	0.028	0.031	0.034	0.037
10	0.005	0.015	0.024	0.031	0.038	0.043	0.049	0.053	0.056
15	0.007	0.022	0.036	0.047	0.057	0.066	0.075	0.083	0.090
20	0.009	0.031	0.052	0.069	0.085	0.099	0.113	0.125	0.137
25	0.013	0.044	0.075	0.101	0.125	0.146	0.167	0.187	0.206
30	0.017	0.061	0.105	0.144	0.180	0.213	0.245	0.275	0.304
35	0.022	0.084	0.147	0.203	0.256	0.305	0.353	0.398	0.442
40	0.028	0.113	0.202	0.284	0.360	0.432	0.502	0.569	0.634
45	0.037	0.151	0.275	0.390	0.499	0.604	0.705	0.803	0.899
50	0.047	0.200	0.371	0.531	0.685	0.833	0.978	1.119	1.258
55	0.059	0.262	0.494	0.715	0.929	1.137	1.341	1.541	1.740
60	0.074	0.340	0.652	0.953	1.247	1.536	1.820	2.101	2.378
65	0.093	0.437	0.852	1.258	1.657	2.053	2.443	2.831	3.218
70	0.114	0.558	1.104	1.645	2.182	2.717	3.250	3.780	4.310

Table 3. Concentration and partial pressure of formaldehyde measured at the boiling points (101.3 kPa) of aqueous formaldehyde solutions

Formaldehyde concentration, wt%		Partial pressure (p_F), kPa
Liquid phase (F_l)	Gaseous phase (F_g)	
3.95	3.68	2.35
8.0	7.3	4.75
12.1	10.6	7.0
15.3	13.2	8.65
20.1	16.95	11.2
25.85	21.45	14.45
30.75	24.9	16.8
35.65	27.4	18.8
42.0	30.5	21.4
47.5	33.1	23.4
49.8	34.0	24.1

concentrations were measured at the boiling points of the solutions at a pressure of 101.3 kPa [21].

Aqueous Formaldehyde–Methanol Solutions. Technical-grade formaldehyde solutions contain a small amount of methanol as a result of the incomplete methanol conversion during formaldehyde production. The amount of methanol present depends on the production process employed. The presence of methanol is often desirable in aqueous solutions containing more than 30 wt% formaldehyde because it inhibits the formation of insoluble, higher mass polymers. Methanol concentrations of up to 16 wt% stabilize the formaldehyde.

The approximate *density* ϱ (in grams per cubic centimeter) of aqueous formaldehyde solutions containing up to 13 wt% methanol at a temperature of 10–70 °C can be calculated by using the following equation [22]:

$$\varrho = a + 0.0030\,(F - b) - 0.0025\,(M - c) - 10^4[0.055\,(F - 30) + 5.4](t - 20)$$

where

F = formaldehyde concentration in wt%
M = methanol concentration in wt%
t = temperature in °C
a, b, and c are constants

The following values can be assumed when F is in the range 0–48: $a = 1.092$, $b = 30$, and $c = 0$. The corresponding values in the range $F = 48$–55 are $a = 1.151$, $b = 50.15$, and $c = 1.61$.

The *boiling points* of pure aqueous solutions containing up to 55 wt% formaldehyde are between 99 and 100 °C at atmospheric pressure [23]. In dilute aqueous solutions, formaldehyde lowers the freezing point of water. If solutions containing more than 25 wt% formaldehyde are cooled, polymer precipitates out before the freezing point is reached. According to NATTA [22], the approximate *refractive index* n_D^{18} of aqueous 30–50 wt% formaldehyde solutions containing up to 15 wt% methanol can be calculated from the following equation:

$$n_D^{18} = 1.3295 + 0.00125\,F + 0.000113\,M$$

where F and M are wt% concentrations of formaldehyde and methanol, respectively.

In close agreement with measurements of commercial solutions, the *dynamic viscosity* η of aqueous formaldehyde–methanol solutions may be expressed by the following equation [24]:

$$\eta(\text{mPa}\cdot\text{s}) = 1.28 + 0.039\,F + 0.05\,M - 0.024\,t$$

This equation applies to solutions containing 30–50 wt% formaldehyde and 0–12 wt% methanol at a temperature t of 25–40 °C.

Detailed studies on chemical reactions, vapor–liquid equilibria and caloric properties of systems containing formaldehyde, water, and methanol are available [216]–[226].

3. Chemical Properties

Formaldehyde is one of the most reactive organic compounds known and, thus, differs greatly from its higher homologues and aliphatic ketones [25], [26]. Only the most important of its wide variety of chemical reactions are treated in this article; others are described in [27]. For a general discussion of the chemical properties of saturated aldehydes, see → Aldehydes, Aliphatic and Araliphatic.

Decomposition. At 150 °C, formaldehyde undergoes heterogeneous decomposition to form mainly methanol and CO_2 [28]. Above 350 °C, however, it tends to decompose into CO and H_2 [29]. Metals such as platinum [30], copper [31], chromium, and aluminum [32] catalyze the formation of methanol, methyl formate, formic acid, CO_2, and methane.

Polymerization. Anhydrous monomeric formaldehyde cannot be handled commercially. Gaseous formaldehyde polymerizes slowly at temperatures below 100 °C, polymerization being accelerated by traces of polar impurities such as acids, alkalis, or water (see paraformaldehyde, Section 11.1). Thus, in the presence of steam and traces of other polar compounds, the gas is stable at ca. 20 °C only at a pressure of 0.25 – 0.4 kPa, or at a concentration of up to ca. 0.4 vol% at ca. 20 °C and atmospheric pressure.

Monomeric formaldehyde forms a hydrate with water; this hydrate reacts with further formaldehyde to form polyoxymethylenes (see Section 2.2). Methanol or other stabilizers, such as guanamines [33] or alkylenebis(melamines) [34], are generally added to commercial aqueous formaldehyde solutions (37 – 55 wt%) to inhibit polymerization.

Reduction and Oxidation. Formaldehyde is readily reduced to methanol with hydrogen over a nickel catalyst [27], [35]. For example, formaldehyde is oxidized by nitric acid, potassium permanganate, potassium dichromate, or oxygen to give formic acid or CO_2 and water [27], [36].

In the presence of strong alkalis [37] or when heated in the presence of acids [38], formaldehyde undergoes a Cannizzaro reaction with formation of methanol and formic acid [39]. In the presence of aluminum or magnesium methylate, paraformaldehyde reacts to form methyl formate (Tishchenko reaction) [27].

Addition Reactions. The formation of sparingly water-soluble sodium formaldehyde bisulfite is an important addition reaction of formaldehyde [40]. Hydrocyanic acid reacts with formaldehyde to give glycolonitrile [*107-16-4*] [27]. Formaldehyde undergoes an acid-catalyzed Prins reaction in which it forms α-hydroxymethylated adducts with olefins [24]. Acetylene undergoes a Reppe addition reaction with formaldehyde [41] to form 2-butyne-1,4-diol [*110-65-6*]. Strong alkalis or calcium hydroxide convert formaldehyde to a mixture of sugars, in particular hexoses, by a multiple aldol condensation which probably involves a glycolaldehyde intermediate [42], [43]. Mixed aldols are formed with other aldehydes; the product depends on the reaction conditions. Acetaldehyde, for example, reacts with formaldehyde to give pentaerythritol, $C(CH_2OH)_4$ [*115-77-5*] (→ Alcohols, Polyhydric).

Condensation Reactions. Important condensation reactions are the reaction of formaldehyde with amino groups to give Schiff's bases, as well as the Mannich reaction [27]. Amines react with formaldehyde and hydrogen to give methylamines. Formaldehyde reacts with ammonia to give hexamethylenetetramine, and with ammonium chloride to give monomethylamine, dimethylamine, or trimethylamine and formic acid, depending on the reaction conditions [44]. Reaction of formaldehyde with diketones and ammonia yields imidazoles [45].

Formaldehyde reacts with many compounds to produce methylol ($-CH_2OH$) derivatives. It reacts with phenol to give methylolphenol, with urea to give mono-, di-, and trimethylolurea, with melamine to give methylolmelamines, and with organometallic compounds to give metal-substituted methylol compounds [27].

Aromatic compounds such as benzene, aniline, and toluidine combine with formaldehyde to produce the corresponding diphenylmethanes. In the presence of hydrochloric acid and formaldehyde, benzene is chloromethylated to form benzyl chloride [*100-44-7*] [46]. The possible formation of bis(chloromethyl)ether [*542-88-1*] from formaldehyde and hydrochloric acid and the toxicity of this compound are reported elsewhere (→ Ethers, Aliphatic).

Formaldehyde reacts with hydroxylamine, hydrazines, or semicarbazide to produce formaldehyde oxime (which is spontaneously converted to triformoxime), the corresponding hydrazones, and semicarbazone, respectively. Double bonds are also produced when formaldehyde is reacted with malonates or with primary aldehydes or ketones possessing a CH_2 group adjacent to the carbonyl group.

Resin Formation. Formaldehyde condenses with urea, melamine, urethanes, cyanamide, aromatic sulfonamides and amines, and phenols to give a wide range of resins.

4. Production

Formaldehyde is produced industrially from methanol [*67-56-1*] by the following three processes:

1) Partial oxidation and dehydrogenation with air in the presence of silver crystals, steam, and excess methanol at 680–720 °C (BASF process, methanol conversion = 97–98%).
2) Partial oxidation and dehydrogenation with air in the presence of crystalline silver or silver gauze, steam, and excess methanol at 600–650 °C [47] (primary conversion

Table 4. Specifications of commercial methanol (grade AA) used for the production of formaldehyde

Parameter	Specification
Methanol content	> 99.85 wt%
Relative density, d_4^{20}	0.7928 g/cm^3
Maximum boiling point range	1 °C
Acetone and acetaldehyde content	< 0.003 wt%
Ethanol content	< 0.001 wt%
Volatile iron content	< 2 µg/L
Sulfur content	< 0.0001 wt%
Chlorine content	< 0.0001 wt%
Water content	< 0.15 wt%
pH	7.0
KMnO$_4$ test, minimum decolorization time	30 min

of methanol = 77–87%). The conversion is completed by distilling the product and recycling the unreacted methanol.

3) Oxidation only with excess air in the presence of a modified iron–molybdenum–vanadium oxide catalyst at 250–400 °C (methanol conversion = 98–99%).

Processes for converting propane, butane [48], ethylene, propylene, butylene [49], or ethers (e.g., dimethyl ether) [50] into formaldehyde are not of major industrial significance for economic reasons. Processes that employ partial hydrogenation of CO [51] or oxidation of methane [52] do not compete with methanol conversion processes because of the lower yields of the former processes.

The specifications of the methanol, used for formaldehyde production according to processes 1–3 are listed in Table 4. However, crude aqueous methanol obtained by high- [54], medium-, or low-pressure [55] synthesis can also be used for process 1. This methanol contains low concentrations of inorganic impurities and limited amounts of other organic compounds. The methanol must be first subjected to purification processes and preliminary distillation to remove low-boiling components.

4.1. Silver Catalyst Processes

The silver catalyst processes for converting methanol to formaldehyde are generally carried out at atmospheric pressure and at 600–720 °C. The reaction temperature depends on the excess of methanol in the methanol–air mixture. The composition of the mixture must lie outside the explosive limits. The amount of air that is used is also determined by the catalytic quality of the silver surface. The following main reactions occur during the conversion of methanol to formaldehyde:

$$CH_3OH \rightleftharpoons CH_2O + H_2 \quad \Delta H = +84 \text{ kJ/mol} \quad (1)$$
$$H_2 + 1/2\, O_2 \rightarrow H_2O \quad \Delta H = -243 \text{ kJ/mol} \quad (2)$$
$$CH_3OH + 1/2\, O_2 \rightarrow CH_2O + H_2O \quad \Delta H = -159 \text{ kJ/mol} \quad (3)$$

The extent to which each of these three reactions occurs, depends on the process data.
Byproducts are also formed in the following secondary reactions:

$CH_2O \rightarrow CO + H_2$ $\Delta H = +12.5$ kJ/mol (4)
$CH_3OH + 3/2\ O_2 \rightarrow CO_2 + 2\ H_2O$ $\Delta H = -674$ kJ/mol (5)
$CH_2O + O_2 \rightarrow CO_2 + H_2O$ $\Delta H = -519$ kJ/mol (6)

Other important byproducts are methyl formate, methane, and formic acid.

The endothermic dehydrogenation reaction (1) is highly temperature-dependent, conversion increasing from 50% at 400 °C to 90% at 500 °C and to 99% at 700 °C. The temperature dependence of the equilibrium constant for this reaction K_p is given by

$$\log K_p = (4600/T) - 6.470$$

For detailed thermodynamic data of reactions (1)–(6) see [56]. Kinetic studies with silver on a carrier show that reaction (1) is a first-order reaction [57]. Therefore, the rate of formaldehyde formation is a function of the available oxygen concentration and the oxygen residence time on the catalyst surface:

$$\frac{dc_F}{dt} = kc_o$$

where

c_F = formaldehyde concentration
c_O = oxygen concentration
k = rate constant
t = time

A complete reaction mechanism for the conversion of methanol to formaldehyde over a silver catalyst has not yet been proposed. However, some authors postulate that a change in mechanism occurs at ca. 650 °C [58]. New insight into the reaction mechanism is available from recent spectroscopic investigations [227]–[229], which demonstrate the influence of different atomic oxygen species on reaction pathway and selectivity. The synthesis of formaldehyde over a silver catalyst is carried out under strictly adiabatic conditions. Temperature measurements both above and in the silver layer show that sites still containing methanol are separated from sites already containing predominantly formaldehyde by only a few millimeters.

The oxygen in the process air is shared between the exothermic reactions, primarily reaction (2) and, to a lesser extent depending on the process used, the secondary reactions (5) and (6). Thus, the amount of process air controls the desired reaction temperature and the extent to which the endothermic reactions (1) and (4) occur.

Another important factor affecting the yield of formaldehyde and the conversion of methanol, besides the catalyst temperature, is the addition of inert materials to the reactants. Water is added to spent methanol–water-evaporated feed mixtures, and nitrogen is added to air and air–off-gas mixtures, which are recycled to dilute the methanol–oxygen reaction mixture. The throughput per unit of catalyst area provides another way of improving the yield and affecting side reactions. These two methods of process control are discussed in [59].

The theoretical yield of formaldehyde obtained from Reactions (1)–(6) can be calculated from actual composition of the plant off-gas by using the following equation:

$$\text{Yield (mol\%)} = 100 \left[1 + r + \frac{(\%CO_2) + (\%CO)}{0.528(\%N_2) + (\%H_2) - 3(\%CO_2) - 2(\%CO)} \right]^{-1}$$

Percentages signify concentrations in vol% and r is the ratio of moles of unreacted methanol to moles of formaldehyde produced [60]. The equation takes into account the hydrogen and oxygen balance and the formation of byproducts.

4.1.1. Complete Conversion of Methanol (BASF Process)

The BASF process for the complete conversion of methanol to formaldehyde is shown schematically in Figure 1 [61]. A mixture of methanol and water is fed into the evaporating column. Fresh process air and, if necessary, recycled off-gas from the last stage of the absorption column enter the column separately [60]. A gaseous mixture of methanol in air is thus formed in which the inert gas content (nitrogen, water, and CO_2) exceeds the upper explosive limit. A ratio of 60 parts of methanol to 40 parts of water with or without inert gases is desired. The packed evaporator constitutes part of the stripping cycle. The heat required to evaporate the methanol and water is provided by a heat exchanger, which is linked to the first absorption stage of the absorption column [62]. After passing through a demister, the gaseous mixture is superheated with steam and fed to the reactor, where it flows through a 25–30 mm thick bed of silver crystals. The crystals have a defined range of particle sizes [63] and rest on a perforated tray, which is covered with a fine corrugated gauze, thus permitting optimum reaction at the surface. The bed is positioned immediately above a water boiler (cooler), which produces superheated steam and simultaneously cools the hot reaction gases to a temperature of 150 °C corresponding to that of the pressurized steam (0.5 MPa). The almost dry gas from the gas cooler passes to the first stage of a four-stage packed absorption column, where the gas is cooled and condensed. Formaldehyde is eluted countercurrent to water or to the circulating formaldehyde solutions whose concentrations increase from stage to stage.

The product circulating in the first stage may contain 50 wt% formaldehyde if the temperature of the gas leaving this stage is kept at ca. 75 °C; this temperature provides sufficient evaporation energy for the feed stream in the heat exchanger. The final product contains 40–55 wt% formaldehyde, as desired, with an average of 1.3 wt% methanol and 0.01 wt% formic acid. The yield of the formaldehyde process is

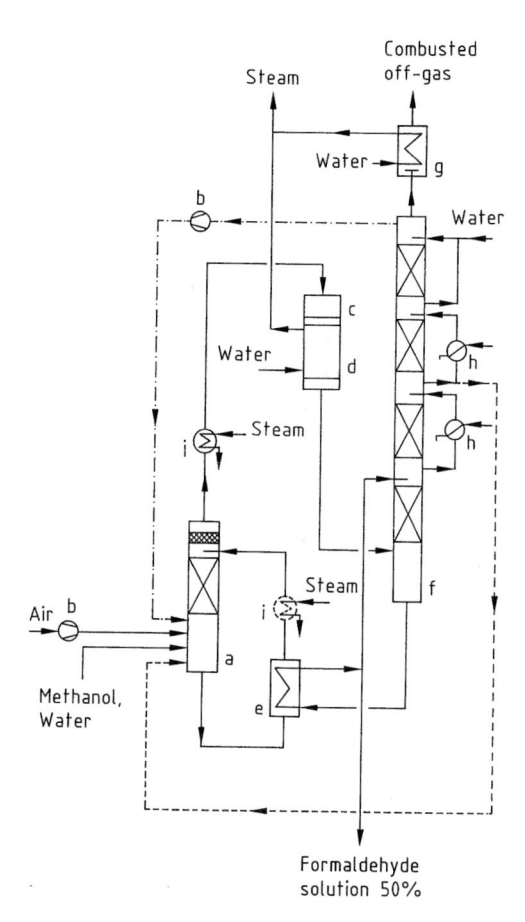

Figure 1. Flowchart of formaldehyde production by the BASF process
a) Evaporator; b) Blower; c) Reactor; d) Boiler; e) Heat exchanger; f) Absorption column; g) Steam generator; h) Cooler; i) Superheater
Recycling schemes: –·–·– off-gas, – – – – formaldehyde solution.

89.5–90.5 mol%. Some of the off-gas is removed at the end of the fourth stage of the column [60] and is recycled due to its extremely low formaldehyde content (Fig. 1, route indicated by dashed-dotted lines). The residual off-gas is fed to a steam generator, where it is combusted [64] (net calorific value = 1970 kJ/m^3). Prior to combustion the gas contains ca. 4.8 vol% CO_2, 0.3 vol% CO, and 18.0 vol% H_2 as well as nitrogen, water, methanol, and formaldehyde. The combusted off-gas contains no environmentally harmful substances. The total steam equivalent of the process is 3 t per ton of 100 wt% formaldehyde.

In an alternative procedure to the off-gas recycling process (Fig. 1, dashed lines) the formaldehyde solution from the third or fourth stage of the absorption tower is recycled to the evaporator; a certain amount of steam is used in the evaporation cycle. The resulting vapor is combined with the feed stream to the reactor to obtain an optimal methanol:water ratio [65]. In this case, the temperature of the second stage of the absorption column is ca. 65 °C.

The yields of the two processes are similar and depend on the formaldehyde content of the recycled streams.

The average life time of a catalyst bed depends on impurities such as inorganic materials in the air and methanol feed; poisoning effects caused by some impurities are partially reversible within a few days. The life time of the catalyst is also adversely affected by long exposure to excessively high reaction temperatures and high throughput rates because the silver crystals then become matted and cause an increase in pressure across the catalyst bed. This effect is irreversible and the catalyst bed must be changed after three to four months. The catalyst is regenerated electrolytically.

Since formaldehyde solutions corrode carbon steel, all parts of the manufacturing equipment that are exposed to formaldehyde solutions must be made of a corrosion-resistant alloy, e.g., certain types of stainless steel. Furthermore, tubes that convey water or gases must be made of alloys to protect the silver catalyst against metal poisoning.

If the throughput and reaction temperature have been optimized, the capacity of a formaldehyde plant increases in proportion to the diameter of the reactor. The largest known reactor appears to be that of BASF in the Federal Republic of Germany; it has an overall diameter of 3.2 m and a production capacity of 72 000 t/a (calculated as 100 wt% formaldehyde).

4.1.2. Incomplete Conversion and Distillative Recovery of Methanol

Formaldehyde can be produced by partial oxidation and distillative recovery of methanol. This process is used in numerous companies (e.g., ICI, Borden, and Degussa) [66] and is based on recent developments. As shown in Figure 2, a feed mixture of pure methanol vapor and freshly blown-in air is generated in an evaporator. The resulting vapor is combined with steam, subjected to indirect superheating, and then fed into the reactor. The reaction mixture contains excess methanol and steam and is very similar to that used in the BASF process (cf. Section 4.1.1). The vapor passes through a shallow catalyst bed of silver crystals or through layers of silver gauze. Conversion is incomplete and the reaction takes place at 590–650 °C, undesirable secondary reactions being suppressed by this comparatively low temperature. Immediately after leaving the catalyst bed, the reaction gases are cooled indirectly with water, thereby generating steam. The remaining heat of reaction is then removed from the gas in a cooler and is fed to the bottom of a formaldehyde absorption column. In the water-cooled section of the column, the bulk of the methanol, water, and formaldehyde separate out. At the top of the column, all the condensable portions of the remaining formaldehyde and methanol are washed out of the tail gas by countercurrent contact with process water. A 42 wt% formaldehyde solution from the bottom of the absorption column is fed to a distillation column equipped with a steam-based heat exchanger and a reflux condenser. Methanol is recovered at the top of the column and is recycled to the bottom of the evaporator. A product containing up to 55 wt% formaldehyde and less than 1 wt% methanol is taken from the bottom of the distillation column and

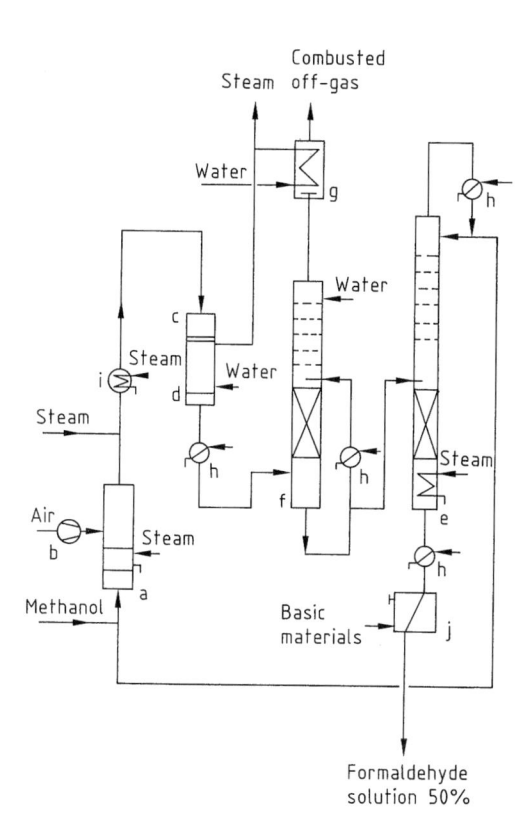

Figure 2. Flowchart of formaldehyde production with recovery of methanol by distillation
a) Evaporator; b) Blower; c) Reactor; d) Boiler; e) Distillation column; f) Absorption column; g) Steam generator; h) Cooler; i) Superheater; j) Anion-exchange unit

cooled. The formaldehyde solution is then usually fed into an anion-exchange unit to reduce its formic acid content to the specified level of less than 50 mg/kg.

If 50–55 wt% formaldehyde and no more than 1.5 wt% methanol are required in the product, steam addition is restricted and the process employs a larger excess of methanol. The ratio of distilled recycled methanol to fresh methanol then lies in the range 0.25–0.5. If a dilute product containing 40–44 wt% formaldehyde is desired, the energy-intensive distillation of methanol can be reduced, leading to savings in steam and power as well as reductions in capital cost. The off-gas from the absorption column has a similar composition to that described for the BASF process (in Section 4.1.1). The off-gas is either released into the atmosphere or is combusted to generate steam, thus avoiding environmental problems caused by residual formaldehyde. Alternatively, the tail gas from the top of the absorber can be recycled to the reactor. This inert gas, with additional steam, can reduce the excess methanol needed in the reactor feed, consequently providing a more concentrated product with less expenditure on distillation. The yield of the process is 91–92 mol%.

Process variations to increase the incomplete conversion of methanol employ two-stage oxidation systems [67]. The methanol is first partly converted to formaldehyde, using a silver catalyst at a comparatively low temperature (e.g., 600 °C). The reaction gases are subsequently cooled and excess air is added to convert the remaining methanol in a second stage employing either a metal oxide (cf. Section 4.2) or a further silver bed as a catalyst.

Formaldehyde solutions in methanol with a relatively low water content can be produced directly by methanol oxidation and absorption in methanol [68]. Anhydrous alcoholic formaldehyde solutions or alcoholic formaldehyde solutions with a low water content can be obtained by mixing a highly concentrated formaldehyde solution with the alcohol (ROH) and distilling off an alcohol–water mixture with a low formaldehyde content. The formaldehyde occurs in the desired solutions in the form of the hemiacetals $RO(CH_2O)_nH$.

4.2. Formox Process

In the Formox process, a metal oxide (e.g., iron, molybdenum, or vanadium oxide) is used as a catalyst for the conversion of methanol to formaldehyde. Many such processes have been patented since 1921 [69]. Usually, the oxide mixture has an Mo:Fe atomic ratio of 1.5–2.0, small amounts of V_2O_5, CuO, Cr_2O_3, CoO, and P_2O_5 are also present [70]. Special conditions are prescribed for both the process and the activation of the catalyst [71]. The Formox process has been described as a two-step oxidation reaction in the gaseous state (g) which involves an oxidized (K_{OX}) and a reduced (K_{red}) catalyst [72]:

$$CH_3OH_{(g)} + K_{OX} \rightarrow CH_2O_{(g)} + H_2O_{(g)} + K_{red}$$
$$K_{red} + 1/2\ O_{2(g)} \rightarrow K_{OX} \qquad \Delta H = -159\ \text{kJ/mol}$$

In the temperature range 270–400 °C, conversion at atmospheric pressure is virtually complete. However, conversion is temperature-dependent because at >470 °C the following side reaction increases considerably:

$$CH_2O + 1/2\ O_2 \rightleftharpoons CO + H_2O \qquad \Delta H = -215\ \text{kJ/mol}$$

The methanol oxidation is inhibited by water vapor. A kinetic study describes the rate of reaction to formaldehyde by a power law kinetic rate expression of the form [230]

$$r = k P_{CH_3OH}^x\ P_{O_2}^y\ P_{H_2O}^z$$

where $x = 0.94 \pm 0.06$; $y = 0.10 \pm 0{,}05$ and $z = -0.45 \pm 0.07$. The rate is independent of the formaldehyde partial pressure. The measured activation energy is 98 ± 6 kJ/mol.

As shown in Figure 3, the methanol feed is passed to a steam-heated evaporator. Freshly blown-in air and recycled off-gas from the absorption tower are mixed and, if necessary, preheated by means of the product stream in a heat exchanger before being fed into the evaporator. The gaseous feed passes through catalyst-filled tubes in a heat-

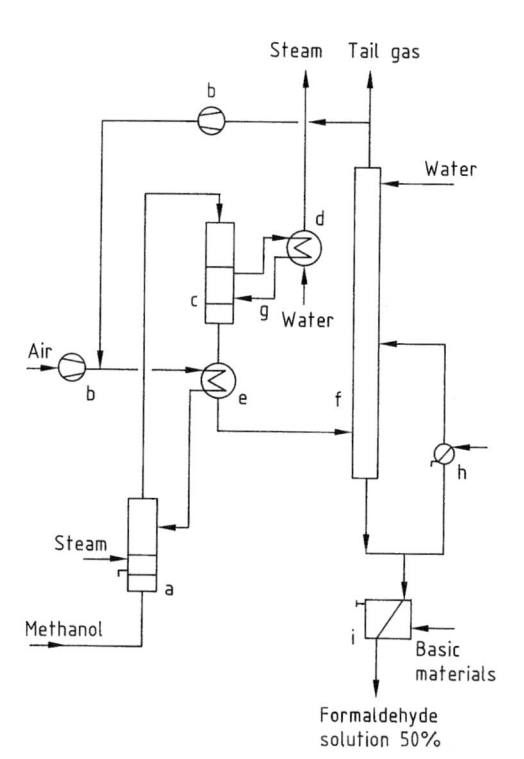

Figure 3. Flowchart of formaldehyde production by the Formox process
a) Evaporator; b) Blower; c) Reactor; d) Boiler; e) Heat exchanger; f) Formaldehyde absorption column; g) Circulation system for heat-transfer oil; h) Cooler; i) Anion-exchange unit

exchanging reactor. A typical reactor for this process has a shell with a diameter of ca. 2.5 m that contains tubes only 1.0–1.5 m in length. A high-boiling heat-transfer oil circulates outside the tubes and removes the heat of reaction from the catalyst in the tubes. The process employs excess air and the temperature is controlled isothermally to a value of ca. 340 °C; steam is simultaneously generated in a boiler. The air–methanol feed must be a flammable mixture, but if the oxygen content is reduced to ca. 10 mol% by partially replacing air with tail gas from the absorption tower, the methanol content in the feed can be increased without forming an explosive mixture [73]. After leaving the reactor, the gases are cooled to 110 °C in a heat-exchange unit and are passed to the bottom of an absorber column. The formaldehyde concentration is regulated by controlling the amount of process water added at the top of the column. The product is removed from the water-cooled circulation system at the bottom of the absorption column and is fed through an anion-exchange unit to reduce the formic acid content. The final product contains up to 55 wt% formaldehyde and 0.5–1.5 wt% methanol. The resultant methanol conversion ranges from 95–99 mol% and depends on the selectivity, activity, and spot temperature of the catalyst, the latter being influenced by the heat transfer rate and the throughput rate. The overall plant yield is 88–91 mol%.

Well-known processes using the Formox method have been developed by Perstorp/Reichhold (Sweden, United States, Great Britain) [74], [75], Lummus (United States) [76], Montecatini (Italy) [77], and Hiag/Lurgi (Austria) [78].

The tail gas does not burn by itself because it consists essentially of N_2, O_2, and CO_2 with a few percent of combustible components such as dimethyl ether, carbon monoxide, methanol, and formaldehyde. Combustion of Formox tail gas for the purpose of generating steam is not economically justifiable [79]. Two alternative methods of reducing atmospheric emission have been developed. The off-gas can be burned either with additional fuel at a temperature of 700–900 °C or in a catalytic incinerator at 450–550 °C. However, the latter system employs a heat exchanger and is only thermally self-sufficient if supplementary fuel for start-up is provided and if an abnormal ratio of oxygen : combustible components is used [80].

4.3. Comparison of Process Economics

Considering the economic aspects of the three formaldehyde processes in practice, it becomes obvious that the size of the plant and the cost of methanol are of great importance. Generally, the Formox process proves to be advantageous regarding the attainable formaldehyde yield. However, in comparison with the silver process this process demands a larger plant and higher investment costs. For the purpose of a cost comparison, a study was undertaken based on the cost of methanol being $ 200 / t and a plant production capacity of 20,000 t / a of 37 % formaldehyde (calculated 100 %) [1]. Table 5 summarizes the economic data.

According to these data the silver process, without the recovery of methanol (cost of formaldehyde $ 378 / t), offers the most favorable production costs, followed by the Formox process ($ 387 / t) and the silver process with recovery ($ 407 / t). The two latter processes produce a product with < 1 % methanol whereas the methanol content in the silver process without recovery lies between 1–5 %.

The study takes into consideration the benefit of the production of steam only in the case of the Formox process. If the production of steam is included in the silver process (3 t per tonne CH_2O without and 1.5 t per tonne CH_2O with methanol recovery) better results than demonstrated in Table 5can be obtained (costs per tonne $ 24 and $ 12 lower, respectively). The proven capacity limits of a plant with only one reactor are about 20,000 t / a (calculated 100 %) with the metal oxide process and about 72,000 t/a with the silver process.

The key feature of the BASF process for the production of 50 wt % formaldehyde is a liquid circulation system in which heat from the absorption unit of the plant is transferred to a stripper column to vaporize the methanol–water feed. Therefore, the process produces excess steam, with simultaneous savings in cooling water.

Plant operation and start-up are simple; the plant can be restarted after a shutdown or after a short breakdown, as long as the temperatures in the stripping cycle remain high. The BASF process has several other advantages. Formaldehyde is obtained from a single pass of the methanol through the catalyst. If a lower formaldehyde concentration is needed (e.g., 40 wt %) the yield can be increased by employing a feedstock of suitably pretreated crude aqueous methanol instead of pure methanol (cf. Section 4.1.1). Deacidification by means of ion exchangers is not necessary. The off-gas does not

Table 5. Comparison of economic factors in formaldehyde production processes

		Complete methanol conversion (BASF process)	Incomplete conversion and methanol recovery	Formox process
Total capital investment, $\$\times10^6$		6.6	8.6	9.6
Methanol consumption, t/t		1.24	1.22	1.15
Raw materials, $/t		255	252	227
	Methanol	250	247	232
	Catalyst and chemicals	5	5	7
	Byproduct credit (steam)	not mentioned	not mentioned	12
Utilities, $/t		12	20	13
	LP Steam	3.4	9.5	
	Power purchased	3.4	4.3	8.0
	Cooling water	2.9	2.8	4.0
	Process water	2.4	3.3	1.0
Variable costs, $/t		267	272	240
Direct fixed costs, $/t		27	29	30
Total allocated fixed costs, $/t		18	20	21
Total cash cost, $/t		312	321	291
Depreciation, $/t		33	43	48
Cost of production, $/t		345	364	339
Return of total capital investment (ROI), $/t		33	43	48
Cost of production and ROI, $/t		378	407	387

present any problems because it is burned as a fuel gas in power stations to generate steam or steam and power. The catalyst can be exchanged within 8–12 h of plant shutdown to restart and can be regenerated completely with little loss. The plant is compact due to the small volume of gas that is used and the low space requirements; both factors result in low capital investment costs.

Formaldehyde production processes based on incomplete methanol conversion employ a final distillation column to recover the methanol and concentrate the formaldehyde. As shown in Table 5, this means that more steam and cooling water is consumed than in the BASF process. The BASF process has a somewhat lower yield but all other aspects are roughly comparable. Other distinctive features of the incomplete conversion of methanol are the relatively large amount of direct steam introduced into the feedstock and the lower reaction temperature, which give a somewhat larger amount of hydrogen in the off-gas with a net calorific value of 2140 kJ/m^3. The additional ion-exchange unit also increases production costs.

The Formox process uses excess air in the methanol feed mixture and requires at least 13 mol of air per mole of methanol. A flammable mixture is used for the catalytic conversion. Even with gas recycling, the process must handle a substantial volume of gas, which is 3–3.5 times the gas flow in a silver-catalyzed process. Thus, the equipment must have a large capacity to accommodate the higher gas flow. The main disadvantage of the Formox process is that the off-gas is noncombustible, causing substantial costs in controlling environmental pollution. To reduce air pollution to the levels obtained in the silver-catalyzed processes, a Formox plant must burn the tail gas with sulfur-free fuel, with or without partial regeneration of energy by means of steam production. Advantages of the process are its very low reaction temperature, which permits high catalyst selectivity, and the very simple method of steam generation. All these aspects mean in easily controlled process. Plants based on this technology can be very small with annual capacities of a few thousand tons. As a result, plants employing Formox methanol oxidation are most commonly encountered throughout the world. However, if higher capacities are required and a small number of reactors must be arranged in parallel, the economic data favor the processes employing a silver catalyst.

Although approximately 70% of existing plants use the silver process, in the 1990s new plant contracts have been dominated by the metal oxide technology [1].

4.4. Distillation of Aqueous Formaldehyde Solutions

Since formaldehyde polymerizes in aqueous solutions, the monomer content and thus the vapor pressure of formaldehyde during distillation are determined by the kinetics of the associated reactions.

Vacuum distillation produces a more concentrated bottom product and can be carried out at a low temperature, an extremely low vapor pressure, and an acid pH value of 3–3.4 [81]. However, the distillation rates are low, making this procedure uneconomical.

High-pressure distillation at 0.4–0.5 MPa and above 130 °C with long columns produces a relatively concentrated overhead product. Efficiency is high, but yields are limited due to the formation of methanol and formic acid via the Cannizzaro reaction [82].

If formaldehyde solutions are subjected to slow *distillation at atmospheric pressure* without refluxing, the distillate has a lower formaldehyde content than the bottom product [21]. If the condensate is refluxed, the ratio of condensate (reflux) to distillate determines the formaldehyde content of the distillate removed [81].

In the case of aqueous formaldehyde solutions that contain methanol, a virtually methanol-free product can be obtained by using distillation columns with a large number of plates and a relatively high reflux ratio. The product is taken from the bottom of the column [83].

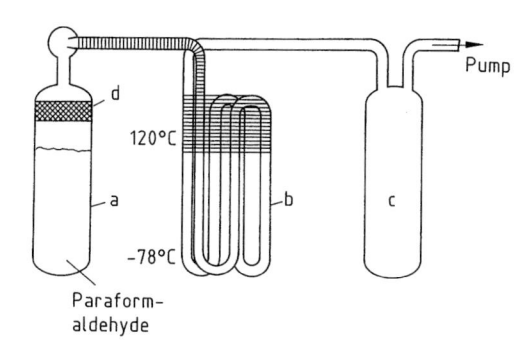

Figure 4. Apparatus for the preparation of liquid monomeric formaldehyde
a) Distillation flask; b) Glass tube with hairpin turns; c) Condenser; d) Glass wool

4.5. Preparation of Liquid Monomeric Formaldehyde

Two methods have been described for the preparation of liquid monomeric formaldehyde from paraformaldehyde, the first was developed by F. WALKER [11] and the second by R. SPENCE [84]. In Walker's method, liquid formaldehyde is prepared by vaporizing alkali-precipitated α-polyoxymethylene. The resultant vapor is then condensed and the crude liquid condensate is redistilled. The process is performed in an apparatus made of Pyrex glass. A vaporizing tube is charged to about one-half its height with the polymer. The thoroughly dried system is then flushed with dry nitrogen. The vaporizing tube is heated to 150 °C in an oil bath and the condensing tube is chilled in a bath of solid carbon dioxide and methanol. The polymer is vaporized in a slow stream of nitrogen by gradually raising the temperature. Formation of polymer on the tube walls is minimized by winding wire round the tubes and heating with electricity. The crude liquid product, which is opalescent due to precipitated polymer, is then distilled in a slow current of nitrogen.

According to the method of SPENCE, paraformaldehyde is dried over sulfuric acid in a vacuum desiccator and introduced into a distillation flask. This flask is connected to a glass condenser via glass tubes with relatively long hairpin turns designed to separate traces of water (Fig. 4). The system is first evacuated by means of a mercury diffusion pump, and the distillation flask is then heated to 110 °C in an oil bath to remove traces of oxygen. The distillate is heated electrically to 120 °C when it flows through the upper parts of the hairpin turns; in the lower parts of the loops, it is cooled to −78 °C by means of a cooling bath. After the valve to the pump is shut and the condenser flask is cooled in liquid air, a colorless solid product condenses. The inlet and outlet tubes of the condenser flask are then sealed with a flame. The contents of the condensing flask liquefy when carefully warmed. The procedure can be repeated to obtain an even purer substance. The liquid formaldehyde that is prepared does not polymerize readily and, when vaporized, leaves only very small traces of polymeric product.

Table 6. Sources emitting formaldehyde into the atmosphere

Emission source	Formaldehyde level
Natural gas combustion	
Home appliances and industrial equipment	2400 – 58 800 µg/m^3
Power plants	15 000 µg/m^3
Industrial plants	30 000 µg/m^3
Fuel-oil combustion	0.0 – 1.2 kg/barrel oil
Coal combustion	
Bituminous	< 0.005 – 1.0 g/kg coal
Anthracite	0.5 g/kg coal
Power plant, industrial, and commercial combustion	2.5 mg/kg coal
Refuse incinerators	
Municipal	0.3 – 0.4 g/kg refuse
Small domestic	0.03 – 6.4 g/kg refuse
Backyard (garden refuse)	up to 11.6 g/kg refuse
Oil refineries	
Catalytic cracking units	4.27 kg/barrel oil
Thermofor units	2.7 kg/barrel oil
Automotive sources	
Automobiles	0.2 – 1.6 g/L fuel
Diesel engines	0.6 – 1.3 g/L fuel
Aircraft	0.3 – 0.5 g/L fuel

5. Environmental Protection

As already stated, formaldehyde is ubiquitously present in the atmosphere [85]. It is released into the atmosphere as a result of the combustion, degradation, and photochemical decomposition of organic materials. Formaldehyde is also continuously degraded to carbon dioxide in processes that are influenced by sunlight and by nitrogen oxides. Formaldehyde washed out of the air by rain is degraded by bacteria (e.g., *Escherichia coli*, *Pseudomonas fluorescens*) to form carbon dioxide and water [86].

The major source of atmospheric formaldehyde is the photochemical oxidation and incomplete combustion of hydrocarbons (i.e., methane or other gases, wood, coal, oil, tobacco, and gasoline). Accordingly, formaldehyde is a component of car and aircraft exhaust fumes and is present in considerable amounts in off-gases from heating plants and incinerators. The main emission sources of formaldehyde are summarized in Table 6.

The formaldehyde in the exhaust gases of motor vehicles is produced due to incomplete combustion of motor fuel. Formaldehyde may be produced directly or indirectly. In the indirect route, the unconverted hydrocarbons undergo subsequent photochemical decomposition in the atmosphere to produce formaldehyde as an intermediate [88]. The concentration of formaldehyde is higher above densely populated regions than above the oceans as shown in Table 7 [89]. According to a 1976 report of the EPA [89], the proportions of formaldehyde in ambient air are derived from the main emission sources as follows:

Table 7. Geographical distribution of formaldehyde in ambient air

Location	Formaldehyde concentration (max.), ppm *
Air above the oceans	0.005
Air above land	0.012
Air in German cities	
normal circumstances	0.016
high traffic density	0.056
Air in Los Angeles (before the law on catalytic combustion of exhaust gases came into effect)	0.165

* 1 ppm = 1.2 mg/m^3

Exhaust gases from motor vehicles and airplanes (direct production)	53–63%
Photochemical reactions (derived mainly from hydrocarbons in exhaust gases)	19–32%
Heating plants, incinerators, etc.	13–15%
Petroleum refineries	1–2%
Formaldehyde production plants	1%

Formaldehyde in confined areas comes from the following sources:

1) Smoking of cigarettes and tobacco products [88], [90], [91]
2) Urea–, melamine–, and phenol–formaldehyde resins in particle board and plywood furniture
3) Urea–formaldehyde foam insulation
4) Open fireplaces, especially gas fires and stoves
5) Disinfectants and sterilization of large surfaces (e.g., hospital floors)

Sources generating formaldehyde must be differentiated into those which release formaldehyde for a defined period, cases (1), (4), and (5) and those which release formaldehyde gas continuously, i.e., decomposition of resins as in cases (2) and (3). In recent years many regulations have been issued to limit pollution of the atmosphere with formaldehyde in both general and special applications [92]. Protection against pollution of the environment with formaldehyde must be enforced with due attention to its sources.

The most effective limitation of atmospheric pollution with formaldehyde is the strict observation of the maximum allowable concentration indoors and outdoors. A maximum workplace concentration of 0.5 ppm (0.6 mg/m^3) has, for example, been established in the Federal Republic of Germany [93]. Other limit values and guide values have been specified for formaldehyde levels in outdoor and indoor air. Emission limits for stationary installations have also been established and regulations for specific products have been formulated. Table 8 gives a survey of regulations valid in some countries of the Western world in 1987.

In the Federal Republic of Germany formaldehyde levels and emissions are subjected to stringent regulations. Plants operating with formaldehyde must conform to the plant emission regulations introduced in 1974 which limit formaldehyde in off-gases to a maximum of 20 mg/m^3 for mass flow rates of 0.1 kg/h or more [94]. This presupposes a closed handling procedure. For example, industrial filling and transfer of formaldehyde solutions is carried out by using pressure compensation between communicating vessels. Discharge of formaldehyde into wastewater in Germany is regulated by law since it endangers water and is toxic to small animals [95]. Formaldehyde is, however, readily degraded by bacteria in nonsterile, natural water [96].

A maximum limit of 0.1 ppm formaldehyde in indoor living and recreation areas has been recommended by the BGA (German Federal Health Office) [97]. To avoid unacceptable formaldehyde concentrations in room air, the German Institute for Structural Engineering has issued guidelines for classifying particle board into emission categories E1, E2, and E3, class E3 having the highest emission [98]. The lowest class (E1) is allowed a maximum formaldehyde emission of 0.1 ppm and a maximum formaldehyde content of 10 mg per 100 g of absolutely dry board (as measured by the DIN EN-120 perforator method) [99]. Furthermore, the uses and applications of urea–formaldehyde foams, which are used to some extent for the heat insulation of cavity walls, have been controlled by DIN 18 159 [99] since 1978. No formaldehyde emission is permitted after the construction has dried.

Cigarette smoke contains 57–115 ppm of formaldehyde and up to 1.7 mg of formaldehyde can be generated while one cigarette is smoked. If five cigarettes are smoked in a 30 m^3 room, with a low air-change rate of 0.1 (i.e., 10%) per hour, the formaldehyde concentration reaches 0.23 ppm [88], [91].

The best protection against accumulation of formaldehyde in confined spaces is, however, proper ventilation. The strong smell of formaldehyde is perceptible at low concentration and thus provides adequate warning of its presence. If all manufacturing and application regulations are strictly observed, possible emission of formaldehyde from consumer products is very low and will not therefore constitute a human health hazard.

Formaldehyde concentrations in cosmetic products have been limited since 1977, they must be appropriately labeled if they contain > 0.05 wt% formaldehyde [100]. Below this level, formaldehyde does not cause allergic reactions even in sensitive subjects.

Table 8. International regulations restricting formaldehyde levels

Country	Emission limit		Product-specific regulations
	Outdoor air, ppm	Indoor air, ppm	
Canada		0.1 (1982)	Urea–formaldehyde (UF) foam insulation prohibited. Voluntary program of particle board manufacturers to reduce emission, no upper limit. Registration of infection control agents
Denmark		0.12 (1982)	Guidelines for particle board: max. 10 mg/100 g of absolutely dry board (perforator value). Guidelines for furniture and in situ UF foam. Cosmetic regulations. Prohibited for disinfecting bricks, wood, and textiles if there is contact with food
Federal Republic of Germany	0.02 (MIK_D, 1966)[a] 0.06 (MIK_K, 1966)[b]	0.1 (1977)	Particle board classification. Guidelines (GefStoffV, Gefahrstoffverordnung) for wood and furniture: upper emission limit 0.1 ppm, corresponding to 10 mg/100 mg of absolutely dry board (perforator value); detergents, cleaning agents, and conditioners: upper limit 0.2%; textiles: compulsory labeling if formaldehyde content >0.15%. Guidelines for in situ UF foam: upper limit 0.1 ppm. Cosmetic regulations
Finland		0.12 0.24 for pre 1983 buildings (1983)	Upper limit for particle board: 50 mg/100 g absolutely dry board (perforator value). Prohibited as an additive in hairsprays and antiperspirants. Guidelines for cosmetics, but as yet (1987) no EEC directives
Great Britain			Upper limit for particle board: 70 mg/100 g of absolutely dry board (perforator value)
Italy		0.1 (1983)	Cosmetic regulations (July 1985)
Japan			Prohibited as an additive in foods, food packaging, and paints. Guidelines for particle board, textiles, wall coverings, and adhesives
The Netherlands		0.1 obligatory for schools and rented accommodation (1978)	Particle board quality standard on a voluntary basis: upper limit 10 mg/100 g of absolutely dry boad (perforator value). Particle board regulations in preparation
Sweden		0.4–0.7 (1977)	Particle board and plywood quality standards: upper limit 40 mg/100 g of absolutely dry board (perforator value)
Switzerland		0.2 (introduced 1984, came into force 1986)	Particle board quality standard on a voluntary basis: upper limit 10 mg/100 g of absolutely dry board (perforator value, Oct. 1985); quality symbol "**Lignum** CH 10"
Spain			Regulations for in situ UF foam (1984): upper limit 1000 µg/m^3 = 0.8 ppm, 7 days after installation; 500 µg/m^3 = 0.4 ppm, 30 days after installation

Table 8. (continued)

Country	Emission limit		Product-specific regulations
	Outdoor air, ppm	Indoor air, ppm	
United States		0.4 (Minnesota, 1984)[c] 0.4 (Wisconsin, 1982)[c]	UF foam insulation prohibited in Massachusetts, Connecticut, and New Hampshire; upper limit for existing UF-insulated houses in Massachusetts 0.1 ppm (1986). FDA limit for nailhardening preparations: 5%. Department of housing and urban development (HUD) guidelines for emission from particleboard and plywood for the construction of mobile houses: upper limit 0.3 ppm.

[a] MIK_D = Maximum allowable concentration for constant immission (mean annual value).
[b] MIK_K = Maximum allowable concentration for short-term immission (30 min or 24 h).
[c] Replaced by HUD product standards, 1985.

Table 9. Typical specifications of commercial formaldehyde solutions

Formaldehyde content, wt%	Methanol content (max), wt%	Formic acid content (max), mg/kg	Iron content (max), mg/kg	Density		Added Stabilizer
				t, °C	g/mL	
30	1.5	150	0.8	20	1.086–1.090	
37	1.8	200	1	20	1.107–1.112	
37	8–12	200	1	20	1.082–1.093	Methanol
37	1.8	200	1	20	1.108–1.112	Isophthalobisguanamine, 100 mg/kg
50	2.0	200	1	55	1.126–1.129	
50	2.0	200	1	40	1.135–1.138	Isophthalobisguanamine, 200 mg/kg

6. Quality Specifications and Analysis

6.1. Quality Specifications

Formaldehyde is commercially available primarily in the form of an aqueous (generally 30–55 wt%) solution, and in solid form as paraformaldehyde or trioxane (cf. Chap. 11). Formaldehyde solutions contain 0.5–12 wt% methanol or other added stabilizers (see Chap. 7). They have a pH of 2.5–3.5, the acid reaction being due to the presence of formic acid, formed from formaldehyde by the Cannizzaro reaction. The solutions can be temporarily neutralized with ion exchangers. Typical product specifications for formulations on the European market are listed in Table 9. Other manufacturers' specifications are described in [102]–[108].

6.2. Analysis

The chemical reactivity of formaldehyde provides a wide range of potential methods for its qualitative and quantitative determination in solutions and in the air.

Qualitative Methods. Qualitative detection of formaldehyde is primarily by colorimetric methods, e.g., [109], [110]. Schiff's fuchsin – bisulfite reagent is a general reagent used for detecting aldehydes. In the presence of strong acids, it reacts with formaldehyde to form a specific bluish violet dye. The detection limit is ca. 1 mL/m^3. Further qualitative detection methods are described in [111].

Quantitative Methods. Formaldehyde can be quantitatively determined by either physical or chemical methods.

Physical Methods. Quantitative determination of pure aqueous solutions of formaldehyde can be carried out rapidly by measuring their specific gravity [27]. Gas chromatography [112], [113] and high-pressure liquid chromatography (HPLC) [114] – [116] can also be used for direct determination.

Chemical Methods. The most important chemical methods for determining formaldehyde are summarized in [111]. The sodium sulfite method is most commonly used. This method was developed by LEMMÉ [117] and was subsequently improved by SEYEWETZ and GIBELLO [118], STADTLER [119], and others. It is based on the quantitative liberation of sodium hydroxide produced when formaldehyde reacts with excess sodium sulfite:

$$CH_2O + Na_2SO_3 + H_2O \rightarrow HOCH_2SO_3Na + NaOH$$

The stoichiometrically formed sodium hydroxide is determined by titration with an acid [27].

Formaldehyde in air can be determined down to concentrations in the µL/m^3 range with the aid of gas sampling apparatus [120], [121]. In this procedure, formaldehyde is absorbed from a defined volume of air by a wash liquid and is determined quantitatively by a suitable method. The quantitative determination of formaldehyde in air by the sulfite/pararosaniline method is described in [122].

A suitable way of checking the workplace concentration of formaldehyde is to take a relevant sample to determine the exposure of a particular person and to use this in combination with the pararosaniline method. The liquid test solution is transported in a leakproof wash bottle [111]. A commercial sampling tube [123], [124] can also be used, in which the formaldehyde is converted to 3-benzyloxazolidine during sampling. Evaluation is carried out by gas chromatography.

Continuous measurements are necessary to determine peak exposures, e.g., by the pararosaniline method as described in [125].

Table 10. Storage temperatures for commercial formaldehyde solutions

Formaldehyde content, wt%	Methanol content, wt%	Storage temperature, °C
30	≤ 1	7–10
37	< 1	35
37	7	21
37	10–12	6–7
50	1–2	45*
50	1–2	60–65

* Stabilized with 200 mg/kg of isophthalobisguanamine

7. Storage and Transportation

With a decrease in temperature and/or an increase in concentration, aqueous formaldehyde solutions tend to precipitate paraformaldehyde. On the other hand, as the temperature increases, so does the tendency to form formic acid. Therefore, an appropriate storage temperature must be maintained (Table 10). The addition of stabilizers is also advisable (e.g., methanol, ethanol, propanol, or butanol). However, these alcohols can be used only if they do not interfere with further processing, or if they can be separated off; otherwise, effluent problems may be encountered.

The many compounds used for stabilizing formaldehyde solutions include urea [126], melamine [127], hydrazine hydrate [128], methylcellulose [129], guanamines [130], and bismelamines [33]. For example, by adding as little as 100 mg of isophthalobisguanamine [*5118-80-9*] per kilogram of solution, a 40-wt% formaldehyde solution can be stored for at least 100 d at 17 °C without precipitation of paraformaldehyde, and a 50-wt% formaldehyde solution can be stored for at least 100 d at 40 °C [32].

Formaldehyde can be stored and transported in containers made of stainless steel, aluminum, enamel, or polyester resin. Iron containers lined with epoxide resin or plastic may also be used, although stainless steel containers are preferred, particularly for higher formaldehyde concentrations. Unprotected vessels of iron, copper, nickel, and zinc alloys must not be used.

The flash point of formaldehyde solutions is in the range 55–85 °C, depending on their concentration and methanol content. According to German regulations for hazardous substances (Gefahrstoffverordnung, Appendix 6) and Appendix 1 of the EEC guidelines for hazardous substances, aqueous formaldehyde solutions used as working materials that contain ≥ 1 wt% of formaldehyde must be appropriately labeled. The hazard classifications for the transport of aqueous formaldehyde solutions with a flash point between 21 and 55 °C containing > 5 wt% formaldehyde and < 35 wt% methanol are as follows [131]:

GGVS/GGVE, ADR/RID	Class 8, number 63 c
CFR	49 : 172.01 flammable liquid
IMDG Code (GGVSee)	Class 3.3
UN No.	1198

Formaldehyde solutions with a flash point > 61 °C and aqueous formaldehyde solutions with a flash point > 55 °C that contain > 5 wt% formaldehyde and < 35 wt% methanol are classified as follows:

GGVS/GGVE, ADR/RID	Class 8, number 63 c
CFR	49 : 172.01 combustible liquid
IMDG Code (GGVSee)	Class 9
UN No.	2209

8. Uses

Formaldehyde is one of the most versatile chemicals and is employed by the chemical and other industries to produce a virtually unlimited number of indispensable products used in daily life [132].

Resins. The largest amounts of formaldehyde are used for producing condensates (i.e., resins) with urea, melamine, and phenol and, to a small extent, with their derivatives. The main part of these resins is usedfor the production of adhesives and impregnating resins, which are employed for manufacturing particle boards, plywood, and furniture. These condensates are also employed for the production of curable molding materials; as raw materials for surface coating and as controlled-release nitrogen fertilizers. They are used as auxiliaries in the textile, leather, rubber, and cement industries. Further uses include binders for foundry sand, rockwool and glasswool mats in insulating materials, abrasive paper, and brake linings. A very small amount of urea – formaldehyde condensates are used in the manufacture of foamed resins that have applications in the mining sector and in the insulating of buildings.

Use as an Intermediate. About 40% of the total formaldehyde production is used as an intermediate for synthesizing other chemical compounds, many of which are discussed under separate keywords. In this respect, formaldehyde is irreplaceable as a C_1 building block. It is, for example, used to synthesize 1,4-butanediol [110-63-4], trimethylolpropane [77-99-6], and neopentyl glycol [126-30-7], which are employed in the manufacture of polyurethane and polyester plastics, synthetic resin coatings, synthetic lubricating oils, and plasticizers. Other compounds produced from formaldehyde include pentaerythritol [115-77-5] (employed chiefly in raw materials for surface coatings and in permissible explosives) and hexamethylenetetramine [100-97-0] used as a cross-linking agent for phenol – formaldehyde condensates and permissible explosives).

The complexing agents nitrilotriacetic acid [139-13-9](NTA) and ethylenediaminetetraacetic acid [60-00-4] (EDTA) are derived from formaldehyde and are components of modern detergents. The demand for formaldehyde for the production of 4,4'-diphenylmethane diisocyanate [101-68-8] (MDI) is steadily increasing. This compound is a constituent of polyurethanes used in the production of soft and rigid foams and, more recently, as an adhesive and for bonding particle boards.

2667

The so-called polyacetal plastics produced by polymerization of formaldehyde are increasingly being incorporated into automobiles to reduce their weight and, hence, fuel consumption. They are also used for manufacturing important functional components of audio and video electronics equipment [232].

Formaldehyde is also a building block for products used to manufacture dyes, tanning agents, dispersion and plastics precursors, extraction agents, crop protection agents, animal feeds, perfumes, vitamins, flavorings, and drugs.

Direct Use. Only a very small amount of formaldehyde is used directly without further processing. In the Federal Republic of Germany, ca. 8000 t/a are used in this way, which corresponds to ca. 1.5% of total production. It is used directly as a corrosion inhibitor, in the metal industry as an aid in mirror finishing and electroplating, in the electrodeposition of printed circuits, and in the photographic industry for film development. However, formaldehyde as such is used mainly for preservation and disinfection, for example, in medicine for disinfecting hospital wards, preserving specimens, and as a disinfectant against athlete's foot.

Modern hygiene requires preservatives and disinfectants to prevent the growth of microorgansims which can produce substances that may be extremely harmful to man. As an antimicrobial agent, formaldehyde displays very few side effects, but has a broad spectrum of action. All alternative agents have unpleasant or even dangerous side effects. Moreover, their toxicity has not been as thoroughly investigated as that of formaldehyde, and their spectrum of action is limited (i.e., they do not provide comprehensive disinfectant protection). Another advantage of formaldehyde is that it does not accumulate in the environment since it is completely oxidized to carbon dioxide within a relatively short time. In the cosmetics industry, formaldehyde is employed as a preservative in hundreds of products, for example, soaps, deodorants, shampoos, and nail–hardening preparations; in some of these items, upper limits have been set for the formaldehyde concentration due to its sensitizing effect (cf. Table 8). Formaldehyde solutions are also used as a preservative for tanning liquors, dispersions, crop protection agents, and wood preservatives. Furthermore, formaldehyde is required in the sugar industry to prevent bacterial growth during syrup recovery.

9. Economic Aspects

Formaldehyde is one of the most important basic chemicals and is required for the manufacture of thousands of industrial and consumer products. It is the most important industrially produced aldehyde.

Formaldehyde can seldom, if ever, be replaced by other products. Substitutes are generally more expensive; moreover, their toxicities have been less thoroughly investigated than that of formaldehyde.

Table 11. Worldwide formaldehyde production capacities in 1996

Country	Total capacity, 10^3 t/a
Western Europe	3119
Germany	1464
Italy	389
Spain	265
United Kingdom	197
France	126
Sweden	124
Netherlands	115
Others	439
Eastern Europe	1850
North America	2008
United States	1772
Canada	236
South America	253
Mexico	65
Chile	63
Brazil	48
Argentina	44
Others	33
Japan	651

According to recent publications [1], [231], the worldwide capacity is approximately 8.7×10^6 t/a (see Table 11; the values are based on 100 % formaldehyde); the five largest manufactures account for ca. 25 % of this capacity:

Borden	0.66×10^6 t/a
BASF	0.444×10^6 t/a
Hoechst Celanese	0.38×10^6 t/a
Georgia Pacific	0.38×10^6 t/a
Neste Resins	0.37×10^6 t/a

The three leading countries with a capacity share of about 45 % are:

United States	1.77×10^6 t/a
Germany	1.46×10^6 t/a
Japan	0.65×10^6 t/a

Formaldehyde consumption is ca. 6×10^6 t/a, although present data about capacity use in Eastern Europe are not available. The demand and the estimated average annual growth rate in the Western hemisphere is summarized in Table 12.

Formaldehyde and its associated products are used in ca. 50 different branches of industry, as described in Chapter 8.

In the mid 1980s the sales of industrial products derived from formaldehyde was more than DM 300×10^9 in the Federal Republic of Germany [132]. At least 3×10^6 people in the Federal Republic of Germany work in factories that use products manufactured from formaldehyde.

Table 12. Consumption of formaldehyde in the United States, Western Europe, and Japan in 1995

	United States	Western Europe	Japan
Consumption, 10^6 t/a	1.37	2.22	0.52
Use, %			
Urea–formaldehyde resin	32	50	27
Melamine–formaldehyde resin	4	6	
Phenol–formaldehyde resin	24	10	6
Polyoxymethylenes	10	10	24
1,4-Butanediol	11	7	
MDI	5	6	4
Others	14	11	39
Average annual growth rate, %	2.5	1.5	2.0

Table 13. Dose–response relationship following human exposure to gaseous formaldehyde

Effect	Exposure level, ppm
Odor threshold	0.05–1.0
Irritation threshold in eyes, nose, or throat	0.2–1.6
Stronger irritation of upper respiratory tract, coughing, lacrimation, extreme discomfort	3–6
Immediate dyspnea, burning in nose and throat, heavy coughing and lacrimation	10–20
Necrosis of mucous membranes, laryngospasm, pulmonary edema	> 50

10. Toxicology and Occupational Health

Formaldehyde toxicity was investigated extensively during the last decades and comprehensive reviews are available [233]–[235]. Formaldehyde is an essential intermediate in cellular metabolism in mammals and humans, serving as a precursor for the biosynthesis of amino acids, purines and thymine. Exogenously administered formaldehyde is readily metabolized by oxidation to formic acid or reacts with biomolecules at the sites of first contact. Inhalation exposure of rats, monkeys and humans to irritant concentrations did not increase blood formaldehyde levels, which were found to be around 80 mM (= 2.4 ppm) in these species.

Formaldehyde gas is toxic via inhalation and causes irritation of the eyes and the mucous membranes of the respiratory tract. Concentration–response relationship following human exposure is given in Table 13. Aqueous formaldehyde solutions cause concentration dependent corrosion or irritation and skin sensitization. There is no evidence for Formaldehyde to cause respiratory allergy [236].

In *chronic inhalation* studies with rats, mice, hamsters, and monkeys no systemic toxicity occurred in irritant concentrations. Upper respiratory tract irritation ceased at concentrations < ca. 1 ppm. At concentrations above 1–2 ppm changes in the nasal mucosa (respiratory epithelium) occur. At high concentrations (15–20 ppm) olfactory epithelium, laryngeal mucosa, and proximal parts of the tracheal epithelium might also be affected. The lesions are characterized by epithelial hyperplasia and metaplasia. Studies using other routes of administration also failed to show systemic toxicity or reproductive effects.

Formaldehyde was *genotoxic* in several in vitro test systems. In animals, there are some indications of in vivo genotoxicity in tissues of initial contact (portal of entry) but not in remote organs or tissues. In workers exposed to formaldehyde no systemic genotoxicity and no convincing evidence of local genotoxicity was found.

No evidence of systemic *carcinogenicity* was found after oral dermal and inhalative administration of formaldehyde. Several chronic inhalation studies in rats showed development of nasal tumors starting at concentrations at or above 6 ppm, causing in addition severe chronic epithelial damage in the nasal epithelium [237]. The nonlinear concentration response curve shows a disproportionately high increase in tumor incidence at concentrations of 10 and 15 ppm. The same nonlinear concentration response was observed for DNA protein cross-link (DPX) formation in nasal mucosa, which is a surrogate of formaldehyde tissue "dose", and for increase in cell proliferation in nasal epithelium. This leads to the suggestion that increased cell proliferation is a prerequisite for tumor development [237]. Chronic inhalation studies in mice failed to show statistically significant increases in tumor incidences at similar concentrations while in hamsters no nasal tumors were found. This may be attributed to differences in local formaldehyde tissue dose or lower susceptibility of the species for nasal tumor formation.

In humans numerous *epidemiological studies* failed to give convincing evidence of carcinogenicity [235]. IARC [234] concluded that the epidemiologic data available represent "limited evidence of carcinogenicity" and classified formaldehyde as "probably carcinogenic to humans (Group 2A)". The European Union categorizes the compound as "possibly carcinogenic to humans (Class 3)".

Current occupational exposure limits in different countries vary between 0.3 and 2 ppm [238]. Proposed limit values for indoor air are in the range of 0.1 ppm [239].

11. Low Molecular Mass Polymers

The ability of formaldehyde to react with itself to form polymers depends directly on the reactivity of formaldehyde as a whole. Two different types of formaldehyde polymers are possible and are based on the following structural elements:

1) –CH_2–O–
2) –CH(OH)–

The polyhydroxyaldehydes consist of the structural unit (2). The highest molecular mass representatives of this group are the sugars. Although these substances can be made by aldol condensation of formaldehyde, they do not revert to formaldehyde on cleavage, and are not discussed in this article.

The representatives of group (1), the real formaldehyde polymers (polyoxymethylenes), revert to formaldehyde on cleavage and, therefore, can be considered as a solid, moisture-free form of formaldehyde. If these linear or cyclic compounds contain no more than eight formaldehyde units, they are defined as low molecular mass polymers. The high molecular mass substances are the real polymers. Chemical and physical analyses of these low and high molecular mass compounds as well as investigation of their chemical reactions led to the elucidation of the molecular structure of polymers in general [135].

11.1. Linear Polyoxymethylenes

Apart from the poly(oxymethylene) glycols, also called poly(oxymethylene) dihydrates or simply polyoxymethylenes, of the formula $HO–(CH_2 O)_n–H$, derivatives such as poly(oxymethylene) diacetates $CH_3COO(CH_2O)_nCOCH_3$ and poly(oxymethylene) dimethyl ethers $CH_3O(CH_2O)_nCH_3$ should be mentioned. Some of their physical properties are given in Table 14. The n values of the real low molecular mass polyoxymethylenes are $2-8$; the n values of paraformaldehyde are $8-100$. However, high polymers with a degree of polymerization $n \geq 3000$ are also obtained. The polyoxymethylenes are also classified as α, β, γ, δ, and ε polymers which are of historical importance. They differ in their degrees of polymerization and in their chemical structures (Table 15). Their toxicology is the same as that of formaldehyde (see Chap. 10).

The lower poly(oxymethylene) glycols are colorless solids with melting points between 80 and 120 °C (Table 14). In contrast to the high molecular mass materials, they dissolve in acetone and diethyl ether without or with only slight decomposition; they are insoluble in petroleum ether. When dissolved in warm water, they undergo hydrolysis to give a formaldehyde solution. The low molecular mass polymers constitute a homologous series, whose properties change continuously with the degree of polymerization.

A freshly prepared, aqueous formaldehyde solution polymerizes to the lower polymers when allowed to stand (see also Section 2.2). Indeed, formaldehyde exists in dilute solution as dihydroxymethylene (formaldehyde hydrate), which in turn undergoes polycondensation to yield low molecular mass poly(oxymethylene) glycols:

$CH_2O + H_2O \Leftrightarrow HO–CH_2–OH + n\ HOCH_2OH \Leftrightarrow HO–CH_2O–(CH_2O)_n–H + n\ H_2O$

Equilibrium is attained under the influence of a hydrogen ion catalyst. At a low temperature and a high concentration, equilibrium favors formation of high molecular

Table 14. Physical properties of low molecular mass poly(oxymethylene) glycols HO–(CH$_2$O)–H and their derivatives

n	Poly(oxymethylene) glycols		Poly(oxymethylene) diacetates			Poly(oxymethylene) dimethyl ethers		
	fp, °C	Solubility in acetone	fp, °C	bp, °C	ϱ, g/m^3 (13 Pa, 24 °C)	fp, °C	bp, °C (101.3 kPa)	ϱ, g/cm^3 (25 °C)
2			−23	39–40	1.128			
3	82–85	Very soluble in the cold	−13	60–62	1.158	−69.7	105.0	0.9597
4	82–85	Very soluble in the cold	−3	84	1.179	−42.5	155.9	1.0242
5	95–105 (decomp.)	Very soluble in the cold	7	102–104	1.195	−9.8	201.8	1.0671
6			17	124–126	1.204	18.3	242.3	1.1003
7		Soluble in the cold	ca. 15	180–190				
8		Soluble in the cold	ca. 15	180–190				
9	115–120 (decomp.)	Soluble when heated	32–34		1.216*			

* Value at 13 Pa and 36 °C.

Table 15. Structure and synthesis of polyoxymethylenes

Polymer	Formula	Synthesis
Paraformaldehyde	$HO(CH_2O)_nH$ $n = 8-100$	from aqueous formaldehyde solution [137]
α-Polyoxymethylene	$HO(CH_2O)_nH$ $n > 100$	from aqueous formaldehyde solution [137]
β-Polyoxymethylene	$HO(CH_2O)_nH$ $n > 200$	by heating paraformaldehyde [138]
γ-Polyoxymethylene	$H_3CO(CH_2O)_nCH_3$ $n = 300-500$	from a methanolic paraformaldehyde solution in the presence of sulfuric acid [139]
δ-Polyoxymethylene	$H_3CO[CH_2OC(OH)HO]_nCH_3$ $n = 150-170$	by prolonged boiling of γ-polyoxymethylene with water [138]
ε-Polyoxymethylene	$HO(CH_2O)_nH$ $n > 300$	by sublimation of 1,3,5-trioxane [138]

mass polymers. However, the major product is of lower molecular mass when the system is heated. The polymers partially separate out, crystallize, and slowly undergo further condensation polymerization [140]. The low molecular mass substances can be further precipitated and isolated by concentrating the solution at low temperature under vacuum conditions; the polymers can be further precipitated by evaporation [141]. The resulting mixture can be separated into the individual substances by exploiting their varying solubilities in different solvents [135].

The transformation of poly(oxymethylene) dihydrates to diacetates and, above all, to diethers produces a remarkable increase in thermal and chemical stability. This is because the unstable hemiacetals at the ends of the chains are eliminated through saturation of the hydroxyl groups. The diethers are stable up to 270 °C in the absence of oxygen and up to 160 °C in the presence of oxygen. These diethers and diacetates are resistant to hydrolysis under neutral conditions, the diethers are also stable in the presence of alkali. Similar to the dihydrates, the properties of the diacetates and diethers also change continuously as the degree of polymerization increases (see Table 14).

Poly(oxymethylene) diacetates are produced by the reaction of paraformaldehyde with acetic anhydride [135]. Pure products are isolated by vacuum distillation, solvent extraction, and crystallization.

The formation of poly(oxymethylene) dimethyl ethers involves the reaction of poly(oxymethylene) glycols or paraformaldehyde with methanol at 150–180 °C in the presence of traces of sulfuric or hydrochloric acid in a closed vessel [135]. Alternatively, they can be synthesized by the reaction of formaldehyde dimethyl acetal with either paraformaldehyde or a concentrated formaldehyde solution in the presence of sulfuric acid. This synthesis can be varied by substituting other formaldehyde dialkyl acetals for the dimethyl compound [142].

Paraformaldehyde [*30525-89-4*] was first produced in 1859. This polymer, at first mistakenly called dioxymethylene and trioxymethylene, consists of a mixture of poly-

Table 16. Vapor pressure of formaldehyde () released from paraformaldehyde

t, °C	10	21	25	33	37
p, kPa	0.112	0.165	0.193	0.408	0.667
t, °C	43	47	51	58	65
p, kPa	0.943	1.096	1.376	1.808	20.8
t, °C	80	90	100	110	120
p, kPa	32.8	44.1	49.6	53.5	78.3

* Values up to 58 °C according to [143] and from 65–120 °C according to [144].

(oxymethylene) glycols HO–(CH$_2$O)$_n$–H with $n = 8-100$. The formaldehyde content varies between 90 and 99 % depending on the degree of polymerization n (the remainder is bound or free water). It is an industrially important linear polyoxymethylene.

Properties. Paraformaldehyde is a colorless, crystalline solid with the odor of monomeric formaldehyde. It has the following physical properties: fp 120–170 °C, depending on the degree of polymerization; heat of combustion 16.750 kJ/kg (product containing 98 wt % formaldehyde); energy of formation 177 kJ/mol formaldehyde (product containing 93 wt % formaldehyde); flash point 71 °C; ignition temperature of dust 370–410 °C; lower explosion limit of dust 40 g/m^3 (the last three values strongly depend on the particle size); minimum ignition energy 0.02 J.

Even at ambient temperature, paraformaldehyde slowly decomposes to gaseous formaldehyde (Table 16), a process which is greatly accelerated by heating. Depolymerization is based on a chain "unzipping" reaction which starts at the hemiacetal end groups of the individual molecules. The rate of decomposition therefore depends on the number of end groups, i.e., on the degree of polymerization.

Paraformaldehyde contains only a few acetone-soluble components (lower diglycols). It dissolves slowly in cold water, but readily in warm water where it undergoes hydrolysis and depolymerization to give a formaldehyde solution. Although this solution is generally cloudy as a result of impurities, these can be removed by filtration. Indeed, the resulting solution is identical with the solution obtained on dissolving gaseous formaldehyde in water. Not only heat, but also dilute alkali or acid increase the solubility of paraformaldehyde. Alkali catalyzes formaldehyde cleavage at the chain ends, acid causes additional splitting at the oxygen bridges. The rate constant of dissolution passes through a minimum between pH 2 and 5, it increases rapidly above and below this pH range [145]. The situation becomes more complex at higher concentrations of formaldehyde in the solution due to the incipient back reaction. Solubility in nonaqueous solvents is also improved in the presence of acidic or alkaline agents as a result of the onset of depolymerization. Paraformaldehyde also dissolves in alcohols, phenols, and other polar solvents, in which depolymerization and solvation can occur.

Production. Paraformaldehyde is prepared industrially in continuously operated plants by concentrating aqueous formaldehyde solutions under vacuum conditions. At first, colloidal, waxy gels are obtained, which later become brittle. The use of a fractionating column through which gases were passed dates from about 1925 [146], [147].

Paraformaldehyde is currently produced in several steps which are carried out at low pressure and various temperatures [148], [149]. Highly reactive formaldehyde is produced under vacuum conditions starting with solutions that contain 50–100 ppm of formic acid and also 1–15 ppm of metal formates where the metals have an atomic number of 23–30 (e.g., Mn, Co, and Cu) [150]. The solutions are processed in thin-layer evaporators [151] and spray dryers [152].

Other techniques such as fractional condensation of the reaction gases in combination with the formaldehyde synthesis process [153] and very rapid cooling of the gases [154] are also applied. Alternatively, formaldehyde-containing gas is brought into contact with paraformaldehyde at a temperature that is above the dew point of the gas and below the decomposition temperature of paraformaldehyde [155]. The product is obtained in the form of flakes when a highly concentrated formaldehyde solution is poured onto a heated metal surface. The hardened product is subsequently scraped off and thoroughly dried [156]. Paraformaldehyde beads are produced by introducing a highly concentrated melt into a cooling liquid (e.g., benzene, toluene, cyclohexane [157]. Acids [158] and alkalis [148], [159] are also added; they apparently accelerate polymerization and lead to the formation of higher molecular mass but less reactive paraformaldehyde.

Highly soluble, highly reactive paraformaldehyde with a low degree of polymerization is very much in demand. It is produced from concentrated, aqueous–alcoholic formaldehyde solutions [160], [161].

Stabilizing agents include hydantoins [162], nicotinamide, and succinamide [163].

Producers. The main producers of paraformaldehyde are Degussa (Federal Republic of Germany), Ticona (owned by Hoechst, United States), Mitsui Toatsu, and Mitsubishi Gas (Japan). Smaller producers include Derivados Forestales (Spain), Synthite (United Kingdom), Alder (Italy), and Lee Chang Yung (Taiwan).

Quality Specifications. An important factor regarding the quality of paraformaldehyde is its tendency to "age", especially under the influence of heat, which results in decreased reactivity. The paraformaldehyde loses its residual water and becomes brittle. This may be prevented by storage below 10 °C; in a moist atmosphere regeneration is possible. After production, the residual moisture in the product should therefore be retained.

Reactivity is determined with the aid of the resorcinol test in which the rate at which paraformaldehyde condenses with resorcinol in an alkaline medium is measured [162]. The formaldehyde content is measured by using the sodium sulfite method (cf. Section 6.2) and the water content by the Karl Fischer method.

Transport and Storage. Paraformaldehyde is stored and transported as granules or a free-flowing coarse powder in bags, containers, and in silos or silo wagons. Cool, dry conditions must be maintained.

Uses. Paraformaldehyde is used in place of aqueous formaldehyde solutions, especially in applications where the presence of water interferes, e.g., in the plastics industry for the preparation of phenol, urea, and melamine resins, varnish resins, thermosets, and foundry resins. Other uses include the synthesis of organic products in the chemical and pharmaceutical industries (e.g., Prins reaction, chloromethylation, Mannich reaction), the production of textile auxiliaries (e.g., for crease-resistant finishes), and the preparation of disinfectants and deodorants.

11.2. Cyclic Polyoxymethylenes

11.2.1. Trioxane

1,3,5-Trioxane [*110-88-3*], trioxymethylene, or 1,3,5-trioxycyclohexane, $C_3H_6O_3$, M_r 90.1, is the cyclic trimer of formaldehyde and was first produced and identified in 1885. A laboratory oddity for many years, this substance is very important today, especially as an intermediate in the production of acetal resins (polyoxymethylenes).

$$\begin{array}{c} CH_2-O \\ O \quad\quad CH_2 \\ CH_2-O \end{array}$$

Physical Properties. Trioxane is a white crystalline solid with a characteristic odor resembling that of chloroform. Other physical properties are as follows (see also [164]):

fp	62–63 °C
bp	115 °C
Density of crystals	1.39 kg/m^3
Density of liquid at 65 °C	1.17 kg/m^3
Refractive index n_D^{20} at 65 °C	1.3891
Specific heat	1.23 kJ kg^{-1}K^{-1}
Heat of fusion	222 kJ/kg
Heat of evaporation	452 kJ/kg
Trouton constant	25
Heat of formation	180 kJ/mol formaldehyde
Heat of combustion	16.850 kJ/kg

Vapor pressure

t, °C	25	37.5	86	87	90	114.5	129
p, kPa	1.69	4.16	37.7	39.5	44.0	101.2	161.8

Flash point	45 °C
Ignition temperature	410 °C
Lower explosion limit (38 °C)	3.57 vol%
Upper explosion limit (75 °C)	28.7 vol%
Dynamic viscosity η	
at 65 °C	2.05 mPa s
at 85 °C	0.91 mPa s
Dielectric constant	
at 20 °C	3.2–3.4
at 70 °C	8
Conductance at 78 °C	<1 µS

For the spectroscopic analytical identification of trioxane, see [164]. The compound forms an azeotropic mixture with 30 wt% water, *bp* 91.3 °C.

Trioxane is soluble in water (0.172 g/cm^3 at 18 °C, 0.211 g/cm^3 at 25 °C and completely soluble at 100 °C), alcohol, ketones, organic acids, ethers, esters, phenols, aromatic hydrocarbons, and in chlorinated hydrocarbons; however, it is only sparingly soluble in aliphatic hydrocarbons.

Like dioxane and other cyclic ethers, melted trioxane, (mixed with water if necessary) is an excellent solvent for organic substances. It forms addition compounds with phenol and 1,3,5-trinitrobenzene.

Studies of vapor-liquid equilibria of systems containing formaldehyde, water, methanol, and trioxane are available [220], [224], [226].

Chemical Properties. Trioxane is stable up to 224 °C. Like acetals, it is hydrolyzed by aqueous solutions of strong acids. However, trioxane is inert under neutral and alkaline conditions. For this reason, the usual detection methods for formaldehyde, such as the sodium sulfite method, cannot be applied directly. Formaldehyde in acetals, and thus trioxane, can be determined by heating in water in the presence of a strong acid (e.g., sulfuric acid) to convert the product to monomeric formaldehyde. The formaldehyde can then be determined by the sulfite method (see Section 6.2).

The acid-catalyzed hydrolysis of trioxane is accelerated by using stronger acids, higher acid concentrations, and nonaqueous solvents. When trioxane is heated in nonaqueous systems in the presence of either strong acids (sulfuric acid, hydrochloric acid, zinc chloride, iron chloride etc.) or comonomers, it is converted to high molecular mass oxymethylene homopolymers or oxymethylene copolymers, respectively. This method is applied today in the large-scale production of acetal resins. If additional substances which react with formaldehyde (e.g., phenols, melamine) are also present during these hydrolytic or moisture-free reactions, then the nascent formaldehyde reacts very vigorously with them.

Production. Trioxane is prepared by the trimerization of formaldehyde. At first, the starting materials used were paraformaldehyde or polyoxymethylenes which were heated with acid, usually sulfuric acid [165]. As the commercial importance of trioxane grew, aqueous formaldehyde solutions and high formaldehyde concentrations [166] were introduced. World production in 1987 was 250 000 t, in 1997 about 380 000 t.

The currently preferred process for trioxane production is illustrated in Figure 5 (see also [167]). Commercial aqueous formaldehyde is concentrated to 60–65 wt% in a column (a) under vacuum conditions. The 60–65% formaldehyde solution is fed into the reactor (b). A dilute (6–10%) formaldehyde solution is removed and, if necessary, can be reconcentrated by pressure distillation and returned to the process. In the reactor (b), formaldehyde is trimerized in the presence of a catalyst (e.g., sulfuric acid, phosphoric acid, ion exchanger [168], or heteropoly acid [169]) which is present at a concentration of up to 25 wt% [170]. Technical and kinetic experiments designed to achieve optimal reactor conditions are described in [171]. A mixture of trioxane,

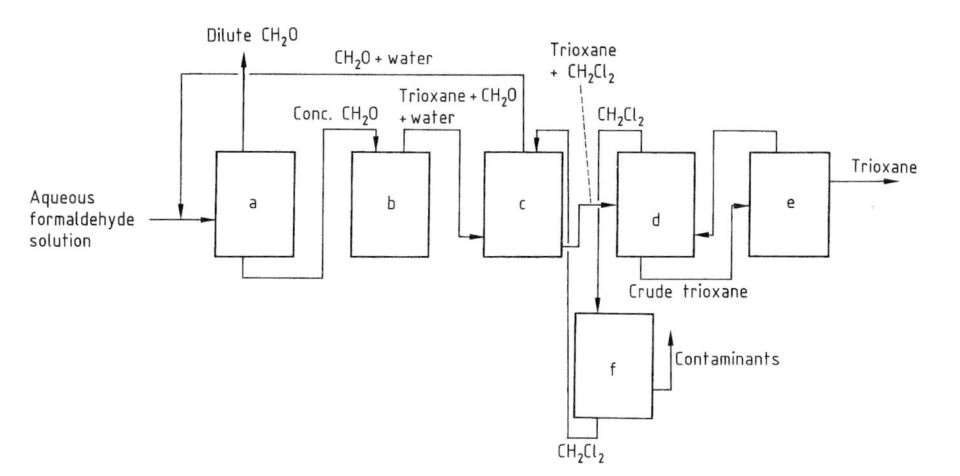

Figure 5. Trioxane production
a) Concentration column; b) Reactor; c) Extraction column; d, e) Distillation columns; f) Solvent purification

formaldehyde, and water is distilled off and, if necessary, concentrated in a distillation column [172]. In the extraction column (c) [173], the trioxane is extracted from the aqueous mixture by using methylene dichloride [75-09-2] [166] or another inert, water-immiscible solvent (benzene [174], 1,2-dichloroethane [175], or nitrobenzene [176]). The formaldehyde–water mixture is returned to the concentrating column (a). The trioxane–methylene dichloride mixture is separated in the distillation column (d). The crude trioxane is obtained at the bottom of the column and the methylene dichloride as overhead product. The trioxane is subsequently purified by distillation in column (e). The solvent methylene dichloride is freed of impurities in the solvent purification step (f) and recycled to the extraction column (c).

Impurities and byproducts (formic acid, methyl formate, dimethylformal, and formaldehyde) can be eliminated by washing with sodium hydroxide solution after the extraction step [177]. Emulsification of the reaction mixture in an inert, nonvolatile liquid, e.g., paraffin oil, may reduce the formation of paraformaldehyde in the reactor [178]. Addition of inert organic substances with low dielectric constants to the trimerization mixture increases conversion [179]. Special processes in a synthesis reactor have also been described [180]. Crystallization of liquid trioxane and contact with an inert gas stream can be used for further purification, especially from higher homologues of bis(methoxymethyl) ether [181].

In the last years a new method has been proposed to remove trioxane from the mixture of trioxane, formaldehyde, and water. After trimerization formaldehyde is separated by pervaporation, e.g., with a polyether block amide membrane [182], [183].

Since water is added with the aqueous formaldehyde solution, the energy requirement for trioxane production is very high (14–17 t of steam per ton of trioxane). This may be reduced by using a process based on the trimerization of gaseous formaldehyde with Lewis acids [184] and catalyzed by vanado- and/or molybdo- and/or wolframo-phosphoric acids [185] – [187].

Trioxane can be produced by recycling of poyoxymethylenes (POM) [188], [189].

Producers. Ticona (owned by Hoechst, Federal Republic of Germany and United States), Ultraform (owned by BASF and Degussa, Federal Republic of Germany and United States), Polyplastics (owned by Hoechst and Daicel, Japan), Mitsubishi Gas Chemical (Japan), Asahi (Japan), Korean Engineering Plastics (owned by Tong Yang Nylon and Mitsubishi Gas Chemical, Korea), Taiwan Engineering Plastics, Tepco (owned by Polyplastics and Chang Chun Group, Taiwan), Formosa Plastics (Taiwan), Tarnoform (Poland). The world capacity is about 400 000 t.

Uses. The production of plastics (polyoxymethylene plastics, POM) is the most important commercial application of trioxane and requires extremely pure material. Since trioxane depolymerizes to produce formaldehyde, it can be used in almost all formaldehyde reactions, especially when anhydrous formaldehyde is desired. Trioxane is used, e.g., as a textile auxiliary; in various cross-linking agents, for instance for pitch fibers in manufacturing of carbon fibers; for stabilization and reinforcement of wood to produce musical instruments; and in distillable binders in ceramic compositions for shaped articles, e.g., in an injection moulding process. It is also employed in the stabilization and deodorization of waxes, as an anticorrosive additive for metals (especially when corrosion is caused by chlorinated hydrocarbons) [190], as an additive in photographic developers, as a solid fuel, in coating removers and, because of its hydrolysis to formaldehyde in presence of acid, in disinfectants.

Storage and Transport. Liquid trioxane is stored and transported in heated tanks. Solid trioxane is transported in drums, trioxane flakes are transported in drums and bags.

Toxicology. Trioxane is not known to display any special toxic reactions; its oral LD_{50} value in rats is > 3200 mg/kg. Hydrolysis in the presence of acid produces formaldehyde (see Chap. 10).

11.2.2. Tetraoxane

1,3,5,7-Tetraoxane [*293-30-1*], tetraoxymethylene, or 1,3,5,7-tetroxocane, $C_4H_8O_4$, M_r 120, is the cyclic tetramer of formaldehyde and is a crystalline solid (*fp* 112 °C). Tetraoxane vapors are stable even at 200 °C (*bp* at 101.3 kPa, 163 – 165 °C).

Tetraoxane is produced by heating a water-insoluble, high molecular mass poly(oxymethylene) diacetate [165]. In the tetramerization of formaldehyde, metal sulfates may be used as catalyst [191].

Tetraoxane accumulates as a high-boiling substance in the distillation residue obtained during the production of trioxane. It is also obtained in the polymerization of trioxane and can be isolated by extraction with, for example, ethylbenzene [192]. The formation of tetraxane by irradiation of carbon monoxide and hydrogen has also been described [193].

Tetraoxane is a more favorable textile auxiliary for achieving crease resistance, e.g., for cellulose fibers, than trioxane because of its higher boiling point. Like trioxane, tetraoxane can also be employed in the production of acetal plastics.

11.2.3. Higher Cyclic Polyoxymethylenes

Higher cyclic polyoxymethylenes can also undergo polymerization. The tendency to polymerize in the melt decreases with increasing ring size.

1,3,5,7,9-Pentoxecane [*16528-92-0*], pentaoxane, pentaoxacyclodecane, or pentaoxymethylene, $C_5H_{10}O_5$, the cyclic pentamer of formaldehyde, is a crystalline product (*fp* 55–56 °C). The polymerization of trioxane with the aid of a tin tetrachloride catalyst produces several cyclic homologues including pentoxecane [194].

The fractional distillation and subsequent preparative gas chromatography of residues from the commercial trioxane polymerization affords 1,3,5,7,9-pentoxecane [195].

Although *hexoxecane* (six formaldehyde units) has not clearly been identified in the residue, *pentadecoxane* (15 formaldehyde units) has been unambiguously detected (*fp* 68–70 °C). Pentoxecane is used in addition to tetraoxane in patent formulations for the shrinkproofing of cotton [196].

12. Formaldehyde Cyanohydrin

Formaldehyde cyanohydrin [*107-16-4*], $CH_2(OH)CN$, M_r 57.05, is also known as hydroxyacetonitrile, glycolonitrile, or glycolic nitrile. It is less commercially important than acetone cyanohydrin (→ Acetone) but of about the same importance as the cyanohydrin derivatives of acetaldehyde, benzaldehyde, and ethylene.

Physical Properties. Formaldehyde cyanohydrin is a colorless liquid with an odor similar to that of hydrogen cyanide. It is soluble in water, ethanol, and diethyl ether but insoluble in chloroform and benzene. For IR spectrum see [197]. Other physical properties follow:

mp	−72 °C
bp at	
101.3 kPa	183 °C (slight decomp.)
3.2 kPa	119 °C
0.4 kPa	76.5 – 78 °C
Relative density d_4^{19}	1.1039
n_D^{20}	1.4112
Electrolytic dissociation constant k at 25 °C	0.843×10^{-5}
Dielectric constant	178

Chemical Properties. Like all cyanohydrins, reactions can occur at both the nitrile and the hydroxyl groups of formaldehyde cyanohydrin. Heating at 100 °C with water yields hydroxyacetamide, acid-catalyzed hydrolysis yields hydroxyacetic acid. Reaction with absolute ethanol in the presence of hydrochloric acid produces ethyl hydroxyacetate [198]. *N*-Substituted amides can be synthesized by heating formaldehyde cyanohydrin with amines in water [199]. Catalytic hydration (Ni – Al catalysts) of the nitrile group gives amines. The resulting mixtures can be separated into primary, secondary, and tertiary *β*-hydroxyethyleneamines [200].

The hydroxyl group of formaldehyde cyanohydrin can be replaced by other electronegative groups. For example, reaction with ammonia gives amino nitriles including nitrilotriacetonitrile $N(CH_2CN)_3$ which is an intermediate in the synthesis of nitrilotriacetic acid (NTA) [201]. Ethylenediaminetetraacetonitrile is similarly synthesized from formaldehyde cyanohydrin and ethylene diamine in the presence of sulfuric acid as a catalyst. Subsequent hydrolysis yields ethylenediaminetetraacetic acid (EDTA) [201]. The reaction of cyanohydrins with ammonia is a step in the Strecker synthesis of amino acids. Hydration of formaldehyde cyanohydrin and subsequent treatment with an aqueous solution of sodium cyanide and ammonia (Bucherer reaction, a variant of the Strecker synthesis) leads to *DL*-serine [202].

Formaldehyde cyanohydrin reacts with hydrogen halides or phosphorus pentachloride to form α-halonitriles. In aqueous solutions at pH 8 and a temperature of 10 °C or less, it trimerizes to form 4-amino-2,5-bis(hydroxymethyl)-pyrimidin-5-ol [203]. Hydantoin can be synthesized by reacting formaldehyde cyanohydrin with carbon dioxide, ammonia, or ammonium carbonate at high pressure [204], [205].

Production. Formaldehyde cyanohydrin is formed in 79.5 % yield when 37 % formaldehyde is mixed with a stoichiometric amount of aqueous hydrogen cyanide for 1 h at 2 °C in the presence of sodium hydroxide as a catalyst [206]. Patents also describe synthesis by (1) heating formaldehyde with hydrogen cyanide at 250 °C [207] or in the presence of sulfur dioxide at pH 1.5 – 2.0 in an aqueous medium [208] and (2) by reacting acetonitrile with oxygen and with or without water at high pressure in the presence of various vanadium oxides [209] – [211].

Formaldehyde cyanohydrin is generally handled as an aqueous solution but it can also be isolated in the anhydrous state by ether extraction, drying, and vacuum distillation [212]. Although the extremely pure product is reported to be very stable [213], addition of a small amount of stabilizer (e.g., monochloroacetic, mercaptoacetic, lactic, sulfonilic, and phosphoric acids) is advisable [214]. Monochloroacetic acid is especially suitable because it codistils with the product. Cyanohydrin with a lower degree of purity is not very stable, especially at high pH values, this should be borne in mind when handling anhydrous formaldehyde cyanohydrin (see also [203]).

Transport and Storage. Formaldehyde cyanohydrin is used almost exclusively as an intermediate and is normally produced at the site where it is to be processed. However, it is also supplied as a 70% aqueous solution and is transported in steel containers and tank cars. Attention should be paid to stabilization during transport and storage. Impure formaldehyde cyanohydrin can decompose to give formaldehyde and hydrogen cyanide, attention should therefore be paid to explosion and fire hazards.

Toxicology. Formaldehyde cyanohydrin, like all cyanohydrins, is extremely toxic when inhaled or ingested and moderately toxic when absorbed via the skin [215]. All forms of contact should be avoided because of the possible formation of hydrogen cyanide or cyanides.

13. References

[1] Chem. Systems Inc.: "Formaldehyde" (April 1996).
[2] J. C. D. Brand, *J. Chem. Phys.* **19** (1951) 377; *J. Chem. Soc.* 1956, 858–872; *J. Chem. & Ind.* 1955, no. 7, 167.G. H. Dieke, G. B. Kistiakowsky, *Proc. Natl. Acad. Sci., U.S.A.* **18** (1932) 367–372; *Phys. Rev.* **45** (1934) 4–28. P. J. Dyne, *J. Chem. Phys.* **20** (1952) 811–818. W. C. Price, *Phys. Rev.* **46** (1934) 529; *J. Chem. Phys.* **3** (1935) 256–259. R. I. Reed, *Trans. Faraday Soc.* **52** (1956) 1194. G. W. Robinson, V. E. Digiorgio, *Can. J. Chem.* **36** (1958) 31–38. S. A. Schou, V. Henri, *C. R. Hebd. Seances Acad. Sci.* **182** (1926) 1612–1614; **186** (1928) 690–692, 1050–1052. *Nature London* **118** (1926) 225; *Z. Phys.* **49** (1928) 774–826. V. Henri, *J. Chim. Phys. Phys. Chim. Biol.* **25** (1928) 665–721; **26** (1929) 1–43. P. Torkington, *Nature London* **163** (1949) 446. R. S. Mulliken, *J. Chem. Phys.* **3** (1935) 514.
[3] Th. Förster, *Fluoreszenz organ. Verbindungen*, Vandenhoeck u. Rupprecht, Göttingen 1951, pp. 49, 97, 125, 143.
[4] H. H. Blau, H. H. Nielsen, *J. Mol. Spectrosc.* **1** (1957) 124–132. D. W. Davidson, B. P. Stoicheff, H. J. Bernstein, *J. Chem. Phys.* **22** (1954) 289–294. K. B. Harvey, J. F. Ogilvie, *Can. J. Chem.* **40** (1962) 85–91. I. C. Hisatsune, D. F. Jr. Eggers, *J. Chem. Phys.* **23** (1955) 487–492. R. H. Pierson, A. N. Fletcher, E. St. Clair Gantz, *Anal. Chem.* **28** (1956) 1218–1239. H. M. Randall, R. G. Fowler, N. Fuson, J. R. Dangl: *Infrared Determination of Organic Structures*, D. Van Nostrand Company, New York 1949, pp. 49, 51, 53, 55, 56. W. G. Schneider, H. J. Bernstein, *Trans. Faraday Soc.* **52** (1956) 13–18. W. E. Singer, *Phys. Rev.* **71** (1947) 531–533.
[5] J. H. Hibben: *The Raman Effect and its Chemical Applications*, Reinhold Publ. Co., 1939, p. 187.
[6] J. K. Bragg, A. H. Sharbaugh, *Phys. Rev.* **75** (1949) 1774–1775. R. B. Lawrance, M. W. P. Strandberg, *Phys. Rev.* **83** (1951) 363. T. Oka, H. Hirakawa, K. Shimoda, *J. Phys. Soc. Japan* **15** (1960) 2265–2279.
[7] R. Spence, W. Wild, *J. Chem. Soc.* 1935 506–509.

[8] J. Legrand, R. Delbourgo, P. Lafitte, *C. R. Hebd. Seances Acad. Sci.* **249** (1959) 1515–1516. M. Vanpée, *Bull. Soc. Chim. Belg.* **64** (1955) 235–263.

[9] National Fire Protection Association: *Fire Protection Guide on Hazardous Materials*, 5th ed., 49 Hazardous Chem. Data, Boston 1973, p. 150.

[10] S. Sapgir, *Bull. Soc. Chim. Belg.* **38** (1929) 392–408.

[11] J. F. Walker, *J. Am. Chem. Soc.* **55** (1933) 2821–2825.

[12] S. Bezzi, A. Iliceto, *Chim. Ind. (Milan)* **33** (1951) 212–217. A. Iliceto, *Gazz. Chim. Ital.* **81** (1951) 915–932, **84** (1954) 536–552. F. J. Walker, *Formaldehyde*, 3rd ed., Reinhold Publ. Co., New York 1964, p. 62. E. Koberstein, K. P. Müller, G. Nonnenmacher, *Ber. Bunsenges. Phys. Chem.* **75** (1971) 549–553.

[13] J. F. Walker, *J. Phys. Chem.* **35** (1931) 1104–1113.

[14] A. Iliceto, *Gazz. Chim. Ital.* **81** (1951) 786–794.

[15] H. Schecker, G. Schulz, *Z. Phys. Chem. (NF)* **65** (1969) 221–224.

[16] A. Iliceto, *Ann. Chim. (Rome)* **39** (1949) 703–716.

[17] E. W. Blair, W. Ledbury, *J. Chem. Soc.*, 1925, 26–40. W. Ledbury, E. W. Blair, *J. Chem. Soc.* 1925, 2832–2839.

[18] P. Skell, H. Suhr, *Chem. Ber.* **94** (1961) 3317–3327.

[19] W. Dankelman, J. M. H. Daemen, *Anal. Chem.* **48** (1976) 401. Z. Fiala, M. Navratil, *Collect. Czech. Chem. Comm.* **39** (1974) 2200–2205. D. A. Young, *unpublished results*, Celanese Research Company, Summit, N. J. 1978.

[20] W. Ledbury, E. W. Blair, *J. Chem. Soc.* 1925, 33–37, 127, 2834–2835. J. F. Walker, T. J. Mooney, *unpublished data*, Du Pont. B. S. Lacy, unpublished data, Du Pont.

[21] E. L. Piret, M. W. Hall, *Ind. Eng. Chem.* **40** (1948) 661–672.

[22] G. Maue, *Pharm. Ztg.* **63** (1918) 197. H. Gradenwitz, *Chem. Ztg.* **42** (1918) 221; *Pharm. Ztg.* **63** (1918) 241. G. Natta, M. Baccaredda, *G. Chim. Ind. Appl.* **15** (1933) 273–281.

[23] S. J. Green, R. E. Vener, *Ind. Eng. Chem.* **47** (1955) 103–109.

[24] *Kirk-Othmer*, **11**, 231–250.

[25] *Organikum*, VEB Deutscher Verlag der Wissenschaft, Berlin 1986.

[26] H. Gault, *Chim. Ind. (Paris)* **67** (1952) 41–64.

[27] J. F. Walker: *Formaldehyde, ACS Monographic Series*, 3rd ed., Reinhold Publ. Co., New York, Amsterdam, London 1967.

[28] J. G. Calvert, E. W. R. Steacie, *J. Chem. Phys.* **19** (1951) 176–182.

[29] W. A. Bone, H. L. Smith, *J. Chem. Soc.*,1905, 910–916.

[30] M. J. Marshall, D. F. Stedman, *Trans. R. Soc. Can.* Sect. 3, 17 (1923) 53.

[31] Y. Miyazaki, J. Yasumori, *Bull. Chem. Soc. Japan*, **40** (1967) 2012.

[32] H. Tropsch, O. Roehlen, *Abh. Kenntnis Kohle* **7** (1925) 15.

[33] BASF, DE 2 365 180, 1973 (H. Diem et al.); DE 2 358 856, 1973 (H. Diem et al.)

[34] Degussa, DE-OS 3 143 920, 1981 (P. Werle et al.).

[35] P. Sabatier, J. B. Senderens, *C. R. Hebd. Seances Acad. Sci.* **137**, (1903) 301–303.

[36] *Beilstein*, **E III, 1**, 2549 ff.

[37] O. Loew, *Ber. Dtsch. Chem. Ges.* **22** (1889) 470–478.

[38] H. Staudinger et al., *Justus Liebigs, Ann. Chem.* **474** (1929) 254–255. H. J. Prins, *Rec. Trav. Chim. Pays Bas* **71** (1952) 1131–1136.

[39] H. S. Fry, *Rec. Trav. Chim. Pays Bas* **50** (1931) 1060–1065.

[40] W. M. Lauer, L. M. Langkammerer, *J. Am. Chem. Soc.* **57** (1935) 2360–2362.

[41] GAF, US 2 232 867, 1941 (W. Reppe, E. Keyssner).

[42] L. Orthner, E. Gerisch, *Biochem. Z.* **259** (1933) 30–52.

[43] E. Katzschmann, *Ber. Dtsch. Chem. Ges.* **77** (1944) 579–585.
[44] H. Gilman: *Org. Synthesis Coll.*, vol. **I**, J. Wiley & Sons Inc., New York 1932, p. 514.
[45] R. Behrend, J. Schmitz, *Justus Liebigs Ann. Chem.* **277** (1893) 338. A. Windus, *Chem. Ber.* **42** (1909) 760.
[46] G. Blanc, *Bull. Soc. Chim. Belg.* **33** (1923) 313–319.
[47] A. R. Chauvel et al., *Hydrocarbon Process.* **52** (1973) 179.
[48] J. V. Hightower, *Chem. Eng. (N.Y.)* **55** (1948) 136–139; Celanese Co., US 2 570 216, 1949 (H. K. Dice, R. L. Mitchell) US 2 570 217, 1949 (H. K. Dice, R. L. Mitchell).
[49] Ruhrchemie, US 3 255 238, 1962 (O. Roelen, W. Rottig).
[50] H. Tadenumar et al., *Hydrocarbon Process.* 45 (1966), 195–196.
[51] V. N. Ipatieff, G. S. Monroe, *J. Am. Chem. Soc.* **67** (1945) 2168–2171.
[52] F. Fischer, *Oel Kohle* **39** (1943) 521–522. Chemische Werke Hüls, DE-OS 2 201 429, 1972 (R. Bröckhaus). G. E. Haddeland, *Formaldehyde*, Stanford Research Institute, Menlo Park, California 1967, p. 56, 107–135. F. J. Walker, *Formaldehyde*, 3rd ed., Reinhold Publ. Co., New York 1964, pp. 25–26.
[53] D. D. Mehta, W. W. Pan, *Hydrocarbon Process.* **45** (1971), 115–120.
[54] BASF, DE 1 277 834, 1966 (V. Gerloff et al.).
[55] BASF, DE 2 034 532, 1970 (H. Hohenschutz et al.).
[56] E. Jones, G. G. Fowlie, *J. Appl. Chem.* **3** (1953) 206–213.
[57] V. I. Atroshchenko, I. P. Kushnarenko, *Int. Chem. Eng.* **4** (1964) 581–585.
[58] V. N. Gavrilin, B. I. Popov, *Kinet. Catal. (Engl. Transl.)* **6** (1965) 799–803.
[59] H. Sperber, *Chem. Ing. Tech.* **41** (1969) 962–966. Heyden Chem. Corp., US 2 465 498, 1945 (H. B. Uhl, I. H. Cooper).
[60] BASF, DE 2 442 231, 1974 (G. Halbritter et al.).
[61] H. Diem, *Chem. Eng. N.Y.* **85** (1978) 83.
[62] BASF, DE 0 150 436, 1984 (A. Aicher et al.).
[63] BASF, DE 2 322 757, 1973 (A. Aicher et al.).
[64] BASF, DE 2 655 321, 1976 (A. Aicher et al.).
[65] BASF, DE E 2 444 586, 1974 (A. Aicher et al.).
[66] *Chem. Week* **105** (1969) 79. D. G. Sleemann, *Chem. Eng. N.Y.* **75** (1968) no. 1, 42–44. M. Weimann, *Chem. Eng. N.Y.* **77** (1970) no. 3, 102–104. *Hydrocarbon Process.* **52** (1973) 135, 179. J. H. Marten, M. T. Butler, *Oil Gas J.* **72** (1974) no. 10, 71–72. Du Pont, FR 1 487 093, 1967.
[67] Du Pont, US 2 519 788, 1947, (W. A. Payne); 3 959 383, 1974 (E. S. Northeimer); 4 076 754, 1978 (G. L. Kiser, B. G. Hendricks). Allied Chemical, US 2 462 413, 1943 (W. B. Meath).
[68] Borden, US 3 629 997, 1970 (C. W. DeMuth).
[69] Barrett Comp., US 1 383 059, 1921 (G. C. Bailey, A. E. Craver).Bakelite Corp., US 1 913 405, 1933 (V. E. Meharg, H. Adkins). Montecatini, US 3 198 753, 1965 (F. Traina). Perstorp AB, GB 1 080 508, 1967 (S. A. Bergstrand).
[70] G. D. Kolovertnov, G. K. Boreskov et al., *Kinet. Catal./Engl. (Tansl.)* **6** (1965) 950–954; **7** (1966) 125–130. Reichhold Chemicals, US 2 973 326, 1954 (T. S. Hodgins, F. J. Shelton). Montecatini, AT 218 513, 1959 (G. Natta et al.).
[71] Lummus Comp., US 3 408 309, 1964 (A. W. Gessner). Reichhold Chem. Inc., US 2 812 309, 1954 (C. L. Allyn et al.); 3 855 153, 1973 (G. M. Chang).
[72] P. Jiru et al., *Proc. 3rd Intern. Congr. Catalysis* Amsterdam, 1 (1965) 199–213.M. Dente et al., *Chim. Ind. (Milan)* **46** (1964) 1326–1336; **47** (1965) 359–367; **47** (1965) 821–829.

[73] Du Pont, US 2 436 287, 1948 (W. F. Brondyke, J. A. Monier Jr.); GB 589 292, 1947 (W. F. Brondyke, J. A. Monier Jr.).
[74] *Chem. Eng. N.Y.* **61** (1954), 109–110.
[75] *Chem. Process Eng. (London)* **51** (1970) 100–111.
[76] Lummus Comp., US 3 277 179, 162 (M. C. Sze).
[77] Montecatini, G. Greco, U. Soldano, *Chem. Ing. Tech.* **31** (1959) 761–765.
[78] W. Exner et al., *Chem. Anlagen + Verfahren*, 1973, 87–92.
[79] H. R. Gerberich et al., Celanese Chemical Comp., Inc., *Kirk-Othmer*, **11**, 240.
[80] C. W. Horner, *Chem. Eng. N.Y.* **84** (1977) no. 14, 108–110.
[81] A. Iliceto, *Chim. Ind. (Milan)* **36** (1954) 523–528. B. Olsson, S. G. Svennson, *Trans. Inst. Chem. Eng.* **53** (1975) 97–105. Cities Service Oil Co., US 2 665 241, 1948 (L. G. Willke et al.).
[82] M. I. Faberov, V. A. Speranskaya, *Zh. Prikl. Khim. (Leningrad)* **28** (1955) 205–208. J. Meissner, FR 1 255 022, 1960.
[83] S. J. Green, R. E. Vener, *Ind. Eng. Chem.* **47** (1955) 103–109. Du Pont, GB 931 688, 1961 (C. H. Manwiller, J. B. Thompson). Sumitomo, GB 869 764, 1959. M. W. Hall, E. L. Piret, *Ind. Eng. Chem.* **41** (1949) 1277–1286.
[84] R. Spence, W. Wild, *J. Chem. Soc.* 1935, 338–340.
[85] *Formaldehyd*, Gemeinsamer Bericht des Bundesgesundheitsamtes, der Bundesanstalt für Arbeitsschutz und des Umweltbundesamtes, Familie und Gesundheit, vol. **148**, Verlag Kohlhammer, Stuttgart 1984.
[86] G. Bringmann, R. Kühn, *Gesund. Ing.* **11** (1960) 337–339.
[87] IARC (Int. Agency f. Res. on Cancer): *Formaldehyde*, vol. **29**, Lyon 1982, pp. 346–389.
[88] US National Academy of Sciences, *Formaldehyde – An Assessment of its Health Effects*, Washington, D.C. 1980.
[89] J. F. Kitchens et al., *Investigation of Selected Potential Environmental Contaminants, Formaldehyde*, EPA 560/2-76-002, US Environmental Protection Agency, Office of Toxic Substances, Washington, D.C. 1976.
[90] A. Weber-Tschopp et al., *Int. Arch. Occup. Environ. Health* **39** (1977) 207–218.
[91] WKI-Report No. 13/82, Fraunhofer Institut für Holzforschung, Wilhelm-Klauditz-Institut, Braunschweig (1982).
[92] W. Lohrer, N. J. Nantke, R. Schaaf, *Staub-Reinhalt. Luft* **45** (1985) no. 5, 239–247.
[93] Technische Regeln für Gefahrstoffe TRGS 900; Oktober 1996; BArb. Bl. Nr. 10/1996, p. 88.
[94] Technische Anleitung zur Reinhaltung der Luft, TA-Luft, Feb. 27, 1986, GMBl. 1986, p. 95.
[95] Verwaltungsvorschrift wassergefährdender Stoffe, April 18, 1996, GMBl. 1996, p. 327.
[96] U. Pagga, *Vom Wasser* **55** (1980) 313–326 Bundesministerium für Forschung und Technologie, Forschungsbericht (02-WA 822), Wassertechnologie, Dec. 1980.
[97] H. Petri, H. L. Thron, J. Wegner, *Grenzwert-Bestimmung für Formaldehyde in der Innenraumluft*, BGA Jahresbericht 1977; Bundesgesundhbl. 28 (1985) no. 6.
[98] Ausschuß für Einheitliche Technische Baubestimmungen: *Richtlinien über die Verwendung und Klassifizierung von Spanplatten bezüglich der Formaldehydabgabe*, Beuth Verlag, Berlin, Köln 1980.
[99] Beuth Vertrieb, DIN EN 120, Perforator-Methode.
[100] *Kosmetikverordnung*, Dec. 16 1977, BGBl. I, p. 2589.
[101] *Formaldehyde Product Bulletin*, Badische Anilin- & Soda-Fabrik AG, Ludwigshafen, Sept. 1975.
[102] *Formaldehyde Product Bulletin*, Celanese Chemical Co., Dallas.
[103] *Formaldehyde Product Bulletin*, Du Pont, Wilmington, Del.
[104] *Formaldehyde Product Bulletin*, Georgia-Pacific Corp., Portland, Oregon.

[105] *Formaldehyde Product Bulletin,* Deutsche Gold- und Silber-Scheideanstalt AG, Frankfurt, Germany.
[106] *Formaldehyde Product Bulletin,* Borden Chemical Co., Columbus, Ohio, Feb. 1978.
[107] *Formox Process for Producing Formaldehyde, Product Bulletin,* Reichhold Chemicals, White Plains, N.Y.
[108] *Formaldehyde Product Bulletin,* ICI, Birmingham, UK, Feb. 1977.
[109] G. Denigès, *J. Pharm. Chim.* **6** (1896) 193.
[110] G. Denigès, *C. R. Hebd. Seances Acad. Sci.* **150** (1910) 529–531.
[111] H. Petersen, N. Petri, *Melliand Textilber.* **66** (1985) 217–222, 285–295, 363–369.
[112] H. L. Gruber, H. Plainor, *Chromatographia* **3** (1970) 490.
[113] L. Gollob, J. D. Wellons: "Analytical Methods for Formaldehyde", *For. Prod. J.* **30** (1980) 27–35.
[114] K. Mopper, W. L. Stahovec, *J. Chromatogr.* **256** (1983) 243–252.
[115] K. Kuwata, H. Uebori, H. Yamasaki, Y. Kuge, *Anal. Chem.* **55** (1983) 2013–2016.
[116] K. Fung, D. Grosgean, *Qual. Chem.* **53** (1981) 168–171.
[117] G. Lemmé, *Chem. Ztg.* **27** (1903) 896; *Chem. Zentralbl.* **II** (1903) 911.
[118] A. Seyewetz, Gibello, *Bull. Soc. Chim. Fr.* **31** (1904) 691–694.
[119] S. S. Stadtler, *Am. J. Pharm.* **76** (1907) 84–87; *Chem. Zentralbl.* **I** (1904) 1176.
[120] W. Leithe: *Die Analyse der Luft und ihre Verunreinigungen,* Wissenschaftl. Verlags GmbH, Stuttgart 1974.
[121] D. Henschler, *Luftanalyse, Analytische Methoden zur Prüfung gesundheitsschädlicher Arbeitsstoffe,* vol. **1,** Verlag Chemie, Weinheim 1976.
[122] Verein Deutscher Ingenieure: 1, *Messen gasförmiger Immissionen, Bestimmen der Formaldehydkonzentration nach dem Sulfit-Pararosanilin-Verfahren,* Richtlinie VDI 3484, Blatt 1, Düsseldorf 1979.
[123] Supelco, Bulletin 794, Supelchem GmbH, 8399 Friesbach, Federal Republic of Germany.
[124] NIOSH Manual of Analytical Methods, vol. **7,** Analytical Method P & CAM 354 E, Cincinatti, Ohio, 1981.
[125] Antechnika, Karlsruhe, Formaldehyd-Monitor TGM 555.
[126] Du Pont, US 2 000 152, 1932 (J. F. Walker).
[127] American Cyanamid, US 2 237 092, 1939 (R. C. Swain, P. Adams).
[128] G. Altieri, IT 509 976, 1954.
[129] Celanese, US 3 137 736, 1959 (R. H. Prinz, B. C. Kerr); US 3 532 756, 1968 (R. H. Prinz, B. C. Kerr).
[130] Süddeutsche Kalkstickstoff-Werke, DE 1 205 072, 1962 (J. Seeholzer); DE 1 205 073, 1964 (P. Bornmann, H. Michaud).Koei Chemical Co., CA 809 691, 1963 (S. Matsuura, Y. Hikata).
[131] G. Hommel: *Handbuch der gefährlichen Güter,* Springer Verlag, Berlin, Heidelberg 1970, 1974, 1980, 1987, Merkblatt 95, 95 a.
[132] Formaldehyd, Porträt einer Chemikalie, BASF, Ludwigshafen 1984.
[133] WHO: *Environmental Health Criteria (EHC) for Formaldehyde. Interantional Program on Chemical Safety (IPS),* Hannover 1987.
[134] D. Henschler: *Formaldehyd, MAK-Nachtrag 1987,*VCH Verlagsgesellschaft, Weinheim1987.
[135] H. Staudinger: *Die hochmolekularen organischen Verbindungen,* Springer Verlag, Berlin 1932. W. Kern et al.: *Angewandte Chemie* **73** (1961) 177.
[136] *Houben-Weyl,* **E 20,** 1390.
[137] *Houben-Weyl,* **XIV/1 a,** 408–410, 413–421.

[138] J. F. Walker, *Formaldehyde*, 3rd ed.,Reinhold Publishing Corp., New York 1964, pp. 52 ff., 158 ff.
[139] Du Pont, US 2 512 950, 1950 (T. E. Londergan).
[140] O. Vogl, *J. Macromol. Sci. Rev. Macromol. Chem.* **C 12** (1975) 109.
[141] *Houben-Weyl*, **7/1,** 429.
[142] Du Pont, US 2 449 469, 1948 (W. F. Gresham, R. E. Brooks).
[143] G. Nordgren, *Acta Pathol. Microbiol. Scand.* **40** (1939) 21.
[144] H. H. Nielsen, E. S. Ebers, *J. Chem. Phys.* **5** (1937) 824.
[145] J. Löbering, *Ber. Dtsch. Chem. Ges. (B. Abhandlungen)* **69** (1936) 1846.
[146] Consortium f. Elektrochem. Ind., DE 489 644, 1925 (M. Mugdan, J. Wimmer).
[147] Du Pont, US 2 527 654, 1947 (C. Pyle, J. A. Lane); US 2 527 655, 1948 (C. Pyle, J. A. Lane), US 2 581 881, 1948 (C. Pyle, J. A. Lane).
[148] Celanese, US 2 568 016 – 18, 1949 (A. F. McLean, W. E. Heinz).
[149] Cities Service Oil Co., US 2 498 206, 1948 (B. W. Greenwald, R. K. Cohen).
[150] Degussa, DE 1 260 143, 1968 (H.-J. Mann, H. Murek).
[151] Société Chimique des Charbonages, FR 2 067 169, 1969 (Y. Moreaux).
[152] Jos. Meissner KG, DE 2 037 759, 1972 (F. M. Deisenroth).
[153] Celanese, GB 682 737, 1950.
[154] Petric V. N. et al., SU 675 051, 1979 (V. N. Petric, N. V. Kudrina, A. E. Obraztsov, G. G. Vidorenkov).
[155] Du Pont, US 2 529 269, 1948 (J. F. Walker); US 2 992 277, 1959 (H. F. Porter).
[156] Sumitomo, US 3 001 235, 1959 (D. Komiyama, T. Takaki, T. Ando, T. Nii).
[157] Degussa, DE 1 795 551, 1963 (H. Junkermann, F. Löffler).
[158] Heyden Newport Chem. Corp., US 2 915 560, 1957 (D. Steinhardt, D. X. Klein, R. H. Barth).
[159] Societa Italiana Resine, US 3 772 392, 1972 (T. Paleologo, J. Ackermann).
[160] Degussa, DE 884 947, 1951 (H. Leyerzapf).
[161] Pan American Petroleum Corp., US 2 823 237, 1954 (J. F. McCants).
[162] Du Pont, US 2 481 981, 1948 (R. L. Craven); US 2 519 550, 1948 (R. L. Craven).
[163] Mitsubishi Gas Chemical, JP 73 21 082, 1973 (M. Katayama, S. Yahara, T. Endo).
[164] *Beilstein,* **19 (3, 4)** 4710.
[165] H. Staudinger, H. Luthy, *Helv. Chim. Acta* **8** (1925) 65.
[166] Du Pont, US 2 304 080, 1940 (C. E. Frank).
[167] J. Mahieux, *Hydrocarbon Process.* **48** (1969) no. 5, 163.
[168] G. Rotta, DE 2 225 267, 1972.
[169] Asahi Kasei Kogyo, DE 3 106 476, 1981 (K. Yoshida, T. Iwaisako, J. Masamoto, K. F. Hamanaka et al.).
[170] BASF, DE 1 543 340, 1966 (H. Buchert, H. Sperber).
[171] E. Bartholomé, W. Köhler, H. G. Schecker, G. Schulz, *Chem. Ing. Tech.* **43** (1971) 597.
[172] ICI, GB 1 012 372, 1963 (W. R. Bamford).
[173] BASF, DE 1 668 867, 1968 (H. Sperber, H. Fuchs).
[174] Houillères du Bassin du Nord, FR 1 459 000, 1965 (E. Comber, H. Montanbric).
[175] BASF, DE 1 668 867, 1968 (H. Sperber, H. Fuchs, H. Libowitzky).
[176] Asahi, JP 95 215 961, 1995 (J. Masamoto, H. Morishita).
[177] Péchiney-Saint-Gobain, FR 1 449 675, 1965 (M. P. Raoul).
[178] Degussa, FR 1 377 169, 1963.
[179] Polyplastics, US 3 732 252, 1971 (H. Komazawa, O. Matsumo).

[180] Hoechst, DE 2 853 091, 1978 (K. F. Mück, G. Sextro, K. H. Burg); DE 2 912 767, 1979 (H. Bär, K. H. Burg, H. Mader, K. F. Mück, Cr. Sextro); DE 2 943 984, 1979 (H. Bär, H. Mader, K. F. Mück, P. Zorner).
[181] Mitsubishi Gas Chem., DE 2 855 710, 1978 (A. O. S. Sugio et al.).
[182] Asahi, JP 95 33 762, 1995 (J. Masamoto).
[183] Hoechst, EP 596 381, 1994 (D. Arnold).
[184] Perstorp, AT 252 913, 1964 (P. G. M. Flodin, P. Komfeldt, J. I. Gardshol).
[185] G. Emig, F. Kern, St. Ruf, H.-J. Warnecke, *App. Catal. A.* **118** (1994) L17–L20.
[186] Hoechst, EP 604 884, 1994 (G. Emig et al.).
[187] Hoechst, EP 691 338, 1996 (M. Hoffmockel, G. Sextro, G. Emig, F. Kern).
[188] D. Fleischer, K. F. Mueck, G. Reuschel, *Kunststoffe* **82** (1992) no. 9, 763–766.
[189] Hoechst, EP 484 786, 1992 (K.-F. Mueck, G. Reusch, D. Fleischer).
[190] Solvay, DE 2 142 920, 1971 (A. Ryckaert).Péchiney-Saint-Gobain, DE 1 793 235, 1968 (Y. Correia).
[191] Toyo Koatsu Industries, US 3 426 041, 1965 (Y. Miyake, S. Adachi, N. Yamanchi, T. Hayashi et al.).
[192] Houillères du Bassin du Nord, FR 1 548 554, 1967 (E. Gombar, J. Mahieux).
[193] S. Sugimoto et al., *Int. J. Appl, Radiat. Isot.* **34** (1983) no. 11, 559.
[194] Miki Tetsuro et al., *J. Polym. Sci. Polym. Chem. Ed.* **6** (1968) no. 11, 3031.
[195] K. H. Burg et al., *Makromol. Chem.* **111** (1968) 181.
[196] Mitsui Toatsu Chem., JP 74 24 836, 1974 (K. Yamamoto, M, Naito, M. Hata).
[197] *American Petroleum Institute Res. Proj 44,* no. 1264 (1951).
[198] V. P. Belikov et al., *Izv. Akad. Nauk SSSR Ser. Khim* **8** (1967) 1862.
[199] Coastal Interchemical Co, US 3 190 916, 1965 (N. B. Rainer).
[200] Du Pont, GB 598984, 1948.
[201] Hampshire Chemical, US 2 855 428 1958 (J. J. Sin-ger, M. Weisberg).
[202] Degussa DE 3 242 748, 1984 (A. Kleemann, B. Lehman, K. Deller).
[203] D. B. Lake, T. E. Londergan, *J. Org. Chem.* **19** (1954) 2004.
[204] Mitsui Toatsu Chem., JP 61 72 761, 1986 (H. Inagaki et al.).
[205] Mitsui Toatsu Chem., JP 61 83 164, 1986 (K. Takeuchi et al.).
[206] Dow Chemical, US 2 890 238, 1959 (A. R. Sexton).
[207] Monsanto, US 2 752 383, 1953 (S. F. Belt).
[208] Mitsubishi Chemical, JP-Kokai 76 100 027, 1976 (Y. Ono).
[209] Standard Oil Co., US 4 634 789, 1987 (R. G. Teller, J. F. Brazdil, L. C. Glaeser).
[210] J. F. Brazdil et al., *J. Catal.* **100** (1986) no. 2, 516.
[211] Standard Oil Co., US 4 515 732, 1985 (J. F. Brazdil, W. A. Marrit, M. D. Ward).
[212] R. Gaudry; *Organic Synthesis, Coll.* vol. 3, J. Wiley & Sons, New York 1955, 436.
[213] Rohm and Haas Co., US 3 057 903, 1962 (J. W. Nemec, C. H. McKeever).
[214] Röhm und Haas, DE 811 952, 1949 (H. Beier); US 2 623 896, 1952 (H. Beier).
[215] H. E. Christensen (ed.): *Registry of Toxic Effects of Chemical Substances,* U.S. Dept. of Health, Education and Welfare, Rockville 1976.
[216] M. Albert, I. Hahnenstein, H. Hasse, G. Maurer: "Vapor-Liquid Equilibrium of Formaldehyde Mixtures: New Data and Model Revision," *AIChE J.* **42** (1996) no. 6, 1741–1752.
[217] I. Hahnenstein, H. Hasse, Y.-Q. Liu, G. Maurer: "Thermodynamic Properties of Formaldehyde Containing Mixtures for Separation Process Design," *AIChE Symp. Ser.* **90** (1994b) no. 298, 141–157.

[218] I. Hahnenstein, H. Hasse, C. G. Kreiter, G. Maurer: "^{1}H- and ^{13}C-NMR Spectroscopic Study of Chemical Equilibria in Solutions of Formaldehyde in Water, Deuterium Oxide, and Methanol," *Ind. Eng. Chem. Res.* **33** (1994b) no. 4, 1022–1029.

[219] I. Hahnenstein, M. Albert, H. Hasse, C. G. Kreiter, G. Maurer: "NMR-Spectroscopic and Densimetric Study of Reaction Kinetics of Formaldehyde Polymer Formation in Water, Deuterium Oxide, and Methanol," *Ind. Eng. Chem. Res.* **34** (1995) no. 2, 440–450.

[220] H. Hasse: "Dampf-Flüssigkeits-Gleichgewichte, Enthalpien und Reaktionskinetik in formaldehydhaltigen Mischungen," PhD Thesis, Universität Kaiserslautern, 1990.

[221] H. Hasse, G. Maurer: "Kinetics of the Poly(oxymethylene) Glycol Formation in Aqueous Formaldehyde Solutions," *Ind. Eng. Chem. Res.* **30** (1991a) no. 9, 2195–2200.

[222] H. Hasse, G. Maurer: "Vapor-Liquid Equilibrium of Formaldehyde-Containing Mixtures at Temperatures below 320 K," *Fluid Phase Equilib* **64** (1991b) 185–199.

[223] H. Hasse, G. Maurer: "Heat of Dilution in Aqueous and Methanolic Formaldehyde Solutions," *Ber. Bunsenges. Phys. Chem.* **96** (1992) no. 1, 83–96.

[224] H. Hasse, I. Hahnenstein, G. Maurer: "Revised Vapor–Liquid Equilibrium Model for Multicomponent Formaldehyde Mixtures," *AIChE J.* **36** (1990) no. 12, 1807–1814.

[225] Y.-Q. Liu, I. Hahnenstein, G. Maurer: "Enthalpy Change on Vaporization of Aqueous and Methanolic Formaldehyde Solutions," *AIChE J.* **38** (1992) no. 11, 1693–1702.

[226] G. Maurer: "Vapor-Liquid Equilibrium of Formaldehyde- and Water-Containing Multicomponent Mixtures," *AIChE J.* **32** (1986) no. 6, 932–948.

[227] H. Schubert, U. Tegtmeyer, R. Schlögl, *Catalysis Letters* **28** (1994) 383–395.

[228] H. Schubert et al., *Catalysis Letters* **33** (1995) 305–319.

[229] G. J. Millar, J. B. Metson, G. A. Bowmaker, R. P. Cooney: *J. Chem. Soc. Faraday Trans.* **91** (1995) no. 22, 4149–4159.

[230] W. L. Holstein, C. J. Machiels, *J. Catal.* **162** (1996) 118.

[231] *Chemical Week* (1996) Feb. 28.

[232] J. Eckenberger, *Kunststoffe* **86** (1996) no. 10, 1514.

[233] WHO International Program on Chemical Safety (IPCS), Environmental Health Criteria Document 89, Formaldehyde, 1989.

[234] WHO international Agency for Research on Cancer (IARC), IARC Monographs on the Evaluation of Carcinogenic Risks to Humans, Volume 62, Wood Dust and Formaldehyde, 1995.

[235] European Centre for Ecotoxicology and Toxicology of Chemicals (ECETOC), Technical Report no.65, Formaldehyde and Human Cancer Risk, 1995.

[236] J. Hilton et al.: "Experimental assessment of the sensitizing properties of formaldehyde," *Food Chem.Toxicol.* **34** (1996) 571–578.

[237] T. M. Monticello et al: "Correlation of regional and nonlinear formaldehyde induced nasal cancer with proliferating population of cells," *Cancer Res.* **56** (1996) 1012–1022.

[238] NIOSH Registry of Toxic Effects of Chemicals, July 1996.

[239] Bundesgesundheitsblatt 9, 1992, p. 482–483; WHO Air Quality Guidelines (in press).

Formamides

HANSJÖRG BIPP, BASF Aktiengesellschaft, Ludwigshafen, Federal Republic of Germany (Chaps. 2–4)
HEINZ KIECZKA, BASF Aktiengesellschaft, Ludwigshafen, Federal Republic of Germany (Chap. 5)

1.	Introduction	2691	3.1.	Physical Properties	2698
2.	Formamide	2692	3.2.	Chemical Properties	2699
2.1.	Physical Properties	2692	3.3.	Production	2700
2.2.	Chemical Properties	2693	3.4.	Environmental Protection	2701
2.3.	Production	2694	3.5.	Quality Specifications and Analysis	2701
2.4.	Environmental Protection	2695	3.6.	Storage and Transportation	2702
2.5.	Quality Specifications and Analysis	2696	3.7.	Uses	2703
2.6.	Storage and Transportation	2696	3.8.	Economic Aspects	2704
2.7.	Uses	2697	4.	Other Formamides	2704
2.8.	Economic Aspects	2697	5.	Toxicology and Occupational Health	2705
3.	N,N-Dimethylformamide	2698	6.	References	2708

1. Introduction

Formamides are derivatives of the smallest of the aliphatic carboxylic acids, formic acid, and are thus structurally and chemically rather straightforward substances. Two of the simplest members of the family—formamide itself [75-12-7], HCONH$_2$, and N,N-dimethylformamide [68-12-2], HCON(CH$_3$)$_2$—have considerable industrial importance because of their chemical bifunctionality and high polarity. Formamide is especially valued for its chemical reactivity; N,N-dimethylformamide is used widely for its solvent properties.

Formamide (methanamide) was first prepared by A. W. HOFFMANN in 1863 by reaction of ethyl formate with ammonia. The compound became commercially important only after the development of high-pressure syntheses based on carbon monoxide and ammonia, to yield formamide directly or via the intermediate methyl formate. N,N-Dimethylformamide (formyldimethylamine) was reported in 1893 by

VERLEY, who obtained it upon distillation of a mixture of calcium formate and dimethylamine hydrochloride. Referred to commonly as "DMF," this compound is a key solvent in numerous operations, especially the spinning of synthetic fibers (a development of the 1950s).

No other member of the formamide family has achieved the industrial importance of these two substances.

2. Formamide

2.1. Physical Properties

Formamide [75-12-7], CH_3NO, M_r 45.04, is a hygroscopic, clear, colorless, odorless, and pH-neutral liquid with a consistency like that of glycerol. It melts at 2.55 °C and boils (decomp.) at 210 °C (101.3 kPa). Below 80 °C, the substance exhibits a greater degree of association than water, although this association diminishes rapidly with increasing temperature. The most important physical properties of formamide may be summarized as follows:

mp	2.55 °C
bp (at 101.3 kPa)	210 °C (decomp.)
Density ϱ (at 20 °C)	1.1334 g/cm³
(at 50 °C)	1.1078 g/cm³
Refractive index n_D^{15}	1.4491
n_D^{20}	1.4468
Dynamic viscosity η (at 15 °C)	4.320 mPas
(at 30 °C)	2.926 mPas
Surface tension σ (at 20 °C)	0.0582 N/m
Coefficient of expansion α	0.000742 cm³/K
Specific heat c_p^{19}	2300 J kg⁻¹ K⁻¹
Thermal conductivity λ (at 40 °C)	0.352 W m⁻¹ K⁻¹
Heat of vaporization	1673 kJ/kg
Heat of combustion	12 530 kJ/kg
Heat of formation (at 18 °C)	−3529 kJ/kg
Electrical conductivity (at 25 °C)	2×10^{-7} Ω⁻¹cm⁻¹
Dielectric constant (at 25 °C) (wavelength 1 – 5 m)	109 ± 1.5
Dipole moment (at 30 °C)	3.37 D
Flash point (Abel Martens)	175 °C
Ignition temperature (DIN 51 794)	> 500 °C
Explosive limits in air	2.7 – 19 vol %

Vapor pressure of formamide at various temperatures

t, °C	70.5	122.5	157.5	193.5	210
p, kPa	0.13	2.67	13.33	53.32	101.3 (decomp.)

Because of its high polarity and dielectric constant, formamide is an effective nonaqueous solvent for many inorganic salts, including chlorides, iodides, nitrates, phosphates, and carbonates. Formamide itself is very soluble in polar solvents such as water, lower aliphatic alcohols, carboxylic acids, esters, glycols, acetone, phenol, chloroform, and ether. Many high molecular mass polymers, natural products, and dyes also exhibit solubility or swelling in formamide. Examples include peptones and proteins (e.g., keratin), saccharides (e.g., cellulose and starch), and poly(vinyl alcohol). By contrast, hydrocarbons, fats, and oils have little solubility in formamide.

2.2. Chemical Properties

Pure formamide is a slightly hygroscopic liquid that is stable to the effects of light and atmospheric oxygen below ca. 100 °C. Above this limit, and especially above 160 °C, significant decomposition occurs. The decomposition rate reaches 0.5 % per minute at the atmospheric pressure boiling point. Principal decomposition products are carbon monoxide and ammonia, along with smaller amounts of hydrogen cyanide and water. Further transformation products are also detectable, including polymeric hydrogen cyanide derivatives. Above 500 °C, and with the aid of catalysts such as aluminum oxides or aluminum silicates, decomposition can be so accelerated that hydrogen cyanide yields may exceed 90 %. Formamide is stabilized by the presence of traces (e.g., 0.001 wt %) of purine bases [7].

Formamide is quite stable to water at room temperature. However, at elevated temperatures it undergoes hydrolysis to formic acid and ammonia, a combination that appears as ammonium formate. Both acid and alkali accelerate this hydrolysis, and the reaction of sulfuric acid with formamide at high temperature produces a nearly quantitative yield of formic acid. Until recently, this process served BASF as the basis for industrial-scale preparation of formic acid; today, however, direct synthesis of formic acid by high-pressure hydrolysis of the formamide precursor methyl formate is more economical, because the formation of ammonium sulfate as a side product is avoided (→ Formic Acid). Alkaline hydrolysis of formamide leads to alkali formates and ammonia, a reaction that can be used for quantitative analysis of formamide.

Formamide has attracted commercial interest not only because of its favorable solvent characteristics but also because of its bifunctional nature, which makes it a valuable component in many synthetic sequences. It is particularly useful as a formylating reagent, e.g., in the N-formylation of primary and secondary amines. Formamide can also serve as the starting point for substituted formamides (e.g., formanilide [103-70-8]), which are obtained by transamidation. It is used in the preparation of nitrogen-containing heterocycles such as theophylline [58-55-9], caffeine [58-08-2], pyrimidine [289-95-2], and imidazole [288-32-4]. Formate esters are produced by the reaction of formamide with the corresponding alcohols in the presence of an acid catalyst. Formamide reacts with acyl chlorides to give triacylamines [8]. Reaction with sodium methoxide produces sodium diformylamide [*18197-26-7*] [9].

2.3. Production

The production of formamide when carbon monoxide and ammonia are heated under pressure and subjected either to dark electrical discharge or to the catalytic effects of ceramic fragments has been known since the beginning of this century. However, the work of R. WIETZEL in 1924 led to the first economical, continuous process for the manufacture of formamide. His results showed that alkyl formates, especially methyl formate, could be produced in satisfactory yield at high pressure (25 MPa) and low temperature (30–140 °C) with the aid of alkali alkoxide catalysts, thus establishing what is now referred to as the two-step formamide synthesis [10]. WIETZEL also introduced a direct synthesis of formamide from carbon monoxide and ammonia in the presence of alcohols as a solvent, again promoted by alkali alkoxides [11].

Modern facilities for the preparation of formamide utilize either this *direct synthesis*,

$$CO + NH_3 \xrightarrow{NaOCH_3} HCONH_2$$

or the indirect *two-step process*:

$$CO + CH_3OH \xrightarrow{NaOCH_3} HCOOCH_3$$

$$HCOOCH_3 + NH_3 \longrightarrow HCONH_2 + CH_3OH$$

Direct Synthesis. In the direct synthesis of formamide, carbon monoxide or a gas mixture containing carbon monoxide is mixed with ammonia (either in stoichiometric amounts or in excess) and subjected to the catalytic action of alkoxides dissolved in alcohol (usually methanol). Other potential catalysts have been examined, including organometallic compounds from groups 4–10 of the periodic table (i.e., titanium to nickel), but these have not proved satisfactory so far [12]. The amount of catalyst required depends on the extent to which catalyst poisons are present, i.e., substances such as water that lead to undesirable side reactions. Poisons of this type render a portion of the catalyst inactive, causing it to precipitate from the reaction mixture in the form of sodium formate. Requisite pressures are 0.8–1.7 MPa at 75–80 °C [13]. Ammonia may be introduced in either liquid or gaseous form. Pure carbon monoxide is suitable for the process, but so is a gas mixture containing carbon monoxide, such as the coke-oven gas or blast-furnace gas obtained from a steel mill [14]. Water, hydrogen sulfide, and carbon monoxide are potential catalyst poisons and must be removed from the starting material, but hydrogen and inert gases pose no problem.

The direct synthesis of formamide actually consists of a series of steps, including synthesis, methanol recovery, catalyst recovery, salt removal, and final product purification. A separate purification step (flushing with water or separation of formamide by distillation) is necessary because impurities collect during salt removal [15], [16]. Because starting materials and product are both corrosive and stringent purity standards exist with respect to iron contamination, exposure to iron must be avoided. According to that, stainless steel is used as material of construction.

Two-Step Process. The two-step approach to formamide synthesis (as practiced, for example, by BASF) begins with ordinary commercial (96%) methyl formate and ammonia. No other starting materials are involved, so this process avoids the problem posed in direct synthesis by residual salts and other impurities. Thus, no overhead distillation of formamide is required. Apart from excess starting materials, the only byproducts are traces of entrained water and inert gas, both of which can be removed easily by distillation to yield formamide of >99.5% purity. The process divides itself naturally into three phases: the reaction itself, the recovery of methanol and other low-boiling materials, and vacuum evaporation of trace residues of volatile substances [17].

The BASF Process. The BASF process utilizes a continuous reactor equipped with a circulating pump and an external heat exchanger. Equimolar amounts of liquid methyl formate and gaseous ammonia [18] are introduced into the reactor [19], and the excess heat that results is removed by circulating the mixture over the external heat exchanger. Temperature is maintained in the range of 40–100 °C at 0.1–0.3 MPa. The gaseous phase is washed to remove excess ammonia and methyl formate. The inert gases methane, nitrogen, and a small amount of carbon monoxide (the latter arising as a decomposition product) are subsequently combusted together with off-gases from the evaporation step that also contain carbon monoxide. The reaction mixture, which consists of formamide, methanol, methyl formate, and ammonia then passes through a methanol recovery unit. This is a distillation system in which volatiles are isolated and separated into two fractions: methanol, and the low-boiling substances ammonia and methyl formate. Methanol accumulates at the bottom of the separator in sufficient purity to permit its use as a starting material in the synthesis of methyl formate. The low-boiling fraction contains virtually all the excess ammonia and methyl formate, together with a small amount of methanol, and this is recycled to the reactor. The residual formamide contains trace amounts of methanol which are removed under vacuum, leaving a product of sufficient purity to meet all standards applicable to both end-product and intermediate uses.

2.4. Environmental Protection

Flammability poses few problems, since formamide has very high ignition and flash points (>500 and 175 °C, respectively), as well as a low vapor pressure (see data in Section 2.1). Thus, the risk of combustion is quite low. If the substance should happen to ignite, the resulting fire may be quenched with water, extinguishing foam, carbon dioxide, or dry chemical extinguishers. Formamide should not be allowed to enter surface or groundwater sources but it can be readily biodegraded in appropriate wastewater treatment systems.

Serious effort should be made to prevent uncontrolled release of formamide from production and storage areas. The preferred method for transferring the liquid within an enclosed area involves the use of sealed pumps, although these should not be the stuffing-box type. Ideally, production should be carried out in the open, with exhaust gases washed or passed through a combustion unit prior to their release into the environment. If production or processing must be conducted within an enclosed facility, exceptionally good ventilation is essential because inhalation of formamide may be hazardous.

2.5. Quality Specifications and Analysis

Market demands require that formamide be available in a purity exceeding 99%. The maximum permissible color number is 10 APHA, which implies a very low content of polymeric hydrocyanic acids and iron compounds. The material supplied by BASF is certified as meeting the following specifications.

Formamide content	min. 99.5%	(by difference)
Methanol	max. 0.1%	(photometric)
Ammonium formate	max. 0.1%	(titrimetric)
Water content	max. 0.05%	(DIN 51 777)
Iron	max. 1 ppm	(photometric)
Hazen color number	max. 10	(DIN 53 409)
pH	6.5	
Freezing point	≥ 2.1 °C	(DIN)

Methanol is determined photometrically after oxidation to formaldehyde, which in turn reacts with chromotropic acid to a purple dye. Ammonium formate content is measured by releasing the ammonia, trapping it with formaldehyde, and titrating the residual formic acid. The Karl Fischer method is employed for measuring water content. Iron is determined photometrically as the bipyridyl complex. The most accurate estimate of total formamide content results from subtracting the combined impurities from 100%. Estimation by chromatography is less precise because of the compound's thermal instability, a consideration which also limits the usefulness of Kjeldahl determinations. Purity determination can also be based on melting point, which is related directly to the content of methanol and water.

2.6. Storage and Transportation

Unalloyed steel, copper, brass, and lead are all subject to corrosion by formamide, especially in the presence of water. Moreover, most protective paints and plastic coatings are attacked by the substance. In applications where iron contamination is undesirable aluminum and stainless steel containers are used instead of iron tanks.

Formamide is shipped in containers made of stainless steel, in polyethylene drums, or in iron drums coated with polyethylene. Typical containers are designed to hold either 60 or 220 kg.

The physical and chemical properties of formamide are such that it is exempt from classification as a hazard with respect to highway, rail, inland waterway, or ocean transport.

When kept in closed containers, formamide is stable almost indefinitely. It is neither explosive nor subject to spontaneous combustion, although it is flammable. Those involved in the storage or transport of formamide should be familiar with the safety considerations discussed in Section 2.4, and Chapter 5.

2.7. Uses

Until relatively recently, the major application of formamide involved its reaction with sulfuric acid to produce formic acid and ammonium sulfate. However, alternative syntheses of formic acid based on methyl formate and water have now largely displaced the formamide method (→ Formic Acid). Currently, the principal application relates to the so-called formamide vacuum synthesis of hydrogen cyanide [74-90-8] [20], in which formamide is pyrolyzed to hydrogen cyanide and water.

Formamide is also used as a synthetic intermediate in the preparation of various pharmaceuticals and fungicides. Examples include caffeine [58-08-2], theophylline [58-55-9], and theobromine [83-67-0] [21] – [24]. It serves as a catalyst in certain synthetic reactions, including carbonylation [25].

Formamide is employed widely as a *solvent*. It is used to dissolve lacquers [26] and plays a role in the spinning of acrylonitrile mixed polymerizates [27]. Together with nitromethane [75-52-5], it is used to dissolve polyacrylonitrile [25014-41-9] [28], and it is also employed in the polymerization of unsaturated amines to produce ion-exchange resins [29]. Formamide is a good solvent for wood stains, as well as for the ink used in felt- and fiber-tip pens. It was formerly employed in conjunction with silicates as a floor sealant, and also in de-icing airport runways, but these uses have been abandoned because of its teratogenic properties.

2.8. Economic Aspects

Only four major producers of formamide can be found worldwide. The leading manufacturer is BASF, with an annual capacity of 100 000 t. Smaller amounts of formamide are produced in Czechoslovakia, the German Democratic Republic (VEB Leuna), and Japan (Nitto). Most of the industrial output is utilized directly by the manufacturers. Despite the potential diversity of applications for formamide, market demand is relatively low, which explains why the number of production sites is so limited.

3. N,N-Dimethylformamide

3.1. Physical Properties

N,N-Dimethylformamide [68-12-2] (DMF), C_3H_7NO, M_r 73.09, is a colorless, high-boiling, nonviscous, highly polar, distillable, hygroscopic liquid with a weak, characteristically amine-like odor. Its most important physical characteristics can be summarized as follows:

mp	−61 °C
bp (at 101.3 kPa)	152.5 – 153.5 °C
Density ϱ (at 20°C)	0.9500 g/cm^3
(at 50 °C)	0.9215 g/cm^3
Refractive index n_D^{20}	1.4310
n_D^{25}	1.4282
Dynamic viscosity η (at 15 °C)	0.87 mPa·s
(at 50 °C)	0.63 mPa·s
Surface tension σ (at 20 °C)	0.0355 N/m
Specific heat c_p^{20}	2062 J kg^{-1} K^{-1}
Thermal conductivity λ (at 40 °C)	0.180 W m^{-1} K^{-1}
Heat of vaporization	573.5 kJ/kg
Heat of combustion	26 197 kJ/kg
Heat of formation (at 18 °C)	−3342 kJ/kg
Electrical conductivity (at 25 °C)	6×10^{-8} Ω^{-1} cm^{-1}
Dielectric constant (at 25 °C) (wavelength 1 – 5 m)	36
Dipole moment (at 20 °C)	3.82 D
Flash point (DIN 51 755)	58 °C
Ignition temperature (DIN 51 794)	410 °C
Explosive limits in air	2.2 – 16 vol %

Vapor pressure of DMF

t, °C	−50	−10	0	10	40	80	120	153
p, kPa	0.0003	0.033	0.081	0.182	1.34	9.67	42.0	101.3

Dimethylformamide is miscible in all proportions with water, alcohols, ethers, ketones, esters, and carbon disulfide, as well as with chlorinated and aromatic hydrocarbons. By contrast, it has only limited solubility or complete insolubilty in aliphatic hydrocarbons. Many high molecular mass natural products and synthetics are quite soluble in DMF. Examples include poly(vinyl alcohol) [9002-89-5], poly(vinyl chloride) [9002-86-2], poly(vinylidene chloride) [9002-85-1], poly(vinyl acetate) [9003-20-7] poly(vinyl ether), polyacrylate, polyacrylonitrile [25014-41-9], polystyrene [9003-53-6], polyester resins, alkyd resins, urea–formaldehyde resins, melamine – formaldehyde resins, phenol – formaldehyde resins, epoxy resins, and natural resins. It is also a good

Table 1. Solubility of organic gases in dimethylformamide (partial pressure 101.3 kPa)

Gas	CAS registry no.	Solubility (in 100 g of DMF), g	
		20 °C	100 °C
Vinylacetylene	[689-97-4]	112	18
Acetylene	[74-86-2]	5.1	0.9
1,3-Butadiene	[106-99-0]	11.2	2.5
Isobutene	[115-11-7]	7.3	1.5
Butane	[106-97-8]	4.7	1.3
Propene	[115-07-1]	0.78	0.46
Propane	[74-98-6]	0.85	0.33
Ethylene	[74-85-1]	0.29	0.14

Table 2. Solubility of inorganic gases in dimethylformamide (partial pressure 101.3 kPa)

Gas	Solubility (in 100 g of DMF), g at 25 °C
Hydrogen bromide	47.55
Hydrogen chloride	39.37
Hydrogen sulfide	1.32
Sulfur dioxide	179.00

solvent for a number of drying oils, plasticizers, and cellulose derivatives, as well as for both natural and chlorinated rubber. The solubility of polyurethanes in DMF is a function of the conditions under which they are prepared. Most halogenated hydrocarbons are also soluble in DMF.

Various inorganic salts are highly soluble in DMF [30]–[38]. For example, the following solubilities (in 100 g DMF) are reported at 25 °C: 11.43 g of $FeCl_3$ [7705-08-0], 13.71 g of KI [7681-11-0], 15.97 g of KSCN [333-20-0], 27.53 g of LiCl [7447-41-8], and 39.6 g of $NaClO_4$ [7601-89-0]. Several organic and inorganic gases are soluble in DMF, as indicated in Tables 1 and 2. Other gases such as ammonia, carbon monoxide, carbon dioxide, oxygen, and hydrogen dissolve in DMF to only a limited extent at atmospheric pressure.

3.2. Chemical Properties

N,N-Dimethylformamide is stable to air and light and, unlike its parent compound formamide, can be distilled without decomposition at atmospheric pressure. Nevertheless, decomposition is observed at temperatures above the boiling point, the products being carbon monoxide and N,N-dimethylamine. The compound also shows less tendency than formamide to hydrolyze. It is hygroscopic, however, and water absorbed from the air will produce traces of hydrolysis products; these produce the faint amine odor that frequently accompanies commercial DMF. The resistance of DMF to hydrolysis is demonstrated by the fact that a refluxing 5% aqueous solution was found to have decomposed only 0.17% after 120 h.

Pure DMF can be handled quite safely. Nevertheless, a few reports exist of reactions in which DMF displays a reactivity bordering on the explosive. Thus, with certain halogenated hydrocarbons, e.g., carbon tetrachloride or hexachlorobenzene, at elevated temperature and in the presence of ionic iron, explosive decomposition has occurred. Explosion during the reaction of DMF with thionyl chloride has also been described [39]. Furthermore elemental bromine and triethylaluminum [97-93-8] react violently with DMF. Vigorous oxidation reactions have been described in DMF systems containing chromic acid, as well as when nitrates are dissolved in DMF. Finally, DMF vapors form a flammable mixture with air.

3.3. Production

The first experiments directed toward the production of *N*-alkylformamides were conducted in the 19th century. Thus, A. BEHAL reported in 1899 that *N*-alkylformamides could be prepared by reacting mixed anhydrides of formic and other aliphatic acids with *N*-alkylamines. Since then, *N*,*N*-dimethylformamide has been synthesized by the reaction of hydrocyanic acid with methanol [40] or methylamines in the presence of water [41]. Other approaches include the reaction of ammonia or formamide with hydrogen and carbon dioxide in the presence of catalysts [42], as well as the reaction of *N*,*N*-dimethylamine with hydrogen and carbon dioxide [43]. Processes have also been described starting with carbon monoxide and *N*,*N*-dimethylamine, either alone [44] or together with trimethylamine [75-50-3] [45].

As with formamide, two methods dominate present-day commercial preparation of DMF: *direct synthesis* and a *two-step process*.

Direct Synthesis. The direct or one-step synthesis [46] of DMF begins with either pure carbon monoxide or a gas stream containing carbon monoxide. This is reacted in a continuous process with *N*,*N*-dimethylamine, by using a solution of alkali alkoxide (usually sodium methoxide) in methanol as catalyst. Methyl formate is presumably formed as an intermediate. The reaction mixture passes over an external heat exchanger to remove the excess heat generated and to ensure thorough mixing of the components [47].

The reaction is conducted between 0.5 and 11 MPa at 50–200 °C. The reaction mixture exits the reactor through a decompression chamber. In addition to *N*,*N*-dimethylformamide, the crude product contains methanol, a certain amount of unreacted *N*,*N*-dimethylamine, dissolved carbon monoxide, and residual catalyst. The addition of acid or water deactivates any catalyst present, resulting in the formation of sodium formate [141-53-7]. Dissolved carbon monoxide, together with inert gases, escapes from the mixture during decompression, and the off-gases are removed by combustion. Preliminary distillation is followed by a second distillation in a separate column; here, dimethylformamide is separated from methanol which contains traces of N,N-dimethylamine. Further distillation results in a product of 99.9% purity.

Two-Step Process. The two-step process [48] for the synthesis of N,N-dimethylformamide differs from direct synthesis because methyl formate is prepared separately and introduced in the form of ca. 96% pure (commercial-grade) material. Equimolar amounts of methyl formate and N,N-dimethylamine are subjected to a continuous reaction at 60–100 °C and 0.1–0.3 MPa. The resulting product is a mixture of N,N-dimethylformamide and methanol. The purification process involves distillation and is analogous to that described for direct synthesis. However, no separation of salts is required because no catalysts are involved in the process. According to the corrosive properties of both starting materials and products, stainless steel has to be used as material of construction for production facilities.

3.4. Environmental Protection

The physical properties of N,N-dimethylformamide (i.e., its autoignition temperature and flash point) are such that the compound is exempt from classification under the ordinance dealing with combustible liquids in the Federal Republic of Germany (VbF, Verordnung für brennbare Flüssigkeiten). If fire involving DMF does occur, it can be extinguished with water, dry chemical agents, or carbon dioxide.

N,N-Dimethylformamide is regarded as a hazardous liquid with respect to water pollution, and direct discharge into streams and lakes must be avoided. Nevertheless, it is easily biodegradable in appropriate wastewater treatment plants.

Regulations governing the use and disposal of N,N-dimethylformamide are essentially the same as those for formamide. The advisory covering atmospheric emissions in the Federal Republic of Germany (TA Luft, Verwaltungsvorschrift zum Bundesimmissionsschutzgesetz, 28th August 1974) places DMF in Category II. This means that DMF emissions exceeding 3 kg/h must be diluted to such an extent that the maximum concentration in air does not exceed 150 mg/m^3. If this limit cannot be met, the corresponding off-gases must be treated with water or combusted. The content of DMF in air can be measured either by gas chromatography or with an infrared detector.

3.5. Quality Specifications and Analysis

Most applications of DMF require that it be available in high purity. Table 3 contains the specifications applicable to DMF marketed by BASF.

The acid and base content of a DMF sample is established easily by potentiometric titration. The method of choice for determining DMF in an aqueous solution depends upon the anticipated concentration: those above 10% may be quantified by gas

Table 3. Sales specifications for -dimethylformamide from BASF

Sales specifications	Limit	Test method
DMF content	min. 99.9%	gas chromatography
Methanol content	max. 100 mg/kg	gas chromatography
Water content	max. 0.03 wt%	DIN 51 777
Boiling range	152.5–153.5 °C	DIN 53 171
Acid content calcd. as HCOOH	max. 20 mg/kg	titrimetry
Base content calcd. as $NH(CH_3)_2$	max. 20 mg/kg	titrimetry
Hazen color no.	max. 10	DIN 53 409

chromatography, while lower concentrations are best measured with a doublebeam infrared photometer. If only traces of DMF are expected, as in wastewater, a combination of chemical and physical methods of analysis should be employed. Thus, any DMF present can be hydrolyzed with acid, after which the solution is made alkaline and the resultant N,N-dimethylamine is distilled. The latter is then converted to the corresponding nitroso compound, which can be determined polarographically. In the absence of impurities, DMF content can be calculated from simple physical measurements such as refractive index or density.

3.6. Storage and Transportation

If it is pure and anhydrous, at ordinary temperatures, N,N-dimethylformamide does not tend to corrode metals during storage and transportation. Even with standard steel (ST 37), the linear corrosion rate for DMF is less than 0.01 mm per year. Exceptions to this generalization include copper, tin and their alloys.

Because DMF is hygroscopic, it readily absorbs moisture from the air, which renders it more corrosive. To prevent iron contamination, DMF should be packaged under an atmosphere of dry nitrogen if iron containers are used. To avoid all risk of contamination, it should be stored in stainless steel containers, although aluminum containers are recommended for DMF used in synthetic fiber production. Suitable gasket materials for pipelines, industrial fittings, and other apparatus exposed to DMF include poly(tetrafluoroethylene) [9002-84-0], high molecular weight polyethylene [9002-88-4], polypropylene [9003-07-0], and high molecular weight poly(vinyl alcohol) [9002-89-5]. N,N-Dimethylformamide may be shipped in tank cars, containers, and tank trucks fabricated from stainless steel. Iron drums with a capacity of 200 kg are also utilized. However, DMF should not be allowed to come in contact with chlorinated hydrocarbons because of the potential for explosive reaction, especially with such compounds as carbon tetrachloride [56-23-5] or hexachlorobenzene [118-74-1] in the presence of dissolved iron.

N,N-Dimethylformamide is subject to the following international transport designations (UN No. 2265):

Rail transport	GGVE/RID	3,32 C
Highway transport	GGVS/ADR	3,32 C
Water transport	GGVSee/IMDG-Gde	3,3
Air carriers	IATA-DGR/ICAO-TI	

Designations applicable to its transport and handling in the Federal Republic of Germany are derived from the ordinance "Verordnung über gefährliche Stoffe (Gefahrstoffverordnung GefStoffV)" of 26th August 1986, in which DMF is among the substances classified as having minimal toxicity. The corresponding designations include

Hazard symbol	Xn
R	20/21-36
S	26-28-36
EEC Guideline	616-001-00X

3.7. Uses

The principal applications of N,N-dimethylformamide are as a solvent and as an extractant, particularly for salts and compounds with high molecular mass. This role is consistent with its interesting combination of physical and chemical properties: low molecular mass, high dielectric constant, electron-donor characteristics, and ability to form complexes. The use of DMF as a component in synthesis is of relatively minor significance, at least commercially.

The most important use of DMF as a solvent for polymers is in the preparation of polyacrylonitrile solutions for the manufacture of fibrous polyacrylonitrile [25014-41-9]. Residual DMF can be recovered from the spun fibers with relatively little expenditure of energy. In addition DMF is used as a solvent in the preparation of high-quality surfaces based on polyurethanes, which are otherwise difficult to solubilize, and it is employed as a solvent for coating surfaces with polyamides.

The high boiling point and high solvating power of DMF lead to its further use as a solvent for wire enamel formulated with polyamides, polyesterimides, or polyurethanes. Conversely, its solubilizing ability also makes it an effective component of paint removers, and it is used as a selective solvent in a wide range of manufacturing processes. For example, DMF is useful in separating acetylene from ethylene; butadiene from C_4 cuts; and hydrogen chloride, hydrogen sulfide, or sulfur dioxide from CO_2-containing gases. N,N-Dimethylformamide or mixtures containing it are used by oil refiners to extract aromatic compounds from hydrocarbon mixtures.

Solutions of salts in specially purified DMF serve as fillers for electrolytic capacitors, a role made possible by the high dielectric constant of DMF. This substance is also used either alone or with other solvents in the recrystallization of aromatics, heterocyclics,

2703

Table 4. Manufacturers of -dimethylformamide

Manufacturer	Country	Annual capacity, t
Air Products	USA	7 000
BASF AG	Fed. Rep. of Germany	60 000
BASF	Brazil	6 000
Celanese	Mexico	6 000
Chinook	Canada	10 000
Du Pont	USA	40 000
Ertisa	Spain	5 000
ICI	UK	15 000
Korea Fertilizer Co.	Republic of Korea	8 000
LCY	Taiwan	10 000
Mitsubishi Gas Chemicals	Japan	20 000
Nitto Chemicals	Japan	25 000
UCB	Belgium	16 000
VEB Leuna	German Democratic Republic	19 000

dyes, fluorescent whitening agents, and pigments. The pharmaceutical industry also utilizes its broad solvent and extractant properties.

N,N-Dimethylformamide could potentially be employed in the synthesis of aldehydes, acetals, amides, amidines, esters, and heterocycles, but thus far its use has been limited [49]–[53].

3.8. Economic Aspects

N,N-Dimethylformamide is produced by a number of manufacturers worldwide; total capacity is estimated at 250 000 t per year. Much of the production is utilized directly by the manufacturers, the most important of whom are listed in Table 4.

The principal market for DMF is as a solvent for polyacrylonitrile, followed by its use in processing polyurethanes. The remaining applications are remarkably diverse but minor relative to the principal uses. Current operating capacity is sufficient to meet worldwide demand.

4. Other Formamides

While many other N-substituted formamides are known, only three have achieved a measure of commercial significance; N-methylformamide, N-formylmorpholine, and formanilide. These are prepared by a two-step process: either the corresponding nitrogen compound is reacted with methyl formate, or a transamidation is performed, typically involving formamide.

N-Methylformamide. N-Methylformamide [123-39-7], HCONHCH$_3$, M_r 59.1, is a colorless, nearly odorless liquid that is infinitely soluble in water and relatively stable to hydrolysis at low temperature. Commercial material is available with a purity of at least 99.7%. Its most important physical properties include the following: mp −3.2 °C; boiling range 199.3–200.5 °C; ϱ (at 20 °C) 1.003 g/cm^3; n_D^{20} 1.432–1.433; flash point 111 °C; ignition temperature 425 °C; explosive limits in air 3.8–17.8 vol%; vapor pressure (at 20 °C) 0.014 kPa (0.14 bar). The substance is used as an intermediate in the synthesis of the insecticide formothion. It also has limited use in oil refineries as a selective solvent for aromatic hydrocarbons.

N-Formylmorpholine. N-Formylmorpholine [4394-85-8], HCON(CH$_2$)$_4$O, M_r 115.13, is a colorless liquid with a faint amine-like odor. It is very resistant to hydrolysis at low temperature and has sufficient thermal stability to be distilled at atmospheric pressure. Commercial material has a minimum purity of 99.5%. The following are important physical properties of this substance: mp 20–21 °C; bp 244 °C; ϱ (at 20 °C) 1.152 g/cm^3; n_D^{20} 1.4870; flash point 125 °C; ignition temperature 370 °C; explosive limits in air 1.2–8.5 vol%; vapor pressure (at 80 °C) 0.17 kPa.

N-Formylmorpholine is used by petroleum refiners as a solvent for the extraction of aromatics and also as an anticorrosive agent in fuel oil.

Formanilide. Formanilide [103-70-8]; HCONHC$_6$H$_5$, M_r 121.14, is used as an intermediate in the pharmaceutical industry and as an antioxidant (anti-ager) in the rubber industry.

Formamides containing various aliphatic substituents on nitrogen have acquired limited significance as solvents and extractants. Thus, N,N-di-n-butylformamide [761-65-9] or N-methyl-N-2-heptylformamide [68018-42-8] can be used to extract organic acids from aqueous media [53].

5. Toxicology and Occupational Health

Formamide. *Acute Toxicity.* Acute toxicity studies demonstrate that single doses of formamide, administered by a variety of routes, have low toxicity. In rats the acute *oral* LD$_{50}$ is >5000 mg per kilogram body weight [55], [56] and the acute dermal LD$_{50}$ is 13.5 g per kilogram body weight [56]. The acute inhalation hazard is low: no mortality was observed in rats exposed for 8 h at 20 °C to an atmosphere enriched in or saturated with formamide [55]. In the rabbit skin and eye the substance is not irritating to slightly irritating [55], [56]. Skin sensitization was not observed in guinea pigs [57].

Subacute, Subchronic Toxicity. Twenty oral doses of 300 µL formamide per kilogram body weight, given to rats via stomach tube, over four weeks were lethal in 50% of the animals. The affected organs were the stomach, spleen thymus, adrenals, and testis. Treatment with 100 µL per kilogram body weight resulted in reduction of body weight gain and a reduced food intake whereas 30 µL per kilogram body weight did not produce any observable adverse effects [58]. Rats were subjected to doses of 30–3000 mg per kilogram body weight, applied dermally for three months (6 h per day, 5 d per week), semi-occlusive on the intact skin. Mortality occurred after treatment with 3000 mg per kilogram body weight. The affected organs were the liver, kidney, adrenals, spleen and testis. Hematological investigation revealed polycytemia, a change in the red blood picture, as the only effect observed. This effect also occurred at doses of 1000 and 300 mg/kg but not at 100 and 30 mg/kg body weight [59].

Mutagenicity. In several mutagenicity tests on bacterial systems (Ames test [60]–[62]), on mammals (dominant lethal test, mouse [64]), or in an in vitro short term test for carcinogenic activity (cell transformation test in rat embryo cultures [64], [65]) formamide did not demonstrate a mutagenic potential.

Teratogenicity. Formamide has been tested in a variety of studies for its teratogenic effects. In rats and mice the compound produced adverse effects on the offspring after oral and dermal administratation. The degree of fetal damage depends on the stage of pregnancy at which the substance was given. Doses that had adverse effects on the dams also had distinct effects on the offspring; at lower doses the effects were less pronounced [56], [66].

TLV Value. The TLV value (1988–1989) was established to be 10 ppm (15 mg/m^3).

Dimethylformamide. *Acute Toxicity.* Dimethylformamide which has been tested in different animal species using various administration methods is slightly toxic. The affected organs, especially after repeated administration are the liver and kidney.

The acute *oral* LD_{50} in rats and mice is 2200–7550 mg per kilogram body weight [67]–[69]. The acute *dermal* LD_{50} in rats is between 800 and 17 000 mg per kilogram body weight and in rabbits between 1000 and 5000 mg per kilogram body weight. The acute inhalative LC_{50} in rats is given as >2500 ppm for a four or six hour exposure and 2475 ppm for a one-hour exposure [69]. In mice LC_{50}-values are 9.4 mg/L (3133 ppm, two-hour exposure) or 686 ppm (two-hour exposure) [56]. Rats survived a four-hour inhalation of an atmosphere enriched or saturated with DMF at room temperature, but deaths occurred after 6 h [67]. In other studies cats, rabbits, guinea pigs, and rats survived an eight-hour exposure to a saturated atmosphere without any adverse effects. Rabbits, cats, and dogs survived a 6 h-inhalation of 1000 ppm [67].

Irritation. DMF slightly irritates the rabbit skin. The liquid is absorbed by the skin and can facilitate the absorption of other substances [56], [67]. The compound has a pronounced irritating effect on the eye of rabbits, the effects being reversible within 14 d [56], [67], [69]. Washing the eye with water immediately after application increased the irritating effect and produced opacity of the cornea [69]. Although DMF has not been shown to cause sensitization [67], a case of allergic contact dermatitis has been reported [70].

Subacute and Subchronic Toxicity. In subacute and subchronic animal studies (rats, mice, rabbits, cats, and dogs) in which repeated doses of DMF were given orally, dermally, or by inhalation, nephrotoxicity and hepatotoxicity were the prominent effects. The doses and concentrations used varied widely depending on the method and duration of administration. Changes in the red blood picture were observed in a few cases (for further details, see [56], [67]).

Mutagenicity. DMF has been investigated in several mutagenicity tests covering all different end points of mutagenic activity and has in all these tests been found to be non-mutagenic [56].

Carcinogenicity. Animal studies published so far have not provided any evidence of carcinogenicity [56]. Rats receiving 150 mg DMF per kilogram body weight in their drinking water for 500 d did not develop tumors. A weekly subacute treatment with doses of 200 or 400 mg per kilogram body weight (total dose 8–20 g per kilogram body weight) did not increase the incidence of tumor rate in the rat. A single intraperitoneal injection of 0.5 g per kilogram body weight to rats with or without simultaneous partial hepatectomy did not cause formation of tumors or "nodules" as early stages. In an insufficiently designed, inadequately described study, rats developed multiple tumors, liver necrosis, and ulceration of the intestinal mucosa after ten weekly intraperitoneal injections of 1 mL per rat. In comparison to all other mentioned studies the latter result seems to be inadequate.

Teratogenicity. DMF has been tested in different animal species for its adverse effects on the offspring using various administration methods. Fetotoxic and embryotoxic effects (fetal death and/or reduced fetal weight) were observed, especially at doses that produced toxic effects in the dams. In some studies DMF increased the incidence of malformations.

Oral doses that are toxic to the dams increased the incidence of malformations in rats and rabbits, whereas lower doses which were not maternally toxic are still embryotoxic [71], [72]. In mice oral administration of DMF leads to embryotoxicity and to an increase in the incidence of spontaneous malformations without detectable maternal toxicity [72]. Dermal application of DMF (200 mg kg^{-1} d^{-1}) in the rabbit caused fetal death; maternal toxicity was not discussed [73]. Other investigators found that dermal administration of a 200 mg kg^{-1} d^{-1} dose of DMF, applied semi-occlusively, did not have any effects in rats but a dose of 400 mg kg^{-1} d^{-1} increased the incidence of skeletal anomalies and showed slight maternal toxicity [72]. In other studies in rats, dermal application of DMF at doses which were slightly toxic to the dams had adverse effects on the offspring [72], [73]. In inhalation studies, embryotoxic effects were observed in rats at doses that produced maternal toxicity; in a few cases these effects already occurred in the absence of maternal toxicity [56].

Dermal administration of 400 mg kg^{-1} d^{-1} caused a higher incidence of anomalies in the offspring of rabbits, parallelled by reduced body weight in the dams. The next lower dose of 200 mg kg^{-1} d^{-1} did not cause any effects [74].

Inhalative exposure of rabbits (6 h/d, 7–19 day of pregnancy) to a DMF concentration of 150 and 450 ppm in the respiratory air produced dose-dependent maternal toxic, embryotoxic, and teratogenic effects. A concentration of 50 ppm did not cause any effects [75]. According to these investigations DMF seems to be teratogenic after inhalation only in higher concentrations which normally also lead to observable maternal toxicity. The threshold concentration for embryotoxic and teratogenic effects after inhalative exposure is between 50–150 ppm.

MAK Value. The MAK value is established to be 20 ppm (60 mg/m^3) refering to the danger of skin resorption.

6. References

General References

[1] The Sigma Aldrich Library of Chemical Safety Data, Edition I, 1985.
[2] Robert Kühn, Karl Birett: *Merkblätter gefährliche Arbeitsstoffe,* Verlag Moderne Industrie, München.
[3] *Beilstein* **2**, 26–27; **2 (1)**, 20–21; **2 (2)**, 36–37, **2 (3)**, 55–59; **2 (4)**, 45–50; system no. 156.
[4] BASF, *Technisches Merkblatt über Formamid,* 1983. (Technical Leaflet Formamide, 1983.) for N,N-Dimethylformamide
[5] *Beilstein* **4**, 58; **4 (1)**, 171–174; **4 (3)**, 122; system no. 335.
[6] BASF, *Technisches Merkblatt über N,N-Dimethylformamid,* 1984.(Technical Leaflet Dimethylformamide, 1986.)

Specific References

[7] Du Pont, US 2 346 425, 1944 (E. C. Kirkpatrik)
[8] Q. E. Thompson, *J. Am. Chem. Soc.* **73** (1951) 5841.
[9] I. C. Gramain, R. Rémuson, *Synthesis* 1982, 264.
[10] I. G. Farben, DE 495 935, 1930 (R. Wietzel).
[11] I. G. Farben, DE 550 749, 1932 (R. Wietzel).
[12] Ethyl Corporation, US 3 099 689, 1961 (H. J. Cragg).
[13] Du Pont, DE 624 508, 1936.
[14] O. Bankowski, M. Köthnig, K. Klemt, DD 12 113, 1953.
[15] VEB Leuna-Werke, DD 13 619, 1954 (K. Smeykal, M. Köthnig).
[16] VEB Leuna-Werke, DD 61 536, 1967 (G. Peinze).
[17] BASF, DE 1 215 130, 1966 (E. Germann, H. Friz)
[18] BASF, DE 2 623 173, 1980 (H. Hohenschutz et al.)
[19] Buss AG, CH 370 057, 1963 (W. Steiner)
[20] I. G. Farben, FR 906 114, 1944.
[21] H. Bredereck, *Angew. Chem.* **56** (1943) 328–329.

[22] H. Bredereck et al., *Chem. Ber.* **83** (1950) 201–211.
[23] H. Bredereck, H. G. v. Schuh, DE 864 870, 1953.
[24] H. Bredereck, et al., *Chem. Ber.* **86** (1953) 333–351.
[25] Esso Research and Engineering Co., US 2 834 812, 1958 (L. Hughes, C. Township, J. Kirshenbaum).
[26] Radio Corp. of America, US 2 737 465, 1956 (L. Pessel).
[27] Allied Chemical and Dye Corp., US 2 776 945, 1957 (F. J. Rahl, H. H. Weinstock).
[28] P. Halbig, CH 295 072, 1953.
[29] State Inst. of the State Florida, US 2 687 382, 1954 (G. B. Butler, R. L. Bunch).
[30] R. Ch. Paul et al., *Indian J. Chem.* **4** (1966) 382.
[31] R. Ch. Paul et al., *Indian J. Chem.* **2** (1964) 97.
[32] W. Gerrard et al., *J. Chem. Soc.* 2 (1960) 2144.
[33] T. S. Piper et al., *J. Am. Chem. Soc.* **76** (1954) 4318.
[34] E. L. Muetterties et al., *J. Am. Chem. Soc.* **75** (1953) 490.
[35] J. Archambault et al., *Can. J. Chem.* **36** (1958) 1461.
[36] R. Ch. Paul et al., *Indian J. Chem.* **3** (1965) 300.
[37] A. J. Jungbauer, *Nature (London)* **202** (1964) 290.
[38] M. Glavas et al., *J. Inorg. Nucl. Chem.* **31** (1969) 291.
[39] S. J. Mukund, *Chem. Eng. News* **64** (1986) no. 14, 2.
[40] Y. Fukuoka, N. Kominami, *Chem. Technol.* 1972, 670–674.
[41] Asahi Chem. Ind. Co., JP 7 208 051, 1968.
[42] Air Products and Chemicals Inc., US 4 558 157, 1985 (J. A. Marsella, G. P. Pez).
[43] Shell Oil Co., US 3 530 182, 1970 (P. Haynes, J. F. Kohnle, L. H. Slaugh).
[44] The Leonard Process Co., *Hydrocarbon Process.* **32** (1983) no. 11, 90.
[45] Nitto Chemical Industry Co., JP 160 075, 1978.
[46] Du Pont, US 2 204 371, 1937 (D. J. Loder).
[47] UCB S. A., DE 2 710 725, 1987 (W. Couteau, J. B. Ramioulle).
[48] I. G. Farben, DE 714 311, 1937 (R. Seidler, H. Klein).
[49] R. S. Kittila: *Dimethylformamide Chemicals, Uses,* Du Pont de Nemours, Wilmington, Del., 1967.
[50] H. Bredereck et al., *Angew. Chem.* **74** (1962) 354.
[51] H. Bredereck et al., *Chem. Ber.* **96** (1963) 3265.
[52] H. Bredereck et al., *Angew. Chem.* **75** (1963) 825.
[53] H. Bredereck et al., *Chem. Ber.* **97** (1964) 61.
[54] BASF, DE 2 545 658, 1985 (H. Hohenschutz, J. E. Schmidt, H. Kiefer).
[55] BASF Aktiengesellschaft, unpublished results 1963.
[56] G. L. Kennedy, Jr., *CRC Crit. Rev. Toxicol.* **17** (1986) no. 2, 129–182.
[57] P. H. Chanh et al., *Therapie* **26** 409–424 (1971).
[58] BASF Aktiengesellschaft, unpublished results 1975.
[59] BASF Aktiengesellschaft, unpublished results 1982–1984.
[60] National Toxicology Program, Fiscal Year 1984, Annual Plan.
[61] National Toxicology Program, *Technical Bulletin* no. 9, April 1983.
[62] K. Mortelmans et al., *Environ. Mutagen.* Suppl. 7, 8 (1986) 1–119.
[63] BASF Aktiengesellschaft, unpublished results 1970.
[64] A. E. Freeman et al., *J. Nat. Cancer Inst.* **51** (1973) no. 3, 799–808.
[65] Ishidate, Kada: *Environmental Mutagens Data Book,* 1980.
[66] BASF Aktiengesellschaft, unpublished results 1963–1969.

[67] Toxikologisch-arbeitsmedizinische Begründung von MAK-Werten der Kommission zur Prüfung gesundheitsschädlicher Arbeitsstoffe der Deutschen Forschungsgemeinschaft, Stand 16.12.1974, Nachtrag 20.12.1977, Verlag Chemie, Weinheim.
[68] J. Lundberg, et al.,*Environ. Res.* **40** (1986) 411–420.
[69] G. L. Kennedy, Jr., H. Sherman, *Drug Chem. Toxicol.* **9** (1986) 2, 147–170.
[70] J. G. Camarasa, *Contact Dermatitis* **16** (1987) no. 4, 234.
[71] J. Merkle et al., *Arzneim. Forsch.* **30** (1980) 1557–1562.
[72] BASF Aktiengesellschaft, unpublished results 1970–1976.
[73] E. F. Stula et al., *Toxicol. Appl. Pharmacol.* **41** (1977) 35–55.
[74] BASF Aktiengesellschaft, unpublished results 1984.
[75] BASF Aktiengellschaft, unpublished results 1988.

Formic Acid

WERNER REUTEMANN, BASF Aktiengesellschaft, Ludwigshafen, Federal Republic of Germany (Chaps. 2–12)

HEINZ KIECZKA, BASF Aktiengesellschaft, Ludwigshafen, Federal Republic of Germany (Chap. 13)

1.	Introduction	2711	7.	Chemical Analysis	2732
2.	Physical Properties	2712	8.	Storage and Transportation	2733
3.	Chemical Properties	2716	9.	Legal Aspects	2734
4.	Production	2718	10.	Uses	2734
4.1.	Methyl Formate Hydrolysis	2719	11.	Economic Aspects	2735
4.1.1.	Kemira–Leonard Process	2723	12.	Derivatives	2735
4.1.2.	BASF Process	2724	12.1.	Esters	2735
4.1.3.	USSR Process	2725	12.1.1.	Methyl Formate	2736
4.1.4.	Scientific Design–Bethlehem Steel Process	2726	12.1.2.	Ethyl Formate	2738
			12.1.3.	Isobutyl Formate	2738
4.2.	Other Processes	2727	12.1.4.	Miscellaneous Esters	2738
4.2.1.	Oxidation of Hydrocarbons	2727	12.2.	Salts	2738
4.2.2.	Hydrolysis of Formamide	2727	12.2.1.	Alkali-Metal and Alkaline-Earth Metal Formates	2738
4.2.3.	Production of Formic Acid from Formates	2728	12.2.2.	Ammonium Formate	2739
4.2.4.	Direct Synthesis from Carbon Monoxide and Water	2729	12.2.3.	Aluminum Formate	2739
4.2.5.	Use of Carbon Dioxide	2729	12.2.4.	Nickel Formate	2739
4.3.	Concentration of Formic Acid	2730	12.2.5.	Copper Formate	2739
4.4.	Construction Materials	2731	13.	Toxicology and Occupational Health	2740
5.	Environmental Protection	2732			
6.	Quality Specifications	2732	14.	References	2741

1. Introduction

Formic acid [64-18-6] HCOOH, M_r 46.03, is a colorless liquid with a pungent odor, which is completely miscible with water and many polar solvents but only partially miscible with hydrocarbons. Formic acid derived its name from the red ant, *Formica rufa*, in which it was discovered around 1670. Formic acid has been detected in the poison or defense systems of ants, bees, and other insects and also of cnidarians.

Formic acid is used primarily in dyeing, in the textile and leather industries; in rubber production; and as an intermediate in the chemical and pharmaceutical in-

Table 1. Density of pure formic acid as a function of temperature

t, °C	ϱ, g/cm^3	t, °C	ϱ, g/cm^3
0	1.244 *	40	1.195
10	1.232	50	1.182
15	1.226	60	1.169
20	1.220	70	1.156
25	1.214	80	1.143
30	1.207	90	1.130
		100	1.117

* Supercooled liquid.

Table 2. Density of aqueous formic acid solutions at 20 °C

Formic acid content, wt %	ϱ, g/cm^3	Formic acid content, wt %	ϱ, g/cm^3
2	1.003	50	1.118
5	1.011	60	1.141
10	1.024	70	1.163
20	1.048	80	1.185
30	1.072	90	1.204
40	1.095	100	1.221

dustries. The use of formic acid as an aid in the ensilage of green forage has increased sharply.

The worldwide production of formic acid is about 260 000 t/a (1987). Formic acid is produced by hydrolysis of methyl formate or formamide or from its salts. In addition, formic acid is a byproduct of acetic acid production by liquid-phase oxidation of hydrocarbons.

2. Physical Properties

Formic acid, mp 8.3 °C [1], bp 100.8 °C (at 101.3 kPa) [2], is a colorless, clear, very corrosive liquid with a pungent odor. With a pK_a of 3.739, formic acid is the strongest unsubstituted fatty acid, about ten times stronger than acetic acid [3]. Pure formic acid is hygroscopic.

The temperature dependence of the *density* of formic acid is given in Table 1. The density of formic acid–water binary mixtures, as a function of formic acid concentration, is shown in Table 2.

The *freezing point* diagram for the formic acid–water binary mixture exhibits a eutectic point (Table 3) at −48.5 °C and 41.0 mol % of formic acid [1]. Formic acid does not increase in volume when it solidifies and has a tendency to undergo supercooling.

Table 4 shows the *vapor pressure* curve of pure formic acid. The vapor of formic acid deviates considerably in behavior from an ideal gas because the molecules dimerize partially in the vapor phase. At room temperature and normal pressure, 95 % of the formic acid vapor consists of dimerized formic acid [7]:

Table 3. Freezing points of various formic acid–water mixtures

Formic acid molar fraction, %	fp, °C	Formic acid molar fraction, %	fp, °C
100.0	8.3	41.0	−48.5
80.0	−5.6	39.8	−46.8
74.8	−8.2	37.0	−43.0
69.5	−13.0	31.7	−33.5
64.5	−17.3	28.0	−29.3
59.0	−23.2	23.1	−22.8
54.2	−28.3	20.0	−19.7
50.0	−35.3	12.0	−10.3
44.6	−42.0	0.0	0.0
42.0	−46.7		

Table 4. Vapor pressure of pure formic acid

Liquid		Solid	
t, °C	p, kPa	t, °C	p, kPa
−5.23	1.083	−5.07	0.664
0.00	1.488	0.00	1.096
8.25	2.392	8.25	2.392
12.57	3.029		
20.00	4.473		
29.96	7.248		
39.89	11.357		
49.93	17.347		
59.98	25.693		
70.04	37.413		
79.93	52.747		
100.68	101.667		
110.62	135.680		

$$\text{H-C} \begin{matrix} \text{O} \cdots \text{H-O} \\ \\ \text{O-H} \cdots \text{O} \end{matrix} \text{C-H}$$

The enthalpy of the gas-phase dimerization is −63.8 kJ/mol [7]. The thermodynamic properties of monomeric and dimeric formic acid have been investigated by WARING [8].

The ring-type dimeric structure exists both in the vapor phase and in solution. Liquid formic acid consists of long chains of molecules linked to each other by hydrogen bonds. Solid formic acid can also be isolated in two polymorphic forms (α and β) [9]:

Table 5. Azeotropic data for the formic acid – water system

p, bar	bp of azeotropic mixture, °C	Formic acid content, wt %	Reference
0.093	48.6	66.2	[11]
0.267	72.3	70.5	[11]
1.013	107.6	77.6	[11]
2.026	128.7	84	[12]
3.140	144	85	[12]

Table 6. Dynamic viscosity of pure formic acid as a function of temperature

t, °C	η, mPa·s
10	2.262
20	1.804
30	1.465
40	1.224
50	1.025

Portion of the chain in an α-type formic acid crystal

Portion of the chain in a β-type formic acid crystal

The thermodynamic properties of monomeric and dimeric formic acid have been investigated [9].

Vapor – liquid equilibrium data for formic acid – water mixtures and for mixtures of formic acid with organic compounds have been collected in [10]. Formic acid and water form a maximum-boiling azeotropic mixture whose boiling point is 107.6 °C at 101.3 kPa; it consists of 77.6 wt% formic acid and 22.4 wt% water. The composition and boiling point of formic acid – water azeotropic mixtures are shown as a function of pressure in Table 5. Formic acid can form azeotropic mixtures with many other substances [13].

The variation of dynamic *viscosity* with temperature is shown in Table 6. The dynamic viscosity of formic acid – water mixtures decreases approximately linearly as the water content of formic acid increases [14]. The *thermal conductivity* of formic acid is markedly higher than that of comparable liquids, because of its pronounced polarity.

Table 7. Specific heat capacity of formic acid as a function of temperature at constant pressure

t, °C	c_p, J g^{-1} K^{-1}	t, °C	c_p, J g^{-1} K^{-1}
Solid		*Vapor*	
−150	0.921	25	1.058
−100	1.114	100	1.192
−75	1.193	200	1.348
−50	1.285	300	1.480
−25	1.411	400	1.589
0	1.800	600	1.757
		800	1.870
		1000	1.953
		1200	2.037
Liquid			
20	2.169		
50	2.202		
100	2.282		

The variation of specific *heat capacity* with temperature is shown in Table 7. The specific heat capacity of formic acid–water mixtures decreases approximately linearly as the concentration of formic acid increases [5].

The physical properties of formic acid are listed below:

Heat of fusion			276 J/g [6]
Heat of vaporization (at *bp*)			483 J/g [6]
Dielectric constant			
	Liquid (at 20 °C)		57.9 [20]
	Solid (at −10.1 °C)		11.7
Refractive index n_D^{20}			1.37140 [18]
Surface tension σ			
	(at 20 °C)		37.67 × 10^{-3} N/m [19]
	(at 40 °C)		35.48 × 10^{-3} N/m
	(at 60 °C)		33.28 × 10^{-3} N/m
	(at 80 °C)		31.09 × 10^{-3} N/m
Heat of formation ΔH_f^0			
	Liquid	(at 25 °C)	−425.0 kJ/mol [21]
	Vapor	(at 25 °C, monomer)	−378.57 kJ/mol [7]
		(at 25 °C, dimer)	−820.94 kJ/mol
Heat of combustion ΔH_c^0			
	Liquid	(at 25 °C)	−254.8 kJ/mol [21]
Entropy S^0			
	Liquid	(at 25 °C)	129.0 J K^{-1} mol^{-1} [9]
	Vapor	(at 25 °C, monomer)	248.88 J K^{-1} mol^{-1} [7]
		(at 25 °C, dimer)	332.67 J K^{-1} mol^{-1} [7]
Heat of neutralization			56.9 kJ/mol [22]
Critical data p_{crit}			7.279 MPa [17]
T_{crit}			581 K [17]
ϱ_{crit}			0.392 g/cm^3 [17]

Thermal conductivity λ			
Liquid	(at 20 °C)		0.226 W m^{-1} K^{-1} [17], [23]
	(at 60 °C)		0.205 W m^{-1} K^{-1}
	(at 100 °C)		0.185 W m^{-1} K^{-1}
Vapor	(at 50 °C)		0.0136 W m^{-1} K^{-1}
	(at 100 °C)		0.0176 W m^{-1} K^{-1}
	(at 200 °C)		0.0267 W m^{-1} K^{-1}
Electrical conductivity (at 25 °C)			6.08 × 10^{-5} Ω^{-1} cm^{-1} [24]
Coefficient of cubic expansion α	(at 30 °C)		0.001 [2]

In 1968, GALLANT reported the important physical properties of formic acid and other low molecular weight fatty acids as a function of temperature [17].

3. Chemical Properties

To some extent, formic acid reacts differently from its higher homologues because it is both a carboxylic acid and an aldehyde.

Aldehyde Reactions. In some reactions, formic acid behaves like an aldehyde. Thus if an aqueous solution of formic acid is heated with a silver nitrate solution, metallic silver is deposited. Mercury, gold or platinum ions are reduced in solution, as are organic compounds. For example, formic acid reduces triphenylmethanol to triphenylmethane. The reduction of imines (Schiff bases) by formic acid has been known for a long time. Primary amines can be prepared from ketones, ammonia, and formic acid (Leuckart reaction). Like other aldehydes, formic acid has bactericidal properties.

Carboxylic Acid Reactions. Magnesium, zinc, and iron dissolve in formic acid with the evolution of hydrogen. Because of its strongly acidic nature, many alcohols can be esterified by formic acid, without the addition of mineral acids. Primary and secondary alcohols are esterified in pure formic acid 15 000 – 20 000 times more rapidly than in pure acetic acid [25]. The rate of esterification of primary, secondary, and tertiary alcohols in formic acid has been determined [26], [27]. Because of its extremely reactive carboxyl group, amines can be formylated by formic acid, with high yields, for example, the yield of N-methylformanilide [93-61-8] is 93 – 97 %. Formic acid adds to the double bonds of olefins to form esters. Acetylenes react with formic acid in the vapor phase to yield vinyl formates [28].

Formic acid may also be used as a source of carbon monoxide. Thus, carboxylic acids can be prepared from olefins and formic acid in the presence of sulfuric or hydrofluoric acid (Koch carboxylic acid synthesis). Formic acid and tertiary organic bases form addition compounds (3:1 and 2:1 ratio of formic acid:base). If, for instance, 1 mole of gaseous triethylamine is passed in 3 moles of anhydrous formic acid, the addition product is obtained in quantitative yield. In these compounds formic acid is present in an extremely active form. Addition compounds of formic acid and trimethylamine or

Table 8. Volume of carbon monoxide produced per liter of formic acid (99 wt%) as a function of storage time and temperature

Time, d	V, L		
	at 20 °C	at 30 °C	at 40 °C
15	0.012	0.064	0.303
30	0.024	0.128	0.597
60	0.049	0.255	1.161
90	0.073	0.381	1.700

Table 9. Volume of carbon monoxide produced per liter of aqueous formic acid as a function of storage time and concentration

Time, d	V, L ($t = 40$ °C)		
	90 wt%	98 wt%	99 wt%
15	0.010	0.110	0.303
30	0.019	0.220	0.597
60	0.039	0.437	1.161
90	0.058	0.653	1.700

triethylamine can be used as liquid reducing agents in many selective reductions. The reduction of sulfur dioxide to sulfur is approximately quantitative [29].

Formic acid reacts with hydrogen peroxide in the presence of an acidic catalyst to form unstable performic acid [*107-32-4*] (HCOOOH); may decompose explosively when heated to 80–85 °C. Performic acid exhibits typical peroxide properties [30]. The reactions of formic acid were summarized by GIBSON in 1969 [31].

Although formic acid is not very stable, it can be distilled at atmospheric pressure in the absence of catalysts. Formic acid may decompose into either carbon monoxide and water (dehydration) or carbon dioxide and hydrogen (dehydrogenation); the nature of the decomposition and the reaction rate depend strongly on the presence of catalysts, the temperature, and the concentration of formic acid. *Dehydrogenation* is catalyzed preferentially by metals (platinum sponge, copper, nickel, silver). *Dehydration*, on the other hand, is catalyzed by aluminum oxide, silicon dioxide, and charcoal. Dehydration is promoted by mineral acids but inhibited by water; it also occurs on activated metal surfaces which must be considered during the preparation and processing of formic acid, as well as during transportation and storage of the highly concentrated acid. Tables 8 and 9 show the formation of carbon monoxide as a function of formic acid concentration and temperature [32]. A summary of the catalytic decomposition of formic acid has been published by MARS [33], along with an extensive list of references. The temperature dependence of formic acid decomposition at low temperature (between 40–100 °C) has been investigated [34].

4. Production

The worldwide installed capacity for producing formic acid was about 330 000 t/a in 1988. The installed formic acid processes can be classified in four groups:

1) methyl formate hydrolysis,
2) oxidation of hydrocarbons,
3) hydrolysis of formamide, and
4) preparation of free formic acid from formates.

The significance of individual processes has shifted considerably in the last decade because the economic data for production and raw products have changed and new processes have been developed. In the past, a large amount of the formic acid utilized was a byproduct in the manufacture of acetic acid. However, the proportion of formic acid produced by dedicated processes has increased in recent years.

Formic acid is a byproduct in the production of acetic acid [64-19-7] by liquid-phase oxidation of butane or naphtha (→ Acetic Acid). For many years, oxidation of hydrocarbons was the most important method of producing acetic acid. Over the past decade, however, the preferred process in the establishment of new acetic acid plants has been the carbonylation of methanol [35], which is expected to remain the preferred method in the future. As a result, the proportion of formic acid produced worldwide, which results as a byproduct in the manufacture of acetic acid, will continue to decrease.

The production of formic acid by hydrolysis of formamide [75-12-7] played an important role in Europe; in 1972, about one-third of the world production was obtained by this process [2]. However, the consumption of ammonia and sulfuric acid, along with the unavoidable production of ammonium sulfate, have in many cases made this process economically inferior. As a result, the direct hydrolysis of methyl formate is currently preferred.

Another industrial production method involves formation of the free acid from its salts. Sodium formate [141-53-7] and calcium formate [544-17-2] are used for this purpose.

The economic disadvantages of the methods discussed led to the development of a process specifically dedicated to the production of formic acid, with no undesirable byproducts. In the 1970s, the hydrolysis of methyl formate [107-31-3] to methanol and formic acid was developed commercially by various firms into an economically feasible method. This process involves carbonylation of methanol and subsequent hydrolysis of the methyl formate produced. The methanol resulting from this process is returned to the first stage. Formic acid plants based on this process were started up at BASF (Federal Republic of Germany) in 1981 and Kemira (Finland) in 1982 [36]. Most of the formic acid plants begun in recent years employ this principle of methyl formate hydrolysis.

Table 10 shows the world capacity for various formic acid production processes in 1986. Although other methods for producing formic acid have been patented, they do

Table 10. Formic acid production capacity in 1988 (without standby plants; BASF assessment)

Process	Capacity, t/a	Percentage of total capacity, %
Hydrolysis of methyl formate	160 000	49
Byproduct in hydrocarbon oxidation	74 000	22
Hydrolysis of formamide	10 000	3
Acidolysis of alkali formates	86 000	26
Total	330 000	

not appear to have been implemented industrially. Direct synthesis of formic acid from carbon monoxide and water cannot be carried out economically because of the unfavorable position of the equilibrium.

In recent years, various ways of producing formic acid from carbon dioxide and hydrogen have been investigated extensively, but this method has not yet been applied industrially. The oxidation of methyl *tert*-butyl ether [1634-04-4] with oxygen to form *tert*-butyl formate [762-75-4] and formic acid has been described [37]. Various methods have been developed for the electrochemical production of formic acid, but they have not been used commercially [2], [38]. The combined production of formic acid and acetic acid from methyl formate, carbon monoxide, and water with methyl iodide [74-88-4] as promoter and anhydrous rhodium chloride [13569-65-8] as catalyst, has been described by BP [39].

4.1. Methyl Formate Hydrolysis

Synthesis of formic acid by hydrolysis of methyl formate is based on a two-stage process: in the first stage, methanol is carbonylated with carbon monoxide; in the second stage, methyl formate is hydrolyzed to formic acid and methanol. The methanol is returned to the first stage:

$$CH_3OH + CO \longrightarrow HCOOCH_3$$
$$CH_3OOCH + H_2O \longrightarrow CH_3OH + HCOOH$$
$$CO + H_2O \longrightarrow HCOOH$$

Although the carbonylation of methanol is relatively problem-free and has been carried out industrially for a long time [2], [40], only recently has the hydrolysis of methyl formate been developed into an economically feasible process. The main problems are associated with work-up of the hydrolysis mixture. Because of the unfavorable position of the equilibrium, reesterification of methanol and formic acid to methyl formate occurs rapidly during the separation of unreacted methyl formate. Problems also arise in the selection of sufficiently corrosion-resistant materials.

Older publications suggest the reaction of methyl formate and a dicarboxylic acid, with subsequent distillation of formic acid from the higher-boiling esters [2]. Degussa suggested hydrolyzing methyl

Table 11. Maximum methyl formate concentration (in weight percent) achievable by carbonylation of methanol (unpublished BASF data)

t, °C	p, MPa			
	1	5	9	17
50	59	95	98	99
60	43	91	96	99
70	31	85	94	98
80	21	78	90	96
90	15	69	85	94
100	10	59	79	91
110	7	49	71	87
120	5	39	62	82

formate at 200 °C above atmospheric pressure and then reducing the pressure suddenly to less than 0.1 MPa. Separation of the methyl formate–methanol mixture flash-evaporated from the aqueous formic acid can then take place in a liquid separator [4].

Industrial methods involving carbonylation of methanol and hydrolysis of methyl formate, followed by isolation of the formic acid, were developed by (1) the Leonard Process Company (United States), (2) BASF (Federal Republic of Germany), (3) Halcon – Scientific Design – Bethlehem Steel (United States), and (4) the Scientific Research Institute for the Chlorine Industry of the Ministry of the Chemical Industry (Soviet Union).

Key patents relating to the carbonylation of methanol and the hydrolysis of methyl formate are listed in [42].

Basic Principles. *Carbonylation of Methanol.* In the four processes mentioned, the first stage involves carbonylation of methanol in the liquid phase with carbon monoxide, in the presence of a basic catalyst:

$$CH_3OH + CO \longrightarrow HCOOCH_3 \quad \Delta H_R^0 = -29 \text{ kJ/mol}$$

This reaction was first described by BASF in 1925 [40]. As a rule, the catalyst is sodium methoxide [124-41-4]. Potassium methoxide [865-33-8] has also been proposed as a catalyst; it is more soluble in methyl formate and gives a higher reaction rate [43]. Although fairly high pressures were initially preferred, carbonylation is carried out in new plants at lower pressure. Under these conditions, reaction temperature and catalyst concentration must be increased to achieve acceptable conversion. According to published data, ca. 4.5 MPa, 80 °C, and 2.5 wt% sodium methoxide are employed. About 95% carbon monoxide, but only about 30% methanol, is converted under these circumstances. Nearly quantitative conversion of methanol to methyl formate can, nevertheless, be achieved by recycling the unreacted methanol. The carbonylation of methanol is an equilibrium reaction, and Table 11 specifies the equilibrium concentrations of methyl formate.

The reaction rate can be raised by increasing the temperature, the carbon monoxide partial pressure, the catalyst concentration, and the interface between gas and liquid:

$$r = C \cdot a \cdot e^{-E/RT} \cdot c \cdot p_{CO}$$

where

- r = reaction rate,
- C = proportionality coefficient,
- a = interface between gas and liquid,
- E = activation energy,
- R = molar gas constant,
- c = catalyst concentration in solution, and
- p_{CO} = CO partial pressure in the gas phase [44].

This equation reveals the dependence of reaction rate on the partial pressure of carbon monoxide. Therefore, to synthesize methyl formate, gas mixtures with a low proportion of carbon monoxide must first be concentrated; this can be achieved by low temperature distillation, pressure-swing adsorption, prism separators (Monsanto), or the Cosorb process (Tenneco).

In a side reaction, sodium methoxide reacts with methyl formate to form sodium formate and dimethyl ether, and becomes inactivated:

$$NaOCH_3 + HCOOCH_3 \longrightarrow HCOONa + CH_3OCH_3$$

The substances used must be anhydrous; otherwise, sodium formate is precipitated to an increasing extent:

$$NaOCH_3 + H_2O + HCOOCH_3 \longrightarrow 2\,CH_3OH + NaOOCH$$

Sodium formate is considerably less soluble in methyl formate than in methanol. The risk of encrustation and blockage due to precipitation of sodium formate can be reduced by adding poly(ethylene glycol) [45]. The carbon monoxide used must contain only a small amount of carbon dioxide; otherwise, the catalytically inactive carbonate is precipitated.

Basic catalysts may reverse the reaction, and methyl formate decomposes into methanol and carbon monoxide. Therefore, undecomposed sodium methoxide in the methyl formate must be neutralized [46].

Hydrolysis of Methyl Formate. In the second stage, the methyl formate obtained is hydrolyzed:

$$HCOOCH_3 + H_2O \rightleftarrows CH_3OH + HCOOH \qquad \Delta H_R^0 = 16.3 \text{ kJ/mol}$$

The equilibrium constant for methyl formate hydrolysis depends on the water : ester ratio [47]. With a molar ratio of 1, the constant is 0.14, but with a water : methyl formate molar ratio of 15, it is 0.24. Because of the unfavorable position of this equilibrium, a large excess of either water or methyl formate must be used to obtain an economically worthwhile methyl formate conversion. If methyl formate and water are used in a molar ratio of 1:1, the conversion is only 30%, but if the molar ratio of water to methyl formate is increased to 5–6, the conversion of methyl formate rises to 60%. However, a dilute aqueous solution of formic acid is obtained this way, and excess water must be removed from the formic acid with the expenditure of as little energy as possible.

Another way to overcome the unfavorable position of the equilibrium is to hydrolyze methyl formate in the presence of a tertiary amine, e.g., 1-(n-pentyl)imidazole [19768-54-8], [48]. The base forms a salt-like compound with formic acid; therefore, the concentration of free formic acid decreases and the hydrolysis equilibrium is shifted in the direction of products. In a subsequent step formic acid can be distilled from the base without decomposition.

A two-stage hydrolysis has been suggested, in which a water-soluble formamide is used in the second stage [49]; this forms a salt-like compound with formic acid. It also shifts the equilibrium in the direction of formic acid.

To keep undesirable reesterification as low as possible, the time of direct contact between methanol and formic acid must be as short as possible, and separation must be carried out at the lowest possible temperature. Introduction of methyl formate into the lower part of the column in which lower boiling methyl formate and methanol are separated from water and formic acid, has also been suggested. This largely prevents reesterification because of the excess methyl formate present in the critical region of the column [50].

The hydrolysis of methyl formate is catalyzed by strong acids, but the efficiency of strong mineral acids [51] is restricted because they also promote the decomposition of formic acid. In the processes described by BASF, Leonhard, and Halcon – Scientific Design – Bethlehem Steel, autocatalysis by formic acid is therefore used. In the Soviet Union, a process has been described in which hydrolysis is carried out in a vertical column reactor divided into two zones [52]; the upper zone is filled with a strongly acidic cation exchanger which partially hydrolyze methyl formate. In the lower part of the reactor, hydrolysis is carried out with autocatalysis by formic acid produced in the upper section.

Dehydration of the Hydrolysis Mixture. Formic acid is marketed in concentrations exceeding 85 wt%; therefore, dehydration of the hydrolysis mixture is an important step in the production of formic acid from methyl formate. For dehydration, the azeotropic point must be overcome. The concentration of formic acid in the azeotropic mixture increases if distillation is carried out under pressure (Table 5), but the higher boiling point at high pressure also increases the decomposition rate of formic acid. At the same time, the selection of sufficiently corrosion-resistant materials presents considerable problems. A number of entrainers have been proposed for azeotropic distillation [42].

BASF describes various energy-saving processes for the dehydration. One process involves extractive distillation with *N*-formylmorpholine [*4394-85-8*] [53]. Another method of increasing formic acid concentration in dilute solutions without a considerable input of energy is to extract formic acid from water by liquid–liquid extraction. Secondary amides have been proposed [54] as extractive agents. Formic acid binds weakly to the amide; therefore, water can be distilled to overcome the azeotropic point. In a subsequent column, formic acid can then be distilled from the extractant in vacuo. The energy consumed can be reduced further by optimizing the exchange of heat [55].

4.1.1. Kemira–Leonard Process

A 20 000-t/a formic acid plant based on a method developed by the Leonard Process Co. [56] was built at Kemira in Finland and put into operation in 1982. The process has been developed further by Kemira, and licences for it have been issued in Korea, India and Indonesia.

In the Kemira–Leonard process, carbonylation is carried out at about 4 MPa and a temperature of approximately 80 °C, with additive-containing alkoxides used as catalyst [57]. Hydrolysis is carried out in two reactors with different operating conditions. Methyl formate and water react in the preliminary reactor in approximately equimolar proportions. The formic acid produced catalyzes the hydrolysis in the main reactor. An excess of methyl formate is employed in the main reactor, and hydrolysis is carried out at ca. 120 °C and 0.9 MPa. The reactor discharge is brought to atmospheric pressure in a flash tank; reesterification is largely prevented by the cooling that occurs during flash-evaporation. Methyl formate and methanol are separated under vacuum. Although the reverse reaction is attenuated further as a result, condensation of the low-boiling methyl formate in vacuo presents a problem. The formic acid is dehydrated by distillation. If 85 wt% formic acid is desired, dehydration must be carried out at ca. 0.3 MPa. Even higher concentrations can be achieved by connecting an additional dehydration column downstream under atmospheric pressure, and formic acid concentrations up to ca. 98 wt% can then be drawn off as the distillate.

Process Description [56]–[59] (Fig. 1). Compressed carbon monoxide and methanol are converted into methyl formate in reactor (a). Catalyst is fed into the reactor in a methanol solution. The amount of methanol introduced in this way makes up for methanol losses in the process. The discharge from reactor (a) is flashed and fed into the methyl formate column (b) from which methyl formate is drawn off as the distillate. Methanol and the dissolved catalyst are returned to the reactor; inactivated catalyst (primarily sodium formate) is crystallized and discharged.

Off-gas from column (b) and waste gas from reactor (a) are burned. Methyl formate reacts partially with water in the preliminary reactor (c), and discharge from the preliminary reactor is fed into the main reactor (d) along with recycled methyl formate, methanol, and water. The contact time in this reactor is largely sufficient for equilibrium to be established. Reactor discharge is flashed to approxi-

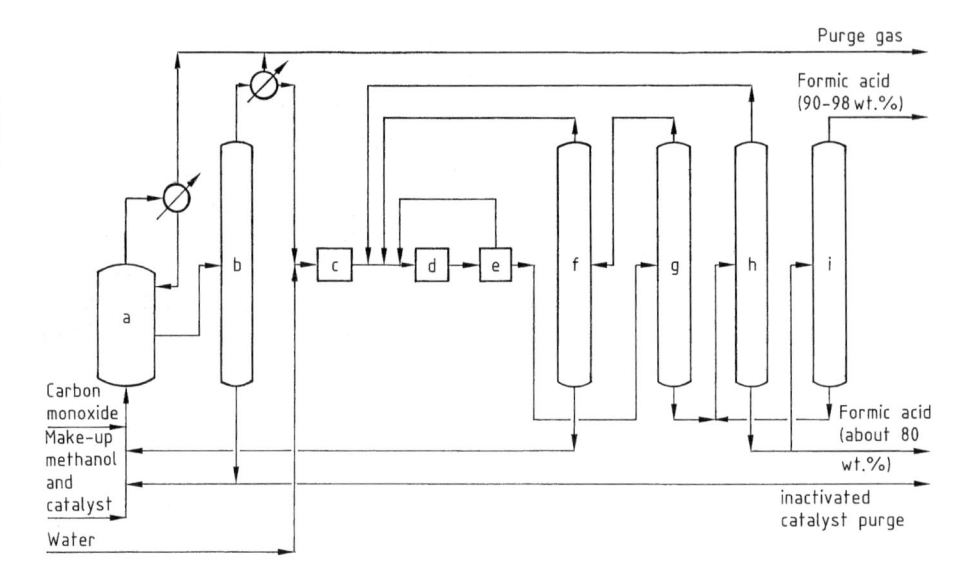

Figure 1. Production of formic acid (Kemira – Leonard process)
a) Methyl formate reactor; b) Methyl formate column; c) Preliminary reactor; d) Main reactor; e) Flash tank; f) Recycle column; g) Acid separation column; h) First product column; i) Second product column

mately atmospheric pressure in the flash tank (e); methyl formate, methanol, and small quantities of formic acid evaporated in this process are recycled to the main reactor (d). Methyl formate and methanol are distilled in vacuo in the acid separation column (g). Contact time is minimized by internals with a small liquid holdup. The distillate is separated into methyl formate and methanol in the recycle column (f). Formic acid can be concentrated under pressure in a column; if the pressure is ca. 0.3 MPa, 85% formic acid is drawn off as the bottom product. Distillation in two columns is more industrially controllable. Water is distilled overhead in the first product column (h). The bottom product is concentrated further in the second product column (i), and formic acid with a maximum concentration of ca. 98 wt% is drawn off as distillate. The bottom product is recycled to the first product column.

4.1.2. BASF Process

A 100 000-t/a formic acid plant began operating in Ludwigshafen (Federal Republic of Germany) in 1981. In this plant, a newly developed technology for the hydrolysis and dehydration was used for the first time.

The production of methyl formate by carbonylation of methanol has been carried out on a large scale for many years at BASF [40], [60]. The carbonylation stage is largely identical to that of the Kemira – Leonard process, but the hydrolysis stage and the dehydration of formic acid are noticeably different.

In the BASF process, hydrolysis is carried out with a large excess of water (about 5 mol of water per mole of methyl formate) to shift the equilibrium in the direction of

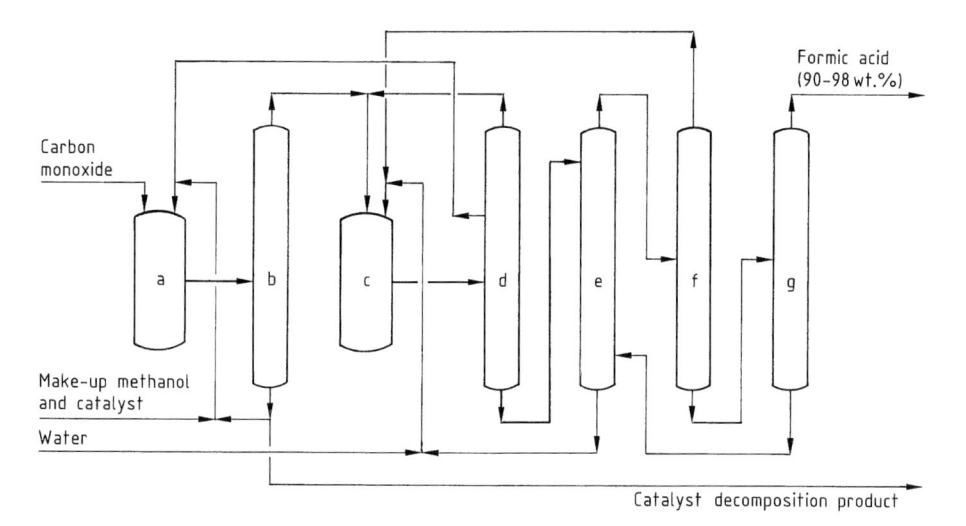

Figure 2. Production of formic acid (BASF process)
a) Methyl formate reactor; b) Methyl formate column; c) Hydrolysis reactor; d) Low-boiler column, e) Extraction unit; f) Dehydration column; g) Pure acid column

formic acid. Much of the water is separated by liquid–liquid extraction with a secondary amide.

Process Description (Fig. 2) Carbon monoxide and methanol react in the methyl formate reactor (a) in the presence of sodium methoxide. Methyl formate is fed, as a distillate, from the methyl formate column (b) into the formic acid reactor (c) together with recycled methyl formate. Methanol and dissolved catalyst are drawn off from the bottom of column (b) and returned to reactor (a); catalyst decomposition products are discharged by crystallization. In reactor (c), methyl formate is hydrolyzed with excess water at elevated temperature and increased pressure. The reaction product is flashed into the low-boiler column (d). Methyl formate is removed as the distillate, with methanol as a side stream, and dilute aqueous formic acid is drawn off from the bottom into the extraction unit (e). Here, the formic acid and some of the water are extracted by the secondary amide. Most of the water (largely free of formic acid) is recycled to reactor (c). The extract—a mixture of extractant, formic acid, and some water—is distilled in the dehydration column (f). Enough water is distilled via the head for the required formic acid concentration to be obtained in the pure acid column (g). This column is operated in vacuo. The extraction agent is recycled from the bottom of column (g) to the extraction unit.

4.1.3. USSR Process

A 40 000-t/a formic acid plant based on a process [52] developed in the Soviet Union is being built in Saratov (Ukraine) [61] and is expected to be operational in 1989.

According to published material, this formic acid process differs from the above-mentioned processes [62], [63] in the hydrolysis stage. Hydrolysis is carried out at 80 °C and 0.6 MPa in a fixed-bed reactor equipped with bubble-cap trays in the upper section.

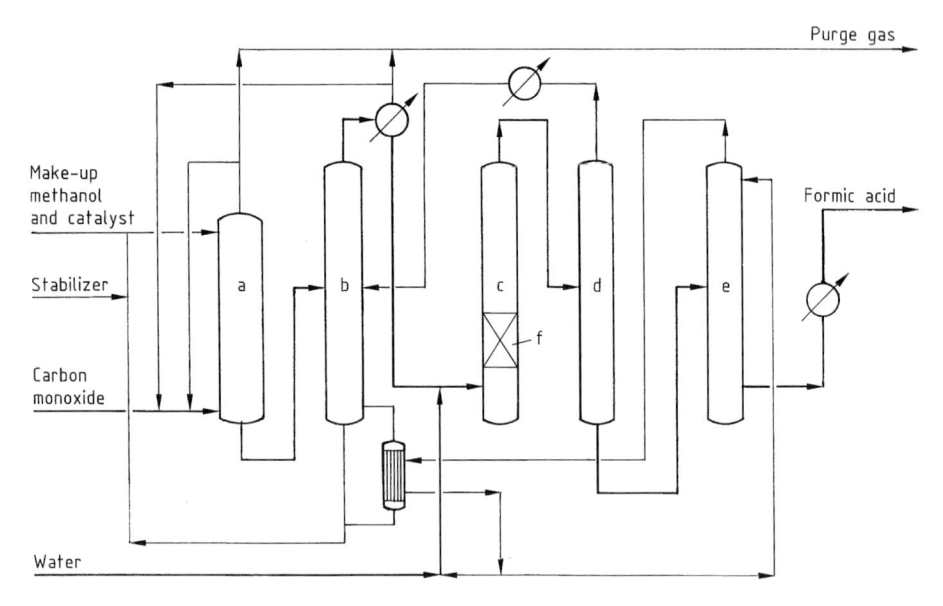

Figure 3. Production of formic acid (USSR process)
a) Methyl formate reactor; b) Methyl formate column; c) Hydrolysis reactor; d) Low-boiler column; e) Formic acid column; f) Fixed-bed catalyst

Reesterification is prevented by decompressing and cooling to about 45 °C. Formic acid is produced in a pressure column as an azeotropic mixture with water which contains 85 wt % formic acid.

Process Description [62], [63] (Fig. 3). Carbon monoxide reacts with methanol in column reactor (a), in the presence of the catalyst and a stabilizer, at about 3 MPa, to yield methyl formate. Separation into methyl formate (distillate) and methanol plus catalyst takes place in the methyl formate column (b) at 0.2 MPa. The exhaust gas produced is partially recycled. In reactor (c), methyl formate is hydrolyzed with water in two stages; in the first stage, packed acidic cation exchanger is used as catalyst (f), which partially hydrolyzes the methyl formate. In the second stage, hydrolysis is carried out with autocatalysis by the formic acid produced in stage one. Methanol and methyl formate are distilled from the hydrolysis product in the low-boiler column (d); these compounds are separated, along with the carbonylation reaction mixture, in the methyl formate column (b). Aqueous formic acid is drawn off the bottom of column (d) and further dehydrated in the formic acid column (e) under pressure. Formic acid with a content of 85 wt % is drawn off in the side stream, and the concentration can be increased to about 98 wt % in the column situated downstream.

4.1.4. Scientific Design – Bethlehem Steel Process

The process developed jointly by Scientific Design and Bethlehem Steel [64] has not yet been implemented industrially on a large scale.

The carbonylation stage is essentially similar to the processes described previously. If the concentration of carbon monoxide is higher than 90 mol%, a single back-mixed reactor is provided, but if the carbon monoxide concentration is between 50 and 90 mol%, a two-stage, countercurrent, back-mixed reactor system is recommended. If the carbon monoxide concentration is low, however, the total pressure must be increased considerably [65]. The hydrolysis is catalyzed by formic acid, and the dehydration is carried out by distillation alone.

4.2. Other Processes

4.2.1. Oxidation of Hydrocarbons

Formic acid is produced as a byproduct in the liquid-phase oxidation of hydrocarbons to acetic acid (→ Acetic Acid). In the United States, butane [106-97-8] is used as the hydrocarbon, and ca. 50 kg of formic acid is produced per ton of acetic acid. In Europe, the oxidation of naphtha is preferred, and up to 250 kg of formic acid is produced per ton of acetic acid in this process. For the mechanism of formic acid production, see [58].

Unreacted hydrocarbons, volatile neutral constituents, and water are separated first from the oxidation product. Formic acid is separated in the next column; azeotropic distillation is generally used for this purpose. The entrainers preferred in this process are benzene or chlorinated hydrocarbons. The formic acid contains about 2 wt% acetic acid, 5 wt% water, and 3 wt% benzene [58]. Formic acid with a content of about 98 wt% can be produced by further distillation.

4.2.2. Hydrolysis of Formamide

In 1972, 35% of the formic acid produced worldwide was still made by the formamide process developed by BASF. However, because of the direct hydrolysis of methyl formate, this process has lost much of its significance (see Table 10).

Formic acid is produced this way in a three-stage process. In the first stage, methanol is carbonylated to yield methyl formate:

$$CO + CH_3OH \rightarrow HCOOCH_3 \quad \Delta H_R^0 = -29 \text{ kJ/mol}$$

In the second stage, formamide is produced by ammonolysis of methyl formate:

$$HCOOCH_3 + NH_3 \rightarrow HCONH_2 + CH_3OH \quad \Delta H_R^0 = -72 \text{ kJ/mol}$$

In the third stage, sulfuric acid is used to hydrolyze formamide to formic acid and ammonium sulfate:

$$HCONH_2 + 1/2\ H_2SO_4 + H_2O \longrightarrow HCOOH + 1/2\ (NH_4)_2SO_4 \quad \Delta H_R^0 = -48\ kJ/mol$$

Methanol is carbonylated as in the first stage of the processes already described (Section 4.1). Conversion to formamide is usually carried out at 0.4 – 0.6 MPa and 80 – 100 °C; methanol is distilled and recycled to the methyl formate stage (see → Formamides). Formamide is hydrolyzed continuously in the third stage by using 68 – 74 % sulfuric acid at temperatures between 85 °C and the boiling point of formic acid [2]. This reaction is carried out preferably in stirred containers and provides the heat for the distillation of formic acid. A hot slurry consisting essentially of ammonium sulfate and formic acid flows out of the stirred container into a rotary kiln where the residual formic acid is distilled so that dry, pure ammonium sulfate is produced at the kiln end. The yield of formic acid exceeds 90 %. The economic efficiency of this process is determined by the commercial value of ammonium sulfate.

Hydrochloric acid, nitric acid, or phosphoric acid can also be used to hydrolyze formamide [2].

4.2.3. Production of Formic Acid from Formates

The reaction of sodium formate [141-53-7] or calcium formate [544-17-2] with strong mineral acids, such as sulfuric and nitric acids, is the oldest known process for producing formic acid commercially. If formates or sodium hydroxide are available cheaply or occur as byproducts in other processes, formic acid can still be produced economically in this manner.

Formates as Byproducts in the Production of Polyhydric Alcohols. Sodium formate and calcium formate occur as byproducts in the production of pentaerythritol 115-77-5], trimethylolpropane [77-99-6], and 2,2-dimethyl-1,3-propanediol [126-30-7] (neopentyl glycol) (→ Alcohols, Polyhydric). The formates unavoidably produced in this manner are used in several plants as feed for manufacturing formic acid.

Formates from Sodium Hydroxide and Carbon Monoxide. The production of sodium formate from sodium hydroxide and carbon monoxide once had considerable significance, but today this process is carried out simply to utilize excess sodium hydroxide solution in low-capacity plants. The economic efficiency of producing formic acid from its salts suffers because one equivalent of a low-value inorganic salt (e.g., sodium sulfate), is produced per mole of formic acid.

Carbon monoxide reacts with alkaline compounds, even in aqueous solution, to yield the corresponding formates:

$$\text{CO} + \text{NaOH} \xrightarrow{\text{H}_2\text{O}} \text{HCOONa}$$

Carbon monoxide is mixed countercurrently with aqueous sodium hydroxide, for example, in a tower reactor at 1.5 – 1.8 MPa and 180 °C [58]. Sodium formate crystallizes and reacts with strong mineral acids, (e.g., concentrated sulfuric acid) at normal pressure in a cooled, stirred reactor at 35 °C:

$$2\,\text{HCOONa} + \text{H}_2\text{SO}_4 \rightarrow 2\,\text{HCOOH} + \text{Na}_2\text{SO}_4$$

The reaction mixture is separated, for example, in a thin-film evaporator at normal pressure and 100 – 120 °C, to produce formic acid and dry sodium sulfate [66]. The suggestion has been made that reaction with sulfuric acid be carried out in two stages. In this case, the first stage takes place in vacuo in a horizontal tubular reactor with rotating internals, and some of the formic acid is evaporated at the same time. In the second stage, the remaining formic acid is evaporated, also in vacuo, at 30 – 60 °C in a smaller tubular reactor [67].

Norsk Hydro describes the production of formic acid from calcium formate by reaction with nitric acid. The calcium nitrate produced may be used, for example, as a fertilizer.

4.2.4. Direct Synthesis from Carbon Monoxide and Water

The simplest theoretically possible method of producing formic acid is to react carbon monoxide with water:

$$\text{H}_2\text{O} + \text{CO} \rightleftarrows \text{HCOOH}$$

As pressure increases and temperature decreases, the equilibrium of this exothermic reaction shifts in favor of formic acid. Below 150 °C, the reaction rate is very slow, and although equilibrium is reached rapidly at higher temperature, the pressure must be increased drastically to obtain acceptable formic acid concentrations. Table 12 lists the calculated equilibrium concentrations at various temperatures and pressures.

Inorganic acids and salts have been described as catalysts in the literature. Unless the reaction rate can be increased markedly by the development of new catalysts, direct synthesis cannot be used for the economical production of formic acid.

4.2.5. Use of Carbon Dioxide

Interesting developments in carbon dioxide chemistry have resulted in new formic acid syntheses, but so far these have not been used industrially [69].

Table 12. Calculated equilibrium concentrations of formic acid at different temperatures and pressures

Carbon monoxide pressure, MPa			Formic acid concentration, wt %
at 25 °C	at 100 °C	at 217.9 °C	
0.027	0.63	6.95	1
0.30	7.15	79.1	10
0.72	16.95	187	20
3.7	85.8	950	50
108	2 540	28 170	90

Zinc selenide [1315-09-9] and zinc telluride [1315-11-3] have been proposed as catalysts for the hydrogenation of carbon dioxide [70]. If homogeneous transition-metal catalysts are employed, carbon dioxide, hydrogen, and water can be converted to formic acid. If alcohol is used instead of water, the corresponding ester is produced. Ruthenium and palladium complexes have proved particularly active; either inorganic bases [71] or organic bases [72] (e.g., aliphatic tertiary amines) can be used as the alkaline material.

BP Chemicals has developed a process in which formic acid is produced from carbon dioxide and hydrogen via several process stages [73]–[75]. In the first stage, a nitrogen base such as triethylamine [121-44-8] reacts with carbon dioxide and hydrogen in the presence of a noble-metal catalyst (e.g., a ruthenium complex) to yield the formate of the nitrogen base. In the second and third stages, the formate is separated from the catalyst and the low-boiling constituents. In the fourth stage, the formate reacts with a high boiling base, for example, 1-(n-butyl)imidazole [4316-42-1], to yield a formate that can be thermally decomposed. Simultaneously, the low-boiling base is liberated and distilled. In the fifth stage, the formate is thermally decomposed, formic acid distilled, and the high-boiling base is returned to the fourth stage.

4.3. Concentration of Formic Acid

Formic acid–water mixtures cannot be concentrated to more than the azeotropic composition by simple distillation. Further dehydration is carried out by ternary azeotropic distillation, extractive distillation, or extraction. Additional concentration can then be achieved by simple distillation. Through suitable choice of distillation conditions, formic acid with a content exceeding 99 wt% can be drawn off as distillate.

Azeotropic Distillation. Propyl and butyl formates have been proposed as entrainers for azeotropic distillation [2]. A particularly economical separation of water from the water–formic acid–acetic acid mixture obtained by oxidizing butane is achieved by azeotropic distillation with ethyl n-butyl ether [628-81-9] [76].

Extractive Distillation. In extractive distillation, formic acid is extracted in a distillation column by means of a basic extracting agent introduced countercurrently and fed into the bottom of the column. If this mixture is heated in a column downstream, the formic acid is liberated from the salt-like compound and distilled. *N*-Formylmorpholine [*4394-85-8*] has been suggested as extractant [53], and various sulfones have been described as auxiliary liquids for extractive distillation [77].

Extraction. A number of extractants have been proposed for extracting acetic and formic acids [78]. According to BASF, secondary amides are efficient extracting agents [54]; Hüls recommends compounds in the series tri-*n*-octylamine [*1116-76-3*] to tri-*n*-dodecylamine [*102-87-4*] [79].

4.4. Construction Materials

The use of sufficiently corrosion-resistant materials in the individual process stages is part of the expertise of formic acid production. Tests on materials under laboratory conditions do not always guarantee that corrosion resistance can be achieved in practice. A low impurity content or a change in concentration or temperature may produce a decisive change in the corrosion resistance of a material toward formic acid.

If austenitic CrNi steels are used, those that contain molybdenum are preferred to molybdenum-free types [80]. Above 20 °C, stainless steel AISI 304 (Cr 18 wt%, Ni 8 wt%) has only adequate resistance, if the formic acid concentration is below 10% or above 90%. Stainless steel AISI 316 (Cr 18 wt%, Ni 10 wt%, Mo 2 wt%) is resistant up to ca. 50 °C to formic acid below 40 wt% or over 80 wt% [81]. For higher temperatures and formic acid concentrations between 40 and 80 wt%, steels with higher Cr, Ni, and Mo contents (for example, AISI 317 or 317 L) or special nickel-rich alloys must be used. Aluminum ist not sufficiently resistant to formic acid, except at concentrations below 30 wt% or above 80 wt% and at temperatures below 20 °C [81]. Titanium is resistant under oxidizing conditions but is attacked very strongly under reducing conditions [80]. The corrosion behavior of titanium is considerably improved by addition of platinum or palladium. The more exotic metals tantalum, niobium, and zirconium are also corrosion resistant [80].

Of the nonmetallic materials, graphite and glass are sufficiently corrosion-resistant, but their use is limited because of their poor mechanical properties. Among the common plastics, polyethylene is resistant up to about 50 °C. Polytetrafluorethylene (PTFE) and poly(vinylidene fluoride) (PVDF) are resistant to formic acid. Hard poly(vinyl chloride) (PVC) is resistant only at room temperature, and soft PVC should be used only for formic acid concentration below 10 wt%.

Liquid-crystal polymers (LCP) are recommended as packing materials for columns because their resistance to formic acid is reportedly much better than that of ceramics. A survey of construction materials used in formic acid production is given in [82].

5. Environmental Protection

Like most other simple organic acids, formic acid can be degraded rapidly and completely by biological methods. For this reason, formic acid is totally mineralized in a short time in purification plants and streams. The ecological equilibrium of the streams can, however, be disturbed by a change in pH. For formic acid, the theoretical biological oxygen demand (BOD) is 350 mg/g. The chemical oxygen demand has been determined to be 86 mg/g after 5 d (COD_5) and 250 mg/g after 20 d (COD_{20}). In the modified OECD screening test (OECD method 30 IE), formic acid is degraded to $>90\%$ [83]. If an additional carbon source is available, degradation proceeds more quickly.

Toxicity to fish was tested on the blue gill sunfish (*Lepomis macrochirus*). The lethal dose (TLm 24) was found to be 175 ppm [84]. Toxicity in invertebrates was determined on *Daphnia magna*. The TLm 48 value was 120 ppm [85]. In the Federal Republic of Germany, formic acid is classified among the substances that slightly endanger water (Class 1 of water endangerment, WGK 1) [86].

Formic acid vapor can be removed from exhaust streams by washing with water. Fog results when formic acid vapor is emitted during periods of high humidity. In the Federal Republic of Germany emission is controlled by the technical guidelines for air (TA Luft) [87]. Formic acid belongs to Class 1, which means that for emission rates of 0.1 kg/h or higher, the formic acid concentration must be <20 mg/m^3 in the exhaust gas. Upon filling of formic acid, the displaced air is recycled via a gas-transfer line or it is emitted after purification (e.g., by water washing or passage through an incinerator).

6. Quality Specifications

Formic acid is marketed in concentrations of 85, 90, 95, 98, and >99 wt%. The impurity content depends on the production process. Formic acid produced by oxidation of hydrocarbons still contains a small proportion of acetic acid. A typical specification for formic acid produced by hydrolysis of methyl formate is given in Table 13; this acid meets the specifications required in the Food Chemicals Codex (FCC) [89].

7. Chemical Analysis

The concentration of formic acid is normally determined by titration with sodium hydroxide solution. Volatile impurities can be determined by gas chromatography on packed columns. Polyesters or Carbowax 20 M, both treated with phosphoric acid and adsorbed on chromosorb as a carrier, have proved successful as the stationary phase. Formic acid content is measured with a thermal-conductivity cell because the substance cannot be detected with flame ionization detectors (FID).

Table 13. Sales specification for formic acid

Specification	Limits	Analytical method
Content	min. 85 wt%	titrimetric,
	min. 90 wt%	ISO 731/II, 5.2
	99 wt%	
Color number	max. 10 APHA	ASTM D 1209-79
Acetic acid	max. 150 mg/kg	ISO 731/VII
Sulfate	max. 1 mg/kg	ISO 731/V
Chloride	max. 1 mg/kg	ISO 731/IV
Iron	max. 1 mg/kg	ISO 731/VI
Other heavy metals	max. 1 mg/kg	FCC III
Evaporation residue	max. 20 mg/kg	DIN 53 172
Quality	the formic acid complies with DAB 7 and FCC III	

Test methods for formic acid have been standardized by the International Organization for Standardization (ISO) [90]. Other standard analytical methods for formic acid are described in [91].

8. Storage and Transportation

Formic acid is stored in tanks made of austenitic CrNi steels (18% Cr, 10% Ni, 2% Mo), for example, AISI 316 L. The use of other materials, such as AISI 304 L, polyethylene-lined carbon steel, glass-lined carbon steel, and polypropylene, is restricted to specific concentrations and low temperatures. Tank cars and tank trucks made of AISI 316 L are used for transportation. Small canisters may be made of polyethylene; glass bottles are used for laboratory quantities. When aqueous formic acid mixtures are stored or transported the melting points at various concentrations (Table 3) should be considered. Tanks are best heated with hot water or subsidiary electrical heating systems. To prevent corrosion, even of AISI 316, wall temperatures above 50 °C must be avoided. Formic acid decomposes to highly toxic carbon monoxide and water. For this reason, formic acid with a concentration of 99 wt% or higher is transported in canisters without gastight seals [2]. Containers of highly concentrated formic acid should be protected against exposure to heat and stored in a well-ventilated place. Before entering ship tanks or storage tanks that have contained highly concentrated formic acid, personnel should monitor the carbon monoxide content of the air. The storage method should ensure that contact with strong alkali solutions (evolution of heat) and strong acids or oxidizing agents (decomposition) does not occur. Table 14 contains important safety data.

Table 14. Safety data

	Formic acid content, wt%				Method
	85	90	94	99	
Flashpoint, °C	59	58	55	48	DIN 51 755
Autoignition point, °C	500	480	480	480	DIN 51 794
Flammable limits in air, vol%	15 – 47	14 – 44	14 – 38	12 – 38	

9. Legal Aspects

Formic acid is marked according to the regulations of the European Economic Community (council directive 76/907/EEC including the latest amendment) and to the regulations of the Federal Republic of Germany (Gef Stoff V) [93] as: Corrosive, R34 (formic acid content 25 – 90 wt%), R35 (formic acid content > 90 wt%), and S 2-23-26. The EEC numbers of formic acid are 607-001-00-0 (formic acid content > 90 wt%) and 607-001-01-8 (formic acid content 25 – 90 wt%). Hazard classifications for the transport of formic acid are listed in the following:

RID/ADR	class 8, item 32 b
Fed. Rep. of Germany: GGVE/GGVS	class 8, item 32 b
ADNR	class 8; item 21 b
IMDG Code (Fed. Rep. of Germany: GGVSee)	8
IATA-DGR	1779
UN Number	1779
DOT Hazard Classification (USA)	corrosive material
HazChem Code (UK)	2 R

10. Uses

Because of its acidity, its aldehyde nature, and its reducing properties, formic acid is used in a variety of fields (Table 15).

In contrast to mineral acids, formic acid evaporates without leaving any residue. The leather and textile industries, in particular, make use of its strongly acidic nature. Animal skins and hides are acidified with a mixture of formic acid, sulfuric acid, and sodium chloride before chrome tanning.

Formic acid is used for adjusting the pH of dye baths in dyeing of natural and synthetic fibers. Use is made of its highly reactive nature in the manufacture of pharmaceuticals and crop-protection agents [31], [42]. Formic acid plays an important role in the coagulation of rubber latex.

In Europe, most of the formic acid is used as a silage aid. If freshly mown, damp grass is sprayed with formic acid to establish a pH of ca. 4, such fodder can be used directly for silage. Formic acid promotes the fermentation of lactic acid and suppresses the harmful formation of butyric acid. If formic acid is used, fermentation processes occur more rapidly and at lower temperature so that fewer nutrients are lost [94], [95].

Table 15. Applications of formic acid (1988, BASF assessment)

Textiles, leather	25%
Pharmaceuticals, crop-protection agents	10%
Latex, rubber auxiliaries	10%
Silage	35%
Miscellaneous	20%

Mixtures of formic acid with formaldehyde or with ammonium formate are also used as silage aids. Silage that utilizes formic acid is restricted largely to Europe, but it should be of interest in all regions where climate makes *green forage* difficult to dry.

New applications of formic acid that have emerged in recent years are used as an additive for cleaning agents (replacement of mineral acid for environmental reasons), in the synthesis of the sweetener aspartame [*22839-47-0*] [96] (→ Amino Acids), and in the desulfurization of flue gas by the Saarberg–Hölter process (pH stabilization of the wash liquid) [97]. A large potential market for formic acid is anticipated by the Leonard Process Co. in steel pickling and wood pulping [57]. In both fields, however, formic acid has not yet replaced the standard techniques (steel pickling with hydrochloric acid or wood pulping by the Kraft process).

11. Economic Aspects

The market for formic acid has increased only slightly in recent years. Nevertheless, several new plants have been built because less formic acid is now produced as a byproduct, and more economic processes have been developed. A breakdown of production capacity for formic acid is given in Table 16.

12. Derivatives

Among the derivatives of formic acid, amides (→Formamides), various esters, and salts are the most important.

12.1. Esters

Esters of formic acid have the typical physical and chemical properties of esters; they are readily volatile and flammable, and have a sweet, often fruity smell. They are only partially soluble in water. A summary of their chemical properties has been published by GIBSON [31].

Table 16. Worldwide formic acid production capacity (BASF assessment)

Company	Basis of process	Capacity, t/a
BASF (Fed. Rep. of Germany)	methyl formate	100 000
BP (UK)	naphtha	50 000
Kemira Oy (Finland)	methyl formate	40 000
Norsk Hydro (Norway)	calcium formate	13 000
Chem. Werke Hüls (Fed. Rep. of Germany)	sodium formate	10 000
Perstorp (Sweden)	sodium formate	10 000
Far (Italy)	sodium formate	10 000
CSSR	formamide	10 000
Celanese (United States)	butane	13 000
PT Pupuk Kujang (Indonesia)	methyl formate	10 000
GNFC (India)	methyl formate	5 000
Korea Fertilizer (Republic of Korea)	methyl formate	5 000
Daicel (Japan)	naphtha	5 000
Others	sodium formate calcium formate formamide	50 000
Under construction or planned (start-up):		
USSR	methyl formate	40 000 (1989)
USSR	methyl formate	40 000 (1990)

12.1.1. Methyl Formate

Methyl formate [*107-31-3*], $C_2H_4O_2$, M_r 60.05, is an intermediate in the production of formic acid (Sections 4.1, 4.2) and formamide (→ Formamides) in large manufacturing plants. With increasing interest in the production of synthesis gas by coal gasification, methyl formate will become important as a versatile key intermediate. It can also be used as a starting material in the production of high-purity carbon monoxide [98]. Physical properties, safety data, and marking regulations for methyl formate are summarized below.

Boiling point (101.3 kPa)	31.7 °C [16]
Melting point	−99 °C [16]
Specific gravity d_4^{20}	0.975 [16]
Refractive index n_D^{20}	1.3433 [98]
Flash point (open cup)	−20 °C [98]
Ignition temperature	450 °C [98]
Flammable limits in air	5.0 – 23 vol % [98]
MAK value	100 mL/m^3; 250 mg/m^3 (category I)
TLV—TWA	100 mL/m^3; 250 mg/m^3
TLV—STEL	150 mL/m^3; 375 mg/m^3
UN number	1243
EEC number	607-014-00-1
Marking according to Gef Stoff V (FRG)	symbol: F R12, S 9-16-33
Hazchem Code (UK)	2 SE
RID/ADR	class 3, item I a
ADNR	class 3, item I a
IMDG Code (F.R. of Germany GGVSee)	3.1
IATA-DGR	1243
DOT Hazard Classification (USA)	flammable liquid

Production. Methyl formate is produced by base-catalyzed carbonylation of methanol (Section 4.1). Processes have been developed for dehydrogenating methanol, but these have not yet been implemented commercially [99]:

$$2\,CH_3OH \longrightarrow HCOOCH_3 + 2\,H_2 \qquad \Delta H_R^0 = 98.9 \text{ kJ/mol}$$

Catalysts containing copper are used.

Mitsubishi Gas Chemical (MGC) has developed highly active catalysts which contain Cu–Zr–Zn or Cu–Zr–Zn–Al. The conversion of methanol is approximately 50%, with a methyl formate selectivity of ca. 90% and a space–time yield of 3000 g L^{-1} h^{-1} [100]. Air Products uses a copper–chromite catalyst for the dehydrogenation of methanol [101]. Methyl formate can also be produced by oxy-dehydrogenation of methanol:

$$2\,CH_3OH + O_2 \longrightarrow HCOOCH_3 + 2\,H_2O \qquad \Delta H_R^0 = -472.8 \text{ kJ/mol}$$

Soluble chromium compounds are used as catalyst [102].

The direct synthesis of methyl formate from synthesis gas has been investigated:

$$2\,CO + 2\,H_2 \longrightarrow HCOOCH_3 \qquad \Delta H_R^0 = -157.2 \text{ kJ/mol}$$

The high pressure of 100–200 MPa (1000–2000 bar) required makes economical implementation impossible [99]. The dimerization of formaldehyde has likewise not been exploited industrially [99]:

$$2\,CH_2O \longrightarrow HCOOCH_3 \qquad \Delta H_R^0 = -146.4 \text{ kJ/mol}$$

Uses. Most of the methyl formate produced is used as an intermediate for the production of formic acid and formamide. Dimethylformamide [68-12-2] is produced by reacting methyl formate with dimethylamine [124-40-3] (→ Formamides). A new use has been found for methyl formate in the production of foundry molds. The mold is formed from granular refractory material and a binder consisting of phenol–formaldehyde resin which is cured by exposure to methyl formate vapor. This process makes possible the production of rapidly cured foundry molds at low temperature [103].

Small quantities of methyl formate are used as solvents and insect control agents. With the anticipated extension of C_1 chemistry, methyl formate may be used as an intermediate for a number of products [98], [99].

12.1.2. Ethyl Formate

Ethyl formate [109-94-4], $C_3H_6O_2$, M_r 74.08, bp 54.1 °C, mp −80 °C, d_4^{20} 0.923, flash point −20 °C, ignition point 440 °C, UN number 1190, has a sharp rumlike smell and taste.

Ethyl formate is produced by carbonylation of ethanol with carbon monoxide or by esterification of ethanol with formic acid.

Ethyl formate is a solvent for acetyl cellulose and nitrocellulose. Ethyl formate is used in alcohol-free drinks, ice cream, chewing gum, and other confectionery as a component of apple-, pineapple-, banana-, and peach-type flavors.

12.1.3. Isobutyl Formate

Isobutyl formate [542-55-2], M_r 102.13, bp 98 °C, mp −96 °C, d_4^{20} 0.88, flash point 5 °C, ignition point 320 °C, UN number 2393, is produced by carbonylation of isobutyl alcohol [78-83-1] with carbon monoxide or esterification of isobutyl alcohol with formic acid. Isobutyl formate is used as a solvent for colors, lacquers, adhesives, and cleansing agents.

12.1.4. Miscellaneous Esters

A number of formates produced by esterification of the corresponding alcohols with formic acid are used as perfumes and flavorings. Some representative products are citronellyl formate [105-85-1], geranyl formate [61759-63-5], phenylethyl formate [104-62-1], benzyl formate [104-57-4], and isopentyl formate [110-45-2].

Orthoformates are used in the preparation of acetals and ketals. Chloroformic esters are used as intermediates in organic syntheses.

12.2. Salts

12.2.1. Alkali-Metal and Alkaline-Earth Metal Formates

Sodium formate [141-53-7] and calcium formate [544-17-2] are byproducts in the synthesis of polyols such as pentaerythritol. However, they are also produced directly from the corresponding hydroxides and carbon monoxide (Section 4.2.3) [2].

Sodium or calcium formate is used to produce formic acid (Section 4.2.3). An important process for manufacturing sodium dithionite [7775-14-6] starts with sodium formate. Oxalic acid [144-62-7] production employs sodium formate as an intermediate.

Sodium formate is used in chrome tanning and as a mordant in the dyeing and printing of fabrics by the textile industry. The reducing power of sodium formate is utilized in electroplating baths and photographic fixing baths. Calcium formate is also used in tanning. Adding calcium formate to concrete reduces the setting time considerably.

12.2.2. Ammonium Formate

Ammonium formate solutions are used as a low-corrosion silage aid; the complex salts ammonium diformate [64165-14-6] and ammonium tetraformate [70179-79-2] are stable in aqueous solution [104].

12.2.3. Aluminum Formate

Stepwise replacement of the hydroxyl groups in aluminum hydroxide [21645-51-2] yields dibasic aluminum formate [18748-09-9], $(HO)_2Al(OOCH)$; monobasic aluminum formate [51575-25-8], $(HO)Al(OOCH)_2$; and aluminum formate [22918-74-7], $Al(OOCH)_3$. Aluminum hydroxy acetate formate [34202-30-7] $(HO)Al(OOCH)(OOCCH_3)$, is also manufactured commercially. The various formates are produced either directly from the metal or by reaction with an aluminum compound, for example, aluminum hydroxide [105].

Aluminum formates are used to impregnate textiles and to make paper water-resistant. Formates are also employed mordants in the textile industry. Because of the low toxicity and antiseptic, astringent, and basic properties of the aluminum salts, aluminum formates are used in pharmaceuticals.

12.2.4. Nickel Formate

Nickel(II) formate [15694-70-9] usually exists in the form of a green crystalline dihydrate $Ni(HCOO)_2 \cdot 2 H_2O$; it is prepared by dissolving nickel(II) hydroxide [12054-48-7] or nickel(II) carbonate [16337-84-1] in formic acid or by reaction of nickel(II) sulfate with sodium formate. Nickel(II) formate is used for the production of nickel hydrogenation catalysts.

12.2.5. Copper Formate

Copper(II) formate [544-19-4] is produced by reacting copper(II) hydroxide [20427-59-2], copper(II) oxide [1317-38-0], or copper(II) carbonate [12069-69-1] with formic acid. It is used as an antibacterial agent for treating cellulose.

13. Toxicology and Occupational Health

Formic acid occurs naturally in small amounts in plants, animal poisons, and higher organisms. In humans, it is formed in toxic amounts during methanol poisoning. After intake of large amounts of methanol, the metabolic capacity to detoxify it to carbon dioxide is insufficient, and the high amount of formic acid that results leads to metabolic acidosis. Under physiological conditions, formic acid which is observed under these conditions in low concentration is a component of human blood and tissues, and plays an important role in the transfer of C_1 compounds during intermediate metabolism. Most of the formic acid taken up by the body is metabolized; a small portion is excreted unchanged in the urine. After oral intake, the biological half-life of formic acid is about 45 minutes in human blood plasma [106].

The most prominent toxic property of this compound is its corrosive effect upon skin and mucous membrane. When applied to the clipped skin of rabbits, formic acid causes necrosis, which heals slowly [107]. Following oral intake of aqueous dilutions of formic acid, corrosion of the oral cavity and esophagus occurs even at concentrations as low as 6% [107]–[109]. The LD_{50} after oral administration is reported to be 1830 mg/kg in rats [110] and 1076 mg/kg in mice [111]. Formic acid, whose solubility in water is good, can be absorbed easily by the skin [106], [107] and causes serious local injuries upon inhalation. In rats, increasing concentrations of aqueous acid solutions, inhaled as saturated vapors (20 °C), lead to death during decreasing exposure times: a 10% aqueous solution of formic acid is tolerated for 7 h. Eyelid closure and watery nasal discharge as a result of pronounced irritation is no longer observed after termination of the exposure period. A 25% aqueous solution is lethal after an exposure of more than 3 h; a 50% aqueous solution, after 30 min; and undiluted (98%) formic acid, after only 3 min inhalation. In rats, the LC_{50} is 7.4 mg/L after inhalation of the vapor [112]. The predominant symptoms of inhalative intake are irritation of the eyes, irritation and corrosion of the nasal mucous membranes, and corneal opacity.

According to earlier investigations [107], formic acid affects the central nervous system (CNS). In rabbits, intravenous administration of buffered formic acid in the range of 0.46–1.25 g per kilogram of body weight leads to CNS depression; higher dosages cause convulsions. Subcutaneously administered doses of 0.8 g/kg in dogs and over 0.4 g/kg in cats lead to staggering, while a dosage of 0.4 g/kg leads to sleepiness in cats. In dogs, lethal doses administered intravenously result in tonic and clonic convulsions [107]. The mechanism of CNS effects was investigated by exposing rats to formic acid vapors (20 ppm) for 6 h a day, five days per week, for two or three weeks. This treatment caused changes in glial cell metabolism, which could be the reason for neurological effects [113].

Rats administered formic acid in concentrations of 0.5 and 1.0% in feed and drinking water, respectively, showed a slight increase in body weight gain. In addition the weights of liver, kidneys, adrenals, and (for the lower dose group) spleen were lower than those of control animals [110]. In rats, the administration of dosages of 8.2, 10, 25, 90, 160, and 360 mg kg^{-1} d^{-1} in drinking water for 2 to 27 weeks led to a lower intake of food and delay in body weight gain only in the highest dose group [114].

The influence of formic acid on biochemical parameters was investigated in rats by exposing them for 6 h a day to vapors of 20 ppm for three or eight days [115]. This treatment led to a decrease in the concentration of glutathione in the liver and kidneys.

Few experimental results are available regarding the mutagenic effects of formic acid. The substance was weakly mutagenic when tested on *Escherichia coli* [116]. A mutagenic effect was also observed on the germ cells of drosophila [117]. However, no primary genotoxic mechanism of action appears to be involved; rather, the mutagenicity is probably associated with the acidity of the compound, which leads to radical formation after inhibition of the enzyme catalase.

The TLV-TWA exposure limit (ACGIH) for formic acid is 5 ml/m^3 (9 mg/m^3); its MAK value is also 5 ml/m^3 (9 mg/m^3) (short term exposure maximum: category I). Formic acid has a PDK (Soviet Union) of 0.5 ml/m^3 (1 mg/m^3).

14. References

[1] M. N. Kuznetsova, A. G. Bergmann, *J. Gen. Chem. USSR (Engl. Transl.)* **26** (1956) 1497–1504; *Chem. Abstr.* **50** (1956) 14337.
[2] *Ullmann*, 4th ed. **7** 362–373.
[3] J. E. Prue, A. J. Read, *Trans. Faraday Soc.* **62** (1966) 1271.
[4] The Texas A & M University System, Table 23-2-1-(1.1210)-d, p 1.
[5] Landolt-Börnstein, *New Series,* Group IV vol. I, Part b, Springer Verlag, Heidelberg 1977, pp. 114, 301.
[6] A. S. Coolidge, *J. Am. Chem. Soc.* **52** (1930) 1874–1887.
[7] J. Chao, B. J. Zwolinski, *J. Phys. Chem. Ref. Data* **7** (1978) no. 1, 363–377.
[8] W. Waring, *Chem. Rev.* **51** (1952) 171–183.
[9] R. J. Jakobsen, Y. Mikawa, J. W. Brasch, *Spectrochim. Acta Part A:* **23 A** (1967) 2199–2209.
[10] J. Gmehling, U. Onken, et al.: *Vapor-Liquid Equilibrium Data Collection,* Dechema Chemistry Data Series, vol. **I,** part 1 (1977), 1 a (1981), 2 d (1982), 3+4 (1979), 5 (1982).
[11] T. Ito, F. Yoshida, *J. Chem. Eng. Data* **8** (1963) 315–320.
[12] M. M. Gil'burd, F. B. Moin et al., *Zh. Prikl. Khim. (Leningrad)* **57** (1984) 915–917; *Chem. Abstr.* **100** (1964) 216, 549.
[13] L. H. Horsley: "Azeotropic Data III," Adv. Chem. Ser. **n116** (1973) 66–69.
[14] P. B. Davis, H. C. Jones, *J. Am. Chem. Soc.* **37** (1915) 1194–1198.
[15] *Landolt-Börnstein,* 6th ed., **II/5 a,** 209.
[16] VDI-Wärmeatlas, 4th ed., VDI-Verlag, Düsseldorf 1984.
[17] R. W. Gallant, *Hydrocarbon Process.* **47** (1968) no. 6, 139–148.
[18] R. R. Dreisbach, R. A. Martin, *Ind. Eng. Chem.* **41** (1949) no. 12, 2875–2878.
[19] J. J. Jasper, *J. Phys. Chem. Ref. Data* **1** (1972) no. 4, 851.
[20] J. F. Johnson, R. H. Cole, *J. Am. Chem. Soc.* **73** (1951) 4536–4540.
[21] G. C. Sinke, *J. Phys. Chem.* **63** (1959) 2063.

[22] W. J. Canady, H. M. Papée, K. J. Laidler, *Trans. Faraday Soc.* **54** (1958) 502–506.
[23] W. Jobst, *Int. J. Heat Mass Transfer* **7** (1964) 725–732.
[24] Th. C. Wehman, A. I. Popov, *J. Phys. Chem.* **72** (1968) 4031.
[25] A. Kailan, N. H. Friedmann, *Monatsh. Chem.* **62** (1933) 284–316.
[26] A. Kailan, F. Adler, *Monatsh. Chem.* **63** (1933) 155–185.
[27] A. Kailan, G. Brunner, *Monatsh. Chem.* **51** (1929) 334–368.
[28] E. N. Rostovskii, J. A. Arbuzova, *Met. Khim. Prom. Kazakhstana Nauchn. Tekhn. Sb.* **71** (1961) *Chem. Abstr.* **58** (1963) 5489.
[29] K. Wagner, *Angew. Chem.* **82** (1970) no. 2, 73–77.
[30] D. Swern, *Chem. Rev.* **45** (1949) 3–5.
[31] H. W. Gibson, *Chem. Rev.* **69** (1969) 673–692.
[32] W. Harder (BASF), private communication, Sept. 1984.
[33] P. Mars, J. J. F. Scholten, P. Zwieterling, *Adv. Catal.* **14** (1963) 35–113.
[34] H. N. Barham, L. W. Clark, *J. Am. Chem. Soc.* **73** (1951) 4638–4640.
[35] S. F. Dickson, J. Bakker, A. Kitai: "Acetic Acid", in *Chemical Economics Handbook*, SRI International, Menlo Park, Calif. March 1982, revised July 1982, pp. 602.5020, 602.5021.
[36] *Eur. Chem.* 1982 no. 5, 64.
[37] Mitsubishi Gas Chem. KK, JP-Kokai 58 167 529, 1982.
[38] *Chem. Eng.*, (1979, July 30), 17.
[39] BP Chemicals Ltd. EP 60 695, 1981 (R. D. J. Mathieson).
[40] IG Farben, GB 252 848, 1925 (R. Witzel).
[41] Deutsche Gold- und Silber-Schneideanstalt, DE 1 035 637, 1956 (L. Hüther, M. Petzold).
[42] R. Hoch, H. W. Scheeline, SRI International, Report No. 156, "Formic Acid," private report by the Process Economics Program, Aug. 1983.
[43] Shell International Petroleum Maatschappij, N. V. GB 1 047 408, 1966.
[44] Union Chimique Belge, US 4 216 339, 1980 (W. Couteau, J. Ramioulle).
[45] BASF, EP 48 891, 1980 (F. J. Müller, W. Steiner, K. H. Ross, O. Kratzer).
[46] D. M. Pond, *Chem. Eng. News* **60** (1982) no. 37, 43.
[47] R. F. Schultz, *J. Am. Chem. Soc.* **61** (1939), 1443–1447.
[48] BASF, EP 1432, 1981 (H. Hohenschutz, H. Kiefer, J. E. Schmidt).
[49] BASF, EP-A 161 544, 1985 (H. Kiefer, H. Hohenschutz, J. E. Schmidt).
[50] Hüls AG, DE-OS 3 428 321, 1984 (M. Zölffel).
[51] Carbide & Carbon Chemicals, US 2 160 064, 1936 (J. F. Eversole).
[52] I. I. Moiseev, O. A. Tagaev, N. M. Zhavoronkov et al., DE 3 220 555, 1982.
[53] BASF US 4 076 594, 1978 (H. Bülow, H. Hohenschutz, W. Sachsze, J. E. Schmidt).
[54] BASF, DE 2 545 658, 1975 (H. Kiefer, H. Hohenschutz, J. E. Schmidt).
[55] BASF, EP 17 866, 1980 (H. Schoenmakers, D. Wolf, K. Bott, et al.).
[56] J. D. Leonard, EP 5 998, 1978.
[57] L. J. Kaplan, *Chem. Eng. (N.Y.)* **89** (1982) no. 14, 71–73.
[58] A. Aguilo, Th. Horlenko, *Hydrocarbon Process.* **59** (1980) no. 11, 120–130.
[59] *Hydrocarbon Process.* **62** (1983) no. 11, 104.
[60] BASF, DE 1 147 214, 1961 (E. Germann).
[61] *Eur. Chem. News* **42** (1984) May 7, 22.
[62] Salzgitter Industriebau GmbH: *Technical Information Formic Acid*, 1982.
[63] Salzgitter Industriebau GmbH: *Technical Information Formic Acid*, 1985.
[64] Bethlehem Steel Corp., US 3 907 884, 1973 (J. B. Lynn, O. A. Homberg, A. H. Singleton).
[65] A. Peltzman, *Oil Gas J.* **79** (1981) no. 46, 103–109.

[66] Skanska Attikfabriken AB (Perstorp), GB 1 049 013, 1964 (E. Lindkvist).
[67] Davy Mc Kee, DE 3 137 356, 1981 (W. Dürkoop, E.-W. Wisskirchen).
[68] Commercial Solvents, US 1 606 394, 1926 (W. C. Arsem).
[69] A. Behr, *Chem. Ing. Tech.* **57** (1985), no. 11, 893–903.
[70] Omsk Poly., SU 1 097 366, 1983.
[71] Adogar Kogyo KK, JP-Kokai 53 46 820, 1978,
[72] Mitsubishi Gas Chem. Ind., JP-Kokai 51 138 614, 1976.
[73] BP Chemicals Ltd., EP-A 0095 321, 1982 (D. J. Drury, J. E. Hamlin).
[74] BP Chemicals Ltd., EP-A 0126 524, 1983 (J. J. Anderson, J. E. Hamlin).
[75] BP Chemicals Ltd., EP-A 0181 078, 1984 (J. J. Anderson, D. J. Drury, J. E. Hamlin, A. G. Kent).
[76] W. Hunsmann, K. H. Simmrock, *Chem. Ing. Tech.* **38** (1966) no. 10, 1053–1059.
[77] L. Berg, An-I Yeh, US 4 642 166, 1986.
[78] P. Eaglesfield, B. K. Kelly, J. F. Short, *Ind. Chem.* **29** (1953) April, 147–151; **29** (1953) June, 243–250.
[79] Hüls AG, DE-OS 3 428 319, 1984 (F. v. Praun, H.-U. Hög, G. Bub, M. Zölffel).
[80] National Association of Corrosion Engineers Publication 5A 174, July 1974, pp. 13–18.
[81] F. F. Berg: *Korrosionsschaubilder, Corrosion Diagrams*, VDI-Verlag, Düsseldorf 1965.
[82] D. Behrens (ed.): *DECHEMA-Werkstoff-Tabelle*, Frankfurt, 1988.
[83] OECD-Guidelines for Testing of Chemicals (1981), Paris.
[84] B. F. Dowden, H. J. Bennett, *J. Water Pollut. Control Fed.* **37** (1965) 1308–1316.
[85] G. Bringmann, R. Kühn, *Gesund. Ing.* **80** (1959) 115–120.
[86] *Katalog wassergefährdender Stoffe*, Gemeinsames Ministerialblatt Ausgabe A, 36. Jahrgang, Nr. 11, Bonn April 1985, pp. 173–186.
[87] T. A. Luft: *Gemeinsames Ministerialblatt* Ausgabe A, 37. Jahrgang, Nr. 7, Bonn Febr. 1986, pp. 93–143.
[88] BASF Data sheet D 027 d, e (Sept. 1986).
[89] *Food Chemicals Codex*, 3rd. ed., National Academy Press, Washington D.C., 1981.
[90] International Standard ISO 731/I-VII, 1977.
[91] R. M. Speights, A. J. Barnard, Jr. in F. D. Snell, L. S. Ettre (eds.): Formic Acid, *Encyclopedia of Industrial Chemical Analysis*, vol. **13,** Interscience Publishers, New York 1971, pp. 117–138.
[92] BASF DIN-Sicherheitsdatenblatt für Ameisensäure 85%, 90%, 95%, 99% (1986).
[93] Verordnung über gefährliche Stoffe (Gefahrstoffverordnung-GefStoffV), 26. 08. 86 (BGBl. I S. 1470), Carl Heymans Verlag KG, Köln 1986.
[94] P. Mc Donald: *The Biochemistry of Silage*, J. Wiley & Sons, New York 1981.
[95] M. P. Czaikowski, A. R. Bayne, *Hydrocarbon Process.* **59** (1980) no. 11, 103–106.
[96] C. A. M. Hough, K. J. Parker, A. J. Vlitos (eds.): *Developments in Sweeteners-1*, Applied Science Publ., London 1979.
[97] W. Kaminsky, *Chem. Ing. Tech.* **55** (1983) no. 9, 667–683.
[98] T. Ikarashi, *CEER Chem. Econ. Eng. Rev.* **12** (1980) no. 8, 31–34.
[99] M. Röper, *Erdöl Kohle Erdgas Petrochem.* **37** (1984) no. 11, 506–511.
[100] Mitsubishi Gas Chemical Ind., DE 2 753 634, 1978 (M. Yoneoka, M. Osugi).
[101] Air Products and Chemicals, EP 26 415, 1979 (J. V. Martinez de Pinillos, G. B. De La Mater, H. Ladenheim).
[102] BP Chemicals Ltd., EP 60 718, 1981 (D. J. Drury, J. Pennington).
[103] Borden, EP 86 615, 1982 (P. H. Lemon, J. D. Rallton, P. R. Ludlam, T. J. Reynolds).
[104] BP Chemicals Ltd., DE-OS 2 653 448, 1975 (J. J. Huitson).
[105] *Kirk-Othmer*, **2**, 202–204

[106] G. Malorny, *Z. Ernährungswiss.* **9** (1969) no. 4, 340–348.
[107] W. F. von Oettingen., *Arch. Ind. Health* **20** (1959) 517–531.
[108] S. Moeschlin: *Klinik und Therapie der Vergiftungen,* Thieme-Verlag, Stuttgart – New York 1980, p. 273.
[109] K. E. von Mühlendahl et al., *Arch. Toxicol.* **39** (1978) 299–314.
[110] A. Sporn et al., Igenia **11** (1962) 507–515, cited in: "Evaluation of the Health Aspects of Formic Acid, Sodium Formate and Ethyl Formate as Food Ingredients," Life Sci. Res. Office, Fed. Am. Soc. Exp. Biol., Bethesda, MD., 1976, NTIS No. 03 266 282.
[111] C. Malorny, *Z. Ernährungswiss.* **9** (1969) no. 4, 332–339.
[112] BASF Aktiengesellschaft, Ludwigshafen, unpublished results, 1979–1981.
[113] H. Savolainen et al., *Acta Pharmacol. Toxicol.* **47** (1980) 239–240.
[114] T. Sollmann, *J. Pharmacol. Exp. Ther.* **16** (1921), 463–474.
[115] A. Zitting et al., *Res. Commun. Chem. Pathol. Pharmacol.* **27** (1980) no. 1, 157–162.
[116] D. M. Demerec et al., *Am. Nat.* **85** (1951) no. 821, 119–136.
[117] B. F. A. Stumm-Tegethoff, *Theor. Appl. Genet.* **39** (1969) 330–334.

Furan and Derivatives

Individual keyword: → *Tetrahydrofuran.*

WILLIAM J. MCKILLIP, QO Chemicals Inc., West Lafayette, Indiana 47 906, United States (Chaps. 2 – 8, 11 [in part])

GERD COLLIN, Rütgerswerke AG, Duisburg, Federal Republic of Germany (Chaps. 9 and 10, 11 [in part])

HARTMUT HÖKE, Rütgerswerke AG, Duisburg, Federal Republic of Germany (Chaps. 9 and 10, 11 [in part])

1. Introduction 2745
2. Furan 2746
3. Furfural 2750
4. Furfuryl Alcohol 2756
5. 2-Methylfuran 2760
6. 2-Methyltetrahydrofuran 2761
7. Tetrahydrofurfuryl Alcohol. . . 2762
8. 2,3-Dihydropyran 2763
9. Benzofuran 2764
10. Dibenzofuran 2765
11. Toxicology and Occupational Health. 2767
12. References 2768

1. Introduction

This article discusses the properties, production, and industrial uses of furan and those derivatives derived commercially from furfural. For a complete compilation of the chemistry, properties, and uses of furan and its derivatives up to 1953, see [1]. For more recent summaries, see [2] and [3]. For information on production, supply and demand by geographic region, and pricing refer to [4].

Radicals derived from furan are named in a similar manner to analogous radicals in the benzene series. Typical furan radicals are furyl (**1**), furfuryl (**2**), furoyl (**3**), and furfurylidene (**4**).

2. Furan

Furan [*110-00-9*], furfuran, oxole, C_4H_4O, M_r 68.07, is the parent compound of the five-membered heterocycles containing one oxygen atom.

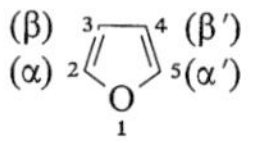

Physical Properties. Furan is a colorless liquid with a strong ether odor; it has a low boiling point and is highly flammable; it is miscible with most common organic solvents but only slightly soluble in water. Some physical properties of furan are listed in Table 1 [5].

Furan possesses some aromatic character arising from the delocalization of the four carbon π-electrons and one of the oxygen electron pairs [6]. The delocalized structure of furan is supported by bond lengths, the dipole moment, the UV spectrum, ^{1}H NMR chemical shifts, and the heat of combustion [6]. A stabilization energy of 96 kJ/mol has been observed for furan, whereas that for benzene is 155 kJ/mol. Furan derivatives are therefore less aromatic than their benzene analogues and less aromatic than other heterocycles. Aromaticity decreases in the order:

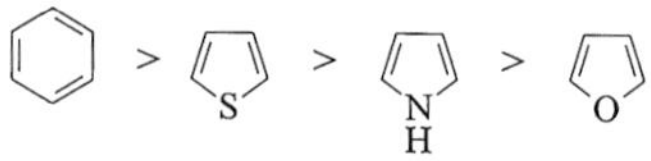

As a consequence, furan compounds undergo a number of addition reactions.

The overall electron distribution in furan is shown by the canonical forms **5–7**:

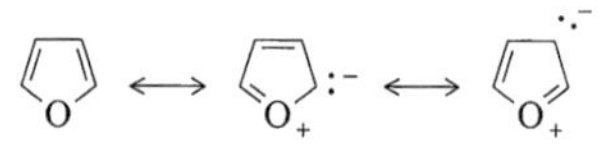

The contribution of these resonance structures to the overall electron distribution decreases in the order **5 > 6 > 7**. The importance of **5** is shown by the tendency of furan to behave as a nucleophilic diene or a nucleophilic vinyl ether. In electrophilic substitution, reaction occurs exclusively at the 2- rather than the 3-position, demonstrating the importance of **6** compared with **7**.

Chemical Properties. Furan is an electron-rich heterocycle, which is more reactive than benzene. When warmed with aqueous mineral acid, furan and alkylfurans are cleaved or polymerize, a reaction typical of vinyl ethers:

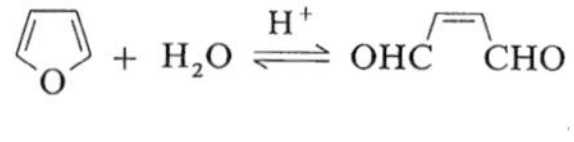

The reverse reaction, cyclodehydration of 1,4-dicarbonyl compounds, has been used to prepare substituted furans.

Table 1. Physical properties of furan derivatives

	CAS registry number	mp, °C	bp, °C	ϱ (20 °C), g/cm³	n_D^{20}	Vapor density (air = 1)	Flash point[a], °C	Explosion limits in air, vol%	Ignition temperature, °C	Solubility in H₂O (25 °C), wt%	Dielectric constant (20 °C)	Viscosity (25 °C), mPa · s
Furfural	[98-01-1]	−36.5	161.6	1.1598	1.5261	3.3	61.7	2.1–19.3	315	8.3[b]	41.9	1.49
Furfuryl alcohol	[98-00-0]	−29[c] −14.6[d]	170	1.1285	1.4868	3.4	77	1.8–16.3	391	∞		5.0
Furan	[110-00-9]	−85.6	31.36	0.9378	1.4214	2.36	−35.5	2.3–14.3		1		
Tetrahydrofurfuryl-alcohol	[97-99-4]	<−80	178	1.0511	1.4520	3.52	83.9[e]	1.5–9.7	282	∞	13.6	6.24
2-Methylfuran	[534-22-5]	−88.7	63.2–65.6	0.913	1.4320		−27			<0.3		
2-Methyltetra-hydrofuran	[96-47-9]	−136	80.2	0.854	1.4025		11.1			15.1		
Dihydropyran	[110-87-2]	−70	84–85	0.923	1.4380		−17.8			1.6		

[a] Tag, closed cup.
[b] At 20 °C.
[c] Metastable crystalline phase.
[d] Stable crystalline phase.
[e] Tag, open cup.

Furan and alkylfurans react as nucleophiles with α, β-unsaturated aldehydes and ketones in the Michael reaction, e.g., with methyl vinyl ketone:

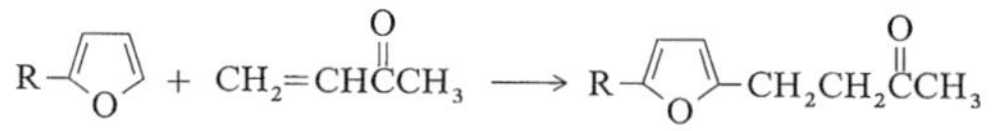

$$R\text{-furan} + CH_2=CHCCH_3 \longrightarrow R\text{-furan-}CH_2CH_2CCH_3$$

Furan undergoes the Diels–Alder reaction, e.g., with maleic anhydride:

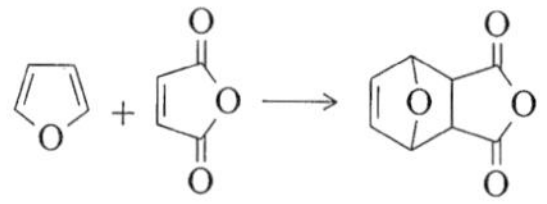

The radical-catalyzed polymerization of furan and maleic anhydride yields a 1:1 furan–maleic anhydride copolymer [7], [8]. The structure of the product, as shown by ^{1}H NMR, is that of an unsaturated alternating copolymer [8], [9].

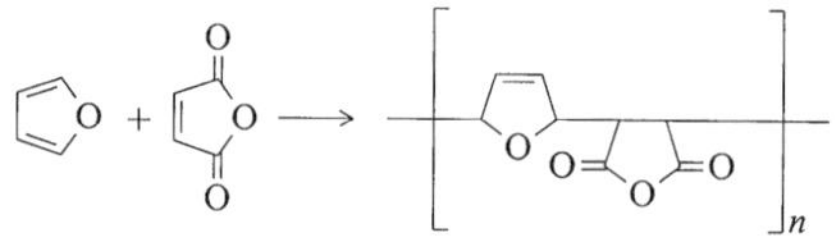

These copolymers are used as complex-forming agents for copper(II) ions.

The balance between aromatic and aliphatic character in substituted furans is markedly affected by substituents. Furan itself, halofurans, alkoxyfurans, furfuryl esters and ethers, and furfural diacetate, for example, react as dienes in the Diels–Alder reaction. On the other hand, furans with strongly electron-withdrawing substituents (e.g., furfural, 2-furoic acid [88-14-2], and nitrofurans) do not react even with the strongest dienophiles. Acid hydrolysis of some furan derivatives yields open-chain dicarbonyl compounds. Furans with electron withdrawing substituents are quite stable to acid cleavage because the substituent deactivates the ring toward electrophiles but renders it vulnerable to nucleophilic attack.

Some apparent substitution reactions occur by an addition–elimination mechanism, especially at low temperature. For example, furan is converted at −5 °C into 2-bromofuran in good yield; at −50 °C, addition products can be detected.

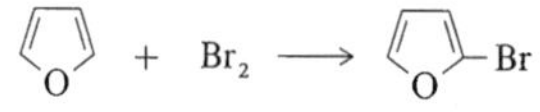

$$\text{furan} + Br_2 \longrightarrow \text{furan-}Br$$

Nitration in acetic anhydride to give 2-nitrofuran also occurs via an addition–elimination process:

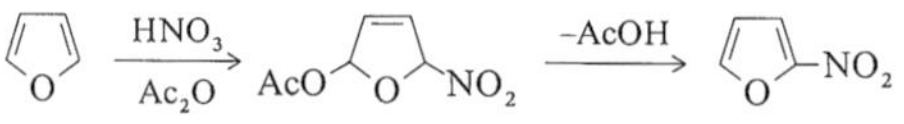

$$\text{furan} \xrightarrow[Ac_2O]{HNO_3} AcO\text{-furan-}NO_2 \xrightarrow{-AcOH} \text{furan-}NO_2$$

Commercially, the most significant reaction is the catalytic hydrogenation of furan to tetrahydrofuran (→Tetrahydrofuran). Reduction occurs under both liquid- and vapor-phase conditions. Nickel catalysts are preferred; noble metals may also be employed, but considerable hydrogenolysis occurs as a side reaction.

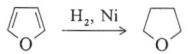

$$\text{furan} \xrightarrow{H_2,\ Ni} \text{tetrahydrofuran}$$

Furan slowly oxidizes in air to form peroxides, whose formation can be inhibited by the addition of hydroquinone.

The oxygen atom of the furan ring can be replaced by nitrogen or sulfur (→Pyrrole; →Thiophene). Passage of furan and ammonia over alumina [1344-28-1] at 400 °C gives pyrrole [109-97-7] [10], [11]:

$$\text{furan} + NH_3 \xrightarrow{Al_2O_3} \text{pyrrole}$$

With hydrogen sulfide, in the presence of alumina at 400–450 °C, furan is converted into thiophene (Yur'ev reaction) [12], [13]. Various furan derivatives, such as alcohols, ketones, esters, and alkylfurans, are similarly converted by hydrogen sulfide into the corresponding substituted thiophenes at room temperature, with yields in excess of 80 % [14]. A variety of substituted thiophenes is thus available:

$$\text{furan-R} + H_2S \xrightarrow{Al_2O_3} \text{thiophene-R}$$

Production. Furan is produced commercially by the decarbonylation of furfural (Chap. 3) over a noble metal catalyst, preferably palladium on charcoal.

Quality Specifications and Analysis. Furan is assayed by gas chromatography. A typical quality specification for a U.S. manufacturer is given below:

Furan (min.), wt %	99.0
Methylfuran (max.), wt %	0.2
Tetrahydrofuran (max.), wt %	0.05
Furfural (max.), wt %	0.05
Water (max.), wt %	0.2
Peroxide* (max.), wt %	0.015
Butylated hydroxytoluene, wt %	0.025–0.040

* Calculated as H_2O_2.

Handling, Storage, and Transportation. Furan is potentially hazardous. Because of its low boiling point, a dangerous concentration of vapor can build up easily unless work areas are ventilated adequately.

Furan is classified as a flammable liquid and in the United States must be packed and shipped according to Interstate Commerce Commission (ICC) regulations. Each container must carry the red ICC label and be identified clearly. In case of fire or explosion involving furan, carbon dioxide or carbon tetrachloride may be used to extinguish it. Furan is shipped in drums, cans, and tank cars. Containers of furan should be able to withstand the vapor pressure exerted during storage under variable climatic conditions.

Furan should be stored in a cool place where the drums will not be heated. In the event of spillage, only persons wearing protective clothing and respiratory equipment should be allowed in the area.

Leaking containers should be removed to a well-ventilated area, and the furan should be transferred to undamaged containers.

Drums of furan may be emptied safely by using hand pumps or electrical pumps with explosion-proof motors. Gravity or pressure pumps must never be used. Drums must be earthed to prevent static discharge.

Uses. Furan is used in the production of pharmaceuticals, agricultural chemicals, stabilizers, and fine chemicals. Its most significant industrial use is in the production of tetrahydrofuran, thiophene, and pyrrole.

3. Furfural

Furfural [*98-01-1*], 2-furancarboxaldehyde, 2-furaldehyde, furfuraldehyde, $C_5H_4O_2$, M_r 96.08, is the starting material for the production of various industrially important furans.

Physical Properties. When freshly distilled, furfural is a colorless liquid with a pungent, aromatic odor reminiscent of almonds. Although it darkens appreciably on exposure to air, furfural has excellent thermal stability in the absence of oxygen. Up to 230 °C, exposure for many hours is required to produce detectable changes in its physical properties, with the exception of color. Some physical properties of furfural are listed in Table 1 (see also [15]).

Furfural has exceptional solvent properties, with low viscosity over a wide temperature range. It is miscible with most common organic solvents but only slightly miscible with saturated hydrocarbons, which makes it a selective solvent, its major industrial use. The solubilities of representative organic compounds and selected thermoplastic resins in furfural are given in [15]. With the exception of zinc chloride and ferric chloride hexahydrate, inorganic compounds are insoluble in furfural.

The vapor pressure of furfural versus temperature is as follows:

t, °C	55	75	95	105	130	150
p, kPa	1.97	4.61	11.18	19.7	39.5	79.0

Chemical Properties. The chemistry of furfural is that of an aromatic aldehyde (→Benzaldehyde), with other reactions due to the dienic character of the furan ring. Furfural can be reduced to furfuryl alcohol (Chap. 4), oxidized to furoic acid, and decarbonylated to furan (see Chap. 2); Cannizzaro reaction with alkali gives furfuryl alcohol and sodium furoate [*17256-62-1*]. Reaction of furfural with ammonia and hydrogen in the presence of nickel or nickel–cobalt catalyst yields furfurylamine [*4795-29-3*].

$$\text{furfural-CHO} + NH_3 + H_2 \xrightarrow{Ni} \text{tetrahydrofurfuryl-NH}_2$$

In the presence of NaCN, furfural dimerizes to furoin [552-86-3], the furan analogue of benzoin. Furfural condenses with compounds possessing active methylene groups, such as aliphatic carboxylic esters, ketones, and nitriles, to give α, β-unsaturated esters, ketones, or nitriles, respectively. With phenols, urea, or furfuryl alcohol, polymers (resins) are obtained [16]–[18].

Catalytic vapor-phase oxidation of furfural yields maleic acid:

$$\text{furfural-CHO} \longrightarrow HOOC\text{-CH=CH-}COOH$$

Furfural can be halogenated or nitrated, under carefully controlled conditions, with introduction of the substituent at the vacant 5-position:

$$\text{furfural-CHO} \longrightarrow X\text{-furan-CHO}$$
$$X = Br, Cl, NO_2$$

The furan ring of furfural is susceptible to attack by atmospheric oxygen, leading to acidic autoxidation products; this reaction is inhibited by the addition of trace amounts of base, for example, a tertiary amine.

Resources, Raw Materials. Furfural, an industrially important chemical, is unusual in that it is produced from renewable agricultural sources such as fibrous residues of food crops. Cereal grasses or byproducts such as corncobs and the hulls of cottonseeds, oats, and rice constitute one large natural feedstock; others include bagasse (a byproduct of sugarcane harvesting), wood, and wood products. The pentosan polysaccharides xylan and arabinan are the major precursors of furfural; they are almost as widely distributed in nature as cellulose.

The chief constituent of the pentosan fraction is xylan, a polysaccharide with backbone chains of β-1,4-linked D-xylopyranose units [19]. Together with a small amount of arabinan, a highly branched polysaccharide of 1,3- and 1,5-linked α-L-arabinofuranose units [20], xylan accounts for 25–30% of cereal straw and grains, ca. 15–25% of deciduous wood, and 5–15% of coniferous wood, based on the dry matter of these plants [21]. Only a few pentosan-containing substances have been used to manufacture furfural. The more important of these are listed in Table 2.

The recoverable yield of furfural is considerably less than the potential values given in Table 2. Viable commercial production of furfural requires a raw material with a minimum pentosan content of 18–20% (dry weight). The raw material should also be available in large amounts within a limited radius of the furfural-producing plant.

The mechanism of formation of furfural from C_5 sugars or their precursors has not been established unequivocally. The potential for side reactions complicates the recovery of furfural and limits its practical yield, which is far from quantitative under any commercial mode of operation. The economics of manufacturing furfural depend on

Table 2. Available furfural raw materials

Raw material	Pentosan content, wt%*	Average furfural content, wt%**
Acacia wood (after tannin extraction)	20	
Bagasse	25–27	17.4
Bagasse pith	25	
Birch wood (after tannin extraction)	25	
Chestnut husks	11	
Chestnut wood (after tannin extraction)	16	
Corncobs (stoned)	30–32	23.4
Cornhusks	30–33	
Cornstalks	24	
Cotton husks	23–28	18.6
Flax stalks	16–19	
Hemp husks	15	
Oak wood (after tannin extraction)	20–21	
Oat husks	40	22.3
Peanut husks	14–17	
Rice husks	16–18	11.4
Sansa	22–25	
Sunflower husks	30–33	
Wheat chaff	18	

* Based on dry weight.
** Method of the Association of Official Analytical Chemists.

the availability, price, and pentosan content of the raw material and on the cost of collecting, transporting, storing, and handling the raw material. Estimated furfural production capacities for various countries, together with the respective feedstock, are given in Table 3.

Production. Furfural is produced commercially in batch or continuous digesters [23]. The pentosans (xylan) are hydrolyzed to pentoses (xylose), which are subsequently cyclodehydrated to furfural :

$$(C_5H_8O_4)_n \xrightarrow{H_3O^+} \text{Pentose} \xrightarrow{-3\,H_2O} \text{Furfural}$$

In the conventional furfural manufacturing process (Fig. 1), the raw material is charged to the digester and treated with strong mineral acid. High-pressure steam is introduced, and after operating temperature and pressure are attained, furfural is recovered by steam distillation. The condensed reactor vapors are fed to a stripping column from which an enriched furfural–water distillate is withdrawn to a decanter, where separation into two layers occurs. The furfural-rich layer, containing about 6% water, is processed further to obtain furfural; the water-rich layer (containing about 8% furfural) is recycled.

Another process, commercialized in 1968 by Rosenlew in Finland, was first used to manufacture furfural from birch chips; the process has now been modified to handle other pentosan-containing raw

Table 3. Feedstocks and estimated annual capacity of furfural

Country	Capacity, 10^3 t	Feedstock
Argentina	1.8	tanning byproduct
Brazil	4–6	bagasse
Dominican Republic	32	bagasse
Mexico	1.8	corncobs
United States	77	bagasse, corncobs, oat hulls, rice hulls
Austria	2	cellulose waste liquor
Federal Republic of Germany	0.5	sulfite waste liquor
Italy	5	chestnut wood, rice hulls
Spain	2.5–5.0	almond shells
Hungary	2	cereal waste
Poland	5	pulping products
USSR	18–20	(not available)
Yugoslavia	2	(not available)
People's Republic of China	30–40	corncobs, cottonseed hulls
India	6	bagasse
South Africa	7.5	bagasse
Kenya	5	corncobs
Total	199.1–213.1	

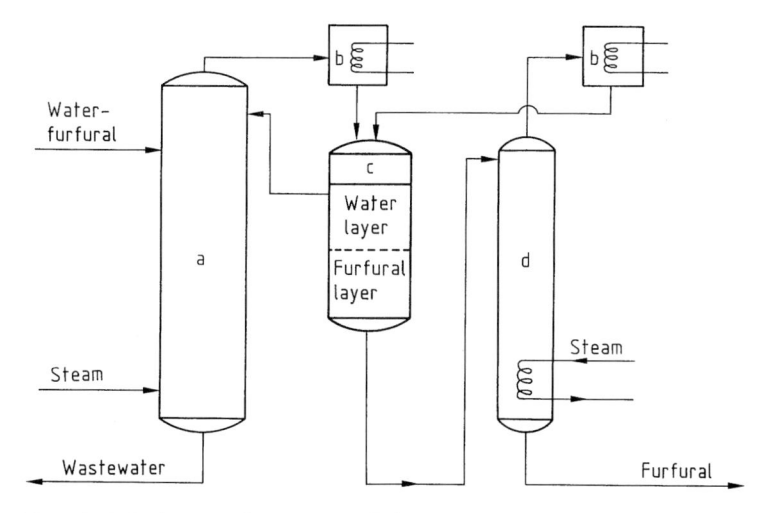

Figure 1. Furfural recovery from aqueous solution
a) Stripping column; b) Condenser; c) Decanter; d) Dehydrating column

materials. Treatment of the wood chips with steam produces acetic acid in situ, which catalyzes hydrolysis of the pentosans to pentoses. This milder acid hydrolysis also produces diacetin (glyceryl diacetate) as a saleable coproduct. At 180 °C and with 90 min reaction, the furfural yield is about 8%, based on the dry weight of birch chips. For each kilogram of furfural approximately 0.6 kg of acetic acid is coproduced. Furfural has been produced by the Rosenlew process in Finland, Spain, Poland, South Africa, and the Philippines, although not all the plants constructed in the past 15 years are currently operating.

Regardless of the process employed, the cellulose in the raw material undergoes only partial degradation and remains with the lignin and ash as a residue that is burned as fuel to support the energy requirements of the furfural plant.

Furfural has been recovered from sulfite wood-pulping liquors in the past, although little is produced this way. Concern over environmental pollution, however, has led to renewed interest in such methods, and pulp producers in Europe are reported to have installed facilities to treat the sulfite liquors and to convert the available pentoses into furfural.

As an alternative to the kraft process [24] of producing pulp, several methods have been reported that use solvents in the manufacture of cellulose pulp for paper production.

Solvent pulping, if commercially viable, could yield products including furfural from the pentoses obtained from hemicellulose and low molecular mass lignin.

The Finnish Sugar Company has reported that it will build a plant in Finland to produce pentose sugars from biomass [25], based on a process developed by Stake Technology (Canada). The pentose sugars can be used to produce either furfural or xylitol. The Finnish Sugar Company is the world's largest producer of the latter.

Quality Specifications and Analysis. Procedures for the quantitative estimation of furfural may be divided into two groups: reactions of the aldehyde group and reactions of the furan ring; none has gained universal acceptance [1]. The method selected depends on the nature of other materials present, many of which can interfere with the determination. Distillation of furfural from an aqueous solution will occasionally eliminate interfering impurities.

The accepted method of the Association of Official Analytical Chemists (AOAC) for the determination of pentosans involves distillation in the presence of hydrochloric acid under carefully controlled conditions, followed by precipitation of the aldehyde with phloroglucinol [*108-73-6*] [26]. Considerable proficiency is required to ensure reproducible results: values obtained range from 98 to 102% of the true value.

A volumetric procedure, based on the reaction of furfural with sodium bisulfite, is useful for determining the aldehyde in the presence of other furan compounds, especially furfuryl alcohol [27].

The Hughes–Acree method is based on the reaction of bromine with the furan ring under carefully controlled conditions of temperature and time [28]; results are accurate to within 1%.

Gas chromatography is used to measure furfural in the presence of other organic compounds and water. Furan derivatives normally present do not interfere. Furfural may also be determined in water by using UV spectroscopy; absorbance is measured at 276 nm. Many UV-absorbing compounds, however, interfere.

A typical quality specification for a U.S. manufacturer is given below:

Furfural (min.), wt%	98.0
Acidity (max.), equivalents/L	0.02
Water content (max.), wt%	0.2
Residue (max.), wt%	0.5

Handling, Storage, and Transportation. Furfural is shipped in steel tank cars, aluminum tank trucks, and steel drums. Above-ground storage (in iron or steel tanks) is satisfactory.

Care must be taken to ensure that all joints are tight and that pump and valve packings are in good condition because furfural is an excellent solvent and penetrant.

Drum lots may be stored for months without appreciable change in physical properties. Change will occur more quickly if the drums have been opened and part of the contents removed, thereby increasing the unit exposure to air. Gradual darkening of color, increase in acidity, and some polymer formation occur when furfural is stored for long periods in contact with air. Autoxidation can be prevented by storing in an oxygen-free atmosphere; storage under nitrogen is recommended.

The flammability of furfural is comparable with that of kerosene or No. 1 fuel oil. Furfural should not be mixed with strong acid or alkali, or stored near strong oxidizing agents.

Uses. Principal uses of furfural are as follows:

1) Production of derivatives: 5-nitrofurfural and 5-nitrofurfural diacetate [29] (to manufacture antimicrobial reagents); furfuryl alcohol [30], furfurylamine [31], furoic acid [32], furan [33], tetrahydrofuran, methylfuran, and methyltetrahydrofuran
2) Selective solvent for separating saturated from unsaturated compounds in petroleum lubricating oil, gas oil, and diesel fuel [34]
3) Extractive distillation of C_4 and C_5 hydrocarbons for the production of synthetic rubber [35] (current availability of butadiene from ethylene production has reduced the use of furfural for this purpose)
4) Decolorizing agent for wood resin
5) Reactive solvent in resins, especially of the phenol–aldehyde type for production of composites with high carbon yield [36], abrasive wheels, and brake linings
6) Reactive solvent in furfuryl alcohol resins for fiberglass-reinforced plastic articles, which are corrosion and flame resistant and generate only low levels of smoke upon ignition [37]

Economic Aspects. The data in Table 4 are world (excluding the Soviet Union) furfural capacity and consumption in 1985, compared with that for 1978. Consumption decreased during this period due to worldwide recession and loss of markets, principally in butadiene and lubricating oil refining. Demand for furfural continues to be significantly less than capacity. Prices for furfural have decreased since 1981; discounting from list price was common in the United States in 1985.

Table 4. Worldwide furfural and furfuryl alcohol capacity and consumption (10^3 t) in 1978 and 1985

	Furfural		Furfuryl alcohol	
	1978	1985	1978	1985
Capacity	230–239	199–213	112–116	96
Consumption	145	133	72–77	79

4. Furfuryl Alcohol

Furfuryl alcohol [*98-00-0*], furanol, 2-furanmethanol, $C_5H_6O_2$, M_r 98.10, is a colorless liquid that slowly darkens on exposure to air.

Physical Properties. Furfuryl alcohol is completely miscible with water and is soluble in most organic solvents with the exception of saturated hydrocarbons. Some physical constants are listed in Table 1. An extensive tabulation of properties is available.

The vapor pressure of furfuryl alcohol versus temperature is as follows:

t, °C	40	60	80	100	120	140
p, kPa	0.24	0.84	2.7	7.1	17.0	36.1

Chemical Properties. Furfuryl alcohol undergoes most of the reactions of a primary alcohol, such as oxidation, esterification, and etherification (→Alcohols, Aliphatic). It is stable in basic media but forms a carbonium ion very readily in acidic media, with resultant condensation or rearrangement reactions. In the presence of strong acid, reaction is highly exothermic and may be violent. The reactivity under acidic conditions has led to commercially important resins. Esterification is generally carried out by transesterification due to the acid sensitivity.

The solubility of some organic compounds and selected thermoplastic resins in furfuryl alcohol is given in [15]. Depending on the conditions, reduction of furfuryl alcohol with hydrogen produces a series of chemicals including tetrahydrofurfuryl alcohol (Chap. 7) [38], 2-methylfuran (Chap. 5) [39], 2-methyltetrahydrofuran (Chap. 6) [40], and open-chain compounds by hydrogenolysis of the furan ring:

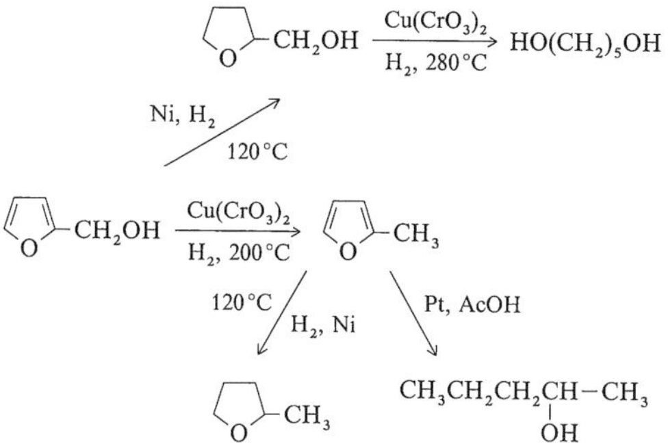

[Reaction scheme: furfuryl alcohol reductions]

- Furfuryl-CH$_2$OH $\xrightarrow[\text{H}_2,\ 280\,°\text{C}]{\text{Cu(CrO}_3)_2}$ HO(CH$_2$)$_5$OH
- Ni, H$_2$, 120 °C → (tetrahydrofurfuryl alcohol)
- Furfuryl-CH$_2$OH $\xrightarrow[\text{H}_2,\ 200\,°\text{C}]{\text{Cu(CrO}_3)_2}$ 2-methylfuran
- H$_2$, Ni, 120 °C → 2-methyltetrahydrofuran
- Pt, AcOH → CH$_3$CH$_2$CH$_2$CH(OH)CH$_3$

Most of these reductions proceed in high yield, so that furfuryl alcohol is a versatile chemical from which a variety of heterocyclic and aliphatic compounds can be produced.

Although polymerization is usually the primary reaction of furfuryl alcohol in acidic media, under controlled conditions other reactions can become important. In dilute aqueous acid, the predominant reaction is hydrolytic ring cleavage – rearrangement to levulinic acid [123-76-2]. In alcoholic media with dilute acid, levulinic esters are formed:

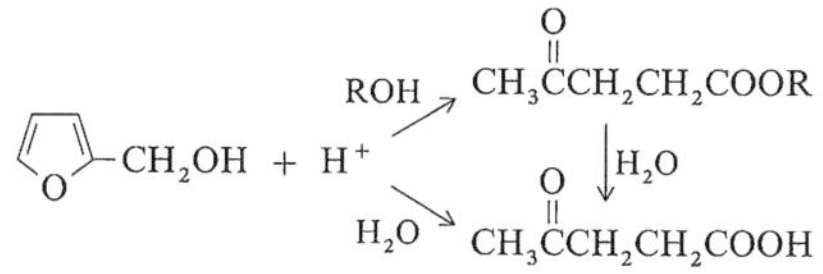

Furfuryl-CH$_2$OH + H$^+$
- $\xrightarrow{\text{ROH}}$ CH$_3$COCH$_2$CH$_2$COOR $\xrightarrow{\text{H}_2\text{O}}$
- $\xrightarrow{\text{H}_2\text{O}}$ CH$_3$COCH$_2$CH$_2$COOH

The reaction in butanol has been used commercially for the production of butyl levulinate, which is then hydrolyzed to levulinic acid. Furfuryl alcohol is hydroxymethylated to 2,5-bis(hydroxymethyl)furan [1883-75-6] in good yield under mild acidic conditions [41].

Furfuryl alcohol also undergoes the Mannich reaction with ammonia or amines and formaldehyde; e.g., with dimethylamine, 5-dimethylaminomethylfurfuryl alcohol [15433-79-1] is produced [42], [43].

By far the most important industrial reaction of furfuryl alcohol is acid-catalyzed resinification.

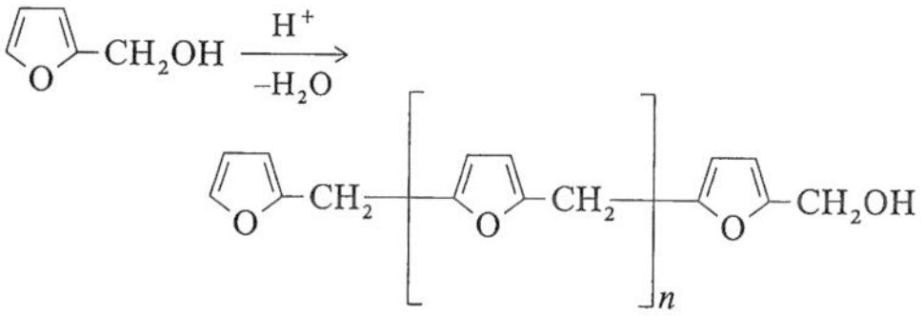

Furfuryl-CH$_2$OH $\xrightarrow[-\text{H}_2\text{O}]{\text{H}^+}$ [–Furfuryl–CH$_2$–Furfuryl–CH$_2$–]$_n$–Furfuryl–CH$_2$OH

Condensation of a methylol group with a second molecule of furfuryl alcohol at the 5-position gives liquid oligomers, consisting mainly of dimers and trimers, with methylene bridges between the furan rings. Mineral acids, strong organic acids, Lewis acids, and acyl halides are active catalysts. The polymerization of furfuryl alcohol is highly exothermic; care must be exercised to keep the reaction under control. This is generally accomplished by regulating the catalyst concentration (i.e., the pH) and by cooling the reaction mixture with refluxing solvent or an external cooling fluid. The homopolymerization of furfuryl alcohol can be carried to a predetermined level as measured by viscosity; the reaction is then stopped by adjusting the pH to 5–8. The liquid resins obtained are stable for a minimum of six months up to 38 °C. Furfuryl alcohol can also be copolymerized with aldehydes such as formaldehyde or furfural, with phenol or urea in the presence of an aldehyde, or with other reactive condensable monomers.

These thermosetting furan polymers are cross-linked by condensation of a terminal methylol group in one oligomer with a methylene bridge in another chain, under the influence of strong acid:

$$\sim\!\!\text{furan}-CH_2-\text{furan}-\sim \;+\; \sim\!\!\text{furan}-CH_2OH$$

$$\xrightarrow[-H_2O]{H^+} \sim\!\!\text{furan}-CH-\text{furan}-\sim$$
$$\underset{\text{furan}}{\overset{|}{CH_2}}$$

With the advent of more advanced spectroscopic and chromatographic analytical techniques, polymerization intermediates have been identified more accurately. Elucidation of the polymer structure and the kinetics of polymerization has resulted in improved understanding of the mechanism of furfuryl alcohol polymerization [44].

Production. Furfuryl alcohol can be produced by both liquid- and vapor-phase hydrogenation of furfural. Selective catalysts based on copper are preferred because they do not promote hydrogenation of the ring.

Quality Specifications and Analysis. A typical quality specification for a U.S. producer is given below:

Furfuryl alcohol (min.), wt%	98.0
Water content (max.), wt%	0.3
Furfural (max.), wt%	0.7
Cloud point * (max.), °C	10.0
* Modification of ASTM D 97-47.	

Gas chromatography is used to determine furfuryl alcohol in the presence of furfural, other organic compounds, and water. The refractive index and specific gravity are also used to determine the quality of furfuryl alcohol.

Handling, Storage, and Transportation. Furfuryl alcohol is shipped in steel tank cars, aluminum tank trucks, and steel drums. The containers must not contain any

Table 5. Estimated annual capacity for furfuryl alcohol

Country	Capacity, 10^3 t
Belgium	40
Brazil	6
France	6
Hungary	4
Japan	13
Turkey	2
United States	25
Total	96

traces of acidic materials because a violent polymerization reaction can occur. Trace amounts of acid may result in polymerization occurring slowly during an induction period, followed by rapid acceleration due to the heat liberated.

Because furfuryl alcohol is a good solvent and penetrant, containers, tanks, lines, and valves used to handle the product should be in excellent condition to avoid possible leakage. The use of hoses made of cross-linked polyethylene, fluorocarbons, ethylene–propylene rubber (EPDM), or styrene–butadiene rubber (SBR) is recommended for transfer of furfuryl alcohol.

Furfuryl alcohol can be stored in ordinary steel tanks above ground.

Steel containers may be unlined or lined with baked phenolic coatings. Containers coated with epoxy resins should not be used because furfuryl alcohol is an excellent solvent for such coatings.

Furfuryl alcohol can be stored up to six months at ambient temperature without appreciable change. When stored longer, the alcohol becomes progressively darker and less water soluble. Exposure to air, acid, or heat causes autoxidation and dehydration. These reactions can be retarded by the addition of small amounts of base or by storage under nitrogen.

The flammability of furfuryl alcohol is comparable with that of kerosene or No. 1 fuel oil; it should not be used or stored near an open flame. If ignited, the fire can be extinguished with water, foam, carbon dioxide, or dry chemicals.

Uses. Uses of furfuryl alcohol include (1) resins for bonding foundry sand to produce cores and molds for metal casting [45]–[47], (2) corrosion-resistant mortar for installing acid-proof brick [48], (3) laminating resins for corrosion-resistant fiberglass-reinforced equipment [32], (4) resins for corrosion-resistant furan polymer concrete, (5) impregnating solutions and carbon binders [49], (6) nonreactive diluent for epoxy resins [50], (7) modifier for phenolic and urea resins, (8) oil-well sand consolidation [51], (9) solvent [52], (10) production of tetrahydrofurfuryl alcohol, and (11) other chemical syntheses.

Economic Aspects. Demand for furfuryl alcohol is significantly less than existing capacity. In 1985, demand was estimated to be approximately 79 000 t. Estimated capacity (1985) is given in Table 5. Consumption is concentrated in the industrialized countries where demand resides primarily in the foundry industry. Furfuryl alcohol resins are the predominant materials in the no-bake sector of foundry resins used to make sand cores and molds for metal casting. The sand bonded with such resins hardens with acid catalysis at ambient temperature, hence the term no-bake.

The recession in the United States in the early 1980s and the shifting of casting production away from the United States to other countries have reduced domestic demand for furfuryl alcohol. Discounting from list price was common in 1985 in the United States.

5. 2-Methylfuran

2-Methylfuran [534-22-5], Sylvan, C_5H_6O, M_r 82.10, is a colorless, mobile liquid.

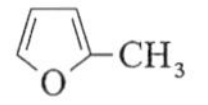

Physical Properties. Some physical properties of 2-methylfuran are listed in Table 1.

Chemical Properties. Generally, the chemical properties of 2-methylfuran are similar to those of furan (see Chap. 2). Alkylfurans are more sensitive to acid-catalyzed polymerization than the parent compound.

2-Methylfuran also undergoes methyl group substitution, e.g., bromination with *N*-bromosuccinimide (NBS) to give furfuryl bromide:

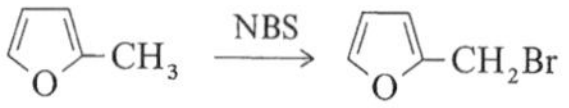

Ring-opening reactions under controlled conditions provide access to 1,4-difunctional compounds. The cleanest ring-opening reactions are those involving hydrogenolysis: 2-pentanol [6032-29-7], is formed in 90% yield by hydrogenolysis of 2-methylfuran in acetic acid over a platinum catalyst [53]; reaction of hydrogen with 2-methylfuran over nickel results in 2-pentanone [107-87-9] [54]:

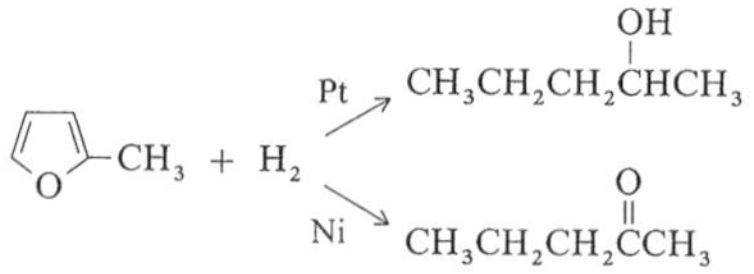

Production. 2-Methylfuran is produced commercially by vapor-phase hydrogenation of furfural over a copper catalyst.

Quality Specifications and Analysis. Commercial 2-methylfuran is analyzed by gas chromatography. Typical quality specifications are given below:

2-Methylfuran (min.), wt%	97.0
Furan (max.), wt%	0.5
Water (max.), wt%	0.25
Others (max.), wt%	0.25
Inhibitor *, wt%	0.01

* Hydroquinone.

Handling, Storage, and Transportation. Procedures used to handle, store, and transport 2-methylfuran are essentially the same as those for furan. 2-Methylfuran is classified as a flammable liquid and must be packed and shipped according to ICC regulations. 2-Methylfuran is shipped in drums.

Uses. 2-Methylfuran is used as a selective solvent and as a feedstock for the production of nitrogen or sulfur heterocycles and functionally substituted aliphatic compounds.

6. 2-Methyltetrahydrofuran

Physical Properties. 2-Methyltetrahydrofuran [96-47-9], $C_5H_{10}O$, M_r 86.13, is a colorless, mobile liquid with high solvent power.

It has the property of inverse water solubility, i.e., solubility decreases with increasing temperature. The two components, water and 2-methyltetrahydrofuran, may be conveniently separated at higher temperature, with 2-methyltetrahydrofuran being recovered from aqueous solution by distillation. Some physical properties of 2-methyltetrahydrofuran are listed in Table 1.

Chemical Properties. 2-Methyltetrahydrofuran has chemical properties similar to those of tetrahydrofuran (→Tetrahydrofuran). Ring cleavage produces difunctional pentane derivatives: reaction with hydrobromic acid yields 1,4-dibromopentane; catalytic dehydrogenation gives 1,3-pentadiene (piperylene).

The 2-methyltetrahydrofuran – lithium hexafluoroarsenate complex has been reported to be a superior electrolyte for the secondary lithium electrode [55].

Quality Specifications and Analysis. 2-Methyltetrahydrofuran is assayed with gas chromatography. Quality specifications are given below:

2-Methyltetrahydrofuran (min.), wt%	99.0
2-Methylfuran (max.), wt%	0.05
Water (max.), wt%	0.10
Pentanols (max.), wt%	0.50

Handling, Storage, and Transportation. 2-Methyltetrahydrofuran requires the usual precautions recommended for volatile, flammable organic compounds.

Uses. 2-Methyltetrahydrofuran is used as a specialty solvent and as a reactant for production of chemicals, e.g., 2-methylpyrrolidine and N-substituted 2-methylpyrrolidines.

7. Tetrahydrofurfuryl Alcohol

Physical Properties. Tetrahydrofurfuryl alcohol [97-99-4], 2-(hydroxymethyl)tetrahydrofuran, 2-tetrahydrofuranmethanol, THFA, $C_5H_{10}O_2$, M_r 102.13, is an industrially impor-tant important solvent with a high boiling point and low toxicity; it is completely miscible with water, is biodegradable, and has a low photochemical oxidation potential. Physical properties are listed in Table 1.

Chemical Properties. The reactions of tetrahydrofurfuryl alcohol are typical of a primary alcohol (→Alcohols, Aliphatic) and a cyclic ether.

Tetrahydrofurfuryl acrylate [2399-48-6], a specialty monomer, has been produced by carbonylation (nickel carbonyl and acetylene) of the alcohol and also by ester interchange.

Tetrahydrofurfuryl alcohol reacts with ammonia to give a variety of nitrogen-containing compounds, depending on the conditions employed. Over a barium hydroxide-promoted nickel–aluminum catalyst, 2-tetrahydrofurfurylamine [4795-29-3] is produced; with palladium on alumina catalyst in the vapor phase (250–300 °C), the principal product at low pressure (275 kPa) is pyridine; at 200 °C over a reduced nickel catalyst, piperidine is obtained in 70 % yield.

Tetrahydrofurfuryl alcohol is oxidized to 2-tetrahydrofurfural [7681-84-7] in good yield by passage with oxygen over silver gauze at 500 °C. When tetrahydrofurfuryl alcohol is passed over alumina-containing catalysts at 200–500 °C, it undergoes dehydration and ring expansion to 2,3-dihydropyran, a major intermediate for 1,5-difunctional open-chain aliphatic compounds.

$$\text{(tetrahydrofurfuryl alcohol)} \xrightarrow{-H_2O} \text{(dihydropyran)}$$

Production. Tetrahydrofurfuryl alcohol is produced commercially by vapor-phase hydrogenation of furfuryl alcohol with a nickel catalyst.

Quality Specifications and Analysis. Tetrahydrofurfuryl alcohol is assayed by gas chromatography. Color (APHA; American Public Health Association) is determined by using a modification of ASTM D 1045-58. A typical specification for tetrahydrofurfuryl alcohol is given below:

Tetrahydrofurfuryl alcohol (min.), wt%	98.0
Furfuryl alcohol (max.), wt%	0.1
1,2-Pentanediol (max.), wt%	1.6
Water (max.), wt%	0.3
Color (max.), APHA	50
Inhibitor	
Polygard, wt%	0.025
Sodium borohydride, wt%	0.005

Handling, Storage, and Transportation. Tetrahydrofurfuryl alcohol is shipped in drums, aluminum tank trucks, or standard steel tank cars. It can be stored in mild steel tanks under ambient conditions, with balloon protectors against vapor escape. Steel piping is normally used because tetrahydrofurfuryl alcohol is an excellent solvent for many resins and organic materials. Skin and eye exposure should be avoided.

Uses. Tetrahydrofurfuryl alcohol is used as a specialty solvent for dyes, resins, lacquers, pesticides, and herbicides, and in commercial and industrial cleaners. The major use of tetrahydrofurfuryl alcohol is in proprietary stripping formulations employed extensively in the automotive and equipment industries to remove protective coatings, paint, and grime prior to final painting and finishing. Stripping formulations to remove epoxy coatings are also industrially important. Tetrahydrofurfuryl alcohol is also used as an intermediate in the synthesis of pharmaceuticals and fine chemicals.

8. 2,3-Dihydropyran

2,3-Dihydropyran [*110-87-2*], 3,4-dihydro-2*H*-pyran, C_5H_8O, M_r 84.11, is a colorless, mobile liquid.

Physical Properties. Some physical properties of dihydropyran are listed in Table 1.

Chemical Properties. Dihydropyran reacts with alcohols under mild acid catalysis to form tetrahydropyranyl ethers, which are useful protecting groups for alcohols.

Tetrahydropyranyl ethers are stable to alkali, organolithium and Grignard reagents, lithium aluminum hydride, acetic anhydride, and chromium trioxide; the alcohol is easily regenerated by treatment with dilute acid.

Dihydropyran self-polymerizes to give a homopolymer or reacts with other unsaturated monomers to form copolymers. No significant use has been found for these polymers.

Production. Dihydropyran is produced by the dehydration of tetrahydrofurfuryl alcohol over activated alumina catalyst at 200–500 °C.

Quality Specifications and Analysis. Dihydropyran is assayed by gas chromatography. Typically, dihydropyran is sold at a minimum purity of 97%, with a maximum water content of 0.2%.

Handling, Storage, and Transportation. Dihydropyran is flammable and an explosion hazard. The vapor may travel a considerable distance to a source of ignition, and flashback may occur. In case of fire, carbon dioxide, dry chemical powder, or polymer foam is used. Water may be effective for cooling but may not effect extinction. Dihydropyran is incompatible with oxidizing agents and strong acids. It is stored and shipped in steel drums.

Uses. Dihydropyran is used as a protecting group for a number of functional groups including hydroxyl, carboxyl, and thiols.

9. Benzofuran

Benzofuran [271-89-6], coumarone, C_8H_6O, was obtained by R. Fittig and G. Ebert in 1882, by calcining coumarilic acid (from coumarin) with lime. In 1890, G. Kraemer and A. Spilker discovered benzofuran in coal tar.

Physical Properties. Benzofuran, M_r 118.14, bp (101.3 kPa) 171.4 °C, mp –28.9 °C, ϱ (20 °C) 1.0948 g/cm^3, is a colorless liquid with an aromatic odor; it is volatile in steam, soluble in ethanol, and insoluble in water.

Chemical Properties. Benzofuran polymerizes slowly at ambient temperature, more rapidly with heat and in the presence of an acidic catalyst; it is resistant to aqueous alkali.

Production. Coal tar contains ca. 0.01 % benzofuran, which can be concentrated through distillation to ca. 15 % in a light oil fraction with a boiling range of 165 – 175 °C. It is recovered either by precipitation as the dibromide or the picrate, or by sulfonation in acetic anhydride, followed by decomposition of the benzofuran derivatives. If the benzofuran fraction is largely free from indene, benzofuran can be concentrated further through either selective extraction or azeotropic distillation with alkylene glycols.

Benzofuran can be synthesized by dehydrogenation and cyclization of 2-ethylphenol.

Uses. Benzofuran, together with indene, is the raw material used for the production of indene – coumarone resins from light oil fractions of coal tar. Normal commercial resins contain up to 10 % benzofuran polymers. No large-scale commercial use has been found for pure benzofuran.

10. Dibenzofuran

Dibenzofuran [*132-64-9*], diphenylene oxide, C$_{12}$H$_8$O, was synthesised for the first time by C. LESIMPLE in 1866 from triphenyl phosphate and lime; its constitution was recognized in 1871 by W. HOFFMEISTER, and it was found in coal tar by G. KRAEMER and R. WEISSGERBER in 1901.

Physical Properties. Dibenzofuran, M_r 168.18, bp (101.3 kPa) 287 °C, ϱ (99.3 °C) 1.0886 g/cm^3, mp 82.8 °C (a higher melting point indicates impurities of fluorene or acenaphthene), consists of colorless crystals with a weak, characteristic odor. It is resistant for a short time in an inert atmosphere to > 600 °C. Dibenzofuran is volatile in steam; sublimable; readily soluble in diethyl ether, benzene, and acetic acid; fairly soluble in ethanol; insoluble in water; its enthalpy of vaporization (130 – 140 °C) is 66.4 kJ/mol, enthalpy of combustion 5882 kJ/mol, dielectric constant (100 °C) 3.0, and dipole moment (liquid) 0.9 D.

Chemical Properties. The skeleton of the dibenzofuran molecule combines that of both biphenyl and diphenyl ether. In comparison with biphenyl, the reactivity of dibenzofuran to electrophilic substitution is lower. The 2- and 2,8-derivatives can be prepared by halogenation, sulfonation, Friedel–Crafts reaction, Gattermann's aldehyde synthesis, and chloromethylation; the 3- and 3,8-derivatives are obtained by nitration; the 4- and 4,6-positions exclusively by metallation and mercuration. In the 1-position, substituted derivatives can be formed only by indirect methods or by radical substitution. If dibenzofuran is already substituted, this substituent determines whether a new substituent is homonuclear, i.e., in the same ring (e.g., hydroxyl, amino, ether groups), or heteronuclear (e.g., halogen, nitro, acyl groups). Examples of opening the furan nucleus include fusion with potassium hydroxide at 280–300 °C to give 2,2'-dihydroxybiphenyl, reaction with potassium metal to form 2-hydroxybiphenyl, or conversion to phenol with aluminum chloride at 140 °C.

Production. Dibenzofuran is present in coal tar at a level of ca. 1%. It is recovered from a wash oil fraction that boils between 275 and 290 °C (containing ca. 30% dibenzofuran). A redistillation step is largely successful in separating dibenzofuran from acenaphthene, which boils 7 °C lower. The technically pure compound is obtained through crystallization of the redistilled fraction. In admixture with ca. 60% acenaphthene, dibenzofuran forms a eutectic and, with fluorene, which boils about 13 °C higher, a continuous series of mixed crystals [56].

Dibenzofuran can be synthesized by dehydrogenation of phenol at 450 °C [57]; by oxidative dehydrogenation of phenol and cyclization of 2-cyclohexenylcyclohexanone [58]; by cyclization of diphenyl ether in the presence of palladium(II) acetate [59]; or by the pyrolysis of phthalic anhydride with furan [60]. As sufficient quantities of dibenzofuran can be produced from coal tar, these syntheses are not industrially applied. Some dibenzofuran derivatives can be synthesized better by ring closure than by substitution of dibenzofuran.

Uses. Due to its high heat resistance, dibenzofuran, together with biphenyl, is a component of heat-transfer oils. When combined with methylnaphthalenes, it is suitable as a carrier for dyeing and printing textiles [61]. Some of its derivatives serve as intermediates for naphtol AS and chromium complex dyes, e.g., 2,2'-dihydroxybiphenyl and 2-hydroxydibenzofuran-3-carboxylic acid (from 2-hydroxydibenzofuran through Kolbe–Schmitt carboxylation).

Derivatives of dibenzofuran phosphonic acid can be used as antioxidants in plastics [62]. A range of polymers can be produced from dibenzofuran, e.g., heat-resistant polyarylacetylene [63] and quinoxaline polymers [64] or photoconductive polymers for electrophotography [65]. Dibenzofuroyl silicon polymers and monomers can be employed as perfume oil bases, in cosmetics, and as fluids for the production of elastomers and resins [66]. Dibenzofuran derivatives have been proposed as pharmaceuticals, e.g., dibenzofuranyl aminoalcohols as sedatives and tranquilizers [67] and dibenzofurylpropionic acid and derivatives as antiphlogistics [68].

11. Toxicology and Occupational Health

Furfural has been reported to be a polytropic toxin with a primary effect on the central nervous system and the liver [69]. Furfural vapor is irritating to the mucous membranes; its low volatility reduces the risk of exposure.

Furfural penetrates intact skin (rabbits) but with no fatal effect up to 500 mg/kg. Several drops of furfural in standard eye tests (rabbit) cause significant irritation but no irreversible damage. Specific data include:

LD_{50} (oral), mg/kg	400 (mouse);
	127 (rat);
	2300 (dog);
	541 (guinea pig)
LD_{50} (i.v.), mg/kg	152 (mouse);
	250 (dog)
LD_{50} (i.m.), mg/kg	78 (rabbit)
LD_{50} (i.p.), mg/kg	102 (mouse)
LD_{50} (inhalation, 1-h exposure), ppm	1037 (rat)
Occupational exposure limit (8 h, TWA), ppm	2

Furfural is metabolized rapidly and excreted in the urine as furoylglycine, with a smaller proportion (0.5 – 5%) as 2-furanacryluric acid [70].

Furfuryl alcohol is toxic, and irritating to the eyes and skin; it may penetrate intact skin. Values of LD_{50} are 160 mg/kg (mouse, oral); 275 mg/kg (rat, oral); 650 mg/kg (rat, i.p.); and 650 mg/kg (rabbit, i.v.). The LC_{50} has been determined as 592 ppm (rat, inhalation, 1 h). The occupational exposure limit is 50 ppm (8 h, TWA) [45].

Some operations may create aerosols containing furfuryl alcohol, and properly engineered ventilation must be provided. Additional information on the toxicity and handling of furfuryl alcohol is available [71].

Furan. Exposure to furan produces a narcotic effect. Its low boiling point requires that adequate ventilation be provided in areas where furan is handled. Liquid furan can be absorbed through the skin. People with circulatory disorders, abnormal liver conditions, or chronic gastrointestinal complaints should not work with furan. Symptoms of furan exposure include fatigue, headache, and gastrointestinal disturbance. Furan is not a poison by DOT definition with regard to inhalation toxicity [70]. The LD_{50} values are 7 mg/kg (mouse, i.p.) and 5200 mg/kg (rat, i.p.); the LC_{50} is 3464 ppm (rat, inhalation, 1 h).

2-Methylfuran is not a poison by DOT definition with regard to inhalation toxicity [72]. Symptoms of overexposure are nausea and temporary lowering of blood pressure. The LD_{50} is 167 mg/kg (rat, oral); the LC_{50} is 1485 ppm (rat, inhalation, 1 h).

Benzofuran showed acute toxic effects in rabbits when administered orally at a dosage of 1 g [73]. The LD_{50} (mice, intraperitoneal) is 500 mg/kg [74]. (A carcinogenicity study on benzofuran in rats and mice was performed in the United States under the National Toxicology Program. The results have not yet been published [75].)

Dibenzofuran caused no adverse effects when fed to rats at high dosage for 200 d (0.025 – 0.4 % of feed) [76]. In skin- and eye-irritant tests (24-h exposure, FDA method), dibenzofuran proved to be nonirritating [77]. On exposure of fish (mullet sp.) to 750 µg/L for 6 and 24 d, dibenzofuran caused weak biochemical stress responses in brain, interrenal gland, and liver. This dosage was one-third the one-week LC_{50} level [78], [79]. Dibenzofuran is degraded by microbes [80], [81]. The microbial fate of benzofuran is unknown [82], although it is metabolized in mammals (rats) by oxidation, followed by conjugation with sulfate and excretion in urine [73]. No TLV has been stipulated for either substance.

Tetrahydrofurfuryl alcohol is not considered an environmentally hazardous substance. It is completely miscible with water and at levels up to 0.5 mg/L (0.5 ppm) is biodegraded. Its vapor density is 3.52 (air = 1), and 1 ppm = 4.18 mg/m^3. Under Los Angeles County Rule 66, it is not considered a photochemical oxidant.

Tetrahydrofurfuryl alcohol is slightly irritating to the skin and severely irritating to the eye of albino rabbits. Young adult albino rats were exposed to air containing 4700 ppm tetrahydrofurfuryl alcohol for 4 h. No deaths occurred during the subsequent 14-day observation period; body weight gains were within normal limits. Massive doses of the vapor produce a narcotic effect in animals. The values of LD_{50} are 2300 mg/kg (mouse, oral); 2500 mg/kg (rat, oral); and 3000 mg/kg (guinea pig, oral).

Dihydropyran is harmful if inhaled, ingested, or absorbed by the skin. The vapor or mist is irritating to the eye, mucous membranes, and upper respiratory tract. The LDLo for dihydropyran is 256 mg/kg (rat, oral) [45]. No TWA exposure data are available.

12. References

General References

Beilstein, **17,** 54; **17(1),** 24; **17(2),** 57; **17(3/4),** 478 – 480.

H.-G. Franck, G. Collin: *Steinkohlenteer,* Springer Verlag, Berlin – Heidelberg – New York 1968, pp. 11, 41, 93.

A. Mustafa: *Benzofurans. Chemistry of Heterocyclic Compounds (Weissberger),* vol. **29,** Wiley Interscience, New York 1974.

P. Cagniant, D. Cagniant in A. R. Katritzky, A. J. Boulton (eds.) *Advances in Heterocyclic Chemistry,* vol. **18,** Academic Press, New York – San Francisco – London 1975, pp. 337 – 482.

Beilstein, **17,** 70; **17(1),** 30; **17(2),** 67; **17(3/4),** 585 – 586.

H.-G. Franck, G. Collin: *Steinkohlenteer,* Springer Verlag, Berlin – Heidelberg – New York 1968, pp. 10 – 12, 21 – 26, 58 – 59, 180 – 181.

M. V. Sargent, P. O. Stransky in A. R. Katritzky (ed.): *Advances in Heterocyclic Chemistry,* vol. **35,** Academic Press, Orlando 1984, pp. 2 – 81.

Specific References

[1]   A. P. Dunlop, F. N. Peters: *The Furans,* Reinhold Publ. Corp., New York 1953.
[2]   *Ullmann,* 4th ed., **12,** 15 – 22.
[3]   *Kirk-Othmer,* 3rd ed., **11,** 499 – 527.
[4]   R. F. Bradley, J. Bakker, K. Tsuchiya: *Furfural, Chemical Economics Handbook,* SRI International, 1986.
[5]   N. Irving Sax: *Dangerous Properties of Industrial Materials,* 6th ed., Van Nostrand Rheinhold, New York 1984.
[6]   M. J. Cook, A. R. Katritzky, P. Linda, *Adv. Heterocycl. Chem.* **17** (1974) 255 – 356.
[7]   N. G. Gaylord, S. Maiti, B. K. Patnaik, A. Takahashi, K. Birendra, *J. Macromol. Sci. Chem.* **6** (1972) no. 8, 1459 – 1480.
[8]   N. G. Gaylord, M. Martan, A. B. Deshpande, *J. Polym. Sci. Polym. Chem. Ed.* **16** (1978) no. 7, 1527 – 1537.
[9]   B. Kamo, I. Morita, S. Horie, S. Furusawa, *Polym. J.* **6** (1974) no. 2, 121 – 125.
[10] E. I. DuPont de Nemours & Co., US 2 600 289, 1952 (C. A. Bordner).
[11] Daicel Chemical Industries, JP-Kokai 58/90548 A2[83/90548] 30 May 1983.
[12] Yu K. Yur'ev, V. A. Tranova, *J. Gen. Chem. USSR Engl. Transl.,* **18** (1940) 31 – 35; *Chem. Abstr.,* **34** 4733.
[13] Yu K. Yur'ev, *Uch. Zap. Mosk. Gos. Univ.* **175,** Org. Khim. 1956 159 – 162; *Chem. Abstr.,* **51** 104 70 b.
[14] V. G. Kharchenko, I. A. Markushina, T. I. Gubina, *Dokl. Akad. Nauk. SSSR* **255** (1980) no. 5, 601 – 602.
[15] E. W. Flick: *Industrial Solvents Handbook,* 3rd ed., Noyes Data Corp. Park Ridge, N.J., 1985.
[16] P. M. Heertjes, G. J. Kok, *Delft Prog. Rep. Ser. A* **1** (1974) no. 2, 59 – 63.
[17] A. Gandini, *Adv. Polym. Sci.* **25** (1977) 47 – 96.
[18] A. Gandini: *Encyclopedia of Polymer Science;* 2nd ed., vol. **7,** John Wiley, New York 1986.
[19] R. L. Whistler, E. L. Richards: *Hemicelluloses, The Carbohydrates,* vol. **IIA,** Academic Press, New York 1970, 447 – 467.
[20] G. O. Aspinall: *Polysaccharides,* Pergamon Press, Oxford 1970, 103 – 115.
[21] R. I. Browne: *Handbook of Sugar Analyses,* John Wiley & Sons Inc., New York 1912.
[22] H. K. Krivanec, *Manufacturing Guide on Furfural,* United Nations Industrial Development Organization, September 11, 1974, p. 33.
[23] Making and Marketing Furfural: Added Value for Agro-Industrial Wastes, International Trade Center UNCTAD/GATT, Geneva, Switz., 1979.
[24] *Chem. Week,* Jan. 4 (1984) 26 – 27.
[25] *Chem. Mark. Rep.* **230** (1986) no. 2, 7.
[26] *AOAC Methods of Analysis,* 10th ed., Association of Official Analytical Chemists, Washington, D.C., 1965, 336.
[27] A. P. Dunlop, F. Trimble, *Ind. Eng. Chem. Anal. Ed.* **11** (1939) 602 – 606.
[28] E. Hughes, S. F. Acree, *Ind. Eng. Anal. Ed.* **6** (1934) 123 – 126.
[29] H. J. Sanders, *Ind. Eng. Chem.* **47** (1955) 358.

[30] B. W. Bycroft: *Dictionary of Antibiotics and Related Substances,* Chapman and Hall, London – New York 1988.
[31] Mitsubishi Petrochemical Co., US 4 598 159, 1986. (T. Ayusawa, S. Mori, T. Aoki, R. Hamana).
[32] Interox Chemicals Ltd., GB 2 188 927, 1987 (K. Rowbottom).
[33] P. Lejemble, A. Gaset, P. Kalck, *Biomass* **4** (1984) 263.
[34] B. Gillespie, L. W. Manley, C. J. DiPerna, *Chem. Technol.* **8** (1978) 750.
[35] W. D. Peters, R. S. Rogers, *Hydrocarbon Process.* **47** (1968) 11, 131.
[36] QO Chemicals, Inc., Technical Bulletin no. 196.
[37] R. H. Leitheiser, M. E. Londrigan, D. W. Akerberg, K. B. Bozer, *Fire Prev. Sci. Technol.* **22** (1979) 21.
[38] Quaker Oats Co., US 2 838 523, 1958 (A. P. Dunlop, H. Schegulla).
[39] Quaker Oats Co., US 3 020 291, 1962 (A. P. Dunlop, D. G. Manly).
[40] Quaker Oats Co., US 3 021 342; 3 021 343 1962 (D. G. Manly).
[41] BHMF: 2,5-Bis(hydroxymethyl)furan, Technical Bulletin No. 194, QO Chemicals, Oak Brook, IL. 1979.
[42] Smith Kline & French Laboratories, US 4 347 191, 1982 (J. Cooper).
[43] Quaker Oats Company, US 4 219 585; 4 219 486; 4 219 487, 1980 (A. P. Dunlop).
[44] J. Milkovic, G. E. Myers, R. A. Young, *Cellul. Chem. Technol.* **13** (1979) 651 – 672.
[45] R. F. Frankenberg, *Castings (Sydney)* **25** (1979) 87.
[46] A. Thomas, P. V. Allen, *Met. Australas.* **10** (1978) 8.
[47] QO Chemical, Inc., Technical Bulletin WBP (1985).
[48] R. H. Leitheiser, M. E. Londrigan, C. A. Rude, "Furan Resinous Cements," Am. Chem. Soc. Symp. Ser. no. 113, ACS Washington, D.C., 1979.
[49] E. I. DuPont de Nemours & Co., EP 0 251 596, 1988 (M. Katz).
[50] L. Manocha, E. Yasuda, Y. Tanabe, S. Kimura, *Carbon* **26** (1988) 3, 333.
[51] Halliburton Co., US 3 237 691, 1966 (R. R. Koch, J. Ramos).
[52] QO Chemicals, Inc., Technical Bulletin 205 D.
[53] H. A. Smith, J. F. Fuzek, *J. Am. Chem. Soc.* **71** (1949) 415 – 419.
[54] C. L. Wilson, *J. Am. Chem. Soc.* **70** (1948) 1313 – 1315.
[55] V. R. Koch, J. H. Young, *Science (Washington, D.C.)* **204** (1979) May 4, 499 – 501.
[56] G.-P. Blümer, G. Collin, *Erdöl Kohle Erdgas Petrochem.* **36** (1983) 22 – 27.
[57] Monsanto, US 4 008 254, 1975 (D. E. Gross, N. A. Fishel); US 4 013 694 (N. A. Fishel).
[58] Anic SpA., DE 2 552 652, 1975 (P. A. Moggi, G. Iori).
[59] Schering, DE 2 418 503, 1974 (B. Akermark, L. Eberson, E. Jonsson, E. Pettersson); Halcon SD Group Inc., US 4 362 883, 1981 (R. J. Harvey); Ashland Oil Inc., US 4 490 550, 1983 (A.B. Goel).
[60] Standard Oil, US 3 514 458, 1967 (E. K. Fields).
[61] Cassella Farbwerke Mainkur, Rütgerswerke, DE 2 236 551, 1972 (G. Weckler, R. Mildenberg).
[62] Sandoz, DE 2 215 544, 1972 (K. Hofer, G. Tscheulin).
[63] Hercules Inc., US 4 026 859, 1975 (L. C. Cessna).
[64] US Dept. of the Air Force, US 3 792 017, 1972 (F. E. Arnold, F. L. Hedberg, R. F. Kovar).
[65] Azoplate Corp., US 3 294 531, 1960 (H. Schlesinger).
[66] General Electric, US 3 715 370, 1970; US 3 855 241, 1972 (E. V. Wilkus, A. Berger).
[67] Aldrich Chemical Co., US 3 701 786, US 3 716 539, US 3 751 390, 1971 (H. B. Hopps, D. Jackman, J. H. Biel).
[68] Merck Patent GmbH, DE 2 261 745, 1972 (J. Gante, W. Mehrhof, A. Wild).

[69] V. B. Danilov, Ye V. Mel'nikova, *Probl. Gig. Organ. Zdravookhr. Uzb.* **5** (1976) 55–58; *Chem. Abstr.* **90** 146 950x.
[70] J. Flek, V. Sedivec, *Int. Arch. Occup. Environ. Health* 41 (1978) no. 3, 159–168.
[71] Criteria for a Recommended Standard: Occupational Exposure to Furfuryl Alcohol, DHEH (NIOSH) Publ. No. 79–142, Government Printing Office, Washington, D.C., 1979.
[72] QO Chemicals Report, Hazelton Laboratories America Inc., August 26, 1987.
[73] F. Böhm, *Hoppe-Seyler's Z. Physiol. Chem.* **269** (1941) 17–23.
[74] M. Fatome, L. Andrieu, J.-D. Laval, J.-M. Clavel et al. *Eur. J. Med. Chem.-Chim. Ther.* **12** (1977) 383–384.
[75] U.S. Department of Health & Human Services: personal communication.
[76] J. O. Thomas, R. H. Wilson, C. W. Eddy, *Food Res.* **5** (1940) 23–30.
[77] H. Willeitner, H. O. Dieter, *Holz Roh-Werkst.* **42** (1984) 223–232.
[78] P. Thomas, H. W. Wofford, J. M. Neff, *Aquat. Toxicol.* **1** (1981) 329–342.
[79] P. Thomas, H. W. Wofford, *Toxicol. Appl. Pharmacol.* **76** (1984) 172–182.
[80] M. D. Lee, J. T. Wilson, C. H. Ward, *Dev. Ind. Microbiol.* **25** (1984) 557–565.
[81] C. E. Cerniglia, J. C. Morgan, D. T. Gibson, *Biochem. J.* **180** (1979) 175–185.
[82] P. W. Trudgill in D. T. Gibson (ed.): *Microbial Degradation of Organic Compounds*, Marcel Dekker, New York 1984, pp. 295–308.

Gluconic Acid

HELMUT HUSTEDE, Joh. A. Benckiser GmbH, Ladenburg, Federal Republic of Germany
HANS-JOACHIM HABERSTROH, Biochemie Ladenburg GmbH, Ladenburg, Federal Republic of Germany
ELISABETH SCHINZIG, Biochemie Ladenburg GmbH, Ladenburg, Federal Republic of Germany

1.	Introduction	2773	4.4.	Methods Involving Biotechnology	2777
2.	Physical Properties	2774	5.	Uses	2780
3.	Chemical Properties	2775	6.	Economic Aspects	2782
4.	Production	2776	7.	Physiology, Product Specifications	2782
4.1.	Chemical Oxidation	2776			
4.2.	Electrochemical Oxidation	2776			
4.3.	Catalytic Oxidation	2777	8.	References	2783

1. Introduction

D-Gluconic acid [526-95-4], 1,2,3,4,5-pentahydroxypentane-1-carboxylic acid, $C_6H_{12}O_7$, M_r 196.16, was discovered in 1870 by HLASIWETZ and HABERMANN during the oxidation of glucose with chlorine. The substance was isolated in the form of its barium and calcium salts. Several authors subsequently reported that gluconic acid could be obtained by treatment of various mono-, di-, and polysaccharides with oxidizing agents such as elemental halogen, copper(II) or hexacyanoferrate(III) salts, or mercury(II) oxide. Depending on the type of sugar and the oxidant employed, byproducts of the reaction include formic acid, glycolic acid, oxalic acid, and carbon dioxide.

As early as 1880 BOUTROUX recognized that gluconic acid was produced, together with acetic acid, by the oxidative action of *Acetobacter aceti* on glucose. This characteristic was also found to be associated with numerous other bacteria [1]. MOLLIARD was the first to report the presence of gluconic acid in cultures of *Sterigmatocystis nigra*, now known as *Aspergillus niger* [2]. The currently preferred method for preparing gluconic acid and its derivatives with the aid of *Aspergillus* strains is based on the work of a number of authors [3]. The catalytic activity of the enzyme glucose oxidase was first described by MÜLLER [4].

During the 1930s anodic oxidation was suggested for the preparation of calcium gluconate [5]; proposals for catalytic oxidation of glucose with the aid of air or oxygen followed somewhat later [6].

Sources. Gluconic acid is a naturally occurring substance in humans and other organisms (cf. Chap. 7); it is found in food products such as wine and honey.

2. Physical Properties

Free D-gluconic acid is difficult to prepare in crystalline form. According to MILSOM and MEERS [7], crystallization of the anhydrous substance is possible below 30 °C, and a monohydrate has been reported to crystallize at 0–3 °C with a characteristic crystalline structure, *mp* ca. 85 °C [8].

Anhydrous gluconic acid is a white, odorless, crystalline powder, specific rotation $[\alpha]_D^{20}$ −6.7° [9], $[\alpha]_D^{25}$ −5.4° [10].

Literature values for the melting point of gluconic acid fall in the range 120–131 °C. The spread in values is due to the formation of intramolecular anhydrides whose presence lowers the melting point.

Two *lactones* of gluconic acid are known. These exist both in the solid state and also in aqueous solution, where they are in equilibrium with each other and with the free acid.

This complex equilibrium is also influenced by the dissociation of the free acid, which has a reported [10] dissociation constant K_A at 25 °C of 1.99×10^{-4} and a pK_A of 3.70 (3.62 according to [11]).

Lactone hydrolysis and formation are both first order with respect to hydrogen ion concentration, with rate constants (k and b, respectively) related as follows [12]:

$$k_{1,5} + b_{1,5} = c_{H^+} \cdot 5.5 \times 10^{-2}\,\text{s}^{-1}$$

$$k_{1,4} + b_{1,4} = c_{H^+} \cdot 4.3 \times 10^{-4}\,\text{s}^{-1}$$

Thus, equilibrium is established with the 1,5-lactone roughly 100 times faster than with the 1,4-lactone. The following empirical equation describing the kinetics of hydrolysis for the 1,5-lactone in the pH range 0.6–8.2 has been reported [12]:

$$k_{1,5} = c_H + \cdot\ 4.7 \times 10^{-2} + c_{OH^-} \cdot 4 \times 10^3 + 2.5 \times 10^{-4}\ s^{-1}$$

This expression indicates that the OH⁻ ion strongly catalyzes the opening of the lactone ring, while the H⁺ ion is considerably less effective. In the region of neutrality, hydrolysis is quite slow. The rate of hydrolysis over the narrower pH range 3–5 is given by

$$-dc_L/dt = kc_L$$

where c_L is the concentration of the 1,5-lactone and $k = 2.3 \times 10^{-4}\ s^{-1}$. The activation energy for the process is reported to be 62.8 kJ/mol [13].

Gluconic acid is quite soluble in water. At 20 °C, commercial 50 % gluconic acid solution has a pH of 1.82 and a density of 1.23 g/cm³. In this concentration region, solution density increases almost linearly with increasing concentration. The acid is only slightly soluble in ethanol and virtually insoluble in nonpolar solvents.

Storing gluconic acid over a desiccant at room temperature or heating it above 50 °C leads to the formation of lactones. Pyrolysis occurs above 200 °C.

Gluconic acid 1,5-lactone [90-80-2] is a white, crystalline substance with a faintly sweet taste, mp 153 °C, $[\alpha]_D^{20}$ +66.2 ° [13]. On standing, both the pH and the specific rotation of a lactone solution decrease as a result of hydrolysis, becoming constant only after equilibrium is established. Approximately 90 g of 1,5-lactone dissolves in 100 mL of water at 20 °C; this increases to about 205 g at 50 °C. Only small amounts dissolve in organic solvents. The *1,4-lactone* [1198-69-2] crystallizes as fine needles, mp 134–136 °C, $[\alpha]_D^{20}$ +67.8 ° [13].

3. Chemical Properties

The action of oxidizing agents such as nitric acid or hydrogen peroxide on either the lactones or the calcium salt of gluconic acid leads under mild conditions to mixtures of oxogluconic acids, with the carbonyl groups at the 2- and the 5-positions. 2-Oxo-D-gluconic acid [669-90-9] is the principal product of anodic oxidation, oxidation with sodium chlorate in acid solution, or fermentation by species such as *Acetobacter suboxydans, A. xylinum,* or *A. gluconicum.* Treatment with concentrated nitric acid or with N_2O_4 results in formation of glucaric acid, HOOC(HCOH)₄COOH [87-73-0].

Hydrogenation of gluconic acid in aqueous solution over a platinum oxide catalyst results in a modest yield of D-glucose, whereas the 1,5-lactone undergoes this reaction in high yield. Refluxing with concentrated hydriodic acid in the presence of red phosphorus causes reduction to hexanoic acid [142-62-1].

The six functional groups of gluconic acid can, in principle, react with a variety of reagents, such as alcohols, acids, etc. Nevertheless, the resulting derivatives tend to be stable only if reaction is complete. Partial reaction leads to nonuniform mixtures that are sensitive to hydrolysis; such reactions have little significance.

In contrast, considerable interest exists in the ability of gluconic acid and its alkali salts to form complexes with polyvalent cations; some of these complexes are very stable (see Chap. 5). Nuclear magnetic resonance spectra suggest that complex formation involves both the carboxyl and the hydroxyl groups.

4. Production

For commercial purposes, D-gluconic acid and its salts are prepared exclusively by the oxidation of glucose or glucose-containing raw materials. The oxidation method may be chemical, electrolytic, catalytic, or biochemical.

4.1. Chemical Oxidation

Chemical methods have the disadvantage of limited specificity even under carefully controlled and optimized reaction conditions, which results in unsatisfactory yields (60–80%) and undesirable byproducts. Isolation and purification of the product are correspondingly difficult. In recent years, hydrogen peroxide [14], ozone [15], and oxygen [16] have been the oxidants of choice, presumably because of ecological problems associated with other oxidizing agents and their reduction products.

4.2. Electrochemical Oxidation

Similar problems accompany electrochemical procedures, most of which are based on the oxidation potential of halogens. The usual approach is to electrolyze a glucose solution containing bromide (ca. 10 mol% relative to glucose) at a current density of 1–20 A/dm^2. Carbonates (preferably calcium carbonate) or hydroxides are added to neutralize the resulting acid [17]. Electrical current yields are 80–86%; the product yield based on glucose is 80–97%. The rapid increase in the cost of electricity in recent years has made the electrolytic process noncompetitive.

4.3. Catalytic Oxidation

In contrast, recent developments have led to great improvements in the technique of oxidizing glucose with oxygen or air and serious consideration has been given to adopting this strategy on an industrial scale. Oxidation is carried out, with the aid of an appropriate catalyst, on a glucose solution whose concentration is 1–2 mol/L. The pH is held between 8 and 11 (preferably 9–10) by continuous addition of alkaline solution.

The catalysts originally employed were finely divided platinum-group metals suspended on activated charcoal, aluminum oxide, or some other carrier [18]. The corresponding reaction mechanism for a Pt–C catalyst has largely been elucidated [19].

The most important observations with respect to the catalytic reaction include a rapid decrease in activity of the catalyst from its initial levels (a decline that cannot be completely reversed despite efforts at reactivation), the need for highly purified glucose solutions, and the formation of various byproducts as a result of insufficient catalyst specificity and the tendency of the substrate to undergo side reactions under the prevailing reaction conditions.

The foregoing disadvantages have recently been circumvented through a series of developments with respect to the catalysts employed. Thus, platinum-group metals have been found to become more active and more selective if they are doped with lead [20], [21], selenium [22], thallium [21], or bismuth [21], [23]. The preferred carrier is activated charcoal.

The catalyst described in [23] permits conversion of a 2 M glucose solution to gluconic acid with oxygen at $50 \pm 1\,°C$ and pH 9.5 ± 0.2 (neutralization with 40% sodium hydroxide solution) in >99.5% yield within 1 h. The process is also highly selective: product purity (based on sodium gluconate) is >99.5%. In addition, no special effort need be taken to purify the glucose solution. Loss of catalytic activity is minimal, permitting the catalyst to be recycled many times without an intervening reactivation step.

The economic viability of the process depends largely on the activity, selectivity, lifetime, and cost of the catalyst, as well as on the measures required for product purification and the energy demands.

Despite earlier suggestions to the contrary, glucose cannot be oxidized photochemically to gluconic acid [24].

4.4. Methods Involving Biotechnology

The principal organisms employed today for large-scale biological preparation of gluconic acid are *Aspergillus niger* and *Gluconobacter suboxydans*.

Aspergillus Niger Process. The biochemistry of gluconic acid formation by *Aspergillus niger* is illustrated in the following diagram [25], [26]:

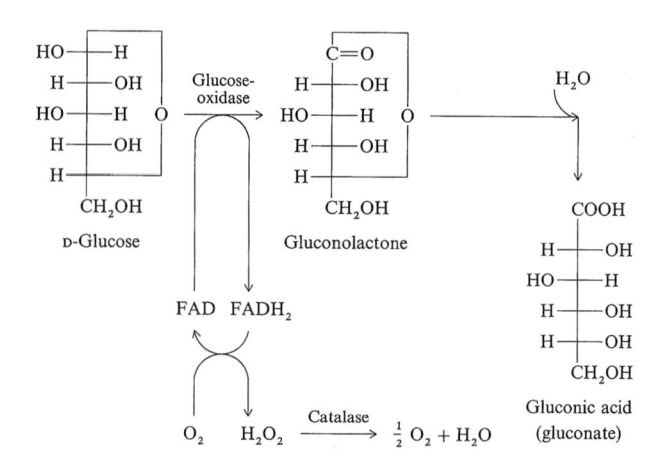

```
HO─┬─H                          C═O
H──┼─OH    Glucose-     H───┬─OH                    H₂O
HO─┼─H  O   oxidase    HO──┼─H   O
H──┼─OH    ─────→       H───┼─OH
H──┤                    H───┤
   │                        │
 CH₂OH                    CH₂OH                    COOH
 D-Glucose              Gluconolactone          H───┬─OH
                                                HO──┼─H
           FAD  FADH₂                           H───┼─OH
              ⤫                                 H───┼─OH
                                                  CH₂OH
                           Catalase           Gluconic acid
           O₂   H₂O₂  ─────────→  ½O₂ + H₂O    (gluconate)
```

where FAD and FADH$_2$ are the oxidized and reduced forms, respectively, of flavine-adenine dinucleotide. β-D-Glucose is oxidized catalytically to glucono-1,5-lactone by the enzyme glucose oxidase. Subsequent hydrolysis of the lactone to gluconic acid is to some extent spontaneous, but the enzyme gluconolactonase also plays a role.

Glucose oxidase (E.C. 1.1.3.4) [9001-37-0], M_r 186 000, contains two FAD moieties as prosthetic groups. These are responsible for the abstraction of hydrogen atoms from glucose (to form FADH$_2$), which ultimately combine with oxygen to produce hydrogen peroxide. This is in turn cleaved to water and oxygen by the enzyme catalase. Both glucose oxidase and catalase are present as endoenzymes in the fungal mycelium [27], [28].

In addition to glucose oxidase, *Aspergillus niger* may have other enzymes that also lead to gluconic acid formation, comparable to the glucose dehydrogenase and phosphomonoesterases found in *A. oryzae* [29], [30].

Metabolic investigation has shown fermentation to be a three-stage process [31]: a brief lag phase, followed by an acceleration phase during which glucose oxidase activity rises nearly logarithmically, and finally a stationary phase, which is characterized by a constant rate of gluconic acid formation. The optimum pH for the reaction lies near 5.6 [32].

Production Parameters. Fermentative synthesis of sodium gluconate requires a sterilized substrate; a typical formulation is as follows: glucose 250–350 g/L, MgSO$_4 \cdot$ 7H$_2$O 0.2–0.3 g/L, KH$_2$PO$_4$ 0.2–0.3 g/L, and (NH$_4$)$_2$HPO$_4$ or urea 0.4–0.5 g/L. The substrate may be sterilized either in batches at 110 °C with a residence time of 45 min or continuously under conditions providing several minutes exposure to a temperature of 135–150 °C. The sterile medium is then transferred to the fermentation vessel, adjusted to a pH of 4.5–5.0, and inoculated with cultured microorganism.

The culture medium contains only about 100 g of glucose per liter, but as much as twice the above amounts of nutrient salts, an increased level of nitrogen compounds, and a growth-stimulating additive consisting of 0.2–0.4 g of corn steep powder (a residue obtained by evaporating the water in which maize has been soaked). A lyophilized permanent culture is used to grow the conidia. This culture is first activated with a specific growth medium in culture tubes. After several subcultures have been prepared, the organism is introduced into a culture flask containing a special medium that encourages the formation of conidia. After an incubation period of 5–10 d the conidia are harvested, either in dry form or as an aqueous suspension, and used to inoculate the preculture.

During the production phase, the system is maintained at 30–32 °C and pH 5.5–6.5. The pH is controlled by continuous neutralization with 30–50% sodium hydroxide solution, the consumption of which is an indication of reaction progress. Depending on the starting concentration of glucose, fermentation continues for a period of 40–100 h. Under conditions of maximum specific rate of product formation, 20–30 mmol of gluconic acid is generated per hour for each gram of biomass (dry weight) [33], [34].

After fermentation is complete, the microorganisms are removed by filtration and washed. The filtrate is decolorized with activated charcoal, subjected to fine filtration, and then either evaporated and crystallized or directly spray-dried.

The fungal mycelium residue may be used for the isolation of glucose oxidase or it may simply be burnt [35].

Various attempts have been made to use mathematical models to optimize the above process [33], [34], [36], [37]. These efforts have shown that both the specific rate of oxygen uptake and the specific rate of formation of gluconic acid increase approximately linearly with increasing concentration of dissolved oxygen or increasing oxygen partial pressure. If the effects of the brief growth phase and retentive metabolism are ignored, the productive phase follows simple Michaelis–Menten kinetics, with K_M 3.6×10^{-4} mol/L at 33 °C and pH 6.6 [34]. At 25 °C and pH 5.6, $K_M = 2.0 \times 10^{-4}$ mol/L [32]. Isolated glucose oxidase under these conditions has K_M 4.8×10^{-4} mol/L.

To ensure economically viable yields of more than 80%, an adequate oxygen supply (0.1 L of oxygen per liter of solution per minute) must be maintained, and gas distribution within the fermentor must be optimized. The oxygen partial pressure may be increased by conducting fermentation at elevated pressure or by employing air enriched in oxygen [38].

Product concentration in the fermentation solution can be increased if glucose is introduced at a higher concentration and pH control is discontinued near the end of the fermentation [39]. These conditions result in formation of a mixture of sodium gluconate and free gluconic acid, which has a considerably higher solubility than the salt alone. Such a solution (e.g., Naglusol) is suitable for commercial use in technical applications as soon as the mycelium has been removed.

Gluconobacter Suboxydans Process. The *Gluconobacter suboxydans* process is rather different, in that conversion of β-D-glucose to glucono-1,5-lactone occurs through the action of two glucose dehydrogenases: a particulate quinoprotein glucose dehydrogenase and a soluble NADP-dependent glucose dehydrogenase [40], [41].

Gluconobacter suboxydans fermentation is distinguished by a high affinity for oxygen relative to the *Aspergillus* process: simple aeration suffices to ensure a high rate of glucose conversion. In addition, the process displays greater tolerance with respect to acidity, permitting the isolation of free gluconic acid directly from the fermentation solution [42].

Other Methods. Another approach that leads directly to gluconic acid uses acidophilic methylotropic bacteria (*Acetobacter methanolicus*) under nonsterile conditions [43]. Other procedures have recently been described that involve immobilized enzymes, microorganisms, or both [44]–[47]. To date, none of these methods is used for commercial production of gluconate, largely because of rapid deactivation of the enzymes or limitations with respect to oxygen diffusion.

Downstream Processing. Ordinary commercial grades of free gluconic acid may be prepared from the corresponding sodium salt either through cation exchange or electrodialysis. Passage of a mixture of sodium gluconate and free gluconic acid over a strong anion-exchange resin results in adsorption of gluconate [48].

Glucono-1,5-lactone crystallizes from a supersaturated aqueous gluconic acid solution between 36 and 57 °C [49]. A two-step continuous crystallization is said to result in increased productivity per unit volume under conditions that are favorable in terms of energy consumption [50]. In the first step, a crystalline suspension (3–10% crystalline material) is produced in a supersaturated gluconic acid solution held at 65–75 °C. The second step involves controlled cooling to 40–45 °C, followed by a period of crystal maturation at constant temperature. A portion of the resulting crystalline suspension is then recycled to the first step.

According to [51], the 1,5-lactone may also be obtained from an aqueous gluconic acid solution by dehydration involving azeotropic distillation with alcohols, followed by crystallization from the alcohol-containing residue.

Crystallization above 70 °C leads to the 1,4-lactone.

5. Uses

The primary applications of gluconic acid follow from its most important characteristic: it is a weak acid capable of dissolving the oxides, hydroxides, and carbonates of polyvalent cations without attacking metallic or nonmetallic surfaces. The value of this property is further enhanced by the fact that the acid forms water-soluble complexes with such cations. Gluconic acid is thus exceptionally well suited for use in removing calcareous and rust deposits from metals or other surfaces, including beer and milk scale on galvanized iron, magnesium alloys, or stainless steel. The compound is also used in the textile industry together with magnesium salts as a stabilizer for peroxide bleach baths.

Because of its physiological properties, D-gluconic acid has also been used in both the food and beverage and the pharmaceutical industries. Thus, low concentrations (0.02–0.1%) of the substance effect the inversion of sucrose without causing the resulting fructose to undergo further reaction. Trace elements are usually administered in the form of gluconate salts because such compounds are readily absorbed into the body and are well tolerated. Potassium gluconate also has certain pharmaceutical applications and may be dispensed either in anhydrous form or as the monohydrate.

In many applications, gluconic acid 1,5-lactone is a convenient substitute for the free acid. The lactone offers particular advantages in circumstances involving acidic conditions over a long period, e.g., in the preparation of pickled goods and frankfurter-type sausages, or in curing fresh sausage. Another example is its use as a leavening agent in baked goods.

The most common gluconate is sodium gluconate, M_r 218.14, $[\alpha]_D^{25}+12.0°$ [10], whose solubility behavior is as follows:

t, °C	0	20	50	80	100
c, g/100 g H$_2$O	43	60	85	133	160

Sodium gluconate, like the free acid, forms complexes with metal cations. The stability of such complexes often increases considerably with increasing pH [13]. This salt also displays cleansing characteristics with respect to surfaces of widely varying type and structure.

Combinations of sodium gluconate with sodium carbonate or sodium hydroxide solution are useful for removing grease and corrosion from aluminum, rust from steel, oxide coatings from copper and copper alloys, and for etching aluminum. Strongly alkaline gluconate baths permit the removal of zinc coatings from metallic objects, castings, and scrap metal. Such baths are characterized by a high zinc capacity and a correspondingly long life.

Alkaline sodium gluconate solutions at 95 – 100 °C are effective agents for the rapid removal of paint and varnish, without damaging underlying surfaces. Gluconate is also useful in the pretreatment of certain surfaces, e.g., in the galvanic deposition of nickel – cobalt brazing surfaces onto aluminum or in the baths required for preparing smooth, shiny surface platings of nickel, tin, and zinc. In the latter application, gluconate is increasingly used to replace the highly toxic cyanide ion.

Mixtures of gelatin and sodium gluconate are used as sizing agents in the paper industry because they result in a product with increased acid resistance. Textile manufacturers employ gluconate (sometimes in combination with polyphosphates) for desizing polyester or polyamide fabrics and for finishing natural cellulose fibers.

Sodium gluconate is a component of commercial cleansers because of its absolute stability to hydrolysis at high temperature and high pH, as well as its sequestering properties with respect to hardening agents in water. For example, sodium gluconate is used in bottle washing and in cleaning aluminum surfaces (e.g., building facades, aircraft, and containers).

Concrete manufacturers have found sodium gluconate to be a highly effective agent for retarding the curing process. Addition of 0.02 – 0.2 wt% of this substance (relative to cement) produces a very homogeneous concrete with high resistance to water, frost, and cracking. This material is also easily worked and shows increased stability upon setting. Other advantages accompanying the introduction of gluconate include better flow properties for the wet concrete mixture, increased wettability with respect to iron reinforcing structures, and maintenance of plasticity despite reduced water content.

Relative to other organic and inorganic complexing agents, gluconic acid, its lactone, and its salts offer one major advantage: they are all destroyed effectively and quickly by biological wastewater treatment, whether adapted or not. The same applies [52] to the soluble complexes of gluconate with aluminum, copper, iron, and zinc, although biodegradation of the chromium gluconate complex occurs more slowly. Metal ions released during gluconate complex degradation are largely removed from the purified wastewater, either by precipitation or by adsorption on the sludge. This circumstance reduces a major threat that might appear to accompany the use of gluconate as a complexing agent: mobilization or remobilization of heavy metals into surface water [53].

Destruction of heavy-metal gluconate complexes in wastewater may occur not only through biodegradation but also by simple hydroxide precipitation [54], in which iron(III) acts as a precipitant at a pH of 9–10; see also [55].

6. Economic Aspects

Current annual worldwide production capacity for gluconic acid and its derivatives is estimated to be 60 000 t. The bulk of production (85%) is in the form of sodium gluconate and other alkali gluconate salts. Present consumption levels are, however, significantly lower, resulting in only 60–70% utilization of existing capacity. Principal manufacturers include: Akzo-Chemie, Amersfoort, Holland; Biochemie Ladenburg, Ladenburg, Federal Republic of Germany; Finnish Sugar Co. Ltd., Espoo, Finland; Fujisawa Pharmaceutical Co., Osaka, Japan; Pfizer, New York, New York; Roquette Frères, Lille, France.

7. Physiology, Product Specifications

D-Gluconic acid and its 1,5-lactone are important intermediates in carbohydrate metabolism. These compounds serve two important functions: (1) they contribute to the synthesis of reduced nicotinamide–adenine dinucleotide phosphate (NADPH), which is required in the biosynthesis of fatty acids and steroids, and (2) they lead to the formation of ribose 5-phosphate, which is used in nucleic acid synthesis [56].

The gluconate salts of sodium [527-07-1] calcium [299-28-5], copper [13005-35-1], iron(II) [299-29-6], and manganese [6485-39-8] are on the list of GRAS substances [57], [58]. The gluconates of potassium [299-27-4], magnesium [3632-91-5], and zinc [4468-

02-4] have not been shown to have teratogenic, mutagenic, or carcinogenic properties [56].

Under the food and beverage law of the Federal Republic of Germany, D-gluconic acid is classified as a foodstuff and not regarded as an additive. Purity standards for gluconate salts are prescribed in the U.S. Food Chemical Codex (FCC III) of 1981, the U.S. Pharmacopeia (1985), and the European Pharmacopeia. Methods of analysis can be found in [32] and [59].

8. References

[1] W. Henneberg: *Handbuch der Gärungsbakteriologie*, 2nd ed., vol. **2,** Springer Verlag, Berlin 1926.
[2] M. Molliard, *C.R. Hebd. Seances Acad. Sci.* **174** (1922) 881.
[3] K. Bernhauer: *Die oxydativen Gärungen,* Springer Verlag, Berlin 1932.
[4] D. Müller, *Biochem. Z.* **199** (1928) 156.
[5] H. S. Isbell, L. Frush, F. J. Bates, *J. Res. Nat. Bur. Stand.* **6** (1931) 1148: **8** (1932) 572.
[6] M. Busch, DE 702 729, 1941.
[7] P. E. Milsom, J. L. Meers in M. Moo-Young (ed.): *Comprehensive Biotechnology,* vol. **3,** Pergamon Press, Oxford-New York-Toronto-Sydney-Frankfurt 1985, pp. 681–700.
[8] Noury and van der Lande, GB 902 609, 1962 (D. W. van Gelder); *Chem. Abstr.* **57** (1962) 13020 g.
[9] F. J. Prescott, J. K. Shaw, J. P. Bilello, G. O. Cragwall, *Ind. Eng. Chem.* **45** (1953) 338–342.
[10] D. T. Sawyer, J. B. Bagger, *J. Am. Chem. Soc.* **81** (1959) 5302–5306.
[11] L. H. Skibsted, G. Kilde, *Dan. Tidsskr. Farm.* **45** (1971) 320–324; *Chem. Abstr.* **76** (1972) 72 722 d.
[12] R. E. Mitchell, F. R. Duke, *Ann. N.Y. Acad. Sci.* **172** (1970) 129.
[13] D. T. Sawyer, *Chem. Rev.* **69** (1964) 633.
[14] M. Boruch, T. Pierzgalski, *Zesz. Nauk. Politech. Lodz. Chem. Spozyw.* **32** (1977) 33; *Chem. Abstr.* **89** (1978) 131 450 g.
[15] Rin Kagaku, Kogyo Co., Ltd., JP 7 330 618, 1973 (T. Nakagawa, Y. Murai, N. Kanai); *Chem. Abstr.* **80** (1974) 71 052 f.
[16] Kao Soap Co., DE-AS 2 344 990, 1973 (Y. Nakagawa, S. Koshigaya, K. Sato, S. Hakozaki, C. Funabashi).
[17] H. V. K. Udupa et al., *Proc. Semin. Elektrochem.* **14th** (1974) 79. R. Ohme, P. Kimmerl, DD 104 779, 1974.
[18] Johnson, Matthey and Co., Ltd., GB 1 208 101, 1970 (G. J. K. Acres, A. E. R. Budd); *Chem. Abstr.* **74** (1971) 14 347 h. Kawaken Fine Chemicals Co., Ltd., JP 8 007 230, 1980 (H. Saito, M. Nozue); *Chem. Abstr.* **93** (1980) 72 211 n. Mitsui Toatsu Chemicals Inc., JP 7 652 121, 1976 (T. Kiyoura, T. Kimura, T. Sugiura); *Chem. Abstr.* **85** (1976) 160 467 r. Kawaken Fine Chemicals Co., Ltd./Kao Corp., JP 5 872 538, 1983; *Chem. Abstr.* **99** (1983) 88 543 g.
[19] J. M. H. Dirkx, H. S. van der Baan, *J. Catal.* **67** (1981) 1.
[20] Asahi Chemical Industry Co., Ltd., DE-OS 2 936 652, 1980 (J. Nishikido, N. Tamura, Y. Fukuoka); *Chem. Abstr.* **93** (1980) 150 595 g. Kao Corp./Kawaken Fine Chemicals Co., Ltd., JP 59/205 343, 1984; *Chem. Abstr.* **102** (1985) 149 721 t.

[21] N. Xonoglou, G. Kokkinidis, *Bioelectrochem. Bioenerg.* **12** (1985) no. 5–6, 485–498; *Chem. Abstr.* **102** (1985) 149 654 y.

[22] Kawaken Fine Chemicals Co. Ltd./Kao Corp., JP 60/92 240, 1985; *Chem. Abstr.* **103** (1985) 88 175 g.

[23] Kawaken Fine Chemicals Co. Ltd./Kao Corp., EP 0 142 725, 1984 (H. Saito, S. Ohnaka, S. Fukuda).

[24] P.-P. Degelmann, G. Ehmig, H. Hofmann, *Zuckerindustrie (Berlin)* **107** (1982) 117.

[25] G. Gottschalk: *Bacterial Metabolism*, 2nd ed., Springer Verlag, New York-Berlin-Heidelberg-Tokyo 1986, p. 172.

[26] H. G. Schlegel: *Allgemeine Mikrobiologie*, Thieme Verlag, Stuttgart-New York 1985, p. 327.

[27] E. E. Bruchmann: *Angewandte Biochemie*, E. Ulmer, Stuttgart 1976, p. 64.

[28] J. P. van Dijken, M. Veenhuis, *Eur. J. Appl. Microbiol. Biotechnol.* **9** (1980) no. 4, 275–283.

[29] H. M. Müller, *Zentralbl. Bakteriol. Parasitenkd. Infektionskr. Hyg. Abt. 2 Naturwiss. Allg. Landwirtsch. Tech. Mikrobiol.* **132** (1977) 14–24; *Chem. Abstr.* **87** (1977) 1598 j.

[30] T.-G. Bak, *Biochim. Biophys. Acta* **139** (1967) 277–293.

[31] B. Drews, H. Smalla, *Branntweinwirtschaft* **109** (1969) 21–27; *Chem. Abstr.* **71** (1969) 79 730 k.

[32] H. U. Bergmeyer: *Methoden der enzymatischen Analyse*, 3rd ed. vol. **1**, Verlag Chemie, Weinheim 1974, p. 486.

[33] T. Takamatsu, S. Shioya, T. Furuya, *J. Chem. Technol. Biotechnol.* **31** (1981) 697.

[34] S. Fröhlich, *Inst. Ferment. u. Brauwesen*, Dipl.-Arbeit, TU Berlin 1981.

[35] L. B. Lockwood in J. E. Smith, D. R. Berry (eds.): *Filamentous Fungi*, vol. **1**, Wiley, New York 1975, pp. 140–157.

[36] Y. Miura, K. Tsuchiya, H. Tsusho, K. J. Miyamoio, *Ferment. Technol.* **48** (1970) 795.

[37] M. Reuss, S. Fröhlich, B. Kramer, K. Messerschmidt, H. Niebelschütz, *3rd Eur. Congr. Biotechnol.*, vol. **2,** 455–460, Verlag Chemie, Weinheim 1984.

[38] L. B. Lockwood in H. J. Peppler, D. Perlmann (eds.): "Production of Organic Acids by Fermentation," *Microbial Technology I*, Academic Press, London–New York 1979, pp. 355–387.

[39] Pabst Brewing Co./Joh. A. Benckiser GmbH, DE 1 817 907, 1979 (J. Ziffer, A. S. Gaffney, S. Rothenberg, T. J. Cairney).

[40] W. Olijve, J. J. Kok, *Arch. Microbiol.* **121** (1979) 283–297.

[41] P. R. Levering, C. Lucas, J. v. Heijst, W. Olijve, *4th Eur. Congr. Biotechnol. Amsterdam*, vol. **3,** 477, Verlag Chemie, Weinheim 1987.

[42] J. B. M. Meiberg, H. A. Spa, *Antonie van Leeuwenhoek* **49** (1983) 89–90.

[43] Akad. der Wissenschaften der DDR, DD 236 754 A1, 1986 (W. Babel, D. Miethe, V. Iske, K. Sattler, H.-P. Richter, J. Schmidt, R. Duresch); *Chem. Abstr.* **106** (1987) 31 387 t.

[44] K. Buchholz in H. Dellweg (ed.): *Fifth Int. Ferment. Sympos. Berlin 1976*, "Abstracts of Papers," Verlag Versuchs- und Lehranstalt für Spiritusfabrikation und Fermentationstechnologie im Institut für Gärungsgewerbe und Biotechnologie zu Berlin 1976, p. 273.

[45] G. Richter et al., *Starch Stärke* **31** (1979) no. 12, 418–422.

[46] Boehringer Mannheim, DE-AS 2 214 442, 1972 (K. U. Bergmeyer, D. Jaworek).

[47] T. Döppner, W. Hartmeier, *Starch Stärke* **36** (1984) no. 8, 283–287.

[48] Roquette Frères, FR 2 054 829, 1971 (M. Huchette, F. Devos).

[49] Daiichi Kogyo Seiyaku Co., Ltd., JP 75 123 674, 1975 (T. Yamauchi, K. Shimizu); *Chem. Abstr.* **85** (1976) 6008 k.

[50] Roquette Frères, EP 0 203 000, 1986 (J. B. Lelen, P. Lemay).

[51] Daicel Chemical Industries, Ltd., JP 60/45 570, 1985; *Chem. Abstr.* **103** (1985) 123 356 y.

[52] G. Magureanu, *Z. Wasser Abwasser Forsch.* **14** (1981) 10.

[53] G. Müller, *Umsch. Wiss. Tech.* **79** (1979) 778.
[54] D. Ott, *Galvanotechnik* **73** (1982) no. 4, 339; no. 5, 453.
[55] D. Ott, C. J. Raub, *Proc. Electrochem. Soc.* **83–12** (1983) (Proc. Symp. Electroplat. Eng. Waste Recycle New Dev. Trends 1982, 399–407).
[56] Federation of American Societies for Experimental Biology (FASEB) PB 288 675 (1978).
[57] Federal Register, vol. **47,** no. 210, 49 028 (1982).
[58] Federal Register, vol. **49,** no. 114, 24 118 (1984).
[59] R. Lenz, G. Zoll in I. Molnar (ed.): "Pract. Aspects Mod. High Perform. Liq. Chromatogr.,"- *Proc., Meeting Date 1981,* De Gruyter, Berlin 1983, pp. 355–361.

Glycerol

GERALD JAKOBSON, Deutsche Solvay-Werke GmbH, Solingen and Rheinberg, Federal Republic of Germany

FRIEDRICH W. KATHAGEN, Deutsche Solvay-Werke GmbH, Solingen and Rheinberg, Federal Republic of Germany

MARTIN KLATT, Deutsche Solvay-Werke GmbH, Solingen and Rheinberg, Federal Republic of Germany

UDO STEINBERNER, Henkel KGaA, Düsseldorf, Federal Republic of Germany

1.	Introduction 2788	5.2.	Production from Epichlorohydrin 2801
2.	Physical Properties 2788	5.3.	Environmentally Important Properties of Glycerol 2802
3.	Chemical Properties........ 2792	6.	Quality Specifications and Analysis 2802
4.	Production 2793	7.	Storage and Transportation .. 2803
4.1.	Glycerol from Fats and Oils .. 2793	8.	Uses 2804
4.1.1.	Fat Splitting 2793	9.	Derivatives 2804
4.1.2.	Pretreatment and Concentration of Crude Glycerol 2794	9.1.	Esters 2804
4.1.3.	Purification and Refining..... 2794	9.2.	Polyglycerols 2805
4.2.	Synthesis from Propene 2797	9.3.	Chlorohydrins 2806
4.2.1.	Production from Allyl Chloride . 2797	10.	Economic Aspects 2806
4.2.2.	Production from Acrolein 2799	11.	Toxicology and Occupational Health................ 2806
4.2.3.	Production from Propene Oxide . 2800	12.	References.............. 2807
4.3.	Other Processes........... 2800		
5.	Environmental Protection ... 2801		
5.1.	Production from Fats and Oils 2801		

1. Introduction

Glycerol [56-81-5], $C_3H_8O_3$, M_r 92.09, 1,2,3-propanetriol, commonly known as glycerin, is the simplest triol. It can be found in all natural fats and oils as fatty esters and is an important intermediate in the metabolism of living organisms.

$$\begin{array}{c} CH_2-OH \\ | \\ CH-OH \\ | \\ CH_2-OH \end{array}$$

Glycerol was discovered in 1779 by SCHEELE through saponification of olive oil with lead oxide. In 1813, CHEVREUL showed that fats are glycerol esters of fatty acids. He also gave glycerol its name, after the Greek word γλυκερος which means sweet. The first industrial use of glycerol was in 1866 when NOBEL produced dynamite, in which the trinitrate of glycerol—nitroglycerin—is stabilized by absorption on diatomaceous earth.

The most important industrial synthesis of glycerol, which uses propene as starting material, was developed in the late 1930s by I.G. Farben in Germany and by Shell in the United States.

Glycerol is now used in a large variety of applications because of its particular combination of chemical and physical properties and because it is physiologically innocuous. Total production is estimated (1986) to be ca. 550 000 t/a; 75% is produced by splitting of natural oils and 25% is synthesized from propene [1].

2. Physical Properties

Glycerol is a sweet-tasting liquid; when pure, it is colorless and odorless. At room temperature it is viscous.

The boiling point of glycerol is 290 °C at atmospheric pressure (101.3 kPa). The corresponding values under reduced pressure are listed in Table 1. With some liquids, (e.g., biphenyl), glycerol forms azeotropes; with others (e.g., water), nonazeotropic mixtures are formed. The vapor pressure curves of glycerol–water solutions at various glycerol concentrations are shown in Figure 1. The liquid–vapor equilibria of glycerol–water solutions, important for distillation and fractionation, have been established experimentally [2]–[4]. The boiling- and condensing curves at 13.3 and 1.33 kPa are shown in Figure 2. Suitable theoretical methods of calculation are also available [5].

The freezing point of pure glycerol is ca. 18 °C. The crystalline state is seldom reached, however, because of a strong tendency toward supercooling. A solid, glassy state can be observed from ca. −70 to −110 °C. Crystallization can be brought about at ca. 0 °C by seeding with glycerol crystals. The freezing point of glycerol–water solutions is shown in Figure 3. At 66.7 wt% glycerol, a eutectic mixture is formed, fp −46.5 °C.

Table 1. Vapor pressure vs. temperature of glycerol

Temperature, °C	Pressure, kPa
290	101.3
266	53.3
222	13.3
204	6.67
175	2.00
152	0.67
130	0.18
100	0.03
20	<0.0001

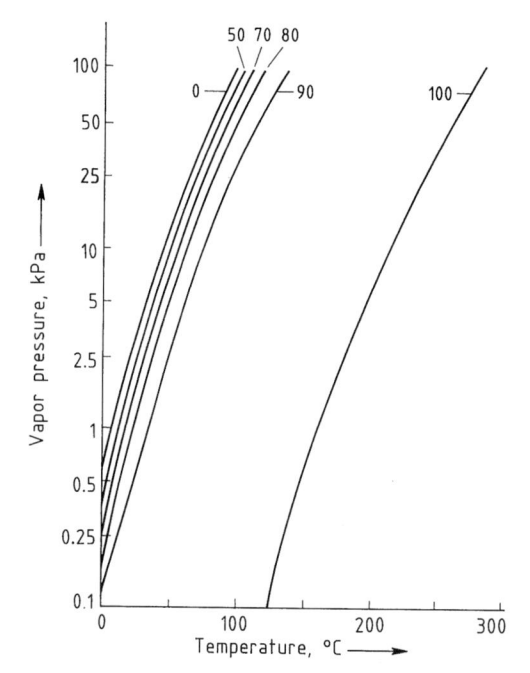

Figure 1. Vapor pressure of glycerol–water solutions (wt% glycerol) [8], [9]

The density of glycerol is 1.261 g/cm^3 (20 °C). Tables of the density of glycerol–water solutions at various concentrations of glycerol are given in [8]. For rough estimations, a mean density change of 0.0007 g/cm^3 per unit change in temperature may be used in the range from 0 to 200 °C. The compressibility of glycerol is only about half that of water.

The refractive index of aqueous glycerol for the range 0–100% glycerol is tabulated in 1% steps in [8]. The viscosity of glycerol–water solutions (0–100 wt% glycerol) from −40 to 100 °C has been determined [8]. Some curves are shown in Figure 4. Aqueous glycerol solutions, at any concentration, will supply moisture until equilibrium with

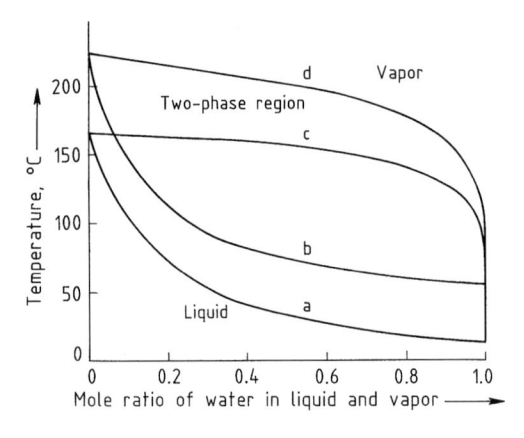

Figure 2. Liquid–vapor equilibria for glycerol–water solutions
a) Boiling curve, 1.33 kPa; b) Boiling curve, 13.3 kPa; c) Condensing curve, 1.33 kPa; d) Condensing curve, 13.3 kPa

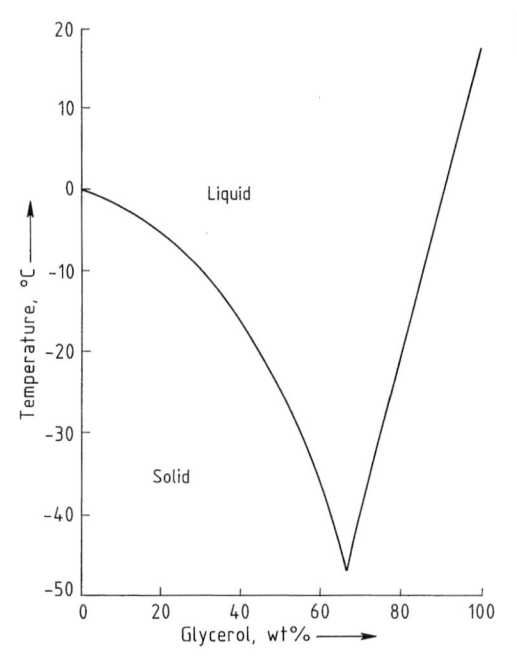

Figure 3. Freezing point of glycerol–water solutions

water vapor in the atmosphere is reached [8]. The relative humidity over glycerol solutions is illustrated in Figure 5.

Glycerol has useful solvent properties, similar to those of water and simple aliphatic alcohols, because of its three hydroxyl groups. It is, accordingly, completely miscible with water, methanol, ethanol, and the isomers of propanol, butanol, and pentanol. It is also fully miscible with phenol, glycol, propanediols, amines, and heterocyclic com-

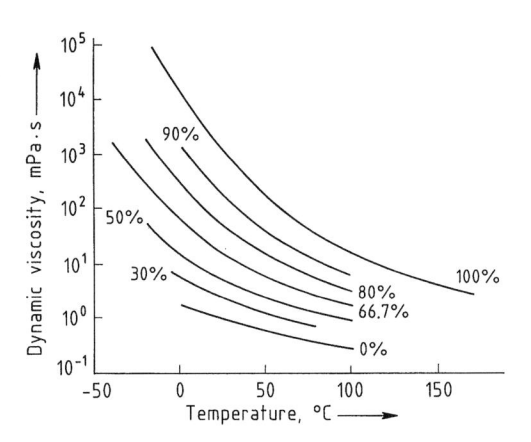

Figure 4. Dynamic viscosity of glycerol–water solutions

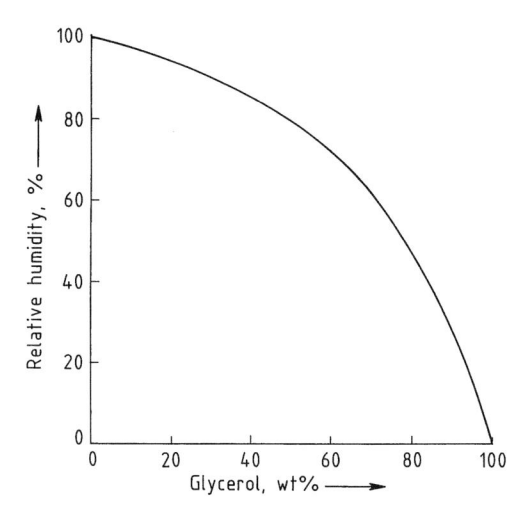

Figure 5. Relative humidity over aqueous glycerol (20–100 °C)

pounds containing a nitrogen atom in the ring (e.g., pyridine, quinoline). Its solubility, however, is limited in acetone, diethyl ether, and dioxane. Glycerol is almost insoluble in hydrocarbons, long-chain aliphatic alcohols, fatty oils, and halogenated solvents such as chloroform. Ternary systems, such as glycerol–water–phenol and the nonaqueous system glycerol–ethanol–benzene, display significant temperature-dependent gaps in miscibility. Glycerol is a useful solvent for many solids, both organic and inorganic, which is particularly important for the preparation of pharmaceuticals. The solubility of gases in glycerol, like other liquids, is temperature- and pressure-dependent [8]. Some important physical properties of glycerol are presented in Table 2; a comprehensive collection of data is available [9].

Table 2. Physical properties of glycerol

M_r	92.09
mp	18.0 °C
bp (101.3 kPa)	290.0 °C
Density (20 °C)	1.261 g/cm³
Refractive index n_D^{20}	1.4740
Dynamic viscosity (20 °C)	1.410 Pa · s
Compressibility (28.5 °C)	2.1×10^{-4} MPa^{-1}
Gravity coefficient of thermal expansion (15–20 °C)	0.000615 K^{-1}
Surface tension (20 °C)	63.4 mN/m
Heat of formation	669 kJ/mol
Heat of combustion	1662 kJ/mol
Heat of vaporization (55 °C)	88.2 kJ/mol
(195 °C)	76.1 kJ/mol
Heat of fusion (18 °C)	18.3 kJ/mol
Heat of solution (infinite dilution)	5.8 kJ/mol
Heat capacity (26 °C)	2.41 kJ kg^{-1} K^{-1}
(−80 °C)	1.91 kJ kg^{-1} K^{-1}
(−108 °C)	0.91 kJ kg^{-1} K^{-1}
Thermal conductivity (0 °C)	0.29 W m^{-1} K^{-1}
Diffusion constant of water into glycerol (20 °C)	1.33×10^{-11} m²/s
Specific electrical conductivity (20 °C)	0.1 µS/cm
Relative dielectric constant (25 °C)	42.48
Flash point	177 °C
Fire point	204 °C
Autoignition temperature on glass	429 °C

3. Chemical Properties

Glycerol is a reactive molecule that undergoes all the usual reactions of alcohols (→ Alcohols, Aliphatic; → Alcohols, Polyhydric). The two terminal primary hydroxyl groups are more reactive than the internal secondary hydroxyl group. Under neutral or alkaline conditions, glycerol can be heated to 275 °C without formation of acrolein. By contrast, in the presence of small amounts of strong mineral acid, the odor of acrolein (odor threshold 0.2–0.4 ppm) is already quite perceptible at 160 °C; by 200 °C, the evolution of acrolein is vigorous [6]. Reactions with glycerol are, therefore, best carried out under neutral or alkaline conditions. At 180 °C, alkaline glycerol begins to dehydrate, forming ether-linked polyglycerols. At room temperature glycerol rapidly absorbs water; when dilute it is attacked by microorganisms. Glycerol is easily oxidized: the terminal carbon atoms to aldehyde or carboxyl groups and the central carbon atom to a carbonyl group [10].

Industrially important reaction products of glycerol include:

1) mono-, di-, and triesters of inorganic and organic acids;
2) mono- and diglycerides of fatty acids formed by transesterification of triglycerides (from fats);

3) aliphatic and aromatic esters, formed by reaction with alkylating or arylating agents, respectively;
4) polyglycerols, formed by the intermolecular elimination of water with alkaline catalysts;
5) cyclic 1,2- or 1,3-acetals or ketals, formed by reaction with aldehydes or ketones, respectively;
6) mono- or diglycerates, formed by the action of alkali or metal alcoholates.

4. Production

Glycerol is obtained as a byproduct in the conversion of fats and oils to fatty acids or fatty acid methyl esters. This type of glycerol is known as natural or native glycerol, in contrast to synthetic glycerol from propene. Other methods of production (e.g., fermentation of sugar or hydrogenation of carbohydrates) are not industrially important.

4.1. Glycerol from Fats and Oils

Although a component of all plant and animal fats and oils, glycerol does not exist in the free form but as fatty acid esters; all three hydroxyl groups are usually esterified. Such esters are called triglycerides ($\rightarrow$ Fats and Fatty Oils; $\rightarrow$ Fatty Acids). The glycerol content of fats and oils varies between 8 and 14 wt%, depending on the proportion of free acid and on the chain-length distribution of the fatty acid esters. To obtain glycerol, the oils and fats must be split.

4.1.1. Fat Splitting

(For a detailed description of processes used in the splitting of fats and oils see $\rightarrow$ Fatty Acids)

The main sources of natural glycerol are now high-pressure splitting and transesterification. Glycerol from the neutral saponification of oils is encountered only in small quantities.

High-Pressure Splitting. Splitting under pressure has been known since 1854; continuous process reactors are now used. Water and fat are fed into a splitting column in countercurrent fashion at 5–6 MPa and 250–260 °C, leading to a 15% solution of glycerol in water, known as sweet water. This glycerol is marketed as 88% saponification- or hydrolysis-crude glycerol. Such glycerol is extremely low in ash: a typical value is ca. 0.1% or less of inorganic salts.

Transesterification. Natural crude glycerol of similar quality can be obtained from the continuous transesterification of oils and fats to their methyl esters. The crude glycerol is obtained directly at a concentration of ca. 90–92%.

Saponification. The splitting of fats by saponification of neutral oils is a traditional method; caustic alkali or alkali carbonates are used, as in the production of soap. The use of calcium hydroxide in the form of milk of lime is also possible.

4.1.2. Pretreatment and Concentration of Crude Glycerol

The pretreatment of glycerol water is an important step in the work-up, which is crucial for trouble-free production and for the quality of the final product. Glycerol water produced by high-pressure hydrolysis is slightly acidic. Dispersed fat and fatty acid components can be largely removed under these conditions simply by settling or by centrifugation. In a subsequent step the concentration is immediately increased to 70–90% glycerol to avoid decomposition by fermentation (resulting in formation of 1,3-propanediol) and to render the crude product stable to storage. A multistage flash evaporator of stainless steel yields the greatest economic advantage. This main pretreatment takes place directly before final purification and refining.

An optimum effect can be obtained in dilute solutions with a glycerol content of ca. 50%. A two-step process is normally employed. Activated carbon and sodium hydroxide are used as auxiliaries, the former for bleaching and absorption of impurities, the latter to saponify any remaining fat components. A separation by filtration follows each step. The so-called soda-lime process has been valued for decades for the pretreatment of crude glycerol contaminated with larger amounts of organic impurities [8]. Further concentration of the treated crude glycerol is necessary with distillative work-up. Single- or multiple-step plants are used with a vacuum of 10–15 kPa. The solubility limits of the salts must be taken into consideration, and the relevant stages suitably equipped for removal of the salt.

4.1.3. Purification and Refining

Until the end of the 1950s the large-scale production of pharmaceutical glycerol was possible only by *distillation*. Batches were distilled under vacuum with fractional condensation. The development of new technologies allowed conversion to continuous production processes, with correspondingly low losses.

A glycerol refining unit used for the thermal work-up of crude pretreated and evaporated glycerol to glycerol of 99.5% purity is shown in Figure 6. For work-up of glycerol with a high salt content, *thin-film distillation* is employed [12], [13]. The glycerol is spread by a rotor in a thin film onto the inner wall of an externally heated

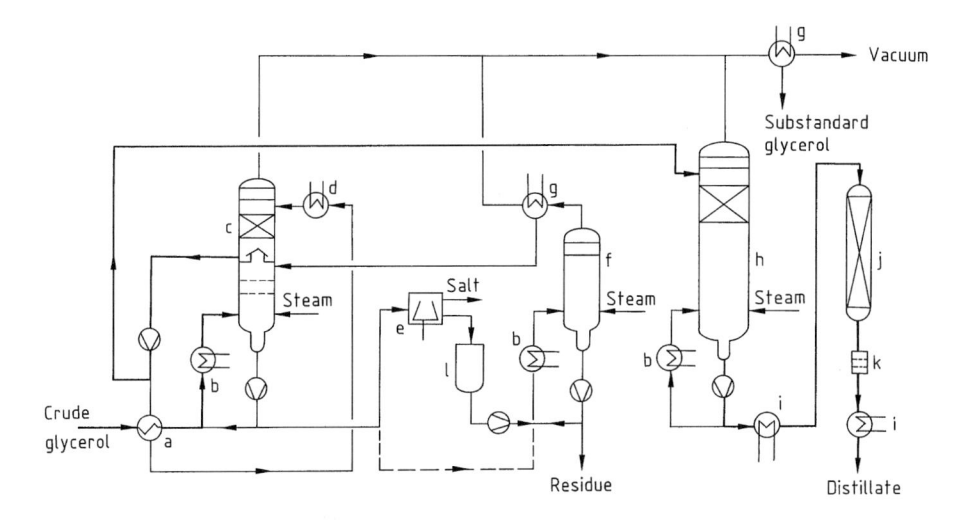

Figure 6. Continuous glycerol distillation (Wurster and Sanger) [7], [11]
a) Feed heater/reflux cooler; b) Reboiler; c) Distillation tower; d) Reflux cooler; e) Centrifuge; f) Pitch still; g) Condenser; h) Deodorizing column; i) Cooler; j) Bleaching stage; k) Polishing filter; l) Intermediate tank

column and vaporized. The residue flows or trickles to the bottom. This equipment, combined with a fractionating column, is shown in Figure 7. The advantage of thin-film distillation is the extremely short residence time; thermal loading of the glycerol is thereby minimized. Refinement of low-salt crude glycerol can also be carried out by *ion exchange;* this process became available with the development of exchange resins suitable for glycerol [14]. Ion exchangers are used in pairs and consist of a cationic and an anionic exchanger. Charged impurities are thereby successively removed and ultimately exchanged for water [15]. The process permits removal of inorganic salts, fat and soap components, colored matter, odor-causing substances, and other impurities. The so-called adsorption resins possess special decolorizing properties; nonionic components such as chromophores can be removed physically by adsorption [16], [17]. The concentration of pharmaceutical glycerol solutions is carried out under special mild conditions to avoid changes in color and odor: a three- or four-step stainless steel flash evaporator with natural circulation is mostly used (Fig. 8).

At 10–15 kPa, a glycerol concentration of ca. 90–95% is attained. The removal of residual water to give a glycerol concentration >99.5% is best carried out in a separate final step by forced circulation in a falling-film evaporator. The vacuum used is 0.5–1.5 kPa. The continuous processes described above are suitable for automation when equipped with the appropriate control equipment.

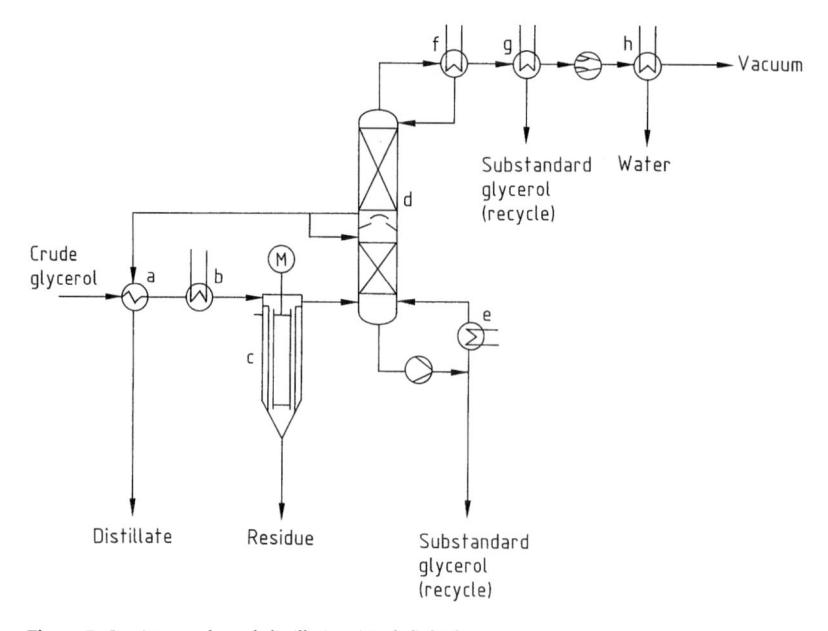

Figure 7. Continuous glycerol distillation (Henkel) [13]
a) Feed heater/distillate cooler; b) End heater; c) Thin-film distillation; d) Fractionating column; e) Reboiler; f) Reflux condenser; g) Glycerol condenser; h) Water condenser

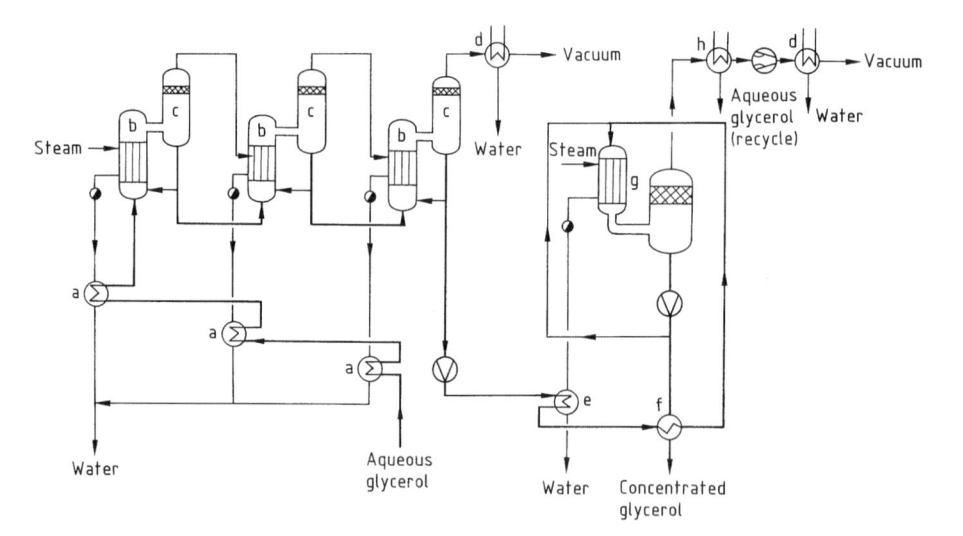

Figure 8. Continuous glycerol concentration
a) Feed heater; b) Evaporator; c) Separator with demister; d) Water condenser; e) Glycerol heater; f) Glycerol heater/final product cooler; g) Falling film evaporator; h) Glycerol condenser

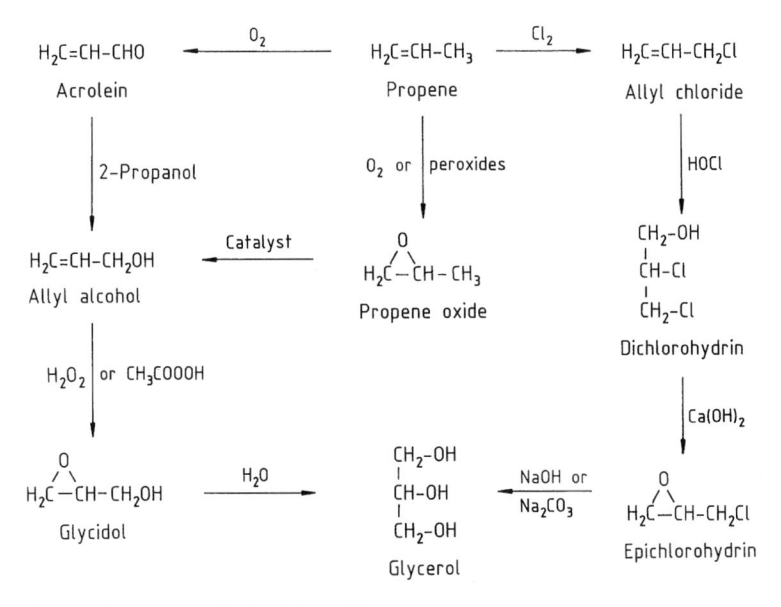

Figure 9. Routes for synthesis of glycerol from propene

4.2. Synthesis from Propene

Three processes are known for the production of glycerol from propene [115-07-1], involving the following intermediate stages (Fig. 9):

1) allyl chloride – epichlorohydrin
2) acrolein – allyl alcohol – glycidol
3) propene oxide – allyl alcohol – glycidol

Only the first of these is important industrially.

4.2.1. Production from Allyl Chloride

The first synthetic glycerol was produced in 1943 by I.G. Farben in Oppau and Heydebreck and in 1948 by Shell in Houston, Texas. This method became available once the high-temperature chlorination of propene to allyl chloride [107-05-1] could be controlled properly (→ Allyl Compounds). The allyl chloride produced is oxidized with hypochlorite to dichlorohydrin, which is converted without isolation to epichlorohydrin [106-89-8] by ring closure with calcium or sodium hydroxide. Hydrolysis to glycerol is carried out with sodium hydroxide or sodium carbonate [18] – [20].
Chlorination of propene to allyl chloride:

$$H_2C=CH-CH_3 + Cl_2 \longrightarrow H_2C=CH-CH_2Cl + HCl$$

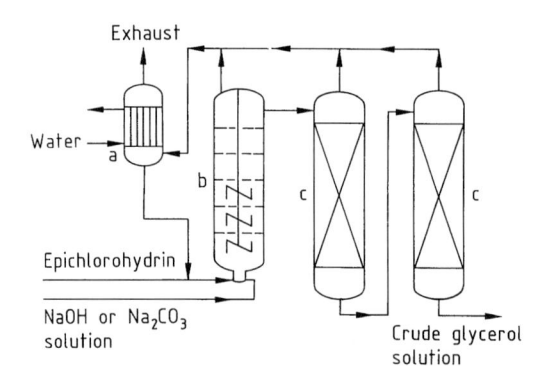

Figure 10. Synthesis of glycerol from epichlorohydrin
a) Exhaust gas cooler; b) Stirred reactor; c) Packed reactor

Hypochlorination:

$$H_2C=CH-CH_2Cl + HOCl \longrightarrow H_2\underset{|}{C}-\underset{|}{CH}-CH_2Cl$$
$$ClOH$$

Dehydrochlorination:

$$2\ H_2\underset{|}{C}-\underset{|}{CH}-CH_2Cl + Ca(OH)_2 \longrightarrow$$
$$ClOH$$

$$2\ H_2C\overset{O}{\overset{/\backslash}{-}}CH-CH_2Cl + CaCl_2 + 2\ H_2O$$

Hydrolysis of epichlorohydrin to glycerol:

$$H_2C\overset{O}{\overset{/\backslash}{-}}CH-CH_2Cl + NaOH + H_2O \longrightarrow$$

$$CH_2-CH-CH_2 + NaCl$$
$$\underset{|}{OH}\underset{|}{OH}\underset{|}{OH}$$

or

$$2\ H_2C\overset{O}{\overset{/\backslash}{-}}CH-CH_2Cl + Na_2CO_3 + 3\ H_2O \longrightarrow$$

$$2\ CH_2-CH-CH_2 + 2\ NaCl + CO_2$$
$$\underset{|}{OH}\underset{|}{OH}\underset{|}{OH}$$

Epichlorohydrin is hydrolyzed at 80–200 °C with a 10–15 % aqueous solution of sodium hydroxide or sodium carbonate at atmospheric or overpressure. The residence time in one or a series of several closed, continuously operating reactors amounts to several minutes or several hours depending on the plant concerned. The yield of dilute (10–25 %) glycerol solution is > 98 %. The solution contains 5–10 % sodium chloride and less than 2 % of other impurities (Fig. 10). This aqueous glycerol solution containing sodium chloride is evaporated in a multistage evaporation plant under vacuum to a glycerol concentration of > 75 %; precipitated sodium chloride is separated at the same time (Fig. 11). The glycerol solution is then distilled under high vacuum (ca. 0.5–1.0 kPa) (Section 4.1.3., Fig. 6 and 7); codistilled water is separated by fractional condensation. Residual inorganic salts and higher oligomers of glycerol remaining after

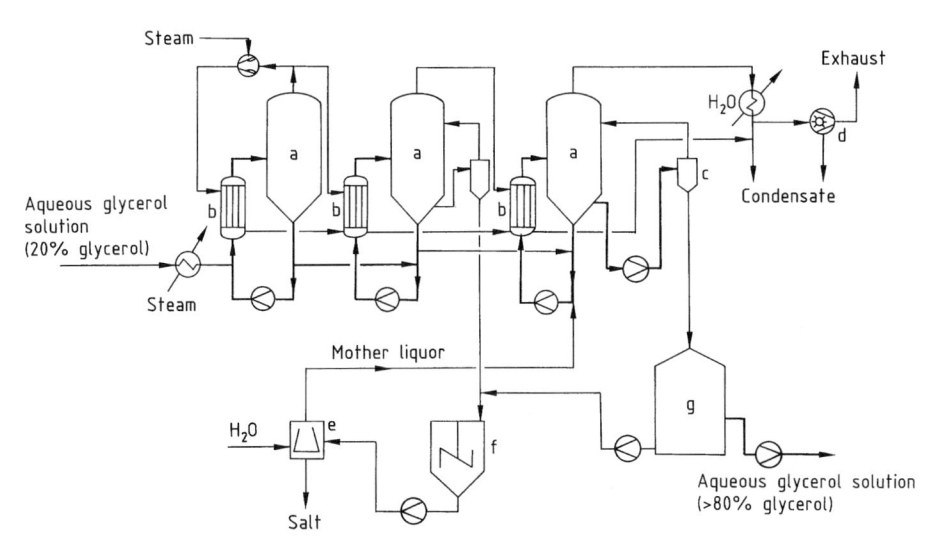

Figure 11. Three-stage evaporation plant for aqueous glycerol solutions: Buss glycerol evaporation process [12]
a) Evaporator vessel; b) Reboiler; c) Cyclone; d) Vacuum pump; e) Centrifuge; f) Stirred tank; g) Storage tank for glycerol solution

the evaporation must be worked up further or discarded. The glycerol, practically free of water, is treated further to remove colored impurities and odorous material; this can be performed, for example, with activated carbon.

4.2.2. Production from Acrolein

The acrolein process for the production of glycerol was developed by Shell and does not require the use of chlorine [21]; the first plant was built in 1958 in Norco, Louisiana. Propene is oxidized to acrolein [107-02-8], which is then reduced to allyl alcohol [107-18-6] (Meerwein–Ponndorf–Verley reduction). The allyl alcohol is epoxidized with hydrogen peroxide, and the resulting glycidol [556-52-5] is hydrolyzed to glycerol.

Oxidation:

$$H_2C=CH-CH_3 + O_2 \longrightarrow H_2C=CH-CHO + H_2O$$

Reduction:

$$H_2C=CH-CHO + (CH_3)_2CHOH \longrightarrow H_2C=CH-CH_2OH + (CH_3)_2CO$$

Epoxidation:

$$H_2C=CH-CH_2OH + H_2O_2 \longrightarrow$$

$$\underset{\triangle}{H_2C-CH}-CH_2OH + H_2O$$

Hydrolysis:

$$\underset{\triangle}{H_2C-CH}-CH_2OH + H_2O \longrightarrow CH_2-CH-CH_2 \;(OH,\,OH,\,OH)$$

4.2.3. Production from Propene Oxide

As with acrolein, preparation from propene oxide does not use chlorine. Propene is epoxidized to propene oxide, which is then isomerized to allyl alcohol by the Progil process [22] (→ Allyl Compounds). A second epoxidation is carried out with peracetic acid, and the resulting glycidol is hydrolyzed to glycerol.

Direct oxidation or oxidation with peroxides:

$$2\,H_2C=CH-CH_3 + O_2 \longrightarrow 2\,\underset{\triangle}{H_2C-CH}-CH_3$$

Isomerization:

$$\underset{\triangle}{H_2C-CH}-CH_3 \longrightarrow H_2C=CH-CH_2OH$$

Epoxidation:

$$H_2C=CH-CH_2OH + CH_3COOOH \longrightarrow$$

$$\underset{\triangle}{H_2C-CH}-CH_2OH + CH_3COOH$$

Hydrolysis:

$$\underset{\triangle}{H_2C-CH}-CH_2OH + H_2O \longrightarrow CH_2-CH-CH_2 \;(OH,\,OH,\,OH)$$

4.3. Other Processes

Fermentation of Sugar. The formation of glycerol by fermentation of alcohol was discovered in 1858 by PASTEUR. Industrial use became possible once the mechanism was understood; the fermentation could be interrupted at the glyceraldehyde 3-phosphate stage with sodium carbonate or with alkali or alkaline earth sulfites. After reduction to glycerol phosphate, glycerol is obtained in yields up to 25% by hydrolysis [23]. The process is economically unimportant.

Hydrogenation of Carbohydrates. Hydrogenation of natural polyalcohols such as cellulose, starch, or sugar leads to mixtures of glycerol and glycols, which can be separated by distillation. Catalysts used in this high-temperature reaction include nickel, copper, cobalt, chromium, and tungsten, as well as oxides of some of the lanthanides [24].

Other Noncommercial Processes. Glycerol can be obtained in yields of ca. 7% by enzymatic splitting of fats and oils with lipases in special reactors [25]–[27] (→ Fatty Acids).

Photoproduction of glycerol along with other biomass is possible by means of solar energy and algae [28], [29]. Another synthetic process involves the catalytic hydrogenation of carbon monoxide [30].

5. Environmental Protection

Ecologically, glycerol presents no special problems. Discussion of environmental protection is limited to those industrial processes based on fats and oils and on epichlorohydrin, because only these processes are economically important.

5.1. Production from Fats and Oils

Production Plants. No special protective measures are necessary for the work-up of natural crude glycerol with the technology described above.

Gaseous Emissions. Certain measures are necessary to prevent atmospheric pollution by odors during storage and purification of native crude glycerol. Exhaust vapor from the respective tanks and process units should be drawn off and treated (e.g., burned) in a closed system (→ Fatty Acids).

Residues. The residues of higher glycerol oligomers and inorganic salts remaining from evaporation and distillation are water soluble and must be treated further or disposed of properly.

5.2. Production from Epichlorohydrin

Extensive protective measures must be taken to allow proper manipulation of epichlorohydrin, because of the hazardous properties of this material (→ Epoxides).

Production Plants. Production plants must be explosionproof to prevent ignition of epichlorohydrin. Manipulation of epichlorohydrin exclusively in closed systems guarantees protection of personnel.

Gaseous Emissions. Epichlorohydrin and accompanying organic material in exhaust gases from the hydrolysis of epichlorohydrin can be removed, for example, by adsorption on activated carbon.

Wastewater. Wastewater contaminated with epichlorohydrin can be treated with alkali, thereby converting the epichlorohydrin to glycerol. Regulations exist in several countries concerning exhaust gas and wastewater purification, as well as measures for the protection of plant personnel [31]–[36].

Residues. Residues from evaporation and distillation should be treated as described in Section 5.1.

5.3. Environmentally Important Properties of Glycerol

Gaseous Emissions. No problems are encountered with gaseous emissions because of the low vapor pressure of glycerol and the lack of odor of the pure material.

Wastewater. Glycerol poses no problems for wastewater because it is completely biodegradable in sewage treatment plants. Glycerol is not regarded as a danger to the water supply. In some countries, however, its presence in wastewater is liable to payment of duty because of its high oxygen demand: COD = 1217 mg of O_2 per gram; BOD_5 = 780 mg of O_2 per gram [37].

6. Quality Specifications and Analysis

Glycerol can be detected qualitatively by heating a small sample with potassium hydrogen sulfate; the presence of glycerol is indicated by the characteristic pungent odor of acrolein.

The most accurate wet method for quantitative determination of glycerol is periodate oxidation. Glycerol is oxidized by sodium periodate to formaldehyde and formic acid:

$$\underset{\underset{CH_2}{|}}{OH} - \underset{\underset{CH}{|}}{OH} - \underset{\underset{CH_2}{|}}{OH} + 2\,NaIO_4 \longrightarrow 2\,HCHO + HCOOH + H_2O + 2\,NaIO_3$$

Formic acid is determined by titration and is a measure of the amount of glycerol; polyols with three or more neighboring hydroxyl groups (e.g., sugars) interfere. For quantitative determination of glycerol, chromatography is also employed: for speed of determination, high performance liquid chromatography (HPLC) is particularly suitable because derivatives need not be prepared [38]. Glycerol can be determined by gas chromatography as the tris(trimethylsilyl) ether [39].

Water in glycerol can be determined quantitatively by Karl Fischer titration. A rapid method for establishing the glycerol content of commercial glycerol–water mixtures is measurement of the refractive index at 20 °C or determination of the density with a digital instrument that electronically measures the period of oscillation of a swinging body suspended in the glycerol [40]. In contrast, measurement of the density with a pycnometer is, for the same degree of accuracy (0.1%), considerably more time consuming. Color is given as Hazen or American Public Health Association (APHA) color number; typical color values would be 5–20, depending on the glycerol quality. For rapid measurement, equipment is available for direct color comparison with standardized slides. Ash, heavy metals, chloride and organic chlorine compounds, pH, and acidic or alkaline substances can be determined by standard analytical methods or by methods given in various pharmacopoeias. Traces of reducing agents, such as aldehydes, sometimes still present in glycerol can be determined quantitatively by special tests, e.g., with silver nitrate.

Glycerol is sold in crude (natural glycerol) or pure form. In the latter, a distinction is made between technical grade:

Technical glycerol at various concentrations
Nobel test glycerol, 99.5%

and pharmaceutical grade (e.g., Pharmacopoea Europaea, Ph. Eur.):

Glycerol Ph. Eur. 86%
Glycerol Ph. Eur. 99.5%.

7. Storage and Transportation

Storage. Glycerol is stable when stored below 100 °C; it is noncorrosive and presents little risk of ignition because of its high flash point. Anhydrous glycerol does not corrode steel, but storage tanks of carbon steel must be protected by surface coating to prevent rusting by residual moisture. Glycerol is therefore usually stored in tanks of stainless steel [41], [42] or aluminum. Regulations exist in some countries to control the construction and arrangement of industrial storage tanks [32].

Transportation. Glycerol is shipped in tank trucks, containers, and drums. The tank trucks and containers are usually made of stainless steel. Galvanized or resin-coated steel is used for drums; for small drums, plastic is also employed.

Table 3. Applications of glycerol (worldwide)

Application	Percentage
Pharmaceuticals, cosmetics	26
Esters	17
Resins	12
Polyols	12
Food	10
Other chemicals	10
Cellulose	5
Nitration	4
Tobacco	4

8. Uses

Because of its particular combination of physical and chemical properties, glycerol is an extremely useful product with a large variety of applications. Of considerable importance are its nontoxicity and lack of color or odor. Consumption of glycerol according to major areas of application, is listed in Table 3. Glycerol is important in pharmaceuticals, cosmetics, and the food industry because it is nontoxic and possesses good solvent properties for many compounds, both organic and inorganic. It keeps preparations moist and hinders unwanted crystallization. It can be used as a softener, it supplies moisture to the skin, and its high viscosity renders solutions syrupy [44].

The polyfunctional reactivity of glycerol is used in the production of alkyd resins and in cross-linked polyesters from mono- and dicarboxylic acids and polyols. Glycerol serves as a lubricant in the textile industry, as well as in equipment and materials that come in contact with food, pharmaceuticals, cosmetics, or skin. In special cases, it is also employed as a hydraulic fluid. Detailed information is available in the literature; over 800 references are cited in [45] alone. The uses of glycerol derivatives are discussed in the following chapter.

9. Derivatives

9.1. Esters

Glycerol forms esters with both inorganic and organic acids; depending on reaction conditions and degree of esterification, these can be mono-, di-, or triglycerides. Of most importance are esters produced from nitric, acetic, and fatty acids [8], [46].

Glycerol trinitrate [55-63-0], nitroglycerin, is produced in a mixture of nitric and sulfuric acids. In addition to its use in form of dynamite, it is also an important therapeutic agent in the treatment of angina pectoris.

The *acetins* are produced industrially by esterification of glycerol with acetic acid, acetic anhydride, or both. They are the most important esters of glycerol from short-chain carboxylic acids. The greatest use is of triacetin, e.g., in the production of vinyl polymers and rubbers. Triacetin is used preferentially as a plasticizer and a stabilizer in the production of cigarette filters.

Glycerol fatty esters, the partial glycerides, are obtained by transesterification of fats or oils with glycerol to a mixture of mono- and diglycerides. The mono- and diesters of fatty acids are edible, as are the triglycerides. They are used in the food industry as emulsifiers for baked goods, ice cream, and dairy products. Mixed fatty acid esters with citric, lactic, or acetic acid are also used for this purpose. The emulsifying properties of partial glycerides are used in the cosmetic and pharmaceutical industries. They are also employed as lubricants for working plastics and serve as components for special lubricant formulations. The range of emulsifying and stabilizing properties can be increased further by replacing some of the glycerol in the fatty acid ester with polyglycerol or ethoxylated glycerol. These esters are used in the production of fat-containing baked goods or as regreasing agents in shampoos and foam baths.

9.2. Polyglycerols

Polyglycerols are ethers of glycerol; they are produced industrially by base-catalyzed condensation of glycerol and as byproducts of epichlorohydrin hydrolysis [47]. Separation and isolation are achieved by treatment with dimethoxypropane or acetone (ketalization), fractional distillation of the diisopropylidene derivatives, and finally, acid-catalyzed hydrolysis [48]:

$$CH_2-CH-CH_2-O\left[CH_2-CH-CH_2-O\right]_n CH_2-CH-CH_2 \xrightarrow{\underset{CH_3}{\overset{CH_3}{\times}}\underset{OCH_3}{\overset{OCH_3}{}}}_{\text{Ketalization}}$$

$n = 0, 1, 2, \ldots$

$$CH_2-CH-CH_2-O\left[CH_2-CH-CH_2-O\right]_n CH_2-CH-CH_2 \xrightarrow[\text{Hydrolysis}]{H^+, H_2O}$$

Diisopropylidene derivatives, $n = 0, 1, 2, \ldots$

$$CH_2-CH-CH_2-O-CH_2-CH-CH_2$$
$OHOHOHOH$

Diglycerol, $n = 0$

$$CH_2-CH-CH_2-O-CH_2-CH-CH_2-O-CH_2-CH-CH_2$$
$OHOHOHOHOH$

Triglycerol, $n = 1$

Diglycerol, as the simplest condensation product, and the higher oligomers, as synthetic building blocks similar to glycerol, can be converted into interesting products. For example, its fatty acid esters offer a still larger range of properties than the simple glycerides [49]. Because of their lipophilic and hydrophilic properties, they have a wide range of applications as emulsifiers in the cosmetic and food industries.

9.3. Chlorohydrins

Chlorohydrins have been treated in a separate article (→ Chlorohydrins); epichlorohydrin appears in → Epoxides.

10. Economic Aspects

Estimated worldwide consumption of glycerol amounts to ca. 550 000 t/a year; almost two-thirds of the total production is in Europe and the United States. Annual production (10^3 t) is as follows:

Western Europe	208
United States	145
Japan	45
Others	152
Total	550

The fraction of synthetic glycerol is ca. 25 %.

The price of glycerol is subject to fluctuation. In times of higher prices, glycerol is partly replaced in certain applications by products such as glycols or sorbitol. Nevertheless, the consumption of glycerol is increasing annually by ca. 1 %.

11. Toxicology and Occupational Health

Glycerol is not harmful to health. Ingestion even of large amounts causes no harm to humans. There is no MAK value for glycerol.

The use of glycerol as a food additive is permitted in most countries, in particular those of the EEC (Directive 74/329/CEE) and in the United States (GRAS § 182.1320).

Slight irritation of the skin or mucous membranes is possible on contact with undiluted glycerol because the strongly hygroscopic glycerol draws water from the skin.

12. References

[1] U. Steinberner, W. Preuss, *Fat Science Technology,* **89** (1987) 297–303.
[2] D. F. Stedmann, *Trans. Faraday Soc.* **24** (1928) 289–298.
[3] I. C. Chu, S. L. Wang, S. L. Levy, R. Paul: *Vapor Liquid Equilibrium Data,* I. W. Edwards, Michigan 1956.
[4] N. M. Sokulov, L. N. Tsygankova, N. M. Zhavornokov, *Khim. Promst. (Moscow)* **48** (1972) 96; *Chem. Abstr.* **76** (1972) 154 052 z.
[5] R. C. Reid, I. M. Prausnitz, T. K. Sherwood: *The Properties of Gases and Liquids,* 3rd ed., McGraw-Hill, New York 1977;
VDI Wärmeatlas, 4th ed., VDI Verlag GmbH, Düsseldorf 1984.
[6] Internal communication, Henkel KGaA, 1987.
[7] Am. Oil Chem. Soc: "Short Course on Fatty Acids", Kings Island, Ohio, Sept. 13–16, 1987.
[8] C. S. Miner, N. N. Dalton: *Glycerol,* Reinhold Pub. Corp., New York 1953.
[9] *Physical Properties of Glycerine and its Solutions,* Glycerine Producers Association, New York 1963.
[10] *Chemical Properties and Derivatives of Glycerine,* Glycerine Producers Association, New York 1965.
[11] *Glycerine Refining,* Wurster & Sanger Co., Chicago, United States; Company brochure.
[12] *Glycerine,* Buss SMS GmbH, Butzbach, Federal Republic of Germany; Company brochure.
[13] Henkel KGaA EP 014 135 8531, 184 (R. Brockmann et al.).
[14] *Lewabit/Lewasorb,* Bayer AG, Leverkusen (FRG); Company brochure.
[15] *Neue Ionenaustauschertechniken (New Ion Exchanger Technology),* Bayer AG, Leverkusen (FRG); Company brochure.
[16] F. Martinola, *Chem. Ing. Tech.* **51** (1979) 728–736.
[17] K. Terelak et al., *Chem. Tech. (Leipzig)* **33** (1981) 315–316.
[18] Shell Dev. Co., US 2 605 293, 1952 (F. T. Tymstra).
[19] Shell Dev. Co., US 2 810 768, 1957 (K. B. Cofer).
[20] Pittsburgh Plate Glass Co., GB 926 804, 1963 (W. B. Graybill, H. J. Vogt, D. E. Wiley).
[21] Shell Dev. Co., US 2 779 801, 1957 (H. D. V. Finch, A. D. Benedictis).
[22] Progil S. A., FR 1 271 563, 1962 (A. Thizy, M. E. Degeorges, E. Charles).
[23] P. Vijaikishore, N. G. Karanth, *Process Biochem.* **21** (1986) no. 2, 54–57.
[24] Daicel Chem. Ind., JP 58 180 444, 1982 (J. Goto).
[25] Kao Corp., DE 3 533 615, 1986 (M. Tanigaki et al.).
[26] M. Bühler, Ch. Wandrey, *Fat Science Technology* **89** (1987) 156–164; 598–605.
[27] J. Hirano, *Chem. Econ. Eng. Rev.* **18** (1986) 9–13.
[28] M. Avron, *Adv. Photosynth. Res. Proc. Int. Congr. Photosynth.* 6th 1983, 745–753.
[29] B. J. Chem., C. H. Chi, *Biotechnol. Bioeng.* **23** (1981) no. 6, 1267–1287.
[30] R. Fonseca et al., *High-Pressure Sci. Technol. AIRAPT Conf.* 6th 1977 **1** (1979) 733–738.
[31] ChemG, Federal Republic of Germany (16th Sep. 1980); Gefahrstoffverordnung (26th Aug. 1986).
[32] Wasserhaushaltsgesetz, Federal Republic of Germany (23rd Sept. 1986).
[33] Erste Allgemeine Verwaltungsvorschrift zum Bundes-Immissionsschutzgesetz (Technische Anleitung zur Reinhaltung der Luft – TA Luft, 27th Feb. 1986).
[34] National Environmental Policy Act, United States, 1st Jan. 1970.
[35] Clean Air Act (based on the Air Quality Act of 1967), United States, 1977 amendment.

[36] Clean Water Act (Federal Water Pollution Control Act), United States, 1977 amendment.
[37] 2. Gesetz zur änderung des Abwasserabgabengesetzes, Federal Republic of Germany (19th Dec. 1987).
[38] K. Aitzetmüller, M. Böhrs, *Z. Lebensm. Unters. Forsch.* **169** (1979) 155–158.
[39] M. R. Sahasrabudhe, *J. Am. Oil Chem. Soc.* **44** (1967) 376–378.
[40] Rechnender Digitaler Dichtemesser für Flüssigkeiten und Gase (nach O. Kratky, H. Leopold und H. Stabinger, Graz); Fa. Paar K.G., A-8054 Graz–Austria.
[41] DIN 17 440: Nichtrostender Stahl (July 1985).
[42] American Society for Testing and Materials (ASTM), no. A 167, Type 304; no. A 167, Type 316.
[43] J. Knaut, H. J. Richtler: "Oleochemicals Outlook until the 90's", *Proc. 2nd World Conf. Detergents*, Montreux (Switzerland), Oct. 5th–10th, 1986, The American Oil Chem. Soc., Campaign, Ill.
[44] The United Kingdom Glycerine Producers' Assoc., London: *Glycerine in pharmacy, glycerine in cosmetics, glycerine-miscellaneous uses.*
[45] A. A. Newman: *Glycerol,* Morgan-Grampian, London 1968.
[46] E. Schlenker: *Das Glycerin,* Wissenschaftliche Verlagsgesellschaft, Stuttgart 1932.
[47] Deutsche Solvay-Werke, DE-OS 3 410 520 A1, 1984 (H. Dillenburg, H. Schafferus).
[48] Deutsche Solvay-Werke, DE 3 600 388, 1986 (G. Jakobson, W. Siemanowski).
[49] G. Jakobson, *Fette Seifen Anstrichm.* **88** (1986) 101–105.

Glyoxal

GEORGES MATTIODA, Société Française Hoechst, C.R.A., Stains, France
ALAIN BLANC, Société Française Hoechst, C.R.A., Stains, France

1.	Introduction	2809	5.	Uses	2813
2.	Physical Properties	2809	6.	Toxicology and Occupational	
3.	Chemical Properties	2809		Health	2814
4.	Production	2812	7.	References	2815

1. Introduction

Glyoxal [107-22-2], ethanedial, diformyl, $C_2H_2O_2$, M_r 58.04, was first prepared in 1856 by DEBUS, through the controlled oxidation of ethanol with nitric acid [1].

2. Physical Properties

Anhydrous glyoxal is a liquid at ambient temperature; it crystallizes at 15 °C in the form of yellow prismatic crystals and boils at 50.4 °C (101.3 kPa), giving off green vapors with a pungent odor. The liquid has a specific gravity of 1.14 at 20 °C; $n_D^{20.5} = 1.3826$. Electron diffraction and infrared spectra show that the molecule is planar; the trans configuration is the more stable [2].

3. Chemical Properties

As the simplest dialdehyde, glyoxal undergoes reactions characteristic of aldehydes (→Aldehydes, Aliphatic and Araliphatic).

Anhydrous monomeric glyoxal can be obtained by heating polyglyoxal in the presence of phosphorus pentoxide. Anhydrous glyoxal polymerizes rapidly under the action of traces of water, forming a series of hydrated oligomers. Glyoxal is used only as an aqueous solution, at concentrations between 30 and 50%. Evaporation of this solution produces a white, solid mass of polyglyoxal, which is infusible, depolymerizes upon heating, and decomposes above 150 °C.

In 40% aqueous solution, glyoxal exists mainly in the form of the hydrated monomer (**1**), together with a dioxolane dimer (**2**) and two bis(dioxolane) trimers (**3**); these components represent ca. 80% of its total composition [3].

In concentrated aqueous solution the most abundant component is the hydrated trimer (**3**), which is responsible for the cloudiness or crystalline precipitates in aqueous glyoxal solutions.

More highly condensed oligomers are also present, as shown by ^{1}H and ^{13}C NMR during acetalization by low molecular mass primary alcohols [4]; the species present have been quantified as a function of concentration and temperature.

With alcohols, under acid catalysis, a mixture of oligomeric acetals is obtained. Glycolates are also formed; glycolate formation predominates in the presence of a strong acid [5]. Glyoxal acetals and thioacetals are used as complex-forming agents [6].

With primary amines glyoxal forms diimines, reduction of which leads to ethylenediamines, [7] which are used industrially to cross-link isocyanates in coatings.

Glyoxal reacts with β-hydroxyalkylamines to give a heterodecalin:

$$\underset{\text{CHO}}{\overset{\text{CHO}}{|}} + \text{RNHCH}_2\text{CH}_2\text{OH} \longrightarrow \text{[cyclic product 1]}$$

$$+ \text{[cyclic product 2]}$$

With diethanolamine [*111-42-2*], the final product is *N,N*-bis(2-hydroxyethyl)glycine [*150-25-4*] (R = –CH$_2$CH$_2$OH):

$$2 \underset{\text{CHO}}{\overset{\text{CHO}}{|}} + 2\ \text{RNHCH}_2\text{CH}_2\text{OH} \longrightarrow \text{[bicyclic intermediate]}$$

$$\text{RN}\overset{\text{CH}_2\text{CH}_2\text{OH}}{\underset{\text{CH}_2\text{COOH}}{\diagdown}} \longleftarrow \text{[morpholinone]} \longleftarrow \text{[hydroxy morpholine]}$$

The latter is used to form complexes with polyvalent cations (see also → Ethylenediaminetetraacetic Acid and Related Chelating Agents) [8].

The corresponding aminoacetamides are formed with secondary amines [9], [10]:

$$2\ \underset{R^2}{\overset{R^1}{\diagdown}}\text{NH} + \underset{\text{CHO}}{\overset{\text{CHO}}{|}} \longrightarrow \underset{R^2}{\overset{R^1}{\diagdown}}\text{NCH}_2\overset{\text{O}}{\underset{}{\overset{\|}{\text{C}}}}\text{N}\underset{R^2}{\overset{R^1}{\diagup}}$$

Controlled oxidation leads successively to glyoxylic acid [*298-12-4*] and oxalic acid [*144-62-7*] [11], [12].

$$\underset{\text{CHO}}{\overset{\text{CHO}}{|}} \xrightarrow{\text{HNO}_3} \underset{\text{CHO}}{\overset{\text{COOH}}{|}} \xrightarrow{\text{HNO}_3} \underset{\text{COOH}}{\overset{\text{COOH}}{|}}$$

With aldehydes and enolizable ketones, glyoxal leads to the expected aldol condensation products [13]; e.g., with acetone or isobutyraldehyde:

$$\underset{\text{CH}_3}{\overset{\text{CH}_3}{\diagdown}}\text{C}=\text{O} + \underset{\text{CHO}}{\overset{\text{CHO}}{|}} \longrightarrow \text{CH}_3\text{COCH}=\text{CHCHO}$$

$$\underset{\text{CH}_3}{\overset{\text{CH}_3}{\diagdown}}\text{CHCHO} + \underset{\text{CHO}}{\overset{\text{CHO}}{|}} \longrightarrow \text{HO}-\overset{\text{CHO}}{\underset{}{\text{CH}}}-\underset{\underset{\text{CH}_3}{|}}{\overset{\underset{}{|}}{\text{C}}}-\text{CHO}$$

$$\downarrow$$

[cyclic hemiacetal] ← [open-chain form with CH$_2$OH, CHO, CH$_3$, CH$_3$]

Chemical Properties

2811

With amides and urethanes, di- and tetramides are obtained, depending on the pH [14], [15].

$$2\ RCONH_2 + \begin{array}{c}CHO\\|\\CHO\end{array} \xrightarrow{NaHCO_3} \begin{array}{cc}RCONH & OH\\ \diagdown\diagup & \\ HO & NHCOR\end{array}$$

$$\Bigg\downarrow 2\ RCONH_2\ \Big|\ H^+$$

$$\begin{array}{cc}RCONH & NHCOR\\ \diagdown\diagup & \\ RCONH & NHCOR\end{array}$$

In alkaline medium, glyoxal disproportionates by an internal Cannizzaro reaction to the glycolate [16]–[18]:

$$\underset{\text{Fast}}{\xrightarrow{OH^-}} \underset{\text{Slow}}{\xrightarrow{}} \longrightarrow HOCH_2COO^-$$

4. Production

Among the numerous processes for producing glyoxal, only those using *acetaldehyde* [75-07-0] and *ethylene glycol* [107-21-1] as starting materials have been developed commercially.

From Acetaldehyde. Oxidation with nitric acid was examined by LJUBOWIN as early as 1875 and patented in 1942 [19]. Reaction takes place at ca. 40 °C and is carried out industrially as a continuous process. Maximum yield is ca. 70%; selectivity is a function of the relative concentrations of reagents. After the removal of excess acetaldehyde, the glyoxal formed, which is contaminated with acetic, formic, and glyoxylic acids, is purified by passage of the aqueous solution through an ion-exchange resin. The solution is then concentrated to a glyoxal content of about 40%.

Selenium oxide [7446-08-4] is more selective than nitric acid, and the yield is ca. 84%; the selenium can be recycled by oxidation with hydrogen peroxide [20]. This process has not been carried out on an industrial scale.

From Ethylene Glycol. The gas-phase oxidation of ethylene glycol by atmospheric oxygen in the presence of dehydrogenation catalysts (metallic copper or silver) represents the basis of the Laporte process [21] and has been used in several industrial production processes. Reaction occurs between 400 and 600 °C; the yield is 70–80%. The main impurity is formaldehyde [50-00-0], whose subsequent separation is difficult. This reaction has also been carried out in the liquid phase and under irradiation.

Other Processes. Ethylene can be oxidized by aqueous nitric acid in the presence of palladium [22], by atmospheric oxygen, or by selenium oxide deposited on silica [23]. Glyoxal may also be formed by oxidation of acetylene [24] or benzene [25] with ozone. Ethylene oxide has been proposed as a substrate for oxidation. Although oxalic acid and its derivatives can be reduced to glyoxal, these processes have not been developed further.

5. Uses

Glyoxal is supplied mainly as a 40% aqueous solution. Various polyhydroxy polymers (e.g., starch or poly(vinyl alcohol)) can be used to stabilize glyoxal.

The bifunctionality of glyoxal has been used to cross-link functionalized macromolecules such as cellulose, polyacrylamides, poly(vinyl alcohol), keratin, and other polycondensates. For example, glyoxal is used as a cross-linking agent for imparting wet strength to coated paper. With cellulose, unstable hemiacetals are obtained in the cold, which irreversibly form acetals when heated in the presence of acid catalysts [26].

The dual functionality and the ability of glyoxal to form heterocyclic compounds have been used in the production of resins for imparting crease resistance to textiles. The reaction proceeds through the intermediate formation of 4,5-dihydroxy-2-imidazolidinone (**4**) [*3720-97-6*] (DHEU) and its methylolated derivatives (DMDHEU).

Ethers of DMDHEU are among the most effective resins for the crease-resistant treatment of cotton-based fabrics [27], [28].

The reducing properties of glyoxal are used in the photographic industry and in glassmaking for the production of silvered glass mirrors.

Glyoxal bisulfite (available as the monohydrate) is used as a resist agent in printing with reactive dyes and as a leveling agent in dyeing polyamide with acid dyes.

An important class of molecules used as cross-linking agents is obtained by nucleophilic substitution of glyoxal. Examples include the bisacrylamide (**5**) [*76843-24-8*] and the tetraallylacetal (**6**) [*16646-44-9*].

$$\underset{5}{\underset{HO}{CH_2=CHCONH}\diagdown\underset{NHCOCH=CH_2}{OH}}$$

$$\underset{6}{\underset{CH_2=CHCH_2O}{CH_2=CHCH_2O}\diagdown\underset{OCH_2CH=CH_2}{OCH_2CH=CH_2}}$$

Glyoxalbisacrylamide is used for the functionalization of ion-exchange resins and in latices for treating textiles. Tetraallyloxyethane is used as a cross-linking agent for polyacrylates and for pressure-sensitive adhesives.

Glyoxal is used in the fine chemicals industry for the production of various heterocyclic compounds including tetraacetylglycoluril [*10543-60-9*] (**7**) (a perborate accelerator), imidazoles such as metronidazole [*443-48-1*] (**8**) that are effective against anaerobic bacteria, and the pyrazine derivatives sulfapyrazine [*116-44-9*] (**9**), thionazine *297-97-2*] (**10**), and pyrazinamide [*98-96-4*] (**11**).

Glyoxal has bactericidal properties comparable with those of glutaraldehyde and is used as a disinfectant.

6. Toxicology and Occupational Health

Glyoxal has low toxicity. The LD_{50} is 3300 mg/kg (oral, rat). No mortality was observed in rats at a dosage of 1.3 mg/L inhaled in a saturated atmosphere or with aerosols. Glyoxal was well tolerated topically in a rabbit skin test but was irritating to the eyes. Glyoxal causes an allergic sensitization in guinea pigs.

Glyoxal gives a positive result in the Ames mutagenesis test, but it does not change cells in the cell transformation test and has no elastogenic effect in the micronucleus test; it is also not mutagenic in the Chinese hamster lung test. All of these tests are generally positive with substances known to be mutagenic.

7. References

[1] H. Debus, *Ann. Chem. Pharm.* **100** (1856) 5; **102** (1857) 20–24.
[2] U. Pincelli, B. Cadioli, D. J. David, *J. Mol. Struct.* **12** (1971) 171.
[3] E. B. Whipple, *J. Am. Chem. Soc.* **92** (1970) 7183–7186.
[4] F. Chastrette, C. Bracoud, M. Chastrette, G. Mattioda, Y. Christidis, *Bull. Soc. Chim. Fr.* 1983 33–40.
[5] J. M. Kliegman, R. K. Barnes, *J. Org. Chem.* **38** (1973) 556–560; *J. Org. Chem.* **39** (1974) 1772–1776.
[6] F. Chastrette, M. Hassamblay, M. Chastrette, *Bull. Soc. Chim. Fr.* **1976,** 601–606; 607–612; 613–615.
[7] Hoechst, DE 2 938 710, 1979 (H. Diery, W. Wagemann).
[8] A. Le Rouzic, D. Raphalen, D. Papillon, M. Kerfanto, *Tetrahedron Lett.* **26** (1985) 1853–1856; Société Française Hoechst, FR 2 561 645, 1984 (A. Blanc); FR 2 561 648, 1984 (A. Blanc).
[9] K. Maurer, E. H. Woltersdord, *Z. Physiol. Chem.* **254** (1938) 18–24.
[10] P. Ferruti, A. Feré, L. Zetta, A. Bettelli, *J. Chem. Soc. C.* 1971, 2984–2985.
[11] BASF, DE 932 369, 1955 (H. Spänig, G. Triem).
[12] Nobel Bozel, FR 1 326 605, 1962 (L. Gandon).
[13] BASF, DE 2 003 600, 1970 (F. Merger).
[14] S. L. Vail, C. M. Moran, R. H. Barker, *J. Org. Chem.* **30** (1965) 1195–1199.
[15] S. L. Vail, A. G. Pierce, *J. Org. Chem.* **37** (1972) 391–393.
[16] P. Salomaa, *Acta. Chem. Scand.* **10** (1956) 311–319.
[17] C. L. Arcus, B. A. Jackson, *Chem. Ind. (London)* 1964, 2022–2023.
[18] A. R. Fratzke, P. J. Reilly, *Int. J. Chem. Kinet.* **18** (1986) no. 7, 757–773.
[19] I. G. Farbenindustrie AG, BF 885 931 (1942).
[20] Air Liquide, FR 2 038 575, 1969 (J. P. Zumbrunn).
[21] Laporte Chemicals, GB 1 272 592, 1963 (B. K. Howe, F. R. Hary, D. A. Clarke).
[22] BASF, DE 1 166 173, 1962 (R. Platz, W. Fuchs); DE 1 231 230, 1964 (R. Platz, H. Nohe).
[23] E. Costa Novella, *Ann. Quim.* **68** (1972) no. 3, 325–332;*Chem. Abstr.,* **77**, 113 760 f.
[24] Imperial Chemical Industries, GB 1 071 902, 1965 (R. A. Rennie).
[25] Inmont Corp., US 3 637 860, 1968 (W. P. Keaveney, J. J. Pappas).
[26] F. S. H. Head, *J. Text. Inst.* **49** (1958) T 345–T 356.
[27] H. Petersen, *Text. Res. J.* (1958) 156–176.
[28] S. L. Vail, G. B. Verburg, A. H. Young, *Text. Res. J.* (1969) 86–93.

Glyoxylic Acid

GEORGES MATTIODA, Société Française Hoechst, C.R.A., Stains, France

YANI CHRISTIDIS, Société Française Hoechst, C.R.A., Stains, France

1.	Introduction 2817	4.	Production 2821
2.	Physical Properties 2817	5.	Toxicology 2821
3.	Chemical Properties 2818	6.	References 2822

1. Introduction

Glyoxylic acid [298-12-4], oxoacetic acid, glyoxalic acid, OHC–COOH, M_r 74.04, is the simplest α-oxocarboxylic acid. It is found in plants and is involved in the metabolic cycle of animals. Glyoxylic acid was discovered by DEBUS, who established its formula in 1856 [1]. This formula was described more precisely as that of the monohydrate by PERKIN in 1868 [2].

In 1957, KRONBERG and KREBS established the basis of the metabolic cycle of glyoxylic acid [3], in which isocitrate is converted to malate anaerobically. This cycle acts as a relay in the tricarboxylic acid cycle under specific conditions of plant life, such as germination or ineffective photosynthesis [4], [5].

2. Physical Properties

Glyoxylic acid crystallizes as the monohydrate upon controlled concentration of its aqueous solution. When dry, the monohydrate has mp 52–53 °C. The dissociation constant in aqueous solution is 4.7×10^{-4}, and the specific heat is 1.80 kJ kg^{-1} K^{-1}. The density of a 50 % aqueous solution of the acid, the only commercial form, is 1.34 g/cm^3 at 20 °C; the refractive index (20 °C) is 1.416.

The use of NMR spectroscopy (^{1}H [6] and ^{13}C [7]) has confirmed the existence of glyoxylic acid in aqueous solution as the dihydroxy acid (**1**) together with a small proportion of the linear dimer (**2**).

$$\underset{1}{\underset{HO}{HO}}CH-COOH \qquad \underset{2}{\underset{HO}{O}}C-\underset{OH}{\underset{|}{CH}}\underset{O}{\overset{OH}{\underset{|}{CH}}}-C\overset{OH}{\underset{O}{\diagdown}}$$

Glyoxylic acid is strongly hydrophilic; it is soluble in alcohols, with which it reacts, and in water-miscible solvents; it has a very low solubility in ether and other organic solvents.

Glyoxylic acid forms complexes with ions of alkali and alkaline-earth metals; it has a chelating power of 30 mg of calcium carbonate per gram of 50% aqueous solution.

3. Chemical Properties

Glyoxylic acid contains two functional groups: the carbonyl group, which undergoes reactions characteristic of aldehydes (→Aldehydes, Aliphatic and Araliphatic), and the carboxylic acid group (→Carboxylic Acids, Aliphatic). The aldehyde group reacts readily with nucleophilic reagents; the hydrated aldehyde and the hemiacetal react similarly. With ambident nucleophiles, the carboxylic group may also react, leading to intramolecular ring formation. Various heterocyclic compounds can be obtained by coupling polynucleophiles with glyoxylic acid: o-phenylenediamine [95-54-5] gives 2-hydroxyquinoxaline [1196-57-2] (3); with urea [57-13-6] and an acid catalyst, allantoin [97-59-6] (4) is obtained in 60% yield [8].

On heating, glyoxylic acid disproportionates to a mixture of glycolic acid [79-14-1] and oxalic acid [144-62-7]; glyoxylic acid is also easily oxidized to oxalic acid by nitric acid.

Reactions that have been used industrially include the Mannich reaction and amidoalkylation. Thus, with phenol and ethylenediamine in alkaline medium, glyoxylic acid leads to the sodium salt of N,N'-ethylenebis[2-(2-hydroxyphenyl)glycine] (EHPG, 5), which forms complexes with iron (III) [9].

$$2 \; \underset{\text{COONa}}{\overset{\text{CHO}}{|}} + H_2NCH_2CH_2NH_2 \longrightarrow$$

$$\underset{\overset{|}{\text{OH}}}{\overset{\text{COONa}}{|}}{\text{CH}}-NHCH_2CH_2NH-\underset{\overset{|}{\text{OH}}}{\overset{\text{COONa}}{|}}{\text{CH}} \xrightarrow{2\,C_6H_5OH}$$

[Structure 5: bis(hydroxyphenyl) compound with NaOOC and COONa groups bridged by CH-NHCH$_2$CH$_2$NH-HC, with OH and HO on the two phenyl rings]

Glyoxylic acid, phenol, and ammonia react to give 4-hydroxyphenylglycine, an intermediate for the semisynthetic penicillin amoxicillin [10].

$$\underset{\text{COOH}}{\overset{\text{CHO}}{|}} + NH_3 + \text{C}_6\text{H}_5\text{OH} \longrightarrow HO\text{-C}_6\text{H}_4\text{-}\underset{\overset{|}{\text{NH}_2}}{\text{CH}}-COOH$$

At slightly alkaline pH, glyoxylic acid reacts with amides such as acrylamide [79-06-1] to form acrylamidoglycolic acid (AGA), which is used as a copolymerizable cross-linking agent [11].

$$\text{CH}_2=\text{CH-C(O)NH}_2 + \underset{\text{COOH}}{\overset{\text{CHO}}{|}} \longrightarrow \text{CH}_2=\text{CH-C(O)-N(H)-CH(OH)-COOH}$$

Several derivatives of AGA have found industrial application; e.g., the methyl ether ester is used in coatings and paint for automobiles.

Glyoxylic acid combines with phenol in alkaline medium to give 4-hydroxymandelic acid with good selectivity [12].

$$\underset{\text{COOH}}{\overset{\text{CHO}}{|}} + \text{C}_6\text{H}_5\text{OH} \longrightarrow HO\text{-C}_6\text{H}_4\text{-CH(OH)-COOH}$$

The reaction with phenols is used in several industrial syntheses of benzaldehydes. For example, guaiacol [90-05-1] is converted to vanillin [121-33-5] by oxidative decarboxylation of the corresponding mandelic acid [13].

Guaiacol (HO-C$_6$H$_4$-OCH$_3$) + $\underset{\text{COOH}}{\overset{\text{CHO}}{|}}$ →

[HO, OCH$_3$-substituted phenyl-CH(OH)-COOH] $\xrightarrow[-CO_2]{O_2,\,Cu^{2+}}$ Vanillin (HO, OCH$_3$-substituted phenyl-CHO)

The mandelic acid intermediate may be modified further: reaction of phenol and glyoxylic acid in the presence of a reducing agent such as phosphorus–iodine leads directly to hydroxyphenylacetic acid [14], a building block for pharmaceuticals such as Atenolol.

$$\text{HO-C}_6\text{H}_5 \xrightarrow{\text{CHO, COOH}} \text{HO-C}_6\text{H}_4\text{-CH(OH)COOH} \xrightarrow{P, I_2} \text{HO-C}_6\text{H}_4\text{-CH}_2\text{COOH}$$

With thiophene, this reaction yields 2-thi-enylacetic acid, which is used to prepare the semisynthetic cephalosporins cephalothin and cefoxitin.

Glyoxylic acid and hydrochloric acid can be used to chloroalkylate aromatic compounds [15].

Glyoxylic acid undergoes the aldol reaction; the corresponding benzoylacrylic acids are obtained with acetophenones [16]:

$$R\text{-C}_6\text{H}_4\text{-COCH}_3 + \text{CHO-COOH} \longrightarrow R\text{-C}_6\text{H}_4\text{-COCH=CHCOOH}$$

Glyoxylic acid is treated with Wittig ylides to form carboxyl functional polyenes of the vitamin A type [17].

$$\begin{array}{c}\text{CHO}\\|\\\text{COOC}_4\text{H}_9\end{array} + \text{CH}_3\text{CH}_2\text{CHO} \longrightarrow \begin{array}{c}\text{CHO}\\\diagdown\\\text{CH}_3\diagup\end{array}\!\!=\!\text{CHCOOC}_4\text{H}_9$$

[Structure: trimethyl-methoxyphenyl-CH=CH-C(CH$_3$)=CH-CH$_2$P$^+$Ph$_3$Br$^-$ + CHO-C(CH$_3$)=CHCOOC$_4$H$_9$]

$\longrightarrow$ [Structure: extended polyene with aryl group and COOC$_4$H$_9$ terminus]

Esters of glyoxylic acid, obtained by dehydrative distillation of the hemiacetal esters in the presence of phosphorus pentoxide, can be polymerized under the action of a base. After saponification, the resulting polymers show useful chelating properties with the advantage of good biodegradability [18], [19]. Low molecular mass polyglyoxylates have been proposed as substitutes for sodium tripolyphosphates (TPP) in detergents to control water hardness.

$$\begin{array}{c}\text{CHO}\\|\\\text{COOH}\end{array} + 2\,\text{ROH} \longrightarrow \begin{array}{c}\text{CH}{<}^{\text{OH}}_{\text{OR}}\\|\\\text{COOR}\end{array} \xrightarrow{P_2O_5} \begin{array}{c}\text{CHO}\\|\\\text{COOR}\end{array}$$

$$M^+R'^- + \begin{array}{cc}\overset{\frown}{\text{CH}{=}\text{O}} & \overset{\frown}{\text{CH}{=}\text{O}}\\|&|\\\text{COOR}&\text{COOR}\end{array} \longrightarrow$$

$$R'\!-\!(\text{CH}\!-\!\text{O})_n\!-\!\text{CH}\!-\!\text{O}^-M^+\\|\phantom{\text{CH}\!-\!\text{O})_n\!-\!}|\\\text{COOR}\text{COOR}$$

$$\downarrow$$

$$R'\!-\!(\text{CH}\!-\!\text{O})_n\!-\!\text{CH}\!-\!\text{OH}\\|\phantom{\text{CH}\!-\!\text{O})_n\!-\!}|\\\text{COONa}\text{COONa}$$

4. Production

Glyoxylic acid is produced industrially by oxidation of glyoxal in aqueous solution with 65% nitric acid in mole ratios of 1:1 to 1:1.5 between 40 and 90 °C. The main byproduct of this process is oxalic acid, which is separated by low-temperature crystallization. The solution is then purified by passage through an anion-exchange resin or by electrodialysis, which removes the residual nitric acid.

Glyoxal can also be oxidized at the anode of a two-compartment electrolytic cell in the presence of chloride ion [20]. Glyoxylic acid may be synthesized by catalytic oxidation of ethylene or acetaldehyde, but the selectivity is low and these routes have not been used industrially.

The cathodic reduction of oxalic acid gives a very good chemical yield (85%), but this technique encounters problems with passivation of the lead electrodes [21], [22].

Another method is oxidative cleavage of maleic acid or its esters by ozone. This process, has been adapted to the preparation of hemiacetal esters [23].

5. Toxicology

The LD_{50} is 2500 mg/kg (oral, rat). Glyoxylic acid is corrosive to the eye (rabbit), does not irritate the skin, but causes an allergic sensitization in guinea pigs. No mutagenic effect was found in the Ames mutagenicity test; glyoxylic acid is completely inactive in the so-called micronucleus mutagenicity test in mice.

Glyoxylic acid is a metabolite in mammalian biochemical pathways.

6. References

[1] H. Debus, *Ann. Chem. Pharm.* **100** (1856) 1.
[2] W. H. Perkin, B. F. Duffa, *Bull. Soc. Chim. Fr.* **10** (1868) 213.
[3] H. L. Kornberg, H. A. Krebs, *Nature (London)* **179** (1957) 988–991.
[4] H. Beevers, *Ann. N.Y. Acad. Sci.* **168** (1969) 313–324.
[5] M. Cioni et al., *Comp. Biochem. Physiol.* **70 B** (1981) 1–26.
[6] G. Ojelund, I. Wadso, *Acta Chem. Scand.* **21** (1967) no. 6, 1408–1414.
[7] F. Chastrette, C. Bracoud, *Bull. Soc. Chim. Fr.* 1985(II) 66–74.
[8] BASF, DE-OS 1 939 924, 1969 (W. Mesch, O. A. Grosskinsky, N. Lösch).
[9] Ciba-Geigy, US 4 130 582, 1977 (H. E. Petree, H. Myatt, A. M. Jelenevsky).
[10] Beecham, GB 978 178, 1962 (J. H. C. Nayler, H. Smith).
[11] Nobel Bozel, FR 1 411 715, 1964.
[12] Soc. Française Hoechst, FR 2 440 350, 1978 (A. Schouteeten, Y. Christidis).
[13] Haarmann & Reimer, DE-OS 2 115 551, 1971 (K. Bauer, W. Steuer).
[14] Soc. Française Hoechst, FR 2 470 127, 1979 (J. C. Vallejos, Y. Christidis).
[15] Soc. Française Hoechst, FR 2 376 112, 1976 (E. Herman, H. Diery, M. Soreau, Y. Christidis).
[16] Roussel-Uclaf, FR 2 504 005, 1981 (Y. Christidis, R. Fournex, C. Tournemine).
[17] Hoffmann La Roche, US 4 224 244, 1980 (W. Bollag, R. Ruegg, G. Ryser).
[18] Monsanto, US 4 144 226, 1977 (M. M. Crutchfield, V. D. Papanu, C. B. Warren).
[19] Procter & Gamble, US 4 284 524, 1980 (L. A. Gilbert).
[20] Chlorine Engineers and Daicel Chem., FR 2 443 517, 1979 (H. Harada, K. Hirao, M. Ichino, T. Mitani).
[21] I. V. Kudryashov et al., *Tr. Mosk. Khim. Tekhnol. Inst.* **49** (1965) 111–115; *Chem. Abstr.* **65** 3328 d.
[22] K. Scott, *Chem. Eng. Res. Des.* **64** (1986) no. 4, 266–272.
[23] Lentia, DE 3 224 795, 1982 (A. Sajtos, M. Wechsberg, E. Roithner).

Guanidine and Derivatives

CLEMENS GRAMBOW, SKW Trostberg AG, Trostberg, Federal Republic of Germany (Chaps. 1 and 3)

STEFAN WEISS, SKW Trostberg AG, Trostberg, Federal Republic of Germany (Chaps. 2 and 3)

RICHARD YOUNGMAN, SKW Trostberg AG, Trostberg, Federal Republic of Germany (Section 1.8.6)

1.	**Guanidine and Guanidine Salts**	2823	1.8.5.	Propellants and Explosives..... 2831
1.1.	**Properties**...............	2824	1.8.6.	Biotechnological Applications of Guanidine Salts 2831
1.2.	**Production**	2826	1.9.	**Economic Aspects** 2832
1.2.1.	Guanidine Salts from Dicyandiamide	2826	2.	**Derivatives** 2832
1.2.2.	Guanidine Salts from Cyanamide	2827	2.1.	**Nitroguanidine** 2832
1.2.3.	Guanidine Salts from Urea	2828	2.2.	**Aminoguanidine** 2833
1.2.4.	Other Processes............	2828	2.3.	**Diamino- and Triaminoguanidine** 2834
1.3.	**Environmental Protection** ...	2829	2.4.	**Organoguanidines** 2835
1.4.	**Quality Specifications**.......	2829	2.4.1.	Alkylguanidines............ 2835
1.5.	**Analysis**	2829	2.4.2.	Arylguanidines 2837
1.6.	**Storage and Transportation** ..	2829	2.5.	**Cyanoguanidines (Dicyandiamides)** 2839
1.7.	**Legal Aspects**.............	2830	2.6.	**Biguanide and Derivatives** 2839
1.8.	**Uses**	2830	3.	**Toxicology and Occupational Health**.................. 2841
1.8.1.	Pharmaceuticals............	2830	4.	**References**............... 2842
1.8.2.	Plant Protection............	2830		
1.8.3.	Cosmetics................	2831		
1.8.4.	Textile Impregnation, Flame Retardants, Paper and Resin Manufacture	2831		

1. Guanidine and Guanidine Salts

Guanidine [113-00-8], CH$_5$N$_3$, M_r 59.08, was first isolated in 1861 by STRECKER as a degradation product of guanine [73-40-5]. Although free

$$H_2N-\underset{\underset{NH_2}{|}}{\overset{NH}{\overset{\|}{C}}}$$

guanidine or guanidine salts occur naturally only in trace amounts [1], many guanidine derivatives are important components of living organisms [2]. Guanidine and its derivatives have a wide field of application, e.g., as building blocks in the synthesis of pharmaceutical and agricultural chemicals, in the manufacture of textiles and plastics, and also in biochemistry [3]–[7].

1.1. Properties

Free guanidine can be isolated from guanidine salts by reaction with a strong base such as an alkali hydroxide or methoxide. After removal of the precipitated salt the free guanidine base is obtained as colorless, waxy, very hygroscopic crystals (*mp* <50 °C).

Guanidine is a strong monobasic Brönsted base, which absorbs water and carbon dioxide from the air and forms strongly alkaline solutions with water and alcohols; the pH of a 20 wt % aqueous solution is approx. 13.5 (25 °C). In aqueous solution at elevated temperature guanidine is hydrolyzed to urea; airtight alcohol solutions are stable. Guanidine is infinitely soluble in water, ethanol, methanol, and dimethylformamide.

Protonation of guanidine gives the mesomeric, highly stabilized guanidinium cation:

$$H_2N-\overset{+NH_2}{\underset{}{C}}-NH_2 \longleftrightarrow H_2N-\overset{NH_2}{\underset{}{C}}=\overset{+}{N}H_2 \longleftrightarrow H_2\overset{+}{N}=\overset{NH_2}{\underset{}{C}}-NH_2$$

Guanidine forms stable, easily crystallizable salts with acids, even with a weak acid such as stearic, carbonic, or silicic acid. Some physical properties of various guanidine salts are listed in Table 1.

Salts of guanidine should, according to IUPAC nomenclature, be named as guanidinium compounds. In this article, however, common (although technically incorrect) names will be used, e.g., guanidine nitrate rather than guanidinium nitrate.

The very stable guanidinium cation is unreactive, so reactions of guanidine with organic compounds are usually carried out under strongly alkaline conditions. Like other amines, guanidine undergoes alkylation, acylation, methylolation, and condensation. Many guanidine derivatives are synthesized from guanylating agents such as cyanamide [420-04-2] (→ Cyanamides) or O-methylisourea [2440-60-0] (see Sect. 2.4) because of the relatively low reactivity of guanidine itself.

The most important reaction of guanidine is the formation of heterocyclic compounds such as pyrimidines, imidazoles and oxazoles. Guanidine condenses with bifunctional compounds such as diketones, acetoacetic acid esters, and malononitrile [109-77-3], eliminating ammonia or water to form 2-aminoheterocycles, for example:

$$R^2-\underset{R^3}{\overset{R^1}{\underset{|}{CH}}}\underset{=O}{\overset{=O}{}} + \underset{HN}{\overset{H_2N}{}}\!\!\!\!>\!\!-NH_2 \xrightarrow{-2H_2O} R^2-\underset{R^3}{\overset{R^1}{\underset{|}{C}}}\underset{-N}{\overset{=N}{}}\!\!\!\!>\!\!-NH_2$$

The formation of nitroguanidine [556-88-7] from guanidine nitrate is also an industrially important reaction. Many guanidine reactions are cited in the general references [1], [3]–[7]. For toxicological properties, see Chapter 3 and references [7], [8].

Table 1. Physical Properties of Guanidine Salts

	Chloride	Nitrate	Carbonate	Dihydrogen phosphate	Hydrogen phosphate	Sulfate	Stearate	Acetate	Thiocyanate	Aminosulfonate
CAS registry number	[50-01-1]	[506-93-4]	[593-85-1]	[5423-22-3]	[5423-23-4]	[594-14-9]	[26739-53-7]	[593-87-3]	[593-84-0]	[51528-20-2]
M_r	95.54	122.09	180.17	157.07	216.22	216.1	343.36	119.13	118.16	156.2
mp, °C	185	215	>190[a]	130	246	>280	78	>227	117	127
d (25 °C)	1.344	1.436	1.251	1.684	1.481	1.29	1.026		1.45	1.68
pH of aqueous solution (20 wt %, 25 °C)	4.8	4.9	11.2	3.9	8.1	9.6		10	5	7
Solubility[b], g per 100 g of										
Water	228	20	50	230	23	ca. 270	[c]	115	160	120
90 °C	450	148	80	410	160	ca. 400		160	∞	
Ethanol	30	1.4	0.03	<0.1	<0.1	0.02	135	8	65	
70 °C	57	8.6	0.1	<0.1	<0.1	0.05		ca. 25	170	
Methanol 20 °C	76	5.5	0.55		0.1				90	
Acetone	0.05	0.1	0.02	<0.05	<0.05	<0.03	0.24	[c]	105	
Benzene	<0.05	0.05	0.01	<0.05	<0.05	<0.05	0.38	[c]	0.4	
DMF[d]	19.5	6.1	0.02			0.05	0.21	15		

[a] Decomposes.
[b] Unless otherwise stated, $t = 30$ °C.
[c] Trace.
[d] Dimethylformamide.

1.2. Production

No practical method exists for the direct synthesis of free guanidine. All major manufacturing processes therefore involve the production of guanidine salts by reaction of the appropriate ammonium salt with compounds containing the NCN group. At present, only *dicyandiamide* [461-58-5] (→ Cyanamides), *cyanamide*, and *urea* [57-13-6] are used for large-scale production of guanidine salts.

1.2.1. Guanidine Salts from Dicyandiamide

Dicyandiamide is by far the most important raw material for the production of guanidine salts. Earlier processes conducted in acidic aqueous solution with guanylurea as an intermediate [3] have been abandoned. At present, guanidine salts are manufactured by fusing dicyandiamide with ammonium salts, with biguanide [56-03-1] as an intermediate:

$$H_2N-\underset{H}{\overset{NH}{\underset{\|}{C}}}-N-CN + NH_4X \longrightarrow H_2N-\underset{H}{\overset{NH}{\underset{\|}{C}}}-\underset{}{\overset{NH}{\underset{\|}{C}}}-NH_2 \cdot HX$$

$$H_2N-\underset{H}{\overset{NH}{\underset{\|}{C}}}-\underset{}{\overset{NH}{\underset{\|}{C}}}-NH_2 \cdot HX + NH_4X \longrightarrow 2\,H_2N-\overset{NH}{\underset{\|}{C}}-NH_2 \cdot HX$$

When carried out in batches in open stainless steel vessels at atmospheric pressure, reaction begins above 130–140 °C and becomes exothermic at 170 °C. The temperature is allowed to rise to 210–230 °C and is maintained within this range until the reaction has reached completion. The molten mass is then dissolved in water or mother liquor; insoluble triazine byproducts are filtered off. The guanidine salt is crystallized by cooling or concentration of the filtrate, centrifuged, washed, and dried.

Large-scale production of guanidine salts such as guanidine nitrate is performed continuously [4] (see Fig. 1).

Dicyandiamide (85 parts) and ammonium nitrate [6484-52-2] (180 parts) are mixed and dosed by a screw into a heated stirring vessel where the mixture is melted down. The melt is then run into a cascade of two further vessels where the main reaction occurs. The hot molten mass is subsequently dissolved in mother liquor to give a warm salt solution, which is pumped through a filter (to remove solid triazines) and cooled in a continuously operating crystallizer. The crystallized guanidine nitrate is centrifuged, dried, and stored. This process is conducted on a technical scale for guanidine nitrate, hydrochloride, and sulfamate. The yield of pure salt is normally ca. 90–95 %. By solidifying the molten mass on a cooling conveyor belt immediately after it leaves the reaction cascade, a crude product can be obtained containing ca. 91–92 % guanidine salt and 3–5 % ammonium salt, together with small amounts of dicyandiamide and triazines.

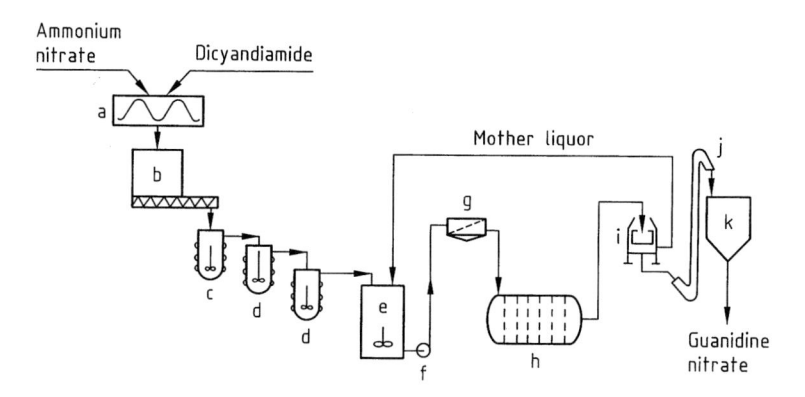

Figure 1. Production of guanidine nitrate
a) Mixer; b) Metering bunker with screw; c) Melting vessel; d) Reaction vessels; e) Dissolving tank; f) Pump; g) Filter press; h) Crystallizer; i) Centrifuge; j) Dryer; k) Storage bunker

Other processes involve the reaction of dicyandiamide, ammonium salts, and ammonia in aqueous solution or in liquid ammonia under pressure [4]. These processes are not industrially important.

1.2.2. Guanidine Salts from Cyanamide

Cyanamide is used mainly for the production of substituted guanidines (see Chap. 2). For simple guanidine salts, calcium cyanamide [156-62-7] is industrially important only in the production of guanidine nitrate [9].

Calcium cyanamide is mixed with mother liquor (from guanidine nitrate crystallization) and concentrated ammonium nitrate solution, and reacted at 120 °C. An aqueous solution of ammonium carbonate [506-87-6] is subsequently added to precipitate calcium carbonate. The solids are separated by concentration and filtration. The guanidine nitrate solution is neutralized with nitric acid, freed of precipitated sulfur by filtration, and of noxious gases such as hydrogen sulfide by stripping, and worked up as described above. Although this process gives good yields, it is technically very complicated and thus susceptible to breakdowns.

In modern cyanamide processes an aqueous solution of an ammonium salt and some free ammonia is heated in a stirred autoclave to 120–140 °C [10]. Aqueous cyanamide solution is then added slowly, and the reaction is carried out at 0.5–1.5 MPa. After being cooled to < 100 °C, the solution can be worked up as usual. This method is used particularly to make guanidine salts that cannot be produced efficiently in a fusion process.

1.2.3. Guanidine Salts from Urea

Many attempts have been made to produce guanidine salts from urea, because of the great cost advantage of urea compared with calcium cyanamide derivatives. As an example, a mixture of molten urea and ammonium nitrate is treated with a silica gel catalyst [11]. Such processes are not industrially important because of safety hazards and the complicated reaction control and purification procedures.

Guanidine carbonate is formed from urea as a byproduct of some melamine [108-78-1] production processes. It can be isolated from the melamine mother liquor by heating (to remove ammonia and carbon dioxide, and precipitate the remaining melamine) and filtering. After further concentration, the liquor is cooled and the crystallized guanidine carbonate is separated [12]. Although the removal of byproducts is rather expensive, significant amounts of guanidine carbonate are isolated in this way by certain melamine producers, mainly to minimize the waste from melamine production, which would otherwise have to be burned or buried.

1.2.4. Other Processes

Many processes for the production of guanidine salts are mentioned in the literature: e.g., from urea and carbon disulfide; from urea, sulfur dioxide, and ammonia; from carbon disulfide and ammonia; from carbon dioxide, ammonia, and sulfur dioxide; from ammonium thiocyanate; or from thiourea. None of these methods is used on an industrial scale. For details, see [1], [3]–[7].

Guanidine salts of weak acids or anions that do not form stable ammonium salts can be manufactured via the free guanidine base. The latter is prepared by reaction of a suitable guanidine salt with a strong base, whose cation can be precipitated with the anion of the guanidine salt. Such reactions are usually carried out in alcohol solution [13]. To obtain alkali-free guanidine solution, guanidine carbonate or phosphate is reacted in water with calcium hydroxide. After removal of the precipitated salt, the free guanidine base can be isolated by careful vacuum distillation of the solvent. A further possibility is addition of acid or acid anhydride to give the desired guanidine salt. As an example, guanidine silicate is produced in this manner from guanidine carbonate, calcium hydroxide, and amorphous silica [14].

A good way to prepare various guanidine salts of strong or moderate acids, such as stearic acid, in smaller amounts is to react guanidine carbonate with acids or anhydrides. This reaction can be carried out in aqueous solution or by direct mixing and controlled heating of the pure compounds [15].

This method gives guanidine salts of very high purity (depending on the raw materials), because the only byproducts are carbon dioxide and water, which are lost in the vapor.

1.3. Environmental Protection

Apart from the ecological aspects of the production of raw material (e.g., calcium cyanamide), guanidine production itself causes few environmental problems. Ammonia, the only volatile byproduct, must be removed by proper ventilation and burned or recycled. Solid byproducts are mainly nontoxic melamine salts that can be worked up (for applications, see → Melamine and Guanamines), deposited, or burned.

At moderate concentration, guanidine and its common salts are nontoxic toward microorganisms, but nitrification is rather slow [16]. Wastewater containing guanidine can therefore best be removed by slow addition to a sewage plant that does not receive too many other nitrogen-containing compounds.

1.4. Quality Specifications

Guanidine salts are normally available in pure (98.5–99.5%) grade as white crystalline substances containing some ammonium salts (0.3–0.6%), triazines 0.2–0.4%), water (0.1–0.3%), and possibly anticaking agent (SiO_2, 0.2–0.4%). Technical grades are also marketed for some purposes as white to yellowish lumps, granules, or powder, with 91–96% guanidine salt, 2–5% ammonium salt, and 1–3% triazines. Some salts are also available as aqueous solutions, e.g., guanidine hydrochloride (60%), sulfamate, dihydrogen phosphate, or silicate (each 50%).

Guanidine hydrochloride solutions are sold in special purity grades, e.g., iron-free (Fe $\leq$ 0.1 ppm) or low-absorbance products ($\varepsilon_{225} \leq 0.1$).

1.5. Analysis

Guanidine is analyzed gravimetrically by precipitation as the picrate [17]. For special purposes, other methods such as colorimetric, spectroscopic, or chromatographic analyses have also been developed; see general references.

1.6. Storage and Transportation

Guanidine salts are packed in paper or polypropylene bags (25 or 50 kg), drums (150 kg) with polyethylene lining, or larger polypropylene bags. If dry storage is available, guanidine salts can be stored for an unlimited period. Solutions must be stored and shipped in plastic or coated steel vessels to avoid corrosion. Cooling of aqueous solutions may cause precipitation of the guanidine salt.

The customary precautions recommended for handling chemicals should generally be observed. When specific data for guanidine salts are unavailable, the precautions recommended are those required to handle the corresponding ammonium or potassium salts.

1.7. Legal Aspects

Guanidine perchlorate [*10308-84-6*] and picrate [*4336-48-5*] are classified in the Federal Republic of Germany as explosives according to the Explosives Act (Sprengstoffgesetz). Handling and storage of these salts are, therefore, subject to special safety regulations prescribed by this act [18].

Although the act does not specifically mention pure guanidine nitrate (or nitroguanidine), some mixtures of these substances with other materials (e.g., potassium nitrate, ammonium dichromate, or chlorinated hydrocarbons) are also classified as explosives [18]. The export of guanidine nitrate or nitroguanidine is controlled by special regulations depending on the country of destination.

1.8. Uses

1.8.1. Pharmaceuticals

Guanidine is a valuable intermediate in the synthesis of a variety of pharmaceuticals. Important industrial processes based on guanidine are, for example, the synthesis of sulfonamides such as sulfaguanidine [*57-67-0*] [19], sulfamethazine [*57-68-1*] [20], sulfadiazine [*68-35-9*] [20], and related compounds like pyrimethamine [*58-14-0*] [21] and trimethoprim [*738-70-5*] [22]. In addition, purine derivatives such as triamterene [*396-01-0*] [23], methotrexate [*59-05-2*] [24], or folic acid [*59-30-3*] are manufactured from guanidine. Many syntheses involve 2-aminopyrimidines as intermediates. The guanidine used is generally prepared in situ by reacting salts such as guanidine nitrate, hydrochloride, or carbonate with a strong base such as sodium methoxide.

1.8.2. Plant Protection

2-Aminopyrimidines derived from guanidine are also used to produce a special group of recently developed herbicides, which are closely related chemically to the sulfonamides.

1.8.3. Cosmetics

Guanidine carbonate is used industrially as a nontoxic, highly soluble component of hair care products.

1.8.4. Textile Impregnation, Flame Retardants, Paper and Resin Manufacture

Guanidine hydrochloride is used for textile impregnation as an antistatic agent. Guanidine dihydrogen phosphate is used as a flame retardant for wood, paper, and textiles. Guanidine hydrogen phosphate, which is only sparingly soluble in water, is used for the same purpose in phenolic resins. Guanidine sulfamate has large application in the manufacture of flame-retardant papers, especially in Japan where paper walls are common. Guanidine stearate is used as an emulsifier and surface-modifying agent in the production of resin coatings and papers.

1.8.5. Propellants and Explosives

Guanidine nitrate is used in large amounts to produce nitroguanidine, which is employed widely in the manufacture of propellants and explosives. This process involves the reaction of guanidine nitrate with concentrated sulfuric acid.

Nitroguanidine was formerly used in the production of aminoguanidine carbonate [2582-30-1]. For economic reasons, however, this method has been abandoned in favor of cyanamide. Triaminoguanidine nitrate [4000-16-2], which has been recommended as a propellant [25], is prepared from hydrazine and guanidine nitrate.

Guanidine perchlorate and picrate have been tested as explosives but are of no industrial importance.

1.8.6. Biotechnological Applications of Guanidine Salts

Guanidine hydrochloride and thiocyanate are among the most potent protein denaturants known [26], readily surpassing urea, the other widely employed denaturing agent, in effectiveness for most applications. As a consequence, the use of guanidine salts for biochemical purposes has increased steadily over the past few years. Their primary application appears to be in the isolation of proteins and enzymes, particularly from genetically modified microorganisms. After successful manipulation of the genetic constitution of a microorganism, the microbial cell begins to produce the protein of interest, which then accumulates in special cellular compartments known as inclusion bodies. To isolate these proteins, they must be released from the inclusion bodies and separated from other cellular constituents.

Traditionally, urea (6–9 M) has been used for this purpose, but guanidine salts, particularly the hydrochloride (2–6 M), are gaining acceptance as substitutes with distinct advantages [27]. For example, guanidine hydrochloride tends to cause less protein inactivation [27], partly because lower concentrations can be used, compared with urea. Furthermore, guanidine hydrochloride is removed from the extracted protein after denaturing simply by dilution [28]. The use of guanidine salts generally results in higher yields of extracted protein than is the case with urea.

Guanidine salts have attained an unrivaled position in the extraction and purification of biologically active RNA [29], [30], because of their strong denaturing properties. Guanidine thiocyanate is often the salt of choice and is used frequently in combination with, or in place of, phenol for the extraction of RNA [31]. The above-mentioned advantages make guanidine salts particularly attractive for the isolation of substances such as interferons [27], [32], interleukins [33], and the blood factors tissue plasminogen activator (tPA) and urokinase.

1.9. Economic Aspects

The worldwide production capacity of guanidine salts is estimated to be several tens of thousands of tons per year. This figure includes significant spare capacity for the production of guanidine nitrate. The exact capacity and output of these plants is not known. In most cases, the amount produced is far less than capacity and is used exclusively as captive raw material for the production of nitroguanidine. Apart from this, worldwide annual consumption of guanidine salts is estimated to be several thousand tons. The main producers are located in the Federal Republic of Germany and Japan. Smaller amounts are produced in Austria, India, and the People's Republic of China.

2. Derivatives

2.1. Nitroguanidine [34]–[36]

Nitroguanidine [556-88-7], $CH_4N_4O_2$, M_r 104.07, is a very weak base that forms salts only with strong acids. Its solubility in water at 20 °C (100 °C) is 0.27 % (7.6 %). It decomposes at 220 °C.

$$H_2N\underset{\underset{NO_2}{N}}{\overset{}{\diagdown}}NH_2$$

Nitroguanidine is made by treating guanidine nitrate with concentrated sulfuric acid, fuming nitric acid, or both. The reaction of nitroguanidine with amines gives substituted nitroguanidines [34].

Nitroguanidine (available from Nigu Chemie) is used as a component in triple-based propellants.

2.2. Aminoguanidine [37]–[40]

Aminoguanidine [79-17-4], CH$_6$N$_4$, M_r 74.09, is a base that forms salts with both inorganic and organic acids [37]. It is soluble in water and ethanol. Preparation of the nitrate [10308-82-4], CH$_6$N$_4$·HNO$_3$, involves the risk of explosion [41].

$$H_2N-\underset{H}{N}-C(=NH)-NH_2$$

Free aminoguanidine is stable for only a short time; therefore, it is marketed in the form of its stable salts. The most common commercial salt is the bicarbonate [2582-30-1], CH$_6$N$_4$·H$_2$CO$_3$ (available from Bayer); the solubility of this salt in water at room temperature is <0.5%. Other salts can be prepared from the bicarbonate by reaction with acid.

Aminoguanidine is prepared from hydrazine and aqueous cyanamide solution. Processes based on the reduction of nitroguanidine have no industrial significance.

$$H_2N-C\equiv N + H_2N-NH_2 \longrightarrow H_2N-\underset{H}{N}-C(=NH)-NH_2$$

Uses. Aminoguanidine can act as both a hydrazine and a guanidine derivative; it is particularly suitable for the synthesis of various nitrogen-containing heterocyclic compounds such as pyrazoles, 1,2,4-triazoles, thiadiazoles, tetrazoles, pyrimidines, 1,2,4-triazines, and tetrazines [40].

Cyclization with formic acid produces 3-amino-1,2,4-triazole [61-82-5]:

$$H_2N-\underset{H}{N}-C(=NH)-NH_2 + HCOOH \xrightarrow{-2H_2O} \text{3-amino-1,2,4-triazole}$$

This compound is an important herbicide (Amitrole, ATA), which has a synergistic action with many nonselective herbicides. 3-Amino-1,2,4-triazole is also used for the preparation of red and yellow cationic triazole dyes such as C.I. Basic Red 22 and C.I. Basic Yellow 25, for poly-(acrylonitrile) fibers.

Aminoguanidine reacts with nitrous acid in aqueous solution to form tetrazene [31330-63-9], which is used as a component of primers that are sensitive to shock and heat.

$$\text{H}_2\text{N}-\underset{\text{H}}{\text{N}}\overset{\text{HN}}{\underset{\|}{\text{C}}}-\text{NH}_2 \xrightarrow{\text{HNO}_2} \underset{\text{N-N}}{\overset{\text{N-NH}}{\|}}\text{C}-\text{N}=\text{N}-\underset{\text{H}}{\text{N}}-\underset{\text{H}}{\text{N}}-\text{C}\underset{\text{NH}}{\overset{\text{NH}_2}{\diagdown}} \cdot \text{H}_2\text{O}$$

Salts and derivatives of aminoguanidine are recommended as stabilizers for poly (vinyl chloride), as flame retardants, and as textile auxiliaries.

2.3. Diamino- and Triaminoguanidine [38], [39], [42]

Diaminoguanidine. Development of the chemistry of N,N'-diaminoguanidine [4364-78-7], CH_7N_5, M_r 89.10, has been hampered by the lack of convenient methods for synthesizing it in reasonable purity; a summary of the available methods is given in [39]. In preparation of the hydrochloride [38360-74-6] by reaction of hydrazine with cyanogen chloride in a molar ratio of 2:1, the formation of byproducts can be suppressed if the reaction is carried out in saturated aqueous sodium chloride solution: yield 71%, mp 170–176 °C [43].

$$\text{H}_2\text{N}-\underset{\text{H}}{\text{N}}-\overset{\text{NH}}{\underset{\|}{\text{C}}}-\underset{\text{H}}{\text{N}}-\text{NH}_2$$

Diaminoguanidine is used in the preparation of robenidine [25875-51-8], whose hydrochloride [25875-50-7] (Cycostat, American Cyanamid) is an anticoccidial agent for poultry.

Robenidine

Triaminoguanidine. Salts of $N,N'N''$-triaminoguanidine [2203-24-9], CH_8N_6, M_r 104.11, are formed by heating (hydrazinolysis) the corresponding guanidine salts with hydrazine (molar ratio 1:3) in aqueous solution or in alcohol. The triaminoguanidine base can be liberated from its salts by treatment with sodium hydroxide and can be crystallized from aqueous dimethylformamide.

$$\text{H}_2\text{N}-\underset{\text{H}}{\text{N}}-\overset{\text{N}-\text{NH}_2}{\underset{\|}{\text{C}}}-\underset{\text{H}}{\text{N}}-\text{NH}_2$$

Triaminoguanidine nitrate (TAGN) [4000-16-2] is an effective explosive [44]. It is useful as a component of propellants.

2.4. Organoguanidines

A large number of organoguanidine derivatives are known. The many methods of preparation are described in detail in a recent, excellent monograph on carbon dioxide derivatives [45].

The addition of amines, polyamines, hydrazines, or hydroxylamine to cyanamides is one of the most common industrial processes for the synthesis of guanidines. Cyanamide [420-04-2] is available as a 50% aqueous solution or in crystalline form; substituted cyanamides can be obtained from amines and cyanogen chloride.

The classical preparation of guanidines by the action of amines on S-alkylisothioureas is versatile and suitable also for the preparation of cyclic compounds. The elimination of alkanethiol byproducts is, however, a disadvantage.

$$H_2N-C(SR^1)=NH + R^2-NH_2 \longrightarrow H_2N-C(NHR^2)=NH + R^1SH$$

A new method for synthesizing di- and trisubstituted guanidines is the reaction of oxidized thioureas with nucleophilic amines [46].

$$R^1-NH-C(=S)-NH_2 \xrightarrow{H_2O_2} R^1-N=C(SO_xH)-NH_2 \xrightarrow{R^2-NH_2} R^1-N=C(NHR^2)-NH_2$$

$$x = 2, 3$$

2.4.1. Alkylguanidines

Monosubstituted and N,N-disubstituted compounds, including guanidinocarboxylic acids (e.g., creatine [57-00-1]), are produced industrially by the reaction of cyanamide with the corresponding amines or aminocarboxylic acids.

$$HN=C(NH_2)-N(CH_3)-CH_2COOH$$

Creatine

For convenient guanylation (amidination) of primary alkylamines and amino acids without formation of methanethiol, S-methylisothiourea sulfate (2:1) [867-44-7] can be replaced by commercially available O-methylisourea sulfate (2:1) [52328-05-9]. Thus, methylguanidine sulfate (2:1) [598-12-9] and ethylguanidine sulfate (2:1) [3482-86-8] are obtained in quantitative yield by the action of methylamine and ethylamine, respectively, on O-methylisourea sulfate in aqueous solution at 50 °C:

$$H_2N-C(OCH_3)=NH + R-NH_2 \longrightarrow H_2N-C(NHR)=NH + CH_3OH$$

For steric reasons, O-methylisourea salts react preferentially with primary amines and are particularly suitable for selective guanylation of the terminal amino groups of polyamines. O-Methylisourea sulfate (available from SKW Trostberg) is a convenient guanylating agent and is also used on an industrial scale for the synthesis of heterocyclic compounds, e.g., benzimidazole anthelmintics and the antibacterial agents pipemidic acid [51940-44-4] and piromidic acid [19562-30-2].

Guanidine Bases. Free alkylguanidines are strong bases that are more stable than unsubstituted guanidine.

N,N,N',N'-Tetramethylguanidine (TMG) [80-70-6], $C_5H_{13}N_3$, M_r 115.18, is a stable, unusually strong base, pK_a 13.6 (25 °C), with a free imino group. It is a colorless to light-yellow liquid, miscible with water and common organic solvents: ϱ 0.92 g/cm^3 (25 °C), bp 160 °C, vapor pressure 0.03 kPa (25 °C). Prepared from cyanogen chloride and dimethylamine in a 1:2 molar ratio, TMG is used as a polyurethane foam catalyst, in the preparation of pigments for color photography, and in steroid synthesis. It is available from Degussa and Lonza.

$$Me_2N-\overset{NH}{\underset{}{C}}-NMe_2$$

Undiluted N,N,N',N'-tetramethylguanidine strongly irritates the eyes, skin, and mucous membranes. Its LD$_{50}$ is ca. 850 mg/kg (rat, oral).

Short-Chain Alkylguanidines as Building Blocks in Synthesis. Some monosubstituted and N,N-disubstituted derivatives are used as intermediates for the preparation of crop-protecting agents.

Pyrimidines can be prepared by condensation of alkylguanidines with a 1,3-difunctional C_3 fragment. Cyclization of alkylguanidines with 2-alkylacetoacetic acid esters produces fungicidally active 5-alkyl-2-alkylamino-4-hydroxy-6-methylpyrimidines. The systemic fungicides ethirimol [23947-60-6], bupirimate [41483-43-6], and dimethirimol [5221-53-4], developed by ICI, belong to this class of compounds.

$$R^3 = n-C_4H_9$$
$$R^1, R^2 = H, C_2H_5; CH_3, CH_3$$

N,N-Diethylguanidine [18240-93-2] and N,N-dimethylguanidine [6145-42-2] are used as an N–C–N building block for the pyrimidine insecticides (ICI) pirimiphos-methyl [29232-93-7], pirimiphos-ethyl [23505-41-1], and pirimicarb [23103-98-2]. Pirimiphos-methyl (Actellic) has proved particularly effective for the control of storage pests.

Biologically Active Compounds. Higher alkylguanidines and polyguanidines, because of their fungicidal and antibacterial action, are economically important as fungicides in agriculture or as antimicrobial agents in disinfectants and industrial biocides [47], [48].

n-Dodecylguanidine acetate [*2439-10-3*], Dodine, used primarily for the control of apple and pear scab, exhibits both protective and curative action. Dodine is also a bacteriostatic agent. Dodecylguanidine derivatives are recommended as effective biocides for controlling biological growth in industrial coolant systems.

Dodigen 180 (Hoechst) is a broad-spectrum microbiocide based on alkyldiguanidine salts. Uses include disinfection, sanitation, preservation, and control of *Desulfovibrio desulfuricans* corrosion [48].

Guazatine, prepared by the guanylation of aliphatic diamines and polyamines, is used as Panoctine (KenoGard) for cereal seed dressing. Under the trade name Kenopel, it is also used as a postharvest dip in citrus growing. Guazatine is effective against wood-discoloring fungi and, together with quaternary ammonium compounds, is a constituent of wood preservatives (Mitrol 48 from KenoGard).

Radam 30 is a new contact fungicide from KenoGard, based on guazatine. In the Federal Republic of Germany, it is being developed and marketed by Shell Agrar [49] as a successor to Captafol [*2425-06-1*]for the control of *Septoria nodorum* in wheat. Fixan, marketed by ICI, also contains guazatine as the active ingredient.

Iminoctadine triacetate (DF 125, Befran) is a new broad-spectrum contact fungicide, developed by Dainippon Ink and Chemicals (DIC); the active ingredient is the triacetate salt of 1,1'-[iminobis(octamethylene)]diguanidine [*57520-17-9*], which is produced by DIC. The product is used in Japan primarily for the control of top fruit (tree fruit) and citrus diseases; in Western Europe, good action has been achieved against leaf spot and glume blotch in winter wheat [50].

2.4.2. Arylguanidines

Monoarylguanidines. Arylguanidines can be prepared by heating aromatic amine hydrochlorides with cyanamide in aqueous solution. Aromatic–aliphatic guanidines are obtained by several methods [45]. A number of compounds have been synthesized as potential fungicides or herbicides, e.g., *N'*-(3-chlorophenyl)-*N,N*-dimethylguanidine (FDN) [*13636-32-3*], which has been tested in Russia against powdery mildew. However, these compounds have not achieved economic importance.

From the group of 2-guanidinoanilides with anthelmintic properties, febantel [*58306-30-2*] (Rintal), supplied by Bayer, has emerged as an effective broad-spectrum anthelmintic; it is used to treat nematode infections in domestic animals.

Febantel

The guanidine derivative linogliride fumarate (1:1) [*78782-47-5*] is a new oral antidiabetic (Johnson & Johnson), developed by McNeil. It can be prepared by the Maryanoff process [46].

Diarylguanidines. N,N'-Diphenylguanidine (DPG) [*102-06-7*] and N,N'-di-o-tolylguanidine (DOTG) [*97-39-2*] are used as vulcanization accelerators. Because of their low activity they do not play a major role as primary accelerators; however, they are very important in combination with other accelerators. Diarylguanidines can be used as secondary accelerators for thiazoles, sulfenamides, thiurams, and dithiocarbamates; the most frequently used combination is with mercapto accelerators. These accelerators and the guanidine produce an extraordinary increase in the degree of cross-linking and the rate of vulcanization; high mechanical values (high tensile stress and elasticity) are achieved. N,N'-Diphenylguanidine is the most important of the three guanidines DPG, DOTG, and OTBG (o-tolylbiguanide; see Section 2.6).

Both DPG and DOTG are prepared by reaction of cyanogen chloride with aniline or o-toluidine, respectively, in a molar ratio of 1:2.

Suppliers include Akzo (Perkacit DPG, Perkacit DOTG); American Cyanamid (DPG, DOTG); Bayer (Vulkacit D/EGC, Vulkacit DOTG/EGC); Mobay (Vulkacit D/C, Vulkacit DOTG/C); Monsanto (DPG); Ouchi Shinko (DPG, DOTG); Sumitomo (DPG, DOTG); Uniroyal (Vulkacit DC/EGC); Vanderbilt (Vanax DPG, Vanax DOTG); and Vulnax (Vulcafor DPG, Vulcafor DOTG).

Cyclic Derivatives. N,N'-o-Phenyleneguanidine [*934-32-7*], 2-aminobenzimidazole, $C_7H_7N_3$, M_r 133.15, mp 231 – 232 °C (decomp.), obtained by cyclization of o-phenylenediamine with cyanogen chloride or cyanamide (available from SKW Trostberg), is the nucleus for a number of compounds of economic interest; for a review of 2-aminobenzimidazoles, see [51].

2-(Methoxycarbonylamino)benzimidazoles are biologically active; carbendazim (R = H) [*10605-21-7*] and benomyl [*17804-35-2*], the latter prepared by reaction of carbendazim with n-butyl isocyanate, are important systemic fungicides.

The following derivatives are used as effective anthelmintics in veterinary medicine:

albendazole [54965-21-8] (R = C_3H_7S-)
fenbendazole [43210-67-9] (R = C_6H_5S-)
flubendazole [31430-15-6] (R = 4-FC_6H_4CO-)
mebendazole [31431-39-7] (R = C_6H_5CO-)
oxfendazole [53716-50-0] (R = C_6H_5SO-)
oxibendazole [20559-55-1] (R = C_3H_7O-)
parbendazole [14255-87-9] (R = C_4H_9-)

Mebendazole and flubendazole are also effective against worm infection in humans. Substituted 2-(methoxycarbonylamino)benzimidazoles can be prepared by cycloguanylation of the corresponding *o*-phenylenediamine derivatives with *N*-methoxycarbonyl-*S*-methylisothiourea [39259-32-0] or *N*-methoxycarbonyl-*O*-methylisourea [40943-37-1].

2.5. Cyanoguanidines (Dicyandiamides)

Cyanoguanidines generally exist in the more stable diamide form, $(NH_2)_2C=N-CN$, in which the cyano group is conjugated with the C=N group.

N-Substituted cyanoguanidines containing a free amino group are prepared easily by heating sodium dicyanamide [1934-75-4], $NaN(CN)_2$ (available from Degussa, ICI, and Lonza), with aliphatic or aromatic amine hydrochlorides in alcohol or water [52], [53]. N-Substituted cyanoguanidines, e.g., 1,6-hexamethylenebis(dicyandiamide) [15894-70-9], are used for the preparation of commercially important biguanides (see Section 2.6).

N,N'-Substituted cyanoguanides can be prepared by stepwise aminolysis of dimethyl *N*-cyanoimidodithiocarbonate [10191-60-3], $(CH_3S)_2C=N-CN$ (available from Ircha and from Robinson Brothers). Use of the O-analogue dimethyl *N*-cyanoimidocarbonate [24771-25-3], $(CH_3O)_2C=N-CN$, avoids the formation of methanethiol. An economically successful N,N'-disubstituted derivative is cimetidine [51481-61-9], developed by Smith Kline & French; it is used worldwide, as Tagamet, for the treatment of duodenal and gastric ulcers.

2.6. Biguanide and Derivatives [54]–[56]

Biguanide [56-03-1], $C_2H_7N_5$, M_r 101.11, mp 130 °C (which decomposes violently at 142 °C), is a moderately strong base (pK_b 2.9). It is readily soluble in water and ethanol, and insoluble in ether, benzene, and chloroform. The free base decomposes gradually in aqueous solution. Biguanide forms stable salts with most acids.

A hydrogen atom in one of the imino (C=NH) groups of biguanide can be replaced by a transition metal to form colored chelate complexes, several of which have been prepared [54], [55].

$$\left[\begin{array}{c} H_2N \\ HN \\ \diagdown NH_2 \\ HN \end{array} \!\!\!\!=\!\! N\!\!\!-\!\!\!M\!\!\!-\!\!\!\begin{array}{c} H_2N \\ \diagdown NH \\ N\!\!=\!\! \\ NH_2 \end{array}\right] \cdot (HX)_2$$

M = bivalent metal atom
HX = univalent acid

Reviews of methods for the preparation of biguanide can be found in [55], [56]. Biguanide salts are prepared by the fusion of dicyandiamide with ammonium salts. [Dibiguanide – copper(II)] sulfate, $[Cu(C_2H_7N_5)_2]SO_4$, is formed by heating dicyandiamide with ammoniacal copper sulfate solution to 90 – 110 °C in a pressure vessel [57]; it can be converted by decomposition with hot 10 % sulfuric acid or hydrogen sulfide to biguanide sulfate (1:1) [6945-23-9], $C_2H_7N_5 \cdot H_2SO_4$, or biguanide sulfate (2:1) [49719-55-3], $2\,C_2H_7N_5 \cdot H_2SO_4$ [57], [58].

Biguanide is used to prepare 6-substituted 2,4-diamino-1,3,5-triazines [56]. The reaction of biguanide with esters to form guanamines is catalyzed by sodium alkoxides.

$$\underset{H_2NNNH_2}{\overset{HNNH}{\|\|}} + RCOOR^1 \longrightarrow \underset{H_2NNNH_2}{\overset{R}{\underset{}{\underset{}{\bigtriangleup}}}}$$

Biguanides substituted at the N-1 position are obtained as the hydrochlorides by heating dicyandiamide with aliphatic or aromatic amine hydrochlorides in water, higher boiling alcohols, or 2-ethoxyethanol.

The N-1-substituted biguanides buformin, phenformin, metformin show hypoglycemic activity, but are used to only a limited extent at present.

o-Tolylbiguanide (OTBG) [93-69-6], $C_9H_{13}N_5$, M_r 191.23, mp 144 °C, is used as a vulcanization accelerator (Vulkacit 1000, Bayer) in the same manner as the guanidines DPG and DOTG (see Section 2.4.2.). An important use is in the preparation of consumer articles that come in contact with food and drinking water (Recommendation XXI of the German Federal Department of Health). *o*-Tolylbiguanide is also used as a curing agent with moderate reactivity for color-stable, highly glossy epoxy resin powder coatings.

The N-1 and N-5 substituted biguanides can be prepared by addition of amines to cyanoguanidines (dicyandiamides). A particularly convenient synthesis of N-1 alkyl N-5 aryl biguanides involves heating aliphatic cyanoguanidines with aromatic amine hydrochlorides [52], [53]: the antimalarial drug proguanil [500-92-5] is obtained as the hydrochloride [637-32-1] (Paludrine, ICI) by reaction of isopropyldicyandiamide with 4-chloroaniline hydrochloride in boiling water [52].

Chlorhexidine [55-56-1], 1,6-bis(4-chlorophenylbiguanido)hexane, $C_{22}H_{30}Cl_2N_{10}$, M_r 505.45, mp 134 °C, is a widely used antiseptic. It can be prepared, for example, from

1,6-hexamethylenebis(dicyandiamide) and 4-chloroaniline hydrochloride [59]. Chlorhexidine is used primarily as its salts (e.g., the dihydrochloride [3697-42-5], diacetate [56-95-1], and digluconate [18472-51-0]) in disinfectants (disinfection of the skin and hands), cosmetics (additive to creams, toothpaste, deodorants, and antiperspirants), and pharmaceutical products (preservative in eyedrops, active substance in wound dressings and antiseptic mouthwashes) [48].

Poly(hexamethylene biguanide) hydrochloride is the active substance in Vantocil IB [32289-58-0], a broad-spectrum industrial biocide developed by ICI. The most important uses are in surface disinfection in hospitals and the food and beverage industry, and in the preservation of aqueous systems (i.e., aqueous solutions of organic products, such as cutting oils and domestic cleaning agents) [48]. Polymeric biguanide hydrochlorides with broad antimicrobial action, very low toxicity, and good skin tolerance are also contained in preservatives for cosmetic formulations such as Arlagard E [70170-61-5] and Cosmocil CQ [91403-50-8] from ICI.

3. Toxicology and Occupational Health

The guanidino group is low in acute toxicity and is not associated with severe health hazards. No mutagenic, carcinogenic, or teratogenic potential has been demonstrated for guanidine.

Free guanidine and its solutions exhibit strong alkaline reactions. They must therefore be handled like alkali hydroxides or alkoxides. Guanidine salts of weak acids (e.g., guanidine silicate) also have alkaline properties. Animal tests (rabbit) show that guanidine hydrochloride is a severe primary skin and eye irritant [7], [60]. In a similar test, guanidine nitrate produced mild to moderate eye irritation [61]. Guanidine hydrochloride did not show sensitizing properties in the Buehler test [62].

Toxicities of some guanidine salts (acute oral LD_{50}, rat) are guanidine hydrochloride 1120 mg/kg, guanidine nitrate 1260 mg/kg, guanidine carbonate 1020 mg/kg, and guanidine stearate 7500 mg/kg. Further toxicological data for guanidine, its salts, and derivatives can be found in [8].

The toxicological properties of individual guanidine derivatives depend primarily on the type of the substitutents. Commercially available guanidines generally have relatively low toxicity. The following are LD_{50} values (rat, oral) for some derivatives: aminoguanidine bicarbonate >5000 mg/kg; nitroguanidine 4650 mg/kg; N,N,N',N'-tetramethylguanidine ca. 850 mg/kg; n-dodecylguanidine acetate 1000 mg/kg; N,N'-diphenylguanidine 507–850 mg/kg; o-tolylbiguanide 1800 mg/kg.

Guanidine derivatives may, however, contain amines as impurities, because amines are used as starting material in the production of guanidines.

With the exception of nitroguanidine, cyano-guanidines, and acylated guanidines, guanidine derivatives are basic; the free bases can cause irritation and corrosion of the skin, eyes, mucous membranes, and respiratory tract. Skin and eye contact with the strong free bases should be avoided. The salts are less irritating than the free bases. Exposure limits (MAK and TLV values) have not been established.

4. References

General References

[1] *Beilstein* System no. 207.
[2] A. Mori, B. D. Cohen, A. Lowenthal (eds.): *Guanidines – Historical, Biological, Biochemical, and Clinical Aspects of the Naturally Occuring Guanidino Compounds*, Plenum Press, New York 1985.
[3] *Ullmann*, 3rd ed., **8**, 328 – 337.
[4] *Ullmann*, 4th ed., **12**, 441 – 420.
[5] *Kirk-Othmer*, 3rd ed., suppl. vol., 514 – 521.
[6] *Gmelin*, **14 D1**, 466 – 475.
[7] SKW Trostberg: "Produktstudie Guanidinsalze" and data sheet "Guanidine Salts."
[8] NIOSH Registry of Toxic Effects of Chemical Substances, 1980 ed., Washington, D.C., 1982.

Specific References

[9] G. Desseigne, A. Audiffren, *Meml. Poudres* **35** (1953) 15 – 38.
[10] Südd. Kalkstickstoffwerke, DE-OS 1 593 472, 1966 (H. Michaud, H. Prietzel).
[11] Pittsburgh Coke & Chem. Co., US 2 949 484, 1957 (J. S. Mackay).
[12] Lentia GmbH, DE-OS 2 234 732, 1974 (A. Schmidt, K.-H. Wegleitner, J. H. Hatzle, R. Sylpra, F. Weinrotter); DE-OS 2 435 167, 1976 (F. Weinrotter, A. Schmidt, K. Wegleitner, A. Garber, J. H. Hatzl, R. Sykora).
[13] G. Desseigne, *Bull. Soc. Chim. Fr.* 1955, 1193 – 1194.
[14] Du Pont, US 3 475 375, 1969 (P. C. Yates).
[15] Goldschmidt, DE-OS 1 922 043, 1970 (E. Ruf).
[16] H. Lees, J. H. Quastel, *Biochem. J.* **40** (1946) 825.
[17] P. Fainer, J. L. Myers, *Anal. Chem.* **24** (1952) no. 3, 515 – 517.
[18] Sprengstoffgesetz, annex I, part 1, no. 33, 34 and part 2, no. 2.5, mixtures 4,5.
[19] American Cyanamid, US 2 233 569, 1941 (P. S. Winnek).
[20] Sharp & Dohme, US 2 407 966, 1946 (J. M. Sprague).
[21] Burroughs Wellcome, US 2 576 939, 1951 (G. H. Hitchings, P. B. Russell, E. A. Falco).
[22] Burroughs Wellcome, US 2 909 522, 1959 (G. H. Hitchings, B. Roth).
[23] Smith Kline & French, US 3 081 230, 1963 (J. Weinstock, V. D. Wiebelhaus).
[24] D. R. Seeger, D. B. Cosulich, J. M. Smith, M. E. Multquist, *J. Am. Chem. Soc.* **71** (1949) 1753 – 1758.
[25] Rockwell I. C., DE-OS 2 453 167, 1976 (V. E. Haury).
[26] Y. Nozaki, C. Tanford, *J. Biol. Chem.* **245** (1970) 1648 – 1672.
[27] Hoffmann-La Roche Inc., US 4 476 049, 1984 (Hsiang Fu Kung).
[28] Hoechst, DE 3 511 011, 1987 (R. Obermeier).

[29] J. R. Chirgwin, A. E. Przybyla, R. J. McDonald, W. J. Rutter, *Biochemistry* **18** (1979) 5294–5299.
[30] J. Logemann, J. Schell, L. Willmitzer, *Anal. Biochem.* **163** (1987) 16–20.
[31] P. Chomczynki, N. Sacchi, *Anal. Biochem.* **163** (1987) 156–159.
[32] Takeda Chem. Ind., EP 150 066, 1985 (Kiyoshi Nara, Susumi Honda).
[33] Hoffman-La Roche & Co., EP 200 986, 1986 (U. Gubler, P. T. Lomedico, S. B. Mizel).
[34] T. Urbanski (ed.): *Chemistry and Technology of Explosives*, vol. **3**, Pergamon Press, Oxford 1967, pp. 22–33.
[35] T. Urbanski (ed.): *Chemistry and Technology of Explosives*, vol. **4**, Pergamon Press, Oxford 1984, pp. 365–367.
[36] R. Meyer: *Explosives*, 3rd ed., VCH Verlagsgesellschaft, Weinheim 1987, pp. 243–245.
[37] *Beilstein*, **3**, H 117, E I 57, E II 95, E III 230, E IV 236.
[38] E. Lieber, G. B. L. Smith, *Chem. Rev.* **25** (1939) 213–271.
[39] F. Kurzer, L. E. A. Godfrey, *Chem. Ind. (London)* 1962, 1584–1595.
[40] F. Kurzer, L. E. A. Godfrey, *Angew. Chem.* **75** (1963) 1157–1175.
[41] H. Koopman, *Chem. Weekbl.* **53** (1957) 97.
[42] *Beilstein*, **3**, H 122, E I 57, E II 97, E III 232, E IV 241/242.
[43] American Cyanamid, US 2 721 217, 1955 (G. A. Peters, D. W. Kaiser).
[44] W. Sauermilch, *Explosivstoffe* **12** (1964) no. 9, 197.
[45] *Houben-Weyl*, E **4**, 608–624.
[46] C. A. Maryanoff, R. C. Stanzione, J. N. Plampin, *Phosphorus Sulfur* **27** (1986) 221. A. E. Miller, J. J. Bischoff, *Synthesis* 1986 no. 9, 777.
[47] H. R. Hudson, I. A. O. Ojo, M. Pianka, *Int. Pest Control* **28** (1986) no. 6, 148–155.
[48] K. H. Wallhäußer: *Praxis der Sterilisation-Desinfektion-Konservierung-Keimidentifizierung-Betriebshygiene*, 4th ed., Thieme Verlag, Stuttgart–New York 1988, pp. 558–564.
[49] W. Wimschneider, A. O. Klomp, K. E. Friedrich, *Gesunde Pflanz.* **39** (1987) no. 6, 263. D. G. Cameron, I. Hylten-Cavallius, E. Jordow, A. O. Klomp, W. Wimschneider, *Proc. Br. Crop Prot. Conf. Pests Dis.* 1986, vol. **3**, p. 1201.
[50] M. Masui, N. Yoshioka, R. I. Harris, H. Roos, A. Cornier, *Proc. Br. Crop Prot. Conf. Pests Dis.* 1986, vol. 1, p. 63.
[51] R. Rastogi, S. Sharma, *Synthesis* 1983, no. 11, 861–882.
[52] F. H. S. Curd, J. H. Hendry, T. S. Kenny, A. G. Murray, F. L. Rose, *J. Chem. Soc.* 1948, 1630.
[53] G. Rembarz, H. Brandner, H. Finger, *J. Prakt. Chem.* **26** (1964) 314.
[54] *Beilstein*, **3**, H 93, E I 44, E II 76, E III 171, E IV 162.
[55] P. Ray, *Chem. Rev.* **61** (1961) 313–359.
[56] F. Kurzer, E. D. Pitchfork, *Fortschr. Chem. Forsch.* **10** (1968) no. 3, 375–472.
[57] K. Rackmann, *Justus Liebigs Ann. Chem.* **376** (1910) 163, 170.
[58] D. Karipides, W. C. Fernelius, *Inorg. Synth.* **7** (1963) 56.
[59] Sogo Pharmaceutical Co., DE-OS 2 932 951, 1980 (S. Uchikuga, M. Itoh, K. Nagahama).
[60] E. W. Morgan, L. Mullen, D. W. Korte, *Gov. Rep. Announce. Index U.S.*, **86** (1986) (15), Abstract no. 633,219; *Chem. Abstr.* **106** (1987) 62 436.
[61] E. W. Morgan, J. W. Bauserman, D. W. Korte jr., *Gov. Rep. Announce. Index U.S.*, **86** (1986) (15), Abstract no. 633,223; *Chem. Abstr.* **106** (1987) 45 362.
[62] G. F. S. Hiatt, E. W. Morgan, D. W. Korte, *Gov. Rep. Announce. Index U.S.*, **86** (1986) (12), Abstract no. 625,904; *Chem. Abstr.* **106** (1987) 151 112.

Hexamethylenediamine

ROBERT A. SMILEY, E. I. du Pont de Nemours & Company, Wilmington, Delaware 19898, United States

1.	Introduction 2845	6.	Storage and Transportation . . 2849
2.	Physical Properties 2846	7.	Economic Aspects 2850
3.	Chemical Properties. 2846	8.	Toxicology and Occupational Health. 2850
4.	Production 2847		
5.	Quality Specifications and Analysis 2848	9.	References. 2851

1. Introduction

CURTIUS published the first synthesis of hexamethylenediamine [124-09-4], 1,6-hexanediamine, 1,6-diaminohexane, $H_2N(CH_2)_6NH_2$, M_r 116.21, in 1900 [1]. The compound did not become important industrially until the late 1930s after the pioneering work at Du Pont on synthetic polymers [2]. One of these polymers was *nylon*, a polyamide produced from adipic acid [124-04-9] (→ Adipic Acid) and hexamethylenediamine. Du Pont subsequently began to produce hexamethylenediamine for manufacture of a synthetic fiber which they named nylon 66 (to distinguish it from nylon 6, a polyamide based on caprolactam). Nylon 66 rapidly gained acceptance for use in hosiery, lingerie, and other types of clothing, for military applications, and later for molded plastic parts. Du Pont was quickly joined by ICI in the production of nylon 66.

The commercial success of nylon led to expanded production of hexamethylenediamine in the postwar years, but it remained a captive product for nylon manufacture and was not generally for sale. This situation changed gradually, particularly after the development of light-stable polyurethanes based on aliphatic diisocyanates and the introduction by Bayer of a polyisocyanate (Desmodur N) made from hexamethylenediamine [3]. Hexamethylenediamine is now a commercial product available as an aqueous solution in several strengths, as anhydrous material in drums, or as a melt in insulated tank cars.

2. Physical Properties [4], [5]

Hexamethylenediamine is a colorless solid with a typical fishlike amine odor. It is very soluble in water, soluble in alcohols and aromatic solvents, and poorly soluble in aliphatic hydrocarbons. The following shows its vapor pressure as a function of temperature [6]:

t, °C	200.6	181.2	154.3	135.1	117.7
p, kPa	101.3	60.0	26.7	13.3	6.7

Some physical properties are:

mp	41 °C
Density (25 °C)	0.854 g/cm^3
Flash point (open cup)	94 °C
Dissociation constant (water, 20 °C)	
pK_1	11.11
pK_2	10.01
Heat of combustion (25°C)	40 208 kJ/kg
Specific heat capacity, c_p (solid)	2.85 kJ kg^{-1} K^{-1}
mp of water eutectic (53% H$_2$O)	−20 °C
Solubility in 100 g water (30 °C)	960 g
Heat of solution in water, J/g	
90% solution	56
85% solution	71
70% solution	112
Infinite dilution	245

Properties of an aqueous solution [7] are as follows:

	Concentration, wt%		
	90	85	70
mp, °C	30	24	6
bp, °C	129.5	122.3	111.5
Density (25 °C), g/cm^3	0.889	0.900	0.929
Flash point (open cup), °C	102	107	116

3. Chemical Properties

Hexamethylenediamine is a strong organic base, e.g., a 10% aqueous solution has a pH of 12.3 (20 °C), and forms stable salts with both organic and inorganic acids. The most important salt is that produced by neutralization with adipic acid (salt strike): the so-called nylon salt or AH salt. This compound is the raw material for the preparation of nylon by thermal dehydration under vacuum.

Other reactions of hexamethylenediamine are typical of aliphatic amines (→ Amines, Aliphatic). The preparation of hexamethylene diisocyanate [822-06-0] by reaction with phosgene is industrially important:

$$H_2N(CH_2)_6NH_2 + 2\ COCl_2 \longrightarrow OCN(CH_2)_6NCO + 4\ HCl$$

The reaction takes place in a chlorinated aromatic solvent with a yield of 95% or more [8]. Hexamethylene diisocyanate can also be produced by converting hexamethylenediamine to an aliphatic urethane, followed by thermal cleavage [9] (→ Isocyanates, Organic).

4. Production

Virtually all hexamethylenediamine is now produced by the *catalytic hydrogenation of adiponitrile* [111-69-3](→ Adipic Acid):

$$NC(CH_2)_4CN + 4\ H_2 \longrightarrow H_2N(CH_2)_6NH_2$$

This process is carried out in the liquid phase in the presence of ammonia, which serves as a heat-transfer medium and which, with some catalysts, improves yields by decreasing the extent of side reactions [10], [11]. A variety of catalysts can be used, most of which are based on promoted (i.e., activated or doped) cobalt [12]–[15], but iron oxide is also effective [16], [17]. Typical reaction temperature is 100–200 °C, and operating pressure is usually high, e.g., 28–41 MPa. Purification is by distillation, with a yield of ca. 85% [18]. Byproducts are 1,2-diaminocyclohexane [694-83-7], hexamethyleneimine [111-49-9], and bis(hexamethylenetriamine) [143-23-7], all of which are recovered for sale.

Adiponitrile itself is produced commercially by at least three processes, each based on a different hydrocarbon. The original process commercialized by Du Pont and still used in a number of adiponitrile plants starts with adipic acid made from cyclohexane:

$$C_6H_{12} \xrightarrow{Air} C_6H_{11}OH + C_6H_{10}O$$
$$C_6H_{11}OH + C_6H_{10}O + HNO_3 \longrightarrow HOOC(CH_2)_4COOH$$

Adipic acid reacts with ammonia at 300–400 °C and atmospheric pressure in the presence of a catalyst to give adiponitrile in yields as high as 95% [19]–[21]:

$$HOOC(CH_2)_4COOH \xrightarrow{NH_3} NC(CH_2)_4CN$$

Du Pont abandoned this method after developing a process for the catalytic addition of hydrogen cyanide to butadiene [22], [23]:

$$CH_2=CHCH=CH_2 \xrightarrow{HCN} NC(CH_2)_4CN$$

This reaction is carried out at ca. 100 °C and sufficient pressure to maintain the reaction in the liquid phase. Adiponitrile is separated from byproducts by distillation.

A third route to adiponitrile is by electrolytic dimerization of acrylonitrile [107-13-1], produced by the ammoxidation of propene [115-07-1] [24]. This process, developed by Monsanto, was commercialized in 1965. The following reaction takes place at the cathode in an electrolytic cell:

$$2\ CH_2=CHCN + 2\ H^+ + 2\ e \longrightarrow NC(CH_2)_4CN$$

Hydrogen ions are formed from water at the anode and pass to the cathode through a membrane. The catholyte consists of a mixture of acrylonitrile, water, and a tetraalkylammonium salt. The anolyte is aqueous sulfuric acid. The yield of adiponitrile based on acrylonitrile is 92 – 95 %. Small amounts of propionitrile and bis(cyanoethyl) ether are formed as byproducts.

A direct route to hexamethylenediamine, not based on adiponitrile, was used by Celanese but discontinued in 1984: hydrogenation of adipic acid esters to 1,6-hexanediol [629-11-8], which then reacted with ammonia at 219 °C and 22.8 MPa.

In Western Europe, hexamethylenediamine is made from adipic acid via adiponitrile by ICI, BASF, and Rhône-Poulenc; BASF also makes adiponitrile by electrolytic dimerization of acrylonitrile. Butachimie, a joint venture of Du Pont and Rhône-Poulenc, uses adiponitrile produced by the Du Pont hydrocyanation route. In Japan, the only hexamethylenediamine producer is Asahi Chemical, which uses electrolytic dimerization of acrylonitrile to manufacture the adiponitrile intermediate. Rhodia, the South American producer, also uses acrylonitrile as raw material.

5. Quality Specifications and Analysis

The color of aqueous hexamethylenediamine solutions and nylon salt solutions is considered an indication of purity. When these solutions are prepared from pure hexamethylenediamine, they are essentially water-white at first but slowly develop color on storage in contact with air. Solutions prepared from diamine containing impurities such as 1,2-diaminocyclohexane, hexamethyleneimine, and tetrahydroazepine develop color more rapidly, depending on the level of impurities and the storage temperature. The color of the polyamide produced is related to the color of the hexamethylenediamine and the nylon salt solution used.

Although no impurity specifications are given for commercial hexamethylenediamine, typical specifications for polymer-grade hexamethylenediamine (maximum content, parts per million) are

1,2-Diaminocyclohexane	50
6-Aminocapronitrile	10
2-Aminomethylcyclopentylamine	100
Hexamethyleneimine	25
Tetrahydroazepine	100
Ammonia	50
Color, Pt–Co scale (APHA)	10

1,2-Diaminocyclohexane and 2-aminomethylcyclopentylamine [21544-02-5] are determined by gas chromatography with a capillary column. Ammonia [7664-41-7], 6-aminocapronitrile [2432-74-8], and hexamethyleneimine are analyzed by a combination of distillation and titration. Ammonia is removed first by simple distillation with methanol as a carrier; the distillate is collected in boric acid solution and titrated with standard acid. After removal of ammonia, the sample is made alkaline with aqueous sodium hydroxide and refluxed to hydrolyze the aminocapronitrile to sodium aminocaproate. The ammonia liberated in this step is distilled with methanol and titrated with acid. The distillation is then continued into an aqueous fraction that carries over the hexamethyleneimine, which forms a binary azeotrope (*bp* 95.5 °C; 50.5 wt % imine) with water. This fraction is collected in neutral water and titrated with standard acid. Tetrahydroazepine can be determined by pulse polarography. Because azepine is a Schiff base, standards may be prepared by adding an aliphatic aldehyde to the hexamethylenediamine sample to give another Schiff base. The color is determined by a standard comparison colorimeter.

6. Storage and Transportation

Hexamethylenediamine should be stored in tightly closed containers in well-ventilated areas.

Shipping Information [5], [7].

United States, other than by air (DOT)

Shipping name	hexamethylenediamine, solid hexamethylenediamine, anhydrous hexamethylenediamine, solution
Hazard class	corrosive material
UN/NA no. (solid)	UN 2280
UN/NA no. (liquid)	UN 1783
DOT hazard classification	corrosive
Freight class	chemicals NOI

International, by water or air (IMO/ICAO)

Hazard class	IMCO 8
UN no. (solid)	UN 2280
UN no. (liquid)	UN 1783
IMO/ICAO label(s) (solid)	corrosive
IMO/ICAO label(s) (solution)	corrosive/poison
Subsidiary risk	6.1
Packaging group	II

7. Economic Aspects

Hexamethylenediamine production capacities (thousand tonnes) in 1986 are given below:

Canada	45
United States	630
South America	13
Western Europe	331
Japan	30
People's Republic of China	21
	1071

The United States and Western Europe account for approximately 90% of the estimated world capacity of hexamethylenediamine. Because more than 95% of production is used for nylon 66 fibers and plastics, the future of hexamethylenediamine remains dependent on the growth of this particular nylon. Additional capacity will probably not be installed before the early or mid-1990s.

8. Toxicology and Occupational Health

Exposure of humans to hexamethylenediamine by inhalation, ingestion, or skin or eye contact may cause eye irritation or blurring of vision, upper respiratory tract irritation, or skin irritation [25]. Higher exposures lead to abnormal liver function, skin burns, or eye corrosion with corneal or conjunctival ulceration. Skin sensitization has been reported in some individuals, and some evidence exists that hexamethylenediamine permeates the skin [26].

The following animal toxicities have been determined [5]: (ALC), LC_{50} 1580 ppm (mouse, inhalation, 10 min); LC_{50} 1100 mg/kg (rabbit, skin absorption); and LD_{50} 792 – 1127 mg/kg (rat, oral). Hexamethylenediamine is corrosive to the skin and eyes of animals but no evidence of skin sensitization exists. On exposure by inhalation, animals have exhibited upper and lower respiratory tract irritation [27]. Other reported effects on animals by ingestion, inhalation, and skin contact include weight loss and liver or kidney damage. The compound is not embryotoxic in animal tests. No mutagenic activity has been observed in bacterial and mammalian cell cultures.

9. References

[1] T. Curtius et al., *J. Prakt. Chem.* **62** (1900) 189.
[2] H. Mark, G. S. Whitby (eds.): *High Polymers*, vol. 1, Interscience, New York 1940.
[3] Products for Surface Coatings, Desmodur/Desmophen, ed. 1.6, Farbenfabriken Bayer AG, 14, 1986.
[4] *Merck Index*, 10th ed., Merck & Co., Rahway, NJ, 1983.
[5] *Hexamethylenediamine (Anhydrous)* Material Safety Data Sheet, Petrochemicals Dept., Du Pont, Wilmington, DE 19803, 1985.
[6] L. S. Scott, *Personal Communication*, Du Pont.
[7] *Hexamethylenediamine (Solution)* Material Safety Data Sheet, Petrochemicals Dept., Du Pont, Wilmington, DE 19803, 1985.
[8] Du Pont, US 2 374 340, 1945 (M. W. Farlow).
[9] BASF, US 4 613 466, 1986 (F. Meyer, G. Nestler, F. Towae, H. Hellbach).
[10] Du Pont, US 3 461 167, 1969 (D. R. Buehler, G. P. Keister, I. F. Long).
[11] BASF, US 3 972 938, 1976 (D. Voges, L. Hupfer, S. Wederl, K. W. Leonhard, H. Hoffman).
[12] BASF, US 3 232 888, 1966 (K. Adam).
[13] Du Pont, US 3 773 832, 1973 (L. D. Brake).
[14] Montedison Fibre, US 3 821 305, 1974 (G. Bartalini, M. Guiggioli).
[15] BASF, US 4 598 058, 1986 (G. Frank, G. Neubauer, P. Duffman, H. J. Wilfinger).
[16] Du Pont, US 3 696 153, 1972 (B. J. Kershaw, M. G. Pounder, K. R. Wilkins).
[17] BASF, US 4 587 228, 1986 (G. Frank, G. Neubauer).
[18] Chemstrand Corp., US 3 017 331, 1962 (C. R. Campbell et al.).
[19] ICI, US 3 325 532, 1967 (J. D. Rushtan, R. A. Williams).
[20] BASF, US 3 671 566, 1972 (M. Decker, J. Schmidt, H. Hoffman, H. J. Pistor).
[21] ICI, US 3 325 531, 1967 (J. G. Mather, R. A. Williams).
[22] Du Pont, US 3 496 215, 1970 (W. C. Drinkard Jr., R. V. Lindsey Jr.).
[23] Du Pont, US 3 536 748, 1970 (W. C. Drinkard Jr., R. V. Lindsey Jr.).
[24] M. M. Baizer, *J. Electrochem. Soc.* **111** (1964) 215.
[25] G. Gallo, L. Ghiringhelli, *Med. Lav.* **49** (1958) 683.
[26] C. Ceresa, *Med. Lav.* **39** (1948) 162.
[27] J. C. Gage, *Br. J. Ind. Med.* **27** (1970) no. 1, 1.

Hydrocarbons

Individual keywords: → *Acetylene;* → *Anthracene;* → *Benzene;* → *Butadiene;* → *Butenes;* → *Cyclododecatriene and Cyclooctadiene;* → *Cyclohexane;* → *Cyclopentadiene and Cyclopentene;* → *Ethylbenzene;* → *Ethylene;* → *Isoprene;* → *Methane;* → *Naphthalene and Hydronaphthalenes;* → *Propene;* → *Styrene;* → *Terpenes;* → *Toluene;* → *Waxes;* → *Xylenes; Propyne;* → *Acetylene*

KARL GRIESBAUM, Universität Karlsruhe (TH), Karlsruhe, Federal Republic of Germany (Chap. 1)

ARNO BEHR, Henkel KGaA, Düsseldorf, Federal Republic of Germany (Chap. 2)

DIETER BIEDENKAPP, BASF Aktiengesellschaft, Ludwigshafen, Federal Republic of Germany (Section 3.1–3.7, Section 6.3 in part)

HEINZ-WERNER VOGES, Hüls Aktiengesellschaft, Marl, Federal Republic of Germany (Section 3.7–3.10, Section 6.3 in part)

DOROTHEA GARBE, Haarmann & Reimer GmbH, Holzminden, Federal Republic of Germany (Section 3.11)

CHRISTIAN PAETZ, Bayer AG, Leverkusen, Federal Republic of Germany (Chap. 4, Section 6.4 in part)

GERD COLLIN, Rüttgerswerke AG, Duisburg, Federal Republic of Germany (Chap. 5)

DIETER MAYER, Hoechst Aktiengesellschaft, Frankfurt, Federal Republic of Germany (Chap. 6)

HARTMUT HÖKE, Rüttgerswerke AG, Castrop-Rauxel, Federal Republic of Germany (Section 6.5)

1.	**Saturated Hydrocarbons**	2854	2.1.4.	Economic Aspects of Higher Olefins	2888
1.1.	**Physical Properties**	2854	2.2.	**Dienes and Polyenes**	2889
1.2.	**Chemical Properties**	2857	2.2.1.	Low Molecular Mass 1,3-Dienes	2889
1.3.	**Production**	2860	2.2.2.	Synthesis of Dienes and Polyenes by Oligomerization	2891
1.3.1.	From Natural Gas and Petroleum	2860	2.2.3.	Synthesis of Dienes and Polyenes by Metathesis	2892
1.3.2.	From Coal and Coal-Derived Products	2861	3.	**Alkylbenzenes**	2893
1.3.3.	By Synthesis and by Conversion of other Hydrocarbons	2862	3.1.	**Trimethylbenzenes**	2893
1.4.	**Uses**	2863	3.2.	**Tetramethylbenzenes**	2895
1.5.	**Individual Saturated Hydrocarbons**	2867	3.3.	**Penta- and Hexamethylbenzene**	2896
2.	**Olefins**	2870	3.4.	**Diethylbenzenes**	2897
2.1.	**Monoolefins**	2870	3.5.	**Triethylbenzenes and more highly Ethylated Benzenes**	2897
2.1.1.	Properties	2870			
2.1.2.	Production of Higher Olefins	2873	3.6.	**Ethylmethylbenzenes (Ethyltoluenes)**	2898
2.1.2.1.	Production from Paraffins	2873			
2.1.2.2.	Oligomerization of Lower Olefins	2876	3.7.	**Cumene**	2899
2.1.3.	Uses of Higher Olefins	2884			

3.8.	Diisopropylbenzenes 2901	5.5.	Phenanthrene 2920
3.9.	Cymenes; C$_4$- and C$_5$-Alkylaromatic Compounds .. 2902	5.6.	Pyrene 2922
3.10.	Monoalkylbenzenes with Alkyl Groups >C$_{10}$ 2903	6.	Toxicology and Occupational Health 2924
3.11.	Diphenylmethane 2904	6.1.	Alkanes............... 2924
4.	Biphenyls and Polyphenyls... 2906	6.2.	Alkenes............... 2925
4.1.	Biphenyl 2906	6.3.	Alkylbenzenes 2925
4.2.	Terphenyls 2911	6.4.	Biphenyls and Polyphenyls... 2926
4.3.	Polyphenyls 2912	6.5.	Hydrocarbons from Coal Tar . 2927
5.	Hydrocarbons from Coal Tar . 2913	6.5.1.	Biological Effects 2927
5.1.	Acenaphthene 2913	6.5.1.1.	Carcinogenicity and Mutagenicity 2927
5.2.	Indene 2915	6.5.1.2.	Mammalian Toxicity and Toxicokinetics 2927
5.3.	Fluorene 2917	6.5.1.3.	Ecotoxicology 2928
5.4.	Fluoranthene 2919	6.5.2.	Safety Regulations 2928
		7.	References............... 2928

1. Saturated Hydrocarbons

The class of saturated hydrocarbons comprises a myriad of individual compounds. A large number of the theoretically possible saturated hydrocarbons is known [18], although only a limited number of individual saturated hydrocarbons are used as raw materials in the chemical industry.

1.1. Physical Properties

Saturated hydrocarbons are colorless nonpolar substances, immiscible with polar solvents, but miscible with many nonpolar organic solvents. Some physical properties are given in Table 1. The boiling points and, starting with C_3, melting points of n-alkanes increase with increasing molecular mass. At 20 °C and atmospheric pressure, C_1 to C_4 n-alkanes are gases, C_5 to C_{16} are liquid, and from C_{17} solid. i-Alkanes do not display a definite correlation between the number of carbon atoms and their boiling- and melting points. The boiling points of i-alkanes all are lower than those of the corresponding n-alkanes; this is also true for many of the melting points. Furthermore, the boiling points of isomeric alkanes decrease with increasing degree of branching. The boiling points and melting points of cycloalkanes are generally higher than those of n-alkanes having the same number of carbon atoms (Table 1).

The density of the saturated hydrocarbons in the liquid state at 20 °C is <1 g/cm^3; it varies from 0.6 g/cm^3 for compounds with low carbon numbers to 0.8 g/cm^3 for

Table 1. Physical properties of saturated hydrocarbons

Compound	Empirical formula	CAS registry number	mp, °C	bp, °C	η (20 °C), mPa·s	ϱ (20 °C), g/cm^3	n_D^{20}
Methane	CH$_4$	[74-82-8]	–182.5 [a]	–161.5			
Ethane	C$_2$H$_6$	[74-84-0]	–183.27 [a]	– 88.6			
Propane	C$_3$H$_8$	[74-98-6]	–187.69 [a]	– 42.1		0.5005 [b]	
Cyclopropane	C$_3$H$_6$	[75-19-4]	–127.4	– 32.8			
n-Butane	C$_4$H$_{10}$	[106-97-8]	–138.4	– 0.5		0.5788 [b]	1.3326 [b]
Cyclobutane	C$_4$H$_8$	[287-23-0]	– 90.7	12.5		0.6943 [b]	1.365 [a]
2-Methylpropane	C$_4$H$_{10}$	[75-28-5]	–159.6	– 11.7		0.5572 [b]	
n-Pentane	C$_5$H$_{12}$	[109-66-0]	–129.7	36.1	0.234	0.62624	1.35748
Cyclopentane	C$_5$H$_{10}$	[287-92-3]	– 93.9	49.3	0.438	0.74538	1.40645
2-Methylbutane	C$_5$H$_{12}$	[78-78-4]	–159.9	27.9	0.224	0.61967	1.35373
2,2-Dimethylpropane	C$_5$H$_{12}$	[463-82-1]	– 16.6	9.5		0.5910 [b]	1.342 [b]
n-Hexane	C$_6$H$_{14}$	[110-54-3]	– 95.3	68.7	0.3117	0.65937	1.37486
2-Methylpentane	C$_6$H$_{14}$	[107-83-5]	–153.7	60.3		0.65315	1.37145
3-Methylpentane	C$_6$H$_{14}$	[96-14-0]		63.3		0.66431	1.37652
2,2-Dimethylbutane	C$_6$H$_{14}$	[75-83-2]	– 99.9	49.7		0.64916	1.36876
2,3-Dimethylbutane	C$_6$H$_{14}$	[79-29-8]	–128.5	58.0		0.66164	1.37495
n-Heptane	C$_7$H$_{16}$	[142-82-5]	– 90.6	98.4	0.4169	0.68376	1.38764
2-Methylhexane	C$_7$H$_{16}$	[591-76-4]	–118.3	90.1		0.67859	1.38485
3-Methylhexane	C$_7$H$_{16}$	[589-34-4]		91.9		0.68713	1.38864
3-Ethylpentane	C$_7$H$_{16}$	[617-78-7]	–118.6	93.5		0.69816	1.39339
2,2-Dimethylpentane	C$_7$H$_{16}$	[590-35-2]	–123.8	79.2		0.67385	1.38215
2,3-Dimethylpentane	C$_7$H$_{16}$	[565-59-3]		89.8		0.69508	1.39196
2,4-Dimethylpentane	C$_7$H$_{16}$	[108-08-7]	–119.2	80.5		0.67270	1.38145
3,3-Dimethylpentane	C$_7$H$_{16}$	[562-49-2]	–134.46 [d]	86.1		0.69327	1.39092
2,2,3-Trimethylbutane	C$_7$H$_{16}$	[464-06-2]	– 24.9	80.9		0.69011	1.38944
n-Octane	C$_8$H$_{18}$	[111-65-9]	– 56.8	125.7	0.5450	0.70252	1.39743
2-Methylheptane	C$_8$H$_{18}$	[592-27-8]	–109.0	117.6		0.69792	1.39494
3-Methylheptane	C$_8$H$_{18}$	[589-81-1]	–120.5	118.9		0.70582	1.39848
4-Methylheptane	C$_8$H$_{18}$	[589-53-7]	–121.0	117.7		0.70463	1.39792
2,2,3-Trimethylpentane	C$_8$H$_{18}$	[564-02-3]	–112.3	109.8		0.71602	1.40295
2,2,4-Trimethylpentane	C$_8$H$_{18}$	[540-84-1]	–107.4	99.2	0.69193	1.39145	
2,3,3-Trimethylpentane	C$_8$H$_{18}$	[564-02-3]	–100.7	114.8		0.72619	1.40750
2,3,4-Trimethylpentane	C$_8$H$_{18}$	[565-75-3]	–109.2	113.5		0.71906	1.40422
2,2,3,3-Tetramethylbutane	C$_8$H$_{18}$	[594-82-1]	–100.7	106.5			
n-Nonane	C$_9$H$_{20}$	[111-84-2]	– 53.5	150.8	0.7139	0.71763	1.40542
n-Decane	C$_{10}$H$_{22}$	[124-18-5]	– 29.7	174.1	0.9256	0.73005	1.41189
n-Undecane	C$_{11}$H$_{24}$	[1120-21-4]	– 25.6	195.9	1.185	0.74024	1.41725
n-Dodecane	C$_{12}$H$_{26}$	[112-40-3]	– 9.6	216.3	1.503	0.74869	1.42160
n-Tridecane	C$_{13}$H$_{28}$	[629-50-5]	– 5.4	235.4	1.880	0.75622	1.42560
n-Tetradecane	C$_{14}$H$_{30}$	[629-59-4]	5.9	253.5	2.335	0.76275	1.42892
n-Pentadecane	C$_{15}$H$_{32}$	[629-62-9]	9.9	270.6	2.863	0.76830	1.43188
n-Hexadecane	C$_{16}$H$_{34}$	[544-76-3]	18.2	286.8	3.474	0.77344	1.43453
n-Heptadecane	C$_{17}$H$_{36}$	[629-78-7]	21.98	302.2	4.196 [c]	0.7779 [c]	1.4368 [c]
n-Octadecane	C$_{18}$H$_{38}$	[593-45-3]	28.2	316.7		0.7818 [c]	1.4389 [c]
n-Nonadecane	C$_{19}$H$_{40}$	[629-92-5]	31.9	330.6		0.7854 [c]	1.4408 [c]
n-Icosane	C$_{20}$H$_{42}$	[112-95-8]	36.4	343.8		0.7886 [c]	1.4425 [c]
n-Triacontane	C$_{30}$H$_{62}$	[638-68-6]	65.8	449.7		0.8096 [c]	1.4535 [c]
n-Tetracontane	C$_{40}$H$_{82}$	[4181-95-7]	81.5	525		0.8204 [c]	1.4592 [c]

[a] For saturation pressure (triple point).
[b] At saturation pressure.
[c] For the super-cooled liquid below the normal melting point.
[d] Value for the melting point of the metastable crystalline form.

Table 2. Molar enthalpies of fusion, vaporization and combustion

	Molar enthalpy, kJ/mol		
	Fusion	Vaporization (25 °C)	Combustion [a]
Methane	0.942		802.861
Ethane	2.86	5.02	1 428.787
Propane	3.53	15.1	2 045.377
Cyclopropane	5.44		1 960.637
n-Butane	4.66	21.1	2 658.827
Cyclobutane	5.78	23.7	2 569.523
2-Methylpropane	4.54	19.1	2 650.454
n-Pentane	8.41	26.8	3 274.287
Cyclopentane	4.89	28.5	3 101.707
2-Methylbutane	5.15	24.8	3 266.248
2,2-Dimethylpropane	3.15	21.8	3 254.735
n-Hexane	13.1	31.7	3 886.81
2-Methylpentane	6.27	30.0	3 879.07
3-Methylpentane		30.4	3 881.71
2,2-Dimethylbutane	0.579	27.8	3 869.78
2,3-Dimethylbutane	0.80	29.2	3 877.86
n-Heptane	14.1	36.6	4 501.44
n-Octane	20.8	41.5	5 115.57
n-Dodecane	35.9	61.3	7 580.076
n-Hexadecane	51.9	81.1	10 040.616
n-Icosane	69.9	100.9 [b]	12 501.198

[a] For the reaction: Hydrocarbon (g) → CO_2 (g) + H_2O (g).
[b] For the supercooled liquid, below the normal melting point.

Table 3. Enthalpy of formation, entropy, and free energy of formation for saturated hydrocarbons as gases (25 °C)

	ΔH^0, kJ/mol	S^0, J mol^{-1} K^{-1}	ΔG^0, kJ/mol
Methane	− 74.898	186.313	−50.828
Ethane	− 84.724	229.646	−32.908
Propane	−103.916	270.090	−23.505
Cyclopropane	+ 53.382	237.810	104.503
n-Butane	−126.232	310.326	−17.166
Cyclobutane	+ 27.256	265.569	110.741
2-Methylpropane	−134.606	294.834	−20.934
n-Pentane	−146.538	349.179	− 8.374
Cyclopentane	− 77.079	293.076	38.895
2-Methylbutane	−154.577	343.820	−14.654
2,2-Dimethylpropane	−166.090	306.599	−15.240
n-Hexane	−167.305	388.661	− 0.293
2-Methylpentane	−174.422	380.789	− 5.024
3-Methylpentane	−171.743	380.036	− 2.135
2,2-Dimethylbutane	−185.685	358.474	− 9.923
2,3-Dimethylbutane	−177.897	366.010	− 4.103
n-Heptane	−187.945	428.058	+ 8.039
n-Octane	−208.586	467.038	16.412
n-Dodecane	−291.066	622.912	50.074
n-Hexadecane	−373.588	778.829	83.820
n-Icosane	−456.068	934.745	117.398

compounds with high carbon numbers (Table 1). Additional physical properties are compiled in Tables 2, 3, 4, 5.

Table 4. Critical data and molar heat capacities of saturated hydrocarbons

Compound	T_c, °C	p_c, MPa	c_p,* J mol^{-1} K^{-1}
Methane	−82.6	4.60	35.74
Ethane	32.28	4.88	52.67
Propane	96.7	4.25	73.56
Cyclopropane	125.1	5.57	56.27
n-Butane	152.0	3.80	97.51
Cyclobutane	187	4.99	72.26
2-Methylpropane	135.0	3.65	96.88
n-Pentane	196.5	3.37	120.3
Cyclopentane	238.5	4.50	82.98
2-Methylbutane	187.2	3.38	118.9
2,2-Dimethylpropane	160.6	3.20	121.7
n-Hexane	234.2	3.01	143.2
2-Methylpentane	224.3	3.01	144.3
3-Methylpentane	231.2	3.12	143.2
2,2-Dimethylbutane	215.6	3.08	142.0
2,3-Dimethylbutane	226.8	3.13	140.6
n-Heptane	267.0	2.74	166.1
n-Octane	295.6	2.49	189.0
n-Nonane	321.4	2.29	211.8
n-Decane	344.4	2.10	234.7
n-Undecane	365.6	1.97	257.6
n-Dodecane	385.1	1.82	280.5
n-Tridecane	402.6	1.72	303.4
n-Tetradecane	418.7	1.62	326.3
n-Pentadecane	433.6	1.52	349.2
n-Hexadecane	447.4	1.42	372.0
n-Heptadecane	460.2	1.32	394.9
n-Octadecane	472.1	1.22	417.8
n-Nonadecane	483	1.12	440.7
n-Icosane	494	1.12	463.6

* Based on 25 °C and ideal gas state.

1.2. Chemical Properties

Alkanes and cycloalkanes are saturated, nonpolar, and lack functional groups; such hydrocarbons can undergo reaction only after cleavage of C–H or C–C bonds. Consequently, the scope of primary reaction steps is essentially limited to dehydrogenation, substitution, and chain- or ring cleavage. Saturated hydrocarbons cannot undergo addition reactions, which represent a versatile synthetic tool for unsaturated hydrocarbons (see Chap. 2). Most industrial reactions involving saturated hydrocarbons are radical reactions, e.g., thermal cracking, oxidation, sulfoxidation, halogenation, sulfochlorination, and nitration. Industrial ionic reactions of saturated hydrocarbons are restricted to acid-catalyzed processes with strong acids. Such reactions are employed mainly in the processing of petroleum by catalytic cracking, isomerization, and alkylation; they yield complex mixtures of products. Other ionic reactions of saturated hydrocarbons can be carried out with super acids (e.g., FSO_3H-SbF_5, $HF-SbF_5$): the alkylation of alkanes by alkanes [20], alkylation of benzene by alkanes [20], and

Table 5. Fuel properties of saturated hydrocarbons

	Flash point,[a] °C	Explosion limits (20 °C),[b] lower	Explosion limits (20 °C),[b] upper	Ignition temperature,[a] °C	Octane number, ^{R}RON
Methane		5.00	15.00	595	
Ethane		3.00	12.50	515	
Propane		2.12	9.35	470	
Cyclopropane		2.40	10.40	495	
n-Butane		1.86	8.41	365	93.8
Cyclobutane		1.8[a]			
2-Methylpropane		1.8[a]	8.5[a]	(460)[d]	
n-Pentane	<−20	1.40	7.80	285	61.7
Cyclopentane	<−20			380	102.5
2-Methylbutane	<−20	1.32	7.6[a]	420	92.3
2,2-Dimethylpropane		1.38	7.50	(450)[d]	85.5
n-Hexane	<−20	1.18	7.40	240	24.8
i-Hexane	<−20	ca. 1[a]	ca. 7.4[a]	ca. 260	
2,2-Dimethylbutane	<−20	1.2[a]	7.0[a]	435	91.8
2,3-Dimethylbutane	<−20	1.2[a]	7.0[a]	415	103.5
n-Heptane	−4	1.10	6.70	215	0
i-Heptane	<−4	ca. 1[c]	ca. 7[c]	ca. 220	
2,3-Dimethylpentane	<0	1.12	6.75	330	91.9
2,2,3-Trimethylbutane	<0			450	112.1
2,3-Dimethylhexane	<12	0.95	6.5[a]	210	
n-Octane	12	0.95	6.5[a]	71.3	
i-Octane	−12	1.0	6.0	410	100
2,2,3-Trimethylpentane	<21			430	109.5
2,3,3-Trimethylpentane	<21			425	106
n-Nonane	31	0.83[d]	5.6[a,d]	205	
n-Decane	46	0.77[d]	5.35[d]	205	
n-Dodecane	74	0.6[a,d]		200	
n-Hexadecane	>100			205	
n-Icosane	>100				

[a] Ref. [24],
[b] Ref. [25],
[c] Ref. [26],
[d] Approximate values.
[e] Values are valid for $t > 20$ °C.

ionic chlorination [22] and bromination [23] of alkanes have been explored, but are not used industrially.

Reactions of saturated hydrocarbons, because of the lack of functional groups, are nonselective, with respect to both the region of attack (regioselectivity) and the number of reaction sites (degree of substitution), unless the molecule possesses specific structural features, e.g., tertiary hydrogen atoms. Such reactions frequently afford mixtures of isomeric or structurally analogous compounds which can only be separated with difficulty or not at all.

On the basis of free enthalpy of formation (see Table 2) most of the saturated hydrocarbons are thermodynamically unstable with respect to the elements carbon and hydrogen. They are, however, kinetically stable at ambient temperature. Thermal decomposition of saturated hydrocarbons proceeds stepwise by loss of hydrogen or

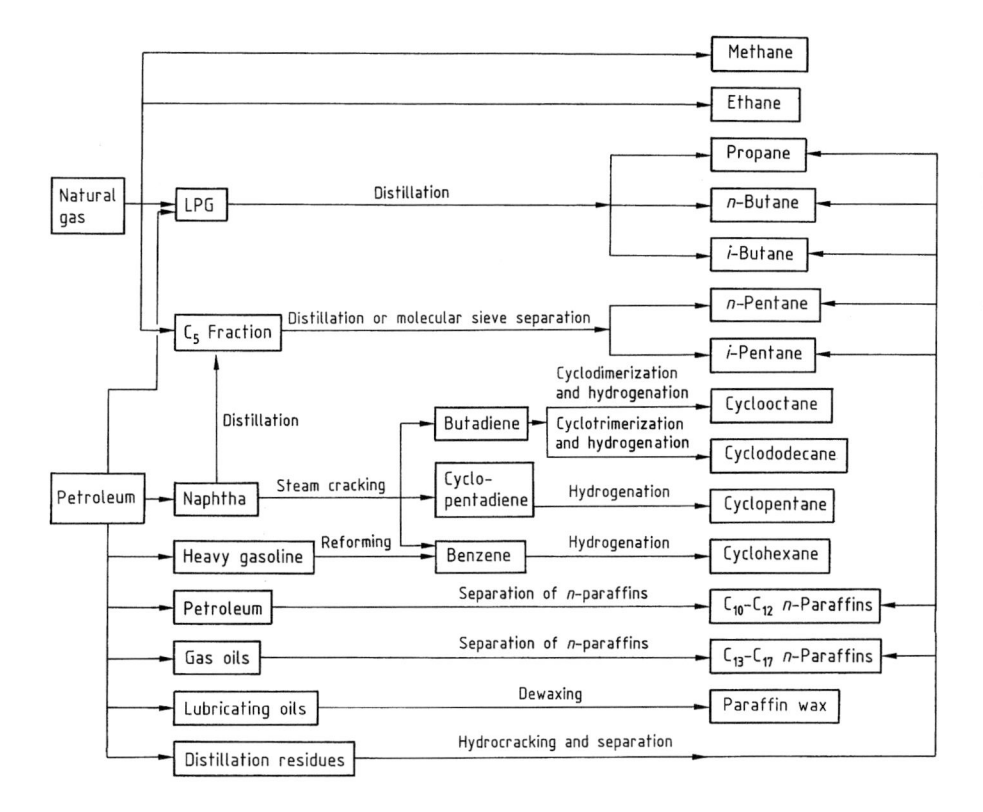

Figure 1. Saturated hydrocarbons from natural gas and from petroleum

hydrocarbon fragments with concomitant formation of industrially useful unsaturated cracking products, such as acetylene, olefins, or aromatic hydrocarbons.

On ignition, mixtures of saturated hydrocarbons and oxygen or air may lead to combustion or explosion, depending on the ratio of hydrocarbon and oxygen. Such reactions can be initiated either by imposed ignition or, if the ignition temperature of the mixture is exceeded, by self-ignition. These reactions are the basis for the use of hydrocarbons as heating- and engine fuel. Pertinent fuel properties of individual saturated hydrocarbons are summarized in Table 4.

1.3. Production

By far the largest amount of saturated hydrocarbons is obtained from the natural sources *natural gas* and *petroleum*, either by isolation or by suitable conversion reactions. Additional sources include various products derived from coal processing. A number of saturated hydrocarbons, unavailable from natural sources, are produced by special synthesis or by conversion processes.

1.3.1. From Natural Gas and Petroleum

Natural Gas contains methane as the single major component (see also →Methane). Depending on the particular source, natural gas may also contain acyclic saturated hydrocarbons up to C_5 in proportions that permit their isolation [27]. The isolation of individual compounds from natural gas (see Figure 1) can be performed either by absorption or by partial condensation at low temperature, followed by distillation.

Petroleum is the most abundant source of saturated hydrocarbons, both with respect to the absolute amount and to the variety of individual compounds. Petroleum is separated into individual fractions by distillation in a refinery. Such fractions contain several (liquefied petroleum gas, LPG) or many individual hydrocarbons (all other fractions); these fractions can be used as such or processed further for use as heating- or engine fuels, as lubricants, as raw materials for the production of petrochemicals (synthesis gas, acetylene, olefins, aromatics), or for the recovery of industrially important alkanes and cycloalkanes.

From the liquefied petroleum gas fraction, propane, *n*-butane, and *i*-butane are isolated by distillation. *n*-Pentane and *i*-pentane can be isolated from the liquid components of natural gas or from light gasoline (naphtha) either by molecular sieve separation or by superfractionation [28]. Steam-cracking of light gasoline provides, among other products, cyclopentadiene (→Cyclopentadiene and Cyclopentene), benzene (→Benzene), and 1,3-butadiene (→Butadiene); these can, in turn, be converted into the C_5-, C_6-, C_8-, and C_{12}-cycloalkanes. Catalytic reforming of heavy gasoline provides C_6- to C_8-aromatics, which can subsequently be converted into cyclohexane (→Cyclohexane), e.g., the C_7 and C_8-aromatics are hydrodealkylated, and the ensuing benzene subsequently hydrogenated. From the higher boiling petroleum, gas-oil, and wax distillate fractions, mixtures of homologous *n*-alkanes are isolated by techniques, such as molecular sieve separation or urea extractive crystallization. Paraffin waxes can also be isolated by precipitation with suitable solvents.

Higher boiling petroleum distillates and distillation residues can be converted into mixtures of lower molecular mass hydrocarbons by hydrocracking [29], [30]. From such mixtures, saturated C_3–C_5-hydrocarbons and a broad spectrum of higher *n*-alkanes can be recovered.

Although petroleum and the various petroleum distillates contain a large variety of other saturated hydrocarbons in addition to those already mentioned, the recovery and subsequent industrial use of individual compounds is largely confined to those depicted in Figure 1; the increasing number of structural isomers with increasing molecular mass poses severe limitations on the economical separation of individual compounds from such distillate mixtures on an industrial scale.

1.3.2. From Coal and Coal-Derived Products

For a long time the most important sources of saturated hydrocarbons were coal and the products derived from the liquefaction, coking, and gasification of coal. These sources became less important when natural gas and petroleum became essential raw materials for organic chemicals [31], [32].

The *liquefaction of coal* provided the greatest variety of saturated hydrocarbons. The Fischer–Tropsch Synthesis can produce alkanes in the range C_1 to C_{30} or higher, depending on the process variant [33]–[35]: the fluidized bed synthesis affords predominantly liquid hydrocarbons in the gasoline range, along with gases from C_1 to C_4. The liquid hydrocarbons contain considerable proportions of branched and olefinic compounds. The fixed bed synthesis provides higher molecular mass hydrocarbons in the range of diesel oil or paraffin wax. These products are rich in *n*-alkanes and are, therefore, suitable raw materials for detergents and for wax products. Suitable fractions may also serve as raw materials for cracking and reforming processes.

The various processes for the *hydrogenation of coal* [36]–[38] or of coal tar afford predominantly liquid products in the boiling range of gasoline, which are rich in aromatics, and minor amounts of C_1- to C_4-alkanes. From the gasoline fractions, benzene can be recovered and hydrogenated to cyclohexane. After the recovery of benzene, the remaining mixtures can be used as feed materials for steam crackers for the production of olefins.

Coking of coal produces methane. In addition, high temperature processes produce benzene, which can be hydrogenated to cyclohexane [39], [40]. Low temperature coking of soft coal provides a tar, from which paraffin wax may be recovered.

Gasification of coal [41] affords mainly carbon monoxide and hydrogen, which can be converted into methane by nickel-catalyzed reactions or into a range of alkanes by the Fischer–Tropsch process. Up to 10% of methane and higher hydrocarbons such as naphtha and benzene are formed in addition to carbon monoxide and hydrogen if the gasification is carried out at elevated pressure [42].

1.3.3. By Synthesis and by Conversion of other Hydrocarbons

The major processing techniques are *cyclization, hydrogenation,* and *isomerization,* forming the basis of the industrial production of most cycloalkanes. Other techniques include *alkylation* of alkenes with alkanes to provide *i*-alkanes, and hydrocracking to give predominantly low molecular mass *i*- and *n*-alkanes, including methane.

Cyclization and Hydrogenation. Among the cycloalkanes only those containing a 5- or a 6-membered ring are directly available from natural sources, i.e., from petroleum or coal-derived products. Steamcracking of naphtha or gas oil produces cyclopentadiene as a byproduct, which can be converted into cyclopentane by catalytic hydrogenation. Similarly, cyclohexane is produced by catalytic hydrogenation of benzene, which in turn is formed by steam cracking of liquid hydrocarbons, by catalytic reforming of heavy gasoline, or by coking of coal.

The skeletons of all other industrially produced cycloalkanes are synthesized from acyclic precursors. Although there is a wide variety of synthetic procedures for such reactions, only a limited number are used industrially. Of prime importance are cycloaddition reactions of acyclic unsaturated compounds followed by hydrogenation of the resulting unsaturated cyclic products: Thus, [4+2]-cycloaddition (Diels–Alder reaction) of conjugated dienes and monoolefins yields saturated 6-membered ring compounds via cyclohexene intermediates, e.g., the synthesis of cyclohexane from 1,3-butadiene and ethylene [43]:

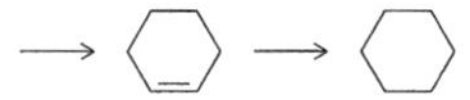

the catalyzed [4+4]-cycloaddition of 1,3-butadiene to give cyclooctane via 1,5-cyclooctadiene;

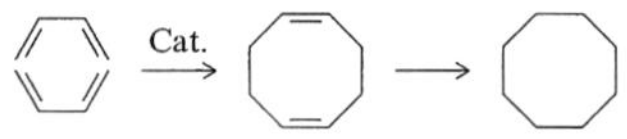

and the catalyzed [4+4+4]-cyclotrimerization of 1,3-butadiene to form cyclododecane via 1,5,9-cyclododecatriene:

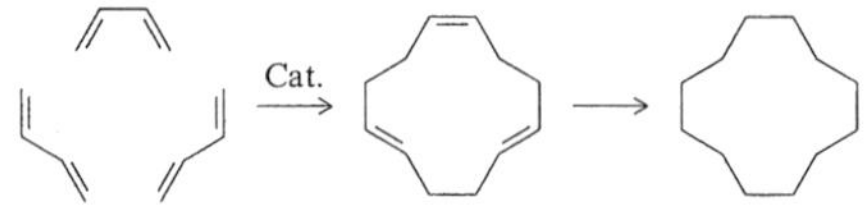

Cyclizations that lead to small (C_3-, C_4-) rings are not industrially important, although various laboratory methods are available [44]. Examples include the coupling of terminal carbon atoms by elimination of suitable α,ω-substituents from acyclic compounds, e.g., α,ω-dihalides, ω-halogenated ketones, esters, or nitriles (Eq. 1), addition of carbenes, to olefins to give cyclopropanes (Eq. 2) and [2+2]-cycloadditions of suitable unsaturated compounds, such as olefins, allenes, or acetylenes, to give cyclobutanes (Eq. 3) [45]:

$$X-CH_2-(CH_2)_n-CH_2-Y \xrightarrow{-XY} \underset{CH_2}{\overset{CH_2}{|}}\!\!>\!(CH_2)_n \quad (1)$$

$$\underset{\diagdown C\diagup}{\overset{\diagup C\diagdown}{\|}} + CX_2 \longrightarrow \underset{-C-}{\overset{-C-}{|}}\!\!>\!CX_2 \quad (2)$$

$$\underset{\diagdown C\diagup}{\overset{\diagup C\diagdown}{\|}} + \underset{\diagdown C\diagup}{\overset{\diagup C\diagdown}{\|}} \longrightarrow \begin{array}{c}-C-C-\\-C-C-\end{array} \quad (3)$$

Most of these synthetic methods yield substituted or unsaturated cyclic compounds in the cyclization step. These primary products can be subsequently converted into the corresponding cycloalkanes by suitable measures.

Isomerization and Alkylation. On account of their specific properties (antiknock activity, possibility for selective reactions), some i-alkanes are in higher demand than can be met by natural sources. i-Alkanes are produced either by acid-catalyzed isomerization of n-alkanes or by acid-catalyzed alkylation of olefins with i-alkanes. The most important isomerizations are the conversion of n-butane to i-butane, n-pentane to i-pentane, and n-hexane to i-hexane [46], [47]. Industrially important alkylations include the reaction of i-butane with propylene or with butenes to give highly branched C_7 or C_8-alkanes, respectively. The latter exhibit high octane numbers [47], [48].

Hydrocracking. A range of acyclic saturated hydrocarbons having low carbon numbers is accessible by catalyzed hydrocracking of heavy distillates or distillation residues. Hydrocracking of distillates can be optimized by adjusting the reaction conditions, so that the products are predominantly in the range of liquefied petroleum gases, including i-butane [30], or in the range of middle distillates [49]. The cracking of petroleum distillates with steam over a nickel catalyst produces methane [50], [51].

1.4. Uses

The major use of saturated hydrocarbons is as a mixture, sometimes with unsaturated or aromatic compounds, as heating fuels, motor fuels and lubricants. Saturated hydrocarbons also serve as raw materials for the production of carbon black. In the chemical industry, the following processes are important (Fig. 2).

Thermal Cracking. Thermal cracking of individual or mixed saturated hydrocarbons is the basis for the petrochemical production of acetylene, olefins, diolefins, and to some extent BTX aromatics: high temperature pyrolysis at >1000 °C affords acetylene

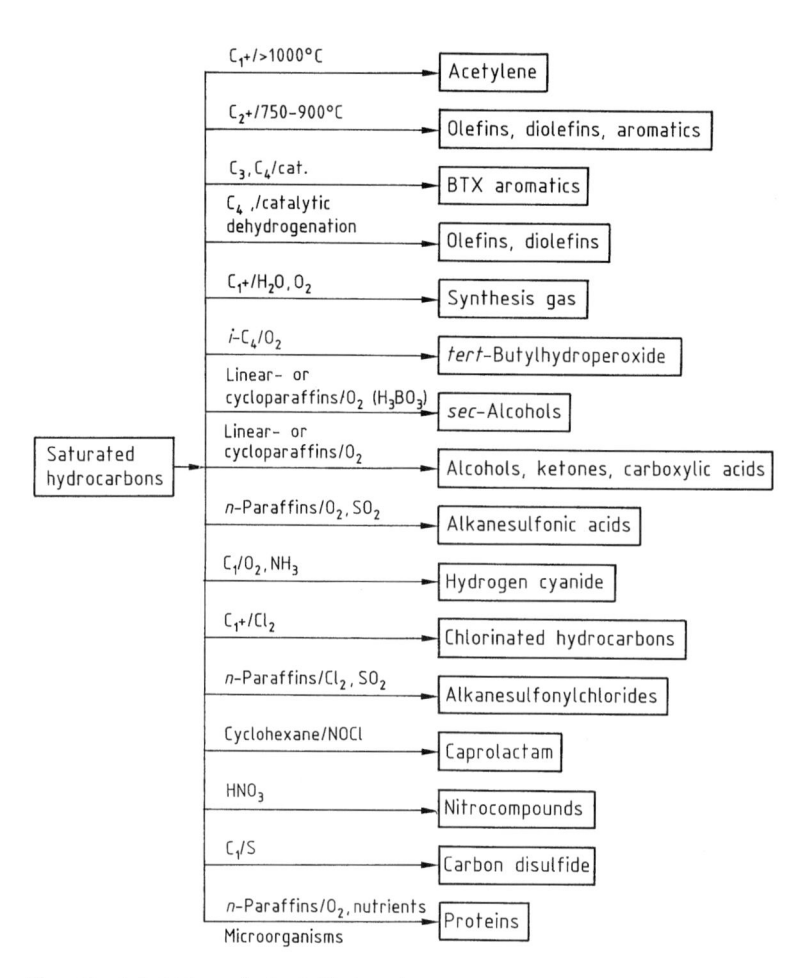

Figure 2. Industrial use of saturated hydrocarbons

(→Acetylene) and, depending on the hydrocarbon feed and the particular process applied, ethylene. The range of hydrocarbon feed materials can vary from methane to crude oil. Medium temperature pyrolysis at 750–900 °C in the presence of inert diluents such as steam (steam-cracking) or hydrogen (hydropyrolysis) yields ethylene (→ Ethylene). A wide range of hydrocarbons, e.g., ethane, liquefied petroleum gas, naphtha, and gas oil may serve as raw material, and increasing proportions of C_3- to C_5-olefins, C_4- and C_5-dienes, and C_6- to C_8-aromatics are coproduced with increasing molecular mass of the feed material. Pyrolysis of petroleum waxes at 500–600 °C affords mixtures of homologous olefins with predominantly terminal double bonds.

Catalytic Dehydrogenation. Catalytic dehydrogenation of acyclic saturated hydrocarbon atoms yields mono- and/or diolefins with the same number of carbon atoms. Industrial processes include the production of butenes and butadiene from *n*-butane, of isoprene from *i*-pentane, and C_6- to C_{19}-monoolefins, mainly with internal double bonds, from the corresponding *n*-alkanes [52]. More recently, improved processes for the production of ethylene, propene, and *i*-butene by catalytic dehydrogenation of the corresponding saturated hydrocarbons have been developed [53]. Catalytic dehydrogenation of cyclohexane and methyl cyclohexanes yields benzene and methylbenzenes, respectively.

Catalytic dehydrocyclooligomerization of propane and *n*-butane over modified zeolite catalysts represents a new method for the production of BTX-aromatics [54]. The process is apparently ready for industrial application.

Partial Oxidation. Partial oxidation of saturated hydrocarbons either by catalyzed reaction with steam (steam reforming) or by noncatalyzed reaction with deficient amounts of oxygen affords a mixture of carbon monoxide and hydrogen, which is used as town gas or as synthesis gas. Desulfurized saturated hydrocarbons from methane to compounds with *bp* ca. 200 °C are suitable feed materials for catalytic processes; for noncatalytic processes all hydrocarbons may be used from methane to distillation residues.

Selective Oxidation. Selective oxidation represents the most important tool for the introduction of oxygen-containing functional groups into saturated hydrocarbons: Liquid-phase autoxidation of *i*-butane with air yields *tert*-butylhydroperoxide; the latter is used for the selective oxidation of propene to propylene oxide [55]. Liquid-phase oxidation of *n*-alkanes or cycloalkanes in the presence of boric acid affords the corresponding secondary alcohols. The reaction is used industrially for the oxidation of C_{10}- to C_{20}-*n*-alkanes, providing raw materials for detergents and for the oxidation of cyclododecane to cyclododecanol as an intermediate for the production of nylon 12. Catalyzed liquid-phase oxidation of *n*-alkanes and cycloalkanes in the absence of boric acid leads to mixtures of secondary alcohols and ketones or to carboxylic acids, depending on the reaction conditions: oxidation of cyclohexane leads to a mixture of cyclohexanol and cyclohexanone (→Cyclohexanol and Cyclohexanone); oxidation of *n*-butane gives acetic acid (→Acetic Acid); oxidation of longer chain *n*-alkanes produces mixtures of fatty acids; and oxidation of cyclohexane gives adipic acid. (→Adipic Acid). Liquid phase oxidation of bicyclo[4.4.0]decane yields the corresponding tertiary hydroperoxide, which is eventually converted into sebacic acid by a sequence of reaction steps [56].

Heterogeneously catalyzed gas-phase oxidation of *n*-butane has been used industrially in recent years as an attractive alternative for the production of maleic anhydride [57]. Selective oxidation of methane to methanol [58] or formaldehyde [59] and of *i*-butane to methacrylic acid [60] has been examined, but is not yet used industrially.

Oxidative Couplings. Oxidative coupling of methane either with itself to give ethane and ethylene [61] (Eq. 4), or with compounds having activated methyl groups (Eq. 5) to give terminal olefinic compounds [62] has been investigated, but is not yet industrially used.

$$CH_4 \xrightarrow[-H_2O]{O_2} CH_3CH_3 + CH_2=CH_2 \qquad (4)$$

$$CH_4 + CH_3-X \xrightarrow[-H_2O]{O_2} CH_2=CH-X \qquad (5)$$

$$X = -CH=CH_2, -C(CH_3)=CH_2, -C_6H_5, -C\equiv N$$

Three-Component Oxidation. Three-component oxidation allows introduction of hetero atoms or groups of hetero atoms other than oxygen into saturated hydrocarbons. The radical-induced action of both oxygen and sulfur dioxide (sulfoxidation) on alkanes yields the corresponding alkanesulfonic acids. The latter, produced from C_{14}- to C_{18}-n-alkanes, are useful as raw materials for detergents. The heterogeneously catalyzed reaction of oxygen and ammonia with methane (ammoxidation) affords hydrogen cyanide. The latter reaction can be also carried out in the absence of oxygen as a two-component codehydrogenation. The reaction of oxygen and ammonia with propane to give acrylonitrile has recently been explored.

Chlorinations. Chlorination of saturated hydrocarbons yields mono- or polychlorinated hydrocarbons, depending on conditions. The following reactions are used industrially: Chlorination of methane to give the four chloromethanes; the chlorination of n-alkanes to give mono- or polychlorinated alkanes having the same number of carbon atoms and chlorinolysis of low molecular mass alkanes or chloroalkanes to give tetrachloroethene and tetrachloromethane. Reaction of chlorine and sulfur dioxide together with alkanes yields alkanesulfonyl chlorides, which in turn can be converted into the corresponding sulfonic acids, sulfonates, sulfonamides, or sulfonic esters.

Photonitrosation. Photonitrosation of saturated alkanes with nitrosyl chloride yields oximes via labile secondary nitroso compounds. The method has been used for the production of caprolactam from cyclohexane (→Caprolactam).

Nitrosation. Nitrosation of saturated hydrocarbons with nitric acid has been carried out with low molecular mass alkanes, particularly with propane in the gas phase. The reaction gave mixtures of nitro compounds with the same number of and fewer carbon atoms than those of the starting alkanes. Such mixtures are useful as specialty solvents, e.g., for polymers.

Reaction with Sulfur. Reaction of methane with sulfur gives carbon disulfide in high yields with hydrogen sulfide as byproduct. The reaction can be performed by thermal or catalytic methods.

Fermentation. Fermentation of certain saturated hydrocarbons by microorganisms (yeast, bacteria) yields single cell proteins [63]–[67]. The microorganisms oxidize the hydrocarbons in the presence of oxygen and nutrients with concomitant formation of protein-containing cells. Certain microorganisms are specific for the fermentation of n-

alkanes. Industrial processes have therefore been developed both for the fermentation of isolated *n*-alkanes and for *n*-alkanes in admixture with branched or cyclic hydrocarbons (e.g., diesel oil). The products obtained are used as animal feed.

1.5. Individual Saturated Hydrocarbons

Methane, Cyclohexane, and *Waxes* are separate keywords. Of the other saturated hydrocarbons only those used industrially will be discussed.

Ethane. Ethane is present in many natural gas sources and in refinery gases; it can be recovered for industrial use. Furthermore, ethane is formed in thermal- and hydrocracking of hydrocarbons and in the liquefaction of coal. The single most important industrial use for ethane is the production of ethylene by steam-cracking (→ Ethylene). Thermally-induced chlorination of ethane yields predominantly monochloro-, 1,1-dichloro-, or 1,1,1-trichloroethane, depending on the conditions used. Combined chlorination and oxichlorination of ethane in a salt melt affords vinyl chloride [68]. Reaction of ethane with nitric acid in the gas phase yields nitroethane and nitromethane.

Propane. Propane is a component of liquefied petroleum gas (LPG), which is derived from natural gas or petroleum. Propane can be recovered from LPG by distillation. The major industrial use of propane is the production of ethylene and propene by the steam cracking process. Chlorinolysis of propane at elevated temperature yields a mixture of tetrachloroethene and tetrachloromethane. The conversion of propane into BTX aromatics by zeolite-catalyzed dehydrocyclooligomerization has been developed recently and may find industrial application. Liquefied propane is used for the deasphalting of petroleum residues.

***n*-Butane.** *n*-Butane can be recovered from LPG by distillation; it has a variety of industrial uses: steamcracking yields ethylene and propene, catalytic dehydrogenation yields butadiene (→ Butadiene), and acid-catalyzed isomerization provides *i*-butane. Oxidation of *n*-butane in the gas phase with heterogeneous catalysis is a modern process for the production of maleic anhydride [57]. Noncatalyzed oxidation of *n*-butane in the liquid or gas phase has been used industrially to produce acetic acid [69] (→ Acetic Acid).

***i*-Butane.** *i*-Butane can be obtained either by recovery from LPG or by isomerization of *n*-butane. Traditional industrial uses include the production of high octane engine fuel by the alkylation of olefins [47], [48] and liquid-phase autoxidation of *i*-butane to give *tert*-butylhydroperoxide. Recent developments comprise improved processes for the catalytic dehydrogenation of *i*-butane [70] to satisfy the increasing demand for *i*-butene as a pecursor for methyl *tert*-butyl ether (MTBE) (→Methyl Tert-Butyl Ether), and the production of methacrylic acid from *i*-butane by oxidation [60].

n-Pentane and i-Pentane. Both isomers are present in light gasoline fractions and can be recovered individually by superfractionation [28], [71], possibly in combination with molecular sieve separation. *n*-Pentane is used as a solvent and for the production of *i*-pentane by acid-catalyzed isomerization [43]. *i*-Pentane is used as a blending component for high octane gasoline and for the production of isoprene by catalytic dehydrogenation [70], [71].

n-Hexane. *n*-Hexane can be isolated from suitable sources (e.g., light gasoline or BTX raffinates) by superfractionation, or by molecular sieve separation. It is used for the extraction of vegetable oils (e.g., from soybeans), as a solvent in chemical reactions (e.g., for coordination complex catalyzed polymerization of olefins) and in adhesive formulations.

Higher n-Alkanes. *n*-Alkanes having more than six carbon atoms are not isolated individually but as mixtures of several homologues. Suitable sources are the appropriate petroleum distillate fractions, from which the *n*-paraffins can be isolated in high isomeric purity ($\geq 95\%$ linearity) by selective separation (e.g., molecular sieve separation or urea extractive crystallization). They are used mainly in applications for which *i*-alkanes are not acceptable for biological reasons, e.g., the production of detergents or proteins. Catalyzed gas-phase dehydrogenation of *n*-alkanes over noble metal catalysts and at low conversion (ca. 10%) yields the corresponding *n*-alkenes with predominantly internal double bonds. The alkenes can be isolated in high purity by selective molecular sieve processes [52], [73]. *n*-Alkenes in the range C_7 to C_{10} are used as raw materials for the production of flexibilizers, by oxo-synthesis with hydrogen and carbon monoxide. *n*-Alkenes in the range C_{10} to C_{19} are used for the production of biodegradable detergents: acid-catalyzed reaction of C_{10}- to C_{13}-*n*-alkenes with benzene, followed by sulfonation and neutralization of the ensuing alkylbenzene sulfonic acids, affords linear alkylbenzenesulfonate (LABS) detergents. Oxo-reaction of C_{10}- to C_{19}-*n*-alkenes yields C_{11}- to C_{20}-alcohols, which are converted into nonionic detergents by reaction with ethylene oxide, or into anionic detergents by reaction with sulfur trioxide or chlorosulfonic acid. Liquid-phase oxidation of *n*-alkanes with oxygen in the presence of boric acid yields predominantly secondary alcohols. The latter can be dehydrated to the corresponding *n*-alkenes. The C_{12}- to C_{19}-alcohols are also used for the production of both ionic and nonionic detergents.

Chlorination of *n*-alkanes yields mono- or polychlorinated alkanes, depending on the conditions. Mixtures of C_{10}- to C_{13}-monochloro-alkanes are used for the production of linear alkylbenzenes as raw materials for LABS-detergents. For this, the chloroalkanes are either directly reacted with benzene or they are dehydrochlorinated to the corresponding *n*-alkenes, which are then used to alkylate benzene. Polychlorinated *n*-alkanes are used in a number of special applications.

n-Alkanes in the range of C_{10} to C_{21} are suitable substrates for the production of single cell proteins by fermentation. The microorganisms used selectively attack *n*-alkanes, so that industrial processes have been developed in which either isolated *n*-alkanes or the appropriate petroleum distillates containing *n*-alkanes (petroleum-, gas oil-, or lube oil fractions) are used as starting materials.

Solid *n*-alkanes (paraffin waxes) are used in a variety of applications, both as such and as feed materials in cracking, oxidation and chlorination.

Cycloalkanes. Of the various accessible cycloalkanes, only cyclohexane and cyclododecane are used industrially on a large scale; cyclooctane is used on a relatively small scale. The strained C_3- and C_4-cycloalkanes are not produced industrially. *Cyclopentane* is present in natural hydrocarbon sources and is formed as a byproduct in the processing of cracking products, e.g., by hydrogenation of cyclopentadiene or cyclopentene in the C_5-fraction of steamcracked products. No important industrial use exists for cyclopentane.

Cyclooctane is accessible by catalytic hydrogenation of 1,5-cyclooctadiene. Liquid-phase oxidation of the former with air yields 1,8-octanedicarboxylic acid, which is used on a moderate scale for the production of nylon 68.

Cyclododecane is produced by liquid-phase hydrogenation of 1,5,9-cyclododecatriene over nickel catalysts. Liquid-phase oxidation of cyclododecane with air in the presence of boric acid yields mixtures of cyclododecanol and cyclododecanone in high selectivity. Further oxidation of such alcohol–ketone mixtures with nitric acid gives 1,12-dodecanedicarboxylic acid, used for the production of polyamides, polyesters, and synthetic lubricating oils. Catalyzed dehydrogenation of alcohol-ketone mixtures in the liquid phase yields cyclododecanone, which is then converted into laurinlactam and nylon 12.

Bicyclo[4.4.0]decane (decalin) is obtained by catalytic hydrogenation of naphthalene. Liquid-phase oxidation of decalin with oxygen gives the corresponding tertiary hydroperoxide, which is eventually converted into 1,10-decanedicarboxylic acid (sebacic acid) by a four-step sequence [56].

$$\text{naphthalene} \xrightarrow{H_2, \text{Cat.}} \text{decalin} \xrightarrow{O_2} \text{decalin-OOH} \xrightarrow{H^+} \text{ketone+OH} \xrightarrow{\text{Cat.}} \text{unsaturated ketone} \xrightarrow{H_2, \text{Cat.}} \text{ketone} \xrightarrow{\text{Oxid.}} HOOC-(CH_2)_8-COOH$$

2. Olefins

Olefins are aliphatic hydrocarbons containing at least one carbon–carbon double bond. Monoolefins (alkenes) contain a single C=C double bond and form a homologous series with the empirical formula C_nH_{2n}. Hydrocarbons with two double bonds are differentiated according to the relative positions of the double bonds: in cumulated dienes (cumulenes, allenes) they are immediately adjacent to each other; in conjugated dienes they are separated by a single bond; and in other diolefins, the double bonds are isolated from each other.

More highly unsaturated polyenes are known, e.g., trienes, tetraenes. The name olefin is derived from the property of these compounds to form oily liquids on reaction with halogens (*gaz oléfiant*, oil-forming gas).

The industrial importance of olefins started in the 1950s when the lower olefins became widely available from thermal cracking of wet natural gas and petroleum fractions, displacing acetylene (ethyne) as the dominant commodity chemical (→Acetylene). In the time since then, ethene (ethylene), propene, 1,3-butadiene, and isoprene (2-methyl-1,3-butadiene) have found wide application, especially for synthetic polymers (→Butadiene; →Ethylene; →Isoprene; →Propene).

2.1. Monoolefins

The low molecular mass monoolefins ethylene, propene, styrene, cyclopentene, and the butenes are separate keywords: (→Butenes; →Cyclopentadiene and Cyclopentene; →Ethylene). Among the higher olefins (C_6–C_{18}), the linear α-olefins are of particular industrial interest.

2.1.1. Properties

Physical Properties. Under normal conditions, the olefins from ethylene to the butenes are gases; from the pentenes to 1-octadecene liquids; and from 1-icosene onwards solids. These thermal phase transitions do not differ significantly from those of the corresponding saturated hydrocarbons (Chap. 1). The density of olefins ranges from 0.63 to 0.79 g/cm^3, only a few percent higher than those of the corresponding alkanes.

The heats of combustion of alkenes and the corresponding alkanes are also nearly identical. Olefins have low solubility in water or are insoluble; they dissolve well in most organic solvents, e.g., alcohols, ethers, and aromatic hydrocarbons. Some physical properties of important monoolefins are listed in Table 6.

Table 6. Physical properties of monoolefins

Olefin	CAS registry number	mp, °C	bp,[a] °C	d^{20}	n^{20}
Ethylene	[74-85-1]	−169.5	−103.8	0.5699 (−103.8 °C)	1.363 (−100 °C)
Propene	[115-07-1]	−185.3	− 47.7	0.6095 (−47.4 °C) 0.5139	1.3640 (−50 °C)
1-Butene	[106-98-9]	−185.3	−6.26	0.6356 (−15.1 °C)	1.3791 (−25 °C)
cis-2-Butene	[590-18-1]	−138.9	3.7	0.6447 (0 °C)	1.3783 (0 °C)
trans-2-Butene	[624-64-6]	−105.5	0.88	0.6269 (0 °C)	1.3696 (−20 °C)
Isobutene	[115-11-7]	−140.4	−6.9	0.6398 (−20 °C)	1.3776
1-Pentene	[109-67-1]	−167.2	29.97	0.6405	1.3718
cis-2-Pentene	[627-20-3]	−151.3	37.1	0.6554	1.3828
trans-2-Pentene	[646-04-8]	−140.2	36.4	0.6482	1.3798
2-Methyl-1-butene	[58718-78-8]	−137.5	31.2	0.6504	1.3778
3-Methyl-1-butene		−168.5	20.1	0.6332	1.3640
2-Methyl-2-butene	[26760-64-5]	−133.6	38.6	0.6625	1.3874
1-Hexene	[592-41-6]	−140.0	63.5	0.6732	1.3878
cis-2-Hexene	[7688-21-3]	−141.8	68.8	0.6969	1.3976
trans-2-Hexene	[4050-45-7]	−133.5	67.5	0.6780	1.3935
cis-3-Hexene	[7642-09-3]	−137.8	66.5	0.6794	1.3951
trans-3-Hexene	[13269-52-8]	−113.4	68.0	0.6778	1.3937
1-Heptene	[592-76-7]	−119.6	93.8	0.6993	1.3998
cis,trans-2-Heptene		−109.5	98.2	0.7034	1.4041
cis-3-Heptene	[7642-10-6]	136.6	95.8	0.7030	1.4059
trans-3-Heptene	[14686-14-7]	−136.3	95.7	0.6981	1.4043
1-Octene	[111-66-0]	−102.7	121.3	0.7160	1.4086
cis-2-Octene	[7642-04-8]	−100.5	125.6	0.7243	1.4146
trans-2-Octene	[13389-42-9]	− 87.8	124.5	0.7184	1.4132
cis-3-Octene	[14850-22-7]	−126	122.7	0.7223	1.4140
trans-3-Octene	[14919-01-8]	−110.0	123.3	0.7149	1.4129
cis-4-Octene	[7642-15-1]	−118.1	122.5	0.7212	1.4150
trans-4-Octene	[14850-23-8]	− 93.8	122.3	0.7136	1.4120
1-Nonene	[124-11-8]	− 81.6	146.8	0.7292	1.4157
2-Nonene	[2216-38-8]		149.4	0.7407	1.4201
3-Nonene	[20063-77-8]		148.2 (102 kPa)	0.7334	1.4191
4-Nonene	[2198-23-4]		148.0	0.7313	1.4200
1-Decene	[872-05-9]	− 66.6	170.6	0.7408	1.4215
2-Decene	[6816-17-7]				
3-Decene	[19348-37-9]		173 – 173.5	0.7447	1.4221
4-Decene	[19689-18-0]		170.6	0.7404	1.4243
cis-5-Decene	[7433-78-5]	−112	169.5 (98.5 kPa)	0.7445	1.4252
trans-5-Decene	[7433-56-9]	− 73	170.2	0.7412	1.4235
1-Undecene	[821-95-4]	− 49.1	192.67	0.75033	1.4261
5-Undecene	[4941-53-1]		192.2	0.7511	1.4289
1-Dodecene	[25378-22-7]	− 33.6	213.3	0.7583	1.4300
6-Dodecene	[7206-29-3]		212.9 (99.3 kPa)	0.7597	1.4328
1-Tridecene	[2437-56-1]	− 22.2	104 (1.33 kPa)	0.7654	1.4335
1-Tetradecene	[1120-36-1]	− 12	125 (1.85 kPa)	0.7726	1.4367
1-Pentadecene	[13360-61-7]	− 3.8	133.5 (1.33 kPa)	0.7763	1.4389
1-Hexadecene	[629-73-2]	4.0	155 (2.0 kPa)	0.7811	1.4412
1-Heptadecene	[6765-39-5]	10.7	159 (1.33 kPa)	0.7860	1.4436
1-Octadecene	[112-88-9]	18.0	176 – 178 (1.85 kPa)	0.7888	1.4450
1-Icosene	[3452-07-1]	28.5	151 (0.2 kPa)		1.4440 (30 °C)
1-Heneicosene	[1599-68-4]	35.5	134 (0.005 kPa)	0.7985	1.4510
1-Triacontene	[18435-53-5]	62 – 63	218 (0.07 kPa)	0.7620 (100 °C)	

[a] At 101.3 kPa, unless otherwise noted.

Table 7. Important reactions of olefins

Reaction	Reagents	Products
Halogenation	Cl_2, Br_2	dihaloalkanes
Hydrohalogenation	HCl, HBr	alkylhalides
Sulfation	H_2SO_4	alkyl sulfates
Hydration	H_2O; catalyst: H^+	alcohols
Hydrogenation	H_2; catalyst: Ni, Pt	alkanes
Epoxidation	a) O_2 catalyst: Ag	ethylene oxide
	b) RCOOOH, ROOH	epoxides
Hydroxylation	H_2O_2; catalyst: H^+	glycols
Wacker oxidation	H_2O; catalyst: Pd–Cu	acetaldehyde, ketones
Ozonolysis	O_3	aldehydes, ketones
Hydroformylation	CO, H_2; catalyst: Co, Rh	aldehydes
Hydrocarboxylation	CO, H_2O; catalyst: Ni, Co	carboxylic acids
Sulfonation	SO_3, NaOH	alkene-sulfonates
Hydroboration	B_2H_6	trialkylboranes
Alkylation	a) of aromatics, with a Friedel–Crafts catalyst	alkylaromatics
	b) of isoalkanes, catalyst: H^+	branched alkanes
Polymerization	Catalyst: radicals, metal complexes	polymers
Oligomerization	Catalyst: metal complexes, BF_3	oligomers

Chemical Properties. The chemistry of monoolefins is dominated by the reactive double bond. Some important reactions of olefins are summarized in Table 7.

Most *addition reactions* proceed by an electrophilic mechanism via carbonium ions. Chlorination and bromination yields the trans adducts. Hydrohalogenation generally follows Markovnikov's rule to give a product containing halogen linked to secondary carbon. Addition of sulfuric acid to olefins affords alkyl sulfates, which are readily hydrolyzed to alcohols; the reaction corresponds to the acid-catalyzed addition of water. Ethanol, isopropylalcohol, and *tert*-butyl alcohol are produced industrially by this method from ethylene, propene, and isobutene, respectively (→Alcohols, Aliphatic).

Hydrogenation of olefins to alkanes requires a catalyst: Raney nickel is most frequently used industrially, although platinum-group metals are also used.

Epoxidation of ethylene may be carried out with oxygen in the presence of a silver catalyst. Higher olefins are converted to epoxides using hydroperoxides or peracids (→Epoxides). A related reaction is the hydroxylation of olefins with hydrogen peroxide in glacial acetic acid: an epoxide is initially formed, which subsequently reacts with water in the acidic medium to form a glycol.

Another olefin oxidation is the *Wacker oxidation* of ethylene to acetaldehyde (→Acetaldehyde). The stoichiometric reaction of ethylene with water and palladium(II) chloride yields acetaldehyde and palladium metal. Addition of copper(II) chloride converts the stoichiometric reaction into a catalytic process, wherein the copper(II) chloride reoxidizes the palladium to palladium(II) chloride the resulting copper(I) chloride is reoxidized to copper(II) chloride by atmospheric oxygen in hydrochloric acid solution.

Ozonolysis is used mainly for structure determination of olefins, but may also be used preparatively. Reaction of olefins with carbon monoxide and additional reagents (e.g., hydrogen, water, alcohols) leads to a plethora of valuable products (e.g., aldehydes, alcohols, carboxylic acids, esters). These reactions proceed under catalysis by transition metals, e.g., hydroformylations (Oxo process) with rhodium- or cobalt carbonyls.

Sulfonation of olefins leads to alkenesulfonic acids, which are used as detergents. Hydroboration is used to generate primary alcohols via trialkylboranes. In contrast to the direct addition of water, this pathway is regioselective to yield the anti-Markovnikov products.

Olefins may undergo a number of reactions resulting in new C-C bonds. With aromatic compounds alkylaromatics are formed by the Friedel–Crafts reaction; with isoalkanes chain-lengthened, branched hydrocarbons, which are used in motor fuels, are formed. Olefins may also react with themselves or each other to form oligomers and polymers.

The industrial use of the higher monoolefins is discussed in Section 2.1.3.

2.1.2. Production of Higher Olefins

Olefins are synthesized on a laboratory scale by a variety of routes: e.g., selective hydrogenation of alkynes; dehydration of alcohols; pyrolysis of alkyl lithium compounds; the Wittig reaction; the Peterson reaction; and deoxygenation of vicinal diols. Only processes starting with cheap raw materials have gained industrial importance for the production of higher olefins. The following discussion considers processes that start either with long-chain alkanes (Section 2.1.2.1) or those that build up longer chain olefins from their lower molecular mass homologs (Section 2.1.2.2).

2.1.2.1. Production from Paraffins

Thermal Cracking of Waxes. Thermal cracking of long chain alkanes yields mainly α-olefins.

$$R^1-CH_2CH_2CH_2CH_2-R^2 \longrightarrow$$
$$R^1-CH=CH_2 + R^2-CH=CH_2 + H_2$$

Thermal cracking involves a radical mechanism. Cleavage of a C-C bond leads to carbon radicals, which are converted into olefins by loss of a hydrogen atom. The latter predominantly combine to form hydrogen; a small fraction reacts with carbon radicals to form alkanes.

Table 8. Composition of fractions (wt%) obtained in a paraffin cracking process (Chevron process)

	C_6–C_7	C_8–C_{10}	C_{11}–C_{12}	C_{13}–C_{14}	C_{15}–C_{18}	C_{18}–C_{20}
α-Olefins	83	83	88	88	89	86
Internal olefins	12	11	5	5	3	1
Diolefins	3	4	5	5	6	4
Paraffins	2	2	2	2	2	9
Aromatics	0.4	0.2	0	0	0	0

To generate linear α-olefins, linear alkanes (waxes) must be used. Suitable sources are the petroleum, diesel oil, and lubricant fractions of paraffin-based crude oils. One possibility of enriching the linear paraffin content of such fractions is to dilute with a suitable solvent, followed by chilling to crystallize a mixture of paraffins. Linear paraffins can be separated from their branched and cyclic analogs by subsequent treatment with molecular sieves. Another method for the separation of *n*-paraffins, especially suitable for long-chain ($>C_{20}$) species, is *extractive crystallization with urea*. This process uses the property of urea to form inclusion compounds with linear paraffins, but not with branched or cyclic alkanes, or aromatics. During crystallization the urea molecules assemble to form cavities having a diameter of 0.53 nm, in which only linear alkanes can be accommodated. The alkanes are then liberated by subsequent decomposition of the complex, e.g., with water at 75 °C. Industrial processes use solid urea (Nurex process) or its saturated aqueous solution (Edeleanu process).

The long-chain paraffins are then cracked by first heating the vapor to 400 °C within a few seconds, followed by thermolysis at 500–600 °C in the presence of steam. Rapid quenching of the reaction mixture reduces side reactions such as isomerization or cyclization.

Using C_{18}–C_{36} alkanes as starting material, α-olefins with even and odd numbers of carbon atoms in the range C_6–C_{20} are obtained; only small amounts of paraffins are present. Byproducts include nonterminal olefins, conjugated and nonconjugated dienes, and traces of aromatics. Typical composition of the olefin fractions from paraffin cracking are given in Table 8.

The dependence of product quality on the starting material and process parameters is discussed in a recent review [74]; catalytic cracking and cracking in the presence of oxygen are also reviewed. The latter processes have not gained acceptance because of the presence of numerous byproducts, e.g., branched olefins and oxidation products.

The first industrial paraffin cracking plant was brought into service in 1941 by Shell at Stanlow (United Kingdom). Other plants were built at Pernis (The Netherlands) and Berre (France). A U.S. paraffin cracking plant owned by Chevron is located in Richmond, California. Future expansion of paraffin cracking processes is unlikely because of the limited flexibility of paraffin cracking plants with respect to carbon number distribution, the limited supply of paraffin-containing crudes, the high percentage ($\geq 10\%$) of byproducts in the α-olefin cut, and because of more efficient competitive processes.

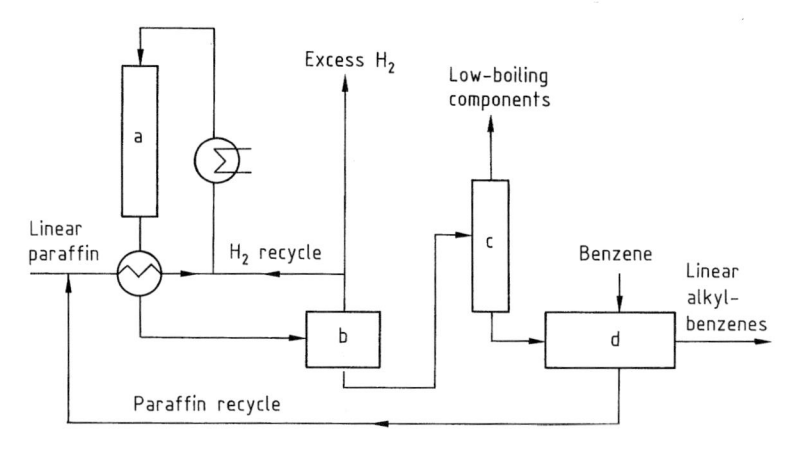

Figure 3. Simplified flow diagram for a combination of the UOP PACOL process and benzene alkylation
a) Pacol reactor; b) Separator; c) Stripper; d) Alkylation reactor

Catalytic Dehydrogenation. Catalytic dehydrogenation of paraffins leads to olefins with the same number of carbon atoms and with random location of the double bond along the chain. Thermally, this reaction cannot be carried out economically because the energy required to cleave a C-H bond (365 kJ/mol) significantly exceeds that for breaking a C-C bond (245 kJ/mol). Cracking of longer chain hydrocarbons is, therefore, favored over dehydrogenation; however, the use of a catalyst facilitates the dehydrogenation of linear paraffins to linear olefins having predominantly the same number of carbon atoms.

The most important industrial process is the Pacol (paraffin catalyst olefin) process of Universal Oil Products (UOP). The reaction is carried out at 450–510 °C and 0.3 MPa using a hydrogen:carbon ratio of 9:1. A supported platinum catalyst, ca. 0.8 wt% Pt, on alumina, activated by addition of lithium and arsenic or germanium, is used. To minimize the formation of byproducts, especially aromatic hydrocarbons, which reduce catalyst activity, relatively low conversion rates of 10–15% are used. This results in paraffin–olefin mixtures that are separated by means of the UOP Olex (olefin extraction) process. The Olex process employs molecular sieves, which absorb olefins more strongly than alkanes. The combination of both processes is known as the Pacol–Olex process [75]. In 1983, there were four Pacol–Olex plants worldwide, as well as 18 other plants where the Pacol process was combined with the alkylation of aromatics [76]. A simplified flow diagram of a Pacol benzene alkylation unit is shown in Figure 3.

Fresh *n*-alkane is added to the circulating paraffin, heated via heat exchangers, and dehydrogenated in the Pacol fixed-bed reactor (a) in the presence of hydrogen and at high throughput velocity. The products are separated into hydrogen and hydrocarbons in the separator (b) and volatile cracking products are removed in the stripper (c); these are then used chiefly as fuel. The stripper sump contains the higher linear monoolefins, which are reacted with benzene under acid catalysis to yield alkylbenzenes. The unreacted *n*-alkanes are returned to the Pacol reactor [77].

Chlorination – Dehydrochlorination. Another method for generating linear olefins from linear paraffins consists of chlorinating the paraffins to chloroalkanes, followed by a catalyzed second step in which hydrogen chloride is eliminated. Chlorination is carried out continuously in the liquid phase at 120 °C; the conversion is limited to a maximum of 30%, to avoid excessive formation of dichlorides. A random mixture of linear olefins results from the dehydrohalogenation because the chlorine substituents are almost randomly distributed along the alkyl chain. The dehydrochlorination of chlorinated paraffins occurs in the presence of iron or iron alloy catalysts at 250 °C. The lower boiling olefins and paraffins are drawn off at the still head; the unreacted, higher boiling chlorinated paraffins remain in the reactor. Separating products in this manner requires that rather narrow cuts be taken, because the boiling points of the *n*-olefins and the chloroparaffins are quite close together [78], [79].

2.1.2.2. Oligomerization of Lower Olefins

All the methods described so far for the production of olefins from paraffins suffer from the disadvantage that crude oil containing linear paraffins with the required distribution of carbon atoms is in short supply. Several companies have, therefore, developed alternative routes to olefins that are based on oligomerization of olefins, especially of ethylene. Such starting materials are readily available from the pyrolysis of liquefied natural gas, naphtha, and gas oil fractions. The worldwide capacity for ethylene alone was ca. 50×10^6 t in 1983. Ethylene supplies appear assured for the forseeable future because it can be generated by the Mobil process from methanol, which is derived from coal-based synthesis gas.

The methods for buildup of longer chain olefins vary greatly. In the following sections syntheses via organoaluminum compounds, transition metal catalyzed oligomerization, and acid-catalyzed processes will be discussed.

Ethylene Oligomerization in the Presence of Organoaluminum Compounds. The first process for the industrial oligomerization of ethylene was discovered by ZIEGLER in the early 1950s [80]. The reaction proceeds in two stages: first a growth step followed by an elimination step.

$$AlEt_3 + 3nCH_2=CH_2 \longrightarrow Al[(CH_2CH_2)_nEt]_3$$
$$Al[(CH_2CH_2)_nEt]_3 + 3\,CH_2=CH_2 \longrightarrow$$
$$AlEt_3 + 3\,CH_2=CH(CH_2CH_2)_{n-1}Et$$

The growth step occurs at ca. 100 °C under 10 MPa ethylene pressure. In the second stage, the high temperature displacement reaction, the alpha-olefins are displaced by ethylene at ca. 300 °C and 1 MPa. The composition of the product mixture corresponds

Table 9. Composition of fractions (wt%) obtained from the Gulf process *

	C_6	C_8	C_{10}	C_{12}	C_{14}	C_{16}	C_{18}
Linear α-olefins	97.0	96.0	95.0	94.0	93.0	92.0	91.0
β-Branched α-olefins **	1.4	2.5	3.4	4.6	5.6	6.6	7.8

* n-Alkane content 1.4%.
** Including traces of olefins with internal double bonds.

to a Poisson distribution. This process suffers from the disadvantage of requiring stoichiometric quantities of reaction components, i.e., industrial plants would require large quantities of aluminum alkyls. Consequently, Gulf Oil and Ethyl Corporation developed two process variants that require less triethylaluminum.

The Gulf Oil Process. In the Gulf process, only catalytic quantities of triethylaluminum are used and growth step and elimination reactions occur simultaneously in the same reactor [81]. The reaction is carried out at <200 °C and ca. 25 MPa. The product mixture consists of a Schulz–Flory distribution of linear olefins. A typical composition of individual fractions from C_6–C_{18} is listed in Table 9.

The Gulf process produces markedly purer α-olefins than can be obtained with paraffin cracking (see Table 8). The major impurities in the α-olefins (trade name: Gulftenes) obtained by the Gulf process are ca. 1.4% paraffins; with increasing carbon number, an increasing amount of β-branched α-olefin is also formed. The occurrence of these vinylidene and 2-ethyl compounds may be rationalized by consecutive reactions of α-olefins with the aluminum alkyls, as follows:

$$Al(CH_2CH_2R^1)_3 + 3\ CH_2=CH-R^2 \longrightarrow$$

$$Al(CH_2\overset{R^2}{\underset{|}{C}}HCH_2CH_2R^1)_3 \xrightarrow{3\ CH_2=CH_2}$$

$$AlEt_3 + 3\ CH_2=\overset{R^2}{\underset{|}{C}}-CH_2CH_2R^1$$

The Gulf Oil Co. uses the Gulf process to produce linear α-olefins in the C_4–C_{30} range at their plant in Cedar Bajou, Texas. Mitsubishi Chemical Industries uses the same process at Mizushima.

Ethyl Process. The Ethyl process combines a stoichiometric and a catalytic stage [82]. This modified Ziegler process is capable of controlling the chain length of the resulting α-olefins much more precisely, because the shorter chain olefins may be subjected to further chain growth.

The Ethyl process is illustrated schematically in Figure 4. Ethylene is first oligomerized using catalytic quantities of triethylaluminum in analogy to the Gulf process. The resulting products are fractionated (b) in Fig. 4) yielding fractions containing C_4, C_6–C_{10}, and C_{12}–C_{18}. The higher boiling fractions may be used directly, whereas the shorter α-olefins, especially C_4, are subjected to transalkylation with long chain aluminum alkyls (c) in Fig. 4). This reaction releases the desired long-chain α-olefins; short-chain trialkylaluminum compounds are simultaneously formed. The aluminum trialkyls are separated in the second distillation step (Fig. 4, (d)) and transformed in

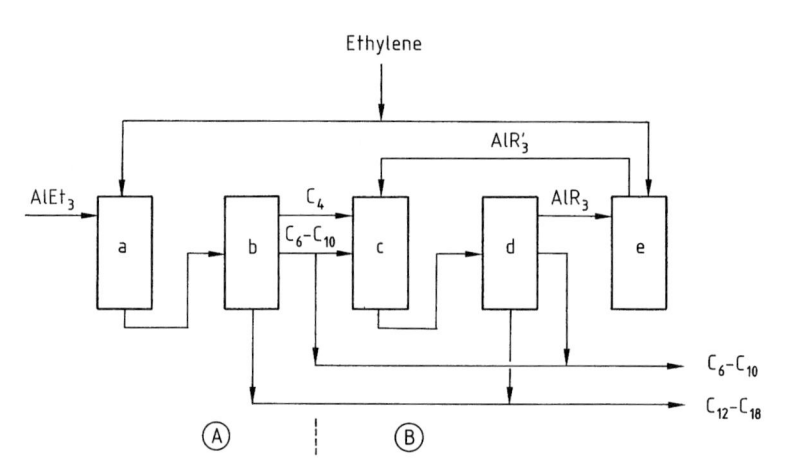

Figure 4. Typical flow diagram of Ethyl process
A) Catalytic step; B) Stoichiometric step;
a) Catalytic reactor; b) First distillation; c) Transalkylation reactor; d) Second distillation; e) Stoichiometric reactor

Table 10. Composition of fractions (wt%) obtained from the Ethyl process

	C_6	C_8	C_{10}	C_{12}	C_{12}–C_{14}	C_{14}–C_{16}	C_{16}–C_{18}
Linear α-olefins	97.5	96.5	96.2	93.5	87.0	76.0	62.7
Internal olefins	0.6	1.2	1.6	1.5	4.2	5.0	8.2
β-Branched α-olefins	1.9	2.3	2.2	5.0	8.8	19.0	29.1
Alkanes	0.1	0.6	0.3	0.4	0.4	0.4	0.8

another reactor (e) in Fig. 4) into longer chain alkyls by stoichiometric reaction with ethylene. The latter are then recycled to the transalkylation.

The Ethyl Corporation has operated a plant in Pasadena, Texas using this process since 1970; the initial capacity was 110 000 t/a. The process permits 95% conversion of ethylene into higher α-olefins, a conversion efficiency not achievable by the single-step catalytic Gulf Oil process. This advantage is however, counterbalanced by a much lower product quality of the long-chain α-olefins. The C_{16}–C_{18} cut consists of only 63% linear α-olefins (Table 10), the remainder being mainly β-branched α-olefins and internal olefins.

Ethylene Oligomerization with Transition-Metal Catalysis. In the presence of nickel, cobalt, titanium, or zirconium catalysts, ethylene may be converted into oligomers [83]. Commercial importance has been attained exclusively by the *Shell higher olefin process (SHOP)*.

Shell Higher Olefin Process. The SHOP process is the most recent development in α-olefin synthesis. It works in the liquid phase using a nickel catalyst and yields α-olefins of unusually high purity [84]–[86]. Monoolefins are formed almost exclusively; analytical data indicate only trace amounts of dienes, aromatics, and alkanes. The

Table 11. Composition of α-olefin fractions (wt%) obtained from the SHOP process

	C_6	C_8	C_{10}	C_{12}	C_{14}	C_{16}	C_{18}
All monoolefins	99.9	99.9	99.9	99.9	99.9	99.9	99.9
Linear α-olefins	97.0	96.5	97.5	96.5	96.0	96.0	96.0
Branched olefins	1.0	1.0	1.0	2.0	2.5	2.5	2.5
Internal olefins	2.0	2.4	1.0	1.5	1.5	1.5	1.5
Dienes	<0.1	<0.1	<0.1	<0.05	<0.05	<0.05	<0.05
Aromatics	<0.1	<0.1	<0.1	<0.05	<0.05	<0.05	<0.05
Alkanes	<0.1	<0.1	<0.1	<0.05	<0.05	<0.05	<0.05

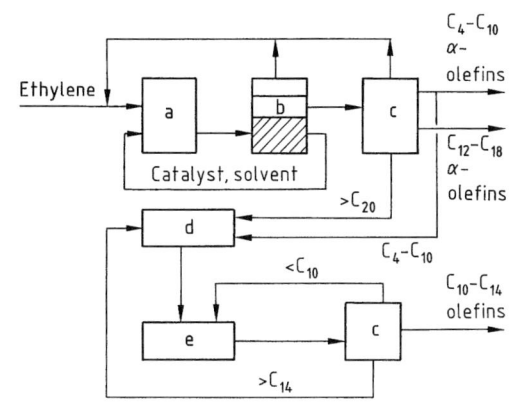

Figure 5. Typical flow diagram of SHOP process
a) Oligomerization reactor; b) Phase separator; c) Distillation; d) Isomerization reactor; e) Metathesis reactor

content of α-olefin for the carbon numbers is in the range 96–97%. A typical product composition is given in Table 11.

The SHOP process involves a combination of three different reactions: *oligomerization, isomerization,* and *metathesis*. This integration of chain lengthening and shortening steps makes the SHOP process very flexible. Whereas all other α-olefin processes permit only limited control over product distribution, with the SHOP process it is possible to achieve almost any olefin cut desired. The SHOP process is shown schematically in Figure 5 [86]–[89].

Oligomerization is carried out in a polar solvent such as ethylene glycol or 1,4-butanediol. The catalyst is produced in situ from a nickel salt, e.g., nickel chloride, sodium borohydride, and a chelating ligand. Suitable ligands are compounds of the general formula $RR^1P-CH_2-COR^2$, e.g., diphenyl phosphinoacetic acid, dicyclohexylphosphinoacetic acid, or 9-(carboxymethyl)-9-phosphabicyclo[3.3.1]-nonane. Ethylene is oligomerized to α-olefins with a Schulz–Flory distribution at 80–120 °C and 7–14 MPa. The olefins formed are immiscible with the polar solvent; product and catalyst phases are thereby readily separated so that the catalyst can be recycled repeatedly. Oligomerization is accomplished in a series of reactors interspersed with heat exchangers to remove the heat of reaction. The reaction rate can be regulated by the amount of catalyst. A high partial pressure of ethylene is required for high product

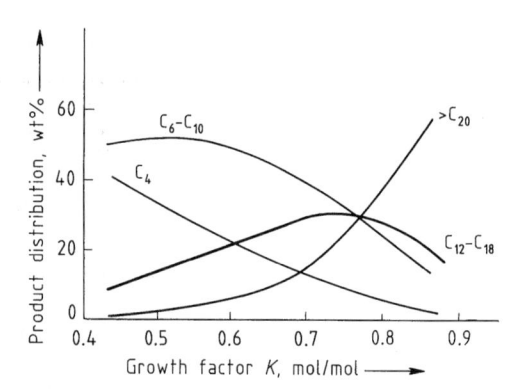

Figure 6. Product distribution as a function of growth factor K

linearity and a suitable rate of reaction. The chain length of the α-olefins is determined by the geometric factor K of molar growth:

$$K = \frac{n(C_{n+2}\text{-olefins})}{n(C_n\text{-olefins})}$$

where n is the number of moles.

The mass distribution of various α-olefins relative to the growth factor K is shown in Figure 6.

Controlling the growth factor is extremely important for the overall process because K determines both the product distribution and the median chain length of the olefins obtained. It is, therefore, advantageous to be able to vary the K-factor by the composition of the catalyst.

After separation of the catalyst and product phase, the latter is washed with fresh solvent to remove the last traces of catalyst. In a subsequent distillation (see Fig. 5) the mixture is separated into the desired C_{12}–C_{18} α-olefins and into low (C_4–C_{10}) and high-boiling ($>C_{20}$) fractions. The C_4–C_{10} fraction can then be further fractionated, if desired, into individual compounds, for example, those used as comonomers for the manufacture of linear low density polyethylene (LLDPE). However, the C_4–C_{10} fraction and the $C \geq 20$ fractions are generally combined for isomerization and subsequent metathesis. These two steps require only moderate reaction conditions: 80–140 °C, 0.3–2 MPa. The heat of reaction of both steps is quite low; because in isomerization only intramolecular migration of double bonds occurs, in metathesis carbon–carbon bonds are simultaneously cleaved and reformed. *Isomerization* takes place in the liquid phase in the presence of magnesium oxide catalyst [90]. About 90% of the terminal olefins are converted to internal olefins in this reaction.

$$R-CH=CH_2 \xrightarrow{MgO} R^1-CH=CH-R^2$$

The subsequent joint *metathesis* of the internal, low, and high molecular mass olefins, mainly over heterogeneous rhenium or molybdenum catalysts, yields a mixture of internal olefins having a new chain length distribution, i.e., a random mixture of

Table 12. Composition of internal olefins (wt%) obtained from the SHOP process

	C_{11}–C_{12}	C_{13}–C_{14}
All monoolefins	>99.5	>99.5
Linear olefins	>96	>96
Branched olefins	3	3
Dienes	<0.1	<0.1
Aromatics	<0.1	<0.1
Paraffins	<0.5	<0.5

olefins of even and odd numbers of carbon atoms.

$$R^1-CH=CH-R^2 + R^3-CH=CH-R^4 \xrightarrow{\text{Re or Mo}} \begin{matrix} R^1-CH \\ \| \\ R^3-CH \end{matrix} + \begin{matrix} CH-R^2 \\ \| \\ CH-R^4 \end{matrix}$$

The products of metathesis are distilled and the C_{11}–C_{14} fraction, representing ca. 10–15% of the mixture, is separated for production of alcohols used in the detergent industry (see Section 2.1.3).

The composition of the internal C_{11}–C_{14} olefins produced by the SHOP process is given in Table 12; more than 99.5% is monoolefins, with only small amounts of dienes, aromatics, and paraffins.

Although the double bonds of the olefins at the beginning of metathesis are distributed almost at random over the carbon backbone, they are shifted in the end product toward the chain end, a result of the high concentration of short-chain olefins (2-butene, 2-hexene, 3-hexene) in the metathesis feed.

Distillation of the metathesis products also yields low- and high-boiling components in addition to the C_{11}–C_{14} fraction. One of the great advantages of the SHOP process is that these fractions do not have to be discarded but can be recycled. The fraction consisting of compounds with less than 11 carbon atoms is returned directly to the metathesis reactor; the fraction containing olefins with more than 14 carbon atoms is again subjected to isomerization before it reaches the metathesis stage. By combining oligomerization, isomerization, and metathesis, the SHOP process is capable of converting almost all the ethylene feed into olefins having the desired number of carbon atoms.

An interesting alternative of the SHOP process described above is the metathesis of the isomerized $C \geq 20$ fraction together with a ten-fold molar excess of ethylene [90]. Olefins, predominantly α-olefins, within the desired carbon range may be prepared in the presence of a rhenium catalyst (20% Re_2O_7 on aluminum oxide).

$$\begin{matrix} C_{10}H_{21}-CH=CH-C_{10}H_{21} \\ + \\ CH_2=CH_2 \end{matrix} \xrightarrow{Re_2O_7-Al_2O_3} 2\ C_{10}H_{21}CH=CH_2$$

The first SHOP processing plant was built in 1977 by Shell Oil Corporation at Geismar, Louisiana (see Section 2.1.4); annual capacity was 200 000 t.

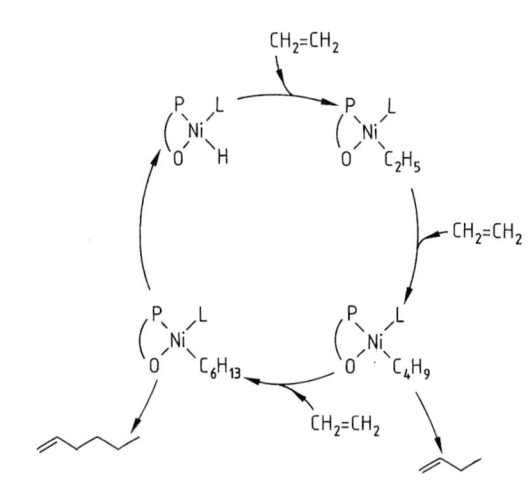

Figure 7. Mechanism of ethylene oligomerization in the presence of nickel P–O-chelate complex catalyst (for P–O ligand see text)

The mechanism of the highly selective ethylene oligomerization with nickel chelate catalysts has been largely elucidated by the research group of W. KEIM [84], [91]–[95]. A chelated nickel hydride complex is formed initially. This complex reacts with ethylene (Fig. 7) to form alkyl nickel complexes. Linear α-olefins are eliminated from the alkyl nickel complexes; the nickel hydride species is formed simultaneously to reenter the catalytic cycle.

The SHOP process is by far the most important method for oligomerization of ethylene involving transition-metal catalysts. To complete the picture, two other processes will be described below, although they have not yet reached commercial importance.

Exxon Process. The Exxon Corporation has developed a process in which ethylene is converted to high molecular mass olefins in the presence of a catalyst consisting of a transition-metal component and a soluble alkyl aluminum chloride [96]. Titanium chloride or an alkoxy titanium chloride are favored as the transition metal components. The catalyst is soluble in saturated hydrocarbons and active at 25–70 °C. This process yields linear α-olefins in the range C_4–C_{100} or higher; it appears to be especially useful for the production of high-melting waxes, i.e., olefins higher than C_{20}.

Alphabutol Process. In the Alphabutol process, developed by the Institut Français du Pétrole (IFP), ethylene is selectively dimerized to 1-butene [97] using a homogeneous titanium catalyst. In contrast to the SHOP process, the mechanism does not involve metal hydrides but a titanium(IV) cyclopentane species that decomposes under the reaction conditions (50–60 °C, slight positive pressure) with β-hydrogen transfer to 1-butene:

The 1-butene, which contains only traces of impurities (hexenes, 2-butene, butanes) may be used as comonomer in the production of LLDPE.

Oligomerization of Propene and Butenes with Transition-Metal Catalysis. Ziegler catalysts consisting of nickel compounds and aluminum trialkyls can be used to dimerize or codimerize propene and butene to give predominantly branched olefins. This process was also developed by the Institut Français du Pétrole and is referred to as the Dimersol process [98]–[103]. IFP research began as early as the mid 1960s. The first commercial Dimersol-G plant was brought into production in 1977 by Total in Alma, Michigan. In the *Dimersol G process*, propene (in admixture with propane) is dimerized mainly (80%) to isohexenes; small amounts of propene trimers (18%) and tetramers (2%) are also formed:

$$2 C_3H_6 \longrightarrow$$
n-Hexenes + 2-methylpentenes + 2,3-dimethylbutenes

The reaction takes place in the liquid phase without solvent. The pressure in the reactor is adjusted so that the C_3 compounds remain liquid at the operating temperature. The product mixture leaving the reactor head is fed to a vessel where the catalyst is neutralized with alkali. The product (trade name: Dimate) is then washed with water and freed of any remaining propane and propene on a stabilizing column.

The *Dimersol X process* is similar, although the starting materials are butenes or a mixture of propene and butenes. The products formed are, accordingly, isooctenes or mixtures of C_6–C_8 olefins. The first Dimersol X plant was started in 1980 by Nissan Chem. in Kashima, Japan. A typical starting material for the Dimersol X process is the raffinate II obtained from the C_4 cut of the steam cracker, after removal of butadiene and isobutene. The resulting C_8 olefins can be converted by Oxo-synthesis into nonanols, which are used for the synthesis of plasticizers (dinonyl phthalate).

A more recent development is the *Dimersol E* process, which employs the exhaust gases of fluid catalytic crackers (FCC) that contain ethylene and propene. A gasoline fraction can be obtained from this product after hydrogenation. The first experiments with cracker exhaust gases from the Elf cracker near Feyzin (France) began in 1984.

By 1982, IFP had licensed 16 Dimersol plants, eight of which were in production. In addition to IFP, other industrial and academic groups have investigated the transition metal-catalyzed oligomerization of lower olefins [104]–[112]. The observation that in the propene dimerization with nickel-based Ziegler catalysts the addition of phosphines may significantly affect the isohexene product distribution is of interest. With the catalyst [NiBr(η^3-C_3H_5)(PCy$_3$)] – EtAlCl$_2$ (Cy = cyclohexyl) a remarkable turnover of 800 000 mol propene per mol nickel h^{-1} could be achieved. A turnover of 60 000 s^{-1} was calculated for operation at room temperature. This value is of an order of magnitude usually achieved only by enzyme reactions.

Oligomerization of Propene and Butene by Acid Catalysis. Mineral acids such as phosphoric or sulfuric acid can also be used as catalysts for the oligomerization of olefins; the major products are branched olefins. Alkylbenzenes were once prepared from tetramers of propenes (tetrapropylene). This reaction is now rarely used in Europe because of the low biodegradability of the alkylbenzenesulfonates obtained.

Oligomerization of isobutene leads to the formation of diisobutene, triisobutene, and tetraisobutene; after hydrogenation, these may be added to gasoline to improve the octane rating (polymer gasoline, oligomer gasoline). In the *cold acid process*, isobutene-containing C_4 fractions are treated with 60–70% sulfuric acid at room temperature. Isobutene dissolves in the sulfuric acid as *tert*-butyl sulfate. This solution is then oligomerized at 100 °C. The products, mainly diisobutenes, separate as a second phase. In the *hot acid process*, the *n*-butenes of the C_4 fraction are also converted, forming *i*-C_4–*n*-C_4 cooligomers.

The oligomerization can also be performed with poly(styrene sulfonic acid) resins as catalysts. This process is used by Erdölchemie–Dormagen using the raffinate I (C_4 fraction after extraction of butadiene). The reaction is performed in loop reactors at 100 °C and 1.5–2.0 MPa with residence times of 20–60 min. The product is composed of 36% diisobutene, 22% C_8 codimers, 38% triisobutene, and 4% tetraisobutene. The product distribution may be changed by recycling specific fractions.

2.1.3. Uses of Higher Olefins

The most important uses of linear olefins are described below.

Synthesis of Oxo-Alcohols. Linear α-olefins can be converted to linear primary alcohols, which are used in the synthesis of plasticizers ($<C_{11}$) or detergents ($>C_{11}$) by hydroformylation (oxo synthesis). The reaction can also be used to give aldehydes as the major products.

$$R-CH=CH_2 \xrightarrow[\text{Cat.}]{CO, H_2} R-CH_2CH_2CH_2OH \text{ or } R-CH_2CH_2CHO$$

The yield of linear products can be increased to >90% if phosphine-modified cobalt- or rhodium catalysts are used.

The described synthesis route for long-chain alcohols has strong competition, however, from the natural fatty alcohols and from the Alfoles, generated from ethylene by an organo aluminum synthesis with subsequent oxidation.

In the United States Monsanto has produced oxo alcohols from C_6–C_{10} α-olefins since 1971. Exxon has used the oxo process since 1983 to produce plasticizer alcohols in the C_7–C_{11} range at Baton Rouge, Louisiana. Shell also converts olefins to alcohols. In recent years, olefins derived from paraffin crackers as starting materials have been replaced by internal olefins from the SHOP process.

The possibility of converting olefin-derived aldehydes to short-chain synthetic fatty acids by oxidation is also notable:

$$R-CH_2CH_2CHO \xrightarrow{\text{Oxid.}} R-CH_2CH_2COOH$$

In 1980 Celanese (now Hoechst-Celanese) built a plant in Bay City, Texas with an annual capacity of 18 000 t to produce heptanoic and nonanoic acids from 1-hexene and 1-octene, respectively, which are used in the production of lubricants. Again, there is strong competition from the natural products, which are generally less expensive. In addition, lubricant production requires highly linear acids. Fatty acids from oxo aldehydes have a linearity of >95%; natural fatty acids have a linearity of almost 100%.

Synthesis of Linear Alkylbenzenes. Currently the most important class of detergents for domestic use is still the linear alkylbenzenesulfonates (LAS or LABS). They are prepared by Friedel–Crafts reaction of benzene with linear olefins followed by sulfonation:

$$\text{benzene} + R-CH=CH_2 \xrightarrow{\text{Cat.}} \text{LAB} \xrightarrow[\text{2. NaOH}]{\text{1. SO}_3} \text{LABS (}-SO_3Na\text{)}$$

Total estimated production of LAS in Western Europe, Japan, and the United States in 1982 was ca. 1.1×10^6 t. In Europe, Asia, South Africa, and South America linear alkylbenzenes are produced from C_{10}–C_{15} *alpha*-olefins. In 1982, 420 000 t of LAB were synthesized in Western Europe from ca. 80 000 t of linear olefins.

Synthesis of α-Alkenesulfonates α-Alkene sulfonates (α-olefin sulfonates, AOS or OS) are obtained by direct sulfonation of C_{14}–C_{18} α-olefins (→Detergents). An ionic mechanism results in a product mixture of 40% isomeric alkenesulfonic acids (with the double bond predominantly in the β- or the γ-position) and 60% 1,3- or 1,4-alkanesulfones. Following hydrolysis with sodium hydroxide, a mixture of alkenesulfonates and hydroxyalkanesulfonates is obtained:

$$R-CH=CH_2 \xrightarrow{SO_3}$$

$$R^1-CH=CH-CH_2-SO_3H + \text{(sultone with } R^2\text{)}$$

$$\downarrow \text{NaOH}$$

$$R^1-CH=CH-CH_2-SO_3Na + R^2-\underset{OH}{CH}-CH_2CH_2CH_2-SO_3Na$$

AOS

The double bonds and the hydroxyl groups contribute to the excellent water solubility of the AOS compounds. They are effective laundry agents even in hard water and at low concentration and are used increasingly in liquid soaps and cosmetics. In order to be used in such products the AOS must be of high quality, especially with regard to color; this is only possible with α-olefins that contain no more than traces of dienes.

Although AOS compounds have been known since the 1930s, they gained commercial importance only in the late 1960s. In 1982, ca. 7000 t of AOS were produced in the United States from α-olefins, with the Stepan Company and Witco Chemical Corporation being the largest U.S. producers. AOS represents an interesting alternative to alkyl sulfates and alkyl ether sulfates in cosmetic products.

Synthesis of Bromoalkanes and Derived Products. An important reaction of α-olefins is radical hydrobromation to give primary bromoalkanes. These compounds are valuable intermediates for, e.g., thiols, amines, amine oxides, and ammonium compounds:

$$R-CH=CH_2 \xrightarrow{HBr} R-CH_2CH_2Br$$

$$R-CH_2CH_2Br \begin{array}{c} \xrightarrow{NaSH} R^1-SH \\ \xrightarrow{HNMe_2} R^1-NMe_2 \end{array}$$

$$R^1-NMe_2 \begin{array}{c} \xrightarrow{H_2O_2} R^1-\overset{+}{N}Me_2 \\ | \\ O^- \end{array} \quad \xrightarrow{R^2X} R^1R^2\overset{+}{N}Me_2 X^-$$

Thiols are produced by this route in Europe and in the United States by Pennwalt Corp. and Phillips Petroleum Co. They are used in the synthesis of herbicides, pesticides, pharmaceuticals, in the textile industry, or as polymer regulators.

In 1982, ca. 15 000 t of alkyldimethylamine (ADA) was produced from linear α-olefins in the United States by Ethyl Corporation and by Procter & Gamble. Most of these amines were directly converted to fatty amine oxides (FAO) or quaternary ammonium salts (Quats). Fatty amine oxides are nonionic surfactants that, because they are biodegradable and do not irritate the skin, are incorporated into numerous products, e.g., into rinsing agents. Quats are produced by alkylation of amines, e.g., with benzylchloride. Benzyl-Quats possess biocidal and antimicrobial properties and are incorporated, for example, into disinfectants and deodorants.

Production of Synthetic Lubricants Oligomerization of α-olefins in the middle C-number range, in particular 1-decene, to isoparaffins (mostly trimers with a few tetramers and pentamers) leads to lubricants known as poly(α-olefins) (PAO) or synthetic hydrocarbons (SHC). They are used industrially, and in the automobile and aircraft sectors [113], [114]. They are in many instances superior to natural mineral oils, especially at lower temperature, because of their favorable viscosity, low volatility, high flash point, and high thermal stability. Oligomerization of the α-olefins can be performed with, e.g., BF_3, to which protonic cocatalysts, e.g., water or alcohols, are added.

In 1982, ca. 25 000 t of PAO were produced in the United States. The major producers are Mobil Oil, Gulf Oil, National Distillers, Burmah-Castrol, and Ethyl Corporation.

Production of Copolymers. Linear α-olefins, in particular 1-butene, 1-hexene, and 1-octene, are used as comonomers in the production of high density polyethylene (HDPE) and linear low density polyethylene (LLDPE). By adding α-olefins the density and other properties of the polymers may be significantly modified.

HDPE copolymers contain only relatively small quantities of α-olefins, generally 0.5–3%. The density of the HDPE homopolymer (0.965–0.955 g/cm^3) is reduced to 0.959–0.938 g/cm^3 for the HDPE copolymer. The most frequently employed comonomer is 1-hexene, followed by 1-butene.

On the other hand, 4–12% α-olefin is generally added to LLDPE to reduce the density to 0.935–0.915. The more α-olefins added to the polymer, the greater the number of short-chain branches and the lower the density. Addition of 1-butene is preferred in many of the gas-phase processes, e.g., in the Unipol process introduced by Union Carbide in 1977 [115], whereas 1-octene and 1-hexene are preferred in liquid-phase processes.

Major United States producers are Dow Chemical, DuPont, Exxon, Gulf Oil, Union Carbide, Soltex Polymer Corporation, Phillips Petroleum Company, and Allied Corporation. An estimated 85 000 t of α-olefins were copolymerized in HDPE and LLDPE in the United States in 1982.

Because Shell began marketing SHOP olefins in Western Europe, the use of HDPE and LLDPE copolymers has increased in these countries. Olefins derived from paraffin cracking plants are unsuitable as comonomers for polyethylene because of the impurities they contain.

Additional Reactions of α-Olefins. The reaction of α-olefins with peracid to form epoxides is an interesting route to bifunctional derivatives [116]. Some examples are given below.

$$R-CH=CH_2 \xrightarrow{R^1COOOH}$$

$$R-\overset{O}{\underset{\diagdown}{CH}}-CH_2 \begin{cases} \xrightarrow{H_2S} R-CH(OH)-CH_2-SH \\ \xrightarrow{HNR_2^2} R-CH(OH)-CH_2-NR_2^2 \\ \xrightarrow{HCN} R-CH(OH)-CH_2CN \\ \xrightarrow{R^3OH} R-CH(OH)-CH_2-OR^3 \\ \xrightarrow{R^4COOH} R-CH(OH)-CH_2-OCOR^4 \end{cases}$$

α-Olefin epoxides have also been used in polymer chemistry, e.g., for the modification of epoxy resins. United States producers of α-olefin epoxides include: Union Carbide, Viking, and Dow Chemical.

Table 13. Linear α-olefin producers (January 1984)

Manufacturer	Location	Process	Annual capacity (10^3 t)
Gulf Oil	Cedar Bayou, Texas	Oligomerization (Gulf process)	90
Mitsubishi	Mizushima, Okayama	Oligomerization (Gulf process)	30
Ethyl Corp.	Pasadena, Texas	Oligomerization (Ethyl process)	450
Chevron	Richmond, California	Paraffin cracking	40
Shell	Berre (France)	Paraffin cracking	90
	Pernis (The Netherlands)	Paraffin cracking	170
	Geismar, Louisiana	SHOP	270
	Stanlow (UK)	SHOP	170
		Total	1310

Another use of α-olefins is conversion into secondary monoalkylsulfates, which are readily degradable surfactants useful in wetting agents and household detergents. They are synthesized by reaction with concentrated sulfuric acid at low temperature (10–20 °C) and with short reaction times (5 min):

$$R-CH=CH_2 \xrightarrow{H_2SO_4} R-\underset{OSO_3H}{CH}-CH_3 \xrightarrow{NaOH} R-\underset{OSO_3Na}{CH}-CH_3$$

Hydrocarboxylation of α-olefins leads to odd-numbered carboxylic acids. This reaction is carried out with carbon monoxide and water at 200 °C and 20 MPa. The formation of linear carboxylic acids is favored in the presence of cobalt carbonyl–pyridine catalysts. Both the free acids and their esters are used as lubricant additives.

$$R-CH=CH_2 \xrightarrow[Co]{CO, H_2O} R-\underset{COOH}{CH}-CH_3 + R-CH_2CH_2COOH$$

The alkenylsuccinic anhydrides (ASA) should also be mentioned. They are obtained by heating maleic anhydride with α- or internal olefins, and are used as lubricant additives, detergents, and in leather and paper production.

Olefins with more than 30 carbon atoms are waxy and can be used as paraffin substitutes. In 1984 ca. 8000 t of α-olefins were used in the United States for the production of candles, crayons, and coatings.

2.1.4. Economic Aspects of Higher Olefins

Production facilities for α-olefins and their capacities (1984) are listed in Table 13.

The α-olefin market is currently subject to fluctuations [117]. Ethyl Corporation, the largest United States α-olefin producer, announced at the end of 1987 the intention to build a large α-olefin plant in Europe with an annual capacity of 250 000 t [118]. This plant would make Ethyl the largest α-olefin producer in Europe.

Table 14. Uses of higher linear olefins (1982) (estimate by the Stanford Research Institute)

Final products	Olefin consumption (10^3 t)	
	United States*	Western Europe
Oxo alcohols	67	145–160
Amines and derivatives	13–17	**
α-Olefinsulfonates (AOS)	7	3
Linear alkylbenzenes (LAB)	1–2	75–80
Copolymers (HDPE, LLDPE)	85	**
Synthetic lubricants (SHC)	25	**
Lubricant additives	15	20

* Figures refer exclusively to α-olefins.
** Unknown.

Shell has also been planning for some time to expand its SHOP processing capacity. The SHOP plant in Geismar, United States, which has been operating since 1977, was expanded in 1982 to an annual capacity of 270 000 t. Construction of a second SHOP unit in Geismar had been announced for 1984 [119]. According to the most recent information [120], this second plant should come on stream in 1989 with an annual capacity of 243 000 t. Additional expansion is to bring the total annual capacity at Geismar to 590 000 t.

Shell has also announced changes in Europe [121]. The plant at Stanlow, United Kingdom, operating since 1982, is to be expanded to an annual capacity of 220 000 t; once this plant is fully operational, the plant at Pernis, The Netherlands which for some time has produced only 60 000–70 000 t/a, is to be closed down.

Mitsubishi Petrochemical has expressed an interest in constructing a SHOP plant in Kashina, Japan in a joint venture with Shell [122]. Other projects under discussion are SHOP plants in New Zealand and in Canada [123].

Uses of linear olefins in the United States and in Western Europe are given in Table 14.

2.2. Dienes and Polyenes

2.2.1. Low Molecular Mass 1,3-Dienes

Industrially, the most important 1,3-dienes are butadiene, isoprene, and cyclopentadiene (→Butadiene; →Cyclopentadiene and Cyclopentene; →Isoprene). The conjugated dienes listed below are of lesser importance.

Piperylene (1,3-pentadiene). Piperylene is present at a level of >10% in the C_5 steam cracker fraction; it is expensive to separate from the C_5 cut. Polymers and copolymers with, for example, butadiene have been produced but have not yet found a technical application.

2,3-Dimethylbutadiene. 2,3-dimethylbutadiene was the first monomer to be converted on an industrial scale to a synthetic rubber, referred to as methyl rubber. It was developed by HOFFMANN in 1910 and produced during 1914–1918. It suffered, however, from several disadvantages (excessive hardness, sensitivity to oxidation) and later lost its importance. The early synthesis of the monomer involved conversion of acetone to pinacol, which is then dehydrogenated to the diene in the presence of aluminum oxide as catalyst. Modern synthesis starts with propene, which is dimerized to 2,3-dimethylbutene [111], [112] and subsequently dehydrogenated:

$$2\ CH_3COCH_3 \longrightarrow (CH_3)_2C(OH)-C(OH)(CH_3)_2 \xrightarrow[-2\ H_2O]{Al_2O_3}$$

$$CH_2=C(CH_3)-C(CH_3)=CH_2$$

$$2\ CH_3CH=CH_2 \xrightarrow{Cat.} CH_2=C(CH_3)-CH(CH_3)_2 \xrightarrow[-H_2]{Cat.}$$

Chloroprene (→Chlorinated Hydrocarbons), $CH_2=CCl-CH=CH_2$, 2-chloro-1,3-butadiene, is used to form polychloroprene (neoprene), which is extremely resistant to oil, abrasion, and ageing. As with all polymers containing a large proportion of halogen, these products are resistant to ignition. Vulcanized polychloroprene is used in, for example, conveyer belts, cable shielding, and protective clothing. Nonvulcanized polymers are used as contact adhesives in plastics manufacturing.

Chloroprene can be produced by two routes. In the older process, acetylene is first dimerized to vinyl acetylene and hydrogen chloride then added to the triple bond; both steps take place in the presence of CuCl as catalyst. The second process involves chlorination of butadiene, at 300 °C with chlorine to yield a mixture of dichlorobutenes. The 1,2-adduct is subsequently dehydrochlorinated to yield chloroprene. The 1,4-adduct can be isomerized to the 1,2-adduct:

$$2\ HC\equiv CH \xrightarrow{CuCl} HC\equiv C-CH=CH_2 \xrightarrow{CuCl/HCl} H_2C=CCl-CH=CH_2$$

$$CH_2=CH-CH=CH_2 \xrightarrow{Cl_2} CH_2Cl-CHCl-CH=CH_2 + CH_2Cl-CH=CH-CH_2Cl$$

$$\downarrow -HCl$$

$$CH_2=CCl-CH=CH_2$$

2.2.2. Synthesis of Dienes and Polyenes by Oligomerization

A multitude of linear and cyclic dienes and polyenes can be synthesized by transition metal-catalyzed oligomerization or by cooligomerization [83]. Only a few examples will be described below, some of which are used commercially.

Butadiene can be converted into six cyclooligomers in a nickel-catalyzed reaction [124]. The synthesis can be steered in the desired direction by a suitable choice of catalyst.

1,5-Cyclooctadiene is produced by Shell and by Hüls and is used, for example, as starting material for suberic acid and poly(octenamer) (→Cyclododecatriene and Cyclooctadiene).

1,5,9-Cyclododecatriene is produced by DuPont, Shell, and Hüls using nickel or titanium as catalyst; it is used in the synthesis of nylon 12, dodecandioic acid, and hexabromocyclododecane.

4-Vinylcyclohex-1-ene is also accessible from butadiene but is, as yet, of little commercial importance. The possibility of using it as a precursor in styrene synthesis has been discussed.

An example of catalytic cooligomerization is the reaction between butadiene and ethylene [125].

Whereas the cyclic 1:1 product vinylcyclobutane and the C_{10} oligomers cyclodecadiene and 1,4,9-decatriene are not yet used commercially, the linear oligomer, 1,4-

hexadiene, is produced as a monomer for the production of ethylene–propene–diene (EPDM) elastomers. Homogenous catalysis with rhodium or palladium complexes leads to *trans*-1,4-hexadiene; cobalt and iron catalysts yield predominantly the *cis*-1,4-product. Rhodium catalysts are used in commercial production.

Ethylidenenorbornene, also a comonomer for EPDM polymers, is also accessible via cooligomerization. One of the routes employs the Diels–Alder reaction starting with butadiene and cyclopentadiene, the other couples norbornadiene and ethylene in the presence of a nickel catalyst. Vinylnorbornene is formed as an intermediate in both routes.

2.2.3. Synthesis of Dienes and Polyenes by Metathesis

Metathesis offers a simple path to α,ω-diolefins that are not easily obtained by other routes [126]. Starting materials are cycloolefins, which can be prepared, for example, by cyclooligomerization of butadiene (see Section 2.2.2) and subsequent partial hydrogenation. Metathetic conversion with ethylene opens the cycloolefin to an α,ω-diene. In this manner, 1,5-cyclooctadiene gives two molecules of 1,5-hexadiene; cyclooctene forms 1,9-decadiene; and cyclododecene leads to 1,13-tetradecadiene.

Heterogeneous rhenium catalysts are preferred because they allow the reaction to proceed under very mild conditions (5–20 °C, 0.1–0.2 MPa ethylene). A plant with an annual capacity of 3000 t using this FEAST (Further Exploitation of Advanced Shell Technology) process was brought on stream by Shell in Berre (France) in 1987. The products are suitable for the synthesis of numerous fine chemicals, e.g., production of long-chain diketones, for dibromides, diepoxides, dialdehydes, dithiols, and diamines.

4-Vinylcyclo-1-hexene undergoes self metathesis. Elimination of ethylene yields 1,2-bis(3-cyclohexenyl)ethylene, which has been used as a starting material for fire retardants.

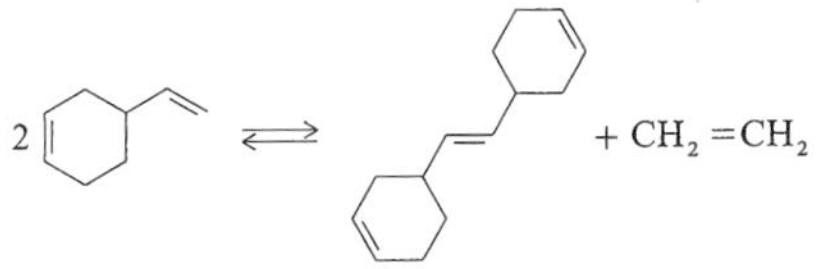

3. Alkylbenzenes

Toluene, Xylenes, and Ethylbenzene are separate keywords.

3.1. Trimethylbenzenes

Properties. The three isomers of trimethylbenzene are colorless liquids. Some of their physical properties are listed in Table 15.

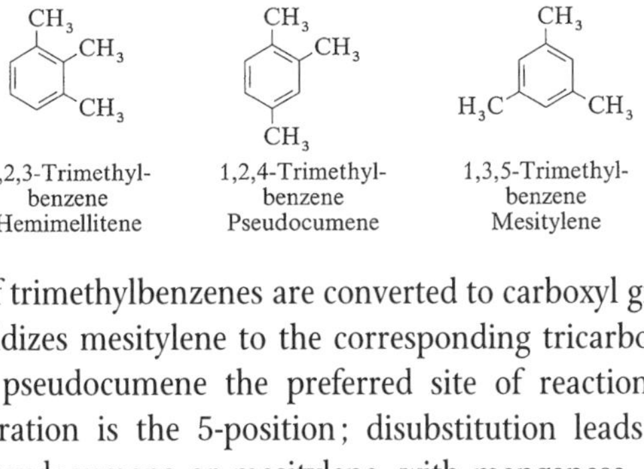

1,2,3-Trimethyl-
benzene
Hemimellitene

1,2,4-Trimethyl-
benzene
Pseudocumene

1,3,5-Trimethyl-
benzene
Mesitylene

The methyl groups of trimethylbenzenes are converted to carboxyl groups on oxidation. Dilute nitric acid oxidizes mesitylene to the corresponding tricarboxylic acid, trimesic acid [554-95-0]. In pseudocumene the preferred site of reaction for halogenation, sulfonation, and nitration is the 5-position; disubstitution leads to 3,5-derivatives. Mild oxidation of pseudocumene or mesitylene, with manganese- or lead dioxide, or electrolytically, leads to 2,4- or 3,5-dimethylbenzaldehyde, respectively.

Production. Hemimellitene, pseudocumene, and mesitylene are found in coal tar oil. The compounds can be isolated by distillation and purified by the differential hydrolysis of their sulfonic acids [128]. More important industrially is the occurrence of trimethylbenzenes in some mineral oils, and their formation during processing of crude oil, especially catalytic cracking and reforming. Pseudocumene is obtained in 99% purity by superfractionation from the C_9 cut of crude oil refining [129]. Pseudocumene and hemimellitene can be obtained by extractive crystallization of the C_9 cut with urea [130]. Trimethylbenzenes can be produced by Friedel–Crafts methylation of toluene or xylene with chloromethane in the presence of aluminum chloride. Trimethylbenzenes, especially pseudocumene and mesitylene, are also formed by liquid-phase dispropor-

Table 15. Physical properties of alkylbenzenes

Compound	CAS registry number	M_r	mp, °C	bp, °C	d_4^{20}, g/cm^3	n_D^{20}
1,2,3-Trimethylbenzene, hemimellitene	[526-73-8]	120.19	−25.4	176.1	0.8944	1.5139
1,2,4-Trimethylbenzene, pseudocumene	[95-63-6]	120.19	−43.8	169.3	0.8758	1.5048
1,3,5-Trimethylbenzene, mesitylene	[108-67-8]	120.19	−44.7	164.7	0.8652	1.4994
1,2,3,4-Tetramethylbenzene, prehnitene	[488-23-3]	134.22	− 6.2	205	0.9052	1.5203
1,2,3,5-Tetramethylbenzene, isodurene	[527-53-7]	134.22	−23.7	198	0.8903	1.5130
1,2,4,5-Tetramethylbenzene, durene	[95-93-2]	134.22	79.2	196–198	0.8380^a	1.4790^a
Pentamethylbenzene	[700-12-9]	148.25	54.5	232	0.917	1.527
Hexamethylbenzene, mellitene	[87-85-4]	162.28	166–167	265	1.0630^b	
1,2-Diethylbenzene	[135-01-3]	134.22	−31.2	183.5	0.8800	1.5035
1,3-Diethylbenzene	[141-93-5]	134.22	−83.9	181.1	0.8602	1.4955
1,4-Diethylbenzene	[105-05-5]	134.22	−42.8	183.8	0.8620	1.4967
1,2,3-Triethylbenzene	[42205-08-3]	162.27				
1,2,4-Triethylbenzene	[877-44-1]	162.27		217.5^c	0.8738	1.5024
1,3,5-Triethylbenzene	[102-25-0]	162.27	−66.5	215.9	0.8631	1.4969
1,2,3,4-Tetraethylbenzene	[642-32-0]	190.33	11.8	251^d	0.8875	1.5125
1,2,3,5-Tetraethylbenzene	[38842-05-6]	190.33	−21	247^d	0.8799	1.5059
1,2,4,5-Tetraethylbenzene	[635-81-4]	190.33	10	250	0.8788	1.5054
Pentaethylbenzene	[605-01-6]	218.38	<−20	277	0.8985^e	1.5127
Hexaethylbenzene	[604-88-6]	246.44	129	298	0.8305^f	1.4736^f
1-Ethyl-2-methylbenzene, 2-ethyltoluene	[611-14-3]	120.19	−80.8	165.2	0.8807	1.5046
1-Ethyl-3-methylbenzene, 3-ethyltoluene	[620-14-4]	120.19	−95.5	161.3	0.8645	1.4966
1-Ethyl-4-methylbenzene, 4-ethyltoluene	[622-96-8]	120.19	−62.3	162	0.8614	1.4959

^{a} At 81 °C.
^{b} At 25 °C.
^{c} 100.4 kPa.
^{d} 97.6 kPa.
^{e} At 19 °C.
^{f} At 130 °C.

tionation of xylene in the presence of aluminum chloride. These reactions can be carried out in the gas phase at high temperature on aluminum silicate catalysts.

A more recent method for the methylation of xylene with methanol or dimethyl ether on fluorine-containing crystalline borosilicate, aluminosilicate, or boroaluminosilicate (ZSM-5) at 300 °C and atmospheric pressure yields pseudocumene and durene almost exclusively, both of which may be isolated at >99% purity [131]. Trimethylbenzenes are also generated in the reaction of toluene with synthesis gas over a dual catalyst consisting of a metal oxide component and an aluminum silicate, at 200–400 °C and up to 20 MPa [132]. Pseudocumene is also generated from methanol as a byproduct of the Mobil MTG (methanol to gasoline) process (see Section 3.2). Fusel oils and other fermentation products may be reacted on the zeolite catalyst HZSM-5 to form aromatic compounds, including trimethylbenzenes [133].

Uses. Mesitylene is used in the synthesis of dyes and antioxidants, and as a solvent. Pseudocumene is a starting material for dyes, pharmaceuticals and, especially, for trimellitic anhydride [552-30-7], which is then converted to heat-resistant polyamideimides and polyesterimides (see also Carboxylic Acids, Aromatic). Hemimellitene is an intermediate in the production of fragrances.

3.2. Tetramethylbenzenes

Properties. Some physical properties of the three tetramethylbenzene isomers are listed in Table 15. Prehnitene and isodurene are colorless liquids; durene forms colorless, monoclinic crystals that sublime slowly at room temperature and that have an odor reminiscent of camphor.

1,2,3,4-Tetramethyl-
benzene
Prehnitene

1,2,3,5-Tetra-
methylbenzene
Isodurene

1,2,4,5-Tetramethyl-
benzene
Durene

Tetramethylbenzenes are readily soluble in aromatic hydrocarbons, acetone, ether, and alcohol. Durene is highly reactive; nitration, chlorination, sulfonation, and chloromethylation occur almost as readily as with phenol. The methyl groups may be successively oxidized to carboxylic acids; the presence of the corresponding mono-, di-, tri-, and tetracarboxylic acids can be demonstrated during liquid-phase oxidation of durene in the presence of cobalt catalyst [134].

Production. The tetramethylbenzenes also occur in coal tar from anthracite and lignite, and in the reformed fraction of oil refineries. The isolation of durene by crystallization is aided by its relatively high melting point; isodurene becomes enriched in the mother liquor. Methylation of xylene leads to an isomeric mixture, from which durene, prehnitene, and isodurene can be obtained [128]. Durene is produced by liquid-phase methylation of pseudocumene with chloromethane in the presence of aluminum chloride [135]. In more recent processes durene is produced by reaction of xylene or pseudocumene with methanol in the gas phase over aluminum silicates; the most successful process uses the zeolites ZSM-5 and TSZ [136], [137]. In the MTG process, commercialized in New Zealand since 1986, the production of automotive fuels from methanol over ZSM-5 yields an unacceptable quantity of durene, which has to be removed from the fuel to avoid carburetor clogging by the high melting point impurity. The durene is currently removed by catalytic isomerization, disproportionation, and reductive demethylation. The possibility of isolating durene from the MTG process is being investigated [138].

Uses. Oxidation of durene leads to the commercially important pyromellitic dianhydride [89-32-7] used as a starting material for heat-resistant polyimides and as a hardener for epoxy resins. Pyromellitic esters serve as migration-resistant plasticizers for poly(vinyl chloride) (→Carboxylic Acids, Aromatic).

3.3. Penta- and Hexamethylbenzene

Properties. Some physical properties are listed in Table 15. Both compounds are crystalline at room temperature; they dissolve easily in benzene and ethanol. The methyl groups are readily oxidized to carboxyl groups. Irradiation with light converts hexamethylbenzene into its valence isomer hexamethyl Dewar benzene (1,2,3,4,5,6-hexamethylbicyclo[2.2.0]hexa-2,5-diene, [7641-77-2]). Electrophilic substitution on the free ring site of pentamethylbenzene proceeds readily.

Production. Penta- and hexamethylbenzene are generated by liquid-phase Friedel–Crafts alkylation of xylene or pseudocumene with chloromethane in the presence of aluminum chloride [135]. If the reaction is carried out at 140–180 °C in o-dichlorobenzene, hexamethylbenzene with a purity of 98 % can be crystallized directly from the reaction mixture [139]. In a more recent method, penta- and hexamethylbenzene can be produced simultaneously by reaction of phenol and methanol in toluene in the presence of a zinc aluminate catalyst at 330–360 °C [140]. The products are then separated by fractional crystallization. Hexamethylbenzene is formed by thermal rearrangement of hexamethyl Dewar benzene, which is readily obtained from 2-butyne [141].

Uses. Pentamethylbenzene is of little commercial importance. Hexamethylbenzene is an intermediate in the fragrance industry. It is of scientific interest in studies of charge transfer complexes and exciplexes, in which, like other methylbenzenes, it may function as donor [142].

3.4. Diethylbenzenes

Properties. For some physical properties see Table 15.

1,2-Diethylbenzene *1,3-Diethylbenzene* *1,4-Diethylbenzene*

Production. The three isomeric diethylbenzenes (DEB) occur as byproducts in the production of ethylbenzene from benzene and ethylene but are, generally, recycled to undergo transalkylation (→Ethylbenzene). In the liquid-phase Friedel–Crafts process with aluminum chloride an increase in the feedstock ratio of ethylene to benzene may raise the DEB concentration in the output to 35%. The DEB fraction consists of ca. 65% 1,3-DEB, 30% 1,4-DEB, and only 5% 1,2-DEB. A similar isomer distribution is found in the product mixture from gas-phase alkylation of ethylbenzene with ethylene on an unmodified HZSM-5 zeolite catalyst. By modifying the catalyst, the zeolite pore may be narrowed to such an extent that, based on geometry alone, only 1,4-diethylbenzene is formed, rather than the bulkier 1,2- and 1,3-isomers. The Mobil gas-phase alkylation, which is based on such a shape-selective zeolite, converts ethylbenzene to DEB at 400 °C and 0.7 MPa with 88% selectivity; more than 99% of the DEB is the 1,4-isomer [143]. Nearly quantitative formation of 1,3-DEB occurs at low temperature in the liquid phase by disproportionation of ethylbenzene in the presence of at least 10 vol% HF and 0.3–0.5 mol BF_3 per mol ethylbenzene.

Uses. Diethylbenzenes may be converted to divinyl benzenes by catalytic dehydrogenation. The latter are useful in the production of crosslinked polystyrenes. DEB is contained in the product of the MTG process (see Section 3.2) and improves the properties of the gasoline.

3.5. Triethylbenzenes and more highly Ethylated Benzenes

Properties. For some physical properties see Table 15.

Production. The more highly ethylated benzenes are produced in the liquid phase in a similar manner as that described for diethylbenzene. 1,3,5-Triethylbenzene may be generated in excellent yield by disproportionation of diethylbenzene with BF_3 and HF at 75 °C. The same result is obtained with $AlCl_3$–HCl as catalyst. 1,3,5-Triethylbenzene is also formed by cyclocondensation of 2-butanone in the presence of sulfuric acid.

Hexaethylbenzene is formed in nearly quantitative yield when ethylene is passed into the AlCl$_3$ complex of aromatics until a mass of crystals is formed, which is then worked up with water.

3.6. Ethylmethylbenzenes (Ethyltoluenes)

Properties. Some physical properties are collected in Table 15.

Production. Ethyltoluene is produced in similar fashion to ethylbenzene, i.e., by ethylation of toluene. Alkylation of toluene with ethylene in the presence of AlCl$_3$ and HCl gives all three isomers of ethyltoluene and more highly ethylated derivatives. A typical reaction mixture contains 48% unreacted toluene, about 12% 4-ethyltoluene, 19% 3-ethyltoluene, 4% 2-ethyltoluene, 14% higher aromatics, and barely 1% tar [143]. According to a Dow process, subsequent distillation gives a mixture of 3- and 4-ethyltoluenes [144]. 2-Ethyltoluene, toluene, and the more highly alkylated toluenes are then recycled. The gas-phase ethylation of toluene on unmodified HZSM-5 zeolite catalyst at 350 °C gives 23% 4-ethyltoluene, 53% 3-ethyltoluene, 8% 2-ethyltoluene, and 10% higher aromatics [145]. Formation of the 2- and 3-isomers in the ethylation of toluene can be almost completely suppressed by the use of a shape-selective ZSM-5 zeolite that has been modified by addition of phosphorus, boron, or other elements, and by suitable physical treatment. In a Mobil process, 4-ethyltoluene of 97% purity is generated in this manner at 350–400 °C [145]. A recent patent granted to Universal Oil Products claims a process for the production of an alkylbenzene mixture, which also contains ethyltoluene, from toluene and synthesis gas [132].

Uses. The mixture of 3- and 4-ethyltoluene obtained by the Dow process is catalytically dehydrogenated to the corresponding mixture of methylstyrenes, known commercially as vinyltoluene [*25013-15-4*], and then polymerized to "poly(vinyltoluene)." Dehydrogenation of the practically pure 4-ethyltoluene from the Mobil process to 4-methylstyrene [*622-97-9*] occurs analogously on the route to "poly(*p*-methylstyrene)". A plant for the alkylation of toluene to 4-ethyltoluene by the Mobil process was opened in 1982 by Hoechst (now Hoechst-Celanese) in Baton Rouge, Louisiana [146]; the plant has an annual capacity of 16 000 t of 4-methylstyrene and has been purchased by Deltech in 1989. The output of the two products vinyltoluene and *p*-methylstyrene is estimated at 25 000 t/a [147].

3.7. Cumene

Properties. Cumene [98-82-8], 2-phenylpropane, isopropylbenzene, C_9H_{12}, M_r 120.19, is a colorless, mobile liquid with a characteristic, aromatic odor.

CH(CH$_3$)$_2$ on benzene ring

Cumene

The effect of cumene in increasing the octane rating of motor fuels has long been known. Large plants for production of cumene, which was used as a component of aviation fuel, were constructed in the United States during the early 1940s. Some physical properties of cumene are listed below [148], [149]:

mp, °C	−96.04
bp, °C	152.39
d^{20}, g/cm^3	0.86179
n_D^{20}	1.49145
Specific heat (25 °C), J g^{-1} K^{-1}	1.96
Heat of vaporization, kJ/mol	
at 25 °C	44.23
at 152.4 °C	37.74
T_{crit}, K	631
p_{crit}, MPa	3.21
V_{crit}, cm^3/mol	428
Flash point, °C	43
Ignition temperature, °C	425

The variation of vapor pressure with temperature is given below:

t, °C	2.9	38.3	66.1	88.1	129.2
Vapor pressure, kPa	0.13	1.33	5.33	15.33	53.33

For more thermodynamic data, see [150]. Explosion limits for the system cumene vapor – oxygen – nitrogen at standard pressure and at 0.56 MPa are given in [151]. For cumene – air mixtures the explosion limits are 0.8 – 6.0 vol% cumene, corresponding to a cumene concentration of 40 – 300 g/m^3 [152]. Cumene forms azeotropes with water, aliphatic carboxylic acids, ethylene glycol ethers, cyclohexanol, and cyclohexanone.

Production. Cumene is produced from benzene [71-43-2] and propene [115-07-1] in the presence of an acidic catalyst:

$$C_6H_6 + C_3H_6 \longrightarrow C_6H_5CH(CH_3)_2 \quad \Delta H = -113 \text{ kJ/mol}$$

Two processes are employed industrially. In the *UOP process* a fixed bed consisting of polyphosphoric acid on a silica support is used as the catalyst [153]. The reaction takes place mainly in the gas phase at ≥ 200 °C and ≥ 2 MPa. Pure propene or propene with a propane content ≥5% is used as raw material, together with benzene (min. purity 99.5%) in up to 12-fold molar excess; this leads to high distillation costs in the product work-up. The yield of cumene is 97% based on benzene and 90% based on propene [154]. The losses are caused by overalkylation and oligomerization of the propene. The high-boiling residue contains 50–60% diisopropylbenzenes (*o-*:*m-*:*p-* ratio 1:2:2–3) and numerous other byproducts. It is not possible to recycle these products to the alkylation reactor, because polyphosphoric acid has no transalkylation properties.

The *second industrial process* works in the liquid phase at 50–80 °C; aluminum chloride or the liquid double compound formed from aluminum chloride, (isopropyl-)benzene, and hydrogen chloride is employed as the catalyst. This complex is separated from the product stream, as a heavy phase and is returned to the alkylation reactor. Except for a small purge stream higher alkylated byproducts are recycled to the alkylation step, whereby essentially complete use of both benzene and propene bound in di- and triisopropylbenzenes is achieved by transalkylation. The yield of cumene, referred to benzene and propene, is 99% and 98%, respectively. The UOP process is used predominantly. Higher selectivity and the lower energy requirements, however, are now making the aluminum chloride process more attractive [154].

Uses. The production capacity for cumene in Western Europe, the United States, and Japan amounts to 2×10^6, 2×10^6, and 4×10^5 t/a, respectively. Cumene is used almost exclusively for the production of phenol [*108-95-2*] and acetone [*67-64-1*] (→Acetone; →Phenol). This process is based on the discovery in 1944 that cumene hydroperoxide [*80-15-9*], obtained by oxidation of cumene with oxygen, can be cleaved under the catalytic action of strong acids:

$$C_6H_5CH(CH_3)_2 + O_2 \longrightarrow C_6H_5C(CH_3)_2OOH$$

$$C_6H_5C(CH_3)_2OOH \longrightarrow C_6H_5OH + CH_3COCH_3$$

Both reactions are exothermic; the enthalpy of reaction amounts to 121 and 255 kJ/mol, respectively. (For cumene quality specifications, → Acetone.)

Pure α-methylstyrene can be obtained as a byproduct of the phenol process. Cumene hydroperoxide is used as a radical initiator for the copolymerization of styrene with butadiene and acrylates, and also for the radical crosslinking of unsaturated polyester resins. Addition of cumene hydroperoxide to α-methylstyrene gives dicumyl peroxide, used for radical crosslinking of polyolefins. Hydrogenation of the aromatic nucleus in cumene gives isopropylcyclohexane (hydrocumene), a cycloaliphatic solvent with a high boiling point (154.5 °C) and a low freezing point (–90 °C).

Table 16. Physical properties of diisopropylbenzene isomers

Property	m-DIPB	o-DIPB	p-DIPB
CAS registry number	[99-62-7]	[577-55-9]	[100-18-5]
mp, °C	−63.13	−56.68	−17.07
bp, °C	203.18	203.75	210.35
d^{20}, g/cm³	0.8559	0.8779	0.8568
n_D^{20}	1.4883	1.4960	1.4893

3.8. Diisopropylbenzenes

1,4-Diisopropylbenzene [100-18-5], p-diisopropylbenzene, p-DIPB, $C_{12}H_{18}$, M_r 162.26, 1,3-diisopropylbenzene [99-62-7], m-diisopropylbenzene, m-DIPB, and mixtures of these isomers, are industrially important. Some physical properties of diisopropylbenzenes are given in Table 16.

The flash point of DIPB mixtures is 77 °C; the ignition temperature is 449 °C. Inhalation of DIPB vapor can lead to the appearance of narcotic effects.

Production. Diisopropylbenzenes are present in the high boiling fractions, in which they may account for 50–70 wt%, of cumene plants, together with numerous other byproducts. Processes for the production of m- and p-DIPB from these fractions have been suggested [155], [156]. The high-boiling fraction of the aluminum chloride process frequently contains 1,1,3-trimethylindan [2613-76-5], bp 204.8–204.9 °C, which cannot be separated from the m- and p-DIPB at a justifiable cost.

For production from benzene and propene alkylation catalysts that possess high isomerization and transalkylation activities are employed. Examples include aluminum chloride–hydrogen chloride [157], [158], silica–alumina [159], [160], boron trifluoride-treated alumina [161], and zeolithes [162]. The process is controlled in such a way that cumene formed at a later stage is alkylated with propene. The desired DIPB isomer is isolated by distillation and the undesired DIPB isomer, together with triisopropylbenzenes, are converted to cumene by reaction with benzene in a transalkylation step. The aim in both the alkylation and the transalkylation steps is to reach thermodynamic equilibrium, at which the o-diisopropylbenzene is present at only ca. 0.1%.

Uses. 1,3-Diisopropyl- and 1,4-diisopropylbenzene are the starting materials for new synthetic routes to resorcinol and hydroquinone (→Hydroquinone), respectively [163], which are used industrially in Japan and the United States [160]. Furthermore, a

Table 17. Physical properties of some alkylbenzenes

Compound	CAS registry number	mp, °C	bp, °C	η_D^{20}
3-Isopropyltoluene	[535-77-3]	−63.75	175.05	1.4929
4-Isopropyltoluene	[99-87-6]	−67.94	177.10	1.4909
2-Isopropyltoluene	[527-84-4]	−71.54	178.15	1.5005
Isobutylbenzene	[538-93-2]	−51.48	172.76	1.4864
tert-Butylbenzene	[98-06-6]	−57.85	169.12	1.4926
4-tert-Butyltoluene	[98-51-1]	−53	192.7	1.4919
5-tert-Butyl-m-xylene	[98-19-1]	−18.3	209	1.4956
tert-Pentylbenzene	[2049-95-8]		192.38	1.4958

process for the production of phloroglucin, 1,3,5-trihydroxybenzene, from 1,3,5-triisopropylbenzene has been developed [164]. The monohydroperoxides of 1,3- and 1,4-diisopropylbenzene are used as initiators for the thermal crosslinking of unsaturated polyester resins, and for the emulsion polymerization of dienes, styrene, and acrylates. Oxidation gives *a,a,a',a'*-tetramethylphenylene-biscarbinols. Dehydrogenation gives diisopropenylbenzenes, which by reaction with phenol or aniline give bisphenols or bisanilines, respectively, which are used as building blocks for polycarbonates, polyamides, polyesteramides, and polyurethanes. Addition of isocyanic acid to diisopropenylbenzenes yields diisocyanates, from which light-stable polyurethanes can be produced [165]. Bis(*tert*-butylperoxyisopropyl)benzene, preferably prepared from the mixture of the corresponding biscarbinols of 1,3- and 1,4-DIPB, is used as a crosslinking agent for polyolefins.

3.9. Cymenes; C_4- and C_5-Alkylaromatic Compounds

The compounds listed in Table 17 are industrially important. Mixtures of 3-isopropyltoluene (*m*-cymene) and 4-isopropyltoluene (*p*-cymene) can be obtained by propylation of toluene in the presence of an isomerization-active Friedel–Crafts catalyst. These two isomers can be separated by distillation only with difficulty. The mixture is used for the production of *m*- and *p*-cresol, (→Cresols and Xylenols), [166]. Treatment of (methyl-)aromatic compounds with isobutene yields *tert*-butyl-substituted compounds, which are used as solvents (*tert*-butylbenzene) and for the production of fragrances, pharmaceuticals, and herbicides.

m-Cymene   *p*-Cymene

The introduction of a *tert*-pentyl group can be achieved by the use of *tert*-pentyl chloride as the alkylating agent and aluminum chloride, iron(III) chloride, or zirconium tetrachloride as the catalyst. 2-*tert*-pentylanthraquinone, obtained from *tert*-pentylbenzene and phthalic anhydride, is employed in the industrial production of hydrogen peroxide. Isobutylbenzene, required for the synthesis of pharmaceuticals and fragrances, is produced in industrial quantities by treatment of toluene with propene over catalysts containing alkali metals, such as potassium–graphite inclusion compounds or sodium–potassium on a support [167].

3.10. Monoalkylbenzenes with Alkyl Groups > C_{10}

Monoalkylbenzenes with secondary alkyl groups C_{10} to ca. C_{14} are used as starting materials for the production of alkylbenzenesulfonates (ABS). Only secondary alkylbenzenesulfonates with unbranched (linear) side chains (LABS) are of practical importance as raw materials for detergents, as they are extensively degraded (>90%) by microorganisms in sewage plants after a relatively short period of time.

Before the introduction of LABS, which took place in the Federal Republic of Germany in 1962, alkylbenzenesulfonates with highly branched C_{12} side chains were used. These compounds, tetrapropylene benzenesulfonates, were produced from the products of alkylation of benzene with propene tetramer, itself obtained by the tetramerization of propene by contact with phosphoric acid. These sulfonates possess excellent detergent properties, but their biological degradeability is insufficient.

Production. Linear secondary alkylbenzenes are obtained industrially by reaction of benzene with secondary alkyl chlorides (chloroparaffins) or with olefins under the influence of a Friedel–Craft catalyst such as aluminum chloride, hydrofluoric acid, or sulfuric acid. The following methods have been described for the production of the required alkyl chlorides and olefins [168]:

1) Dehydrogenation of C_{10}–C_{13}-*n*-alkanes, according to the Pacol process of UOP (see Section 2.1.2.1), on a fixed bed at 400–500 °C, with a conversion of 10–15%; the ensuing mixture of internal olefins and unreacted alkanes is used for the alkylation of benzene. The *n*-C_{10}–C_{13}-alkanes used in the dehydrogenation are obtained from the kerosene fraction of crude oil by adsorption on molecular sieves, according to well known processes such as IsoSiv of Union Carbide, or Molex of UOP.
2) By chlorination of C_{10}–C_{13}-alkanes at ca. 100 °C to a conversion of ca. 30%, and use of the resulting mixture of alkanes and secondary monochloroalkanes for alkylation of benzene.
3) By chlorination of C_{10}–C_{13}-*n*-paraffins as in 2) above, followed by dehydrochlorination (elimination of hydrogen chloride) of the monochloroalkanes at 200–300 °C

over an iron catalyst to give linear alkenes with internal double bonds; these alkenes, in admixture with unreacted paraffins, are then used to alkylate benzene.
4) Metathesis of higher ($>C_{18}$) with lower ($<C_{10}$) internal olefins from the Shell SHOP process (see Section 2.1.2.2) to give olefins of intermediate molecular mass.

A $C_{10}-C_{13}$-α-olefin fraction suitable for use in detergents can also be obtained by thermal cleavage of $C_{20}-C_{40}$-n-paraffins at 500–600 °C, according to the older wax-cracking process of Shell (see Section 2.1.2.1).

The alkylation of benzene with chloroparaffin–paraffin mixtures takes place in the presence of aluminum chloride at ca. 50 °C. A relatively high proportion of 2-phenyl-alkanes is characteristic of secondary alkylbenzenes obtained in this manner [169], [170]. Dichloroalkanes, always present in the mixture of chloroparaffins even at low levels of chlorination, can lead to the formation of diphenylalkanes, 1,4-dialkyltetralins, and 1,3-dialkylindans. For formation and analytical detection see [171]. The removal of these compounds from monoalkylbenzenes is difficult. Formation of these byproducts can be largely avoided by carrying out the alkylation of benzene with olefins from the dehydrochlorination, where dichloroalkanes are dehydrochlorinated to give tar-like products that are easily separated. In industrial production the reaction of $C_{10}-C_{13}$-olefins with benzene is carried out with a 1:10 molar ratio in the presence of anhydrous hydrofluoric acid at 10 °C, a process that dates back to 1949 [172]. The industrial process using the steps paraffin chlorination–dehydrochlorination–alkylation in the presence of hydrofluoric acid was first realized by Chemische Werk Hüls in 1962 [170]. This process is shown in Figure 8. In this way a quality standard for alkylbenzenes for use in detergents was created; in ecological terms, this standard is still valid.

3.11. Diphenylmethane

Diphenylmethane [*101-81-5*] benzylbenzene, $C_{13}H_{12}$, M_r 168.24, was first prepared in 1871 by ZINCKE from benzyl chloride and benzene in the presence of zinc dust or copper (I) oxide.

Physical Properties. Diphenylmethane forms colorless prismatic needles with a harsh herbaceous odor reminiscent of geranium leaves. It is insoluble in water, soluble in alcohol, ether, chloroform, and benzene.

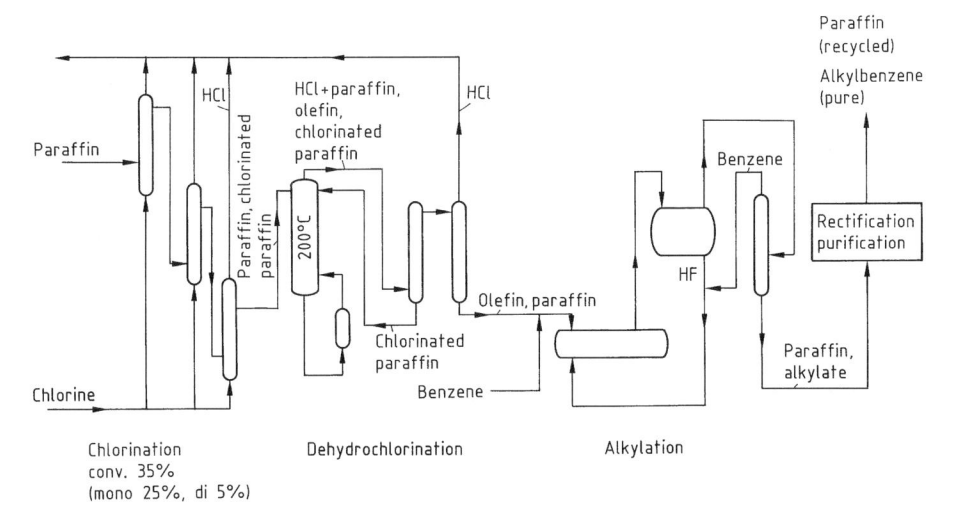

Figure 8. Production of alkylbenzenes, process of Hüls AG [173]

mp	26–27 °C
bp	
101.3 kPa	261–262 °C
1.33 kPa	125.5 °C
d_4^{20}	1.006
Flash point ca.	130 °C
Enthalpy of formation (room temperature)	109.6 kJ/mol
Enthalpy of combustion (20 °C)	41.2×10^3 kJ/kg

Chemical Properties. As an araliphatic hydrocarbon, diphenylmethane displays the chemical properties of both aromatic and aliphatic compounds. It is difficult to limit substitution in the phenyl radicals to one of the rings; therefore, controlled syntheses of particular unsymmetrically substituted diphenylmethanes must be carried out by different paths. Under suitable reaction conditions the methylene group can also be substituted, e.g., by halogen or nitro groups. Benzophenone can be produced by oxidation, for example, with oxygen over various catalysts, chromic acid or dilute nitric acid. Hydrogenation in ethanol over a nickel catalyst leads to dicyclohexylmethane. When diphenylmethane is passed over platinum–charcoal at 300 °C, cyclization to fluorene takes place.

Production. Of the many known methods to synthesize diphenylmethane only a few are used commercially.

The Friedel–Craft reaction of benzyl chloride with benzene in the presence of aluminum chloride is an important method for production of diphenylmethane:

$$C_6H_5CH_2Cl + C_6H_6 \xrightarrow{AlCl_3} C_6H_5CH_2C_6H_5$$

Other known catalysts of this process are elemental aluminum, iron(III) chloride, antimony pentachloride, zinc chloride – hydrogen chloride and boron fluoride dihydrate. When sulfuric acid is used as a condensation agent diphenylmethane yields are 83% [174] and with polyphosphoric acid yields are over 90% [175].

Condensation of benzene with formaldehyde in presence of 85% sulfuric acid yields 79% diphenylmethane [176].

$$2\ C_6H_6 + HCHO \xrightarrow{H_2SO_4} C_6H_5CH_2C_6H_5$$

Quality Specifications. Commercial diphenylmethane should be at least 97% pure (GC). It must be free of halogens and have a minimum melting point of 24 °C.

Uses. Diphenylmethane is used in the fragrance industry both as a fixative and in scenting soaps. It can serve as a synergist for pyrethrin in pesticides and insecticides [177], [178]. Diphenylmethane is recommended as a plasticizer to improve the dyeing properties [179], as a solvent for dyes [180], and as a dye carrier for printing with disperse dyes [181]. The addition of diphenylmethane to saturated, linear polyesters improves their thermal stability [182], and its addition to jet fuels increases their stability and lubricating properties [183].

Substituted diphenylmethanes are used as solvents for pressure-sensitive recording materials [184].

4. Biphenyls and Polyphenyls

In the series of aromatic hydrocarbons with polyphenyls or polyaryls of the general formula $C_6H_5(C_6H_4)_nC_6H_5$, only the lowest members ($n = 0, 1$) are economically important. Higher polyphenyls ($n > 2$) are also known and may be isolated from coal tar or from aryl halides by the Ullmann reaction.

4.1. Biphenyl

[*92-52-4*], phenylbenzene, diphenyl, $C_{12}H_{10}$, $M_r 154.2$, was discovered in 1862 by FITTIG by the reaction of bromobenzene with sodium metal. BERTHELOT obtained the compound in 1867 by passing benzene vapor through a heated tube. In 1875 BÜCHNER demonstrated the presence of biphenyl in coal tar [185].

Table 18. Temperature dependence of some physical properties of biphenyl

t, °C	100	200	300	350
Vapor pressure, kPa		25.43	246.77	558.06
Liquid density, g/cm^3	0.97	0.889	0.801	0.751
Heat capacity, J/g	1.786	2.129	2.468	2.640

Physical Properties. Pure biphenyl is a white solid; when only slightly impure it has a yellow tint. It crystallizes from solution results as glossy plates or monoclinic prisms. Biphenyl is almost insoluble in water, but is easily soluble in organic solvents such as ethanol, diethyl ether, and benzene. The dipole moment and X-ray measurements demonstrate that the two benzene rings are coplanar in the solid state. In the melt, in solution, and in the vapor free rotation about the central C-C bond can take place; in substituted biphenyls, however, this rotation may be severely restricted, e.g., in dinitrodiphenic acid, leading to the existence of optically active isomers. The most important physical properties are listed below:

mp	69.2 °C [186]
bp (101.3 kPa)	255.2 °C [186]
d_4^{20}(solid)	1.041
d_4^{77}(liquid)	0.9896 [187]
n_D^{77}	1.5873
T_{crit}	515.7 °C [188]
p_{crit}	4.05 mPa
Flash point	113 °C [189]
Ignition temperature	570 °C
Heat of combustion	6243.2 kJ/mol
Heat of vaporization	53.9 kJ/mol
Heat of fusion	18.60 kJ/mol

The variation of some physical properties with temperature is given in Table 18.

Chemical Properties. Biphenyl is a very stable organic compound; in an inert atmosphere it remains unchanged even at a temperature >300 °C. The compound sublimes, distills with steam, and undergoes various substitution reactions; the reactivity is similar to, but less than that of benzene. Substitution reactions, e.g., chlorination, nitration, and sulfonation, occur at the 2- and 4-positions, the latter being the preferred site of attack in Friedel–Crafts reactions.

Production. *Hydrodealkylation of Benzene.* Biphenyl is currently obtained mainly as a byproduct in the production of benzene by thermal or catalytic hydrodealkylation of toluene (→Benzene) [190]–[193]. The reaction is carried out in a hot tube reactor at 700 °C with a hydrogen pressure of 4 MPa; the ratio of hydrogen to toluene is 4:1. Toluene conversion ranges from 35 to 85%, with methane being formed at 2.5 mol% above the stoichiometric.

$$\text{C}_6\text{H}_5\text{CH}_3 + \text{H}_2 \longrightarrow \text{C}_6\text{H}_6 + \text{CH}_4 + \text{C}_6\text{H}_5\text{-C}_6\text{H}_5$$

Biphenyl is obtained as the pot residue, following removal of gas and distillation of benzene and toluene; about 1 kg of biphenyl is obtained per 100 kg benzene. The biphenyl can be enriched to a purity of 93–97% by distillation.

Thermal Dehydrocondensation of Benzene. Until the early 1970s, biphenyl was produced exclusively by specific thermal dehydrocondensation of benzene [194]. This process yields a higher quality product than the toluene route, e.g., distilled material with a purity >99.5%; varying quantities of terphenyl isomers and several higher homologs are also obtained.

$$\text{C}_6\text{H}_6 \longrightarrow \text{C}_6\text{H}_5\text{-C}_6\text{H}_5 + \text{H}_2 + \text{[terphenyl/polyphenyl]}_n \quad n = 1, 2$$

The slightly endothermic dehydrocondensation ($\Delta H = +8$ kJ/mol [195]) of benzene to biphenyl and terphenyl has been investigated intensively [196], [197]. The formation of biphenyl from benzene starts at ca. 480 °C and increases with increasing temperature and residence time.

Industrially, the reaction is carried out in a multitube reactor, heated electrically [198] or by heat exchanger, at 700–850 °C using residence times of the order of seconds. An optimal tube length of 45–75 m and diameter of 100–120 mm have been reported [199]–[202]. None of the published procedures work in equilibrium because carbon black formation increases with increasing temperature and longer residence time.

The preferred residence time is ca. 10 s, which results in a benzene conversion of ca. 10% [200], [203]. By recycling the benzene, 1 kg of starting material may yield 0.8–0.85 kg biphenyl, 0.07–0.12 kg terphenyl, and some higher polyphenyls. The major technical problem in the production of biphenyl by dehydrocondensation of benzene is carbonization at the high temperature required [204]. The carbon deposit on the reactor tube walls and the release of large carbon flakes at bends or narrow passages within the tubes may lead to blockages. Countermeasures include addition of about 0.1% oxygen or sulfur compounds, such as methanol, ethanol, acetone, or carbon disulfide [205]–[207] to the benzene feedstock, the use of special alloys [208], [209], uniform heating of the reactor by means of circulating gases, and high feedstock velocity, and the use of turbulent flow [200]. A proposal for further improving the process is to carry out the reaction in fluidized beds. Quartz sand [210] or sintered corundum [211] is used as energy carrier in this process; these materials are recycled after the carbon deposits have been removed by burning them outside of the reactor.

Other Synthetic Routes. In addition to these two industrial processes for the production of biphenyl, other synthetic routes have been described:

1) Ullmann reaction: heating halogenobenzenes with copper powder [212].
2) Pyrolysis of coal tar in the presence of molybdenum oxide – alumina catalyst [213], [214].
3) Reduction of phenyldiazonium chloride solution with copper and zinc [215].
4) Heating phenylmagnesium halides with copper(II) salts [216].
5) Oxidative dehydrocondensation of benzene [217] – [222]. Benzene is oxidized to biphenyl by platinum- or palladium salts in glacial acetic acid at 120 °C; the salts are reduced to the pure metals, and stoichiometric quantities of the salts are required.
6) Dehydrocondensation of benzene in the presence of catalytic quantities of noble metal salts requires reoxidation of the metals during the course of the reaction. This may be done with pure oxygen, but a partial pressure of 6 MPa is required [223]. Additives such as β-diketones [224], heteropolyacids [225], or organic copper salts [226] are incapable of significantly lowering the required oxygen pressure, so that a potentially explosive regime cannot be avoided [225]. Yields of biphenyl are ca. 260 mol %, calculated on the palladium salt charge.
7) Hydrodimerization of benzene to phenylcyclohexane, followed by dehydrogenation to biphenyl [227] – [231]:

$$2\,C_6H_6 + 2\,H_2 \xrightarrow[\text{2–10 MPa}]{\text{250 °C}} \text{Ph–C}_6H_{11} \longrightarrow \text{Ph–Ph}$$
ca. 85 %

Suitable catalysts for this dimerization are noble metals on molecular sieve, alkali metals on alumina or carbon, and transition metals such as nickel, tungsten, or zinc on the zeolites faujasite or mordenite. A selectivity of up to 87 % has been claimed. Subsequent dehydrogenation is carried out at about 400 °C over $Pt-Al_2O_3$ [231], $Cr_2O_3-Na_2O-Al_2O_3$ [230], [232], $Cr_2O_3-K_2CO_3-Fe_2O_3$ [233], or $Cr_2O_3-MgO-Al_2O_3$ [234] giving yields of up to 99 %. Data concerning residence times over the catalyst are not available, either for the hydrodimerization or the dehydrogenation processes.

Analysis. Gas chromatography is used to analyze biphenyl and its derivatives.

Storage and Transportation. Biphenyl is marketed as a solid in the form of plates, flakes, or pellets, containing <10 ppm benzene and with a biphenyl content of >99.5 %; it is transported in bags or fiber drums; in the molten state (ca. 120 °C) in tank cars. It is not a regulated material for transportation, on account of its toxicity; there are no restrictions for transportation by sea, land, or air. The greatest hazards in handling biphenyl are the risk of dust explosions and the ignition of biphenyl vapor – air mixtures over the molten product.

Table 19. Physical properties of terphenyls

Property	o-Terphenyl [84-15-1]	m-Terphenyl [92-06-8]	p-Terphenyl [92-94-4]	Reference
mp, °C	59	87	212	[241]
bp (101.3 kPa), °C	332	365	376	[242]
Flash point, °C	171	206	210	[244]
T_{crit}, °C	613	603	621	[245]
P_{crit}, MPa	3.903	3.503	3.330	[246]
Heat of vaporization (at bp), kJ/mol	58.3	64.2	62.6	[242]
Vapor pressure, kPa				
93 °C	0.01172	0.00165		[241], [247]
204 °C	2.834	0.8274	0.789	
315.6 °C	64.4	27.3	24.56	
426.7 °C	439.9	240.6	174.4	

Uses. Biphenyl is an important heat transfer agent, because of its high thermal stability. It is used as a composite with diphenyl ether, especially as a melting point depressant. The eutectic mixture of 26.5 % biphenyl and 73.5 % diphenyl ether is marketed under the trade names Diphyl (Bayer), Dowtherm A (Dow), Thermic (ICI), Gilotherm (Rhône-Poulenc), Therm S 300 (Nippon Steel), Santotherm VP and Therminol VP-1 (both Monsanto) for use at up to 400 °C. The eutectic begins to boil at 256 °C; the vapor pressure at the maximum operating temperature (400 °C) is ca. 1.1 MPa.

A further large quantity of biphenyl is used as a dye carrier [235], [236]. Until a few years ago, polychlorinated biphenyls (PCBs), generated by chlorination of biphenyl, were produced in large quantities for use as nonflammable hydraulic fluids and as transformer dielectrics. Production has largely ceased on ecological considerations (→Chlorinated Hydrocarbons). Among the hydroxy derivatives, o- and p-hydroxybiphenyl and p,p'-dihydroxybiphenyl are used industrially. o-Hydroxybiphenyl is used as a preservative and fungicide. It is synthesized mainly from cyclohexanone (→Phenol Derivatives). p-Hydroxybiphenyl and p,p'-dihydroxybiphenyl are generated by sulfonation of biphenyl, followed by fusion with alkali.

Economic Aspects. According to a 1976 estimate, 67 % of the annual United States production of 40 000 t biphenyl was produced by dealkylation of toluene and 33 % by thermal dehydrocondensation of benzene [237].

The annual production for 1984 is estimated at 15 000 t, with an increasing proportion from dealkylation of toluene. The production capacities of the major producers, Monsanto, (United States), Bayer (Federal Republic of Germany), and Monsanto (United Kingdom) have not been disclosed. Based on the legal restrictions imposed on the use of PCBs, capacity should far exceed current demand.

Table 20. Terphenyl substitution pattern

	Position for substitution			
	First	Second	Third	Fourth
o-Terphenyl	4-	4''-	4'-	5'-
m-Terphenyl	4'-	4-	4''-	
p-Terphenyl	4-	4''-	2'-	

4.2. Terphenyls

Physical Properties. The three isomeric terphenyls, $C_{18}H_{14}$, M_r 230.29, are colorless to pale yellow crystalline solids [238]–[243]. Some physical properties are listed in Table 19.

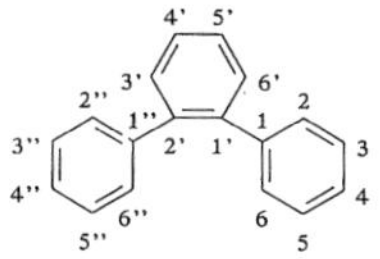

o-Terphenyl
1,1':2', 1''-Terphenyl

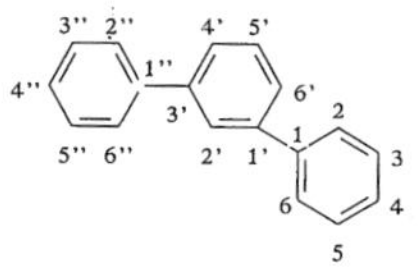

m-Terphenyl

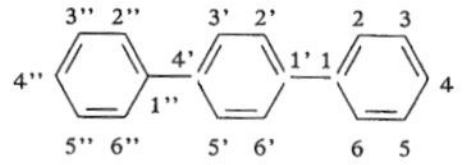

p-Terphenyl

Chemical Properties. Terphenyls, like biphenyls, are thermally extremely stable organic compounds, again offering a potential for use as heat transfer media. Terphenyls are sublimable, steam volatile, and show the typical substitution reactions of aromatic hydrocarbons, such as bromination and nitration. The substitution pattern for the preferred positions of the three terphenyls is shown in Table 20.

Production. Terphenyls are byproducts in the production of biphenyl by dehydrocondensation of benzene; they are found in the high-boiling fraction of the pyrolysis products. The pot residue has the following approximate composition [248]:

  3–8%   o-Terphenyl
  44%   m-Terphenyl
  24%   p-Terphenyl
  1.5%   Triphenylene
  22–27% Higher polyphenylenes and tar.

When, in the early 1960s, terphenyl was proposed to be used as coolant and moderator in nuclear reactors [249], numerous reports were published of attempts to enhance the yield of terphenyl in dehydrocondensation reactions. Increasing the temperature or residence time was ruled out because of increased carbonization. These procedures also enhanced the biphenyl fraction, so that the latter had to be recycled to the reactor [201], [205], [250]–[252]. Recycling a concentration of up to 30% biphenyl has been described [201]. A practical level appears to be using benzene with 10% biphenyl [218]. Attempts have also been made to recycle the extremely high melting *p*-terphenyl, after isolation from the reaction mixture, to obtain a lower melting terphenyl mixture [250]. The *o*-derivative is readily separated from the terphenyl isomer mixture by distillation; *m*- and *p*-terphenyl distill together and the pure isomers can be obtained by zone refining [253]. Terphenyl isomers, including hydrogenated terphenyl, can be analyzed by gas chromatography. The sole producer of terphenyl in the United States is Monsanto. The market amounts to several thousand tons per year. Terphenyl is available in Europe from Bayer.

Storage and Transportation. Solid terphenyl is shipped as flakes in laminated bags or in fiber drums, liquid, hydrogenated terphenyl in tank cars, steel drums, or barrels. Terphenyls and hydrogenated terphenyls possess a low toxicity, and are therefore not regarded as dangerous goods for the purpose of transportation.

Uses. Terphenyl is used as a heat transfer agent because of its excellent thermal stability; a mixture with a composition of 2–10% *o*-terphenyl, 45–49% *m*-terphenyl, 25–35% *p*-terphenyl, and 2–18% higher polyphenyls is used. This composition approximates the pot residue from the distillation of biphenyl. This mixture has *bp* ca. 360 °C; a disadvantage is the high *mp*, 145 °C. The largest fraction is partially hydrogenated to yield a clear oil, miscible with hydrocarbons and chlorinated hydrocarbons. Hydrogenation lowers the setting point and the viscosity but unfortunately also reduces the thermal stability. Partially hydrogenated terphenyl [254] is used mainly as a dye carrier for pressure sensitive recording materials and copy paper, and as a heat transfer oil. Heat transfer oils consisting of partially hydrogenated terphenyl isomers are marketed under the following trade names:

Santotherm 66 or 88, Therminol (Monsanto, United States), Gilotherm (Rhône-Poulenc, France), and Therm S 900 (Nippon Steel, Japan). Their usage range extends to 340 °C.

4.3. Polyphenyls

Polyphenyls with four or more phenyl residues have generally been synthesized as *p*-linked isomers and are used as scintillators [255]. They are economically unimportant.

A symmetrical hydrocarbon within this class is 1,3,5-triphenylbenzene, $C_{24}H_{18}$, *mp* 174 °C, prepared by condensation of three molecules of acetophenone.

5. Hydrocarbons from Coal Tar

Hydrocarbons from coal tar are generally recovered in large-scale operations by distillation and crystallization, after the polar, co-boiling components, phenols and nitrogen bases, have been separated by extraction (→Tar and Pitch). Supplies of coal tar as a raw material have been sufficient to date to meet the demand for these hydrocarbons, so that no other production processes, such as syntheses, have yet been developed to the stage of commerciality. Hydrocarbons from coal tar are used predominantly in the synthesis of dyes, plastics, and pharmaceuticals.

5.1. Acenaphthene

Acenaphthene [*83-32-9*], $C_{12}H_{10}$, M_r 154.21, was discovered in coal tar in 1867 by BERTHELOT.

Physical Properties. Acenaphthene forms colorless needles; readily soluble in chloroform, toluene, and hot ethanol; soluble in cold ethanol, hot diethyl ether, and benzene; insoluble in water.

mp	95.4 °C
bp (101.3 kPa)	277.5 °C
ϱ (25 °C)	1.2195 g/cm^3
n_D^{100}	1.6048
T_{crit}	530 °C
Heat of fusion	1.344×10^5 J/kg
Heat of vaporization (40 °C)	3.48×10^5 J/kg
Heat of combustion (25 °C)	4.033×10^7 J/kg

Chemical Properties. Acenaphthene reacts with halogens, preferentially at 3-, 5-, and 6- positions, or at the 1-position when irradiated with visible light. Nitration and sulfonation occur at the 3-, 5-, or 6-positions. Catalytic hydrogenation leads to tetra- and decahydroacenaphthene. Decahydroacenaphthene can be isomerized in the presence of Lewis acids to 1,3-dimethyladamantane. Oxidation of acenaphthene gives naphthalic acid or its anhydride, acenaphthenequinone, acenaphthenol, and acenaphthenone.

Production. High-temperature coal tar contains, on average, 0.3% of acenaphthene; in addition, it is formed under the conditions of coal tar distillation by hydrogenation of acenaphthylene, which is found in crude tar at a level of 2%. Acenaphthene is concentrated to ca. 25% in the coal tar fraction boiling between 230 and 290 °C (wash oil), which is recovered at ca. 7% in continuous tar distillation. The readily crystallizable acenaphthene fraction, boiling at 270–275 °C, is obtained from the wash oil by redistillation, or by removing it directly during primary tar distillation. From this, technical acenaphthene (95–99% pure) is produced by crystallization in agitated coolers and centrifuges, or by continuous counter-current crystallization [256], [259]. The pure compound is obtained by further distillation and recrystallization.

Uses. Acenaphthene is used on a large-scale for synthesis of naphthalic anhydride by gas-phase with air oxidation at 300–400 °C, in the presence of vanadium containing catalysts [258], [259]. Liquid-phase oxidation with chromate or air in the presence of cobalt- or manganese acetate at 200 °C [260], or cobalt resinate at 120 °C [261] produces naphthalic acid (→Carboxylic Acids, Aromatic). Production of naphthalene-1,4,5,8-tetracarboxylic acid from acenaphthene by condensation with malonic acid dinitrile [262] is at present not industrially important. A technically feasible procedure has been developed using 5-methyl-*peri*-acenaphthindene-7-one [*42937-13-3*] (by reaction of acenaphthene with diketene in HF), [263]. Naphthalic anhydride and naphthalene-1,4,5,8-tetracarboxylic acid are used as intermediates for production of perylene and perinone pigments [262]. Other valuable intermediates, particularly for the production of dyes and optical brighteners, and also for pharmaceuticals and pesticides, are produced from, e.g., 4-bromacenaphthene [264] (for 4-bromonaphthalic anhydride or naphthalic anhydride-4-sulfonic acid), from 5,6-dichloracenaphthene (for 4,5-dichloronaphthalic anhydride) [265] and from 3,5,6-trichloracenaphthene [266]. Nitration and oxidation of acenaphthene produce 4-nitronaphthalic anhydride, from which the fluorescent pigment Solvent Yellow 44 can be made [267]. Reaction of acenaphthalene with sodium cyanate in hydrofluoric acid gives acenaphthene-5,6-dicarboxylic acid imide, which can be used as an intermediate for dyes [268].

Bromination of acenaphthene in the presence of, e.g., iron(III)- or aluminum chloride, followed by further bromination by the addition of a radical initiator, such as azobisisobutyronitrile (AIBN), and subsequent dehydrobromination produces condensed bromacenaphthene, which is particularly well-suited for rendering plastics nonflammable and radiation-resistant [269]. Thermally stable plastics can be obtained from acenaphthene in the following ways: 1) by condensation with formaldehyde; 2) oxidation and reaction with an aromatic polyamine [270]; by cocondensation with phenol and formaldehyde [271]; and by condensation with terephthaloyl chloride [272].Acenaphthene also serves as a feedstock for the production of acenaphthenequinone, acenaphthene, and dimethyladamantane [273].

Derivatives.

Acenaphthenequinone   Acenaphthylene

Acenaphthenequinone [82-86-0], $C_{12}H_6O_2$, M_r 182.17; mp 261 °C; yellow needles; slightly soluble in ethanol; soluble in hot benzene and toluene. It is formed by liquid-phase oxidation of acenaphthene with hydrogen peroxide or dichromate in acetic acid, by chlorination of dibromacenaphthylene followed by acid hydrolysis, or by condensation of acenaphthene with ethyl nitrite and separation of the dioxime. Acenaphthenequinone is suitable for use as an insecticide and fungicide and can be converted into vat dyes.

Acenaphthylene [208-96-8], $C_{12}H_8$, M_r152.18; mp 92–93 °C; yellow prisms; soluble in ethanol, diethyl ether, and benzene; insoluble in water. It is found in coal tar at a concentration of 2% and can be produced industrially by catalytic dehydrogenation of acenaphthene in the gas phase. Thermal, ionic, radical, or radiochemically induced polymerization produces polyacenaphthylene [274]. Polyacenaphthylene and copolymers with other olefins are characterized by high thermal and mechanical stability together with good electrical insulation properties. The thermal stability of polystyrene can be significantly increased by copolymerisation with acenaphthylene [274]. Polymerisation of acenaphthylene with acetylene in the presence of a Lewis acid catalyst gives electrically conductive polymers; they are used for electronic engineering and for the antistatic finishing of plastics [275]. Thermosetting resins with good resistance to heat and chemicals can be obtained by cocondensation with phenol and formaldehyde [271]. Acenaphthylene possesses excellent properties as an antioxidant in crosslinked polyethylene and ethylene–propylene rubber [274]. Thermal trimerization of acenaphthylene leads to decacyclene, which can be further processed to sulfur dyes [276].

5.2. Indene

Indene [95-13-6], C_9H_8, M_r116.16, was discovered in coal tar in 1890 by KRAEMER and SPILKER.

Physical Properties. Indene is a colorless liquid; insoluble in water; soluble in ethanol; it is miscible with diethyl ether, naphtha, pyridine, carbon tetrachloride, chloroform, and carbon disulfide.

mp	−1.5 °C
bp (101.3 kPa)	183 °C
ϱ (20 °C)	0.9966 g/cm^3
n_D^{20}	1.5763
Heat capacity (25 °C)	1.610 × 10^3 J/kg
Heat of fusion	8.33 × 10^4 J/kg
Heat of combustion (25 °C)	4.132 × 10^7 J/kg
Dipole moment (in CCl$_4$)	0.85 D

Chemical Properties. Indene polymerizes readily at ambient temperature and in the dark. Polymerization is accelerated by heat, acid, and Friedel–Crafts catalysts. Depending on reaction conditions, oxidation leads to dihydroxyindan, homophthalic acid, or phthalic acid. Hydrogenation produces indan. Halogenation occurs at the 2,3-double bond. Diene adducts are formed with maleic anhydride and hexachlorocyclopentadiene.

Production. Indene is found in high-temperature coal tar at an average concentration of 1%. Pyrolysis residual oils from olefin production contain varying amounts of indene [277], which can be separated by extractive distillation with N-methyl-2-pyrrolidone [278] or by crystallization [279]. Indene is recovered industrially from dephenolated and debased tar light oil by rectifying distillation followed by crystallization. When the phenol extraction is omitted a highly concentrated indene fraction can also be obtained by azeotropic distillation of an indene–phenol fraction with water, whereby the phenol is separated as a bottom product [280].

Uses. Indene is used as a comonomer in coal-tar derived indene–coumarone and petroleum-derived aromatic hydrocarbon resins. Technical pure indene is used for the production of indan and its derivatives. Esters of indene-1-carboxylic acid, which are effective as acaricides, can be synthesized from indene for use in, e.g., tick collars [281].

Derivatives.

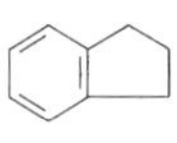

Indan 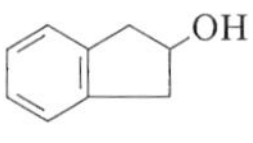   2-Indanol 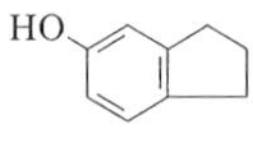   5-Indanol

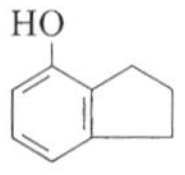

4-Indanol

Indan [496-11-7], hydrindene, C$_9$H$_{10}$, M_r 118.18;

mp	−51.4 °C
bp (1013 kPa)	176 °C
ϱ (20 °C)	0.9644 g/cm^3

Indan is a colorless liquid; insoluble in water; soluble in ethanol and diethyl ether. It is found at a concentration of 0.1% in coal tar, but is mainly produced by catalytic hydrogenation of indene. Alkylated indans are used in the production of synthetic lubricants. Indan is used for the synthesis of indanols (→Phenol Derivatives).

2-Indanol [*4254-29-9*], $C_9H_{10}O$, M_r 134.18; *mp* 69 °C; colorless crystals. It is obtained by hydrogenation of 2-indanone prepared by reaction of indene with hydrogen peroxide and boiling in sulfuric acid over a platinum catalyst in ethanol. 2-indanol is used in the synthesis of the coronary therapeutic, Aprindine [*37640-71-4*], *N*-(3-diethylaminopropyl)-*N*-indan-2-ylaniline; [282] and can be employed as an intermediate for other pharmaceuticals, e.g., for antihypertonics [283].

5-Indanol [*1470-94-6*], $C_9H_{10}O$, M_r 134.18; *mp* 55 °C; colorless crystals. It is produced by fusion of indan-5-sulfonic acid with potassium hydroxide [284]. 5-Indanol is used in the synthesis of the multi-purpose antibiotic, Carindacillin [*26605-69-6*] (6-[2-phenyl-2-(5-indanyloxycarbonyl)acetamido]penicillanic acid; [285]. 5-Indanylesters of substituted picolinic acids can be used as antihypertonic [286]. 5-Indanol itself may be used as an antidandruff additive in hair shampoos [287].

4-Indanol [*1641-41-4*], $C_9H_{10}O$, M_r 134.18; *mp* 49 – 50 °C; colorless crystals. Analogous to 5-indanol it is synthesized by potassium hydroxide fusion of indan-4-sulfonic acid, which is produced as a byproduct in indan sulfonation. Catalytic gas-phase amination of 4-indanol produces 4-aminoindan [288], which is required as an intermediate, e.g., for pharmaceuticals. Clorindanol (7-chloro-4-indanol) is used as an antiseptic.

5.3. Fluorene

Fluorene [*244-36-0*], $C_{13}H_{10}$, M_r 166.22, was discovered in coal tar in 1867 by BERTHELOT.

Physical Properties. Fluorene forms colorless flakes with slight violet fluorescence; sublimes readily; insoluble in water; soluble in benzene, diethyl ether, carbon disulfide, and hot ethanol; soluble with difficulty in cold ethanol.

mp	115 °C
bp (101.3 kPa)	298 °C
ϱ (20 °C)	1.181 g/cm^3
Heat of fusion	1.21×10^5 J/kg
Heat of combustion (25 °C)	3.994×10^7 J/kg

Chemical Properties. Fluorene is capable of numerous chemical reactions, through both the aromatic rings and especially the reactive methylene group. Oxidation produces fluorenone. Reaction with dialkyl carbonate and alkali metal hydride or alcoholate, followed by neutralization and saponification of the resulting ester produces fluorene-9-carboxylic acid; the latter is also produced by metalation with butyl lithium followed by carboxylation. Nitration leads predominantly to 2-nitrofluorene, chloromethylation to 2-chlormethylfluorene. Halogenation, depending on the reaction conditions, gives the mono-, di-, or tri-haloderivatives, substitution taking place at the 2-, 2,7-, or 2,4,7-positions; in the presence of catalysts waxy products containing ca. 50 wt% chlorine are formed.

Production. High-temperature coal tar contains on average, 2% fluorene. The latter is recovered industrially by redistillation of coal tar wash oil or by direct removal of a fluorene fraction during primary distillation of coal tar, followed by recrystallization of the fluorene fraction (dephenolated and debased if necessary) e.g., from solvent naphtha or naphtha [289]. Fluorene can be isolated from a $\geq 60\%$ fluorene fraction by continuous counter-current crystallization [290]. It is possible to refine the technical product further via the sodium compound (by reaction with sodium or sodium amide), or by sulfuric acid purification and crystallization from methanol.

Uses. Fluorene is the starting material for production of fluorenone and fluorene-9-carboxylic acid. Fluorene-9-carboxylic acid is an intermediate in the production of antidiabetic and antiarrhythmic drugs, and plant growth regulators [291]. Plant growth regulators are also obtained by reaction of fluorene with phthalic anhydride [292]. Fluorene-2-acetic acid (by acylation of fluorene) can be used as a precursor in the synthesis of anti-inflammatory drugs [293]. Reaction of fluorene with sulfur produces electrically conductive polymers [294]. The azine dye Sirius Light Violet FRL is produced from 2-aminofluorene. The fluorene analog of malachite green dye is obtainable from 3,6-bis-(dimethylaminofluorene) [295]. 2,7-Diiodofluorene can be employed in the production of styryl dyes for optical brighteners [296]; heptabromofluorene can be used as a fire-retardant for plastics [297]. Fluorene itself can be used in admixture with biphenyl and phenyltoluene as a carrier for dyeing polyester fibers [298].

Derivatives.

9-Fluorenone [486-25-9], $C_{13}H_8O$, M_r 180.21;

mp	85–86 °C
bp 101.3 kPa)	341.5 °C
ϱ (99.4 °C)	1.130 g/cm^3
n_D (99.4 °C)	1.6369

9-Fluorenone forms yellow crystals; insoluble in water; soluble in hot sulfuric acid; very readily soluble in ethanol and diethylether; volatile in steam. Fluorenone can be produced by catalytic oxidation of fluorene [261], [299]–[302], or of fluorene fractions in the presence of a quarternary ammonium salt [303], or by catalytic oxidative cracking (oxicracking) of suitable aromatic [304].

Di- and trinitrofluorenones are synthesized from fluorenone for use as electron acceptors in electrophotography, e.g., in copying systems [305]. 2,4,7-Trinitrofluorenone is used as an analytical reagent for polynuclear aromatic hydrocarbons. Fluorenone can be used as starting material for the synthesis of imidazolylfluorene salts with antimycotic activity [306]. 9,9-Bis(4-hydroxylphenyl-)fluorene (from fluorenone by reaction with phenol [307]) can be used as starting material for thermally stable plastics [308]. 9,9-Bis(4-methylaminophenyl)fluorene (from fluorenone by reaction with phenyl methyl amine) has been suggested as a hardener for epoxy resins [309]. Fluorenone is also suitable as an oxidizing agent in the Oppenauer reaction.

5.4. Fluoranthene

Fluoranthene [206-44-0], $C_{16}H_{10}$, M_r 206.26, was isolated from coal tar by FITTIG and GEBHARDT in 1878.

Physical Properties. Colorless crystals with light blue fluorescence; insoluble in water; readily soluble in diethyl ether, boiling ethanol, chloroform, carbon disulfide, and glacial acetic acid.

mp	110 °C
bp (101.3 kPa)	384 °C
ϱ (20 °C)	1.236 g/cm^3
Heat of vaporization	3.29×10^5 J/kg
Heat of combustion (25 °C)	3.913×10^7 J/kg

Chemical Properties. Oxidation of fluoranthene with chromic acid leads to fluorenone-1-carboxylic acid via fluoranthene-2,3-quinone; hydrogenation leads to 1,2,3,10 b-tetrahydrofluoranthene, and to perhydrofluoranthene via decahydrofluoranthene. 2,3-Dihydrofluoranthene is obtained by metalation with sodium. Halogenation, nitration, and sulfonation take place predominantly at the 4-position. Condensation with phthalic anhydride in the presence of aluminum chloride gives a mixture of 7,8- and 8,9-phthaloyl fluoranthene.

Production. Fluoranthene belongs to the main constituents of high-temperature coal tar, which contains, on average, 3.3% of fluoranthene. It is recovered from the fluoranthene fraction, which boils at 373–385 °C, and is obtained from the high-boiling anthracene oil II and from pitch distillate by redistillation. Subsequent recrystallization of the fluoranthene fraction from solvent naphtha gives 95% pure fluoranthene. Further refining is carried out by recrystallization from xylene with simultaneous partial sulfonation of the impurities with ca. 1% of concentrated sulfuric acid, or by recrystallization from a pyridine–water mixture.

Uses. Fluoranthene is used in the production of fluorescent dyes. Yellow vat dyes are obtained by condensation with phthalic anhydride, via phthaloyl fluoranthenes, whereas bromination [310] and condensation with 1-amino-4-benzoyl-amino anthraquinone produces olive dyes [311]. 2-Benzylfluoranthene (from 2,3-dihydrofluoranthene [312] and benzaldehyde) can also be used for the production of dye intermediates [313]. Fluoranthene and alkyl fluoranthenes can be used as stabilizers in epoxy resin adhesives [314], as additives in electrically-insulating epoxy resin compositions [315], and in electrically-insulating oils [316]. 2,3-Dihydrofluoranthene and 1,2,3,10 b-tetrahydrofluoranthene [317] are used for numerous derivatives, some of which (e.g., their derivatives of phenylethyl amine) are of pharmaceutical interest. 7-Fluoranthenyl aminoalcohols (for cytostatics, anthelmintics, or bactericides [318]) and bis-aminoketones, e.g., 3,8-bis(4-piperidinobutyryl)fluoranthene (as antiviral agents [319]) can be synthesized from fluoranthene.

5.5. Phenanthrene

Phenanthrene [85-01-8], $C_{14}H_{10}$, M_r 178.24 was discovered in coal tar by FITTIG and OSTERMAYER in 1872.

Physical Properties. Colorless flakes with weak blue-violet fluorescence; readily sublimable; insoluble in water; slightly soluble in ethanol; readily soluble in diethyl ether, benzene, carbon disulfide, and glacial acetic acid.

mp	101 °C
bp (101.3 kPa)	338.5 °C
ϱ (20 °C)	1.172 g/cm^3
Heat capacity (25 °C)	1.361×10^3 J/kg
Heat of fusion	1.04×10^5 J/kg
Heat of combustion (25 °C)	3.982×10^7 J/kg

Chemical Properties. Phenanthrene can be oxidized to phenanthrenequinone, diphenic acid, or to phthalic acid, depending on the oxidant. Hydrogenation produces 9,10-di-, tetra-, octa-, or perhydrophenanthrene. Phenanthrene is halogenated predominantly in the 9,10-positions, and nitrated in the 9-position. Sulfonation leads to mixtures of 2-, 3-, and 9-phenanthrene sulfonic acids.

Production. Phenanthrene, at a concentration of 5%, is the second most important coal tar constituent in terms of quantity after naphthalene. During primary distillation of coal tar, it is concentrated in the anthracene oil fraction. After crystallization of the anthracene residues the phenanthrene is recovered as a fraction from the filtrate of this crystallization, or from the top fraction of crude anthracene distillation, by redistillation. Technically pure grades of phenanthrene are obtained by sulfuric acid refining and recrystallization from methanol, or by repeated rectification of the phenanthrene fraction. The accompanying substances can be separated either by partial sulfonation, or by partial condensation with formaldehyde and hydrogen chloride. The most persistent accompanying substance, diphenylene sulfide, can be completely removed by treating the melt with sodium and maleic anhydride.

Uses. Phenanthrene forms the basis for production of 9,10-phenanthrenequinone and 2,2'-diphenic acid (see Chap. 4) [320]. It can be used to synthesize anthracene, via the isomerization product of *sym*-octahydrophenanthrene. Electrically conductive substances, e.g., for use in batteries and solar cells, can be produced by the electrochemical conversion of phenanthrene diazonium salts in a solvent containing a conductive salt, and subsequent doping with various ions (e.g., Na^+, Ba^{2+}, H^+, etc.) [321]. Liquid-crystalline 7-*n*-alkyl-9,10-dihydrophenanthrene-2-carboxylic acid ester, used for optical-electronic applications, can be synthesized from 9,10-dihydrophenanthrene [322]. By crosslinking with *p*-xylene glycol and 4-toluenesulfonic acid, polycondensed thermosetting resins are obtained for composites or temperature-resistant, electrically insulating coatings [323]. A polyamide–polyimide resin can be produced by oxidation of phenanthrene to phenanthrene-9,10-quinone and -9,10-diol, condensation with formaldehyde, oxidation to the polycarboxylic acid, formation of the anhydride and finally reaction with an aromatic diamine. This resin is suitable for use in high-temperature insulators, printed circuit boards, and laminates [324]. Phenanthrene has been proposed as a plasticizer for plastics and molding compounds [325], phenanthrene and alkylphenanthrenes have been suggested as stabilizers for mineral oil products [326], [327]. Deep khaki vat dyes are prepared by heating a mixture of phenanthrene and anthracene with sulfur [328]. [(Phenanthrylmethyl)amino]-propandiol, which possesses antitumor activity, can be synthesized via phenanthrene-9-carboxaldehyde [329].

Derivatives.

Phenanthrene-9-10-quinone [*84-11-7*], $C_{14}H_8O_2$, M_r 208.91; *mp* 217 °C; orange needles; poorly soluble in water; slightly soluble in ethanol; soluble in boiling water and in diethyl ether; readily soluble in hot glacial acetic acid. Phenanthrene-9,10-quinone is produced by liquid-phase oxidation of phenanthrene with chromates [330], catalytically with oxygen [331], with *tert*-butylhydroperoxide [332], [333] or via phenanthrene-9,10-oxide with nitric acid or hypochlorite [334], [335]. Further oxidation transforms the quinone into 9-hydroxyfluorene-9-carboxylic acid, which is used in the form of its salts or esters (in combination with the herbicide methyl chlorophenoxyacetic acid (MCPA), trade name: Aniten) or as the 2-chloro derivative (trade name: Maintain) as plant growth regulators [336], [337]. Phenanthrene-9,10-quinone has been suggested as an additive for photographic or electrophotographic applications [308], [338], [339], in UV-curable coatings and adhesives [340], [341] and in production of cellulose by wood pulping [342]. An intermediate for azo pigments can be obtained by condensation with hydrazinobenzoic acid [343]. Condensation with aromatic amines gives intermediates for pharmaceuticals (e.g., immuno-suppressives) or fungicides [344], [345].

5.6. Pyrene

Pyrene [*129-00-0*], $C_{16}H_{10}$, M_r 202.26, was found in coal tar by GRAEBE in 1871.

Physical Properties. Colorless crystals with blue fluorescence; insoluble in water; very readily soluble in diethyl ether, benzene, and carbon disulfide; poorly soluble in ethanol.

mp	150 °C
bp (101.3 kPa)	393.5 °C
ϱ (23 °C)	1.271 g/cm^3
Heat capacity (18 °C)	1.126×10^7 J/kg
Heat of vaporization	3.21×10^5 J/kg
Heat of combustion (25 °C)	3.878×10^7 J/kg
Dielectric constant (17 °C)	3.21

Chemical Properties. Pyrene is oxidized by mild oxidants such as chromic acid to pyrene-1,6- and 1,8-quinone; further oxidation leads to perinaphthenone [548-39-0] and napthalene-1,4,5,8-tetracarboxylic acid. Hydrogenation produces, depending on the reaction conditions, ditetra-, hexa- or decahydropyrenes; halogenation products substituted in the 1-, 1,6-, 1,8-, 1,3,6- and 1,3,6,8-positions. The substitution pattern for halogenation can be changed in favor of the 4-position by Diels–Alder reaction with hexachlorocyclopentadiene [77-47-4] in the presence of iron powder or thallium acetate, followed by retro-Diels–Alder reaction. With nitration and sulfonation the 1-, 1,6-, and 1,3,6,8-substitution products are formed. Pyrene can be condensed with phthalic anhydride by Friedel–Crafts reaction to give diphthaloyl pyrenes, and with benzoyl chloride to the dye pyranthrone [128-70-1].

Production. High-temperature coal tar contains an average of ca. 2% of pyrene. It is recovered from a fraction crystallizing above 110 °C, which is obtained by redistillation of the high-boiling anthracene oil II or pitch distillate. Pure pyrene is produced by recrystallization, e.g., from solvent naphtha [346] or by fractional crystallization from the melt [256], dephenolation and debasing, and by refining with 80% sulfuric acid. As an alternative to sulfuric acid refining, the pyrene-accompanying brasane (2,3-benzodiphenylene oxide) can also be separated by recrystallization from xylene by adding iron(III) chloride. Traces of tetracene are removed by reaction with maleic anhydride.

Uses. The dye intermediate naphthalene-1,4,5,8-tetracarboxylic acid [128-97-2] can be produced by halogenation of pyrene [347], [348] (halogenation of the pyrene fraction is also possible [349]) followed by oxidation (→ Carboxylic Acids, Aromatic), [350]. Other pyrene-based dyes are Sirius Light Blue F 3 GL from 3-aminopyrene and the green fluorescent dye, pyranin (Solvent Green 7), from pyrene-1,3,6,8-sulfonic acid [351]. Anthraquinone dyes (e.g., pyranthrone) may be obtained from dibenzoyl pyrene or diphthaloyl pyrene. Optical brighteners may be synthesized by reaction of pyrene with a complex of cyanuric chloride and aluminum chloride [352]. By analogy to fluoranthene, pyrene and alkylpyrenes can be used as additives in electro-insulating oils [316], and in epoxy resins for electrical-insulation [315]. In a similar manner to phenanthrene, thermosetting resins may be formed with *p*-xylene glycol and 4-toluenesulfonic acid [323]. Condensation products of 1-bromopyrene or nitropyrene with formaldehyde are used as photoconductive substances in electrophotography [305], [353], [354]. 1,2,3,6,7,8-Hexahydropyrene is effective as a synergist for dialkylsulfide antioxidants in lubricants [355]. On the basis of pyrene, pyrenyl aminoalcohols can be synthesized as cytostatics, anthelmintics, or bactericides [356]. Pyrene itself can serve as an electron donor to increase the blackness in pencil leads [357].

6. Toxicology and Occupational Health

6.1. Alkanes

Methane is toxicologically virtually inert. In very high concentrations (80–90 vol%) it causes respiratory arrest [358]. The bulk of an inhaled dose is exhaled unchanged [359].

Ethane easily causes dyspnea in laboratory animals, beginning at concentrations of approximately 2–5 vol% in the air inhaled [360]. At high concentration respiratory arrest occurs [361]. A slight sensitization of the myocard to catecholamines has been described at a concentration of 15–19 vol% [362].

Propane has narcotic properties [361]. It is neither a skin nor a mucosal irritant. At very high concentration it also causes respiratory arrest [363]. In dogs adverse effects on circulatory function were found with inspiratory concentrations of 1% or greater. A negative inotropic effect, a drop in the mean aortic pressure, and an increase in pulmonary vascular resistance were found [364].

MAK (1987): 1000 ppm.

Butane causes slight drowsiness beginning at a concentration of 10 000 ppm in the air inhaled [365]. In the mouse the inhalatory LC_{50} is 680 g/m^3 [366]. It has been proven in the dog that butane also sensitizes the myocardium to the effects of catecholamines [364].

MAK (1987): 1000 ppm.

n-Pentane has a less pronounced narcotic effect than the C_1–C_4 hydrocarbons. In humans the lowest lethal dose after acute inhalatory exposure is apparently 130 000 ppm, the lowest toxic dose 90 000 ppm [367]. Pentane apparently possesses a neurotoxic effect, although it is not very pronounced [368], [369].

n-Hexane possesses a marked neurotoxic effect that distinguishes it from the other alkanes. The causal factor for peripheral neuropathies is the oxidation product (cytochrome P 450-dependent oxidases) 2,5-hexadione. Toxicity manifests itself clinically in polyneuropathies [370], [371]. Hexane is also a skin irritant. The LC_{50} (inhalation mice) is 120 g/m^3 [372]. In rats the acute oral LD_{50} is 24–49 mL/kg [373]. Hexane is absorbed through the skin. Dermal exposure can lead to poisoning (LD dermal in rabbits ca. 5 mL/kg) [374]. In rats subchronic exposure to a concentration of 400–600 ppm leads to neuropathies [375], which are characterized by degeneration of both myelin sheaths and axons [376]–[378].

MAK (1987): 50 ppm.

n-Heptane has narcotic properties. In humans 0.1% in the inhaled air leads to dizziness and 0.5% to equilibrial disturbances with loss of motor coordination [365]. Within 3 min 4.8 vol% in the air inhaled leads to asphyxia [379]. Like hexane, heptane is a skin and mucosal irritant [365]. The biotransformation of *n*-heptane, as with hexane, is by oxidation. Heptane apparently also possesses neurotoxic properties [380]. Myocardial sensitization to catecholamines has also been found [381].

MAK (1987): 500 ppm.

n-Octane is similar to *n*-heptane in regard to its narcotic effect; however, it does not seem to cause any other effects on the nervous system [372]. Octane is also metabolized by oxidation [382].

MAK (1987): 500 ppm

Only limited toxicological data are available on the higher molecular mass alkanes.

6.2. Alkenes

The alkenes under consideration are toxicologically not very active. Higher molecular mass compounds possess narcotic properties. The alkenes are not neurotoxic. The α-olefins generally seem to be more reactive and toxic than the β-isomers.

6.3. Alkylbenzenes

Trimethylbenzenes. The toxicological data available on trimethylbenzenes is mainly rather old. The acute inhalatory toxic dose for the 1,2,4- and the 1,3,5-isomers is in the range 7000–9000 ppm [383], [384].

As a result of their occurrence in automotive fuel and heating oil, low concentrations of trimethylbenzenes can be demonstrated in air and water. For example, measurements in 1974 in a tunnel in Rotterdam indicated a level of 0.015 ppm pseudocumene; see [385] for more details. An analysis of the drinking water of Cincinnati, Ohio, in 1980, revealed a concentration of 45 ng/L hemimellitene, 127 ng/L pseudocumene, and 36 ng/L mesitylene [386]. Trimethylbenzenes are toxic to aquatic organisms; LC_{50} values between 10 and 100 ppm have been reported [387].

The TLV (TWA) level for trimethylbenzene is 25 ppm [388]. The responsible commission of the Federal German Research Council (Deutsche Forschungsgemeinschaft) has, so far, refrained from publishing an MAK value for mesitylene because of insufficient experience with humans or animal experiments [389]. The United States toxicity assigment ranges from low to moderately toxic for pseudocumene, to highly toxic for the isomer mixture; disturbances in the central nervous system and abnormal blood pictures have been reported [387].

Tetramethylbenzenes. The oral toxicity of tetramethylbenzenes is low; LD_{50} > 5000 mg/kg (rat, oral). Durene is additionally classified as intravenously highly toxic; LD_{50} 180 mg/kg (i.v., mice). Tetramethylbenzenes cause mild skin reactions [387]. Two values are cited for the odor threshold concentration of durene: 0.083 and 0.087 mg/m^3 [385].

Hexamethylbenzene. Hexamethylbenzene has low oral toxicity; LDLo 5000 mg/kg (rat). It is, however, suspected of causing neoplastic effects. Addition of nitromethane to hexamethylbenzene can lead to an explosive reaction [387].

Diethylbenzenes. Diethylbenzene is more toxic than monoethylbenzene ($\to$ Ethylbenzene) [390]. DEB is emitted into the environment from engine fuel and heating oil [386]. Diethylbenzenes are toxic to aquatic organisms; the LC_{50} is between 10 and 100 mg/L [387]. In the cell multiplication inhibition test with protozoa, the toxicity threshold is <10 mg/L [385]. In the Federal Republic of Germany DEB is classified as an aquatic hazard; aquatic hazard class 2. DEB causes mild to moderate irritation of the eye and mucous membranes. The oral toxicity is low: LDLo 5000 mg/kg (rat) [387]. The vapor possesses an anesthetic action and may cause headache, vertigo, or vomiting; the odor threshold concentration is <10 ppm [391].

Triethylbenzene. The LC_{50} of triethylbenzene (aquatic organisms) is between 100 and 1000 mg/L. The oral toxicity is low: LDLo 5000 mg/kg (rat). Implant experiments with hexaethylbenzene in mice have raised the suspicion that the compound may be carcinogenic [387].

Ethyltoluene. Ethyltoluene can, as with aromatic hydrocarbons, be detected in the environment [385]. The oral toxicity of 2- and 4-ethyltoluene is low: LDLo 5000 mg/kg (rat) [387].

Cumene. Cumene appears to be orally as nontoxic as propylbenzene, it is even less so by inhalation [383], [392], [393]. Cumene has a prolonged depressant effect on the central nervous system [394].

The MAK for cumene has been set at 245 mg/m^3, corresponding to a cumene vapor concentration in air of 50 mL/m^3. The vapor pressure of cumene at 20 °C is 0.5 kPa. There is a risk of cumene being absorbed through the skin [395]. Inhalation of cumene vapor leads to the delayed appearance of a long lasting narcotic effect [396].

6.4. Biphenyls and Polyphenyls

Biphenyls. Biphenyl dust or vapor is irritating to the eye and mucous membrane at a concentration as low as 3–4 ppm [390], and to the skin after extended exposure. A concentration of >5 mg/m^3 for long periods is considered a health hazard; systemic toxic effects were elicited in humans by a concentration with inhalatory exposure maxima of 128 ppm [400], [401]. The MAK is 0.2 ppm; the olfactory threshold is ca. 0.06–0.3 mg/m^3. Some toxicity data are listed below [397]–[399]:

LD_{50} (rat, oral)	3280 mg/kg
LD_{50} (rabbit, oral)	2400 mg/kg
LD_{50} (rabbit, dermal)	2500 mg/kg
TL_m (fathead minnow, 96 h)	1.5 mg/L

Triphenyls are only minimally toxic [402].

Terphenyls. Toxicologically, terphenyl isomers should be treated like biphenyl. The oral LD_{50} is ca. 4.6 – 4.7 g/kg [403]. The toxicity of partially (40%) hydrogenated terphenyl (LD_{50} 17.5 g/kg) is much less than that of the fully aromatic terphenyls [404] – [407].

6.5. Hydrocarbons from Coal Tar

The aromatic hydrocarbons under consideration are usually found in complex mixtures of various other polycyclic aromatic hydrocarbons (PAH), which originate from combustion emissions and industrial processes. For this reason toxic phenomena possibly caused by exposure to humans are often unlikely to be attributed to one single aromatic compound. Incorporation of these compounds into the body occurs predominantly by way of the respiratory tract. Skin contact is important with occupational exposure.

6.5.1. Biological Effects

6.5.1.1. Carcinogenicity and Mutagenicity

The aromatic compounds under discussion are considered as noncarcinogens, but pyrene and fluoranthene are able to enhance the carcinogenic potential of benzo[a]pyrene [50-32-8] when applied simultaneously to mouse skin (cocarcinogenic effect) [408]. Phenanthrene resulted in just one positive tumor-initiation test in a series of carcinogenicity studies [408], [409]. Acenaphthene, known as an inhibitor of mitosis in plant cells [410], induced the initial phase of bronchocarcinoma in rats after an inhalation period (4 months) of 4 hours per day with 0.5 – 1.25 mg/m^3 [411], but was negative in repeated topical application to mouse skin [412]. With their low activity in in vivo and in vitro assays, there is limited evidence that phenanthrene and pyrene are mutagenic [408], [409], [413] – [415]. As for fluoranthene, several more recent positive results indicate a mutagenic potential [408], [416] – [419].

So far, tests on fluorene have given negative results [408].

6.5.1.2. Mammalian Toxicity and Toxicokinetics

The acute toxicity of the aromatic compounds under consideration is low: LD_{50} (oral and dermal) is 700 – 2700 mg/kg body weight in rodents [408], [409], [411], [413]; LD_{50} (rat, oral) for acenaphthene is 10 000 mg/kg [421]. The LC_{50} (inhalation) of pyrene is reported to be 170 mg/m^3 [413]. No fatalities occurred in rats exposed to indene vapor (800 – 900 ppm) for six inhalation periods of 7.5 h each; however, systemic pathological changes in the vascular system and several organs were observed [412]. Longterm inhalation (4 months) of acenaphthene (0.5–1.25 mg/m^3) [411], [420] or pyrene (0.3–3.6 mg/m^3) [408], [409], [413] caused local irritation and systemic pathological effects in several organs and the blood of rodents. In direct contact at high concentration, the compounds under consideration are irritating to the skin and mucous membranes; the tri- and polycyclic aromatics especially are photosensitizing [408]–[410], [412], [413]. The compounds under consideration are absorbed on inhalation or oral exposure, and partly leave the body unchanged in the feces and by exhalation [408], [409], [412], [413]. After metabolic conversion to more water-soluble intermediates (hydroxylation and conjugation), they are excreted into the bile or the urine. The half-life of pyrene is estimated to be 24–48 h in rats and swine [413].

6.5.1.3. Ecotoxicology

Several studies have shown that the aromatic compounds, in question are microbially decomposed [422]–[424], although the microbial conversion of indene has not yet been described. These compounds tend to be enriched in sediments, sludges, and aquatic organisms because of their low water solubility. As a result of microbial degradation, biotransformation, and photochemical decomposition, however, accumulation is limited and checked. Nonetheless, a concentration in water of the order of their low solubility (<0.2–4.0 mg/L) may have adverse effects on aquatic life, as shown for acenaphthene [425], phenanthrene, and pyrene [426], [427].

6.5.2. Safety Regulations

A threshold limit value (TLV) at the work place exists only for indene: 10 ppm (45 mg/m^3) (ACGIH, United States 1980). No other TLVs have so far been established. The United States standard for exposure to coal tar volatiles (0.2 mg/m^3) was recommended to be used in case of pyrene-containing PAH mixtures, but a maximum level of 0.1 mg/m^3 should be applied in the case of exposure to pure pyrene [413]. With the exception of indene all aromatics are included in the recommended United States list of priority pollutants (United States EPA, 1977) [428].

7. References

General References

[1]   F. Asinger: *Die Petrolchemische Industrie,* vol. 1 (Gewinnung der Olefine), vol. 2 (Verwendung der Olefine) Akademie-Verlag, Berlin 1971.

[2]   J. Falbe (ed.): *Methodium Chimicum,* vol. **4**, Thieme Verlag, Stuttgart 1980, pp. 35, 111.

[3]   *Houben-Weyl,* 4th ed., vol. **5/1 b**, Thieme Verlag, Stuttgart 1972.

[4] S. Patai, J. Zabicky: *The chemistry of alkenes*, vol. **1** (1964), vol. **2** (1970), Interscience Publishers, New York.
[5] E. S. Strauss, G. Hollis, O. Kamatari: *Chemical Economics Handbook Marketing Research Report* "Linear alpha-olefins," Stanford Research Institute, Menlo Park, CA, 1984.
[6] B. Fell, *Tenside Deterg.* **12** (1975) 3.
[7] H. Isa, *CEER Chem. Econ. Eng. Rev.*, **9** (1977) 26.
[8] A. H. Turner, *J. Am. Oil Chem. Soc.* **60** (1983) 623.
[9] A. Behr, W. Keim, *Arabian J. Sci. Eng.* **10** (1985) 377.
[10] H.-G. Franck, G. Collin: *Steinkohlenteer*, Springer Verlag, Berlin – Heidelberg – New York 1968.
[11] H.-G. Franck, J. W. Stadelhofer: *Industrielle Aromatenchemie*, Springer Verlag, Berlin – Heidelberg – New York – London – Paris – Tokyo 1987.
[12] *Beilstein*, **5**, 586; **5 (1)**, 274; **5 (2)**, 494; **5 (3)**, 1776; **5 (4)**, 1834.
[13] *Beilstein*, **5**, 515; **5 (1)**, 248; **5 (2)**, 410; **5 (3)**, 1355; **5 (4)**, 1532.
[14] *Beilstein*, **5**, 625; **5 (1)**, 300; **5 (2)**, 531; **5 (3)**, 1936; **5 (4)**, 2142.
[15] *Beilstein*, **5**, 685; **5 (1)**, 344; **5 (2)**, 609; **5 (3)**, 2276; **5 (4)**, 2463.
[16] *Beilstein*, **5**, 667; **5 (1)**, 327; **5 (2)**, 579; **5 (3)**, 2136; **5 (4)**, 2293.
[17] *Beilstein*, **5**, 693; **5 (1)**, 343; **5 (2)**, 609; **5 (3)**, 2279; **5 (4)**, 2467.

Specific References

[18] F. D. Rossini, B. J. Mair, *Adv. Chem. Ser.* **5** (1952) 341.
[19] *Selected values of properties of hydrocarbons and related compounds*, Am. Petrol. Inst. Research Project **44**, Supplementary vol. A-72, 1975.
[20] G. A. Olah et al., *J. Am. Chem. Soc.* **91** (1969) 3261.
[21] G. A. Olah et al., *J. Am. Chem. Soc.* **97** (1975) 6807.
[22] G. A. Olah et al., *J. Am. Chem. Soc.* **95** (1973) 7686.
[23] G. A. Olah et al., *J. Am. Chem. Soc.* **95** (1973) 7680.
[24] K. Nabert, G. Schön, *Sicherheitstechnische Kennzahlen brennbarer Gase und Dämpfe*, Deutscher Eichverlag, Berlin 1965.
[25] R. C. Weast, ed., *Handbook of Chemistry and Physics*, 55th ed., CRC Press, Cleveland 1974.
[26] C. Zerbe, *Mineralöle und Verwandte Produkte*, Springer, Berlin 1969.
[27] Deutsche Shell AG, *Erdöl-Nachrichten*, March 1976, p. 4.
[28] F. Schmerling, H. Heneka, *Erdöl Kohle Erdgas Petrochem.* **21** (1968) 705.
[29] K. Hedden, J. Weitkamp, *Chem. Ing. Tech.* **47** (1975) 505.
[30] J. W. Ward, *Hydrocarbon Process.* **54** (1975) no. 9, 101.
[31] R. Oberkobusch, *Erdöl Kohle Erdgas Petrochem. Brennst. Chem.* **28** (1975) 558.
[32] J. Schulze, *Chem. Ing. Tech.* **46** (1974) 925, 976.
[33] H. Pichler, H. Schulz, *Chem. Ing. Tech.* **42** (1970) 1162.
[34] C. D. Frohning, B. Cornils, *Hydrocarbon Process.* **53** (1974) no. 11, 143.
[35] J. C. Hoogendorn, *Gas Wärme Int.* **25** (1976) 283.
[36] K. C. Hellwig, S. B. Alpert, E. S. Johanson, R. H. Wolk, *Brennst. Chem.* **50** (1969) 263.
[37] F. W. Richardson: *Oil from Coal*, Noyes Data Corp., Park Ridge, N.J., 1975.
[38] K. F. Schlupp, H. Wien, *Angew. Chem.* **88** (1976) 348.
[39] G. Nashan, *Erdöl Kohle Erdgas Petrochem. Brennst. Chem.* **28** (1975) 562.
[40] G. Collin, *Erdöl Kohle Erdgas Petrochem. Brennst. Chem.* **29** (1976) 159.
[41] H.-W. v. Gratkowski, *Erdöl Kohle Erdgas Petrochem. Brennst. Chem.* **28** (1975) 81.
[42] H. Hiller, *Erdöl Kohle Erdgas Petrochem. Brennst. Chem.* **28** (1975) 74.
[43] P. D. Bartlett, K. E. Schueller, *J. Am. Chem. Soc.* **90** (1968) 6071.

[44] *Houben-Weyl*, 4th ed., vol. **IV**, parts 3 and 4.
[45] J. H. Lucas, F. Baardman, A. P. Kouwenhoven, *Angew. Chem.* **87** (1975) 740, **88** (1976) 412.
[46] W. C. van Ziill Langhout, *Proc. World Pet. Congr.* 1975, vol. **5**, 197.
[47] W. L. Lafferty, Jr., W. R. Stokeld, *Adv. Chem. Ser.* **103** (1971) no. 7, 130.
[48] L. F. Albright, *Chem. Eng.* **73** (1966) no. 14, 119; **73** (1966) no. 19, 205; **73** (1966) no. 21, 209.
[49] L. Alcock, R. Martin, K. H. Bourne, W. A. Peet, J. Winsor, *Oil Gas J.* **72** (1974) no. 27, 102.
[50] D. P. Thornton, D. J. Ward, R. A. Erickson, *Hydrocarbon Process.* **51** (1972) no. 8, 81.
[51] H. Jockel, B. E. Triebskorn, *Hydrocarbon Process.* **52** (1973) no. 1, 93.
[52] *Hydrocarbon Process.* **52** (1973) no. 11, 145.
[53] P. R. Pujado, B. V. Vora, *Energy Prog.* **4** (1984) no. 3, 186.
[54] M. S. Scurrell, *Appl. Catal.* **32** (1987) 1.
[55] P. Leprince, A. Chauvel, L. Castex: *Procédés de Pétrochimie*, Editions Technip, Paris 1971, p. 269.
[56] K. Ito, I. Dogane, K. Tanaka, *Hydrocarbon Process.* **64** (1985) no. 10, 83.
[57] S. C. Arnold, G. D. Sucui, L. Verde, A. Neri, *Hydrocarbon Process.* **64** (1985) no. 9, 123.
[58] Japan. Chemical Reporter, 1981, October, p. 2.
[59] F. Solymosi, I. Tombácz, G. Kutsan, *J. Chem. Soc. Chem. Commun.* 1985, 1455.
[60] R. V. Porcelli, B. Juran, *Hydrocarbon Process.* **65** (1986) no. 3, 37.
[61] W. Hinsen, M. Baerns, *Chem. Zt.* **107** (1982) 9.
[62] Kh. E. Khcheyan, O. M. Revenko, A. N. Shatalova: "Oxidative Methylation, a new route to petrochemicals through alternative feedstock," Special Paper No. 16, World Petroleum Congress 11th, 1983 (London).
[63] G. de Pontanel: *Proteins from Hydrocarbons*, Academic Press, London 1972.
[64] S. Gutcho: *Proteins from Hydrocarbons*, Noyes Data Corp., Park Ridge, N.J., 1973.
[65] P. J. Rockwell: *Single Cell Proteins from Cellulose and Hydrocarbons*, Noyes Data Corp., Park Ridge, N.J., 1976.
[66] G. Voss, *Erdöl Kohle Erdgas Petrochem. Brennst. Chem.* **26** (1973) 249.
[67] W. Dimmling, R. Seipenbusch, *Hydrocarbon Process.* **54** (1975) no. 9, 169.
[68] H. Riegel, H. D. Schindler, M. C. Sze, *Erdöl Kohle Erdgas Petrochem. Brennst. Chem.* **26** (1973) 704.
[69] R. P. Lowry, A. Aguilo, *Hydrocarbon Process.* **53** (1974) no. 11, 103.
[70] B. V. Vora, T. Imai, *Hydrocarbon Process.* **61** (1982) no. 4, 171.
[71] L. B. Phillips, *Hydrocarbon Process.* **42** (1963) no. 6, 159.
[72] A. A. Di Giacomo, J. B. Maerker, J. W. Schall, *Chem. Eng. Prog.* **57** (1961) no. 5, 35.
[73] Universal Oil Products, US 3 649 177, 1972 (D. H. Rosback).
[74] K. G. Mittal, *J. Chem. Technol. Biotechnol.* **36** (1986) 291–299.
[75] D. B. Broughton, R. C. Berg, *Chem. Eng.* **77** (1970) 86.
[76] B. V. Vora, P. R. Pujado, J. B. Spinner, T. Imai, Proc. World Tenside Congr. **1984** (Munich) vol. II, 16–26.
[77] R. C. Berg, B. V. Vora in McKetta, Cunningham (ed.): *Encyclopedia of Chemical Processing and Design*, vol. **15**, Marcel Dekker, New York 1982, 266–284.
[78] H. D. Wulff, *Fette Seifen Anstrichm.* **69** (1967) 32.
[79] R. Stöbele, *Chem. Ing. Tech.* **36** (1964) 858.
[80] K. Ziegler et al., *Angew. Chem.* **64** (1952) 323, 67 (1955) 542.
[81] Gulf Res. Dev. Co., DE 1 443 927, 1961.
[82] Ethyl Corp., US 3 906 053, 1973.
[83] W. Keim, A. Behr, M. Röper in G. Wilkinson (ed.): *Comprehensive Organometallic Chemistry*, vol. **5**, Pergamon Press, Oxford 1982, 371.

[84] W. Keim, *Chem. Ing. Tech.* **56** (1984) 850.
[85] Shell Development Comp., US 3 635 937, 1972; 3 637 636, 1972; 3 644 564, 1972; 3 644 563, 1972; 3 647 914, 1972; 3 647 915, 1972; 3 661 803, 1972; 3 686 159, 1972 (W. Keim et al.).
[86] E. R. Freitas, R. C. Gum, *Chem. Eng. Progress.* **75** (1979) no. 1, 73.
[87] E. L. T. M. Spitzer, *Seifen Öle Fette Wachse* **107** (1981) 141–144.
[88] M. Sherwood, Chem. Ind. (London) (1982) 994–995.
[89] W. Keim, F. H. Kowaldt, *Erdöl Kohle Erdgas Petrochem.* Compendium (1978/79) 453–462.
[90] Shell Oil Comp., US 3 647 906, 1972 (F. F. Farley).
[91] M. Peuckert, W. Keim, *Organometallics* **2** (1983) 594.
[92] W. Keim, F. H. Kowaldt, R. Goddard, C. Krüger, *Angew. Chem.* **90** (1978) 493.
[93] W. Keim, A. Behr, B. Limbäcker, C. Krüger, *Angew. Chem.* **95** (1983) 505.
[94] W. Keim et al., *J. Mol. Catal.* **6** (1979) 79.
[95] W. Keim et al., *Organometallics* **5** (1986) 2356.
[96] *Chem. Eng. News* (1972), April 10, 16–17.
[97] C. Commereuc, Y. Chauvin, J. Gaillard, J. Léonard, J. Andrews, *Hydrocarbon Process.* **63** (1984) no. 11, 118–120.
[98] W. J. Benedek, J.-L. Mauleon, *Hydrocarbon Process.* **59** (1980) no. 5, 143.
[99] J. Léonard, J. F. Gaillard, *Hydrocarbon Process.* **60** (1981) no. 3, 99.
[100] J. F. Boucher, G. Follain, D. Fulop, J. Gaillard, *Oil Gas J.* **80** (1982) 84–86.
[101] Y. Chauvin, J. Gaillard, J. Léonard, P. Bonnifay, J. W. Andrews, *Hydrocarbon Process.* **61** (1982) no. 5, 110–112.
[102] D. Commereux, Y. Chauvin, G. Leger, J. Gaillard, *Rev. Inst. Fr. Pet.* **37** (1982) 639–649.
[103] J. Gaillard, *Pet. Tech.* **314** (1985) 20–27.
[104] B. Bogdanović, *Adv. Organomet. Chem.* **17** (1979) 105.
[105] W. Keim, A. Behr, G. Kraus, *J. Organomet. Chem.* **251** (1983) 377.
[106] A. Behr, V. Falbe, U. Freudenberg, W. Keim, *Isr. J. Chem.* **27** (1986) 277–279.
[107] W. Keim, *Ann. N.Y. Acad. Sci.* **415** (1983) 191–200.
[108] B. Bogdanović, B. Spliethoff, G. Wilke, *Angew. Chem. Int. Ed. Engl.* **19** (1980) 622.
[109] G. Speier, *J. Organomet. Chem.* **97** (1975) 109.
[110] M. G. Barlow, M. J. Bryant, R. N. Haszeldine, A. G. Mackie, *J. Organomet. Chem.* **21** (1970) 215.
[111] S. Datta, M. B. Fisher, S. S. Wreford, *J. Organomet. Chem.* **188** (1980) 353.
[112] R. Schrock, S. McLain, J. Sancho, *Pure Appl. Chem.* **52** (1980) 729.
[113] J. A. Brennan, *Ind. Eng. Chem. Prod. Res. Dev.* **19** (1980) 2–6.
[114] R. L. Shubkin, M. S. Baylerlan, A. R. Maler, *Ind. Eng. Chem. Prod. Res. Dev.* **19** (1980) 15–19.
[115] F. J. Karol, *CHEMTECH* 1983 (April) 222.
[116] B. W. Werdelmann, *Fette Seifen Anstrichm.* **76** (1974) 1.
[117] *Chem. Eng. News*, (1988) May 30, 10.
[118] *Eur. Chem. News*, (1987) Dec. 21/28, 21.
[119] *Eur. Chem. News*, (1981) July 6, 23.
[120] *Eur. Chem. News*, (1987) Jan. 5/12, 19.
[121] *Eur. Chem. News*, (1987) Jan. 19, 8.
[122] *Eur. Chem. News*, (1981) Jan. 26, 27.
[123] *Eur. Chem. News*, (1982) July 5, 26.
[124] G. Wilke, *Angew. Chem. Int. Ed. Engl.* **2** (1963) 105.
[125] A. C. L. Su, *Adv. Organomet. Chem.* **17** (1979) 269.
[126] S. Warwel, *Erdöl Erdgas Kohle* **103** (1987) 238.

[127] R. C. Weast, M. J. Astle (eds.): *CRC Handbook of Data on Organic Compounds*, CRC Press, Boca Raton, Florida, 1985.
[128] S. Coffey (ed.): *Rodd's Chemistry of Carbon Compounds*, III A, Elsevier, Amsterdam 1971, pp. 228–231.
[129] A. D. Rudkovskii et al., *Chem. Technol. Fuels Oils (Engl. Transl.)* **1** (1983) 387–391.
[130] F. P. McCandless, *Ind. Eng. Chem. Prod. Res. Dev.* **19** (1980) 612–616; **21** (1982) 483–488.
[131] Idemitsu Kosan Co. Ltd., EP 0 104 607, 1983 (H. Sato).
[132] Universal Oil Products, US 4 487 984, 1984 (T. Imai).
[133] O. A. Anunziata et al., *Appl. Catal.* **15** (1985) 235–245.
[134] J. Hanotier, M. Hanotier-Bridoux, *J. Mol. Catal.* **12** (1981) 133–147.
[135] L. I. Smith, *Org. Synth.*, coll. vol. **2** (1948) 248–253.
[136] Toyo Soda Manufacturing, DE-OS 3 330 912, 1983 (T. Sakamoto, T. Hironaka, K. Sekizawa, Y. Tsutsumi).
[137] Toa Nenryo, EP 0 101 651, 1983 (T. Maejima, N. Tagaya, S. Sakurada).
[138] J. Haggin, *Chem. Eng. News*, (1987) June 1, 22–28; (1987) June 22, 22–25.
[139] International Flavors and Fragrances, US 3 869 524, 1974 (K. K. Light, J. B. Anglim).
[140] Phillips Petroleum, US 4 568 784, 1985 (R. L. Cobb).
[141] S. A. Shama, C. C. Wamser, *Org. Synth.* **61** (1983) 62–64.
[142] H. F. Davis, S. K. Chattopadhyay, P. K. Das, *J. Phys. Chem.* **88** (1984) 2798–2803.
[143] W. W. Kaeding, G. C. Barile, M. M. Wu, *Catal. Rev. Sci. Eng.* **26** (1984) 597–612.
[144] Dow Chemical, US 2 763 702, 1956 (J. L. Amos, K. E. Coulter).
[145] W. W. Kaeding, L. B. Young, C.-C. Chu, *J. Catal.* **89** (1984) 267–273.
[146] W. W. Kaeding, G. C. Barile, *Polym. Sci. Technol. (Plenum)* **25** (1984) 223–241.
[147] *Eur. Chem. News*, Oct. 17 (1988) p. 4.
[148] Physical Constants of Hydrocarbons C_1 to C_{10}, "*ASTM Spec. Techn. Publ. STP*" **106** (1963).
[149] Handbook of Chemistry and Physics, R. C. Weast (ed.), 63rd ed., CRC Press, Boca Raton 1982–1983, p. D-213.
[150] D. R. Stull, E. F. Westrum, G. G. Sinke: *The Chemical Thermodynamics of Organic Compounds*, J. Wiley & Sons, New York 1969, p. 370.
[151] J. C. Butler, W. P. Webb, *Chem. Eng. Data Series* **2** (1957) no. 1, 42–46.
[152] G. Sorbe: *Sicherheitstechnische Kenndaten chemischer Stoffe*, vol. **2,** ecomed-Verlagsgesellschaft mbH, Landsberg 1983, pp. 15–16.
[153] P. R. Pujado, J. R. Salazar, C. V. Berger, *Hydrocarbon Process.* **55** (1976) no. 3, 91.
[154] R. C. Canfield, T. C. Unruh, *Chem. Eng. NY* **90** (1983) no. 3, 32.
[155] Toyo Soda Manufacturing Co., DE-AS 2 418 038, 1974 (T. Yanagihera et al.).
[156] Hitachi Chemical Co., JP-Kokai 77 139 030, 1974 (E. Kudo et al.); *Chem. Abstr.* **88** (1978) 120757 r.
[157] BASF, DE-OS 2 141 491, 1971 (F. Neumayr).
[158] Mitsui Petrochemical Ind., EP 0 046 678, 1981 (M. Hisaya).
[159] J. P. Fortuin, M. J. Waale, R. P. Oosten, *Pet. Refiner* **38** (1959) no. 6, 189.
[160] A. H. Olzinger, *Chem. Eng. NY* **82** (1975) no. 6, 50.
[161] Universal Oil Products, US 3 763 259, 1973 (G. L. Hervert).
[162] Mobil Oil Corp., EP-A 0 149 508, 1985 (W. W. Kaeding).
[163] J. Ewers, H. W. Voges, G. Maleck, *Erdöl Kohle Erdgas Petrochem. Compendium* **74/75** (1975) 487.
[164] K. Ito, I. Dogane, K. Tanaka, *Hydrocarbon Process.* **65** (1986) no. 11, 89.
[165] H. A. Colvin, J. Muse, *CHEMTECH* 1986, no. 500.

[166] K. Ito, *Hydrocarbon Process.* **52** (1973) no. 8, 89.
[167] Phillips Petroleum Co., US 3 291 847, 1966 (P. F. Warner); US 3 316 315, 1967 (W. A. Jones).
[168] K. Weissermel, H.-J. Arpe: *Industrielle Organische Chemie*, 3rd ed., VCH Verlagsgesellschaft, Weinheim 1987, pp. 72–79.
[169] A. C. Olson, *Ind. Eng. Chem.* **52** (1960) no. 10, 833.
[170] R. Ströbele, *Chem. Ing. Tech.* **36** (1964) 858.
[171] H.-D.Wulf, T. Böhm-Gößl, L. Rohrschneider, *Fette Seifen Anstrichm.* **69** (1967) 32.
[172] California Research Corp., US 2 477 382, 1949 (A. H. Lewis, A. C. Ettling); *Chem. Abstr.* **44** (1950) 1137 e.
[173] H. Hartig, *Chem. Ztg.* **99** (1975) 175.
[174] K. Ichikawa, H. Shingu, *Kogyo Kagaku Zasshi* **54** (1951) 183–184; *Chem. Abstr.* **47** (1953) 8662 b.
[175] S. Murohashi, *Nagaoka Kogyo Koto Senmon Gakko Kenkyu Kiyo* **9** (1973) 103–106; *Chem. Abstr.* **80** (1974) 132917 r.
[176] J. G. Matveev, D. A. Drapkina, R. L. Globus, *Tr. IREA* **21** (1956) 83–89, *Chem. Abstr.* **52** (1958) 15474 e.
[177] A. D. Harford, H. W. Vernon, US 2 897 112, 1956.
[178] British Petroleum Co., Ltd., DE-AS 1 094 035, 1957 (A. Douglas, H. W. Vernon).
[179] N. K. Moshinskaya, V. S. Olifer, L. I. Zhuoiev, SU 137 993, 1960.
[180] E. Siggel, US 2 710 849, 1952.
[181] BASF, DAS 1 142 338, 1960 (W. Braun, E. Hartwig, R. Krallmann).
[182] FMC Corp., US 3 539 640, 1967 (M. I. Stewart, O. K. Carlson).
[183] Ashland Oil & Refining Co., US 3 529 944, 1967 (A. M. Leas).
[184] Japan Petrochemicals Co., Ltd., DE-OS 2 210 133, 1972 (A. Sato, Y. Aida, I. Shimizi).
[185] H. Winkler: *Der Steinkohlenteer*, Verlag Glückauf, Essen 1951, p. 143.
[186] J. Chipman, S. B. Peltier, *Ind. Eng. Chem.* **21** (1929), 1106.
[187] M. McEwen: *Organic Coolant Handbook*, sponsored by U.S. Atomic Energy Comm., publ. by Monsanto Co., St. Louis, Mo., 1958.
[188] J. M. Cork, H. Mandel, N. Ewbank (North American Aviation Inc.), U.S. At. Energy Comm. Sci. Rept. 5129 (1960) 1–28. Recalculated by A. N. Syverud, The Dow Chemical Co. Thermal Laboratory, Midland, Mich.
[189] M. McEwen, E. Wiederhold, *Nucl. Eng.* Part IV Symp. Ser. 5 (1959) no. 22, 9–15.
[190] M. J. Foole, *Chem. Eng. Prog.* **58** (1962) no. 4, 37–40.
[191] *Hydrocarbon Process.* **46** (1967) no. 11, 187.
[192] S. Feigelman, C. B. O'Connor, *Hydrocarbon Process.* **45** (1966) no. 5, 140–144.
[193] *Hydrocarbon Process. Pet. Refin.* **42** (1963) no. 11, 204.
[194] Bayer, BIOS Final Rep. Nr. 1787.
[195] R. Egbert, Trans. AICHE vol. 34 (1938) 435–488.
[196] R. Dasgupta, B. R. Maiti, *Ind. Eng. Chem.* **25** (1986) 381–386.
[197] Otkrytiya/Itobret, SU 1 165 677, 1985.
[198] BIOS Final Rep. Nr. 893.
[199] Soc. Progil, US 3 227 525, 1964 (M. E. Degeorges, M. Jaymond).
[200] Soc. Progil, US 3 112 349, 1961 (M. E. Degeorges, M. Jaymond).
[201] Soc. Progil, DE-AS 1 249 244, 1961 (M. Jaymond).
[202] Dow Chemical Co., US 1 938 609, 1933 (J. H. Reilly).
[203] Monsanto Co., US 3 228 994, 1962 (W. D. Robinson, W. Groves).
[204] Hawker Siddeley Nuclear Power Co., FR 1 272 068, 1961.

[205] Montecatini, BE 586 826, GB 865 302, IT Anm. 1144/59, 1960, IT Prior. 1959 (C. Ferry, B. Ponzini).
[206] Monsanto Co., US 2 143 509, 1939 (C. Conover, A. E. Huff).
[207] Montecatini, IT 586 113, 1962.
[208] Kambarov YU G; SU 6 79 565, 1977, *Chem. Abstr.* **92** (1980) 58407 d.
[209] Monsanto Co., US 3 228 994, 1962 (W. D. Robinson).
[210] N. M. Indyukov, R. I. Gasanova, *Khim. prom-st. (Moscow)* 1975, no. 7, 555.
[211] Farbenfabr. Bayer AG, DE-AS 1 012 595, 1955 (H. Holzrichter, O. Tegtmeyer).
[212] F. Ullmann, *Justus Liebigs Ann. Chem.* **332** (1904) 48.
[213] E. I. El'bert, SU 215 911, 1968.
[214] R. G. Ismailow, D. A. Aliev, *Azerb. Neft. Khoz.* **44** (1965) no. 1, 30.
[215] L. Gattermann, R. Ehrhardt, *Ber. Dtsch. Chem. Ges.* **23** (1890) 1226.
[216] I. H. Gardner, P. Bergstrom, *J. Am. Chem. Soc.* **51** (1962) 3376.
[217] Asahi Kasei Kogyo K. K., DE-AS 2 039 904, 1970.
[218] Shell Oil Co., US 3 145 237, 1962 (R. van Helden, B. Balder).
[219] R. van Helden, G. Verberg, *Rec. Trav. Chim. Pays-Bas* **84** (1965) 1263.
[220] Gulf Research & Developm. Co., US 3 401 207, 1966 (C. M. Selwitz).
[221] E. S. Rudakov, A. I. Lutsyk, R. I. Rudakova, *Kinet. Katal.* **18** (1977) 525; *Kinet. Catal. (Engl. Transl.)* **18** (1977) 441.
[222] M. K. Starchevskii, M. N. Vargaftik, I. I. Moiseev, *Izv. Akad. Nauk SSSR* 1979, 242; (Engl. Transl.) 227.
[223] Teijin Ltd., US 3 963 787, 1973 (Y. Ichikawa, T. Yamaji).
[224] Ube Industries, JP 55 079 324, 1978.
[225] L. N. Rachkovskaya et al., *Kinet. Katal.* **18** (1977) 792; *Kinet. Catal. (Engl. Transl.)* **18** (1977) 660.
[226] Ube Industries, JP 80 141 417, 1979, *Chem. Abstr.* **94** (1981) 156 561 k.
[227] British Petroleum, DE-OS 1 909 045, 1969 (D. Fiolka, C. R. Jentsch).
[228] L. H. Slaugh, *Tetrahedron* **24** (1968) 4523.
[229] L. H. Slaugh, J. A. Leonard, *J. Catal.* **13** (1969) 385.
[230] G. N. Koshel et al., *Dokl. Akad. Nauk SSSR* **237** (1977) no. 1, 164 (Chem. Tech.); *Chem. Abstr.* **88** (1978) 50 394 u.
[231] Phillips Petroleum, US 4 123 470, 1977 (T. P. Murths).
[232] Farberov M I, SU 416 339, 1972; *Chem. Abstr.* **88** (1978) 62 124 u.
[233] Standard Oil, US 4 218 572, 1978 (S. R. Dolhyj, L. J. Velenyi).
[234] Y. Yamazaki, T. Kawai, A. Yagishita, *Sekiyu Gakkaishi* **21** (1978) no. 4, 229.
[235] Koppers Co., US 3 059 987 1962 (H. P. Baumann).
[236] N. Hendrich, *Melliand Textilber.* **43** (1962) 258.
[237] W. M. Meylan, P. H. Howard: Chemical Market Input/Output Analysis of Biphenyl and Diphenyl Oxide to Assess Sources of Environmental Contamination, Syracuse NY, EPA Contract No. 68-01-3224, 1976.
[238] W. E. Bachmann, H. T. Carke, *J. Am. Chem. Soc.* **49** (1927) 2089–98.
[239] L. W. Pickett, *Proc. R. Soc. London A* **142** (1933) 333–346.
[240] H. G. Römer, E. Hertel, *Z. Phys. Chem. Abt. B* **21** (1933) 292–296.
[241] R. J. Good, E. E. Hardy, A. M. Ellenburg, H. B. Richards, Jr., *J. Am. Chem. Soc.* **75** (1953) 436.
[242] J. A. Ellard, W. H. Yanko: *Thermodynamic Properties of Biphenyl and the Isomeric Terphenyls*, Final Report IDO-11,008, Monsanto Research Corp., U.S. At. Energy Comm. Contract AT (10-1)-1088, Oct. 31, 1963.

[243] R. O. Bolt, R. O. Johnson, R. S. Kirk: *Organic Materials for Use as Reactor Coolant Moderators*, California Research Corp., June 13, 1956.
[244] P. L. Geiringer: *Handbook of Heat Transfer Media*, Reinhold Publ., New York 1962, p. 173.
[245] W. H. Hedley, M. V. Mines, *J. Chem. Eng. Data* **15** (1970) 122.
[246] H. Mandel, N. Eubank: *Critical Constants of Diphenyl and Terphenyl*, Report No. NAA-SR 5129, Atomics International, Dec. 1960, p. 18.
[247] N. Andrews, R. Ubbelohde, *Proc. R. Soc. London* **228 A** (1955) 435–37.
[248] E. Pajda, J. Rusin, *Chemik Wkladka* 1972, 422.
[249] R. Balent, *Proc. Am. Power Conf.* **21** (1959) 120.
[250] Soc. Progil, DE-AS 1 272 281, 1962 (M. Jaymond).
[251] Farbenfabr. Bayer AG, GB 957 065 = FR 1 297 408, 1961.
[252] Hawker Siddeley Nuclear Power Co., FR 1 272 068, 1961.
[253] D. B. Orechkin, SU 570 584, 1977.
[254] Monsanto Co., DE-OS 2 306 454, 1973 (J. K. Sears).
[255] M. He, *Huaxue Shiji* **8** (1986) no. 1, 60–61; *Chem. Abstr.* **105** (1986) 208 532.
[256] Rütgerswerke, DE 3 203 818, 1982 (K. Stolzenberg, K. H. Koch, R. Marrett).
[257] Nippon Steel, JP 82 98 219, 1980.
[258] IG Farbenindustrie, DE 428 088, 1921.
[259] Selden, GB 318 617, 1929.
[260] Standard Oil, US 2 966 513, 1957 (E. K. Fields).
[261] Rütgerswerke, DE 1 262 268, 1959 (E. Marx, H.-G. Franck, E. A. Dierk, K. Fiebach).
[262] *Ullmann*, 4th ed., **18,** 685.
[263] Farbwerke Hoechst, DE 2 209 692, 1972 (K. Eiglmeier); DE 2 343 964, 1973 (F. Röhrscheid).
[264] Institute of Chemical Sciences, Academy of Sciences/Kazakh S.S.R., SU 1 084 275, 1982 (R. U. Umarova, A. K. Ergalieva, D. K. Sembaev, B. V. Suborov).
[265] BASF, DE 2 721 640, 1977 (A. Hettche, M. Patsch); DE 2 653 346, 1976 (W. Eisfeld, W. Disteldorf, A. Hettche).
[266] BASF, DE 2 721 639, 1977 (A. Hettche, M. Patsch).
[267] *Ullmann*, 4th ed., **18,** 693.
[268] Farbwerke Hoechst, DE 2 304 873, 1973 (K. Eiglmeier, H. Lübbers).
[269] Toyo Soda Mfg., JP 85 94 928, JP 85 139 630, 1983 (M. Kubo, H. Satsuka, Y. Tsutsumi); EP 217 316, 1986 (M. Kubo, K. Kawabata, Y. Tsutsumi).
[270] Coal Industry, EP 47 069, 1981 (J. G. Robinson, S. A. Brain).
[271] Nippon Steel, JP 83 176 210, 1982.
[272] Institute of Heteroorganic Compounds, Academy of Sciences/USSR, DE 3 526 155, 1985 (V. V. Korshak et al.).
[273] Sun Research and Development, DE 1 593 398, 1966 (A. Schneider, E. J. Janoski, R. N. McGinnis).
[274] G. Collin, M. Zander, *Erdöl Kohle Erdgas Petrochem.* **33** (1980) 560.
[275] BASF, DE 3 044 111, 1980 (H. Naarmann, K. Penzien, D. Nägele, J. Schlag).
[276] *Ullmann*, 4th ed., **21,** 70.
[277] R. Heering, P. Lobitz, K. Gehrke, *Plaste Kautsch.* **31** (1984) 412–413; J. Stein, G. Liebisch, S. Helling, *Chem. Tech. (Leipzig)* **31** (1979) 302–305.
[278] Gulf Research and Development, US 4 280 881, 1977 (J. C. Montagna, R. D. Galli).
[279] Dow Chemical, US 4 093 672, 1977 (Y. Ch. Sun).
[280] Rütgerswerke, DE 2 443 114, 1974 (K. Rühl, G. Storch, H. Steinke, B. Charpey).
[281] American Cyanamid, US 4 145 409, 1976 (N. R. Pasarela).

[282] Manufacture de Produits Pharmaceutique A. Christiaens, DE 2 060 721, 1970 (P. Vanhoff, P. Clarebout).
[283] Takeda Chemical Industries, JP 82 123 157, 1981.
[284] Rütgerswerke, DE 2 208 253, 1972 (E. Pastorek, H. Miels, W. Orth).
[285] Pfizer, DE 2 244 556, 1973 (D. P. O'Shea, R. C. Adams, S. Nakanishi).
[286] Meiji Seika Kaisha, DE 2 911 492, 1979 (Y. Sekizawa et al).
[287] Wella, DE 3 215 341, 1982 (E. Konrad, H. Mager).
[288] Rütgerswerke, DE 2 635 401, 1976 (P. Stäglich, D. Hausigk).
[289] H. J. V. Winkler: *Der Steinkohlenteer und seine Bedeutung*, Verlag Glückauf, Essen 1951, p. 156.
[290] Nippon Steel, JP 82 98 220, 1980.
[291] Rütgerswerke, DE 3 243 981, 1982 (W. Orth, E. Pastorek, W. Fickert).
[292] Celamerck, DE 2 816 388, 1979 (J. Gante et al.).
[293] Squibb, US 3 856 977, 1974 (E. T. Stiller, S. D. Levine, P. A. Principe, P. A. Diassi).
[294] Bayer, DE 3 530 819, 1985 (J. Hocker, L. Rottmeier).
[295] *Ullmann*, 4th ed., **23,** 415.
[296] Nippon Kayaku, DE 2 325 302, 1973 (D. Matsunaga, M. Sumitani).
[297] Chemische Fabrik Kalk, DE 2 725 857, 1977 (H. Jenkner).
[298] Sandoz, DE 2 225 565, 1972 (H. P. Baumann).
[299] Rütgerswerke, DE 2 704 648, 1977 (C. Finger).
[300] Agency of Industrial Sciences and Technology, JP 81 82 430, 1979.
[301] Magyar Tudomanyos Akademia, HU 36 764, 1983 (L. Simandi et al.).
[302] Nippon Shokubai Kagaku Kogyo, JP 85 233 028, 1984 (T. Yoneda, S. Nakahara).
[303] Dow Chemical, US 4 297 514, 1980 (W. M. King).
[304] Hydrocarbon Research, DE 3 141 645, 1981. (D. Huibers, R. Barclay, R. C. Shah).
[305] Fuji-Xerox, JP 82 144 567, 1981; Ricoh, JP 79 109 438, 1978 (E. Akutsu, A. Kojima); Oce-van der Grinten, DE 2 525 381, 1975 (J. Hagendorn, H. Knibbe); EP 24 762, 1981 (M. T. J. Peters, J. A. Verbunt).
[306] Bayer, DE 2 430 476, 1974 (W. Draber, K. H. Büchel, M. Plempel).
[307] Röhm, DE 3 439 484, 1984 (A. Riemann, W. Ude).
[308] Isovolta, EP 41 496, 1981 (W. Rieder, M. Fehrle).
[309] Minnesota Mining and Mfg., EP 203 828, 1986 (W. J. Schultz, G. B. Portelli, J. P. Tane).
[310] BASF, DE 3 209 426, 1982 (M. Patsch, H. Flohr).
[311] BASF, DE 3 536 259, 1985 (M. Patsch et al.).
[312] Rütgerswerke, DE 2 114 219, 1971 (C. Finger, H.-G. Franck, M. Zander).
[313] Rütgerswerke, DE 2 166 862, 1971 (C. Finger, H.-G. Franck, M. Zander).
[314] Kureha Chemical Industry, JP 75 140 535, 1974 (M. Takahashi, A. Ito, A. Nobuo).
[315] Kureha Chemical Industry, DE 2 447 894, 1973 (Y. Shisakaiso, A. Ito, M. Takahashi).
[316] Kureha Chemical Industry, DE 2 446 591, 1974 (M. Takahashi et al.).
[317] Rütgerswerke, DE 1 931 946, 1969 (C. Finger, H.-G. Franck, M. Zander).
[318] Wellcome Foundation, EP 125 702, 1984 (K. W. Bair).
[319] Richardson Merell, US 3 907 791, 1975 (W. L. Albrecht, R. W. Fleming).
[320] Rütgerswerke, DE 1 126 372, 1958 (J. Altpeter, H. Sauer).
[321] BASF, DE 3 218 229, 1982 (H. Naarmann, J. Schlag, G. Koehler).
[322] VEB Kombinat Mikroelektronik, DE 3 131 625, 1981 (H.-J. Deutscher et al.).
[323] Fuji Standard Research, EP 196 169, 1986 (S. Otani, Y. Nagai).
[324] Coal Industry, EP 36 710, 1981 (J. G. Robinson, D. I. Barnes).
[325] Hercules, US 3 926 897, 1972 (C. L. Cessna).

[326] Licentia, DE 2 406 540, 1974 (M. Saure, H. Wagner, J. Wartusch).
[327] Union Carbide, US 3 834 166, 1974 (R. A. Cupper, G. S. Somckh).
[328] Bayer, DE 3 232 063, 1982 (H. Herzog, W. Hohmann).
[329] Burroughs Wellcome, US 4 511 582, 1983 (K. W. Bair).
[330] Rütgerswerke, DE 1 240 065, 1965 (E. Egerer, W. Fickert, L. Rappen, G. Grigoleit); DE 3 305 528, 1983 (D. Tonnius, W. Orth, H. Miele, E. Pastorek).
[331] US Steel, US 2 956 065, 1958 (C. W. De Walt, K. A. Schowalter).
[332] Mitsubishi Metal, JP 60 238 432, 1984 (H. Yoshida, M. Morikawa, T. Iwamura).
[333] Nihon Joryu Kogyo, Nippon Shokubai Kagaku Kogyo, JP 85 255 747, 1984 (Y. Saito et al.).
[334] Nippon Shokubai Kagaku Kogyo, JP 85 233 029, 1984 (I. Maeda, H. Mitsui, T. Nakamura).
[335] Nippon Shokubai Kagaku Kogyo, JP 8 112 335, 1979.
[336] Merck, US 3 506 434, 1970 (E. Jacobi et al.).
[337] *Ullmann*, 4th ed. **12**, 614.
[338] Agfa-Gevaert, EP 2 546, 1979 (F. C. Heugebaert, W. Janssens).
[339] Fuji Photo Film, GB 2 009 955, 1979 (M. Kitajima, H. Tachikawa, F. Shinozaki, T. Ikeda).
[340] PPG Industries, US 4 054 683, 1976 (G. W. Gruber).
[341] National Starch and Chemical, US 4 069 123, 1972 (M. M. Skoultchi, I. J. Davis).
[342] Canadian Industries, US 4 012 280, 1976 (H. H. Holton).
[343] Ciba-Geigy, CH 580 673, 1972 (W. Müller).
[344] Eli Lilly, DE 2 631 888, 1977 (J. B. Campbell).
[345] EGIS, DE 3 618 715, 1986 (P. Benko et al.).
[346] H. J. V. Winkler: *Der Steinkohlenteer und seine Bedeutung*, Verlag Glückauf, Essen 1951, p. 220.
[347] Rütgerswerke, DE 2 914 670, 1979 (G.-P. Blümer, M. Zander).
[348] Hoechst, DE 3 532 882, 1985 (S. Schießler, E. Spietschka).
[349] Rütgerswerke, DE 2 709 015, 1977 (J. Omran, M. Zander).
[350] Hoechst, DE 3 134 131, 1981; DE 3 324 881, 1983 (S. Schießler, E. Spietschka).
[351] *Ullmann*, 4th ed., **17**, 471–472.
[352] Bayer, EP 20 817, 1979 (H. Harnisch).
[353] Ricoh, JP 7 908 527, 1977 (A. Kojima, M. Hashimoto, M. Ohta).
[354] Hoechst, DE 2 734 288, 1977 (W. Wiedemann); DE 2 726 116, 1977 (H. J. Schlosser, E. Lind).
[355] British Petroleum, GB 1 399 814, 1971 (A. J. Burn, T. A. Lighting).
[356] Wellcome Foundation, EP 125 701, 1984 (K. W. Bair).
[357] Pentel, JP 8 632 352, 1976 (K. Takemura).
[358] W. R. von Oettingen: "Toxicity and Potential Dangers of Aliphatic and Aromatic Hydrocarbons," *Public Health Bull.* (1940) no. 255.
[359] E. A. Wahrenbrock, E. I. Eger, R. G. Laravuso, G. Maruschak, *Anesthesiology* **40** (1974) no. 1, 19.
[360] A. H. Nuckolls, *Underwriters Laboratory Report* (Nov. 13, 1933) no. 2375.
[361] Y. Henderson, H. W. Haggard: *Noxious Gases*, 2nd ed., Reinhold Publ. Co., New York 1943.
[362] J. C. Krantz, Jr., C. J. Carr, J. F. Vitcha, *J. Pharmacol. Exp. Therap.* **94** (1948) 315.
[363] T. Ikoma, *Nichidai Igaku Zasshi* **31** (1972) no. 2, 71.
[364] D. M. Aviado, S. Zakhari, T. Watanabe: *Non-fluorinated Propellants and Solvents for Aerosols*, CRC Press, Cleveland, Ohio 1977, pp. 49–81.
[365] F. A. Patty, W. P. Yant: "Odor Intensity and Symptoms Produced by Commercial Propane, Butane, Pentane, Hexane and Heptane Vapor," U.S. Bur. Mines, Rep. Invest. **1929** no. 2979.
[366] B. B. Shugaev, *Arch. Environ. Health* **18** (1969) 878.

[367] *Documentation of the Threshold Limit Values for Substances in Workroom Air*, 3rd ed., American Conference of Governmental Industrial Hygiene, Cincinnati 1971.
[368] M. G. Rumsby, J. B. Finean, *J. Neurochem.* **13** (1966) 1513.
[369] D. A. Haydon, B. M. Hendry, S. R. Levinson, J. Requena, *Biochim. Biophys. Acta* **470** (1977) 17.
[370] N. Battistini et al., *Riv. Patol. Nerv. Ment.* **95** (1974) 871.
[371] G. W. Paulson, G. W. Waylonis, *Arch. Intern. Med.* **136** (1976) 880.
[372] N. W. Lazarew, *Arch. Exp. Pathol. Pharmakol.* **143** (1929) 223.
[373] E. T. Kimura, D. M. Elery, P. W. Dodge, *Toxicol. Appl. Pharmacol.* **19** (1971) 699.
[374] C. H. Hine, H. H. Zuidema, *Ind. Med.* **39** (1970) no. 5, 215.
[375] H. H. Schaumburg, P. S. Spencer, *Brain* **99** (1976) 182.
[376] K. Kurita, *Jpn. J. Ind. Health Exp.* **9** (1974) 672.
[377] A. K. Ashbury, S. L. Nielsen, R. Telfer, *J. Neuropathol. Exp. Neurol.* **33** (1974) 191.
[378] E. G. Gonzales, J. A. Downey, *Arch. Phys. Med. Rehabil.* **53** (1972) 333.
[379] H. E. Swann, Jr., B. K. Kwon, G. H. Hogan, W. M. Snellings, *Am. Ind. Hyg. Assoc. J.* **35** (1974) no. 9, 311.
[380] W. R. F. Notten, P. W. Henderson, *Biochem. Pharm.* **24** (1975) 1093.
[381] C. F. Reinhardt et al., *Arch. Environ. Health* **22** (1971) 265.
[382] A. Y. H. Lu, H. W. Strobel, M. J. Coon, *Mol. Pharamcol.* **6** (1970) 213.
[383] N. W. Lazarew, *Arch. Exp. Pathol. Pharmakol.* **143** (1929) 223.
[384] *Criteria for a Recommended Standard – Occupational Exposure to Xylene*, U.S. Dept. Health, Education and Welfare, Natl. Inst. Occup. Safety and Health, Washington, D.C. (1975).
[385] O. A. Anunziata et al., *Appl. Catal.* **15** (1985) 235–245. K. Verschueren: *Handbook of Environmental Data on Organic Chemicals*, 2nd ed.,Van Nostrand Reinhold, New York 1983.
[386] W. E. Coleman et al., *Arch. Environ. Contam. Toxicol.* **13** (1984) 171–178.
[387] N. I. Sax: *Dangerous Properties of Industrial Materials*, 6th ed., Van Nostrand Reinhold, New York 1984.
[388] *Threshold Limit Values and Biological Exposure Indices for 1987–1988*, American Conference of Governmental Industrial Hygienists. Cincinnati, Ohio, 1987.
[389] DFG: *Maximale Arbeitsplatzkonzentrationen und biologische Arbeitsstofftoleranzwerte 1987*, VCH Verlagsges., Weinheim 1987.
[390] H. W. Gerarde: *Toxicology and Biochemistry of Aromatic Hydrocarbons*, Elsevier Publishing Co., London 1960.
[391] G. Hommel: *Handbuch der gefährlichen Güter*, Springer Verlag, Berlin Heidelberg 1987.
[392] H. W. Gerarde, *Am. Med. Assoc. Arch. Ind. Health* **19** (1959) 403.
[393] M. A. Wolf et al., *Am. Med. Assoc. Arch. Ind. Health* **14** (1956) 387.
[394] H. W. Werner, R. C. Dunn, W. F. von Oettingen, *Ind. Health Toxicol.* **26** (1974) 264.
[395] *Maximale Arbeitsplatzkonzentrationen und biologische Arbeitsstofftoleranzwerte*, Dt. Forschungsgemeinschaft, Verlag Chemie, Weinheim 1984.
[396] N. I. Sax: *Dangerous Properties of Industrial Materials*, 6th ed., van Nostrand Reinhold. New York 1984, p. 819.
[397] H. E. Christensen, T. T. Luginbyhl: *Registry of Toxic Effects of Chemical Substances*, U.S. Dept. of H. E. W., U.S. Government Printing Office, Washington, D.C., 1975.
[398] J. Opdyke, *Food Cosmet. Toxicol.* **12** (1974) 707.
[399] J. M. Haas, H. W. Earhart, A. S. Todd: *Book of Papers*, 1974 AATCC National Technical Conference, American Association of Textile Chemists and Colorists, Research Triangle Park, N. C., p. 427.

[400] I. Hakkinen et al., *Arch. Environ. Health* **26** (1973) 70.
[401] A. M. Seppalainen, I. Hakkinen, *J. Neurol. Neurosurg. Psychiatry* **38** (1975) 248.
[402] *Patty's Industrial Hygiene and Toxicology*, vol. 2 B, Toxicology, 3rd ed.
[403] Z. F. Khromenko, *Gig. Tr. Prof. Zabol.* 1976, 44.
[404] Y. R. Adamson, J. L. Weeks, Arch. Environ. Health **27** (1973) no. 8, 68.
[405] Y. R. Adamson, J. M. Furlong, *Arch. Environ. Health* **28** (1974) no. 3, 155.
[406] R. J. Hawkins, J. L. Weeks, *At. Energy Can. Ltd. (Rep.)* **4439** (1973); *Chem. Abstr.* **80** (1974) 65 963.
[407] J. S. Henderson, J. L. Weeks, *Ind. Med. Surg.* **40** (1973) 10.
[408] International Agency for Research on Cancer (ed.): *IARC Monographs on the Evaluation of the Carcinogenic Risk of Chemicals to Humans*, vol. **32**, IARC, Lyon 1983.
[409] *Phenanthrene, Dangerous Prop. Ind. Mat. Rep.* **6** (1986) 68–69.
[410] E. E. Sandmeyer in G. D. Clayton, F. E. Clayton (eds.): *Patty's Industrial Hygiene and Toxicology*, 3rd ed., vol. **2 B**, J. Wiley & Sons, New York 1981, p. 3353.
[411] Y. S. Rotenberg, F. D. Mashbits, *Gig. Tr. Prof. Zabol.* **9** (1965) 53–54.
[412] H. W. Gerarde: *Toxicology and Biochemistry of Aromatic Hydrocarbons*, Elsevier Publishing, Amsterdam–London–New York 1960.
[413] S. D. Keimig, R. W. Pattillo, Brookhaven Natl. Lab., Rep. BNL-51885, Upton 1984.
[414] Z. Matijasevic, E. Zeiger, *Mutat. Res.* **142** (1985) 149–152.
[415] M. Sakai, D. Yoshida, S. Mizusaki, *Mutat. Res.* **156** (1985) 61–63.
[416] J. E. Rice, T. J. Hosted, E. J. Lavoie, *Canc. Lett. (Shannon Irel.)* **24** (1984) 327–333.
[417] F. Palitti et al., *Mutat. Res.* **174** (1986) 125–130.
[418] J. R. Babson et al., *Toxicol. Appl. Pharmacol.* **85** (1986) 355–366.
[419] R. P. Bos et al., *Mutat. Res.* **187** (1987) 119–125.
[420] A. L. Reshetyuk, E. J. Talakina, P. A. En'yakova, *Gig. Tr. Prof. Zabol.* **14** (1970) 46–47.
[421] K. Knobloch, S. Szendzikowski, A. Slusarczyk-Zalobna, *Med. Pr.* **20** (1969) 210–222.
[422] D. T. Gibson, V. Subramanian in D. T. Gibson (ed.): *Microbial Degradation of Organic Compounds*, Marcel Dekker, New York–Basel 1984, pp. 181–252.
[423] H. H. Tabak, S. A. Quave, C. J. Mashni, E. F. Barth, *J. Water Pollut. Control Fed.* **53** (1981) 1503–1518.
[424] W. S. Hall, T. J. Leslie, K. L. Dickson, *Environ. Contam. Toxicol.* **36** (1986) 286–293.
[425] G. W. Holcombe, G. L. Phipps, J. T. Fiandt, *Ecotoxicol. Environ. Saf.* **7** (1983) 400–409.
[426] K. Verschueren: *Handbook of Environmental Data on Organic Chemicals*, 2nd ed., Van Nostrand, New York–London 1983.
[427] B. R. Parkhurst in R. A. Conway (ed.): *Environmental Risk Analysis for Chemicals*, Van Nostrand, New York–London, pp. 399–411.
[428] M. Sittig: *Priority Toxic Pollutants*, Noyes Data Corp., Park Ridge, N.J., 1980.

Hydroquinone

PHILLIP M. HUDNALL, Tennessee Eastman Company, Kingsport, Tennessee 37662, United States

1.	Introduction 2941	4.4.	Other Processes 2947
2.	Physical Properties 2942	5.	Environmental Protection ... 2948
3.	Chemical Properties 2942	6.	Quality Specifications and
4.	Production 2945		Analysis 2948
4.1.	Hydroperoxidation of *p*-Diiso-	7.	Uses 2949
	propylbenzene 2945	8.	Toxicology and Occupational
4.2.	Hydroxylation of Phenol ... 2946		Health 2950
4.3.	Oxidation of Aniline 2946	9.	References 2951

1. Introduction

Hydroquinone (**1**) [*123-31-9*] (1,4-benzenediol, 1,4-dihydroxybenzene, *p*-dihydroxybenzene, hydroquinol, quinol), $C_6H_6O_2$, M_r 110.11, first described in 1844 by WOEHLER, was obtained by the addition of hydrogen to 1,4-benzoquinone.

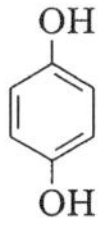

OH
⬡
OH
1

Hydroquinone occurs naturally as hydroquinone β-D-glucopyranoside (arbutin) in the leaves of several plants, including bearberry, cranberry, cowberry, and some varieties of pear [1], usually accompanied by its methyl ether, methylarbutin. Arbutin is easily hydrolyzed to hydroquinone and glucose in hot, dilute aqueous acid; hydrolysis of methylarbutin produces hydroquinone monomethyl ether and glucose.

2. Physical Properties

Hydroquinone is a colorless crystalline solid when pure. Three crystalline modifications exist: the stable α-form, *mp* 173.8 – 174.8 °C, is obtained as hexagonal needles by crystallization from water; the labile γ-form, *mp* 169 °C, is obtained as monoclinic prisms by sublimation; the labile β-modification is obtained as needles or prisms by crystallization from methanol or isopropyl alcohol.

Commercial grades of hydroquinone are typically white to off-white crystalline materials; solutions of hydroquinone are discolored by oxidation on contact with air. Physical constants and solubility data are listed in Table 1. Various spectra of hydroquinone are readily available: IR (prism) [2a], IR (grating) [2b], ^{13}C NMR [2c], proton NMR [2d], UV [2e], and FTIR [3].

3. Chemical Properties

Hydroquinone is easily converted by most oxidizing agents to *p*-benzoquinone [106-51-4] (→ Benzoquinone); even neutral aqueous solutions of hydroquinone darken on exposure to air. The rate of oxidation of hydroquinone by air is accelerated in alkaline solution. The oxidation product can add water to give 1,2,4-benzenetriol; subsequent oxidation may result in the formation of humic acids.

The redox potential E_0 of hydroquinone is 699 mV [5]; the half-wave potential $E_{1/2}$ is 560 mV at pH 0 [5] and 234 mV at pH 5 – 6 [6]. Oxidation of hydroquinone involves loss of an electron and elimination of a proton to give the relatively stable semibenzoquinone radical (**2**). The ability of hydroquinone to act as an antioxidant is a consequence of the formation and stability of this radical. A second one-electron transfer and proton elimination results in the formation of *p*-benzoquinone (**3**).

Hydroquinone and *p*-benzoquinone form the equimolar charge-transfer complex quinhydrone. The quinhydrone complex is typically dark greenish-black, *mp* 171 °C.

Hydroquinone is a reducing agent, whose reduction potential is suitable for reducing silver halide. Grains of silver halide that have been exposed to light are reduced much faster than unexposed or underexposed grains. This property is the basis for the use of hydroquinone as a developing agent in photography. The selectivity is a result of changes in the exposed silver halide crystals, which reduce the energy barrier for reaction with hydroquinone to form elemental silver and *p*-benzoquinone. Sodium sulfite is added to the developing solution to prevent formation of quinhydrone by

Table 1. Physical properties of hydroquinone

mp, °C	172
bp (101.3 kPa), °C	285
d_4^{15}	1.332
Flash point (closed cup), °C	165
Autoignition temperature (closed cup), °C	516
Flammability limits, vol%	1.6 – 15.3
Critical temperature, °C	549
Critical pressure, MPa	7.45
Critical volume, m^3/mol	3.0×10^{-4}
Liquid molar volume, m^3/mol	1.226×10^{-4}
Heat of formation, kJ/mol	−261.71
Gibbs free energy of formation, kJ/mol	−176.13
Absolute entropy, J mol^{-1} K^{-1}	344.17
Heat of fusion at melting point, kJ/mol	27.1
Standard net heat of combustion, kJ/mol	-2.74×10^3
Dipole moment, D	1.40
Van der Waals volume, m^3/mol	5.998×10^{-5}
Van der Waals area, m^2/mol	8.46×10^5
Dissociation constant (30 °C),	
$\quad K_1$	1.22×10^{-10}
$\quad K_2$	9.28×10^{-13}
Vapor pressure (132.4 °C), kPa	0.133
Solubility parameter, (J/m^3)$^{0.5}$	2.3753×10^4
Solubility*, g per 100 g solvent (30 °C)	
$\quad$ Ethanol	46.4
$\quad$ Acetone	28.4
$\quad$ Water	8.3
$\quad$ Benzene	0.06
$\quad$ Tetrachloromethane	0.01

* Reference [4].

converting the *p*-benzoquinone to sodium 2,5-dihydroxybenzenesulfonate (**4**), a weaker developing agent than hydroquinone.

$$\text{hydroquinone} + 2\,\text{AgX} + \text{Na}_2\text{SO}_3 \longrightarrow \underset{\mathbf{4}}{\text{2,5-dihydroxybenzenesulfonate}} + 2\,\text{Ag} + \text{HX} + \text{NaX}$$

The reactivity of hydroquinone is generally similar to that of phenol (→ Phenol). One or both hydroxyl groups may be converted to an ether or ester. Commercially important oxygen derivatives include hydroquinone monomethyl ether (**5**) [*150-76-5*], hydroquinone dimethyl ether (**6**) [*150-78-7*], and hydroquinone bis(2-hydroxyethyl) ether (**7**) [*104-38-1*].

5: 4-methoxyphenol (OCH₃, OH)
6: 1,4-dimethoxybenzene (OCH₃, OCH₃)
7: 1,4-bis(2-hydroxyethoxy)benzene (OCH₂CH₂OH, OCH₂CH₂OH)

Hydroquinone esters undergo the Fries rearrangement to give acyl-substituted hydroquinones.

Hydroquinone and its ethers may be C-alkylated under Friedel–Crafts conditions to produce a variety of mono- and disubstituted products. Among the commercially important alkylated derivatives are 2-*tert*-butylhydroquinone (**8**) [*1948-33-0*], 2,5-bis[2-(2-methylbutyl)]hydroquinone (**9**), and 2-*tert*-butyl-4-methoxyphenol (BHA) (**10**) [*121-00-6*].

8: 2-*tert*-butylhydroquinone
9: 2,5-bis[2-(2-methylbutyl)]hydroquinone
10: 2-*tert*-butyl-4-methoxyphenol (BHA)

Hydroquinone reacts with alkylamines or arylamines to form substituted arylamines. Commercially important products include *p*-N-methylaminophenol (**11**) [*150-75-4*], usually marketed as the sulfate salt [*55-55-0*], and *N,N'*-diphenyl-*p*-phenylenediamine (**12**) [*74-31-7*].

11: *p*-N-methylaminophenol (NHCH₃, OH)
12: *N,N'*-diphenyl-*p*-phenylenediamine (NHPh, NHPh)

Hydroquinone may be chlorinated with chlorine or sulfuryl chloride to provide derivatives that range from mono- to tetrachlorinated. Kolbe–Schmitt carboxylation of hydroquinone with carbon dioxide leads to 2,5-dihydroxybenzoic acid (gentisic acid). Quinizarin [*81-64-1*] (1,4-dihydroxyanthraquinone) can be prepared by the condensation of hydroquinone with phthalic anhydride. Catalytic hydrogenation of hydroquinone gives 1,4-cyclohexanediol (quinitol). Sulfonation of hydroquinone results in the

Table 2. Production of hydroquinone

Process	Capacity, t/a	Major producers/location
Oxidation of p-diisopropylbenzene	25 000	Eastman Chemical Products/United States Goodyear/United States Mitsui/Japan
Hydroxylation of phenol	14 000	Rhône-Poulenc/France Enichem/Italy Ube/Japan
Oxidation of aniline	4 000	People's Republic of China
Total world capacity	43 000	

formation of hydroquinone mono- or disulfonic acid, usually isolated as the corresponding potassium salt.

4. Production

Three processes are used for the industrial production of hydroquinone: hydroperoxidation of p-diisopropylbenzene [100-18-5], hydroxylation of phenol [108-95-2], and oxidation of aniline [62-52-3]. The processes, production capacities, and major producers are listed in Table 2. Although other processes are known, they are not industrially important.

4.1. Hydroperoxidation of p-Diisopropylbenzene

p-Diisopropylbenzene (p-DIPB) is produced by Friedel–Crafts alkylation of benzene with propene. Purified p-DIPB (**13**) is subsequently converted to the dihydroperoxide (DHP) by air oxidation under slightly alkaline conditions at 80 – 90 °C. The DHP (**14**) is separated from the reaction mixture

either by extraction or by crystallization, and is then cleaved to hydroquinone (**1**) and acetone by an acid-catalyzed Hock rearrangement; the DHP solution is treated with sulfuric acid catalyst (0.2–1.0%) at 60–80 °C. The hydroquinone is crystallized and isolated. The overall yield of (**1**) (based on *p*-DIPB) is ca. 80%.

4.2. Hydroxylation of Phenol

The catalyzed hydroxylation of phenol at ca. 80 °C with 70% aqueous hydrogen peroxide produces a mixture of hydroquinone and catechol (**15**). The catalyst may be a strong mineral acid, iron(II), or a cobalt(II) salt. Depending on catalyst selection, the ratio of catechol to hydroquinone can be varied from 3:1 to 0.1:1; in practice, the ratio is typically 1.5:1, i.e., catechol is the major product. The reaction proceeds by an ionic mechanism in which hydrogen peroxide is polarized by the strong acid catalyst and phenol is subsequently hydroxylated the resulting isomers are separated by a series of extractions and solvent-stripping operations.

The ratio of the products hydroquinone and catechol may be influenced by the presence of superacids or shape-selective zeolites. Thus, use of a vanadium–modified Nafion perfluorosulfonate polymer for the oxidation of phenol gave a hydroquinone–catechol ratio of 12.5:1 [7]. The use of a shape–selective zeolite in the hydroxylation of phenol has yielded hydroquinone with 99% selectivity [8].

4.3. Oxidation of Aniline

Oxidation of aniline [62-53-3] is the oldest process used for the production of hydroquinone. Aniline is oxidized with manganese dioxide (15–20% excess) in aqueous sulfuric acid at 0–5 °C to produce *p*-benzoquinone (**3**). This intermediate is removed from the reaction mixture by steam stripping and collected. A byproduct, manganese sulfate, may be retrieved from the depleted reaction mixture and sold for agricultural applications.

Hydroquinone is obtained from the intermediate *p*-benzoquinone by reduction with iron at 55–65 °C or by catalytic hydrogenation. The product (usually technical grade) is crystallized, isolated from the typically aqueous stream by centrifugation, and dried in a vacuum dryer. The overall yield of hydroquinone from aniline is ca. 85%.

$$2 \; C_6H_5NH_2 + 4\,MnO_2 + 5\,H_2SO_4 \longrightarrow 2\;\mathbf{3}\;(\text{p-benzoquinone}) + (NH_4)_2SO_4 + 4\,MnSO_4 + 4\,H_2O$$

$$\mathbf{3} \xrightarrow[\text{or [H]}]{Fe} \mathbf{1}\;(\text{hydroquinone})$$

Aniline oxidation is a batchwise process and, therefore, relatively labor intensive. The use of ground manganese ore and finely divided iron leads to extremely abrasive processing conditions; a great deal of maintenance is usually required. Disposal of the inorganic coproducts (amounting to ca. 85% of the total weight of products) is of environmental concern.

4.4. Other Processes

Carboxylate esters of aromatic diols can be prepared by Baeyer–Villiger oxidation of a 4-hydroxy-substituted aromatic ketone with 97% conversion and 97% efficiency [9]. Microbiological oxidation of either benzene or phenol produces hydroquinone with very high selectivity [10]. Copper-catalyzed air oxidation of phenol provides p-benzoquinone in greater than 90% selectivity [11]. Benzene has been oxidized to hydroquinone in the presence of copper(I) chloride [12] or titanium [13]. Oxidation of benzene in aqueous solution with ozone gives p-benzoquinone and hydroquinone [14].

Other preparative methods include the reaction of p-isopropenylphenol with 30% aqueous hydrogen peroxide under acidic conditions [15]; p-isopropenylphenol may be obtained from bisphenol A [80-05-7] by alkaline cracking. Hydroquinone can also be produced by the carbonylation of acetylene [16], [17]; catalytic hydrogenation of nitrobenzene in acidic solution [18]; acid hydrolysis of nitrobenzene [19] or p-nitrosophenol [20]; and electrochemical oxidation of benzene or phenol in dilute sulfuric acid, followed by reduction [21], [22].

5. Environmental Protection

An estimate of environmental effects can be obtained from the data available [23]–[29]. Hydroquinone has a high BOD and can cause oxygen depletion in aqueous systems. It has high potential for affecting some aquatic organisms and moderate potential for affecting microorganisms used in secondary waste treatment. Hydroquinone has a moderate potential to affect the germination or early growth of some plants. It is readily biodegraded and has a low tendency to concentrate in biological systems. Direct, instantaneous discharge of hydroquinone to a large body of water at a final concentration of 0.005 mg/L or less, followed by secondary waste treatment, is not expected to cause adverse environmental effects.

6. Quality Specifications and Analysis

Five grades of hydroquinone are available: photographic, technical, U.S.P., inhibitor, and polyester. The most important grades (in terms of volume) are photographic and technical. The quality requirements for photographic-grade hydroquinone are defined in American National Standards Institute (ANSI) Specification PH 4.126-1972. Manufacturers specifications for photographic and technical grades are given in Table 3. Specifications for U.S.P.-grade hydroquinone, used in pharmaceutical applications, are described in U.S.P.-XXI [30]. Inhibitor and polyester grades are used in special applications, and the specifications for each grade are provided by the producer. Hydroquinone may be analyzed by a variety of techniques, including spectroscopic, chromatographic, titrimetric, and electrochemical methods.

Table 3. Specifications of hydroquinone

Property	Specifications	
	Photographic [32]	Technical [31]
Appearance	free-flowing white crystals	white to offwhite crystals
mp, °C (min.)	171	169
Assay, wt% (min.)	99.4	99.0
Ash content, wt% (max.)		0.04
pH (5% aqueous solution)	4.1–4.7	
Iron, ppm (max.)	10	
Water, % (max.)	0.6	1.0
Resorcinol, ppm (max.)	50	
Solubility in water		complete
Molar absorptivity* (max.) mL g^{-1} cm^{-1} (λ_{max}, nm)	0.07 (700) 0.12 (520) 0.19 (455)	1.0 (455)

* In 0.1 M aqueous hydrochloric acid.

Table 4. Worldwide demand for hydroquinone and its derivatives by market segment (1987)

Market segment	Demand	
	t	% of total
Photography	11 600	37
Rubber industry	8 100	26
Monomer inhibitors	3 100	10
Dyes and pigments	2 800	9
Antioxidants	650	2
Agricultural chemicals	650	2
Others	4 400	14
Total	31 300	100

7. Uses

Hydroquinone and its derivatives are used in photographic applications, the rubber industry, monomer inhibitors, dyes and pigments, antioxidants, agricultural chemicals, and other diverse and special applications. Demand for hydroquinone and its derivatives is shown in Table 4, according to market segment. Consumption (in tonnes) of hydroquinone and derivatives by geographical area is as follows:

United States	13 100
Europe	9 600
Japan	3 800
Other	4 800
Total	31 300

The largest demand for hydroquinone is as a photographic developer, principally for black-and-white film, lithography, photochemical machining, microfilm, and X-ray film. Many derivatives of hydroquinone are used in photographic applications, e.g., the sulfate salt of *p-N*-methylaminophenol (**11**) and potassium 2,5-dihydroxybenzenesulfonate.

The second largest consumer of hydroquinone is the rubber industry, which requires hydroquinone for the production of antioxidants and antiozonants. Hydroquinone derivatives used in this area include *N,N'*-diaryl-*p*-phenylenediamines (e.g., *N,N'*-diphenyl-*p*-phenylenediamine), dialkylated hydroquinones, *N*-alkyl-*p*-aminophenols, dialkyl-*p*-phenylenediamines, and aralkyl-*p*-phenylenediamines.

Hydroquinone, hydroquinone monomethyl ether, and *p*-benzoquinone are used extensively in the vinyl monomer industry to inhibit free-radical polymerization during both processing and storage. Hydroquinone, *p*-benzoquinone, 2-methylhydroquinone (toluhydroquinone), 2-*tert*-butylhydroquinone, and 2,5-di-*tert*-butylhydroquinone are used as stabilizers for unsaturated polyester resins. Food-grade antioxidants include 2-*tert*-butylhydroquinone (**8**) and 2-*tert*-butyl-4-methoxyphenol (**10**).

Hydroquinone dimethyl ether (**6**) is used as a starting material for a family of dyes and pigments based on the 2-amino- and 2-amino-5-chloro derivatives. Quinizarin is an intermediate for textile dyes. The fungicide Chloroneb is produced from (**6**) as raw material. The herbicide ethofumesate is produced from *p*-benzoquinone. Fluazifop-butyl [33] is representative of a new family of herbicides based on the *o*-alkylation of hydroquinone with 2-halopropionic acid derivatives [34] – [37].

Hydroquinone bis(2-hydroxyethyl) ether (**7**), the product of the reaction of hydroquinone with ethylene oxide, is used as a chain extender in thermosetting urethane polymers [38].

Hydroquinone (U.S.P. grade) and several derivatives are used in topical formulations as skin bleaching and depigmenting agents.

Several new applications for hydroquinone and its derivatives are emerging. The oxygen-scavenging properties of hydroquinone are being exploited for use in boiler water treatment [39]. Hydroquinone and certain C-alkylated or Carylated derivatives are useful monomers for the preparation of a variety of polymers, including liquid crystal polyesters for high-performance plastics, composites, and fibers. In addition to high tensile and impact strengths, these materials exhibit good weatherability, solvent resistance, flame retardance, transparency to microwave radiation, and retention of strength at elevated temperature [40], [41].

8. Toxicology and Occupational Health

Acute toxicity data are shown in Table 5. In a two-year study of rats receiving a diet containing up to 1% hydroquinone, no effects were observed on final body weight, hematology, or pathology; at 5% in the diet, a 46% weight loss together with toxic effects on the bone marrow and liver was found within nine weeks [43]. When applied to the skin of mice, hydroquinone was not carcinogenic and did not promote the carcinogenic activity of benzo[*a*]pyrene [44]. In a skin-painting study in rats, croton oil failed to promote any carcinogenicity from hydroquinone [45]. Hydroquinone produced a negative response in the *Salmonella* bacterial mutagenesis assay [46]. Positive results were observed in the micronucleus test [47] and in an in vitro test for sister chromosome exchanges in human lymphocytes [48].

Maternal toxicity and fetotoxicity were observed in pregnant rats given 300 mg/kg of hydroquinone daily, but no compound-related teratogenic effects occurred; no adverse effects were observed at doses up to 100 mg/kg per day [23]. Rats given a single dose of 200 mg/kg of (^{14}C) hydroquinone excreted 91.9% in the urine within 2 – 4 d, 3.8% in the feces, and 0.39% in expired air, with 1.25% remaining in the body; radiolabeled compounds identified in the urine included hydroquinone monoglucuronide (56 – 66%), hydroquinone monosulfate (25 – 42%), and hydroquinone (1.1 – 8.6%)

Table 5. Toxicity data for hydroquinone

Test	Species	Result
Acute oral LD$_{50}$	rat	400 mg/kg
Acute oral LD$_{50}$	mouse	100–200 mg/kg
Dermal LD$_{50}$	guinea pig	> 1000 mg/kg
Skin irritation	guinea pig	slight
Eye irritation	rabbit	moderate erythema, clearing by day 14
Sensitization (guinea pig)		
Freund's adjuvant		3/10 slight
Modified Buhler		7/15 low, 2/10 moderate
Maximization [42]		70% sensitized, strong

[49]. The percutaneous absorption rate of (^{14}C) hydroquinone in beagles was ca. 1.1 μg cm^{-2} h^{-1}, indicating that hydroquinone is poorly absorbed through the skin [23].

The current OSHA Permitted Exposure Limit (PEL) and ACGIH Threshold Limit Value (TLV) for hydroquinone is 2 mg/m^3 TWA. Effects of human exposure are well documented (see below). Direct contact of hydroquinone dust and quinone vapor with the eye may cause irritation. Chronic exposure at high levels has caused brownish discoloration of the cornea and conjunctiva, as well as distortion of the cornea, leading in some cases to decreased visual acuity and even blindness. Exposure to levels below the TLV of 2 mg/m^3 has not produced these effects [50].

A low incidence of allergic contact dermatitis has been reported from handling large amounts of hydroquinone, where exposure to dust occurred, and from repeated contact with alkaline photographic developer solutions [23]. Skin sensitization was noted in 2 members of a group of 48 middle-aged males who used a depigmentation ointment containing 5% hydroquinone; 38 subjects treated with a 2% ointment and 26 subjects treated with a 3% ointment showed no sensitization [51], [52].

In reported poisonings, the lethal dosage for adults is 5–12 g (70–170 mg/kg). Lower doses cause no adverse effects: two men ingested 500 mg/d for five months and 17 men and women ingested 300 mg/d for three to five months without abnormal symptoms or any abnormalities in blood and urine [43].

9. References

[1] *Beilstein*, VI$_3$, 4374.

[2] *Sadtler Standard Spectra*, Sadtler Research Laboratories, Division of Bio-Rad Laboratories, a. Infrared Prism Spectrum nos. 153, 13206; b. Infrared Grating Spectrum no. 47; c. Carbon-13 NMR Spectrum no. 38; d. Proton NMR Spectrum no. 10350; e. Ultraviolet Spectrum no. 60.

[3] C. J. Pouchert (ed.), *The Aldrich Library of FTIR Spectra*, Aldrich Chemical Company, Milwaukee, Wisconsin 1984, vol. **1**, no. 1107D.

[4] A. Seidell: *Solubilities of Organic Compounds*, 3rd ed., vol. **2**, Van Nostrand Publishing, New York 1941, p. 396.
[5] H. Musso, H. Doepp, *Chem. Ber.* **100** (1967) 3267.
[6] P. J. Elving, A. F. Krivis, *Anal. Chem.* **30** (1958) 1645.
[7] FMC, EP-A 132 783, 1985 (R. A. Bull).
[8] Mobil Oil, US 4 578 521, 1986 (C. D. Chang, S. D. Hellring).
[9] Celanese, EP-A 178 929, 1986; US-A 661 552, 1984 (H. R. Gerberich).
[10] Agency of Industrial Sciences & Technology, JP-Kokai 8 678 394, 1986 (A. Yoshikawa).
[11] J. E. Lyons, C-Y. Hsu, Biological and Inorganic Copper Chemistry, in K. D. Karlin, J. Zubieta (eds.): *Proc. Conf. Copper Coord. Chem. (July 23–27, 1984)*, vol. **2**, Adenine Press, Guilderland, New York 1986, pp. 57–76.
[12] J. Van Gent, A. A. Wismeijer, A. W. P. G. Peters-Rit, H. Van Bekkum, *Tetrahedron Lett.* **27** (1986) 1059–1062.
[13] Mitsui Petrochemicals Industries, JP-Kokai 859 889, 1985.
[14] C. H. Kuo, H. S. Soong, *Chem. Eng. J. (Lausanne)* **28** (1984) 163–171.
[15] Upjohn, DE 2 214 971, 1972, and GB 1 344 602, 1974 (P. S. Charleton).
[16] Ajinomoto, US 3 420 895, 1969 (H. Wakamatsu).
[17] Lonza, GB 1 215 568, 1970 (P. Pino).
[18] Koppers, US 3 953 509, 1974 (N. P. Greco).
[19] Mitsui Toatsu, JP-Kokai 77 102 235, 1976 (F. Matsuda).
[20] Upjohn, US 3 676 503, 1972 (W. v. E. Doering, W. J. Farrisey, Jr.).
[21] M. Ya. Fioshin, *Elektrokhimiya* **13** (1977) 381.
[22] M. Fremery, H. Hoever, G. Schwarzlose, *Chem. Ing. Technol.* **46** (1974) 635.
[23] Unpublished results, Health & Environment Laboratories, Eastman Kodak, Rochester, New York.
[24] K. Verschueren: *Handbook of Environmental Data on Organic Chemicals*, 2nd ed., Van Nostrand Reinhold, New York 1983, pp. 746–748.
[25] *Environmental Effects of Photoprocessing Chemicals*, vols. **1** and **2**, National Association of Photographic Manufacturers, and Hydroscience, Harrison, New York 1974.
[26] G. M. De Graeve, D. L. Geiger, J. S. Meyer, H. L. Bergman, *Arch. Environ. Contam. Toxicol.* **9** (1980) 557.
[27] J. E. McKee, H. W. Wolf (eds.): *Water Quality Criteria*, State of California, 1963, Publication no. 3-A.
[28] I. Juhnke, D. Luedemann, *Z. Wasser Abwasser Forsch.* **11** (1978) 161.
[29] A. J. Leo, C. Hansch (eds.): *Chemical Parameter Data Base*, Medicinal Chemistry Project, Pomona College, Seaver Chemistry Laboratory, Claremont, Calif. 1985.
[30] The United States Pharmacopeia XXI, The United States Pharmacopeial Convention, Rockville, Maryland, 1985.
[31] *Hydroquinone, Technical Grade, Sales Specification* no. 3503-9, Tennessee Eastman Company, Kingsport, Tenn.
[32] *Hydroquinone, Photographic Grade, Sales Specification* no. 4305-1, Tennessee Eastman Company, Kingsport, Tenn.
[33] Ishihara Sangyo Kaisha, DE 2 812 571, 1979 (R. Nishiyama et al.).
[34] Dow Chemical Co., DD 217 694, 1985 (H. Johnson, L. H. Troxell).
[35] Dow Chemical Co., EP-A 97 460, 1984 (H. Johnson, L. H. Toxell).
[36] Bayer AG, DE 3 219 789, 1983 (H. Foerster et al.).
[37] Zoecon Corp., US 4 529 438, 1985 (S. Lee).

[38] Farbenfabriken Bayer Aktiengesellschaft and Mobay Chemical Company, US 3 016 364, 1962 (E. Muller).
[39] S. Romaine, I. J. Cotton, *Proceedings American Power Conference* **48** (1986) 1066.
[40] D. G. Baird, *Fiber Producer* (October 1984) 66.
[41] A. S. Wood, *Modern Plastics* (April 1985) 78.
[42] B. F. J. Goodwin, R. W. R. Crevel, A. W. Johnson, *Contact Dermatitis* **7** (1981) 248.
[43] A. J. Carlson, N. R. Brewer, *Proc. Soc. Exp. Biol. Med.* **84** (1953) 684.
[44] B. L. van Duuren, B. M. Goldschmidt, *J. Nat. Cancer Inst.* **56** (1976) 1237.
[45] *IARC Monographs on the Evaluation of the Carcinogenic Risk of Chemicals to Humans*, vol. **15**, International Agency for Research on Cancer (IARC), Lyon, France 1977, pp. 155–175.
[46] I. Florin, L. Rutberg, M. Curvall, C. R. Enzell, *Toxicology* **15** (1980) 219.
[47] E. Gocke, M.-T. King, K. Eckhardt, D. Wild, *Mutat. Res.* **90** (1981) 91.
[48] K. Morimoto, S. Wolff, A. Koizumi, *Mutat. Res.* **119** (1983) 355.
[49] G. D. Divincenzo, M. L. Hamilton, R. C. Reynolds, D. A. Ziegler, *Toxicology* **33** (1984) 9.
[50] F. L. Oglesby, R. L. Raleigh: *Eye Injury Associated with Exposure to Hydroquinone and Quinone*, 2nd ed., Laboratory of Industrial Medicine, Tennessee Eastman Co., Kingsport, Tenn. 1973, p. 2.
[51] *NIOSH Criteria for a Recommended Standard for Hydroquinone*, National Institute for Occupational Safety & Health (NIOSH), U.S. Government Printing Office, Wahington, D.C., 1978.
[52] M. C. Spencer, *J. Am. Med. Assoc.* **194** (1965) 962.

Hydroxycarboxylic Acids, Aliphatic

Individual keywords: → *Butyrolactone;* → *Citric Acid;* → *Lactic Acid;* → *Tartaric Acid.*

KARLHEINZ MILTENBERGER, Hoechst Aktiengesellschaft, Gersthofen, Federal Republic of Germany

1.	Introduction	2955	5.1.1.	Preparation	2965
2.	General Characteristics	2956	5.1.2.	Uses	2965
2.1.	Physical Properties	2956	5.1.3.	Derivatives	2966
2.2.	Chemical Properties	2956	5.2.	Hydroxypropionic Acids and β-Propiolactone	2967
3.	Preparation	2960			
4.	Analysis	2964	5.3.	Hydroxybutyric Acids	2968
5.	Specific Aliphatic Hydroxycarboxylic Acids of Commercial Significance	2964	5.4.	(R,S)-Malic Acid [(R,S)-Hydroxysuccinic Acid, (R,S)-Hydroxybutanedioic Acid]	2969
5.1.	Glycolic Acid (Hydroxyacetic Acid, 2-Hydroxyethanoic Acid)	2964	6.	Toxicology	2970
			7.	References	2971

1. Introduction

Hydroxycarboxylic acids are widely distributed in nature, playing an important role as metabolic intermediates in both plants and animals. Examples can be found in the tricarboxylic acid cycle leading to the degradation (β-oxidation) of fatty acids, as well as in the fermentation of sugars. Hydroxycarboxylic acids are also excreted in the urine, especially in metabolic disorders. Thus, diabetics are incapable of further metabolizing the β-hydroxybutyric acid and acetoacetic acid derived from butyric acid, so these materials are simply eliminated.

One of the first reports of a naturally occurring hydroxycarboxylic acid occurred in 1780 when SCHEELE described the isolation of (R)-(−)-lactic acid from soured milk. The substance is formed during the fermentation of lactose by lactobacilli. SCHEELE also discovered S-(−)-malic acid (in 1785), S-(+)-tartaric acid (in 1769), and citric acid (in 1784) isolating each as its calcium salt. Another member of the group, glycolic acid, is a constituent of unripe fruit, especially grapes. Related natural substances include tartronic, hydroxystearic, mandelic (see → Hydroxycarboxylic Acids, Aromatic), and numerous other fruit- or sugar-derived acids.

Definition In contrast to the purely aromatic hydroxycarboxylic acids (phenolic carboxylic acids), which contain phenolic hydroxyl groups on the aromatic nucleus and thus display enhanced acidity, aliphatic hydroxycarboxylic acids are characterized by purely aliphatic alcohol hydroxyl groups.

Aliphatic hydroxycarboxylic acids may have one or several hydroxyl groups, together with one or more carboxyl groups. In the simplest case, the monohydroxymonocarboxylic acids, the compounds can be conveniently subdivided into three categories:

1) 2- or α-hydroxycarboxylic acids,
2) 3- or β-hydroxycarboxylic acids, and
3) hydroxycarboxylic acids with the hydroxyl group located in the 4- (i.e., γ-) position, or even more remote from the carboxyl group.

Table 1 lists several commercially important hydroxycarboxylic acids and related compounds, grouped in accordance with the above scheme.

2. General Characteristics

The overall characteristics of a hydroxycarboxylic acid are a function of the number of hydroxyl and carboxylic acid groups present as well as their relative placement.

2.1. Physical Properties

Most naturally occurring hydroxycarboxylic acids are chiral and thus exhibit optical activity. In pure form, they are usually crystallizable. Hydroxycarboxylic acids have exceptionally high boiling points relative to corresponding unsubstituted acids with the same chain length. This is a consequence of association among the hydroxyl groups. For this reason, many of these compounds cannot be distilled without decomposition, even under vacuum.

Table 2 summarizes the physical properties of the commercially important hydroxycarboxylic acids and their derivatives listed in Table 1.

2.2. Chemical Properties

Hydroxycarboxylic acids readily form the expected salts at the carboxyl group. The acidity of α- and β-hydroxycarboxylic acids is enhanced relative to the corresponding unsubstituted acids because of the proximity of the polar hydroxyl group. Compounds in this category may be esterified at either the carboxyl or the hydroxyl function by reaction with other alcohols or acids, respectively. The actual course of such reactions

Table 1. Commercially important hydroxycarboxylic acids and derivatives

Type	Trivial name	CAS registry no.	Formula	Molecular mass, M_r
Monocarboxylic acids				
One −OH group in the α-position	glycolic acid	[79-14-1]	$HOCH_2-COOH$	76.03
	n-butyl glycolate	[7397-62-8]	$HOCH_2-COO-C_4H_9$	132.16
	(R,S)-mandelic acid	[611-72-3]	$C_6H_5-CHOH-COOH$	152.15
	(R,S)-lactic acid	[598-82-3]	$CH_3-CHOH-COOH$	90.08
	(R,S)-α-hydroxybutyric acid	[600-15-7]	$CH_3CH_2-CHOH-COOH$	104.11
One −OH group in the β-position	hydracrylic acid	[503-66-2]	$HOCH_2-CH_2-COOH$	90.08
	β-propiolactone	[57-57-8]	$\begin{array}{l}CH_2-C=O\\ \,\,\,\,\vert\,\,\,\,\,\,\,\,\,\,\vert\\ CH_2-O\end{array}$	72.06
	(R,S)-tropic acid	[552-63-6]	$HOCH_2-CH(C_6H_5)-COOH$	166.18
	(R,S)-β-hydroxybutyric acid	[625-71-8]	$CH_3-CHOH-CH_2-COOH$	104.11
One −OH group in the γ-position (or more remote)	γ-hydroxybutyric acid	[591-81-1]	$HOCH_2-CH_2-CH_2-COOH$	104.11
	γ-butyrolactone	[96-48-0]	$\begin{array}{l}CH_2-CH_2-O\\ \,\,\,\,\vert\,\,\,\,\,\,\,\,\,\,\,\,\,\,\,\,\,\,\,\vert\\ CH_2\text{———}C=O\end{array}$	86.09
	(R,S)-ricinoleic acid	[141-22-0]	$\begin{array}{l}CH-CH_2-CHOH-C_6H_{13}\\ \parallel\\ CH-(CH_2)_7-COOH\end{array}$	298.45
Two −OH groups	(R,S)-glyceric acid	[600-19-1]	$HOCH_2-CHOH-COOH$	
Dicarboxylic acids				
One −OH group	tartronic acid	[80-69-3]	$HOOC-CHOH-COOH$	120.06
	(R,S)-malic acid	[617-48-1]	$HOOC-CH_2-CHOH-COOH$	134.09
Two −OH groups	(R,S)-tartaric acid	[133-37-9]	$\begin{array}{l}\,\,\,\,\,\,\,\,\,\,\,\,\,\,\,OH\\ \,\,\,\,\,\,\,\,\,\,\,\,\,\,\,\vert\\ HOOC-CH-CH-COOH\\ \,\vert\\ \,OH\end{array}$	150.09
	meso-tartaric acid	[147-73-9]	$\begin{array}{l}HOOC-CH-CH-COOH\\ \,\vert\,\,\,\,\,\,\,\,\vert\\ \,OH\,\,\,OH\end{array}$	150.09
Tricarboxylic acids	citric acid	[77-92-9]	$\begin{array}{l}\,COOH\\ \,\vert\\ CH_2\text{—}C\text{—}CH_2\\ \,\vert\,\,\,\,\,\,\,\,\,\,\,\vert\,\,\,\,\,\,\,\,\,\vert\\ COOH\,\,OH\,\,COOH\end{array}$	192.13

depends once again on the placement of the hydroxyl group relative to the carboxylic acid group.

Both α- and β-hydroxycarboxylic acids readily form esters with other alcohol groups but the latter also have a tendency to eliminate water under the reaction conditions. The γ- and δ-hydroxycarboxylic acids undergo intramolecular esterification to form lactones, this reaction being faster than intermolecular esterification. Compounds in which the hydroxyl group is more remote from the acid moiety are again subject to normal esterification.

α-Hydroxycarboxylic acids dimerize with elimination of water to produce six-membered cyclic diesters called 1,4-dioxane-2,5-diones according to IUPAC. These are commonly referred to as lactides, because the phenomenon was first encountered with lactic acid.

Hydroxycarboxylic Acids, Aliphatic

Table 2. Physical properties of the hydroxycarboxylic acids listed in Table 1.

Hydroxycarboxylic acid	mp, °C	bp, °C	Density, d_4^{20}	Refractive index, n_D^{20}	$pK_{A_1}^{25}$, (25 °C)
Glycolic acid	78–80	100 (decomp.)	1.49 (at 25 °C)		3.81
n-Butyl glycolate		178–186 or 82–97 (at 2.7 kPa)	1.015–1.023	1.423–1.4246	
(R,S)-Mandelic acid	118–121		1.300		3.37
(R,S)-Lactic acid	23–33	119–123 (at 1.6–2 kPa)	1.206 (25 °C)	1.4392	3.86
(R,S)-α-Hydroxybutyric acid	43–44	138 (at 1.9 kPa) or 260 (decomp.)	1.125		
Hydracrylic acid	syrup	decomp.	1.0474	1.4489	4.51
β-Propiolactone	−33.4	51 (at 1.4 kPa) or 162 (decomp.)	1.1460	1.4135	
(R,S)-Tropic acid	118	160 (decomp.)			
(R,S)-β-Hydroxybutyric acid	48–50 or syrup	94–96 (at 0.01 kPa) or 130 (at 1.5–1.9 kPa)		1.4424	4.39
γ-Hydroxybutyric acid	−17	decomp.			4.71
γ-Butyrolactone	−43.5	206	1.13	1.4352	
(R,S)-Ricinoleic acid	5	230–235 (at 1.2 kPa)	0.94 (at 15 °C)	1.46393 (at 45 °C)	
(R,S)-Glyceric acid	syrup				
Tartronic acid	156–158 (decomp.)				2.31
(R,S)-Malic acid	131		1.601		3.4
(R,S)-Tartaric acid	205				3.03
meso-Tartaric acid	159–160				3.11
Citric acid	153		1.665		3.13

$$\underset{\substack{R\\H}}{>}C\underset{OH}{\overset{\overset{O}{\|}}{-}}C-OH + \underset{\substack{HO\\HO-C}}{>}C\underset{\overset{\|}{O}}{<}\underset{R}{\overset{H}{}}$$

$$\longrightarrow \underset{\substack{R\\H}}{>}C\underset{O-C}{\overset{\overset{O}{\|}}{-}}\underset{\overset{\|}{O}}{<}C\underset{R}{<}\overset{H}{} + 2\,H_2O$$

Lactide

The process may be accompanied by intermolecular esterification, which produces linear polyesters.

$$\underset{R}{HOCH-COOH} + n\,\underset{R}{HOCH-COOH}$$

$$\longrightarrow \underset{R}{HOCH-}\left[\underset{R}{COOCH}\right]_n-COOH + n\,H_2O$$

Note that the latter type of esterification, known as *estolide formation*, is not restricted to α-hydroxycarboxylic acids. Another important characteristic of both α- and β-hydroxycarboxylic acids is that the proximity of the two functional groups weakens the intervening C–C bond(s). Thus, treatment of such compounds with sulfuric acid results in the elimination of formic acid, which decomposes in the presence of concentrated sulfuric acid to carbon monoxide and water.

$$\underset{\substack{OH\\|\\RCH-COOH}}{} \begin{array}{c} \nearrow +\,H_2SO_4\ \text{dil.} \longrightarrow R-CH=O + HCOOH \\ \\ \searrow +\,H_2SO_4\ \text{conc.} \longrightarrow R-CH=O + CO + H_2O \end{array}$$

The rest of the chain is converted into aldehydes (in the case of α-hydroxycarboxylic acids) or ketones (in the case of β-hydroxycarboxylic acids) with one fewer carbon atom.

This elimination reaction proceeds so smoothly that it is used for quantitative determination of certain hydroxycarboxylic acids. Thus, lactic acid is converted easily to acetaldehyde, and β-hydroxybutyric acid to acetone. Oxidative cleavage is also possible. For example, treatment of α-hydroxycarboxylic acids with hydrogen peroxide in the presence of iron(II) ions leads to easy elimination of carbon dioxide to produce again aldehydes containing one fewer carbon atom in the chain.

β-Hydroxycarboxylic acids readily undergo intramolecular elimination of water to produce α,β-unsaturated carboxylic acids; e.g.,

$$\underset{OH}{CH_2-CH_2-COOH} \xrightarrow{-H_2O} CH_2=CH-COOH$$

Thus, β-hydroxypropionic acid (hydracrylic acid) can be dehydrated to acrylic acid, and β-hydroxybutyric acid to crotonic acid.

The γ- and δ-*hydroxycarboxylic acids* undergo dehydration even at room temperature, yielding γ- and δ-lactones. The corresponding free acid may be regenerated in aqueous alkaline medium, but often it is stable only in the form of its salts. Lactones are internal esters of hydroxycarboxylic acids.

$$\begin{array}{c} H_2C\text{—}CH_2 \\ | \quad\quad | \\ H_2C \quad C=O \\ | \quad\quad | \\ OH \quad OH \end{array} \xrightarrow{-H_2O} \begin{array}{c} H_2C\text{—}CH_2 \\ | \quad\quad | \\ H_2C \quad C=O \\ \quad\backslash \;\; / \\ \quad\; O \end{array}$$

Treatment of a lactone with ammonia produces the corresponding *lactam*.

The related derivatives of α-hydroxycarboxylic acids, the three-membered α-lactones, are unstable and are considered only as reaction intermediates. Four-membered lactones (β-lactones from β-hydroxycarboxylic acids) are known, but special methods are required to prepare them [1], [2]. The exceptional reactivity of these compounds is a consequence of their highly strained rings. For example, they react with ammonia to form amides of the corresponding hydroxycarboxylic acids, which are subject to further conversion to amino acids.

Because of the properties described above, the hydroxyl group of a hydroxycarboxylic acid can be converted into various derivatives. Thus, treatment of lactic acid with hydrogen bromide results in α-bromopropionic acid. Hydrogen iodide can be used to reduce a hydroxycarboxylic acid to the corresponding unsubstituted acid, which is particularly easy with acids from the sugar family.

Saccharic acid lactones are converted easily to polyhydroxyaldehydes (aldoses).

Phosphorus pentachloride leads to replacement of the hydroxyl group in a hydroxycarboxylic acid by chlorine, with the carboxyl group simultaneously converted into an acid chloride moiety. Treatment with *thionyl chloride* results in an unstable sulfurous ester of the acid chloride, which will react, with an alcohol to give a hydroxyester or with ammonia to yield a hydroxyamide. If the hydroxyl group is protected by acetylation, thionyl chloride treatment leads only to the corresponding acid chloride.

3. Preparation

In addition to their isolation from natural sources, hydroxycarboxylic acids can also be obtained by a number of synthetic routes. One general approach involves hydrolysis of a halocarboxylic acid:

$$X\text{–}R\text{–}COOH + OH^- \longrightarrow HO\text{–}R\text{–}COOH + X^-$$

This route is in most cases quite straightforward for the preparation of α-hydroxycarboxylic acids (e.g., glycolic acid), but β- and γ-halocarboxylic acids lead instead to the corresponding lactones.

Certain hydroxycarboxylic acids may also be prepared by acid- or base-catalyzed hydration of an unsaturated carboxylic acid.

$$\begin{array}{c} CH-COOH \\ \parallel \\ CH-COOH \end{array} + H_2O \longrightarrow \begin{array}{c} CHOH-COOH \\ | \\ CH_2-COOH \end{array}$$

For example, either maleic or fumaric acid can serve as a source of racemic tartaric acid, and acrylic acid can be converted in this way to β-hydroxypropionic acid (hydracrylic acid).

α-Hydroxycarboxylic acids may be prepared by the action of zinc and an alkyl iodide on an oxalate ester. Cyanohydrin synthesis is another synthetic route to α-hydroxycarboxylic acids, one example being the synthesis of (R,S)-lactic acid from acetaldehyde via lactonitrile:

$$R-CH=O + HCN \longrightarrow \begin{array}{c} RCH-CN \\ | \\ OH \end{array}$$

$$\xrightarrow{\text{Hydrolysis}} \begin{array}{c} RCH-COOH \\ | \\ OH \end{array}$$

Another interesting preparative method involves diazotization as a means of deaminating an amino acid:

$$\begin{array}{c} R-CH-COOH \\ | \\ NH_2 \end{array} \xrightarrow{HNO_2} \begin{array}{c} R-CH-COOH \\ | \\ OH \end{array} + N_2 + H_2O$$

α-Olefins react with liquid dinitrogen tetroxide at 15–20 °C to produce nitrate esters of α-hydroxycarboxylic acids, compounds whose electrochemical reactivity is described in [3].

β-Hydroxycarboxylic acids can be prepared from epoxides by treatment with hydrogencyanide, followed by hydrolysis of the intermediate nitriles:

$$\begin{array}{c} CH_2-CH_2 \\ \diagdown O \diagup \end{array} + HCN \longrightarrow \begin{array}{c} CH_2-CH_2-CN \\ | \\ OH \end{array}$$

$$\xrightarrow{\text{Hydrolysis}} \begin{array}{c} CH_2-CH_2-COOH \\ | \\ OH \end{array}$$

An alternative is the Kolbe nitrile synthesis involving addition of hypochlorous acid to an olefin. This results initially in a chlorohydrin, which can be converted with cyanide ion to a hydroxynitrile, which is then hydrolyzed:

$$\begin{array}{c} CH_2 \\ \parallel \\ CH \\ | \\ R \end{array} + \begin{array}{c} Cl \\ | \\ OH \end{array} \longrightarrow \begin{array}{c} CH_2-Cl \\ | \\ CHOH \\ | \\ R \end{array} \xrightarrow[-NaCl]{+NaCN} \begin{array}{c} CH_2-CN \\ | \\ CHOH \\ | \\ R \end{array}$$

$$\xrightarrow{\text{Hydrolysis}} \begin{array}{c} CH_2-COOH \\ | \\ CHOH \\ | \\ R \end{array}$$

β-Hydroxycarboxylic acids can also be prepared from α-halocarboxylic acid esters by Reformatsky reaction with an aldehyde or ketone in the presence of activated zinc [4]. Another approach utilizes β-lactones, which are readily accessible through ketene treatment of aldehydes or ketones [5]. Thus, acetaldehyde reacts with ketene in the presence of zinc salts to give β-butyrolactone. Finally, esters of β-ketoacids are subject to catalytic hydrogenation, a method that is also applicable to the synthesis of γ- and δ-hydroxycarboxylic acids from the corresponding γ- and δ-ketoesters.

$$CH_3-\underset{\underset{O}{\parallel}}{C}-CH_2-COOR \xrightarrow[\text{Catalyst}]{H_2} CH_3-\underset{\underset{OH}{|}}{CH}-CH_2-COOR$$

For example, catalytic hydrogenation of acetoacetic esters give the esters of (R,S)-β-hydroxybutyric acid.

Both γ- and δ-*hydroxycarboxylic acids* can be prepared by alkaline hydrolysis of the corresponding halo acids. In this case, the products are normally isolated in the form of alkali-metal salts.

Dihydroxycarboxylic acids may be made by reaction of unsaturated fatty acids (e.g., oleic acid) with hydrogen peroxide or peracid, in which epoxides are formed as intermediates. *Polyhydroxy acids* are also commonly prepared from unsaturated carboxylic acids, usually by permanganate oxidation.

Many of the methods described above lend themselves equally well to the synthesis of *polybasic hydroxycarboxylic acids*. For instance, *tartronic acid* has been prepared both by hydrolysis of monobromomalonic acid or by sodium amalgam reduction of mesoxalic acid (ketomalonic acid):

$$\begin{array}{c} COOH \\ | \\ HC-Br \\ | \\ COOH \end{array} \xrightarrow[-AgBr]{+AgOH} \begin{array}{c} COOH \\ | \\ HC-OH \\ | \\ COOH \end{array} \xleftarrow[Na-Hg]{2H} \begin{array}{c} COOH \\ | \\ C=O \\ | \\ COOH \end{array}$$

Bromomalonic acid   Tartronic acid   Mesoxalic acid

Another approach starts with glycerol, which upon oxidation with potassium permanganate yields tartronic acid, whereas treatment with nitric acid yields glyceric acid.

$$\begin{array}{c} COOH \\ | \\ CHOH \\ | \\ COOH \end{array} \xleftarrow[\text{Oxidation}]{KMnO_4} \begin{array}{c} CH_2OH \\ | \\ CHOH \\ | \\ CH_2OH \end{array} \xrightarrow[\text{Oxidation}]{HNO_3} \begin{array}{c} COOH \\ | \\ CHOH \\ | \\ CH_2OH \end{array}$$

Tartronic acid   Glycerol   Glyceric acid

As in this case, careful selection of oxidizing agent often enables various hydroxycarboxylic acids to be obtained by starting with the same precursor, a point well illustrated by the oxidation of sugars (aldoses and ketoses). Mild oxidizing agents such as aqueous bromine or dilute nitric acid attack only the aldehyde functions of aldoses, producing glyconic acids (e.g., gluconic acid from glucose). Such glyconic acids (also called "onic acids") are stable only in alkaline solution, where they exist as alkali-metal salts. The free acids rapidly cyclize to γ- or δ-lactones, with the γ-lactones preferred.

More powerful oxidizing agents such as concentrated nitric acid result in polyhydroxy dibasic acids (saccharic acids), which immediately cyclize to lactones. Thus, glucose leads to glucaric acid; mannose to mannaric acid; and galactose to mucic acid (galactaric acid). Even more vigorous oxidation causes cleavage of the carbon chain, yielding lower hydroxydicarboxylic acids.

$$\begin{array}{c} H \\ C=O \\ H-C-OH \\ HO-C-H \\ H-C-OH \\ H-C-OH \\ CH_2OH \end{array} \xrightarrow{\text{mild Oxidation}} \begin{array}{c} COOH \\ H-C-OH \\ HO-C-H \\ H-C-OH \\ H-C-OH \\ CH_2OH \end{array}$$

Glucose → Gluconic acid

$$\downarrow HNO_3 \text{ Oxidation}$$

$$\begin{array}{c} COOH \\ H-C-OH \\ HO-C-H \\ H-C-OH \\ H-C-OH \\ COOH \end{array}$$
Glucaric acid

$$\xrightarrow{-H_2O} \begin{array}{c} O \\ \parallel \\ C \\ H-C-OH \\ HO-C-H \\ H-C---O \\ H-C-OH \\ CH_2OH \end{array}$$
Lactone form

Numerous branched and unbranched hydroxycarboxylic acids may be prepared by cleavage or oxidation (with alkaline permanganate, chromic acid, dilute nitric acid, or concentrated nitric acid) of a variety of starting materials [6]–[9]. Enzymatic production by fermentation is becoming increasingly important.

4. Analysis

The literature describes an array of color reactions for *qualitative determination* of hydroxycarboxylic acids. These include the yellow color obtained in the presence of iron(II) chloride, characteristic complexation phenomena with copper salts, and the distinctive color produced with ammonium vanadate. Nevertheless, none of these reactions can be regarded as truly specific.

The most satisfactory approach to *quantitative analysis* involves acidimetric titration of carboxyl groups. Because of the nearly universal tendency to lactone formation in equilibrium with the free acid, free and total acids must be determined separately. Thus, lactones are treated in the same way as other esters: Hydrolysis with excess NaOH is followed by back titration with acid. Other recommended methods for quantitative analysis rely on acetylation of hydroxyl groups or esterification of carboxyl groups, with subsequent determination of the resulting water.

Instrumental methods of analysis are playing an increasingly important role, including IR, NMR and MS spectroscopy and, especially, chromatography (e.g., thin-layer chromatography and HPLC). Nevertheless, such methods prove reliable only if they are carefully adapted to suit the specific problem. A typical procedure involves methylation or silylation, followed by identification and quantification with GC–MS [7].

5. Specific Aliphatic Hydroxycarboxylic Acids of Commercial Significance

5.1. Glycolic Acid (Hydroxyacetic Acid, 2-Hydroxyethanoic Acid) [10]

The chemical structure, molecular mass, and CAS registry number of glycolic acid are given in Table 1. Its physical properties are listed in Table 2 and in the following material:

mp:
    α-Modification   78–80 °C
    β-Modification, metastable   63 °C
Heat of combustion   –697.23 kJ/mol
Heat of solution in water
    (infinite dilution)   –11.71 kJ/mol
Dissociation constant (at 25 °C) 1.54×10^{-4}
pH (1 M solution)   2.4

Solid glycolic acid forms colorless, monoclinic, prismatic crystals. The acid is freely soluble in water, methanol, ethanol, acetone, and ethyl acetate. Glycolic acid is only slightly steam volatile and cannot be destilled under vacuum; attempts result in self-esterification with loss of water, leading to di- and polyglycolides. However, newer results show that glycolic acid can be vaporized without decomposition. In the vapor phase it can be dehydrogenated catalytically to glyoxylic acid [11].

5.1.1. Preparation

Glycolic acid is usually produced by hydrolysis of molten monochloroacetic acid with 50% aqueous sodium hydroxide at 90–130 °C. The resulting glycolic acid solution has a concentration of ca. 60% and contains 12–14% sodium chloride. The salt may be removed by evaporative concentration, followed by extraction of the acid with acetone [12]. Attempts have also been made to conduct the hydrolysis with acid catalysts at 150–200 °C with water or steam under pressure [13]. In this case, the byproduct is hydrogen chloride, rather than sodium chloride, which can be removed by distillation. The principal disadvantage of the method is the need for relatively large volumes of water.

Glycolic acid is produced commercially in the United States (Du Pont) by treating formaldehyde or trioxymethylene with carbon monoxide and water in the presence of acid catalysts at >30 MPa [14]. Another process, previously utilized by Degussa, involves electrolytic reduction of oxalic acid [15]. The compound can also be prepared in 90% yield by hydrolysis of the corresponding nitrile, obtained by reaction of formaldehyde with hydrocyanic acid [16].

5.1.2. Uses

Glycolic acid is available commercially as either a 57% (Hoechst) or a 70% (Du Pont) aqueous solution. Total annual consumption worldwide is ca. 2000–3000 t of solution.

Glycolic acid is used in textile dyeing, printing, and creaseproofing. The fact that it can form a chelate with calcium(II) ions makes it well-suited to hide deliming in the leather industry, as well as to inclusion in alum and chrome mordants and in fur-processing operations. The compound has little tendency to cause corrosion, and this characteristic, coupled with its bactericidal properties, makes it suitable for incorporation into acidic cleansing agents. It is especially well-adapted to cleansing operations involving milk containers, milk-processing equipment, and drinking fountains, as well as rust and scale removal in heat exchangers and pipelines. Glycolic acid inhibits the growth of iron-oxidizing bacteria. The use of glycolic acid eliminates the need for simultaneous addition of chelating agents and bactericides. The effectiveness of glycolic acid as a complexing agent also contributes to its use in copper polishes, as an etching agent for lithographic plates, and in the preparation of electropolishing and galvanizing baths.

Safety Considerations. According to section 1.4 of the hazardous materials regulation, Gef Stoff V (Federal Republic of Germany), and to the EEC guidelines, glycolic acid solutions are "corrosive."

Analysis. Glycolic acid may be detected qualitatively by the violet color formed with 2,7-dihydroxynaphthalene [17]. The preferred method of quantitative analysis (in the absence of other acidic or hydrolyzable substances) is acidimetric titration. Because of the tendency of lactide formation free and total acid must be determined separately.

5.1.3. Derivatives

The glycolic acid esters *methyl glycolate* [96-35-5], bp 147 – 149 °C, and *ethyl glycolate* [623-50-7], bp 158 – 159 °C, both serve as starting materials for the laboratory preparation of pure glycolic acid [18] and were formerly employed as solvents for resins and for nitro- or acetyl cellulose. Apart from these two materials, only carboxymethyl cellulose [9004-32-4], (→ Cellulose Ethers and *n*-butyl glycolate have commercial significance).

***n*-Butyl Glycolate** (see Table 1). *Physical Properties.* *n*-Butyl glycolate is a colorless liquid, which is miscible with most organic solvents. Its physical properties are summarized in Table 2. The solubility in water is limited to 8 wt% (20 °C), although the compound itself may contain up to 25 wt% water.

Production. *n*-Butyl glycolate is produced by treatment of sodium chloroacetate with *n*-butyl alcohol at 125 – 160 °C, followed by vacuum distillation [19].

Applications. *n*-Butyl glycolate is used primarily as a varnish additive, valued because of its low volatility. Trade names include Polysolvan-O (Hoechst) and GB-Ester (Wacker). It confers smooth spreading properties and high gloss on nitrocellulose varnish. For acetyl cellulose, it is effective as a blush inhibitor under conditions of high humidity. Due to its good blending properties, *n*-butyl glycolate is also used as an additive in alkyd resins and oil-based paints.

Safety Considerations. *n*-Butyl glycolate is not regarded as hazardous.

5.2. Hydroxypropionic Acids and β-Propiolactone

Of the two isomeric hydroxypropionic acids, only α-hydroxypropionic acid (→ Lactic Acid) is optically active.

β-Hydroxypropionic Acid (Hydracrylic Acid, 3-Hydroxypropanoic Acid) [20] (see Table 1). *Physical Properties.* The most important physical properties of β-hydroxypropionic acid are listed in Table 2); its dissociation constant at 25 °C is 3.11×10^{-5}. β-Hydroxypropionic acid forms a highly acidic syrup, which loses water upon heating.

Preparation. β-Hydroxypropionic acid can be prepared by alkaline hydration of acrylic acid or by treatment of ethylene chlorohydrin with sodium cyanide, followed by hydrolysis of the resulting β-hydroxypropionitrile. It is also readily obtained by hydrolysis of the commercially important and easily accessible compound β-propiolactone.

β-Propiolactone (Oxetan-2-One) [21] (see Table 1).

$$\begin{array}{c} CH_2 - C^{\nearrow O} \\ | \quad\quad | \\ CH_2 - O \end{array}$$

Physical Properties [1]. The physical properties of β-propiolactone are summarized in Table 2 and in the following material:

Standard heat of formation   −330.0 kJ/mol
Standard heat of combustion   −1422.0 kJ/mol

β-Propiolactone is a colorless, highly reactive liquid, that is soluble in water, alcohol, acetone, and chloroform (solubility in water at 25 °C, 37 vol%).

Chemical Properties. β-Propiolactone reacts with alcohols, acid chlorides, ammonia, and water to yield β-substituted propionic acid derivatives. The most important characteristic of the substance is its ability to polymerize. This highly exothermic process occurs simply by warming, although it is also catalyzed by both acid and base.

Preparation. β-Propiolactone is synthesized by passing equimolar amounts of ketene and formaldehyde into either acetone or β-propiolactone itself. The reaction is carried out at low temperature (<20 °C) with a yield of ca. 90% [5]. Both aluminum chloride and zinc chloride have been employed as catalysts, and the use of methyl borate has also been suggested recently.

Applications. As late as 1974, β-propiolactone was used in the United States in the preparation of acrylic acid and acrylate esters [22]. Today, its principal significance is as a reactive intermediate in organic syntheses; a small amount is treated with ammonia to provide β-alanine. β-Propiolactone was also used as a disinfectant. It appeared to be an attractive replacement for formaldehyde due to its 25-fold greater disinfecting power, but it has since been abandoned because of its carcinogenic properties.

Safety Considerations. In the EEC hazardous materials guideline and in the GefStoffV (Federal Republic of Germany) β-propiolactone is classed as a strong carcinogen and is considered extremely dangerous at concentrations ≥ 1 wt%.

5.3. Hydroxybutyric Acids

Of the three structural isomers α-, β-, and γ-hydroxybutyric acid, only the α - and β- acids are optically active. Racemic mixtures of each can be separated into optical antipodes by way of the corresponding strychnine or brucine salts. All three acids are soluble in water, ethanol, diethyl ether, and numerous organic solvents, and decompose readily on dry distillation. β-Hydroxybutyric acid easily loses water to give crotonic acid, and γ-hydroxybutyric acid is transformed even at low temperature to γ-butyrolactone.

(R,S-)-α-Hydroxybutyric Acid [(R,S-)-2-Hydroxybutanoic Acid] [23], (see Table 1). The physical properties of (R,S)-α-hydroxybutyric acid are listed in Table 2.

Preparation. (R,S)-α-Hydroxybutyric acid results as main product from treating butyraldehyde with alkaline sodium hypochlorite solution or from heating α-bromobutyric acid with formamide at 150 °C.

(R,S-)-β-Hydroxybutyric Acid [(R,S-)-3-Hydroxybutanoic Acid] [24], (see Table 1). *Physical Properties.* The *mp* of racemic β-hydroxybutyric acid is 48–50 °C; it is commonly described in the literature as a syrupy liquid (see Table 2). Pure (R)-(–)-β-hydroxybutyric acid [625-72-9] is a solid with *mp* 49–50 °C.

Preparation. (R,S)-β-Hydroxybutyric acid results from hydrolysis of β-butyrolactone, which in turn can be produced by the addition of ketene to acetaldehyde in the presence of boron trifluoride or zinc salts [5]. The compound is also prepared by reduction of acetoacetic acid with sodium amalgam in alkaline solution.

γ-Hydroxybutyric Acid (4-Hydroxybutanoic Acid) [25] (see Table 1).
Physical Properties. The most important physical properties of γ-hydroxybutyric acid are listed in Table 2. The dissociation constant at 25 °C is 1.93×10^{-5}. At ambient temperature, the acid is a liquid and consists in part of lactone.

Preparation. γ-Hydroxybutyric acid is obtained from γ-butyrolactone by alkaline hydrolysis and isolated as the alkali-metal salt. γ-Butyrolactone is produced by dehydrocyclization of 1,4-butanediol.

Applications. Hydroxybutyric acids were utilized in the manufacture of plasticizers, but their current applications are quite limited. The related lactones are some times used as starting materials or intermediates in organic synthesis.

Optically active *β-hydroxybutyric acid* is an important pharmacological agent [26]. Biodegradable *poly(–(R)-(–)-β-hydroxybutyric acids* (named PHB) have recently been used in the context of controlled long-term parenteral application of drugs through implantation. These materials offer a number of advantages relative to other implantation polymers [e.g., poly(methacrylic acid) derivatives, poly(glycolic acid), poly(lactic acid)] [27]. Poly(β-hydroxybutyric acid) has also been considered as a potential textile material.

γ**-Hydroxybutyric acid** has been used recently as an anesthetic and pain killer. It has a stabilizing effect on blood pressure and induces a state of unconsciousness, characteristics that have encouraged its application especially in geriatrics [28].

5.4. (R,S)-Malic Acid [(R,S)-Hydroxysuccinic Acid, (R,S)-Hydroxybutanedioic Acid] [29], [30] (see Table 1.)

Physical Properties (see Table 2). Malic acid is a colorless, odorless, crystalline substance. It is highly soluble in water (100 g of water dissolves 126.3 g of malic acid at 20 °C), methanol, ethanol, acetone, ether, and other polar solvents. Because it is a dibasic acid, it has two dissociation constants: at 20 °C, $K_{A_1} = 3.9 \times 10^{-4}$ and $K_{A_2} = 1.4 \times 10^{-5}$.

Preparation. (R,S)-Malic acid is prepared commercially in the United States and Canada by hydration of maleic anhydride ($\rightarrow$ Maleic and Fumaric Acids) [31]. The sole manufacturer in the United States is Alberta Gas, with an annual capacity of ca. 5000 t. In this process maleic acid is heated at ca. 180 °C (under a pressure of ca. 1 MPa), malic acid is yielded as the main product. Byproducts are maleic and fumaric acids. The latter can be separated by filtration and returned to the process stream because of its low water solubility.

$$\begin{array}{c} CH-C\begin{subarray}{c}\diagup O \\ \diagdown O\end{subarray} \\ \| \\ CH-C\begin{subarray}{c}\diagup \\ \diagdown O\end{subarray} \end{array} + 2\,H_2O \longrightarrow \begin{array}{c} CHOH-COOH \\ | \\ CH_2-COOH \end{array}$$

The filtrate is then concentrated; this causes separation of the malic acid, which is purified by multiple washings, evaporation, and recrystallization until the contents of fumaric and maleic acids are reduced to 7.5 and <500 ppm, respectively. Additional purification is required to prepare pharmacological-grade material [32], [33].

Applications. (R,S)-Malic acid closely resembles both citric and tartaric acids in its physical and chemical properties. However, the compound has a more neutral flavor, so it is often preferred when the flavor of citric acid is considered objectionable. Examples include the impregnation of packing material for foodstuffs such as cheese or the acidification required during preparation of baked goods. It is widely used in the food industry as an acidulant and, to a lesser degree, as an acidity regulator. Relative to citric or tartaric acid, the flavor imparted by malic acid is not only more neutral but also of longer duration. Thus, citric acid imparts a strongly acidic taste to the tongue almost immediately, but the effect is brief. With malic acid, the sensation is less intense, but it lasts longer and is more consistent. Soft-drink manufacturers have discovered that this property helps mask the aftertaste associated with certain nonnutritive sweeteners and leads to a more balanced flavor profile. Moreover, synergistic effects are observed between these sweeteners and malic acid, which permit reductions of up to 20% in the amount of sweetener required and 10% in malic acid. Consequently, significant economic benefits may be realized. The economic picture is also influenced favorably by the fact that malic acid is available as an anhydrous powder.

Malic acid is used primarily as an ingredient in hard candies and other sweets, jams, jellies, and various canned fruits and vegetables. Most countries authorize its use as an additive in foodstuffs.

(R)-(+)Malic acid [636-61-3] and (S)-(–)-malic acid [97-67-6] are isolated from natural sources by resolution of (R,S)-malic acid with the aid of an optically active base. (S)-(–)-Malic acid is also available through microbiological fermentation of fumaric acid [33]. The levorotatory S-enantiomer is widely distributed in fruit and displays the following characteristics that differ from those of the racemate: mp, 100 °C; d_4^{20}, 1.595; α_D^{18}, –2.3° (7 wt% in H_2O). The compound is soluble to the extent of 36.4 g in 100 g of water at 20 °C.

6. Toxicology

As previously noted, most important aliphatic hydroxycarboxylic acids occur as natural products. Because they serve as intermediates in plant and animal metabolic pathways, they do not exhibit many exceptionally toxic properties. Thus, only a few compounds listed in Table 1are included in the *Register of Toxic Effects of Chemical Substances* prepared by NIOSH [35]. Certain hydroxycarboxylic acids (e.g., lactic, malic, tartaric, and citric) are even accepted as food additives or preservatives in most countries.

The following toxicity data for *citric acid* are taken from the NIOSH list [35, p. 945]:

LD_{50} (oral, rat)	11 700 mg/kg
LD_{50} (oral, mouse)	5 040 mg/kg
LD_{50} (oral, rabbit)	7 000 mg/kg

Glycolic acid was also permitted as a food preservative but later came to be considered suspect [36]. Today, it is described as moderately toxic, with the following data [35, p. 386]:

LD_{50} (oral, rat)	950 mg/kg
LD_{50} (oral, guinea pig)	920 mg/kg

A derivative of glycolic acid *n-butyl glycolate* is described as displaying narcotic properties. It is also an irritant to both skin and mucous membranes. Nevertheless, it is markedly less toxic than glycolic acid, having an LD_{50} (oral, rat) of 4595 mg/kg.

The lowest reported lethal dose for *β-hydroxy propionic acid (hydracrylic acid)* derives from intravenous injection into rats: LD_{50} (i.v., rat) = 50 mg/kg [35, p. 447].

A derivative of *β-hydroxy* propionic acid *β-propiolactone* is of considerable interest because of its viricidal and bactericidal properties, and was used medically in sterilization and disinfection. Nevertheless, it has since been found to be a mutagen [37]. Animal studies have provided incontrovertible evidence that the compound displays carcinogenic properties irrespective of whether it is delivered orally, topically on the skin, or by intravenous, intraperitoneal, or subcutaneous injection [38], [35, p. 1326]. β-Propiolactone is a powerful irritant to the skin and eyes. The following toxicity data have been reported:

LC_{50} (inhalation, rat)	25 ppm, 6 h
LD_{50} (i.v., mouse)	345 mg/kg

Because of its cytotoxic properties, β-propiolactone has been added to Category A 2 of the official list of carcinogenic materials [39].

By contrast, *γ-butyrolactone* is completely free of these cytotoxic characteristics. Despite numerous animal studies, no evidence has been obtained to suggest carcinogenicity. Nonetheless, the compound is moderately toxic, having an LD_{50} (oral, rat) of 1800 mg/kg [35, p. 946]. The narcotic effects observed when γ-butyrolactone is administered orally or intraperitoneally have been ascribed to γ-hydroxybutyric acid, from which it is derived.

From animal studies, the sodium salt of γ-hydroxybutyric acid is known to possess narcotic properties [40]. A dose of ca. 40 mg/kg leads to nearly natural sleep in humans, with essentially no side effects.

7. References

[1] H. E. Zaugg: "β-Lactones," *Org. React. (N.Y.)* **8** (1954) 305 ff.
[2] A. Griesbeck, D. Seebach, *Helv. Chim. Acta* **70** (1987) 1320, 1326.
[3] Jaroslawl. Technological Institute, DE-OS 2 326 297, 1974 (M. S. Rusakowa, et al.).
[4] R. L. Shriner: "The Reformatsky Reaction,"*Org. React. (N.Y.)* **1** (1942) 1 ff.
[5] Goodrich Co., US 2 356 459, 1941 (F. E. Kung); US 2 424 589, 2 424 590, 1944 (T. R. Steadman).
[6] Escambia Chem. Corp., US 2 811 545, 1956 (T. R. Steadman).
[7] K. Niemela, E. Sjostrom, *Carbohydr. Res.* **144** (1985) no. 1, 87–92.
[8] K, Niemela, E. Sjostrom, *Holzforschung* **40** (1986) no. 1, 9–14.

[9] K. Garves in J. F. Kennedy, (ed.): *Cellulose Its Derivatives,* Ellis Horwood, Chichester 1985, pp. 487–494.
[10] *Beilstein,* **3**, 228; **3 (1)**, 88; **3 (2)**, 167; **3 (3)**, 367; **3 (4)**, 571.
[11] Hoechst, DE-OS 2 904 754, 1980 (H. Baltes, E. I. Leupold, F. Wunder).
[12] Hoechst, DE-OS 2 812 682, 1979 (E. I. Leupold, B. Wojtech, H. J. Arpe).
[13] Hoechst, DE-OS 2 810 975, 1979 (H. Klug, H. Woppert, H. Korbanka).
[14] DuPont, US 2 152 852, 1939 (D. J. Loder); US 2 153 064, 1939 (A. T. Larson); US 2 443 482, 1948 (M. T. Shattuck).
[15] Degussa, DE 194 038, 1903.
[16] I. G. Farbenind., DE 459 602, 1925 (O. Schmidt).
[17] E. J. E. Eegriwe, *Z. Anal. Chem.* **89** (1932) 124.
[18] DuPont, US 2 331 094, 1940 (D. J. Loder).
[19] I. G. Farbenind., DE 522 786, 1929 (W. Bülow).
[20] *Beilstein,* **3**, 295; **3 (1)**, 122; **3 (2)**, 212; **3 (3)**, 522; **3 (4)**, 689.
[21] *Beilstein,* **17 (1)**, 130; **17/5 (3/4)**, 4157; **17/9 (5)**, 3.
[22] *Ullmann,* 4th ed., **19**, p. 452.
[23] *Beilstein,* **3**, 302; **3 (1)**, 114; **3 (2)**, 216; **3 (3)**, 561; **3 (4)**, 755.
[24] *Beilstein,***3**, 308; **3 (1)**, 116; **3 (2)**, 220; **3 (3)**, 571; **3 (4)**, 761.
[25] *Beilstein,***3**, 311; **3 (2)**, 222; **3 (3)**, 581; **3 (4)**, 774.
[26] D. Seebach, M. Züger, *Helv. Chim. Acta* **65** (1982) 495–503.
[27] W. Korsatko, B. Wabnegg, G. Braunegg, H. M. Tillian et al., *Pharm. Ind.* **45** (1983) no. 5, 525–527; no. 10, 1004–1007; *Pharm. Ind.* **46** (1984) no. 9, 952–954.
[28] W. Bushart, P. Rittmeyer: *Anästhesie mit γ-Hydroxybuttersäure,* Springer-Verlag, Berlin 1973.
[29] *Beilstein,* **3**, 435; **3 (1)**, 154; **3 (2)**, 289; **3 (3)**, 918; **3 (4)**, 1124.
[30] H. Rudy: *Fruchtsäuren, Wissenschaft und Technik,* Hüthig-Verlag, Heidelberg 1967.
[31] T. Yoshida, O. Hibino, S. Morita, *Nippon Kagaku Kaishi (1921–47)* **53** (1950), 399–401; *Nippon Chem. Ind.* JP 4 360, 1950.
[32] Chemie Linz, FR 2 193 810, 1973.
[33] Allied Chem. Corp., DE-AS 1 518 522, 1965 (Winstrom, O. Leon, Ingleman, Miltin Russel).
[34] Hüls, DE-OS 3 434 880 A1, 1986; Supplement to DE-OS 3 310 849.
[35] Registry of Toxic Effects of Chemical Substances, 1983–84. Cumulative Supplement to the 1981–82 Edition NIOSH.U.S. Department of Health and Human Services Public Health Service, Centers for Disease Control. National Institute for Occupational Safety and Health.
[36] W. F. Riker, G. Gold, *J. Am. Pharm. Assoc. Sci. Ed.* **31** (1942) 306.
[37] H. H. Smith, A. M. Srb, "Induction of mutations with β-propiolacton," *Science* **114** (1951) 490.
[38] D. J. Brusick, *Mutat. Res.* **39** (1977) no. 3–4, 241–255.
[39] D. Henschler, Vors. d. Arbeitsst.-Komm. d. Dtsch. Forschungsgemeinsch.: *Gesundheitsschädliche Arbeitsstoffe, Toxikologisch-arbeitsmedizinische Begründung von MAK-Werten,* vol. **IV**, 14. Lieferung, VCH-Verlagsgesellschaft Weinheim, 1988.
[40] H. Laborit et al., *JAMA J. Am. med. Assoc.* **175** (1961) 520.

Hydroxycarboxylic Acids, Aromatic

Individual keywords; → *Stalicylic Acid. Naphtholcarboxylic Acids see* → *Naphthalene Derivatives*

EDWIN RITZER, Bayer AG, Leverkusen, Federal Republic of Germany
RUDOLF SUNDERMANN, Bayer AG, Leverkusen, Federal Republic of Germany

1.	Introduction 2973	4.1.2.	4-Hydroxybenzoic Acid	2978
2.	General Properties......... 2974	4.1.3.	Cresotic Acids	2979
3.	Production 2975	4.2.	Dihydroxybenzoic Acids.....	2980
4.	Individual Aromatic Hydroxy-	4.3.	Trihydroxybenzoic Acids	2981
	carboxylic Acids........... 2977	4.4.	Mandelic Acid	2982
4.1.	Monohydroxycarboxylic Acids 2977	5.	Toxicology................	2983
4.1.1.	3-Hydroxybenzoic Acid 2977	6.	References................	2984

1. Introduction

Aromatic hydroxycarboxylic acids (i.e., hydroxycarboxylic acids with phenolic hydroxyl groups) occur widely in nature in the form of their esters, ethers, intermolecular esters (depsides), or cyclic ether–esters (depsidones). Structural formulas of the most important depsides and of depsidone are given below:

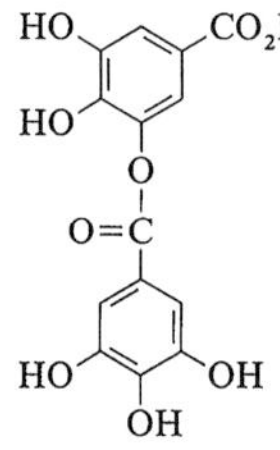

m-Digallic acid
[536-08-3]

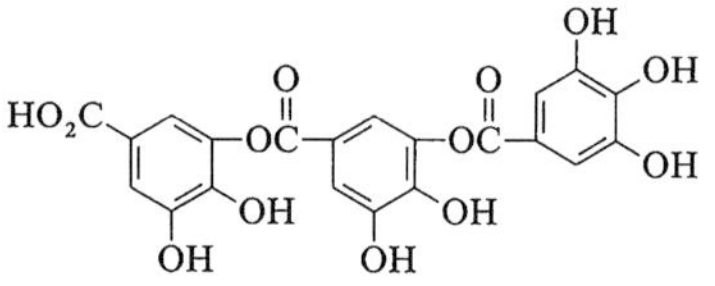

m-Trigallic acid
[2131-66-0]

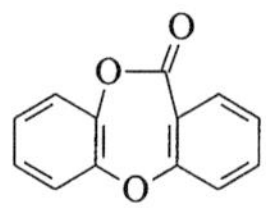

Depsidone
[3580-77-6]

Hydrolyzable tanning agents such as tannin (ester of gallic acid, tannic acid) [5424-20] are esters of aromatic hydroxycarboxylic acids with polyhydric alcohols and sugars. Aromatic hydroxycarboxylic acids are used mainly for pharmaceutical and cosmetic purposes: as disinfectants, preservatives, emulsifiers, analgesics, and antirheumatic and antipyretic drugs. Their use in the production of dyes and plastics is increasingly important, especially in color developing, and in liquid crystal polymers.

2. General Properties

Monohydroxybenzoic acids are dibasic acids. Thus, one mole of alkali forms the carboxylate and a second mole converts the phenolic hydroxyl group into a phenolate. The acidity of each group depends on its relative position (see Table 1). The carboxyl and hydroxyl group acidities of 3- and 4-hydroxybenzoic acids are comparable to those of benzoic acid and phenol. However, the corresponding acidities of 2-hydroxybenzoic acid (salicylic acid [69-72-7]) are different. The higher acidity of the carboxylate group in this case is the result of intramolecular chelate formation.

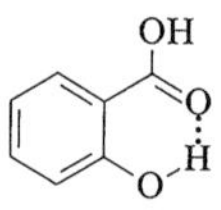

This accounts for its high steam volatility and ease of sublimation. 2-Hydroxy- and 4-hydroxybenzoic acids decarboxylate above their melting point; 2,4- and 2,6-dihydroxybenzoic acids decarboxylate even when heated in an aqueous solution. This process is catalyzed by strong acid. The carboxyl group is esterified with alcohols under acid catalysis without detectable etherization of the phenolic hydroxyl group. Ether–esters can be obtained easily from the dialkali salts by using alkyl halides or dialkyl sulfates.

Table 1. Dependence of dissociation constants for aromatic hydroxycarboxylic acids on relative position of hydroxyl and carboxyl groups

		$K_1(-COOH), \times 10^{-5}$	$K_2(-OH), \times 10^{-10}$
Benzoic acid	[65-85-0]	6.3	
Phenol	[108-95-2]		9.9
2-Hydroxybenzoic acid	[69-72-7]	105	0.0004
3-Hydroxybenzoic acid	[99-06-9]	8.3	1.17
4-Hydroxybenzoic acid	[99-96-7]	2.9	4.8

Alkaline hydrolysis of ether–esters produces alkoxycarboxylic acids. Electrophilic substitution (e.g., nitration, sulfonation, chlorination) leads to attack of the aromatic nucleus at the 2- or 4-position, with the 4-position preferred.

Analysis. Methods of analyzing carboxylic acids and phenols are also applicable to hydroxycarboxylic acids. In practice, alkalimetric titration (in special cases, potentiometric titration) is used. With potentiometric determination of the equivalence point, the hydroxyl group can also be determined. Trace analyses are carried out by colorimetric methods with iron(III) chloride [4]. Thin-layer chromatography is a suitable method for checking purity. Here, individual acids are made visible by means of alkaline (Na_2CO_3) coupling with diazotized 4-nitroaniline, by spraying with iron(III) chloride, or by fluorescence in UV light. For quantitative analysis, HPLC methods are suitable. Alternatively, the components can be derivatized and capillary gas chromatographic methods used to separate them [5].

3. Production

Kolbe–Schmitt or Marassé Carboxylation. Yields according to the Kolbe reaction of solid anhydrous alkali phenolates in a stream of carbon dioxide were improved by SCHMITT who carried out the reaction under pressure [6]. The Marassé method involves the conversion of phenol with potassium carbonate and does not work with less expensive sodium carbonate [7]. The Kolbe–Schmitt synthesis is a gas-solid reaction; thus the phenolate must be present as a fine powder rather than a melt to achieve better contact. For the same reason, inert high boilers, such as higher alkanes, are used for thorough mixing [8]. The alkali phenolates are converted into alkali-metal salts of the corresponding hydroxycarboxylic acids, generally by stirring, in pressurized vessels at 120–180 °C and a carbon dioxide pressure of 0.5 MPa. Conditions up to 260 °C and 120 MPa are required in the presence of electron-withdrawing groups or in the case of sterically hindered phenols. 3-Nitrophenol [555-84-7] is carboxylated with a yield of only 19% [9]. Carboxylation occurs easily when the aromatic compounds are activated by electron-donating substituents, as in the case of aminophenols or di- and trihydroxyaromatic compounds. Reaction then occurs even in an aqueous alkali-metal carbonate solution.

The yield of dicarboxylic acid increases with increasing temperature and pressure above 1.0 MPa. At 10.0 MPa and 220 °C the equilibrium is shifted mainly toward the product [10]. The type of alkali metal is important for the course of the reaction. Sodium and lithium phenolates are carboxylated at medium temperature (150 – 180 °C) to form mainly 2-hydroxycarboxylic acids. Potassium phenolates give 2 – 4 mixtures at medium temperature, but above 220 °C the 4-hydroxycarboxylic acid is favored.

Exchange of a Sulfonic Acid, a Halide, or a Diazonium Group for the Hydroxyl Group. The most common method for producing 3-hydroxybenzoic acid [*99-06-9*] involves the alkali melt of *3-sulfobenzoic acid* [*121-53-9*] [11]. In some cases, if the carboxyl group is activated by electron-donating substituents, decarboxylation takes place during melting. Nucleophilic substitution of *halogen* by the hydroxyl group is particularly effective if the halogen atom is activated by electron-withdrawing groups in the ortho or para position [12]. The conversion of *aminobenzoic acids* into hydroxycarboxylic acids is possible via the diazonium salt [13].

Hydroxycarboxylic Acids via Oxidation. Oxidation of *alkylphenols* is another method of producing aromatic hydroxycarboxylic acids. In general the phenolic hydroxyl group must be protected (e.g., by esterification with sulfuric acid). The alkyl moiety of the sulfuric acid ester is then oxidized easily by using potassium permanganate or chromic acid [14]. Aromatic hydroxycarboxylic acids can also be produced by oxidative alkali melt of appropriately substituted phenols. Alkyl groups ortho to the phenolic hydroxyl group are oxidized most easily [15]. The oxidative alkali melt of *aromatic hydroxyaldehydes* gives the corresponding hydroxycarboxylic acids in good yield [16]. In the presence of silver catalysts, oxidation is possible even in aqueous alkali solution [17]. *Hydroxyacetophenones* can be converted into hydroxycarboxylic acids by using iodine in pyridine (haloform reaction), followed by alkaline hydrolysis of the intermediate pyridinium iodide [18].

Direct introduction of a *hydroxyl* group into the aromatic nucleus by oxidation is possible in the case of ammonium benzoates by using hydrogen peroxide [19]. Thus, the first stage of the Dow process (production of phenols from benzoic acids) involves hydroxylation by air of the metal benzoate ortho to the carboxyl group [20].

Introduction of a *carboxyl group* is achieved by reaction of phenols with tetrachloromethane in the presence of alkali and powdered copper or copper salts [21]. Alkyl ethers of hydroxycarboxylic acids can be obtained directly by carboxylation of alkoxyphenyl metal halides (e.g., magnesium or lithium compounds) with carbon dioxide. Amides and thioamides of alkoxycarboxylic acids are formed by conversion of phenol ethers with carbamoyl chlorides, isocyanates, or isothiocyanates in the presence of aluminum chloride [22]. Rearrangement of arylurethanes in the presence of Lewis acids (e.g., aluminum chloride) may give good yields of the amides of the corresponding hydroxycarboxylic acids [23].

4. Individual Aromatic Hydroxycarboxylic Acids

4.1. Monohydroxycarboxylic Acids

4.1.1. 3-Hydroxybenzoic Acid

Physical Properties. 3-Hydroxybenzoic acid [99-06-9], $C_7H_6O_3$, M_r 138.12, crystallizes as white needles (from water) or rhombic prisms (from alcohol). Important physical properties follow: mp 203 °C; d_{25} 1.473; dissociation constants K_1 8.71×10^{-5}, K_2 1.18×10^{-10} (at 19 °C); solubility (in 100 g of solution): water 6.11 g (69 °C), 98% ethanol 39.6 g (65 °C), and n-butanol 20.7 g (36.5 °C).

Chemical Properties. 3-Hydroxybenzoic acid does not undergo decarboxylation at elevated temperature and is stable up to 300 °C. This is in contrast to 2- and 4-hydroxybenzoic acids. Electrophilic substitution (e.g., nitration, halogenation, sulfonation) occurs ortho or para to the hydroxyl group. Thus, nitration with 62% aqueous nitric acid gives 2-nitro-3-hydroxybenzoic acid as the main product, along with 4-nitro- and 6-nitro-3-hydroxybenzoic acids [24].

Production. 3-Hydroxybenzoic acid can be obtained by alkali melt of the sodium salt of 3-sulfobenzoic acid at 210 – 220 °C. The crude acid precipitates on acidification. It can be purified by recrystallization from water with the addition of activated carbon [11]. 3-Hydroxybenzoic acid can also be obtained from 3-nitrobenzoic acid esters. This occurs via catalytic hydrogenation followed by diazotization of the aminobenzoic acid ester formed. The diazonium salt is then heated [25] with aqueous sulfuric acid. Another production method is oxidation of 3-hydroxybenzaldehyde [100-83-4] by bubbling air through a hot aqueous sodium hydroxide suspension [26].

Uses. 3-Hydroxybenzoic acid is used as an intermediate in the manufacture of pharmaceuticals and pesticides. The 3-hydroxybenzoic acid ester of tropine is used in ophthalmology to dilate the pupils. The sodium salt of 3-hydroxybenzoic acid is a cholepoietic agent. Esters and metal salts have germicidal and preservative properties, and are used in food preservation. Ether – esters are suitable as plasticizers for vinyl and cellulose resins.

Table 2. Properties of 4-hydroxybenzoic acid esters

	Methyl [99-76-3]	Ethyl [120-47-8]	n-Propyl [94-13-3]	n-Butyl [94-26-8]	Benzyl [94-18-8]
Physical properties					
M_r	152.12	166.17	180.2	194.22	228.24
mp, °C	131	118	97	70	111
Solubility at 25 °C in 100 g of					
Water	0.25	0.17	0.05	0.02	0.006
Water (80 °C)	2.0	0.86	0.30	0.15	0.09
Methanol	59	115	124	220	79
Ethanol	52	70	95	210	72
Acetone	64	84	105	240	102
Benzene	0.7	1.65	3.0	40	2.6
Diethyl ether	23	43	50	150	42
Tetrachloromethane	0.1	0.9	0.8	1.0	0.08
Bacteriostatic properties					
Phenol = 1.0	3.8	8.0	17.0	32.0	109.0

4.1.2. 4-Hydroxybenzoic Acid

Physical Properties. 4-Hydroxybenzoic acid [99-96-7] $C_7H_6O_3$, M_r 138.12, is a white granular crystalline powder consisting of monoclinic prisms. The physical properties are summarized here: mp 216.3 °C; d_{20} 1.497; dissociation constants K_1 3.3 × 10^{-5}, K_2 4.8 × 10^{-10} (19 °C); solubility (in 100 g of solution): water 0.49 g (20 °C), 33.5 g (100 °C); 99 % ethanol 38.75 g (67 °C); and n-butanol 19.5 g (32.5 °C). 4-Hydroxybenzoic acid is sparingly soluble in chloroform.

Table 2 lists properties of the most commonly used esters.

Chemical Properties. Above 200 °C, 4-hydroxybenzoic acid undergoes decarboxylation to produce phenol and carbon dioxide. Electrophilic substitution (e.g., nitration, sulfonation, or halogenation) occurs ortho to the hydroxyl group. With iron(III) chloride, no color reaction occurs but a yellow precipitate is formed.

Production. 4-Hydroxybenzoic acid is produced by carboxylation of potassium phenolate above 220 °C and with a carbon dioxide pressure of 0.45 MPa. The yield of 4-hydroxybenzoic acid (partly present as the dipotassium salt) is 80 % of the theoretical amount in which the conversion of phenolate is 60 %. 4-Hydroxybenzoic acid can also be obtained by isomerization of dipotassium salicylate [27] at temperatures up to 240 °C and under carbon dioxide pressure (slightly above atmospheric pressure) [28], or by oxidation of the alkali-metal salt of p-cresol [106-44-5] on metal oxides at 260–270 °C [29]. Reaction of phenol with tetrachloromethane [56-23-5] in the presence of alkali produces salicylic acid [69-72-7] as the main product and smaller amounts of 4-hydroxybenzoic acid [99-96-7] [21].

Uses. *4-Hydroxybenzoic acid* is an intermediate in the manufacture of dyes, pharmaceuticals, pesticides [30], and plastics. It is used as an emulsifier and also as a corrosion-protection agent [31]. Its esters and corresponding sodium salts are antimicrobial substances with a high no-effect level (1200 mg per kilogram of body weight) [32]. They are, therefore, used under weakly acidic to weakly basic conditions as food preservatives (e.g., in fruit juice), cosmetics, and pharmaceuticals, as well as in technical products (e.g., in lubricants, anti-freezing agents) [33]. Benzyl esters are relatively soluble in preparations containing sugar. Table 2 lists observed bacteriostatic effects [34] compared to phenol.

4-Hydroxybenzoic acid is also used in polymers as a component in the manufacture of polyester and as a constituent of *liquid crystal polymers* [35].

Esters. The food preservative "PHB-Ester" is a mixture of the ethyl and *n*-propyl esters of 4-hydroxybenzoic acid. According to the preservatives regulations in the different countries [*EEC guideline* 64/52 date: 5.11.63, 24th amendment 85/585 date: 20.12.85; *United States:* Food and Drug Administration, HHS, 21 CFR Ch.I, 4.1.86, § 172.145, § 184.1490, § 184.1670; *Federal Republic of Germany:* Verordnung über die Zulassung von Zusatzstoffen zu Lebensmitteln, date: 22.12.81 (BGBl. I, p. 1625/33) 1. amendment date: 20.12.84 (BGBl. I, p. 1652)], PHB-Ester in the amount of 0.1 – 10 g/kg is allowed in the preservation of foodstuffs. In Europe, PHB-Esters are identified as E 214 – E 219, in the United States as heptylparaben, methylparaben and propylparaben. A mixture of 60% ethyl and 40% propyl ester is particularly effective. Use of such preservatives is, however, restricted by law in selected foodstuffs (e.g., wine) [36]. Some higher esters are used for termite control [37].

The PHB-Esters that are solids as pure components give eutectic mixtures which are liquid at room temperature and are used as oil-in-water emulsions [38]. See Table 2.

4-Hydroxybenzoic acid esters, particularly PHB-benzyl ester, are also employed as color developers in pressure-sensitive [39] and thermosensitive [40] writing systems. Here, purification and decoloration of the PHB-Ester are achieved through distillation with the addition of sodium hydrogen sulfite [41].

4.1.3. Cresotic Acids

Of the ten isomeric cresotic acids, $C_8H_8O_3$, M_r 152.15, *o*- and *p*-cresotic acids are of most interest.

[83-40-9]   [50-85-1]   [89-56-5]

o-Cresotic acid [83-40-9] (2-hydroxy-3-toluic acid, 2-hydroxy-3-methylbenzoic acid), *mp* 165 °C, dissociation constant K_1 6.76 × 10^{-4}, is produced by carboxylation of the sodium salt of *o*-cresol [95-48-7] at 150–180 °C. The acid is sparingly soluble in water and has properties similar to salicylic acid. It is used as an intermediate in the preparation of pharmaceuticals and dyes. Its methyl ester is used as an aid in textile dyeing.

m-Cresotic Acid [50-85-1] (2-hydroxy-4-toluic acid, 2-hydroxy-4-methylbenzoic acid, 4-methylsalicylic acid), *mp* 177 °C, dissociation constant K_1 6.84 × 10^{-4}, is produced by carboxylation of the sodium salt of *m*-cresol [108-39-4]. It is used as an additive in galvanic baths [42].

p-Cresotic Acid [89-56-5] (2-hydroxy-5-methylbenzoic acid, *p*-homosalicylic acid), *mp* 153 °C, is produced by carboxylation of the sodium salt of *p*-cresol [106-44-5] with carbon dioxide at 150–180 °C. The acid is used as an intermediate in dye production.

4.2. Dihydroxybenzoic Acids

3,4-Dihydroxy- and 2,3-dihydroxybenzoic acids are both obtained by heating catechol [120-80-9] in aqueous ammonium carbonate solution at 140 °C.

[99-50-3]   [303-38-8]   [89-86-1]   [490-79-9]

3,4-Dihydroxybenzoic acid [99-50-3] (protocatechuic acid), $C_7H_6O_4$, M_r 154.1, *mp* 203 °C, undergoes decomposition to catechol [120-80-9] and carbon dioxide. The acid crystallizes as a monohydrate from aqueous solution.

The most important derivative of this acid is the monomethyl ether, *4-hydroxy-3-methoxybenzoic acid* [121-34-6] (vanillic acid), $C_8H_8O_4$, M_r 168.14, *mp* 211 °C (from water). It is obtained by careful alkali melt of vanillin [121-33-5] or oxidation of vanillin with silver oxide [43]. Vanillic acid and its esters are constituents of preservatives and disinfectants.

2,3-Dihydroxybenzoic acid [303-38-8], $C_7H_6O_4$, M_r 154.1, *mp* 206 °C, undergoes decomposition to catechol and carbon dioxide. The acid crystallizes from water as a dihydrate. Among other uses, it serves as an iron-complexing agent in drugs [44].

2,4-Dihydroxybenzoic acid [89-86-1] (*p*-hydroxysalicylic acid, *β*-resorcylic acid), $C_7H_6O_4$, M_r 154.1, *mp* 226 °C (decomp.), is sparingly soluble in water. The dissociation constant K_1 is 6.05 × 10^{-4} (30 °C). The acid is produced by conversion of resorcinol [108-46-3] with aqueous potassium carbonate solution at 100 °C and a carbon dioxide pressure of 0.45 MPa. In this case, 2,6-dihydroxybenzoic acid [303-07-1] (*γ*-resorcylic acid) is formed as byproduct and can be separated by recrystallization [45]. Pure 2,4-dihydroxybenzoic acid can be produced by oxidation of 2,4-dihydroxyacetophenone [89-84-9] with iodine in pyridine [18].

2,4-Dihydroxybenzoic acid is used as an intermediate in the manufacture of dyes, pharmaceuticals and of additives in photography, and cosmetics. Dyes made from resorcylic acids differ from those made from salicylic acid in their deeper and more intense colors.

The acid is used to complex titanium(IV) fluoride [46], in the passivation of aluminum [47], and as an additive in drilling fluids [48].

3,5-Dihydroxybenzoic acid [99-10-5] (α-resorcylic acid), $C_7H_6O_4$, M_r 154.1, mp 237 °C (decomp.), is obtained by alkali melt of 3,5-disulfobenzoic acid [49].

2,5-Dihydroxybenzoic acid [490-79-9] (gentisic acid, 2,5-DHBA), $C_7H_6O_4$, M_r 154.1, mp 205 °C, is soluble in water and alcohols but insoluble in benzene and chloroform. The acid is stable in boiling water and gives a deep blue color with iron(III) chloride. Here, the acid decomposes to form *p*-benzoquinone [106-51-4] and carbon dioxide.

2,5-Dihydroxybenzoic acid is obtained by carboxylation of hydroquinone [123-31-9] with aqueous potassium hydrogen carbonate solution at 130 °C. It can also be made by oxidation of salicylic acid with potassium peroxodisulfate [7727-21-7] in the presence of iron sulfate [50]. 2,5-Dihydroxybenzoic acid is not obtained by reaction of carbon dioxide with the dialkali salt of hydroquinone: instead hydroquinone-2,5-dicarboxylic acid is formed.

2-5-Dihydroxybenzoic acid is used as an antirheumatic drug and, more recently, as a color developer for thermal paper [51].

4.3. Trihydroxybenzoic Acids

3,4,5-Trihydroxybenzoic Acid [149-91-7]. *Physical Properties.* 3,4,5-Trihydroxybenzoic acid (gallic acid), $C_7H_6O_5$, M_r 170.12, is a white to pale yellow crystalline powder, which crystallizes from water in the form of silky needles as the monohydrate (*mp* 258–263 °C) and undergoes decomposition to pyrogallol [87-66-1] and carbon dioxde: d_{25} 1.694, dissociation constants K_1 4.63 × 10^{-3} (at 30 °C) and K_2 1.41 × 10^{-9}. 3,4,5-Trihydroxybenzoic acid is soluble in warm water, ethanol, diethyl ether, and acetone but insoluble in benzene and chloroform.

Chemical Properties. Solutions of gallic acid, particularly of the alkali-metal salts, absorb oxygen and turn brown in air like pyrogallol. The acid is a strong reducing agent that can reduce gold or silver salts to the metal. With iron(III) salts, an intensive blue complex is formed that is used for ink dyes. Gallnut ink, which consists of gallic acid and iron(II) sulfate, produces the blue iron(III)–gallic acid complex on paper in air.

[82-12-2]   [476-66-4]

When gallic acid is heated with concentrated sulfuric acid, hexahydroxyanthraquinone [82-12-2] (rufigallic acid) is produced by condensation. Reaction with *p*-nitrosodimethylaniline hydrogen chloride produces oxazine derivatives. Oxidation with arsenic acid, permanganate, persulfate, or iodine yields ellagic acid [476-66-4], as does reaction of methyl gallate with iron(III) chloride. The carboxyl group undergoes azeotropic esterification with alcohols, or, by the Fischer method, with alcoholic hydrochloric acid. The ether – ester of gallic acid is obtained by conversion of dialkyl sulfates or alkyl halides in the presence of alkali, or by using diazomethane [334-88-3]. The hydroxyl group in the 4-position is the most reactive. 3,4,5-Trimethoxybenzoic acid [118-41-2] undergoes partial hydrolysis in strong acid to form 4-hydroxy-3,5-dimethoxybenzoic acid [530-57-4] (syringic acid).

Production. Gallic acid, a component of many tanning agents is an endogenous product in plants. The acid occurs free or bound to tannin [1401-55-4] (e.g., in divi-divi, oak bark, gallnuts, pomegranate roots, sumac, and tea). The acid is produced from tannin-rich aqueous gallnut extracts by acidic or alkaline hydrolysis. It is also obtained by using the enzyme tannase [9025-71-2] or molds (*Penicillium glaucum*, *Aspergillus niger*) to cleave tannin by fermentation. Both the metabolism of gallic acid and its impact on plant growth enzymes have been studied [52].

Uses. Gallic acid is used for the manufacture of iron gallnut ink (see above) and as an intermediate in the production of dyes (e.g., anthragallol [620-64-2], gallocyanine [1562-85-2], galloflavin [568-80-9], and rufigallic acid [82-12-2]). Gallate esters, particularly methyl gallate [99-24-1] (gallicin), *mp* 157 °C, and propyl gallate [121-79-9], *mp* 150 °C, are used as antioxidants and food preservatives (e.g., for fats) [53]. Gallic acid is also employed as a reducing agent for pharmaceutical purposes (Dermatol, Airol, bismuth salt of gallic acid) [54] and for the production of photographic developers and pyrogallol.

4.4. Mandelic Acid

Racemic (R,S)-Mandelic Acid [611-72-3]. *Physical Properties.* (R,S)-Mandelic acid (α-hydroxyphenylacetic acid, phenylglycolic acid), $C_6H_5-CH(OH)-COOH$, $C_8H_8O_3$, M_r 152.15, *mp* 118 – 121 °C, is a colorless crystalline powder. It is readily soluble in ethanol and diethyl ether, less soluble in chloroform, and insoluble in petroleum ether. Its water solubility is 15 g per 100 mL of water. In alkaline solution, mandelic acid dissolves to form salts.

Chemical Properties. Mandelic acid turns brown when exposed to light. It can be esterified easily with alcohols in the presence of hydrogen chloride.

Production. (R,S)-mandelic acid is produced by hydrolysis of mandelic acid nitrile [532-28-5] (*mp* 10 °C) with hydrochloric acid. Mandelic acid nitrile is obtained by the conversion of benzaldehyde [100-52-7] and hydrogen cyanide in the nascent state

(NaCN + HCl) under refrigeration. Another method of producing mandelic acid is the hydrolysis of α,α-dichloroacetophenone with alkali.

Uses. Mandelic acid esters have analgesic, antirheumatic, and spasmolytic effects. For examples, *cis*-3,5,5-trimethylcyclohexyl mandelate [456-59-7] (cyclandelate, Spasmocyclon, Clan-dilon), $C_{17}H_{24}O_3$, M_r 276.38, mp 54 – 58 °C, is used as a spasmolytic and to stimulate blood circulation. Mandelic acid tropine ester [51-56-9] (homatropine – HBr), $C_{16}H_{22}BrNO_3$, M_r 356.26, mp 212 – 215 °C, is readily soluble in water and ethanol, and is used in ophthalmology to dilate the pupils. Hexamethylenetetramine mandelate [587-23-5] (mandelamine, Mandropine, diuramine, hexydaline) $C_{14}H_{20}N_4O_3$, M_r 292.34, mp 127 – 130 °C, forms colorless crystals that are readily soluble in diethyl ether. This salt is used to treat urinary tract infections.

Levorotatory (R)(–)-Mandelic Acid [611-71-2]. (R)-Mandelic acid forms colorless crystals, mp 132 – 135 °C, specific rotation $[\alpha]_D^{20}$ (c = 5) –154 to 157 °C. It is produced by resolution of racemic (R,S)-mandelic acid with cinchonine [118-10-5] in aqueous solution, whereby the salt of (S)-(+)-mandelic acid crystallizes. If resolution is performed with S-(+)-aminodiol [3306-06-7], (R)-(–)-mandelic acid crystallizes, while (S)-(+)-mandelic acid [17199-29-0] remains in solution. (R)-(–)- and (S)-(+)-mandelic acid can again be subjected to racemic resolution after isomerization.

5. Toxicology

Toxicological and physiological properties of the different aromatic hydroxycarboxylic acids are summarized in the following material.

3-Hydroxybenzoic Acid. LD_{50} 3700 mg/kg (rat, i.p.), TDLo 400 mg/kg (rat, s.c., 11 d pregnant); classification: drug, teratogen [55; date: 1/84].

4-Hydroxybenzoic Acid. LD_{50} 2200 mg/kg (mouse, oral), LD_{50} 210 mg/kg (mouse, i.p.), LD_{50} 1050 mg/kg (mouse, s.c.); classification: drug, mutagen [55; date: 8/83].

o-Cresotic Acid. LD_{50} 1000 mg/kg (mouse, oral). LD_{50} 345 mg/kg (mouse, i.v.); classification: drug [55; date: 6/85].

m-Cresotic Acid. LD_{50} 1800 mg/kg (mouse, oral); classification: drug [55; date: 6/85].

p-Cresotic Acid. LD_{50} 1000 mg/kg (mouse, oral); classification: drug [55; date: 6/85].

4-Hydroxy-3-methoxybenzoic Acid. Classification: mutagen [55; date: 8/82].

2,4-Dihydroxybenzoic Acid. LDLo 642 mg/kg (rat, s.c., 11 d pregnant); classification: teratogen [55; date: 3/84].

3,5-Dihydroxybenzoic Acid. LD_{50} 2000 mg/kg (mouse, i.v.) [55; date: 6/84].

2,5-Dihydroxybenzoic Acid. 2,5 Dihydroxybenzic acid is an intermediate in the metabolism of salicyclic acid [56]: LD_{50} 800 mg/kg (rat, oral) LD_{50} 3000 mg/kg (rat, i.p.), LD_{50} 374 mg/kg (mouse, i.v.), LDLo 380 mg/kg (rat, s.c., 9 d pregnant),

LDLo 642 mg/kg (rat, s.c., 11 d pregnant); classification: teratogen, natural product [55; date: 5/84].

3,4,5-Trihydroxybenzoic Acid. LD_{50} 320 mg/kg (mouse, i.v.), LD_{50} 5000 mg/kg (rabbit, oral), LDLo 800 mg/kg (mouse, i.p.), LDLo 5 mg/kg (mouse, s.c., 1 d pregnant); classification: agricultural chemical, mutagen, teratogen [55; date: 3/85].

6. References

General References

[1] *Beilstein* **10,** 43, 134, 149, 192, 214, 375, 464, **10 (1),** 20, 63, 68, 83, 95, 173, 232, **10 (2),** 25, 79, 88, 112, 248, 331, **10 (3),** 87, 422, 1363, 2057, **10 (4),** 125, 315, 345, 1414, 1971.

[2] E. Rodd: *Chemistry of Carbon-Compounds,* vol. **3,** Elsevier, Amsterdam-London-New York 1956, pp. 756–815.

[3] A. S. Lindey, H. Jeskey, *Chem. Rev.* **57** (1957) 583–620.*Houben Weyl* **8,** 372–377.

Specific References

[4] J. E. Heestermann, *Chem. Weekbl.* **32** (1935) 463.

[5] C. G. Barroso, R. C. Torrijos, J. A. Perez-Bustamante, *Chromatographia* **17** (1983) no. 5, 249–252.R. Shinohara, *Suishitsu Odaku Kenkyu* **7** (1984) no. 9, 549–552; *Chem. Abstr.* **102** (1985) no. 26, 225 717 p.

[6] H. Kolbe, *Justus Liebigs Ann. Chem.* **113** (1860) 125–127. H. Kolbe, E. Lautermann, *Justus Liebigs Ann. Chem.* **115** (1860) 157–206. H. Kolbe, *J. Prakt. Chem.* **10** (1874) no. 2, 89–111. O. Hartmann, *J. Prakt. Chem.* **16** (1877) no. 2, 35–59. H. Kolbe, DE 426, 1877. F. W. v. Heyden, DE 29 939, 1884; 31 240, 1884 (R. Schmitt). F. W. v. Heyden, DE 33 635, 1885 (R. Schmitt). F. W. v. Heyden, DE 38 742, 1886 (R. Seifert). S. Marassé, DE 73 279, 1893.

[7] D. Cameron, H. Jeskey, O. Baine, *J. Org. Chem.* **15** (1950) 233–236. O. Baine, *J. Org. Chem.* **19** (1954) 510–514.

[8] Rhône-Poulenc, DE-OS 27 54 239, 1976 (J. C. Choulet, J. Nouvel).

[9] F. Wessely, *Monatsh. Chem.* **81** (1950) 1071–1091.

[10] S. E. Hunt, J. I. Jones, A. S. Lindsey, *Chem. Ind. (London)* 1955, 417–418.

[11] M. F. Clarke, L. N. Owen, *J. Chem. Soc.* 1950, 2108–2115.

[12] K. W. Rosenmund, H. Harms, *Ber. Dtsch. Chem. Ges.* **53** (1920) 2226–2240.

[13] F. F. Blicke, F. D. Smith, J. L. Powers, *J. Am. Chem. Soc.* **54** (1932) 1460–1471.

[14] B. Heymann, W. König, *Ber. Dtsch. Chem. Ges.* **19** (1886) 704–706.

[15] O. Jakobson, *Ber. Dtsch. Chem. Ges.* **11** (1878) 376, 570.

[16] G. Lock, *Ber. Dtsch. Chem. Ges.* **62** (1929) 1177–1188.

[17] I. A. Pearl, *J. Am. Chem. Soc.* **67** (1945) 1628–1629. I. A. Pearl, *J. Am. Chem. Soc.* **68** (1946) 429–432.I. A. Pearl, *J. Org. Chem.* **12** (1947) 79–84.

[18] L. C. King, M. McWhirter, D. M. Barton, *J. Am. Chem. Soc.* **67** (1945) 2089–2095.

[19] H. D. Dakin, M. D. Herter, *J. Biol. Chem.* **3** (1907) 419–434.

[20] G. Navazio, A. Scipioni, *Chim. Ind. (Milan)* **50** (1968) 1086–1090.

[21] J. Zelter, M. Landau, DE 258 887, 1912.

[22] L. Gattermann, *Justus Liebigs Ann. Chem.* **244** (1888) 29–76.

[23] G. J. Gershzon, *J. Gen. Chem. USSR (Engl. Transl.)* **13** (1943) 82.

[24] H. Schäfer, *Memorial des Poudres* **27** (1937) 153, 156, 157.
[25] H. E. Ungnade, A. S. Henick, *J. Am. Chem. Soc.* **64** (1942) 1737–1738.
[26] I. G. Farben, DE 506 438, 1926 (F. Hans); *Friedländer* **17**, 472.
[27] H. Kupferberg, *J. Prakt. Chem.* **16** (1877) no. 2, 424–447.
[28] C. A. Buehler, W. E. Cate, *Org. Synth. Coll.* **2** (1943) 341–343. Dow Chemical, US 1 937 477, 1932 (L. E. Milk, W. W. Allen). S. E. Hunt, *J. Chem. Soc.* **1958**, 3152–3160. A. J. Rostron, A. M. Spivey, *J. Chem. Soc.* **1964**, 3092–3096.
[29] P. Friedländer, O. Löw-Beer, DT 170 230, 1905.
[30] G. W. K. Cavill, J. M. Vincent, *J. Chem. Soc. Ind.* **66** (1947) 175–182.
[31] BASF Wyandotte, US 4 234 440, 1979 (St. T. Hirozawa, E. F. O'Brien, J. C. Wilson).
[32] G. M. Cramer et al., *Food, Cosmet. Toxicol.* **16** (1978) 255.
[33] A. J. Maga, K. Lorenz, *Cereala Sci. Today* **18** (1973) no. 10, 326–328, 350; *Chem. Abstr.* **84** (1976) no. 23, 163 164 e. H. P. Fiedler: *Lexikon der Hilfsstoffe für Pharmazie, Kosmetik und angrenzende Gebiete*, 2nd ed., Editio Cantor Aulendorf 1981, pp. 480 ff.
[34] F. J. Bandelin, *Drug Cosmet. Ind.* **64** (1949) 430.
[35] F. Hoffmann-La Roche, DE-OS 2 447 098, 1974 (A. Boller, H. Scherrer). Hitachi Ltd., EP 110 299, 1983 (S. Hattori, K. Iwasaki, T. Kitamura, A. Muko et al.). Chisso Corp., DE-AS 2 751 403, 1977 (T. Inukai, H. Sato, K. Yokohama).
[36] C. S. Ough, *Chem. Eng. News* **65** (1987) 19–28.
[37] Yoshitomi Pharmaceutical Industries, JP 62 76 651, 1987 (T. Inoue, M. Horie, T. Sato).
[38] Mallinckrodt Inc., US 4 309 564, 1980 (D. F. Loncrini, J. J. Taylor).
[39] Mitsubishi Paper Mills Ltd., JP 58-138 689, JP 58-151 292, JP 58-8 686, EP-AP 86638, 1983 (H. Tsukahara, T. Torii).
[40] Ueno Pharma Co., JP 58-158 289, 1983 (R. Keno, H. Tsuchiya, S. Itou, T. Tsuchida). Honshu-paper Co., JP 58-158 290, 1983 (M. Jsoda, S. Nishimatsu, K. Kuniyama). Fuji Photo Film Co., JP 58-87 089, 1983 (M. Watanabe). Konzaki Paper Mfg Co. Ltd., JP-P 210 057-82, 1982 (H. A. T. Suzuki, R. O. Hayashi, N. I. O. Arai, Y. I. N. Oeda).
[41] Mitsui Toatsu Kagaku KK, JP 49-19 275, 1974 (R. Fujihara, K. Nitta).
[42] Siemens, DE-OS 3 049 417, 1980 (H. Januschkowetz, H. Laub).
[43] I. A. Pearl, *J. Am. Chem. Soc.* **68** (1946) 2180–2181. I. A. Pearl, *J. Am. Chem. Soc.* **68** (1946) 1100–1101.
[44] A. Cerami, J. Graziano, R. W. Grady, C. M. Peterson, *Proc. Symp.: Dev. Iron Chelaters Clin. Use 1975*, 261–273; Chem. Abstr. 87 (1977) no. 7, 47 789 a.
[45] K. Brunner, *Justus Liebigs Ann. Chem.* **351** (1907) 313–331.
[46] Beecham Inc., US 4 601 898, US 4 601 899, 1985 (R. E. Stier, W. H. Dunn, J. D. Vidra).
[47] Kaiser Aluminium & Chemical Corp., US 3 714 000, 1971 (G. A. Dorsey, Jr.).
[48] Texaco Inc., US 3 537 992, 1967 (J. H. Kolaian).
[49] L. Barth, K. Senhofer, *Justus Liebigs Ann. Chem.* **159** (1871) 216–230.
[50] Chem. Fabr. AG vorm. E. Schering, DE 81 297, 1894. F. Mauthner, *J. Prakt. Chem.* **156** (1940) 150–153.
[51] Fuji Photo Film Co., GB 1 360 581, 1971 (E. Katsusuke, I. Sadao).
[52] E. A. Haddock, *Phytochemistry* **21** (1982) no. 5, 1049–1062. E. Hasalam, *Fortschr. Chem. Org. Naturst.* **41** (1982) 1–46. S. Kobayashi, M. Ariga, T. Ozawa, H. Imagawa, *Agric. Biol. Chem.* **48** (1984) no. 2, 389–395.
[53] J. W. Thompson, E. R. Sherwin, *J. Am. Oil Chem. Soc.* **43** (1966) no. 12, 683–686.
[54] P. Pfeiffer, E. Schmitz, *Pharmazie* **5** (1950) 517–520.
[55] Cis. Inc., 7215 York Road, Baltimore, MD 21212, CIS-Data bank (RTECS and OHMTADS).

[56] D. R. Boreham, B. K. Martin, *Br. J. Pharmacol.* **37** (1969) no. 1, 294–300; *Chem. Abstr.* **71** (1969) 121 980 u.

Imidazole and Derivatives

KLAUS EBEL, BASF Aktiengesellschaft, Ludwigshafen, Federal Republic of Germany (Chaps. 2–6)

HERMANN KOEHLER, BASF Aktiengesellschaft, Ludwigshafen, Federal Republic of Germany (Chaps. 2–6)

ARMIN O. GAMER, BASF Aktiengesellschaft, Ludwigshafen, Federal Republic of Germany (Section 7.1)

RUDOLF JÄCKH, BASF Aktiengesellschaft, Ludwigshafen, Federal Republic of Germany (Section 7.2)

1.	Introduction 2987	6.	Uses 2991
2.	Physical Properties 2987	6.1.	Intermediates for Herbicides and Pharmaceuticals 2991
3.	Chemical Properties. 2988		
4.	Production 2989	6.2.	Other Uses 2992
4.1.	The Radziszewski Reaction. . . 2989	7.	Toxicology and Occupational Health 2992
4.2.	Dehydrogenation of Δ^2-Imidazolines 2989		
4.3.	Functionalized Imidazoles by Modification of Simple Imidazoles. 2990	7.1.	Imidazoles without Nitro Substituents. 2992
		7.2.	Nitroimidazoles. 2995
5.	Quality Specifications and Analysis 2991	8.	References. 2998

1. Introduction

Imidazole [288-32-4], $C_3H_4N_2$, M_r 68.10, was first synthesized in 1858 by DEBUS from ammonia and glyoxal; it was originally named glyoxalin. The name imidazole was introduced by HANTZSCH. Industrial production of imidazole began in the 1950s; a wide range of derivatives is now available in industrial quantities.

2. Physical Properties

Some important physical properties of imidazole are listed in Table 1. Molecular masses, boiling points, melting points, and CAS registry numbers of the most important imidazole derivatives are listed in Table 2. Substitution at nitrogen generally causes a reduction in the melting and boiling points, because the N–H group is no longer

Table 1. Physical properties of imidazole

Flash point	154 °C
d_4^{110}	1.0257
Bulk density	0.55 – 0.63 g/cm^3
Viscosity (100 °C)	2.696 mPa · s
Solubility in water*	241
Enthalpy of fusion	12.96 kJ/mol
Enthalpy of vaporization (220 – 265 °C)	57.12 kJ/mol

* In grams per 100 g of solvent at 20 °C.

Table 2. Physical properties of imidazoles

	CAS registry no.	M_r	mp, °C	bp, °C
Imidazole	[288-32-4]	68.1	90	256
1-Methylimidazole	[616-47-7]	82.1	−1	198.5
2-Methylimidazole	[693-98-1]	82.1	145	270
4(5)-Methylimidazole	[822-36-6]	82.1	46	271
2-Ethylimidazole	[1072-62-4]	96.1	77	268
2-Isopropylimidazole	[36947-68-9]	110.1	134	278
2-Ethyl-4(5)-methylimidazole	[931-36-2]	110.1	45	260
2-Phenylimidazole	[670-96-2]	144.2	145	336
1-Vinylimidazole	[1072-63-5]	94.1		80*
4(5)-Methyl-5(4)-hydroxymethylimidazole	[29636-87-1]	112.1	138	
4(5)-Nitroimidazole	[3034-38-6]	113.1	305**	
2-Methyl-4(5)-nitroimidazole	[696-23-1]	127.1	244**	
2-Isopropyl-4(5)-nitroimidazole	[13373-32-5]	155.2	184	
1-Methyl-5-nitroimidazole	[3034-42-2]	127.1	59	
1,2-Dimethyl-5-nitroimidazole	[551-92-8]	141.1	139	
1-Methyl-2-isopropyl-5-nitroimidazole	[14885-29-1]	169.2	56	
1-(2-Hydroxyethyl)-2-methyl-5-nitroimidazole	[443-48-1]	171.2	155	

* At 1.3 kPa.
** Decomposes.

available for intermolecular hydrogen bonding. Distillation of nitroimidazoles is accompanied by the risk of spontaneous decomposition.

3. Chemical Properties

Imidazole is a moderately strong base (pK_b = 7.0 [1]), and a weak acid (pK_a = 14.9 [1]). Imidazoles substituted with electron-withdrawing groups are stronger acids than imidazole itself; e.g., 4(5)-nitroimidazole has a pK_a of 9.3 [1].

Imidazole is stable at 400 °C, possesses considerable aromatic character, and undergoes the usual electrophilic aromatic substitution reactions. Nitration and sulfonation require, however, far more drastic conditions than the corresponding reactions with benzene. Other substitution reactions of imidazole include halogenation, hydroxymethylation, coupling with aromatic diazonium salts, and carboxylation. Catalytic hydrogenation has not been described [1], [2].

Table 3. Compounds used to prepare simple imidazoles by the Radziszewski reaction

Imidazole	Dicarbonyl compound	Aldehyde	Amine
Imidazole	glyoxal	formaldehyde	ammonia
1-Methylimidazole	glyoxal	formaldehyde	ammonia, methylamine
2-Methylimidazole	glyoxal	acetaldehyde	ammonia
2-Ethylimidazole	glyoxal	propionaldehyde	ammonia
2-Isopropylimidazole	glyoxal	isobutyraldehyde	ammonia
4(5)-Methylimidazole	methylglyoxal	formaldehyde	ammonia
2-Ethyl-4(5)-methylimidazole	methylglyoxal	propionaldehyde	ammonia

4. Production

Of the many known methods for producing imidazoles, only the following are of industrial importance.

4.1. The Radziszewski Reaction

In the generally applicable Radziszewski reaction, a 1,2-dicarbonyl compound is condensed with an aldehyde and ammonia ($R^1 = H$) in a molar ratio of 1:1:2, respectively. Replacement of a molar equivalent of ammonia with a primary amine ($R^1 = $ alkyl or aryl) leads to the corresponding 1-substituted imidazoles.

The reaction is usually carried out in water or a water–alcohol mixture at 50–100 °C. Work-up may involve the usual processes (e.g., distillation, extraction, and crystallization). Distillation leads to imidazole with a purity >99%. The yield is generally 60–85% [3]. This process allows the most important simple imidazoles to be prepared from the starting materials listed in Table 3.

4.2. Dehydrogenation of Δ^2-Imidazolines

Δ^2-Imidazolines can be obtained by several routes from 1,2-diamino compounds and carboxylic acid derivatives:

1) by reaction of a diamine with a carboxylic acid over an acidic heterogeneous catalyst (e.g., alumina–phosphoric acid) in the gas phase [4];
2) by reaction of a diamine with a carboxylic acid nitrile in the presence of sulfur [5] or copper salts [6] in the liquid phase; or
3) by preparation of diformyl derivatives from a diamine and formic acid esters, followed by conversion to imidazoline in the gas phase at 200–350 °C over a heterogeneous catalyst (e.g., zinc oxide–alumina) [7].

The Δ^2-imidazoline is then dehydrogenated in the gas phase over a precious metal at ca. 300 °C [8] or on alumina–zinc oxide at 400–500 °C [9]. In some cases this synthesis gives even better yields than the Radziszewski reaction (e.g., for 2-aryl-substituted imidazoles).

4.3. Functionalized Imidazoles by Modification of Simple Imidazoles

Nitroimidazoles. Reaction of imidazole with 60–80 % nitric acid in the presence of concentrated sulfuric acid at 120–150 °C leads to 4(5)-nitroimidazoles [10], [11]; nitration at C-2 has not been observed. 4(5)-Nitroimidazole, 2-methyl-4(5)-nitroimidazole, and 2-isopropyl-4(5)-nitroimidazole have been prepared by this method.

1-Substituted Imidazoles. Imidazole and alkylimidazoles can be alkylated with alcohols in the gas phase at 300 °C over acidic catalysts such as alumina–phosphoric acid or silica–phosphoric acid [12], [13].

In the liquid phase, imidazoles may be alkylated with dialkyl sulfates or alkyl halides in polar solvents such as alcohols and in the presence of a base, e.g., alkali-metal hydroxides or alkoxides [14].

Of particular importance for the preparation of 1-alkyl-5-nitroimidazoles is the treatment of 4(5)-nitroimidazoles with dialkyl sulfates in formic acid as solvent [15]. This method has been used for 1-methyl-, 1,2-dimethyl-, and 1-methyl-2-isopropyl-5-nitroimidazoles.

1-Vinylimidazole is produced by base-catalyzed addition of acetylene to imidazole in the liquid phase in an autoclave [16].

1-(2-Hydroxyethyl)-2-methyl-5-nitroimidazole (metronidazole) is obtained by addition of ethylene oxide to 2-methyl-4(5)-nitroimidazole in formic acid at 25–40 °C [17].

Michael addition of imidazole to acrylonitrile followed by catalytic hydrogenation of the intermediate 1-(2-cyanoethyl)imidazole [23996-53-4] leads to 1-(3-aminopropyl)imidazole [5036-48-6] [18].

Hydroxymethylimidazoles. 4(5)-Methyl-5(4)-hydroxymethylimidazole can be obtained by treatment of 4(5)-methylimidazole with aqueous formaldehyde under acidic [19] or basic [20] conditions.

Quaternary Salts. Treatment of 1-alkylimidazoles with benzyl chloride at 100 °C in solvent-free melt results in formation of 3-benzyl-1-alkylimidazolium chloride [21].

5. Quality Specifications and Analysis

Simple imidazoles are generally available in >99% purity, which is particularly important for their use in pharmaceutical production. Purity is usually determined by reverse-phase HPLC; volatile derivatives may be analyzed by capillary column gas chromatography. A typical quality specification for imidazole is shown below:

Imidazole (min.)	99.5%
2-Methylimidazole (max.)	0.2%
Water (max.)	0.2%
Other compounds (max.)	0.1%

6. Uses

6.1. Intermediates for Herbicides and Pharmaceuticals

The imidazole ring is a constituent of several important natural products, including purine, histamine, histidine, and nucleic acids. Therefore, the bulk of imidazole produced is used in the preparation of biologically active compounds.

1-Substituted imidazoles such as N-propyl-N-[2-(2,4,6-trichlorophenoxy)ethyl]imidazole-1-carboxamide (prochloraz) [67747-09-5], 1-[2-(2,4-dichlorophenyl)-2-(2-propenyloxy)ethyl]-1H-imidazole (imazalil) [35554-44-0], and 1-(4-chlorophenoxy)-1-(1H-imidazolyl-1-yl)-3,3-dimethylbutanone (climbazole) [38083-17-9] are used as fungicides in crop protection [22]. Clotrimazole [23593-75-1] and bifonazole [60628-96-8] are used as antimycotics.

4(5)-Methylimidazole and 4(5)-methyl-5(4)-hydroxymethylimidazole are used as intermediates for the H_2-blocker cimetidine [51481-61-9].

Nitroimidazoles are important chemotherapeutic agents [23]: e.g., 1,2-dimethyl-5-nitroimidazole (dimetridazole), 1-methyl-2-isopropyl-5-nitroimidazole (Ipronidazole), and 1-methyl-2-carbamoyloxymethyl-5-nitroimidazole (ronidazole) [7681-76-7] are used as veterinary products. Human chemotherapeutics based on 4(5)-nitroimidazole or 2-methyl-4(5)-nitroimidazole include metronidazole [443-48-1], tinidazole [19387-91-8], nimorazole [6506-37-2], ornidazole [16773-42-5], secnidazole [3366-95-8], and carnidazole [42116-76-7].

Quaternary imidazolium derivatives possess algicidal and bactericidal properties [21].

6.2. Other Uses

1-Vinylimidazole is employed as a copolymer in the production of cationic polymers for various uses. Alkylimidazoles are used as hardeners for epoxy resins and for polyurethanes [24]. Less important uses for alkyl- and arylimidazoles include photography and dyes.

7. Toxicology and Occupational Health

7.1. Imidazoles without Nitro Substituents

The prominent feature in imidazole toxicity is the corrosive nature of the compounds. This property is eliminated by 2-phenyl-substitution. In terms of the labelling directives of the EEC the acute systemic toxicity ranges from toxic (T) to harmful to health (X_n).

Acute poisoning results in excitement of the central nervous system and convulsions; therefore, imidazole has been tested as an antidepressant [25]. Interaction of imidazole with the dopaminergic and noradrenergic systems in the brain was postulated [25]. Repeated application of some of the compounds revealed no specific toxic effects. The liver appears to be affected after treatment with high doses. None of the available in vitro short-term tests revealed a mutagenic potential for nonnitroimidazole compounds [26]–[29].

Imidazole. Acute toxicity data (see Table 4) show only slight toxic effects on animals exposed via oral or parenteral routes. The main symptoms of intoxication at high doses are seizures, tremors, loss of balance, opisthotonus (spasm of muscles in the back), and restlessness. Single oral exposure of pregnant rats to a dose of 240 mg/kg of imidazoles on the 12th or 13th day of gestation produced no harmful effect in the progeny [30]. Topically, imidazole is corrosive to the skin and eye, resulting in necrosis and irreversible damage to these organs (Table 4).

Repeated administration to rats for 28 days by gavage (through a stomach tube) revealed no observable effect at a level up to $62.5 \text{ mg kg}^{-1} \text{ d}^{-1}$. Higher doses up to 500 mg/kg led to swelling of the liver and other unspecific toxic effects without mortality. The alteration of the liver may be due to the enhancement of metabolic activity in this organ [26].

1-Alkyl-Substituted Imidazoles. These compounds become more noxious with increasing length of the side chain: 1-n-butylimidazole is fairly toxic, regardless of exposure route. Further increase in chain length reduces the toxicity to that of the parent compound (see Table 4).

2-Methylimidazole behaves like 1-methylimidazole and imidazole (Table 4). Its irritant properties seem to be slightly less than those of the other two compounds but are nevertheless pronounced.

A 28-day gavage study in rats revealed at the lowest dose of $100 \text{ mg kg}^{-1} \text{ d}^{-1}$ only a slight decrease in blood protein content. Higher doses (up to 800 mg/kg) led to unspecific clinical symptoms as well as changes in clinicochemical parameters and organ weights without mortality [26].

4-Methylimidazole is the most toxic of the methyl-substituted imidazoles and is thought to be the toxic agent in ammoniated molasses and hay, which may be responsible for postnatal losses in ruminants fed with these products [31], [32]. It is fairly toxic via the oral and dermal routes. 4-Methylimidazole is the most potent convulsive within this group of compounds. In addition to the data in Table 4, further studies in rabbits [31] and calves [32] demonstrate that single oral doses of 120–150 or 200 mg/kg (rabbit) and 400 mg/kg (calf) lead to convulsion and death. No clinical, clinicochemical, or pathological effects were observed in the study with calves after four repeated doses of 25 or 50 mg/kg within 96 h. At a level of 100 mg/kg, mild behavioral changes appeared.

Intraperitoneal dosage of 43 mg/kg to rats for up to 10 weeks led to enlargement of the liver in some animals, but with no effect on the blood or clinicochemical parameters [33].

1-Vinylimidazole. Resorption of 1-vinylimidazole through the skin in fatal amounts is possible in a relatively short exposure time. The symptoms of intoxication differ with dose and species but excitation and convulsion are a common feature at high doses.

Table 4. Toxicity of imidazoles containing saturated substituents

Substance	Acute toxicity, LD_{50}, mg/kg		Effect on skin and eye	References
Imidazole	Rat, oral	970	Corrosive	[25], [26], [31], [35]
	Mouse, oral	880		
	Mouse, oral (male)	1880		
	Rat, s.c.	626		
	Mouse, s.c.	560; 817		
	Rat, i.p.	620		
	Mouse, i.p.	520		
	Mouse, i.v.	475		
1-Methylimidazole	Rat, oral	1139	Corrosive	[26], [31]
	Mouse, oral (male)	1400		
	Rabbit, dermal	>400 <640		
	Mouse, i.p.	445		
1-Ethylimidazole	Rat, oral	900*	Corrosive	[26]
	Mouse, i.p.	300*		
1-Propylimidazole	Rat, oral	400*	Corrosive	[26]
	Rabbit, dermal	>200*		
	Mouse, i.p.	90*		
1-n-Butylimidazole	Rat, oral	80	Strongly irritant	[26]
	Rabbit, dermal	<200		
	Rat, dermal	>50 <200		
	Rat, inhalation 4 h (aerosol)	0.11 mg/L		
	Mouse, i.p.	40		
1-Decylimidazole	Rat, oral	470	Corrosive	[26]
	Mouse, i.p.	150		
1-Dodecylimidazole	Rat, oral	1100*	Corrosive	[26]
	Rabbit, dermal	>400		
	Mouse, i.p.	125*		
2-Methylimidazole	Rat, oral	1300	Strongly irritant to corrosive	[26], [31]
	Mouse, oral (male)	1400		
	Rat, dermal	ca. 2000		
	Rabbit, dermal	>200		
	Mouse, i.p.	350		
2-Ethylimidazole	Rat, oral	1400	Corrosive	[26]
	Rabbit, dermal	>200		
	Mouse, i.p.	400		
2-Isopropylimidazole	Rat, oral	>1000 <2000	Irritant	[26]
	Rabbit, dermal	>400		
	Mouse, i.p.	>50 <200		
4-Methylimidazole	Rat, oral	350	Strongly irritant to corrosive	[26], [31], [36]
	Mouse, oral (male)	370		
	Rabbit, dermal	440		
	Mouse, i.p. (male)	165		
1,2-Dimethylimidazole	Rat, oral	1300	Strongly irritant to corrosive	[26]
	Rabbit, dermal	>200		
	Rat, inhalation 4 h (aerosol of 50% solution)	> 3 mg/L		
	Mouse, i.p.	300		

Table 4. (continued)

Substance	Acute toxicity, LD_{50}, mg/kg		Effect on skin and eye	References
2-Ethyl-4(5)-methyl-imidazole	Rat, oral	731; 1000	Strongly irritant to corrosive	[26]
	Rabbit, dermal	>400		
	Mouse, i.p.	200		
1-Vinylimidazole	Rat, oral	1100	Strongly irritant to corrosive	[26]
	Mouse, s.c.	650		

* LD_{50}, µL/kg.

After repeated application (rat, rabbit, cat, oral) for up to 7 weeks, unspecific clinical symptoms (weight loss, changes in blood count) and pathological findings (sporadic incidence of liver damage) were reported. The level of no observable effect was between 10 and 100 µL/kg [26].

No toxic effects were observed (mouse, rat, rabbit, cat) after repeated inhalation (five 6-h exposures) of an atmosphere saturated with vinylimidazole vapor at room temperature [26], [34].

N-Methyl-2-vinylimidazole and *N*-butyl-2-vinylimidazole showed a sensitizing effect on the skin (guinea pig) in addition to the corrosive action [34].

Phenylimidazoles. 2-Phenyl-substituted imidazoles display reduced irritancy but no change in acute systemic toxicity compared with imidazole. The acute oral LD_{50} for 2-phenylimidazole [26], 1-methyl-2-phenylimidazole [26], and 1-vinyl-2-phenylimidazole [26] is ca. 1000 mg/kg; the LD_{50} (mouse, i.p.) is ca. 200 mg/kg.

The symptoms of intoxication indicate depression of the central nervous system by 2-phenylimidazole and a convulsive action for the other two compounds.

Inhalation Hazard Tests. With the exception of imidazole, 2-isopropylimidazole, and 1-vinylimidazole, all of the compounds listed in Table 4 have also been subjected to seven- or eight-hour practice-oriented inhalation hazard tests in which rats were exposed to an atmosphere highly enriched in or saturated with the vapor or the volatile components of the test substance [26]. Of the compounds tested, only 1-*n*-butylimidazole resulted in mortality; three out of six rats died after exposure for 8 h [26].

7.2. Nitroimidazoles

Nitroimidazoles are a well-established group of antiprotozoal and bactericidal agents. Due to their aromatic nitro group, they may undergo metabolization to give electrophilic intermediates under certain conditions [37], [38] and thus entail bacteriotoxic, but also mutagenic and potentially carcinogenic properties. Metabolization of the nitro group may be achieved under anaerobic conditions by nitroreductases of intestinal bacteria or of the liver [39]. Recent reports indicate that bone marrow cells also have the potential to reduce the nitro group even under aerobic conditions [40].

Acute Toxicity. The acute toxicity data for nitroimidazoles indicate a moderate to considerable acute toxicity depending on the substituents. In some cases the pH may also play a role. The materials so far investigated are mostly eye irritants, but not skin irritants.

The acute toxicities (LD_{50}, mg/kg) for some nitroimidazoles are as follows:

4(5)-Nitroimidazole [26]	1660 (rat, oral)
1,2-Dimethyl-5-nitroimidazole hydrochloride (dimetridazole hydrochloride) [26]	3000 (rat, oral)
	200–2000 (cat, oral)
1-Methyl-5-nitroimidazole [26]	ca. 200 (rat, oral)
1-(2-Hydroxymethyl)-2-methyl-5-nitroimidazole (metronidazole) [41]	3800 (mouse, oral)
α-Methoxymethyl-2-nitroimidazole-1-ethanol (misonidazole) [42]	2130 (rat, oral)
2-Methyl-4(5)-nitroimidazole [26]	1540 (rat, oral)

Four-hour dust inhalation tests on rats with 4(5)-nitroimidazole and dimetridazole gave LC_{50} values of 6.56 and > 5.3 mg/L, respectively. Dimetridazole and 1-methyl-5-nitroimidazole have an irritant action on the eyes of rabbits, while 4(5)-nitroimidazole is non-irritating [26]. None of the three compounds exhibits irritancy towards the skin of rabbits [26].

Mutagenicity. Several nitroimidazoles have been subjected to bacterial point mutation assays (such as the Ames test), and, in most cases, were found to exert mutagenicity [37], [43]–[51]. Among these were the following:

1-Methyl-5-nitroimidazole
2-Methyl-4(5)-nitroimidazole
1,2-Dimethyl-5-nitroimidazole (dimetridazole)
1-(2-Hydroxymethyl)-2-methyl-5-nitroimidazole (metronidazole)
2-Isopropyl-4-nitroimidazole
1-Methyl-2-isopropyl-5-nitroimidazole (ipronidazole)
1-(3-Chloro-2-hydroxypropyl)-2-methyl-5-nitroimidazole (ornidazole)
(1-Methyl-5-nitroimidazole-2-yl)methyl carbamate (ronidazole)

α-Methoxymethyl-2-nitroimidazole-1-ethanol (misonidazole)

Nitroimidazoles also induce streptomycin-resistant mutants (forward mutations) in *Klebsiella pneumoniae* [44], [45].

Conflicting results were earlier reported with 4(5)-nitroimidazole, 2-methyl-4(5)-nitroimidazole, and 2-methyl-4(5)-nitroimidazole-1-yl-acetate [43], [46], [47]. Negative results may be due to the inadequate metabolic activation achieved in the test systems. Materials bearing a carboxylate or sulfate group may in fact be non-mutagenic [43], [45], but the data are insufficient for a firm conclusion.

Urine and other body fluids of patients treated with metronidazole in doses of 750 mg/d showed mutagenic activity in the Ames test [52], [53]. Patients who received 200–1200 mg/d for several months were found to have two- to four-fold increases in chromosomal abnormalities in peripheral leukocytes [54].

Efforts have been undertaken to lower the mutagenicity of nitroimidazoles by derivatisation and substitution [47], [48], [55]. Note, however, that mutagenicity in vitro is not quantitatively extrapolable to in vivo situations. Furthermore, a lower mutagenicity does not necessarily indicate a decrease in the carcinogenic potential as well.

Carcinogenicity. The carcinogenic potential has been investigated for some nitroimidazoles used or foreseen as drugs. However, the study designs were not fully in accordance with present standards. The carcinogenic responses obtained so far were moderate or low. Human experience with nitroimidazole drugs has not revealed cancer cases due to therapeutic usage. Metronidazole is listed as a carcinogen by IARC [56]. Mice receiving 0.06, 0.15, 0.3, or 0.5 % (ca. 90, 225, 450, or 750 mg/kg) of metronidazole in the diet throughout their life-span showed a dose-dependent increase in the incidence of lung tumors and lymphomas. The lowest dose did not significantly increase the incidence of tumors [57].

No carcinogenic effects were found in rats in a feeding experiment with 0.135 % (ca. 200 mg kg^{-1} d^{-1}) of metronidazole in the diet for 66 weeks and a post-observation period of 10 weeks [58]. This administration period is, however, too short to fully rule out carcinogenic effects.

More recent investigations with metronidazole in the diet of mice (75, 150, or 600 mg kg^{-1} d^{-1} for 92 weeks) and rats (0.06, 0.3 or 0.6 %, equivalent to 30, 150 or 300 mg kg^{-1} d^{-1} for 150 weeks) showed dose-related increases in tumor incidence (liver and mammary tumors, Leydig cell and pituitary tumors) [59], [60].

After gavage administration of metronidazole to mice during pregnancy and lactation (2 mg per animal per day for 5 d per week) an increased tumor incidence was observed in the offspring [61].

Metronidazole (0.15 or 0.3 %) in the diet of hamsters did not increase the incidence of tumors; however, published details of the study are insufficient for a reliable assessment of the carcinogenic effect [62].

A group of patients who had been treated with metronidazole during the period 1960–1969 did not show increased tumor rates in a follow-up study published in 1988 [63].

Dimetridazole was found to have tumorigenic activity in rats receiving 0.2 % in the diet for 46 weeks. Of 35 Sprague–Dawley rats, 25 developed a benign, but possibly premalignant mammary tumor (fibroadenoma) compared to 4 from 35 rats in the control group [58]. The administration period was, however, too short to fully determine the tumorigenic potential.

Ornidazole (25, 100, or 400 mg kg^{-1} d^{-1}) was reported to cause no tumors in a two-year feeding experiment in rats [64].

The overall conclusion from all experiments is that the carcinogenic potential of nitroimidazoles has to be taken into account, but the amount and quality of the available data do not allow firm conclusions.

Neurotoxic effects. There are several case reports on peripheral neuropathy due to Metronidazole treatment. The symptoms appear to be reversible upon ceasing treatment [65].

8. References

General References

[1] M. R. Grimmett in K. T. Potts: *Comprehensive Heterocyclic Chemistry*, vol. **5**, Pergamon Press, 1984, pp. 345–456.
[2] K. Hofmann: *The Chemistry of Heterocyclic Compounds, Imidazole and Derivatives*, Part I, Intersci. Publ. Inc., New York 1953.

Specific References

[3] BASF, DE-OS 2 360 175, 1979 (A. Frank, H. Karn, H. Spänig).
[4] BASF, EP 0 012 371, 1982 (T. Dockner, U. Kempe, H. Krug, P. Magnussen, W. Prätorius, H. Szymanski).
[5] BASF, DE 2 512 513, 1976 (A. Frank, T. Dockner).
[6] BASF, EP 0 111 073, 1987 (A. Frank, T. Dockner).
[7] BASF, EP 0 036 521, 1983 (T. Dockner, U. Kempe, H. Krug, P. Magnussen, W. Prätorius).
[8] Houdry Process Corp., US 2 847 417, 1958 (W. E. Erner).
[9] BASF, EP 0 000 208, 1979 (T. Dockner, A. Frank).
[10] C. Corsar et al., *Arzneim. Forsch.* **16** (1966) 23.
[11] BASF, DE 2 645 172, 1982 (M. Wetzler, T. Dockner).
[12] BASF, DE 1 670 293, 1976 (H. Spänig, A. Steimmig, J. Sand).
[13] BASF, DE-AS 2 233 908, 1977 (T. Dockner, H. Krug).
[14] M. Häring, *Helv. Chim. Acta* **42** (1959) 1845–1846.
[15] A. Grimison, J. H. Ridd, B. V. Smith, *J. Chem. Soc.* 1960, 1357–1362.
[16] W. Reppe, *Justus Liebigs Ann. Chem.* **601** (1956) 81–138.
[17] Rhône-Poulenc, DE-OS 1 470 200, 1970 (C. Podesva, K. Vagi).
[18] Y. Masashige, M. Masaichiro, *Chem. Pharm. Bull.* **24** (1976) 1480–1484.
[19] BASF, EP 0 015 516, 1982 (U. Kempe, T. Dockner, F. Frank, H. Karn).
[20] Delalande S.A., DE-OS 2 934 925, 1978 (Y. A. Hubert-Brierre).
[21] E. R. Shepard, H. A. Shonle, *J. Am. Chem. Soc.* **69** (1947) 2269–2270.
[22] W. Perkow: *Wirksubstanzen der Pflanzenschutz- und Schädlingsbekämpfungsmittel*, 2nd ed., Verlag Paul Parey, Berlin – Hamburg 1985.

[23] M. D. Nair, K. Nagarajan, *Prog. Drug Res.* **27** (1983) 163–252.
[24] Shell Oil Co., US 3 538 039, 1970 (W. L. Lantz, J. P. Manasia).
[25] F. Ferrari, *Arch. Int. Pharmacodyn. Ther.* **227** (1985) 303–312.
[26] BASF Aktiengesellschaft, unpublished results.
[27] L. E. Voogd et al., *Mutat. Res.* **66** (1979) 207–221.
[28] A. Momii et al., *Iyakuhin Kenkyu* **10** (1979) 351–357.
[29] E. Grzybowska et al., *Acta Microbiol. Pol.* **34** (1985) 111–120.
[30] J. A. Ruddick et al., *Teratology* **13** (1976) 263–266.
[31] K. Nishie et al., *Toxicol. Appl. Pharmacol.* **14** (1969) 301–307.
[32] T. E. Fairbrother et al., *Vet. Hum. Toxicol.* **29** (1987) 312–315.
[33] M. H. Idaka, *Okayama Igakkai Zasshi* **88** (1976) 673–680.
[34] *Ullmann*, 4th ed., **23**, 617.
[35] Niosh: Registry of Toxic Effects of Chemicals.
[36] M. Hidaka, *Okayama Igakkai Zasshi* **88** (1976) 653–657.
[37] D. G. Lindmark et al., *Antimicrob. Agents Chemother.* **10** (1976) 476–482.
[38] D. I. Edwards, M. Dye, H. Carne, *J. Gen. Microbiol.* **76** (1973) 135.
[39] J. L. Blumer et al., *Cancer Res.* **40** (1980) 4599–4605.
[40] M. Isildar et al., *Toxicol. Appl. Pharmacol.* **94** (1988) 305–310.
[41] B. Cavalleri et al., *J. Med. Chem.* **20** (1977) 1522.
[42] National Technical Information Service USA, PB 81-121212.
[43] C. E. Voogd et al., *Mutat. Res.* **86** (1981) 243–277.
[44] C. E. Voogd et al., *Mutat. Res.* **26** (1974) 483–490.
[45] C. E. Voogd et al., *Mutat. Res.* **66** (1979) 207–221.
[46] H. S. Rosenkranz et al., *Mutat. Res.* **38** (1976) 203–206.
[47] J. B. Chin et al., *Mutat. Res.* **58** (1978) 1–10.
[48] J. S. Walsh et al., *J. Med. Chem.* **30** (1987) 150–156.
[49] M. Suzangar et al., *Biochem. Pharmacol.* **36** (1987) no. 21, 3743–3749.
[50] P. G. Wislocki et al., *Chem. Biol. Interactions* **49** (1984) 27–38.
[51] H. S. Rosenkranz et al., *Biochem. Biophys. Res. Commun.* **66** (1975) 520–525.
[52] M. S. Legator et al., *Science* **188** (1975) 1118–1119.
[53] E. Mohtashamipur et al., *Mutagenesis* **1** (1986) 371–374.
[54] F. Mitelman et al., *Lancet*, 1976, 802.
[55] G. Cantelli-Forti et al., *Teratog. Carcinog. Mutag.* **3** (1983) 51–63.
[56] IARC Monograph **13**, 113–121.
[57] M. Rustia, P. Shubik, *J. Nat. Cancer Inst.* **48** (1972) 721–729.
[58] S. M. Cohen et al., *J. Nat. Cancer Inst.* **51** (1973) 403–417.
[59] J. H. Rust, An assessment of metronidazole tumorigenicity studies in the mouse and the rat, in S. M. Finegold (ed.): Metronidazole, Proc. Intern. Metronidazole Conf. Montreal, Excerpta Medica, Amsterdam (1977) 138–144.
[60] M. Rustia, P. Shubik, *J. Nat. Cancer Inst.* **63** (1979) 863–868.
[61] M. Chacko et al., *Cancer Res. Clin. Oncol.* **112** (1986) 135–140.
[62] F. C. Roe, Proc. Intern. Metronidazole Conf. Montreal, Excerpta Medica, Amsterdam (1977) 132–137.
[63] C. M. Beard et al., *Mayo Clin. Proc.* **63** (1988) 147–153.
[64] R. Richle et al., *Arzneim. Forsch.* **28** (1978) 612–625.
[65] W. G. Bradley et al., *Br. Med. J.* **2** (1977) 610–611.

Indole

GERD COLLIN, Rütgerswerke AG, Duisburg-Meiderich, Federal Republic of Germany (Chaps. 2–5)
HARTMUT HÖKE, Rütgerswerke AG, Castrop-Rauxel, Federal Republic of Germany (Chap. 6)

1.	Introduction	3001	5.	Uses	3003
2.	Properties	3001	6.	Toxicology	3004
3.	Production	3002			
4.	Derivatives	3002	7.	References	3005

1. Introduction

Indole [120-72-9], 1-benzo[b]pyrrole, C_8H_7N, was discovered by A. v. BAEYER and C. A. KNOP in 1866 as the basic structure of the natural dye, indigo, from which it was obtained. In 1910, R. WEISSGERBER found indole in coal tar.

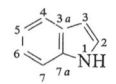

2. Properties

Physical Properties. Indole, M_r 117.15, mp 52.5 °C, bp 254.7 °C (101.3 kPa), d 1.22 g/cm^3 (18 °C), forms colorless, shiny flakes with a slight jasmine aroma. It is readily soluble in ethanol, diethyl ether, and benzene, soluble in hot water, slightly soluble in cold water, and volatile in steam. The heat of combustion is 3.650 kJ/kg (25 °C), the enthalpy of vaporization 597.5 kJ/kg (10.3–27.4 °C), and the dipole moment 2.11 D (benzene).

Chemical Properties. As a secondary amine, indole has a hydrogen atom which can be substituted by alkali metals. Oxidation leads to indigo, mild hydrogenation to 2,3-dihydroindole (indoline). Diindole, triindole, and resinification products are obtained upon treatment with acids.

3. Production

High-temperature coal tar contains on average just under 0.2% indole. In coal tar distillation, the indole is concentrated in a biphenyl–indole fraction which boils between 245 and 255 °C. Following extraction of phenols and bases, the indole is isolated from this fraction and separated from the other major component biphenyl, whose boiling point is only 0.3 °C higher. This is achieved by melting with potassium hydroxide to give the potassium salt of indole, by azeotropic distillation with diethylene glycol, or by extraction with selective solvents (e.g., glycols, aqueous dimethyl sulfoxide, or monoethanolamine). The crude indole can then be purified by crystallization from aliphatic hydrocarbon solvents.

In addition to its recovery from coal tar, indole is also synthesized in technical quantities. Methods used include Madelung synthesis of formyltoluidine (from o-toluidine and formic acid), followed by cyclization [8]–[10], dehydrogenating cyclization of 2-ethylaniline [11]–[20], cyclocondensation of aniline and ethylene glycol in the liquid or gas phase [21]–[23], and cyclization of 2-(2-nitrophenyl)ethanol [24].

4. Derivatives

3-Methylindole [83-34-1], skatole, C_9H_9N, M_r 131.18, mp 95 °C, bp 266.4 °C (101.3 kPa), forms colorless flakes which are soluble in ethanol and slightly soluble in water. It can be recovered from coal tar or synthesized by Fischer reaction from the phenylhydrazone of propionaldehyde or, together with indole, by catalytic dehydration of 2-(2-aminophenyl)-1,3-propanediol [25].

3-(Dimethylaminomethyl)indole [87-52-5], gramine, $C_{11}H_{14}N_2$, M_r 174.25, mp 138–139 °C, forms colorless needles. The compound is soluble in ethanol, diethyl ether, and chloroform and insoluble in water. It can be synthesized from indole by the Mannich reaction with dimethyl-amine and formaldehyde. Gramine is used to synthesize D,L-tryptophan, the plant growth regulators indole-3-acetic acid and indole-3-butanoic acid as well as the intermediate tryptamine.

Indole-3-aldehyde [487-89-8], C_9H_7NO, M_r 145.16, mp 194 °C, forms colorless crystals. It is readily soluble in ethanol and slightly soluble in cold water. It can be produced by Vilsmeier reaction of indole with formylmethylaniline or with dimethylformamide and phosphorus oxychloride. Indole-3-aldehyde can be used to make D,L-tryptophan and cyanine dyes.

2,3-Dihydroindole [495-15-1], indoline C_8H_9N, M_r 119.16, bp 229 – 230 °C (101.3 kPa), is a colorless liquid, which is volatile in steam and soluble in diethyl ether, acetone, and benzene, but only slightly soluble in water. Indoline is obtained by hydrogenation of indole or by catalytic cyclodehydration of 2-(2-aminophenyl)ethanol. A range of pharmaceuticals, as well as fungicides and bactericides, can be produced from indoline.

3-(2-Aminoethyl)indole [61-54-1], trypta-mine, $C_{10}H_{12}N_2$, M_r1 60.22, mp 118 °C, forms colorless needles. It is readily soluble in ethanol and acetone, and slightly soluble in water, diethyl ether, chloroform, and benzene. It can be produced by reacting indole with ethylene imine, by catalytic hydrogenation of indole-3-acetonitrile, or by reacting phenylhydrazine with 4-aminobutyraldehyde diethyl acetal. Tryptamine is used to synthesize the vasodilator and antihypertensive, Vincamine.

3-(2-Hydroxyethyl)indole [526-55-6], tryptophol, $C_{10}H_{11}NO$, M_r 161.20, mp 59 °C, bp 174 °C, (2 kPa) forms colorless prisms or flakes that are readily soluble in methanol, ethanol, diethyl ether, chloroform, and acetone, soluble in benzene, and slightly soluble in water. It is produced by reacting methyl-1H-indole-3-acetate with lithium aluminum hydride, or through reduction of methyl-1H-indole-3-glyoxylate with sodium borohydride. The antihypertensive Indoramin is produced from tryptophol.

5. Uses

In line with its natural occurrence as a component of jasmine oil and orange-blossom oil, indole has been used for many years to fix fragrances and is therefore found in many perfumes [26].

A further important application is the production of the essential amino acid tryptophan (→ Amino Acids). Tryptophan can be synthesized from indole by chemical reactions as a racemate (via the indole derivatives gramine and indole-3-aldehyde). It is

3003

also produced by biotechnology in the enantiomerically pure L-form (e.g., from indole and L-serine).

Indole is also used as a feedstock in the synthesis of plant growth regulators, such as indole-3-acetic acid [6250-86-8] (heteroauxin) and indole-3-butanoic acid [3127-44-4]. Fungicidal and bactericidal plant protectives are synthesized from indoline [27].

Derivatives of indole-3-acetic acid find application as mild analgesics, e.g., indomethacin. Indole and its derivatives are basic building blocks for many other pharmaceuticals. Thus, for example, indoline is used to synthesize analgesics [28] and antihypertensives [29], and tryptamine [61-54-1] [3-(2-aminoethyl)-indole] for the synthesis of the vasodilator and antihypertensive, Vincamine [30]. Tryptophol [526-55-6] [3-(2-hydroxyethyl)indole] is used to prepare the antihypertensive Indoramin [31]. (Pyridylalkyl)indoles can be produced from indole and can be used as antidepressants, antihistamines, and antihypertensives [32]–[34]. 3-Piperidinylindoles can be obtained via the Grignard reaction of 4-chloroindole with 3-piperidinone, and can be used in the prevention of anoxia [35]. Indole-2-carboxylic acid derivatives are useful, e.g., as antihypertensives [36]–[38]. 5-Chloroindole [17422-32-1] can be synthesized from indoline via the intermediates 1-acylindoline, 1-acyl-5-chloroindoline, and 5-chloroindoline [39]. It can be used as an intermediate in the production of 5-chlorotryptamine (for the production of tranquilizers and blood pressure lowering agents), as well as in the production of antidepressives, antiemetics, and drugs for the treatment of Parkinson's disease [40]–[42]. 2-Methylindole [95-20-5] (as the heterocyclic coupling component) is a starting material for commercially important cationic diazo dyes. Cyanine dyes are synthesized from 2-methylindole, 1-methyl-2-phenylindole, 2-methylindole-3-aldehyde, and indole-3-aldehyde. In addition, indolylmethane dyes are produced from 1-methyl-2-phenylindole. Reacting indole with 3,6-bis-(2-carboxybenzoyl)-carbazole affords carbazole phthalein dyes, which are used as optical filter agents in photographic emulsions [43].

6. Toxicology

Apart from contact during production and manufacture of indole, human exposure results from smoking and intestinal degradation of nutritional L-tryptophan. Indole is readily absorbed from the gastrointestinal tract and rapidly converted by the liver, e.g., to indican and oxindole, which are excreted mainly into the urine [44], [45]. The acute oral LD_{50} is 1000–1100 mg/kg in rats [46], [47] and about 500 mg/kg in mice [48]. Autopsy after acute exposure revealed hemorrhages and hyperemia of inner organs and tissues. Upon application to the skin and eyes of rabbits, slight temporary irritation was observed [49], [50]. Unlike 3-methylindole (skatole), indole does not induce pulmonary injury in cattle after repeated oral application [51]. At high chronic dosage (20–200 mg kg^{-1} d^{-1} orally and subcutaneously in mice, rats, rabbits, and dogs for more than three months), indole resulted in a shift in the blood cell differential picture (anemia, leukocytosis, or leukopenia); with persistently high intoxication, the development of leukemia cannot be ruled out [52]–[54]. In cattle, hemolysis and hemoglobinuric nephrosis were observed after several oral doses of indole (100 or 200 mg/kg) [51]. Although there is no other experimental evidence for its cancerogenic potential [55], indole may influence the formation of tumors caused by other agents. Thus, very high oral doses (800 mg kg^{-1} d^{-1}) have been reported to accelerate the development of tar-induced skin tumors in mice [53] and 2-acetylaminofluorene-induced bladder carcinomas in hamsters and rats [56]. On the other hand, it simultaneously seemed to suppress 2-acetylaminofluorene-induced hepato-carcinogenesis [56]. A genotoxic effect of indole was confirmed in the Ames test in combination with other mutagenic agents [57]. However, indole itself was neither mutagenic in the Ames test [57], [58], nor in a cell transformation test [59], although in a DNA repair assay with *Bacillus subtilis*, indole caused reparable DNA damage [59].

7. References

General References

[1] *Beilstein*, **20** 304, **20(1)** 121, **20(2)** 196, **20(3)** 3176–3180.
[2] H. J. V. Winkler: *Der Steinkohlenteer und seine Aufarbeitung*, Verlag Glückauf, Essen 1951, pp. 146–148, 151–152.
[3] Rütgerswerke, *Erzeugnisse aus Steinkohlenteer*, 1958, Frankfurt, Federal Republic of Germany.
[4] H.-G. Franck, G. Collin: *Steinkohlenteer*, Springer Verlag, Berlin 1968, pp. 60–61, 180–181.
[5] H.-G. Franck, J. W. Stadelhofer: *Industrial Aromatic Chemistry*, Springer Verlag, Berlin 1988, pp. 417–418.
[6] R. J. Sundberg: *The Chemistry of Indoles*, Academic Press, New York 1970.
[7] W. J. Houlihan: *Indoles*, John Wiley & Sons, New York 1972.

Specific References

[8] Mitsui Toatsu, JP 74 20 587, 1970 (R. Fujiwars, Y. Yamada, N. Kuroda).
[9] Kyowa Fermentation Industry, JP 73 76 864, 1972 (T. Yamauchi, S. Yada, S. Kudo).
[10] Ube Industries, JP 78 95 965, 1977 (M. Nakai, T. Komata, M. Oda).
[11] Rütgerswerke, DE 2 224 556, 1972; DE 2 401 017, 1974 (G. Grigoleit, R. Oberkobusch, G. Collin).
[12] Houdry, US 2 891 965, 1956 (S. E. Voltz, J. H. Krause, W. E. Erner).
[13] Snam Progetti, DE 2 148 961, 1971 (M. M. Mauri, P. Moggi).
[14] Teijin, JP 73 76 862, 1972 (T. Fukada, K. Tanaka).
[15] E. I. du Pont de Nemours, US 2 409 676, 1944 (W. F. Grasham, W. M. Brunner).
[16] Socony Mobil Oil, NL 6 505 911, 1965.
[17] Lonza, DE 2 441 439, 1974 (C. O'Murchu).
[18] Société Nationale Elf Aquitaine, DE 2 328 284, 1973 (M. Petinaux, J. Metzger, J.-P. Aune, H. Knoche).
[19] Sumitomo Chemical, JP 7 831 660, 1976 (M. Tamura).

[20] Mitsui Toatsu, JP 76 113 868, 1975 (K. Yamamoto, K. Nitta, S. Ichikawa).
[21] Mitsui Toatsu, EP 86 239, 1982 (T. Honda, K. Terada).
[22] Ube Industries, JP 83 128 371, 1982 (N.N.).
[23] Keishitsu Ruibun Shinyoto Kaihatsu Gijutsu Kenkyu Kumiai, JP 86 44 862, 1984 (M. Imanari, T. Seto).
[24] Bofors, DE 2 052 678, 1970 (J. M. Bakke, H. E. Heikmann).
[25] Rütgerswerke, DE 2 328 330, 1973 (G. Grigoleit, R. Oberkobusch, G. Collin).
[26] G. T. Walker, *Seifen öle Fette Wachse* **100** (1974) 375–377.
[27] Pfizer, GB 1 394 373, 1975; GB 1 394 374, 1975 (R. J. Bass, R. C. Koch, H. C. Richards, J. E. Thorpe).
[28] Hoechst-Roussel Pharmaceuticals, US 4 448 784 784, 1982 (E. J. Glamkowski, J. M. Fortunato).
[29] Science Union, FR 2 003 311, 1968 (L. Beregi, P. Hugon, M. Laubie).
[30] Roussel-UCLAF, DE 2 115 718, 1970 (J. Warnant, A. Farcilli, E. Toromnoff).
[31] Wyeth, GB 1 399 608, 1971 (J. L. Archibald).
[32] Pharmindustrie, EP 65 907, 1982 (F. Audiau, G. R. Le Fur).
[33] Boehringer Ingelheim, EP 58 975, 1982 (K. Freter, V. Fuchs, J. T. Oliver).
[34] Wyeth, EP 2 886, 1979 (J. L. Archibald, T. J. Ward).
[35] Roussel-UCLAF, EP 21 924, 1981 (J. Guillaume, L. Nedelec, C. Dumont, R. Fournex).
[36] Hoechst, DE 3 151 690, 1981 (H. Urbach, R. Henning, V. Teetz, R. Geiger, R. Becker).
[37] McNeilab, US 4 529 724, 1983 (C. Y. Ho).
[38] American Cyanamid, US 4 619 941, 1984 (W. B. Wright, J. B. Press).
[39] Rütgerswerke, DE 3 035 403, 1980 (M. Maurer, W. Orth, W. Fickert).
[40] MARPHA, DE 2 618 152, 1977 (A. Champseix, C. Gueremy, G. Le Fur).
[41] Roussel-UCLAF, DE 2 719 294, 1977 (F. Clemence, D. Humbert).
[42] Roussel-UCLAF, DE 2 738 646, 1978 (C. Dumont, J. Guillaume, L. Nedelec).
[43] Polaroid, US 3 932 455, 1976 (R. C. Bilofsky, R. D. Cramer).
[44] A. H. Beckett, D. M. Morton, *Biochem. Pharmacol.* **15** (1966) 937–946.
[45] L. J. King, D. V. Parke, R. T. Williams, *Biochem. J.* **98** (1966) 266–277.
[46] National Institute for Occupational Safety and Health (NIOSH): *Registry of Toxic Effects of Chemical Substances,* US Government Printing Office, Washington D.C. 1987, p. 12992.
[47] Huntingdon Research Centre, Rütgerswerke Order No. 338 a, 1979, Münster, Federal Republic of Germany.
[48] G. J. Martin, E. H. Rennebaum, M. R. Thompson, *Ann. Intern. Med.* **18** (1943) 57–71.
[49] Huntingdon Research Centre, Rütgerswerke Order No. 338 b, 1979, Münster, Federal Republic of Germany.
[50] Huntingdon Research Centre, Rütgerswerke Order No. 338 c, 1979, Münster, Federal Republic of Germany.
[51] A. C. Hammond, J. R. Carlson, *Vet. Rec.* **107** (1980) 344–346.
[52] W. T. Taylor, *Med. Ann. D. C.* **8** (1939) 362–363.
[53] K. Kaiser, *Z. Krebsforsch.* **59** (1953) 448–495.
[54] H. Ehrhart, W. Stich, *Klin. Wochenschr.* **35** (1957) 504–511.
[55] M. J. Shear, J. Leiter, *J. Natl. Cancer. Inst. (US)* **2** (1941) 241–258.
[56] M. Matsumoto, M. L. Hopp. R. Oyasu, *Invest. Urol.* **14** (1976) 206–209.
[57] T. Matsumoto, D. Yoshida, S. Mizusaki, *Mutat. Res.* **56** (1977) 85–88.
[58] M. Curvall, J. Florin, T. Jansson, *Toxicology* **23** (1982) 1–10.
[59] T. Osawa, M. Namiki, K. Suzuki, T. Mitsuoka, *Mutat. Res.* **122** (1983) 299–304.

Iodine Compounds, Organic

PHYLLIS A. LYDAY, US Bureau of Mines, Washington, DC 20241, United States

1. The Individual Compounds .. 3007   2.   References............. 3008

1. The Individual Compounds

As with inorganic iodides, organic iodides have higher specific gravities and indices of refraction than the corresponding chloro and bromo derivatives, and higher melting and boiling points. Compounds containing an iodine atom attached to a primary carbon atom are the most stable. Secondary compounds are less stable, and tertiary are the least stable of all. Aliphatic iodo compounds can be prepared by reacting alcohols with a mixture of iodine and red phosphorus. Other methods for preparing aliphatic iodo compounds include: the addition of iodine halides or iodine to olefins; the substitution reactions between an alkyl bromide or chloride and an alkali metal iodide; and the reaction of alcohols with triphenyl phosphite and methyl iodide.

Organic iodo compounds are highly reactive. The alkyl iodides are generally 50 – 100 times more reactive than the corresponding chlorides. Reactivity of unsaturated organic iodides is affected little if iodine is attached to one of the carbon atoms involved in the double bond, but reactivity is promoted if iodine is attached to the carbon atom in the α position to the double bond. Iodine in organic compounds can be determined after being converted into an inorganic form by heating the compound with one of the following: (1) fuming nitric acid in a sealed strong glass tube, the Carius method; (2) sodium peroxide in a combustion container; or (3) a mixture of chromic and sulfuric acids, followed by addition of phosphoric acid and distillation into a solution of sodium arsenite [7].

Iodoform [75-47-8], triiodomethane, CHI_3, *mp* 119 °C, crystallizes from acetone as yellow hexagonal plates, which have a characteristic penetrating odor and are insoluble in water. The compound can be prepared by the reaction of iodine chloride with acetone in the presence of potassium hydroxide, or by reaction of chloroform with methyl iodide. Iodoform is used as a disinfectant.

Methylene iodide [75-11-6], CH_2I_2, M_r 267.85, *mp* 5 °C, *bp* 180 °C (101.3 kPa), is a colorless liquid when pure. It has a relative density of 3.325 at 20 °C and can be mixed with benzene (d_{20}^{20} 0.879) to produce liquids with different densities. These mixtures are used in the determination of the refractive indices of minerals and for the separation of minerals according to density.

Ethyl iodide [75-03-6], C_2H_5I, M_r 155.98, *bp* 72.3 °C, is a colorless liquid, which is completely miscible with ethanol and ether, but only partially miscible with benzene. Ethyl iodide is prepared by reaction of ethanol with iodine and red phosphorus.

Methyl iodide [74-88-4], CH_3I, M_r 141.95 *mp* −66.1 °C, *bp* 42.5 °C, is a colorless, transparent liquid, which turns brown upon exposure to light. Methyl iodide is prepared from methyl alcohol, iodine, and red phosphorus. Methyl iodide is used as a methylating agent.

Iodobenzene, [591-50-4], M_r 204.02, *mp* −30 °C, *bp* 188.7 °C, is a colorless liquid that rapidly turns yellow and has a characteristic odor. Iodobenzene is prepared by diazotization of aniline followed by treatment of the solution with an aqueous solution of potassium iodide, or by the action of nitric acid on a benzene and iodine mixture.

2. References

General References

[1] G. L. Clark (ed.): *The Encyclopedia of Chemistry,* 2nd ed., Van Nostrand, New York 1966, pp. 556–559.
[2] D. M. Considein et al. (ed.): *Van Nostrand's Scientific Encyclopedia,* 6th ed. Van Nostrand, New York 1983, pp. 1550–1652.
[3] *McGraw-Hill Encyclopedia of Science & Technology,* 5th ed., vol. **7**, McGraw-Hill, New York 1982, pp. 321–325.
[4] *Kirk-Othmer,* 3rd ed., **13**, 649–676.
[5] M. Windholz (ed.): *The Merck Index,* 9th ed., Merck & Co., Rahway, N.J., 1976.
[6] C. A. Hampel: *Encyclopedia of Chemical Elements,* Reinhold Book Corp., New York 1968.

Specific References

[7] J. W. Thomas et al., *Anal. Chem.* **22** (1950) 726–747.

Isocyanates, Organic

Henri Ulrich, Dow Chemical Company, North Haven, Connecticut 06473, United States

1.	Introduction 3009	4.3.	Non-Phosgene Processes. 3021
2.	Physical Properties 3010	5.	Environmental Protection . . . 3023
3.	Chemical Properties. 3010	6.	Quality Specifications and Analysis 3024
4.	Production 3015		
4.1.	Phosgenation of Free Amines . 3017	7.	Storage and Transportation . . 3024
4.2.	Other Phosgenation Procedures 3019	8.	Uses 3025
		9.	Economic Aspects 3026
4.2.1.	Phosgenation of Amine Hydrochlorides 3019	10.	Toxicology and Occupational Health. 3027
4.2.2.	Phosgenation of Carbamate Salts 3020		
4.2.3.	Phosgenation of Ureas 3020	11.	References. 3030

1. Introduction

Organic isocyanates are esters of isocyanic acid, characterized by the formula R–N=C=O. The first synthesis of an organic isocyanate was reported by A. Wurtz in 1848. Prominent chemists, e.g., A. W. von Hofmann and T. Curtius, investigated the chemistry of organic isocyanates in the nineteenth century. The commercially important synthesis of isocyanates from amines and phosgene was discovered by Hentschel in 1884 [1]. Diisocyanates became commercially important in the 1930s when the addition polymerization of difunctional isocyanates and polyols to produce polyurethanes was discovered by Bayer and co-workers at the I.G. Farben laboratories in Leverkusen [2]. More than 270 isocyanates were synthesized between 1934 and 1949 [3], but only toluene diisocyanate (TDI), 4,4′-diphenylmethane diisocyanate (MDI), hexamethylene diisocyanate (HDI), and triphenylmethane triisocyanate found significant use. After 1945, German technical information became available, and polymeric isocyanates (PMDI) were made from oligomeric amines obtained in the reaction of aniline with formaldehyde. Polymeric isocyanates were first developed in the 1950s by The Carwin Chemical Company (now Dow Chemical) in the United States under a license from Goodyear [4].

2. Physical Properties

Organic isocyanates are colorless liquids or low-melting solids; they are normally purified by distillation. Heating above ca. 100–120 °C should be avoided because thermal degradation (evolution of carbon dioxide and carbonization of the residue) can occur. The more volatile derivatives are strong lacrimators with high inhalation toxicity; exposure should therefore be avoided. Polymeric MDI products are dark liquids; the viscosity depends on the diisocyanate content. Some physical properties of commercially available mono- and diisocyanates are listed in Tables 1 and 2, respectively. The boiling and melting points of some diisocyanates have been tabulated [5].

3. Chemical Properties

For general information, see [5]–[7].

Reaction with Nucleophiles. The reactivity of isocyanates is evident from the electronic resonance structures shown below:

$$R-NCO \longleftrightarrow R-\bar{N}-\overset{+}{C}=O \longleftrightarrow R-N=\overset{+}{C}-\bar{O}$$

As a result, nucleophilic reagents rapidly attack the electrophilic carbon of the isocyanate group; aromatic isocyanates react more rapidly than aliphatic. The rate of reaction is increased by electron-withdrawing substituents on the aromatic ring and reduced by electron-donating substituents. Steric hindrance also influences the rate of reaction. Aliphatic isocyanates react in the rate order primary > secondary > tertiary. Aromatic isocyanates with substituents adjacent to the isocyanate group also react more slowly because of steric hindrance. For example, 2,4-toluene diisocyanate reacts with nucleophiles preferentially at the 4-position. Activation energies for the reaction of the 2- and 4-isocyanate groups in toluene 2,4-diisocyanate with 2-ethylhexanol are 33.9 and 33.1 J/mol, respectively [8]. Steric hindrance in the attacking nucleophile similarly influences reaction rate: alcohols react in the order primary > secondary > tertiary. The reactivity of an attacking nucleophile is proportional to its nucleophilicity: aliphatic amines > aromatic amines > alcohols > thiophenols > phenols. The reactivity of the slower reacting phenols and thiophenols can be increased considerably by using the corresponding anions.

Commercially, the most important nucleophilic reaction of isocyanates gives addition or condensation polymers. Reaction with polyether or polyester polyols produces polyurethanes; reaction with polyamines gives polyureas; reaction with dicarboxylic acids yields polyamides; and reaction with carboxylic acid dianhydrides results in polyimides.

Table 1. Industrially important monoisocyanates

Compound	Structure	CAS registry number	bp (101.3 kPa), °C	Producer
Methyl isocyanate	CH_3NCO	[624-83-9]	38	Union Carbide, Bayer, Mitsubishi
n-Butyl isocyanate	C_4H_9NCO	[111-36-4]	115	Dow, BASF, Bayer, Rhône-Poulenc
Phenyl isocyanate	NCO–C₆H₅	[103-71-9]	165	Bayer
3-Chlorophenyl isocyanate	3-Cl-C₆H₄-NCO	[2909-38-8]	201	Bayer
3,4-Dichlorophenyl isocyanate*	3,4-Cl₂-C₆H₃-NCO	[102-36-3]	238	Bayer
p-Toluenesulfonyl isocyanate	4-CH₃-C₆H₄-SO₂NCO	[4083-64-1]	142**	Upjohn

* $Mp = 41 °C$; ** At 1.73 kPa.

The reaction of isocyanates with alcohols to form carbamates is catalyzed by tertiary amines and organotin compounds. The carbamates dissociate on heating:

$$R^1N=C=O + R^2OH \rightleftharpoons R^1NHCOOR^2$$

The thermal stability of polyurethanes is therefore limited by the reversibility of this reaction. Reaction of diisocyanates with phenol has, however, been used to reversibly block the isocyanate function; blocked isocyanates are used in coatings [9].

The reaction of isocyanates with water gives a carbamic acid, which decarboxylates to produce carbon dioxide and an amine; the latter immediately reacts with more isocyanate to yield a disubstituted urea:

$$RN=C=O + H_2O \longrightarrow [RNHCOOH] \longrightarrow RNH_2 + CO_2$$

$$RN=C=O + RNH_2 \longrightarrow RNHCONHR$$

Reaction with water is used in the production of flexible polyurethane foam from toluene diisocyanate and polyether triols. The carbon dioxide generated acts as a blowing agent [9], [10]. 4-Toluenesulfonyl isocyanate undergoes very fast reaction with nucleophiles; this isocyanate is used as a water scavenger; 4-toluenesulfonamide is formed, which does not undergo reaction with a second molecule of isocyanate.

$$RSO_2N=C=O + H_2O \longrightarrow RSO_2NH_2 + CO_2$$

Table 2. Industrially important diisocyanates

Name	CAS registry number	mp, °C	bp, °C (kPa)	Producer
p-Phenylene diisocyanate, PPDI	[104-49-4]	94–96	110–112 (1.6)	Akzo
Toluene diisocyanate, TDI[a]	[1321-38-6]	13.6	121 (1.33)	BASF, Dow, Olin, Bayer, Montedison, Rhône-Poulenc, ICI, Takeda, Mitsui
Methylenediphenyl diisocyanate, MDI	[101-68-8]	39.5	208 (1.33)	BASF, Dow, Bayer, ICI, Montedison, Mitsui
Polymeric diisocyanate, PMDI	[9016-87-9]			BASF, Dow, Bayer, ICI, Montedison, Mitsui
Naphthalene 1,5-diisocyanate, NDI	[3173-72-6]	130–132	183 (1.33)	Bayer, Mitsui
	[91-97-4]	71–72		Upjohn
m-Xylylene diisocyanate, XDI	[3634-83-1]		159–162 (1.6)	Takeda

Table 2. continued

Name	Structure	CAS registry number	mp, °C	bp, °C (kPa)	Producer
Tetramethyl-*m*-xylylene diisocyanate, *m*-TMXDI		[58067-42-8]	−10	150 (0.4)	American Cyanamid
Hexamethylene diisocyanate, HDI	OCN(CH$_2$)$_6$NCO	[822-06-0]		127 (1.33)	Bayer, Rhône-Poulenc, Mitsui
TMDI[b]		[83748-30-5]		149 (1.3)	Veba
		[15646-96-5]			
Cyclohexane diisocyanate, CHDI[c]		[2556-36-7]		122–124 (1.6)	Akzo
1,3-Bis(isocyocyanatomethyl)-cyclohexane, H$_6$XDI[d]		[38661-72-2]		98 (0.05)	Takeda
Isophorone diisocyanate, IPDI		[4098-71-9]		158 (1.33)	Veba
1,1-Methylenebis[4-isocyanato]-cyclohexane, H$_{12}$MDI[d]		[5124-30-1]		179 (0.12)	Bayer

[a] Mixture of 2,4-isomer [584-84-9] (80%) and 2,6-isomer [91-08-7] (20%). [b] Mixture of 1,6-diisocyanato-2,2,4,4-tetramethylhexane and 1,6-diisocyanato-2,4,4-trimethylhexane. [c] The trans isomer. [d] Mixture of stereoisomers

The reaction of isocyanates with amines to give disubstituted ureas is very fast and quantitative. For example, the reaction with excess di-*n*-butylamine is used for determination of isocyanates (see Chap. 6) [11]:

$$R^1N=C=O + R^2_2NH \longrightarrow R^1NHCONR^2_2$$

The rate of reaction of isocyanates with amines increases with increasing basicity of the amine. Sterically hindered or electronically deactivated aromatic diamines are used as chain extenders in segmented polyurethane elastomers to achieve suitable reaction rates [9].

Carboxylic acids react with isocyanates to give mixed anhydrides, which decarboxylate to amides at elevated temperature:

$$R^1N=C=O + R^2COOH \longrightarrow \left[R^1NH-\overset{O}{\underset{\|}{C}}-O-\overset{O}{\underset{\|}{C}}R^2 \right]$$

$$\longrightarrow R^1NH-\overset{O}{\underset{\|}{C}}-R^2 + CO_2$$

Base catalysis is required to avoid formation of disubstituted urea byproducts. The reaction of suitable carboxylic acid anhydrides, such as phthalic, trimellitic, pyromellitic, and benzophenone tetracarboxylic acid dianhydride, with isocyanates to give imides is catalyzed by nucleophiles [12].

Hydrogen halides form labile products with isocyanates:

$$RN=C=O + HX \rightleftharpoons RNH-\overset{O}{\underset{\|}{C}}-X$$

X = halogen

Carbamoyl chlorides (X = Cl) are intermediates in the production of isocyanates from amines and phosgene (see Section 4.1)

Organic isocyanates undergo rapid addition reactions with molecules containing active hydrogen atoms (e.g., esters, ortho esters, and ketones).

$$R^1N=C=O + CH_2(COOR^2)_2 \longrightarrow R^1NH-\overset{O}{\underset{\|}{C}}-CH(COOR^2)_2$$

Insertion reactions occur with a variety of substrates containing metal–oxygen, metal–sulfur, or metal–nitrogen bonds. This reaction is observed with elements of groups 1, 2, 12, and 13 [6].

Cycloaddition Reactions. Isocyanates undergo cycloaddition across the carbon–nitrogen bond with a variety of substrates containing double or triple bonds. Additions across the C=O bond are rare. Dimerization and trimerization are examples of cycloaddition:

$$2\ RN{=}C{=}O \rightleftharpoons \underset{\underset{O}{\|}}{RN}\overset{\overset{O}{\|}}{\underset{}{C}}NR$$

$$3\ RN{=}C{=}O \rightleftharpoons \begin{array}{c} O \\ \| \\ RNNR \\ O{=}\underset{R}{N}{=}O \end{array}$$

Slow dimerization occurs during storage [13]. Aromatic isocyanates dimerize and trimerize more readily than aliphatic isocyanates. Dimerization is catalyzed by pyridine and phosphines; strong bases (such as tertiary amines or potassium salts of carboxylic acids) catalyze trimerization. The trimerization of PMDI is used in the manufacture of rigid polyurethane foams with higher thermal stability and better combustibility characteristics [9]. The cycloaddition of isocyanates with carbodiimides is used in the production of carbodiimide-modified liquid MDI products (see Chap. 8). The conversion of isocyanates to carbodiimides is catalyzed by phospholene oxides [14]:

$$2\ RN{=}C{=}O \longrightarrow RN{=}C{=}NR + CO_2$$

Isocyanates also undergo anionic homopolymerization below room temperature in dimethylformamide solution. Catalysts used include sodium cyanide or aqueous solutions of alkali-metal salts [15]:

$$RN{=}C{=}O \longrightarrow \left[\underset{}{N}{-}\overset{R}{\underset{}{}}\overset{}{\underset{\|}{C}}\overset{O}{\underset{}{}} \right]_n$$

Isocyanate homopolymers have not found commercial use because of their tendency to undergo rapid depolymerization. A review of isocyanate-derived polymers other than urethanes has been published [16].

4. Production

For general information, see [3], [5], [10], [17].

A number of methods for synthesis of isocyanates are described in the literature, but only the reaction of phosgene with amines or amine salts is used commercially [1]. Attempts are being made to develop routes to isocyanates that do not require the use of highly toxic phosgene. On a laboratory scale, liquid trichloromethyl chloroformate (diphosgene) [18] or solid bis(trichloromethyl) carbonate (triphosgene) [19] can be used as substitutes for phosgene to synthesize isocyanates from amines. In addition, novel processes that do not use phosgene and are not based on amines as starting materials are also being developed. Two aliphatic diisocyanates are currently made by procedures that do not require phosgene. Tetramethylxylylene diisocyanate (m-TMXDI) is made by addition of cyanic acid and hydrogen chloride to diisopropenylbenzene [20] and Hofmann degradation of amides is used for the production of cyclohexane diisocyanate (CHDI) [21]. Du Pont has developed a process for the production of methyl isocyanate based on the oxidation of *N*-methylformamide.

The Curtius, Lossen, and Hofmann rearrangements are used on a laboratory scale to prepare isocyanates; the reaction of alkyl halides or sulfates with a cyanate salt, the classical synthesis of isocyanates, is used to synthesize organometallic isocyanates [5], [6].

Dinitration of toluene produces mixtures of the 2,4- and 2,6-dinitro isomers. Catalytic reduction of these derivatives leads to the corresponding diamines, which are treated with phosgene to give TDI. The distillation residue can be hydrolyzed to recover the diamines or incinerated.

The production of PMDI involves condensation of aniline with formaldehyde in the presence of hydrochloric acid to give oligomeric amines that are then phosgenated without purification (→Aniline). The coproduct, MDI, is obtained from PMDI by continuous thin-film distillation. The residual crude oligomeric isocyanate product (PMDI) is used mainly in the production of rigid polyurethane or polyisocyanurate foams. Yields in the production of PMDI–MDI are quantitative. Isomers are also produced in the production of 4,4′-methylenedianiline [*101-77-9*] (MDA), the precursor of MDI, but because 2,4′- and 2,2′-methylenedianiline are more reactive towards formaldehyde, they are converted mainly to triamines or higher oligomers. The amount of diamine formed in this reaction is determined by the ratio of aniline to formaldehyde; temperature and acid concentration also affect product distribution. The production of aromatic amines used in the manufacture of TDI and PMDI is described in [24].

The most common aliphatic diisocyanates are hexamethylene diisocyanate HDI and hydrogenated MDI, the former being obtained by reaction of hexamethylenediamine with phosgene. The vapor pressure of HDI is high; less volatile derivatives such as the biuret (obtained from HDI and water) or the trimer are used in coatings. Hydrogenated MDI is produced by catalytic hydrogenation of methylenedianiline on a rhodium or ruthenium catalyst to yield a mixture of *trans,trans*-, *cis,trans*-, and *cis,cis*-perhydromethylenedianiline (H_{12}MDA), which is then phosgenated. For conversion to pure *trans,trans*-H_{12}DMA, see [25]. To obtain a liquid diisocyanate in the subsequent phosgenation, the amount of the trans,trans isomer must be reduced to ca. 20%, usually by crystallization.

Isophorone diisocyanate [*4098-71-9*] (IPDI), obtained by reaction of isophorone diamine [*2855-13-2*] (1-amino-3-aminomethyl-3,5,5,-trimethylcyclohexane) with phosgene, is also commercially available. Although IPDI has a higher boiling point than HDI, it is also often used as the trimer. Because of the difference in reaction rates between the two isocyanate groups in IPDI, selective trimerization of the primary isocyanate group can be achieved.

4.1. Phosgenation of Free Amines

The reaction of free amines with phosgene is the most important method for production of isocyanates. The amine, dissolved in an inert organic solvent (see below), is added continuously to excess phosgene in the same solvent at low temperature; the resulting slurry of carbamoyl chlorides and amine hydrochlorides is then heated at elevated temperature with excess phosgene until a clear solution is obtained. The overall reaction is shown below:

$$RNH_2 + COCl_2 \xrightarrow{-HCl} \left[RNH-\overset{O}{\underset{\|}{C}}-Cl \right] \xrightarrow{-HCl} RN=C=O$$

The reaction is conducted at ca. 20% solution. Excess phosgene (50–200%) is necessary to reduce the extent of side reactions. Disubstituted ureas are formed as byproducts by reaction of the isocyanate with unreacted amine. In the case of an aliphatic amine, the urea reacts further with phosgene to form allophanoyl chlorides, which can be thermally converted to isocyanates [26]:

$$RNH-\overset{O}{\underset{\|}{C}}-NHR + COCl_2 \xrightarrow{-HCl} RNH-\overset{O}{\underset{\|}{C}}-NR-\overset{O}{\underset{\|}{C}}-Cl$$

$$RNH-\overset{O}{\underset{\|}{C}}-NR-\overset{O}{\underset{\|}{C}}-Cl \xrightarrow{-HCl} 2\,RN=C=O$$

For aromatic amines, the urea byproducts are dehydrated by phosgene to form carbodiimides, which then react with more phosgene to give N-chloroformylchloroformamidines; cycloadducts of carbodiimides with isocyanates are also formed.

$$ArNH-\overset{O}{\underset{\|}{C}}-NHAr + COCl_2 \xrightarrow[-2\,HCl]{-CO_2} ArN=C=NAr$$

$$ArN=C=NAr + COCl_2 \longrightarrow ArN=\overset{Cl}{\underset{|}{C}}-NAr-\overset{O}{\underset{\|}{C}}-Cl$$

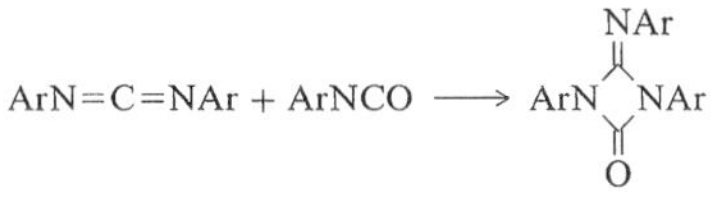

$$ArN=C=NAr + ArNCO \longrightarrow \underset{O}{\overset{NAr}{ArN \bigtriangleup NAr}}$$

The cycloadducts dissociate during distillation of the aromatic isocyanates. Phosgene is sometimes generated in the distillation by thermal dissociation of the N-chloroformylchloroformamidines.

The byproducts formed during phosgenation remain in the PMDI. These byproducts influence the reactivity of the crude isocyanate because hydrolyzable chlorine compounds (e.g., *N*-chloroformylchloroformamidines) deactivate the amine catalysts used in polyurethane production.

Solvents used in the phosgenation of free amines include toluene, xylene, decalin (decahydronaphthalene), chlorobenzene, and *o*-dichlorobenzene; the last two are preferred because of their higher polarity, which aids in dissolution of the amine hydrochlorides. Two-phase phosgenation with an inert solvent and an aqueous phase containing a base (as the hydrogen chloride scavenger) is also feasible for aliphatic amines. Vapor-phase phosgenation of volatile aliphatic amines is also being used.

Batchwise phosgenation is conducted in chlorobenzene or *o*-dichlorobenzene, by using an excess (50 – 200%) of phosgene. Mixing of the amine solution with phosgene is carried out at room temperature. The reaction mixture is then heated until a clear solution is obtained. Excess phosgene and hydrogen chloride are removed by flushing with nitrogen, and the gases are recovered by fractional condensation or by washing with cold solvent. The solvent is removed by distillation and the pure isocyanate is obtained by crystallization, fractional distillation, or sublimation.

Continuous processes are conducted similarly. Higher temperature and pressure are necessary, however, to achieve rapid reaction rates. The exothermic reaction of the amine solution with liquid phosgene or with a solution of phosgene in the same solvent is performed at room temperature with thorough mixing to obtain a fine slurry of the amine hydrochloride. This slurry is pumped into one reactor or a series of reactors, where the reaction is completed at a higher temperature. The first reactor is normally held at 65 – 80 °C to dissociate carbamoyl chloride, and a subsequent reactor is held at 160 – 180 °C to complete the reaction. Hydrogen chloride and excess phosgene are vented at the higher temperature to prevent recombination of hydrogen chloride with the isocyanate. Phosgene and hydrogen chloride are recovered (see Chap. 5). The waste gas, containing some phosgene, is hydrolyzed with water; traces of organic solvents are adsorbed on activated carbon.

Isocyanates are purified by fractional vacuum distillation; the stripped solvents are reused. In the production of PMDI, the coproduct MDI is recovered by rotary film evaporation. In this manner, a mixture consisting of ca. 98% 4,4'-MDI and 2% 2,4'-MDI is obtained. For some applications, production of MDI with higher amounts of the 2,4'-isomer is preferred. This is best accomplished by conducting the aniline – formaldehyde condensation at higher temperature, which favors formation of the 2,4'- and 2,2'-isomers.

A high-pressure process for the production of PMDI is shown in Figure 1 [27]. A 20% solution of the oligomeric amine in chlorobenzene is treated with liquid phosgene at 80 – 120 °C and 1 – 4 MPa. The initial reaction takes place in the in-line mixer at 40 passes per hour and is completed in the reactor at up to 180 °C. Newer single-stage continuous high-pressure processes are being developed to reduce capital investment, increase yields, and prevent blockages resulting from the formation of insoluble byproducts at the higher temperature used (160 °C, 9.5 MPa). Blockage of the in-line mixer requires costly shutdown of the plant.

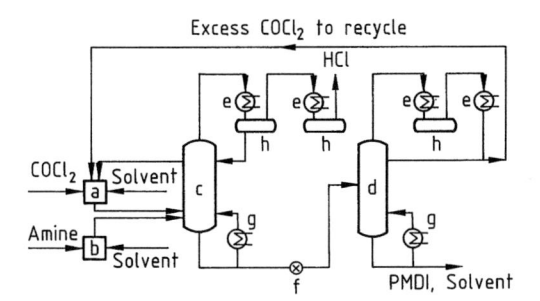

Figure 1. High-pressure production of PMDI
a) Tank for phosgene solution; b) Tank for amine solution; c) Reactor; d) Column; e) Condenser; f) Safety release valve; g) Evaporator; h) Collecting vessel

Low-pressure reactions with phosgene are conducted in a similar manner. A 20% solution of the amine in chlorobenzene is mixed with excess liquid phosgene, and the reaction takes place in several stages at 100–200 kPa. Initial mixing is conducted from 5 °C to room temperature, and the reaction is completed in two reactors at 60–80 °C and 130–180 °C.

For the production of aliphatic isocyanates, continuous free amine phosgenation processes are unsuitable because of the slower reaction rates observed for aliphatic amine hydrochlorides and side reactions resulting from the higher basicity of the aliphatic amines. Aliphatic diisocyanates are produced either by phosgenating a slurry of the corresponding amine hydrochlorides (see Section 4.2.1) or by phosgenating a slurry of the carbamate salts obtained in the reaction of the aliphatic diamines with carbon dioxide (see Section 4.2.2). Exceptions are longer chain aliphatic isocyanates, such as octadecyl isocyanate, 1,12-dodecane diisocyanate, and the diisocyanate obtained from dimer acid (→Dicarboxylic Acids, Aliphatic). Longer alkyl chains increase the solubility of the corresponding hydrochlorides in hydrocarbon solvents. The reaction rates necessary for continuous phosgenation of the free amine can then be obtained.

4.2. Other Phosgenation Procedures

4.2.1. Phosgenation of Amine Hydrochlorides

One method used to avoid side reactions is the phosgenation of the amine hydrochlorides, rather than the free amine, exemplified by the conversion of isophorone diamine dihydrochloride to isophorone diisocyanate [28]. Isophorone diamine in decahydronaphthalene solution flows downward against a hydrogen chloride gas stream. The fine slurry of amine salts obtained is heated to 150 °C and reacted with phosgene in a series of agitated autoclaves to give a clear solution of isophorone diisocyanate. Excess phosgene is removed by purging with nitrogen. Isophorone diisocyanate is isolated by stripping of the solvent and fractional vacuum distillation of the residue.

In the phosgenation of aliphatic diamine dihydrochlorides, the reaction must be conducted below 150 °C to avoid the formation of ω-chloro monoisocyanates because of cleavage of the NCO group by hydrogen chloride [3].

4.2.2. Phosgenation of Carbamate Salts

Carbamate salts can be used instead of aliphatic amine dihydrochlorides to achieve faster reaction [29]. For example, a solution of hexamethylenediamine in chlorobenzene is treated with carbon dioxide to form a suspension of the carbamate salt. This suspension subsequently reacts with phosgene in three reactors. The first two are held at 30 °C with retention times of 12 h each, and the reaction is completed in the third reactor at 100 °C for 6 h. The recovery of phosgene and hydrogen chloride and work-up of the reaction mixture are similar to other phosgenation processes.

4.2.3. Phosgenation of Ureas

Phosgenation of disubstituted ureas is used in the production of arenesulfonyl isocyanates. Direct phosgenation of sulfonamides proceeds in polar solvents only above 200 °C; significant amounts of arenesulfonyl chlorides are formed as byproducts. In contrast, 1-arenesulfonyl-3-alkylureas undergo rapid reaction with phosgene at ca. 150 °C [30]. In the commercial production of arenesulfonyl isocyanates, reaction of the sulfonamides is conducted in the presence of a catalytic amount of n-butyl isocyanate to achieve rapid rates at 150 °C [31]. The rate-determining step is the intermediate formation of the urea. The n-butyl isocyanate catalyst is regenerated in the phosgenation of the urea:

$$RSO_2NH_2 + C_4H_9NCO \longrightarrow RSO_2NH-\overset{\overset{O}{\|}}{C}-NHC_4H_9$$

$$RSO_2NH-\overset{\overset{O}{\|}}{C}-NHC_4H_9 + COCl_2 \longrightarrow RSO_2NCO + C_4H_9NCO$$

The reaction of disubstituted ureas with phosgene can also be used to synthesize aliphatic isocyanates which contain groups such as esters or ethers that are sensitive to prolonged heating with phosgene. The lower boiling aliphatic isocyanate R^2NCO is continuously removed from the reaction mixture [32]:

$$R^1SO_2NCO + R^2NH_2 \longrightarrow R^1SO_2NH-\overset{\overset{O}{\|}}{C}-NHR^2$$

$$R^1SO_2NH-\overset{\overset{O}{\|}}{C}-NHR^2 + COCl_2 \longrightarrow R^1SO_2NCO + R^2NCO$$

Arenesulfonyl isocyanates are used as intermediates for pharmaceuticals and herbicides, and as water scavengers in the formulation of coating systems.

4.3. Non-Phosgene Processes

Attempts have been made in recent years to develop methods for the production of isocyanates that do not require the use of phosgene, because of its volatility and toxicity. Non-phosgene routes developed for TDI and PMDI–MDI have not been adopted commercially. Processes based on the reaction of carbon monoxide with aromatic nitro compounds have been examined [22]. Direct conversion of dinitrotoluene to TDI gives good results [33], but this approach was abandoned because of severe reaction conditions and difficulties in the recovery of the homogeneous noble-metal catalysts, which are expensive. A two-step process, using cheaper but toxic selenium catalysts, was developed by Arco and Mitsui-Toatsu [34], [35]. Cleavage of the intermediate biscarbamates, by using high-boiling organic solvents in the presence of catalysts, proved to be rather difficult [35]. Efficient cocatalysts for the one-step reaction have also been developed [36]. Non-phosgene routes for TDI are shown below:

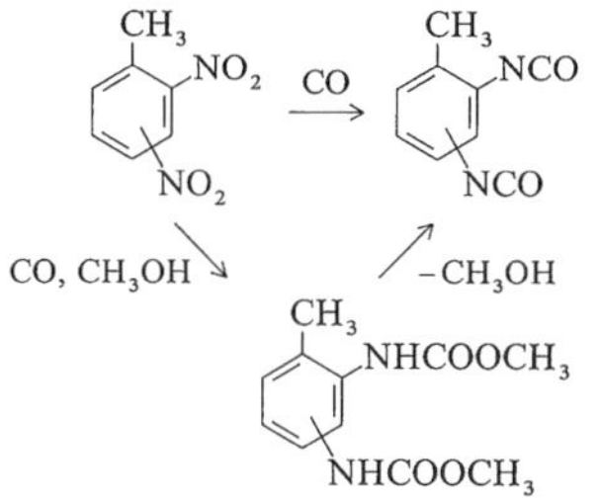

A similar route to PMDI–MDI starting with nitrobenzene was developed by Arco and by Asahi. This process has an additional complication in that unreactive methyl phenylcarbamate must be condensed with formaldehyde [23], [37], [39]:

[Scheme: nitrobenzene → methyl carbamate via CO/CH₃OH, then condensation with CH₂O to oligomeric methyl carbamates, then thermolysis (−CH₃OH) to PMDI]

Another non-phosgene process for PMDI–MDI is based on the reaction of aniline or oligomeric amines derived from aniline and formaldehyde with dimethyl carbonate [38] or with methanol, carbon monoxide, and oxygen [39]. These processes also involve thermolysis of the methyl carbamates as the final step. Reaction with unconverted carbamate cannot be prevented because the reactive isocyanates are formed at high temperature. Carbodiimides are also formed as byproducts.

[Scheme: oligomeric polyamine + (CH₃O)₂CO, −CH₃OH → oligomeric methyl carbamate → Δ, −CH₃OH → PMDI]

Nevertheless, attempts are still underway to find routes to aliphatic diisocyanates that are not based on phosgene. American Cyanamid produces tetramethylxylylene diisocyanates by reaction of 4- or 3-diisopropenylbenzene with hydrogen chloride and isocyanic acid [20]:

[Scheme: diisopropenylbenzene + HCl → dichloride + HNCO, −H₂NCOCl → diisocyanate]

In the second step of this reaction, carbamoyl chloride (H$_2$NCOCl) is formed as a byproduct and reacts with diisopropylbenzene to form the intermediate dichloride and simultaneously regenerate isocyanic acid. The isocyanic acid required is produced in situ by air oxidation of hydrocyanic acid [40].

A nonphosgene process for *trans*-1,4-diisocyanatocyclohexane has been developed by Akzo [21]. Scrap poly(ethylene terephthalate) fibers are hydrolyzed to terephthalic acid, which is converted in several steps to CHDI. Terephthalic acid is converted in a similar way to *p*-phenylene diisocyanate.

$$R = C_2H_5$$

5. Environmental Protection

Hydrogen chloride is a byproduct in the production of organic isocyanates. It is collected and sold as hydrogen chloride gas, absorbed in water and neutralized with sodium hydroxide, or sold as aqueous hydrochloric acid. Some companies oxidize the hydrogen chloride to chlorine by electrolysis; the chlorine is reused in the production of phosgene. Excess phosgene from isocyanate production is separated from hydrogen chloride and recycled. Ammonia or sodium hydroxide scrubbers are available in case of accidental phosgene release.

The inert solvents used in the production of organic isocyanates (e.g., toluene, xylene, chlorobenzene, 1,2-dichlorobenzene, or decahydronaphthalene) are removed by distillation and recycled. Distillation residues and preliminary cuts containing volatile impurities are incinerated. In particular, trace amounts of the volatile phenyl isocyanate formed in PMDI – MDI production, resulting from trace amounts of aniline in the starting amines, must be removed because phenyl isocyanate is a strong lacrimator.

6. Quality Specifications and Analysis

The use of organic isocyanates as monomers and as intermediates for pharmaceutical and agricultural chemicals requires a high degree of purity; this is usually achieved by fractional distillation. Solid isocyanates are sometimes purified by crystallization or sublimation. Exceptions to this are PMDI and crude TDI. Crude TDI is produced from distillation residues by partial trimerization; PMDI is an undistilled liquid product. The composition of polymeric isocyanates (PMDI) varies. The main component is 40 – 60 % 4,4'-MDI; the remainder is composed of other isomers of MDI, triisocyanates, and higher molecular mass oligomers. Important product variables are functionality and acidity. The 4,4'-MDI content varies because some of this coproduct is removed and sold separately, depending on market demand. Samples containing higher amounts of 2,4'-MDI are also produced for specialty elastomer applications. The commodity isocyanate TDI is produced mainly as an 80:20 mixture of the 2,4- and 2,6-isomers. Pure 2,4-TDI and a 65:35 mixture of 2,4- and 2,6-TDI are also available.

Organic isocyanates are characterized by their NCO content, determined analytically by reaction of a sample of the isocyanate with excess di-*n*-butylamine in chlorobenzene. Unreacted amine is determined by back titration with dilute hydrochloric acid [11]. Total and hydrolyzable chlorides are determined for all isocyanates. Iron, which acts as a catalyst in polyurethane reactions, is determined by atomic absorption spectroscopy. Vapor-phase chromatography, high-pressure liquid chromatography, and nuclear magnetic resonance spectroscopy are also used to determine isomer distribution in TDI, MDI, and H_{12}MDI, and to determine the composition of PMDI. Analytical methods used in the determination of organic isocyanates and derived polyurethanes are summarized in [11].

7. Storage and Transportation

The highly reactive organic isocyanates are hazardous materials, and handling of these chemicals is regulated in all industrial countries. Some crystallization of 4,4'-MDI from PMDI may occur on prolonged storage at low temperature. Exposure to atmospheric moisture must be prevented. The reaction of isocyanates with water leads to the formation of carbon dioxide, which results in a buildup of pressure. The disaster in Bhopal, India, involving massive release of highly volatile and toxic methyl isocyanate was caused by the addition of water to the methyl isocyanate storage tank [41].

Isocyanates can undergo slow cyclodimerization on storage [13]. To minimize dimerization MDI, for example, must be stored under refrigeration. The formation of dimers in MDI causes turbidity on melting. Specific storage recommendations for individual isocyanates are available from the producers. Thermal degradation of isocyanates occurs on heating above 100 °C. This reaction is exothermic, and above 175 °C a runaway reaction can occur with charring and evolution of carbon dioxide. In view of the heat sensitivity of isocyanates, solid isocyanates such as MDI must be melted with caution; supplier's recommendations should be followed. Isocyanates are also flammable and their vapors are toxic (see Chap. 10).

Isocyanates are transported in railroad tank cars, tank trucks, tanks in ships, containers, and drums. They are stored in steel tanks and processed in steel equipment. For long-term storage, stainless steel is recommended. To avoid contamination from atmospheric moisture, a dry air or inert gas blanket is essential.

Decontamination of spilled isocyanates and disposal of isocyanate waste are best conducted by using aqueous ammonia (3–8% concentrated ammonia solution in 90–95% water with 0.2–5% liquid detergent) or aqueous sodium carbonate (5–10% sodium carbonate in 90–95% water and 0.2–5% liquid detergent). An alcoholic solution (50% ethanol, isopropyl alcohol, or butanol; 45% water; and 5% concentrated ammonia) may be preferred because of the low miscibility of isocyanates with water.

8. Uses [9], [10]

Monofunctional organic isocyanates (e.g., methyl, *n*-butyl, phenyl, and halogenated phenyl isocyanates) are used as intermediates in the production of carbamate and urea insecticides and fungicides. *n*-Butyl isocyanate is also employed in the manufacture of sulfonyl urea antidiabetic drugs. Arenesulfonyl isocyanates are used in the production of sulfonylurea herbicides and as drying agents for coating systems.

Di- and polyfunctional isocyanates are used as monomers for addition and condensation polymers. For example, the major consumption of the aromatic isocyanates TDI, MDI, and PMDI is in the production of polyurethanes. Water-blown flexible poly(urethane urea) foams are produced from TDI and trifunctional polyols; rigid polyurethane and polyisocyanurate foams are made from PMDI and polyester polyols. The open-cell flexible foams are used mainly in bedding and seating; the fluorocarbon-blown closed-cell rigid foams are used as insulation in building and construction. Also PMDI is used as a wood chip binder for particle board, and the coproduct MDI is used as a monomer in the production of linear thermoplastic and crosslinked thermoset polyurethane elastomers, primarily in the automotive industry. To achieve faster reaction rates and better green strength in reaction injection molding (RIM), aromatic diamine extenders are used to produce poly-(urethane urea) elastomers. Amine-terminated polyethers have also been used recently as comonomers to yield polyureas. Spandex-type elastomeric fibers are made from MDI. Emerging applications for MDI include production of thermoplastic polyamide elastomers [42] and polyurethane [43] or polyamide [42] engineering thermoplastics. The aliphatic diisocyanates HDI, IPDI, and H_{12}MDI are used in the formulation of light-stable polyurethane coatings.

Diisocyanates are often used in modified form. In reaction injection molding, liquid diisocyanates are preferred. Liquid modified MDI is made by conversion of some isocyanate groups to carbodiimide groups [44]; the latter undergo cycloaddition with unreacted MDI to form a cycloadduct. The small amount of cycloadduct depresses the melting point of MDI to give a product that is liquid at room temperature. The cycloadduct dissociates at elevated temperature, thereby maintaining difunctionality in most applications.

Liquid products are also made by reaction of MDI with small amounts of glycols. These isocyanate-terminated oligomers reduce the vapor pressure of diisocyanates or the heat of reaction in molding applications. Biuret and isocyanurate derivatives of aliphatic diisocyanates are also used to reduce the vapor pressure of these relatively volatile compounds.

9. Economic Aspects

Monoisocyanate producers include Mobay, Union Carbide, Dow, and Upjohn (United States), and Bayer, BASF, Hoechst, and Tolochimie (Rhône-Poulenc) (Europe). Methyl isocyanate is produced in Japan by Mitsubishi. The major PMDI–MDI and TDI producers are listed in Table 3 [9].

Organic isocyanates are sold either by their chemical names or under trade names. The latter are used mainly for the commodity isocyanates. Formulated products and systems containing isocyanates are also sold under a variety of trade names. The trade names of isocyanates and formulated isocyanates are listed in Table 4.

The major producers of monoisocyanates are Bayer (Europe) and Union Carbide (United States). The total consumption of monoisocyanates in the United States, Western Europe, and Japan was ca. 40 000 t in 1985. Methyl isocyanate represents nearly 50 % of the total monoisocyanates consumed. In contrast, over 1.1×10^9 t of diisocyanates was consumed in 1985 in these three areas (Table 5).

The market for flexible foams, which accounts for almost 90 % of the TDI used, is estimated to grow by only 1.4 % a year (1985 – 1990). By contrast, the use of PMDI in

Table 3. Worldwide PMDI–MDI and TDI production capacity (1987), 1000 t

Area, Company	PMDI–MDI	TDI	Area, Company	PMDI–MDI	TDI
Western Hemisphere			*Western Europe*		
BASF	55	65	BASF, Belgium	50	
Dow	135	50	Bayer, Federal Republic of Germany	147[a]	152
Mobay (Bayer)	125	110	Bayer-Shell, Belgium	36	30
Olin		72	Bayer-Spain	10	18
Rubicon (ICI)	61[b]	27	Dow	50	
Bayer do Brazil		20	ICI	105	
Industrias Cydsa Bayer Mexico		12	Montedison	40	60
			Rhône-Poulenc		120
Eastern Europe			*Japan and the Far East*		
Sodaso Hemijski, Yugoslavia		18	Mitsui-Nisso	18	28
Yugoslavia		20	Nippoly	31	15
German Democratic Republic (DDR)	23	8	Sumitomo Bayer	24	13
			Takeda		22
USSR	22		Chin Yang (Korea)		10
BVK-Hungary	25[c]				
World total	957	870			

[a] Capacity will be increased by 60 000 t to 1988.
[b] The U.S. plant will be expanded to 136 000 t by 1989.
[c] Planned for 1988.

rigid insulation foams, which represents ca. 77% of the total consumption of PMDI–MDI, will grow at a faster rate because rigid foams based on PMDI have become an important insulation material. Worldwide reduction in fluorocarbon production may affect the growth of this market (fluorocarbons are used as blowing agents in the production of rigid insulating foams). Most of the market growth in pure MDI will result from increased use in automotive RIM (e.g., fascia, body panels, and window gaskets). A recent trend is the gradual replacement of TDI with PMDI–MDI; the latter has a lower vapor pressure and faster rate of reaction.

10. Toxicology and Occupational Health

Isocyanates are classified as dangerous substances (EEC Guidelines); they are usually labeled "toxic" and should therefore be handled with care. The potential for exposure depends on the vapor pressure. Volatile short-chain aliphatic isocyanates (e.g., methyl isocyanate) must be handled with particular caution. Exposure limits for TDI and MDI were set in the United States by OSHA at 0.02 ppm, 8 h TWA. The acute toxicity in rats of several diisocyanates is listed in Table 6 [9].

Isocyanates, Organic

Table 4. Trade names of isocyanates and isocyanate-containing systems or formulations

Trade name	Producer	Country[a]	Product[b]	Trade name	Producer	Country[a]	Product[b]
Adiprene	Uniroyal	USA	TDI Systems, E	Multrathane	Mobay	USA	I
Basonat	BASF	FRG	A, C	Olin-TDI	Olin	USA	TDI
Baygen	Bayer	FRG	IH	Plastidur	Trorion	BR	I, C, A
Baymidur	Bayer	FRG	A, C	Polidur	Polymer Chem	FRG	I, C, A
Baytec	Mobay	USA	E	Polurene	Sapic	I	I, C, A
Beckocoat	Reichold	USA	C	Polylite	Reichold	USA	PU
Beckothane	Hatrick	AUS	C	Polymir	Polymir	USA	PU
Beckuran	Beck	USA	C	Poronat	Thanex	FRG	PMDI Systems
Bermonat	Berol	S	PMDI Systems	Prepol	Interchem	UK	I, C, A
Burnock	Dainippon	J	C	Q-Thane	Quinn	USA	PU
Caradate	Shell	NL	I	Quasilan	Lancro	UK	I, E
Coronate	Nippon	J	C	Reactane	Reacta	E	I, C, A
Cortume	Cariosa	UK	I	Resaminè	Dainichi Saika	J	I, C
Desmocap	Bayer	FRG	BI	Rhenodur	Rheinchemie	FRG	PU
Desmodur	Bayer	FRG	I	Roxdur	C.R.P.	NL	I, C
Duranate	Asahi	J	C	Rubinate	Rubicon	USA	I
Duroderm	Reichold	FRG	A	Scuranate	Sodethane	F	I
Ekanate	BASF	USA	I	Scurane	Rhône-Poulenc	F	I, C
Elastofix	Elastogran	FRG	A	Solithane	Thiokol	USA	I, E
Elastonat	Elastogran	FRG	E	Sothanate	Dainippon	J	I, C
Elastostic	Elastogran	FRG	A	Spenkel	Spencer-Kellog	USA	I, C
Elate	Akzo	USA	CHDI	Spenlite	Spencer-Kellog	USA	I, C
Eldur	Koutlis Elrit	UK	C, A	Stahurane	Stabilital	I	I
Hetrofoam	Hooker	USA	RF	Sumidur	SBU	J	I
Hexacal	ICI	UK	RF	Suprasec	ICI	UK	I, C, A
Hi-Prene	Mitsui	J	PU	Synthane	Synthesia	E	I, C
Hiprox	Dainippon	J	PU	Syntodur	Synres	NL	I
Hysol	Dexter	USA	E	Systanat	Schwarzheide	DDR	I
Hythane	Croda	UK	C	Takenat	Takeda	J	I
Irathane	Irathane	UK	E	Tedimon	Montedison	I	I
Irodur	Iroquois	CAN	A	Tolonate	Rhône-Poulenc	F	I
Isocon	Lancro	UK	I	Trixene	Baxenden	UK	I, C, A
Isofoam	Baxenden	UK	PU	Ucopol	Sapici	I	I, C
Isonate	Dow	USA	PMDI-MDI	Unithane	Gray Valley	UK	I, C
Isopol	Siric	I	C	Uradur	Scado	NL	I, C
Jordathane	Jordan	AUS	C, A	Urestyl	Scholten	NL	I
Lilene	PCUK	F	I	Uronal	Galstaff	I	I, C
Lupranat	BASF	FRG	I	Voranate	Dow	USA	TDI
Millionate	Hodogaya	J	I	Witcobond	Witco	USA	C
Mondur	Mobay	USA	I				

[a] Country codes: AUS, Australia; BR, Brazil; CAN, Canada; DDR, German Democratic Republic; E, Spain; F, France; FRG, Federal Republic of Germany; I, Italy; J, Japan; NL, The Netherlands; S, Sweden; UK, United Kingdom; USA, United States. [b] A, adhesives; BI, blocked isocyanate; C, coatings; CHDI, cyclohexane diisocyanate; E, elastomers; I, isocyanates; IH, isocyanate hardener; MDI, methylene diisocyanate; PMDI, polymeric methylene diisocyanate; PU, polyurethanes; RF, rigid foam; TDI, toluene diisocyanate.

Table 5. Consumption of diisocyanates (1985), 10^3 t

	TDI	PMDI–MDI	Aliphatic isocyanates
United States	224	246	13.5
Western Europe	182	305	20
Japan	63	79	3
Total	469	630	36.5

Table 6. Acute toxicity of diisocyanates in rats

Isocyanate	LC_{50} (oral), μg/kg	LC_{50} (inhalation[a]), mg/m^3	Concentration in air, ppm (STP)
HDI	710	310	6.8
IPDI	> 2 500	260	0.34
TDI	5 800	58–66	19.6
MDI	> 31 600	490	0.1
NDI[b]	> 10 000		0.02[c]

[a] Aerosol, 4 h.
[b] Naphthalene 1,5-diisocyanate.
[c] Vapor pressure at 50 °C.

The more volatile isocyanates are strong lacrimators, but their primary physiological effect is respiratory irritation. An estimated 20 000 – 30 000 people in the United States may have been exposed to TDI; a few developed bronchial asthmatic hypersensitivity. Excessive exposure can result in respiratory irritation. Diisocyanates generally cause irritation when applied directly to the skin or instilled into the eye (rabbits). Repeated skin application leads to an allergic reaction. Studies of subacute TDI toxicity in animals indicate that repeated inhalation causes tracheobronchitis, bronchitis, emphysema, and bronchopneumonia, depending on the degree and frequency of exposure. None of the animal studies showed evidence of sensitization or the asthma-like response reported in humans. A screening test is being developed for workers who come into contact with isocyanates. A tolyl isocyanate antigen has been developed and incorporated into a radioimmunoassay for detection of IgE antibodies in workers with TDI hypersensitivity [45]. The development of hypersensitivity in guinea pigs exposed to tolyl and hexyl isocyanate – ovalbumin aerosols has been demonstrated. The sensitization of guinea pigs to TDI was recently shown by detection of tolyl-specific antibodies [46]. Inhalation of aliphatic diisocyanates (HDI, IPDI) retards growth (rats, mice) [9].

11. References

[1] W. Hentschel, *Chem. Ber.* **17** (1884) 1284.
[2] O. Bayer, *Angew. Chem.* **59** (1947) 257–272.
[3] W. Siefken, *Justus Liebigs Ann. Chem.* **562** (1949) 75–136.
[4] B. F. Goodrich, US 2 683 730, 1954 (N. V. Seeger).
[5] A. A. R. Sayigh, H. Ulrich, W. J. Farrissey, Jr. in J. K. Stille, T. W. Campbell (eds.): *Condensation Monomers*, J. Wiley & Sons, New York 1972 pp. 369–476.
[6] R. Richter, H. Ulrich in S. Patai (ed.): *The Chemistry of Cyanates and Their Thio Derivatives, Syntheses and Preparative Applications of Isocyanates*, J. Wiley & Sons, London 1977, part 2, pp. 619–818.
[7] H. Ulrich: *Cycloaddition Reactions of Heterocumulenes*, Academic Press, New York 1967, pp. 122–213.
[8] N. Onodera, *Kogyo Kagaku Zasshi* **65** (1962) 1964; *Chem. Abstr.* **61** (1964) 6886.
[9] H. Ulrich, Urethane Polymers, in *Kirk-Othmer*, 3rd ed., pp. 576–608.
[10] G. Oertel: *Polyurethane Handbook*, Hanser Publishers, Munich 1985.
[11] D. J. David, H. B. Staley: *Analytical Chemistry of Polyurethanes*, Wiley-Interscience, New York 1969, pp. 85–112.
[12] P. S. Carleton, W. J. Farrissey, J. Rose, *J. Appl. Polym. Sci.* **16** (1972) 2983–2989.
[13] R. B. Wilson et al., *J. Am. Chem. Soc.* **105** (1983) 1672–1674.
[14] J. J. Monagle, T. W. Campbell, H. F. McShane, *J. Am. Chem. Soc.* **84** (1962) 4288–4295.
[15] V. E. Shashoua, W. Sweeny, R. F. Tietz, *J. Am. Chem. Soc.* **82** (1959) 866–873.
[16] H. Ulrich: "Isocyanate-Derived Polymers," *Encyclopedia of Polymer Science and Engineering*, 2nd ed., vol. **8**, J. Wiley & Sons, New York 1987, pp. 448–462.
[17] *Houben-Weyl*, **E-4**, pp. 738–784.
[18] K. Kurita, Y. Iwakura, *Org. Synth.* **59** (1979) 195–201.
[19] H. Eckert, B. Forster, *Angew. Chem.* **99** (1987) 922–923.
[20] K. A. Henderson, Jr., V. A. Alexanian, *Org. Prep. Proced. Int.* **18** (1986) 149–155.
[21] Akzo, AU 523 314, 1982 (P. Hentschel, H. Zengel, M. Bergfeld). *Chem. Abstr.* **98** (1983) 72893.
[22] F. J. Weigert, *J. Org. Chem.* **38** (1973) 1316–1319.
[23] M. Chono, S. Fukuoka, M. Kono, *J. Cell. Plast.* **19** (1983) 385–391.
[24] H. J. Twitchett, *Chem. Soc. Rev.* **3** (1974) 209–230.
[25] Upjohn, US 4 020 104, 4 026 943, 1977 (R. H. Richter).
[26] Upjohn, US 3 275 669, 1966 (H. Ulrich, J. N. Tilley, A. A. R. Sayigh).
[27] BASF, DE-OS 3 403 204, 1985 (R. Ohlinger et al.)
[28] Scholven Chemie, DE 1 202 785, 1965 (K. Schmitt et al.).
[29] Bayer, DE 947 471, 1956 (J. Pfirschke).
[30] H. Ulrich, B. Tucker, A. A. R. Sayigh, *J. Org. Chem.* **31** (1966) 2658–2661.
[31] Upjohn, US 3 371 114, 1968 (A. A. R. Sayigh, H. Ulrich).
[32] Upjohn, US 3 410 887, 1968 (A. A. R. Sayigh, H. Ulrich).
[33] VEB Schwarzheide, DD 123 807, 1975.
[34] Arco, DE-OS 2 343 826, 1974 (J. G. Zajacek, J. J. McCoy, K. E. Fuger).
[35] Mitsui-Toatsu, DE-OS 2 635 490, 1977 (R. Tsumura, U. Takaki, A. Takeshi).
[36] K. Schwetlick, K. Unverferth, H. Tietz, *SYSpur Rep.* **19** (1981) 3–12; *Chem. Abstr.* **97** (1982) 39416.
[37] Arco, US 4 146 727, 1979 (E. T. Shaw, J. G. Zajanek).

[38] Dow Chemical, US 4 268 683, 1981 (A. E. Gurgiolo).
[39] S. Fukuoka, M. Chono, M. Kohno, *J. Org. Chem.* **49** (1984) 1458–1460.
[40] American Cyanamid, US 4 364 913, 1983 (D. S. Katz).
[41] T. D. J. D'Silva et al. *J. Org. Chem.* **51** (1986) 3781–3788.
[42] Upjohn, US 4 420 603, 1983 (R. G. Nelb, R. W. Oertel).
[43] Upjohn, US 4 376 834, 1983 (D. J. Goldwasser, K. Onder).
[44] Upjohn, US 3 384 653, 1968 (W. Erner, A. Odinak).
[45] M. Karol et al. *J. Occup. Med.* **20** (1978) 383; *J. Occup. Med.* **21** (1979) 354.
[46] M. Karol et al. *Toxicol. Appl. Pharmacol.* **51** (1979) 73; **53** (1980) 260.

Isoprene

Hans Martin Weitz, BASF Aktiengesellschaft, Ludwigshafen, Federal Republic of Germany (Chaps. 2–7)

Eckhard Loser, Bayer AG, Wuppertal, Federal Republic of Germany (Chap. 8)

1.	Introduction	3033	5.	Storage and Transportation	3047
2.	Properties	3034	6.	Uses	3048
3.	Production	3036	6.1.	Polymer Synthesis	3048
3.1.	Synthetic Methods	3036	6.2.	Terpene Synthesis	3049
3.2.	Dehydrogenation Procedures	3040	7.	Economic Aspects	3050
3.3.	Recovery from C_5 Cracking Fractions	3041	8.	Toxicity and Occupational Health	3052
4.	Quality Specifications	3046	9.	References	3055

1. Introduction

Isoprene [78-79-5] (**1**), C_5H_8, M_r 68.118, is named according to IUPAC rules as 2-methyl-1,3-butadiene. Its isomers include the commercially less important compounds *cis*- and *trans*-1,3-pentadiene (**2**) [504-60-9] (piperylene).

$$\underset{1}{\underset{\underset{CH_3}{|}}{H_2C=C-CH=CH_2}} \qquad \underset{2}{H_2C=CH-CH=CH-CH_3}$$

Free isoprene has been observed in nature only in very low concentration [1], [2]. The compound is present in roasted coffee and in the gas phase of tobacco smoke [3], and can be regarded as a precursor of polycyclic aromatics [4]. Isoprene is the structural unit of countless natural products (the terpenes, which include natural rubber and camphor) as well as biologically important substances such as vitamin A and the steroid sex hormones (cf. the "isoprene rule" of O. Wallach and L. Ruzicka [5]). It is now known, however, that the biosynthesis of rubber and other natural products containing the isoprene skeleton proceeds not via isoprene itself, but rather via mevalonic acid [150-97-0] (**3**), 3,5-dihydroxy-3-methylpentanoic acid [6].

$$\underset{3}{\underset{\underset{OH}{|}}{\underset{\underset{CH_3}{|}}{HOCH_2-CH_2-C-CH_2-COOH}}}$$

Isoprene was first synthesized in 1860 by C. E. Williams by the pyrolysis of natural rubber [7]. The reverse reaction—polymerization of isoprene to poly(cis-1,4-isoprene), with a structure corresponding to that of natural rubber—was long the subject of intensive effort [8]. The first successful attempts were reported in 1954 and 1955 by the Goodrich Gulf (Al–Ti Ziegler catalyst [9]) and Firestone companies (alkyl lithium catalyst [10]).

Isoprene was commonly prepared on a laboratory scale by thermolysis of turpentine oil (the so-called isoprene lamp of C. HARRIES [11]). Pyrolysis of dipentene [5989-27-5], limonene, was even used in the United States early in the Second World War as a commercial source of isoprene (Bibb process [12]). The reverse of this reaction—synthesis of terpenes from isoprene—has been discussed in a vast number of scientific publications, but has so far achieved little commercial significance.

Isoprene itself was commercially unimportant until after the Second World War because it could not be offered at a price that was sufficiently attractive for its principal potential market, the manufacture of synthetic rubber. This situation changed with the development of improved methods for obtaining isoprene from petrochemical sources, as well as of polymerization techniques for the generation of poly(cis-1,4-isoprene), a synthetic rubber whose valuable material characteristics have been widely confirmed.

2. Properties

Physical Properties. Under normal conditions, isoprene is a colorless, volatile liquid. Important properties of isoprene follow (see also [15]):

mp (101.3 kPa)	−145.95 °C
bp (101.3 kPa)	34.059 °C
Critical data [13]	
Temperature	483.3 K
Pressure	3.74 MPa
Volume	266 cm^3/mol
Density (293 K)	0.68095 g/cm^3
Viscosity (293 K)	0.216 mPa·s
Surface tension (293 K)	18.22 mN/m
Refractive index n_D^{20}	1.42194
Vapor pressure [14] at −20 °C	9.8 kPa
0 °C	26.4 kPa
20 °C	60.7 kPa
34.059 °C	101.3 kPa
40 °C	123.8 kPa
60 °C	229.1 kPa
80 °C	392.1 kPa
100 °C	629.5 kPa
Specific heat	
Vapor (298 K)	102.69 J mol^{-1} K^{-1}
Liquid (298 K)	151.07 J mol^{-1} K^{-1}
Heat of formation	
Vapor (298 K)	75.75 kJ/mol
Liquid (298 K)	49.36 kJ/mol
Free energy of formation	
Vapor (298 K)	146.23 kJ/mol
Liquid (298 K)	145.57 kJ/mol

Entropy of formation	
Vapor (298 K)	314.56 J mol^{-1} K^{-1}
Liquid (298 K)	228.3 J mol^{-1} K^{-1}
Heat of combustion (298 K)	−3186.58 kJ/mol
Heat of hydrogenation (298 K)	−230.20 kJ/mol
Heat of fusion (−145.95 °C)	4.88 kJ/mol
Heat of vaporization (at 25 °C)	26.39 kJ/mol
at 34.059 °C	25.87 kJ/mol
Heat of polymerization (298 K)	−75 kJ/mol
Entropy of polymerization (298 K)	−101 J mol^{-1} K^{-1}
Constants for the Antoine equation	
$\log p = A - B/(t+C)$, p in hPa, t in °C [14]	A 6.05329
	B 1092.997
	C 236.002
Flash point	−48 °C
Ignition temperature	220 °C
Explosive limits in air (total pressure 13 kPa, 25 °C)	
Upper limit	7–9.7 vol%
	200–275 g/m^3
Lower limit	1–1.5 vol%
	28–40 g/m^3
Maximal explosion pressure	0.66 MPa

Spectra are to be found in the literature: IR [16], UV [17], NMR [18], Raman [19].

The *solubility* of isoprene in water at 20 °C is 0.029 mol%. It is miscible in all proportions with organic solvents such as ethanol, diethyl ether, acetone, and benzene. Regarding solubility in high-boiling, polar solvents, see chap.4. The Bunsen absorption coefficient α for oxygen in isoprene is 0.4065 m^3 m^{-3} mbar^{-1} at 0 °C and 0.4557 at 20 °C.

Binary *azeotropic mixtures* of isoprene are listed in Table 1. The azeotrope of isoprene with *n*-pentane has recently acquired commercial significance in the isolation of pure isoprene. For information regarding ternary azeotropes of isoprene see [19].

Data for vapor–liquid equilibria of binary and ternary mixtures of isoprene and other organic compounds are provided in [20] and [21]. Liquid–liquid equilibrium data for binary and multicomponent systems containing isoprene are listed in [22]. Activity coefficients at infinite dilution are given in [23]. Curves describing the temperature dependence of various physical quantities may be found in [24] and [25].

Chemical Properties. Spectroscopic analysis has revealed that most isoprene molecules in the ground state at 50 °C are in the s-trans conformation. Only 15 % of the material has the s-cis form, which is higher in energy by 6.3 kJ/mol.

s-*cis*-Isoprene ⇌ s-*trans*-Isoprene
15%   85%

Equilibrium concentrations for the isomer system isoprene–pentadiene (a total of seven isomers) are presented graphically in [26] for temperatures between 200 and 1600 K. Reaction enthalpies for the corresponding isomerizations are provided in [27].

Table 1. Binary azeotropes of isoprene

Component	bp, °C	Isoprene content, wt%
Methanol	29.57	94.8
n-Pentane	33.6	72.5
Carbon disulfide	<34.15	<93
Methyl formate	22.5	50
Bromoethane	32	>65
Ethanol	32.65	97
Dimethyl sulfide	32.5	65
Acetone	30.5	80
Propylene oxide	31.6	40
Ethyl formate	<32.5	>76
Isopropyl nitrite	33.5	72
Methylal	32.8	70
Diethyl ether	33.2	52
Perfluorotriethylamine	30.2	82
Acetonitrile	33.5–33.6	97.5
Isopropylamine		>72.4

Isoprene exhibits the typical characteristics of a conjugated diene. Its methyl group makes the compound react more readily than butadiene with electrophiles and Diels–Alder dienophiles. Moreover, the reactivity of isoprene is more varied than that of butadiene, which only possesses two types of hydrogen atoms in contrast to the four of isoprene. Isoprene can participate in a wide range of reactions (including substitution, addition, ring formation, complexation, and telomerization reactions). For a summary of the general reactivity of isoprene see [19].

The only chemical reaction of isoprene of practical importance is its conversion to terpenes. Isoprene is mainly used for polymer synthesis (Chap. 6).

3. Production

3.1. Synthetic Methods

Laboratory methods for the synthesis of isoprene from a wide variety of starting materials are summarized in [19].

Industrial syntheses of isoprene utilize the following four principles in creating the C_5 skeleton:

$C_1 + C_4 \longrightarrow C_5$
$C_2 + C_3 \longrightarrow C_5$
$C_3 + C_3 \longrightarrow C_6 \longrightarrow C_5 + C_1$
$C_4 + C_4 \longrightarrow C_8 \longrightarrow C_5 + C_3$

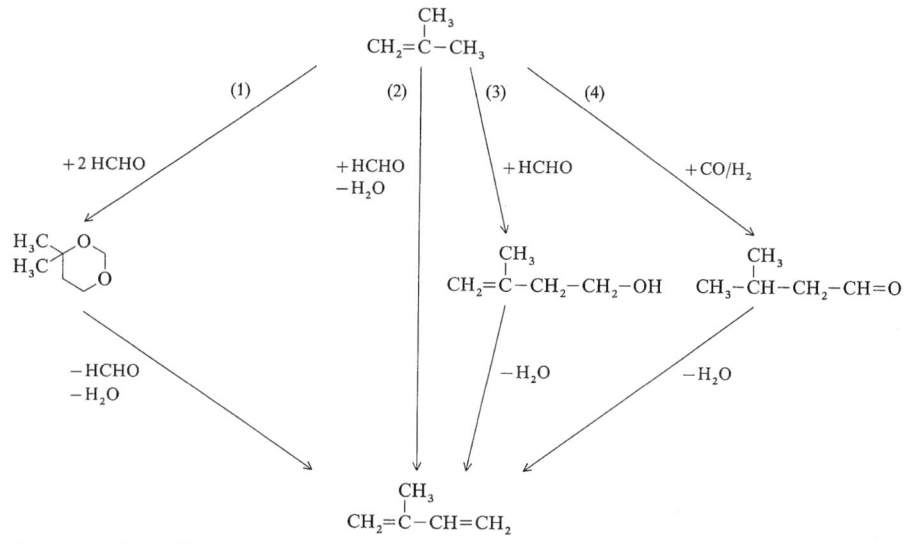

Figure 1. Synthesis of isoprene from isobutene

$C_1 + C_4 \longrightarrow C_5$. Isobutene [115-11-7] is the C_4 building block in several syntheses of this type (Fig. 1). The C_1 component may be formaldehyde [50-00-0] (reactions 1, 2, and 3 in Fig. 1), one of its derivatives, or methanol [67-56-1] [28]. The C_1 molecule carbon monoxide also plays a role in synthesis; this route (4 in Fig. 1) involves hydroformylation of isobutene to 3-methylbutanal [30].

The most frequently used synthetic procedure is the acid-catalyzed addition of formaldehyde to isobutene (Prins reaction, route 1 in Fig. 1). This produces a 1,3-dioxane, which is then cleaved in the gas phase (at 200–300 °C) over an acid catalyst (e.g., H_3PO_4 on a suitable carrier) to give isoprene, with recovery of half of the reacted formaldehyde. This reaction was first suggested for the synthesis of isoprene in 1938 [31], and it became widely known through the work of the Institut Français Pétrol [32], [33]. The method became the subject of further development at several locations as a result of the ready availability of the starting materials [34]–[36]. Individual procedures differ primarily in terms of catalysts employed and engineering details. Production facilities of this type are currently in operation in Japan and the Soviet Union (cf. Chap. 7).

The ultimate goal of developmental efforts in the synthesis of isoprene from isobutene and formaldehyde has been simplification and increased competitiveness. Takeda Chemical in Japan has proposed a one-step gas-phase procedure (cf. Fig. 1, route 2), in which isobutene and formaldehyde are passed over oxide catalysts at 300 °C [36]. Much preliminary work toward this approach was contributed by British Hydrocarbon Chemicals.

An important aspect of the above syntheses with respect to their economic viability is the problem of byproduct formation from formaldehyde. Efforts have therefore been made to use formaldehyde derivatives as a way of suppressing resin buildup. For example, both Idemitsu and Sun Oil have suggested the use of methylal [109-87-5] (**4**), dimethoxymethane [37], [38]. Marathon Oil has proposed a two-step process based on monochlorodimethyl ether [107-30-2] (**5**) [39], [40], while Sumitomo has discussed the use of dioxolane [646-06-0] (**6**), 1,3-dioxacyclopentane [41].

$$\underset{4}{CH_2(OCH_3)_2} \qquad \underset{5}{Cl-CH_2-O-CH_3} \qquad \underset{6}{\underset{\underset{CH_2}{\diagdown\;\diagup}}{\underset{O\quad O}{H_2C-CH_2}}}$$

Consideration has also been given to integrating formaldehyde production directly into the process, i.e., by mixing methanol and oxygen directly with isobutene [42]; a modification of this approach is the use of methyl *tert*-butyl ether (MTBE) and oxygen [43].

Recent Japanese patent applications from Kuraray, Nippon Zeon, and Sumitomo Chemical use *tert*-butyl alcohol [75-65-0] as a C_4 building block in isoprene synthesis, while a number of Soviet publications describe a variety of methods based on cleavage of 1,3-dioxane. Probably the greatest disadvantage of the one-step gas-phase process is the low space–time yield of the catalysts, together with their rapid deactivation under the prevailing reaction conditions [44]–[46]. It is not known whether any of the above-described variants have been carried to production scale.

In addition to the Prins reaction with all its modifications, there is one other economically interesting route to the synthesis of isoprene from isobutene and formaldehyde via 2-methyl-1-buten-4-ol (**7**) (Fig. 1, reaction 3) [47]:

$$H_2C=\underset{\underset{CH_3}{|}}{C}-CH_3 + HCHO \xrightarrow{25\,\text{MPa},\,280\,°C}$$

$$H_2C=\underset{\underset{CH_3}{|}}{\underset{\textbf{7}}{C}}-CH_2-CH_2OH \quad \Big\downarrow\, H^+,\,100\,°C$$

$$H_2C=\underset{\underset{CH_3}{|}}{C}-CH=CH_2 + H_2O$$

One advantage here is that the synthesis (ene reaction) of the intermediate **7** is a purely thermal process occurring in the liquid phase, which permits use of an aqueous formaldehyde solution [48]. Dehydration of **7** is conducted in aqueous NaCl/HCl solution [49] that can be easily regenerated (low catalyst cost).

2-Butene [107-01-7] is the C_4 building block for an entirely different synthesis of isoprene based on the $C_1 + C_4 \rightarrow C_5$ principle via the intermediate 2-methylbutanal [96-17-3] (Fig. 2). 2-Methylbutanal is readily prepared by rhodium-catalyzed hydroformylation of 2-butene [30], [50], [51]. An alternative route to this C_5 aldehyde is the catalytic hydrogenation of 2-ethylacrolein which is easily prepared from *n*-butyraldehyde and formaldehyde [52], [62].

Catalytic dehydration of 2-methylbutanal to isoprene has been the subject of numerous studies [53]. A patent issued to Erdölchemie [54] describes the use of a β-phosphate catalyst, which is also discussed in most of the subsequent publications dealing with

$$CH_3-CH=CH-CH_3$$

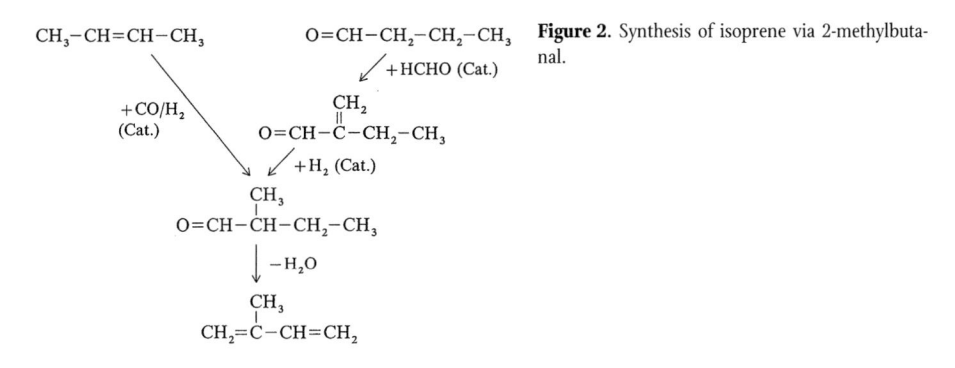

$$\xrightarrow[\text{(Cat.)}]{+CO/H_2}$$

$$O=CH-CH_2-CH_2-CH_3$$

$$\xrightarrow{+HCHO \text{ (Cat.)}}$$

Figure 2. Synthesis of isoprene via 2-methylbutanal.

$$O=CH-\underset{\underset{CH_3}{\|}}{C}-CH_2-CH_3$$

$$\xrightarrow{+H_2 \text{ (Cat.)}}$$

$$O=CH-\underset{\underset{CH_3}{|}}{CH}-CH_2-CH_3$$

$$\downarrow -H_2O$$

$$CH_2=\underset{\underset{CH_3}{|}}{C}-CH=CH_2$$

this reaction [30], [50], [55]–[57]. Other patents for the dehydration of 2-methylbutanal describe the use of magnesium ammonium phosphate [58], molecular sieve [59], as well as zeolite catalysts [60]. Nevertheless, all the catalysts cited are subject to rapid loss of activity [30], or else display low activity from the outset. Zeolites that have been silanized or doped with small amounts of cesium display longer lifetimes. For example, a zeolite containing 0.4 % cesium showed no decrease in yield (51 %) or isoprene selectivity (88 %) after 120 h of use [61].

Cleavage of 3-methylbutanal over a boron phosphate catalyst (Fig. 1, route 4) gave poorer results than were observed with the 2-isomer [30], [62]. At the present time there are still no industrial facilities for the synthesis of isoprene from n-butenes or butanal via 2-methylbutanal.

$C_2 + C_3 \longrightarrow C_5$. Codimerization of ethylene and propene in the presence of triethylaluminum leads to 2-methyl-1-butene, which can be dehydrogenated to isoprene. However, this method has not been carried beyond the experimental stage [63], [64].

By contrast, another isoprene synthesis based on C_2 and C_3 starting materials (acetylene and acetone) was carried to production scale by SNAM in Italy, whose manufacturing facilities in Ravenna had been in operation for several years [65]–[68]; the plant is, however, not currently in operation. A plant based on this principle has been built in the Republic of South Africa (see Chap. 7). The isoprene produced by this method is extremely pure but the process is relatively expensive.

$$H_3C-\underset{\underset{}{\overset{O}{\|}}}{C}-CH_3 + HC\equiv CH \xrightarrow[10-40\,°C]{\text{KOH in liquid NH}_3} H_3C-\underset{\underset{OH}{|}}{\overset{\overset{CH_3}{|}}{C}}-C\equiv CH$$

$$H_3C-\underset{\underset{OH}{|}}{\overset{\overset{CH_3}{|}}{C}}-C\equiv CH + H_2 \xrightarrow[80\,°C,\,0.3\,\text{MPa}]{\text{Pd/Carrier}} H_3C-\underset{\underset{OH}{|}}{\overset{\overset{CH_3}{|}}{C}}-CH=CH_2$$

$$H_3C-\underset{\underset{OH}{|}}{\overset{\overset{CH_3}{|}}{C}}-CH=CH_2 \xrightarrow[250\,°C]{Al_2O_3} H_2C=\underset{}{\overset{\overset{CH_3}{|}}{C}}-CH=CH_2 + H_2O$$

$C_3 + C_3 \longrightarrow C_6 \longrightarrow C_5 + C_1$. The first step of the familiar Goodyear – Scientific Design isoprene synthesis consists of the dimerization of propene to 2-methyl-1-pentene. This is then isomerized to 2-methyl-2-pentene, which is subsequently cracked with loss of methane to give isoprene [64], [69], [70].

$$2\,H_3C-CH=CH_2 \xrightarrow{\text{Trialkylaluminum}} H_2C=\underset{CH_3}{\overset{|}{C}}-CH_2-CH_2-CH_3$$

$$H_2C=\underset{CH_3}{\overset{|}{C}}-CH_2-CH_2-CH_3 \xrightarrow[150-300\,°C]{H^+} H_3C-\underset{CH_3}{\overset{|}{C}}=CH-CH_2-CH_3$$

$$H_3C-\underset{CH_3}{\overset{|}{C}}=CH-CH_2-CH_3 \xrightarrow[650-800\,°C]{H^+} H_2C=\underset{CH_3}{\overset{|}{C}}-CH=CH_2 + CH_4$$

The first commercial synthesis of isoprene was based upon this principle (Beaumont, Texas). Production was discontinued after a fire in 1975 because it was no longer competitive due to sharp increases in the price of propene.

$C_4 + C_4 \longrightarrow C_8 \longrightarrow C_5 + C_3$. Olefin metathesis represents a very interesting approach to isoprene manufacture from readily available petrochemicals. Thus, a butene fraction containing 2-butene and isobutene yields 2-methyl-2-butene and propene:

$$\begin{array}{c} H_3C-CH \\ \parallel \\ H_3C-CH \end{array} + \begin{array}{c} CH_3 \\ | \\ C-CH_3 \\ \parallel \\ CH_2 \end{array} \longrightarrow \begin{array}{c} CH_3 \\ | \\ H_3C-CH=C-CH_3 \\ + \\ H_3C-CH=CH_2 \end{array}$$

2-Methyl-2-butene can then be dehydrogenated to isoprene by standard methods (cf. Section 3.2) [71] – [75]. However, one disadvantage of the metathesis reaction is that all of the olefins in the reactor can react with one another yielding a broad spectrum of byproducts. This is especially true if the starting material is a technical-grade butene mixture containing 1-butene, which represents the only economically feasible starting material [76].

3.2. Dehydrogenation Procedures

The production of isoprene by dehydrogenation of isopentane or the isopentenes (methylbutenes) has been the subject of many investigations [77], [78]. Such reactions closely resemble the analogous dehydrogenations of n-butane and the n-butenes to butadiene ($\rightarrow$ Butadiene).

One-step dehydrogenation of isopentane to isoprene can be carried out according to the Houdry–Catadiene procedure (Cr_2O_3/Al_2O_3 catalyst, ca. 600 °C, and ca. 7 kPa) with a yield of 52 % [79]–[81]. Isoprene is prepared commercially in this way in the Soviet Union [82]. A two-step process for the dehydrogenation of isopentane is described in [83].

Isoprene production plants based on the dehydrogenation of methylbutenes exist in the United States and the Netherlands but are not currently in operation. The requisite starting material comes largely from cat-cracker off-gases [84], [85]. Methylbutenes are extracted in the form of semiesters from the appropriate distillation fraction of these off-gases by dissolution in sulfuric acid, followed by back extraction with paraffinic hydrocarbons (Sinclair procedure). As far as is known, dehydrogenation is effected with a Shell catalyst ($Fe_2O_3-K_2CO_3-Cr_2O_3$) at 600 °C and dilution of the methylbutenes with steam; the yield is 85 % [86]. A Sr–Ni–phosphate catalyst (Dow Type S catalyst, a modification of the well-known Dow Type B catalyst) has also been reported to be suitable for this reaction [87].

Oxidative dehydrogenation of isopentane [88]–[90] and the methylbutenes [91]–[95] has also been widely investigated. The behavior of oxide catalysts in the oxidative dehydrogenation of the methylbutenes is described in [45]. The yield and selectivity for isoprene are reported to be 60 % and 95 %, respectively.

Although many recent publications and patents discuss the preparation of isoprene by dehydrogenation, little significant progress is apparent. Two Soviet reports will be mentioned as examples. The oxidative dehydrogenation of methylbutenes to isoprene with aluminum phosphate catalysts is described in [96]; the reactivity of the olefins decreases in the order 3-methyl-1-butene, 2-methyl-2-butene, and 2-methyl-1-butene. Oxidative dehydrogenation of methylbutenes to isoprene can also be achieved with a silicate catalyst doped with alkaline-earth oxides or the oxides of nickel, iron, or cobalt [97].

Conversion of isopentane and methylbutenes to isoprene by co-oxidation is claimed in [98].

3.3. Recovery from C_5 Cracking Fractions

The C_5 cracking fractions obtained as a byproduct in the pyrolysis of hydrocarbons to ethylene (→ Ethylene) have come to play a major role in isoprene manufacture [99]. Workup of the cracking products from naphtha cleavage yields so-called "crack gasoline", which contains only low concentrations of isoprene and other C_5 hydrocarbons, the major constituents being C_6-C_8 aromatics. Nevertheless, this material may still be distilled into C_5 and aromatic fractions.

The yield of isoprene is typically 2–5 wt % based on ethylene, although it may be increased by starting with a heavier raw material such as gas oil. Table 2 demonstrates

Table 2. Composition (in weight percent) of the C_5 fraction from the pyrolysis of hydrocarbons

Component	bp at 101.3 kPa, °C	Starting material					
		1	2	3	4	5	6
Isopentane	27.85	1.99	11.02	0.99	24.1	7.8	0.8
n-Pentane	36.07	0.6	2.54	2.72	26.1	14.3	20.8
2-Methyl-1-butene	31.16	4.95	5.21	3.80	11.8	2.9	0.8
2-Methyl-2-butene	38.57	2.61	2.13	3.01	11.8	2.4	0.8
1-Pentene	29.97	24.10	3.84	4.08	4.2	2.9	3.1
trans-2-Pentene	36.35	4.74	2.14	2.57	4.2	5.8	
cis-2-Pentene	36.94	3.47	1.54	1.36	4.2	5.8	1.6
trans-1,3-Pentadiene	42.03	14.32	4.95	6.84	8.8	7.3	10.2
cis-1,3-Pentadiene	44.07	8.00	22.13	3.65	8.8	4.7	5.5
Isoprene	34.06	6.71	6.65	4.34.	13.7	16.1	13.3
Cyclopentadiene	42.50	16.46	2.40	17.77	7.5	20.4	37.0
Cyclopentene	44.24	12.50	5.76	6.16	1.6	5.9	5.5

1) Paraffin wax (*mp* 43.5 °C). 2) Low octane gasoline (high severity cracking). 3) Sulfur-containing petroleum (high severity cracking). 4) Naphtha. 5) Romaskino straight-run gasoline (boiling range ≈ 40 – 160/180 °C). 6) Synthetic gasoline from Fischer – Tropsch synthesis (boiling range ≈ 40 – 150 °C).

the influence of the nature of the cracker feed on the C_5 product distribution. Figure 3 shows that the yield of isoprene decreases with increasing crack severity (temperature and residence time). Nevertheless, this overall decrease is accompanied by an increase in isoprene concentration in the C_5 crack fraction (cf. Table 3). Table 3 also provides the product distribution in a typical methylbutene dehydrogenation fraction. Based on energy consumption, recovery of isoprene from crack fractions is considerably more efficient than synthesizing the compound chemically [101].

Increasingly heavier raw materials will be used in Europe and the United States for ethylene production in crackers [100]. The quantity of the isoprene byproduct will therefore also increase. In some crackers only an aromatic fraction is recovered from crack gasoline. The residue (i.e., the C_5 hydrocarbons) is often combined with the C_4 fraction and recycled with the cracker feed.

Pure isoprene cannot be isolated from the C_5 crack fraction simply by distillation, because the numerous components of the mixture differ only slightly in boiling point (cf. Table 2), and isoprene forms an azeotropic mixture with *n*-pentane (cf. Table 1). For the separation of cyclopentadiene from the C_5 crack fraction by dimerization, see → Cyclopentadiene and Cyclopentene.

Distillative enrichment of isoprene in C_5 fractions to a concentration of 25 – 50 % is of some commercial interest because it reduces the transportation and separation costs associated with the isolation of pure isoprene. Examples of special isoprene enrichment processes are described in [103].

Goodyear has developed a process for separating isoprene from C_5 crack fractions in the form of its azeotrope with *n*-pentane [104], [105]. This procedure was carried to production scale in France, although that facility is no longer in operation. The resulting mixture is free of components that might adversely affect the polymerization

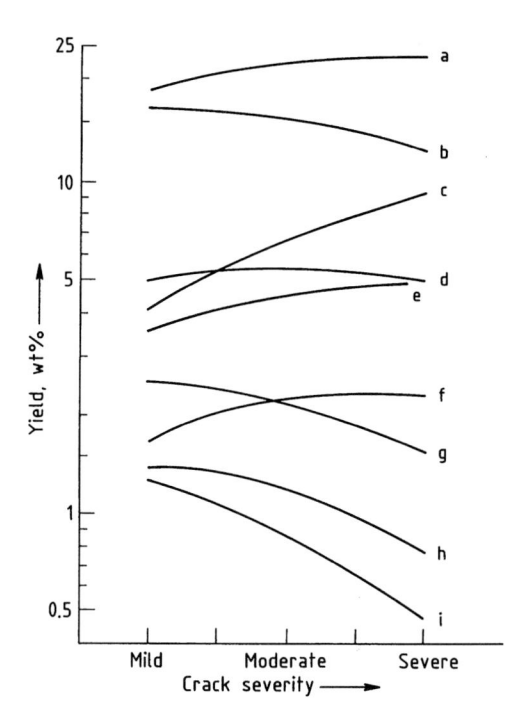

Figure 3. Influence of cracking severity on the product distribution [102]
a) Ethylene; b) Propene; c) Benzene; d) Butadiene; e) Toluene; f) Cyclopentadiene; g) Isobutene; h) Isoprene; i) Piperylenes

Table 3. Typical composition (in weight percent) of C_5 crack fractions as a function of cracking severity

Component	Crack severity			Methylbutene dehydrogenation fractions
	Mild	Normal	Severe	
Pentanes	26.70	34.36	14.12	65–75
n-Pentenes	5.15	6.60	4.21	65–75
Methylbutenes	6.95	9.17	5.76	65–75
Pentadienes	10.90	9.40	19.46	traces
Isoprene	14.20	18.00	23.25	25–35
Cyclopentane and cyclopentene	4.40	3.30	5.76	traces
Cyclopentadiene (including dimer)	16.70	15.97	24.91	traces
2-Butyne	0.40	0.4	1.03	traces
Other hydrocarbons	balance	balance	balance	

of isoprene, and is thus suitable for polymerization processes that employ a solution of isoprene in an inert hydrocarbon.

"Chemical" isolation procedures involving copper(I) compounds have also been considered for the recovery of isoprene from C_5 hydrocarbon mixtures [106]. In contrast to the situation with butadiene (→ Butadiene) this process has never been adopted commercially for isoprene.

Table 4. Relative volatilities * of hydrocarbons to be separated from isoprene in various solvents, based on

Hydrocarbon	bp, °C	Isoprene solvent **						
		ACN	DMF	NMP	MOPN	DMAC	DMSO	γ-BL
Isopentane	27.9	2.92	3.02	3.00	2.62	2.77	2.72	2.79
n-Pentane	36.1	2.37	2.39	2.40	2.27	2.20	2.07	2.13
3-Methyl-1-butene	20.1	2.58	2.63	2.65	2.49	2.53	2.51	2.56
1-Pentene	30.0	1.89	1.90	1.88	1.78	1.82	1.73	1.75
2-Methyl-1-butene	31.2	1.71	1.73	1.75	1.62	1.67	1.62	1.61
trans-2-Pentene	36.3	1.56	1.55	1.56	1.46	1.50	1.45	1.45
cis-2-Pentene	36.9	1.49	1.54	1.53	1.35	1.47	1.41	1.40
2-Methyl-2-butene	38.1	1.38	1.39	1.40	1.32	1.34	1.29	1.28
Isoprene	34.1	(1.00)	(1.00)	(1.00)	(1.00)	(1.00)	(1.00)	(1.00)
trans-1,3-Pentadiene	42.0	0.77	0.76	0.78	0.77	0.77	0.77	0.77
cis-1,3-Pentadiene	44.2	0.70	0.70	0.71	0.70	0.72	0.70	0.70
Cyclopentadiene	41.3	0.62	0.62	0.62	0.63	0.63	0.62	0.63
2-Butyne	26.7	0.96	0.90	0.997	0.97	0.99	1.08	1.08
3-Methyl-1-butyne	26.3	1.04	0.89	0.95	1.10	0.93	0.95	0.94
2-Methyl-1-buten-3-yne	32.5	0.62	0.49	0.48	0.67	0.51	0.55	0.53
1-Pentyne	40.2	0.58	0.55	0.54	0.64	0.54	0.55	0.56
2-Pentyne	56.1	0.42	0.44	0.67	0.43	0.46	0.40	0.41

* The figures indicate the ratio of the volatility of the specified hydrocarbon to that of isoprene measured at 50 °C in a mixture containing 12.5 wt% of the hydrocarbon, 12.5 wt% of isoprene, and 75 wt% of solvent.
** ACN = acetonitrile; DMF = dimethylformamide; NMP = N-methylpyrrolidone; MOPN = β-methoxypropionitrile; DMAC = dimethylacetamide; DMSO = dimethyl sulfoxide; γ-BL = γ-butyrolactone

Selective organic solvents are used in "physical" processes for separating isoprene by extractive distillation with or without liquid – liquid extraction. The principles involved are similar to those developed for separation of butadiene from C_4 crack fractions. The most appropriate solvents for production-scale separation of isoprene from other hydrocarbons include N-methylpyrrolidone [107], [108], dimethylformamide [109], [110], and acetonitrile [111]. The relative volatilities of the hydrocarbons in various polar solvents (i.e., the selectivities of the solvents for the isolation of isoprene from C_5 hydrocarbon mixtures) are listed in Table 4. Figure 4 shows the relationship between the overall concentration and the relative volatilities of 2-methyl-2-butene – isoprene mixtures in various solvents; these data are of considerable importance for estimating the size of a proposed production facility. 2-Methyl-2-butene is the key low-boiling component under the conditions prevailing in this extractive distillation procedure. Additional data relevant to selective solvents for separation of isoprene are presented in [113]. Isolation of isoprene concentrates from C_5 hydrocarbon mixtures by extraction with selective organic solvents is discussed in [114].

Figure 5 is a flow diagram for the isolation of isoprene from a C_5 cracking fraction. The first step in the separation is a combination of a liquid – liquid extraction with an extractive distillation, an approach particularly suited to C_5 fractions with a low isoprene concentration. Preconcentration of isoprene occurs in extractor (a). Crude isoprene is removed as a sidestream from the extractive distillation column (b), after which it is freed from piperylenes and cyclopentadiene in a second extractive distillation

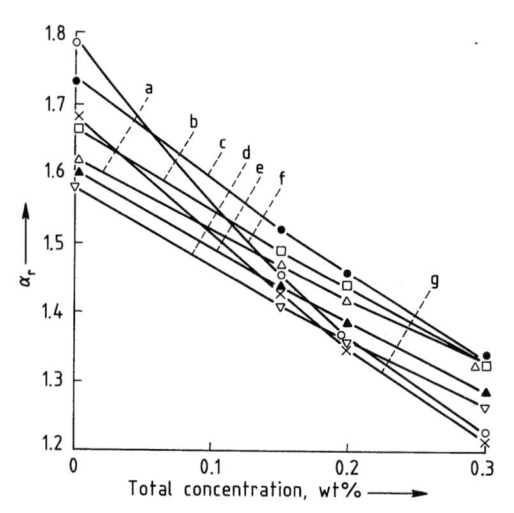

Figure 4. Relationship between the total concentration and the relative volatilities α_r of 2-methyl-2-butene – isoprene mixtures in various solvents [112]
a) Acetonitrile; b) Dimethylformamide; c) N-Methylpyrrolidone; d) β-Methoxypropionitrile; e) Dimethylacetamide; f) Dimethyl sulfoxide; g) γ-Butyrolactone

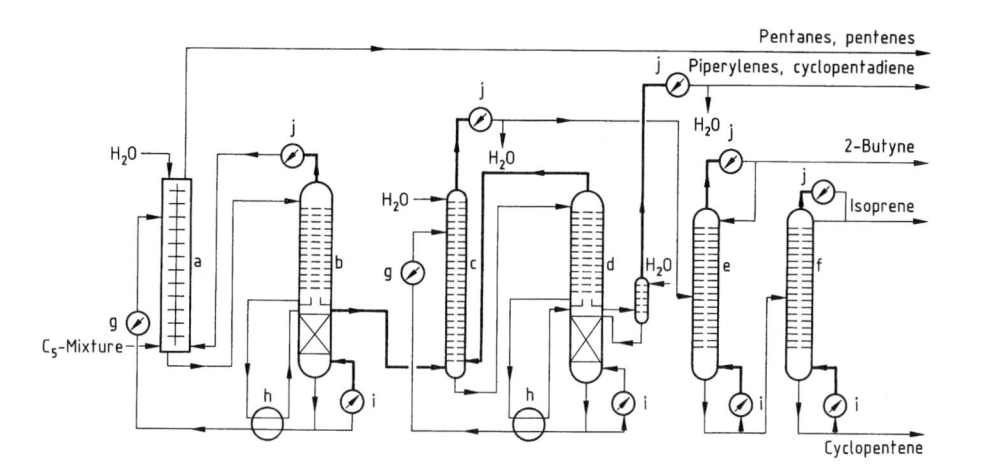

Figure 5. Recovery of isoprene from C_5 hydrocarbon mixtures by a combination of extraction and extractive distillation [107]
a) Liquid – liquid extraction column; b)–d) Columns for extractive distillation; e) Butyne column; f) Isoprene column; g) Cooler; h) Heat exchanger; i) Reboiler; j) Condenser
——— Vapor phase; ——— Liquid phase

step incorporating columns (c) and (d). Final purification of isoprene to polymerization grade occurs in distillation columns (e) and (f) [115]. C_5 fractions that contain cyclopentadiene can be used in this process because the cyclopentadiene is removed together with the piperylenes.

Incorporation of relevant equipment in this process allows the isolation of cyclopentadiene or the commercially valuable cyclopentene (→ Cyclopentadiene and Cyclopentene). Introduction of a supplementary column permits recovery of pure piperylenes.

3045

Table 5. Monomer specifications for isoprene used in the production of poly(-1,4-isoprene)

Component	Alkyl lithium catalyst	Ziegler catalyst (Al/Ti)
Isoprene, wt%	99.5	99.5
Monoolefins, wt%	0.4	0.5
1,3-Pentadiene, ppm	<80	<200
Cyclopentadiene, ppm	< 1	< 1
Acetylene, ppm	< 3	< 50
Allene, ppm	< 5	< 50
Carbonyl compounds, ppm	< 5	< 10
Sulfur (as H_2S), ppm	< 5	< 5
Water, ppm	< 5	< 5

The high quality specifications that are required for isoprene used in the preparation of poly(*cis*-1,4-isoprene) are met with the aid of modern separation facilities. It should be noted, however, that suggestions have also been made for the chemical removal of specific impurities (e.g., cyclopentadiene, acetylene) [116].

Economic comparison of the various techniques for separating isoprene from C_5 cracking fractions is difficult because a large number of criteria must be considered and each must be assigned a proper weighting which depends on the location of the plant. Various selective solvents are compared in [112] with respect to their suitability for isoprene separation. Energy consumption data are analyzed in [117]. A detailed cost analysis for one specific solvent is provided in [118].

4. Quality Specifications

Isoprene is used primarily for the synthesis of polymers, including poly(*cis*-1,4-isoprene) and increasingly in recent years block polymers containing styrene (SIS polymers, see Chap. 6) [119].

Table 5 provides typical specifications for isoprene used as a monomer for the production of poly(*cis*-1,4-isoprene). Purity specifications for polymerization with Ziegler catalysts are less rigorous than for polymerization with alkyl lithium catalysts. The latter are also used for preparing SIS polymers.

Isoprene for use as a monomer in the manufacture of butyl rubber (isobutene – isoprene rubber, IIR) may contain higher levels of the common impurities, as indicated in Table 6. Table 6also contains information about technical-grade isoprene.

In most polymerization processes, isoprene is used in dilution with other inert hydrocarbons. In the production of IR and SIS polymers, it may therefore be more economical to use an isoprene product (isoprene concentrate) in which isoprene of the required purity (content of cyclopentadiene, acetylene, etc.) is diluted with hydrocarbons that do not interfere with polymerization. Thus, in the synthesis of isoprene rubber a starting material containing 60–85% isoprene is sufficient, whereas an

Table 6. Specifications for technical-grade isoprene

Component	Technical-grade isoprene	Isoprene for IIR
Isoprene, wt%	99	>92
Diolefins (total), wt%		>95
C_3 olefins, wt%	1	
Cyclopentadiene, wt%	0.05	<1
α-Acetylene, wt%		<0.35
Peroxides (as H_2O_2), ppm	100	<10
Carbonyl compounds		
(as acetone), ppm	150	<500
Sulfur (as H_2S), ppm	5	<100
Distillation residue, wt%		<1.5

Table 7. Dimerization of isoprene as a function of temperature

Temperature, °C	Dimerized isoprene, wt%/h
20	0.000017
40	0.00019
60	0.0021
80	0.023
100	0.25

isoprene content of 80–90% is required for SIS polymers. For the synthesis of butyl rubber, however, the isoprene cannot be diluted with other hydrocarbons.

The purity of isoprene is normally verified by gas chromatography. Data for sulfur, peroxides, carbonyls, distillation residues, and inhibitors are determined by the usual standard analytical methods, preferably those based on ASTM or ISO/DIN recommendations.

5. Storage and Transportation

Appropriate regulations must be observed during construction and operation of tank storage facilities for isoprene (e.g., the VbF rules in the Federal Republic of Germany). Reference [123] should be consulted regarding special problems, such as prevention of the formation of peroxides (→ Butadiene). Other conditions being equal, the rate of reaction of oxygen with isoprene is significantly greater than that with butadiene. Data regarding the rate of thermal dimerization of isoprene are given in Table 7; this reaction cannot be suppressed by inhibitors. Thermal dimerization at low temperature is a Diels–Alder reaction, producing only cyclohexene derivatives, while at higher temperatures cyclooctadienes are also formed [124]. Transport regulations are summarized in Table 8 [125].

Table 8. Guidelines for the labeling and transport of isoprene

Means of transport	Region	Guideline
Road	Europe	ADR class 3, no. 2 a
	FRG	GGVS class 3, no. 2 a
	USA	DOT regulations
Rail	Europe	RID class 3, no. 2 a
	FRG	GGVE class 3, no. 2 a
	USA	DOT regulations
Inland shipping	Europe	cat. K 1 n
	FRG	ADNR class 3, no. 1 a
Sea freight	FRG	GGV See class 3.1, UN no. 1218
	International	IMDG Code class 3.1, UN no. 1218
Air freight	International	IATA class 3, UN no. 1218

6. Uses

6.1. Polymer Synthesis

Most of the isoprene that is produced is utilized for the synthesis of poly(*cis*-1,4-isoprene) (isoprene rubber, IR), a material (particularly the titanium type) that closely resembles natural rubber in both structure and properties. This IR is largely used for the production of vehicle tires. By contrast, poly(*trans*-1,4-isoprene) has the properties of guttapercha or balata, and has found no major commercial application. Most of the trans polymer is limited to the manufacture of cable insulation and golf balls [124, p. 499], [126].

The second largest market for isoprene in recent years has been the manufacture of styrene–isoprene–styrene (SIS) block copolymers [119], [120, p. 837], [124, p. 532], [127]; 0.79–0.88 t of isoprene are required per tonne of polymer. These products are especially useful as thermoplastic rubbers and as pressure-sensitive or thermosetting adhesives.

Smaller amounts of isoprene are used in the production of butyl rubber (isobutene–isoprene rubber, IIR), a copolymerizate with isobutene [128]. The isoprene content of butyl rubber is 0.5–3.0 mol%. The distinctive features of IIR include its low gas permeability, leading to its use in the construction of hoses and as a liner in tubeless tires.

Finally, mention should also be made of the hydrocarbon resins (petroleum resins), which result from the copolymerization of isoprene from cyclopentadiene-free C_5 crack fractions with other unsaturated C_5 compounds [129].

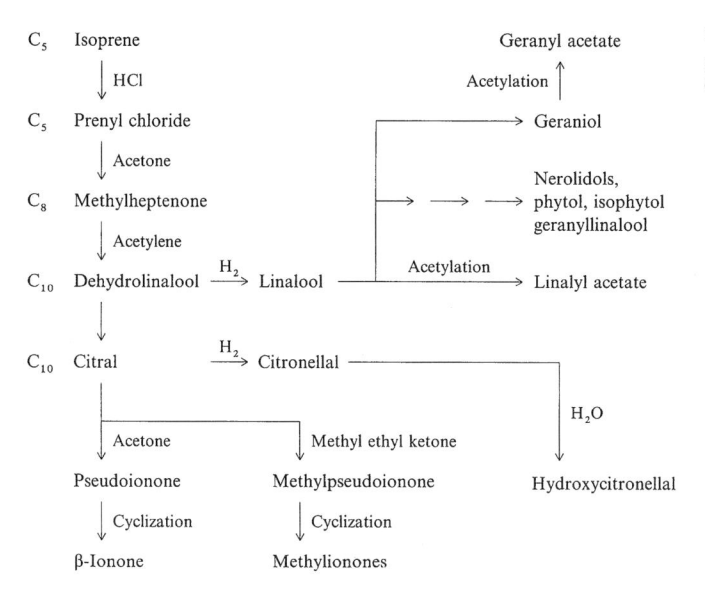

C_5	Isoprene		Geranyl acetate	**Figure 6.** Synthesis of terpenes from isoprene via the $C_5 + C_3 + C_2$ route
	↓ HCl		Acetylation ↑	
C_5	Prenyl chloride	→	Geraniol	
	↓ Acetone		Nerolidols, phytol, isophytol geranyllinalool	
C_8	Methylheptenone	→ → →		
	↓ Acetylene		Acetylation	
C_{10}	Dehydrolinalool —H_2→ Linalool		→ Linalyl acetate	
	↓			
C_{10}	Citral —H_2→ Citronellal			
			↓ H_2O	
	↓ Acetone ↓ Methyl ethyl ketone			
	Pseudoionone Methylpseudoionone		Hydroxycitronellal	
	↓ Cyclization ↓ Cyclization			
	β-Ionone Methylionones			

6.2. Terpene Synthesis

The chemical reactions of isoprene are very varied [19], but they have limited industrial significance. Only terpene synthesis is important and is the subject of intensive investigation.

$C_5 + C_3 + C_2$ Routes. In 1972, Rhodia Incorporated (United States) began to develop a route for the industrial synthesis of C_{10} terpenes and their derivatives from isoprene (C_5), acetone (C_3), and acetylene (C_2) (Fig. 6). The isoprene is first reacted with hydrochloric acid to give prenyl chloride which is then converted to dehydrolinalool in two steps [7], [130]. Rhodia used the dehydrolinalool as a starting compound for the industrial synthesis of most of the compounds shown in Figure 6, but the plant was decommissioned. Kuraray in Japan, however, produces the compounds shown in Figure 6, squalane, and other substances from isoprene [131].

$C_5 + C_5$ Routes. Rhodia has also used a *Grignard synthesis* to produce the C_{10} terpene alcohol lavandulol [1845-51-8] from two molecules of isoprene [132].

Lavandulol

Isoprene can react in a variety of ways to form terpenes via oligomerization or telomerization (→ Terpenes) [133] – [137]. The syntheses of myrcene and *N,N*-diethylnerylamine (C_{10} terpenes) are described below as examples of scientifically and industrially important linear oligomerization and telomerization reactions, respectively.

All naturally occurring terpenes are composed of isoprene units arranged in a "head-to-tail" fashion [133]–[137]. Terpenes synthesized from isoprene must also possess this structure and their double bonds must be correctly positioned [138], [139]. On account of this problem C_{10} terpenes and their derivatives are not generally synthesized from isoprene but by other routes [47], [138].

Variation of the quantitative composition of a Pd-complex catalyst can influence the structure of dimethyloctadienes synthesized from isoprene [140].

Myrcene [123-35-3] has been synthesized by *oligomerization* of isoprene with a sodium/dialkylamine catalyst by TAKABE et al. [141] (see also [142]–[144]).

Myrcene

Nissan Chemical Industries have developed a route for the industrial production of myrcene and have announced that it will be implemented industrially [145].

Telomerization of isoprene proceeds with a variety of compounds, e.g., ammonia or amines [146] (see also [135], [136]). Interestingly, reactions occur in which the regioselectivity can be highly modified by varying the amount and concentration of Brönsted and Lewis acids added to the catalyst [147], [148].

Isoprene can react with diethylamine to form N,N-diethylnerylamine [40137-00-6] [149]; this telomerization reaction is catalyzed by butyl lithium. N,N-Diethylnerylamine can be further reacted to give linalool, geraniol, nerol, hydroxycitronellal, citronellal, and menthol (Fig. 7) [149]–[152].

The Takasago Perfumery Company in Japan has developed an industrial synthesis for enantiomerically pure L-menthol [2216-51-5] from optically active citronellal. The reaction involves the asymmetric allylamine–enamine isomerization of N,N-diethylgeranylamine or N,N-diethylnerylamine with a Rh–BINAP [2,2′-bis(diphenylphosphino)-1,1′-binaphthyl] catalyst [153]–[156]. Geranylamine is obtained by addition of diethylamine to myrcene [22]. The synthesis of nerylamine is described above [149]. Takasago produces 1500 t/a of L-menthol by this route.

7. Economic Aspects

Isoprene is used mainly as a starting material for the manufacture of synthetic rubber, the market for the compound therefore depends heavily upon the widely varying price of natural rubber (cf. [157] and corresponding publications from previous years). As noted in Section 3.3, extraction from C_5 crack fractions is currently the most economical route to isoprene [101]. Nevertheless, a key factor in the calculation is the location of the separation facility, since separation is only profitable if C_5 fractions (possibly enriched in isoprene) from several cracking plants can be transported eco-

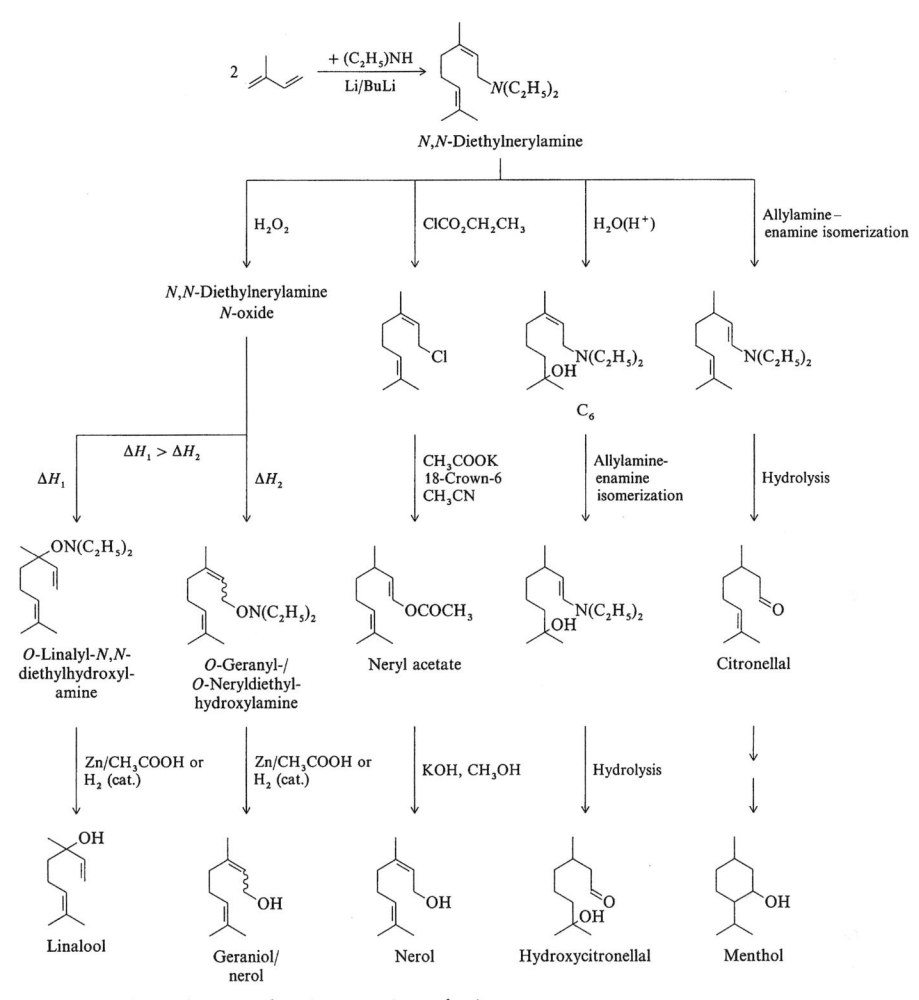

Figure 7. Synthesis of terpenes from isoprene via nerylamine

nomically to a separator of sufficient capacity [158]. Utilization of the large quantities of residual C_5 hydrocarbons is also a major consideration.

The total production of C_5 diolefins in Western Europe in 1987 was 83 000 t [159]. It is of some interest that more (dimeric) cyclopentadiene was produced (44 000 t) than isoprene (23 000 t). The remainder (15 000 t) was accounted for by piperylenes. As shown in Figure 8, an increasing demand for cyclopentadiene relative to isoprene is also anticipated in the United States in the coming years. The growing interest in cyclopentadiene may be attributed to the latest developments in the field of metathetic polymerization of the compound [161]. Combining this technology with the reaction injection molding method opens the way to very economical injection castings with good mechanical properties.

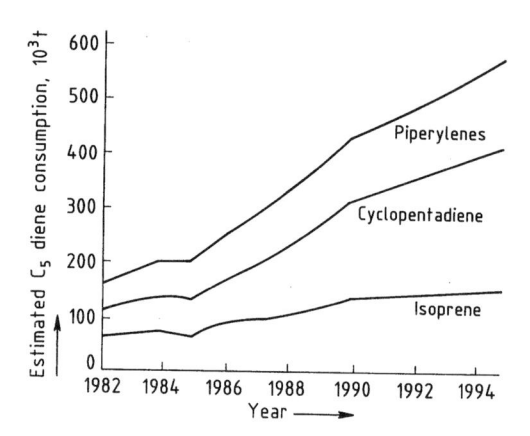

Figure 8. Estimated demand for C_5 dienes in the United States [160]

Table 9 provides an overview of the known production facilities for pure isoprene and isoprene concentrates. The isoprene concentrates are used either directly or, if necessary, after additional purification. Table 10 summarizes production capacities and actual production of polyisoprene (IR), butyl rubber (IIR), and synthetic rubber for the years 1987 and 1989, subdivided according to country. Corresponding totals for the period 1979–1989 are given in Table 11. The obvious decrease in the production of polyisoprene is largely a result of declines in the United States. Consumption data for isoprene used in the manufacture of SIS copolymers in 1986 follow [119]:

United States	25 000 t
Western Europe	18 000 t
Japan	250 t

8. Toxicity and Occupational Health

Acute Toxicity. At high levels of exposure isoprene is an anesthetic in animals, finally resulting in paralysis and death. Single 2-h inhalation exposures of mice to 56 000 mg isoprene per cubic meter of air did not produce toxic effects. Levels of 98 000 – 126 000 mg/m^3 resulted in deep narcosis; death occurred after 2 h of exposure to 140 000 mg/m^3. The LC$_{50}$ for a 2-h exposure was 180 000 mg/m^3 in the rat. The threshold concentration for irritative effects in cats is reported as 800 mg/m^3. An oral LD$_{50}$ of 2100 mg liquid isoprene per kilogram of body weight and an intraperitoneal LD$_{50}$ of 1400 mg/kg in the male rat have been determined. A single dermal exposure of rats to a 1-mL dose of liquid isoprene per kilogram body weight did not cause adverse symptoms or mortality [164], [165].

Table 9. Production facilities for isoprene

Country	Company	Location	Capacity as 100 wt% isoprene, t/a[a]	Product[b]	Process[c]
United States	Goodyear	Beaumont, Texas	84 000	P	refining of isoprene concentrate (extractive distillation?)
	Arco Lyondell	Channelview, Texas	36 000	C	extractive distillation (ACN) of C_5 fractions
	Chevron	Cedar Bayou, Texas	9 000	C	extractive distillation (?)
	Dow	Freeport, Texas	16 000	C	extractive distillation (?)
		Plaquemine, Louisiana	16 000	C	extractive distillation (?)
	Exxon	Baton Rouge, Louisiana	14 000	C	extractive distillation (ACN) of C_5 fractions
		Baytown, Texas	14 000	C	extractive distillation (ACN) of C_5 fractions
	Shell	Dear Park, Texas	40 000	C	extractive distillation (ACN) of C_5 fractions
Netherlands	Shell	Pernis	25 000	C	extractive distillation (ACN) of C_5 fractions
Republic of South Africa	Sentrachem (Karbochem)	New Castle/ Natal	45 000	P	synthesis from acetone and acetylene, plant not yet on line
Japan	Nippon Zeon	Mizushima	45 000	P	extractive distillation (DMF) of C_5 fractions
	Japan Synthetic Rubber	Kashima	30 000	P	extractive distillation (ACN) of C_5 fractions
	Kuraray	Kashima	30 000	P	synthesis from isobutene and formaldehyde
Soviet Union		several locations	ca. 800 000	C	dehydrogenation of isopentane extractive distillation of C_5 fractions synthesis from isobutene + formaldehyde
Standby Capacities					
United States	Shell	Port Neches, Texas	45 000	P	dehydrogenation of isopentene (plant bought from BF Goodrich in 1982 and modified for production of pure isoprene)
Netherlands	Shell	Pernis	50 000	C	dehydrogenation of isopentene
Italy	Enichem (ANIC)	Ravenna	30 000	P	synthesis from acetone and acetylene

[a] Capacities in 1987.
[b] C = concentrate; P = pure isoprene.
[c] ACN = acetonitrile; DMF = dimethylformamide.

Toxicity after Repeated Exposure. Fischer 344 rats and B6C3F1 mice were exposed by inhalation for two weeks to atmospheric isoprene concentrations of 0, 438, 875, 1750, 3500, and 7000 ppm. The rats did not show any exposure-related changes (survival, clinical symptoms, body weight, biochemistry, hematology, macro- and micropathology). Mice at the 7000 ppm level only showed lower body weight gain, but the mice in all groups suffered from anemia, testicular atrophy, olfactory epithelial degeneration, and epithelial hyperplastic changes in the stomach. As in the case of 1,3-butadiene, mice therefore seem to be more susceptible than rats [166].

Table 10. Production capacities (in 10^3 t/a) for polyisoprene (IR), butyl rubber (IIR), and total synthetic rubber

Country	IR		IIR		Synthetic rubber	
	1987	1989	1987	1989	1987	1989
Belgium			85	85	150	150
Germany (FRG)					591	592
France			48	48	719	719
Great Britain			60	60	485	485
Netherlands	40	40			719	719
Western Europe	*40*	*40*	*193*	*193*	*2970*	*3044*
Canada			120	120	284	284
USA	68	68	219	219	2733	2796
North America	*68*	*68*	*339*	*339*	*3017*	*3080*
Brazil		13			350	427
Latin America		*13*			*594*	*689*
Japan	67	67	75	75	1587	1627
Republic of South Africa	45	45			136	136
Countries with market economies	220	233	607	607	8844	9195
Countries with centrally planned economies	1045	1045	120	210	3579	3860
World Total	**1265**	**1278**	**727**	**817**	**12423**	**13055**

* Year-end data, 1989 being based on available projections.

Table 11. Production and production capacity (in 10^3 t) for polyisoprene (IR), butyl rubber (IIR), and total synthetic rubber

		1979	1980	1981	1982	1983	1984	1985	1986	1987	1988	1989
IR	capacity	281	289	289	297	397	289	285	255	225	220	213
IR	production	242	219	168	129	120	146	110	132	113		
IIR	capacity	493	498	446	533	539	583	580	583	585	607	615
IIR	production	457	395	426	370	372						
Synthetic rubber	capacity	7975	8144	8379	8878	8944	8710	8639	8772	8730	8844	9108
Synthetic rubber	production	5830	6167	5347	5281	4623	4854	5397	5211	5486		

* Excluding countries with centrally planned economies.

Repeated dermal exposure (2 applications of 500 µL on each of 5 consecutive days) to the ear of the rabbit caused only limited, reversible irritation [165]. No data on long-term inhalation exposure are available.

Mutagenicity. Isoprene and its monoepoxides do not induce point mutations in the Ames test. The dioxide of isoprene (2-methyl-1,2,3,4-diepoxybutane) was found to be mutagenic in this test system [167] – [170]. Due to some similarities with 1,3-butadiene in the formation of mutagenic metabolites the possible carcinogenic effects of isoprene need further evaluation.

Inhalative exposure of B6C3F1 mice to isoprene levels from 438 to 7000 ppm in air for 6 h/d on 12 days resulted in elevated frequencies of sister chromatid exchange indicating cytogenetic effects [171].

Reproduction. No data on reproductive or teratogenic effects are available.

Metabolism. Rodent mitochondrial fractions transformed isoprene to its monoepoxides 3,4-epoxy-3-methyl-1-butene and 3,4-epoxy-2-methyl-1-butene, leading finally to the corresponding diols. Minor oxidation of the more stable metabolite (3,4-epoxy-2-methyl-1-butene) to a diepoxide has been described [172], [173]. Isoprene is metabolized significantly in the respiratory tract. Up to atmospheric isoprene concentrations of about 300 ppm, metabolism is directly proportional to the exposure concentration [174], [175]; a saturation effect occurs at levels above 300–500 ppm. Mice metabolize isoprene at a higher rate than rats [174].

Isoprene is synthesized endogenously. The synthesis rate in unexposed mice and rats has been calculated to be 0.4 $\mu mol\, h^{-1}\, kg^{-1}$ and 1.9 $\mu mol\, h^{-1}\, kg^{-1}$, respectively [174].

Effects in Humans. Isoprene is narcotic in humans at very high exposure levels. It may also cause irritation of the skin, eye, mucous membranes, and respiratory tract.

Hygienic Standards. No TLV or MAK values have been established for isoprene. A threshold of 40 mg/m^3 has been set in the Soviet Union.

9. References

[1] W. C. Meuly, *Am. Perfum. Cosmet.* **85** (1970) 123.
[2] Y. Yokouchi, Y. Ambe: *J. Geophys. Res. D. Atmos.* **93** (1984) no. 04, 3751; *Chem. Abstr.* **109** (1988) 26 887 p.
[3] T. Dalhamn, R. Rylander, *Arch. Environ. Health* **20** (1970). J. L. Egle, Jr., B. J. Gochberg, *Am. Ind. Hyg. Assoc. J.* 1975 (May) 369.
[4] E. Gil-Av, J. Shabtai, *Nature* **197** (1963) 1065.
[5] L. Ruzicka, *Experientia* **9** (1953) 357.
[6] F. Lynen, U. Henning, *Angew. Chem.* **72** (1960) 820.
[7] C. G. Williams, *Proc. Roy. Soc.* **10** (1860) 516; *Philos. Trans. R. Soc. London* 1860, 241.
[8] F. Hofmann, *Chem. Ztg.* **60** (1936) 693.
[9] Goodrich Gulf, US 3 114 734, 1954 (S. E. Horne).
[10] Firestone Tire and Rubber Comp., US 3 208 988, 1955 (L. E. Forman et al.).
[11] C. Harries, K. Gottlob, *Justus Liebigs Ann. Chem.* **383** (1911) 228.
[12] Newsport Ind., US 2 386 537, 1945 (C. H. Bibb).
[13] K. H. Simmrock, R. Janowsky, A. Ohnsorge: "Critical Data of Pure Substances" in D. Behrens, R. Eckermann (eds.): *Chemistry Data Series*, DECHEMA, Frankfurt 1986, vol. II/1, p. 289.
[14] T. Boublik, V. Fried, E. Hála: *The Vapour Pressure of Pure Substances*, Elsevier Science Publ., Amsterdam, New York 1984, p. 329.

[15] *Selected Values of Physical and Thermodynamic Properties of Hydrocarbons and Related Compounds,* American Petroleum Research Project 44, Carnegie Press, Pittsburg 1953, p. 361.
[16] *The Aldrich Library of FT-IR Spectra,* I/1 ed., 30 D Milwaukee, Wisconsin 1985.
[17] L. C. Jones, L. W. Taylor, *Anal. Chem.* **27** (1955) 228.
[18] Sadtler Research Laboratories: *Nuclear Magnetic Resonance Spectra,* Spectrum no. 3434 M.
[19] W. J. Bailey in E. C. Leonard (ed.): *Vinyl and Diene Monomers,* vol. **2**, Wiley, New York 1971, Chap. 5.
[20] J. Gmehling, U. Onken, W. Arlt, B. Kolbe: "Vapor–Liquid Equilibrium Data Collection," in D. Behrens, R. Eckermann (eds.): *Chemistry Data Series*, DECHEMA, Frankfurt, vol. **I/6 a,** 1980, p. 38; vol. **I/6 c,** 1983, p. 40.
[21] *Landolt–Börnstein,* NS IV/3, 1975, 367.
[22] J. M. Sørensen, W. Arlt: "Liquid–Liquid Equilibrium Data Collection," in D. Behrens, R. Eckermann (eds.): *Chemistry Data Series,* DECHEMA, Frankfurt vol. **V/1,** 1979, p. 582; vol. V/3, 1980, p. 472.
[23] D. Tiegs et al.: "Activity Coefficients at Infinite Dilution," in D. Behrens, R. Eckermann (eds.): *Chemistry Data Series,* DECHEMA, Frankfurt 1986, vol. **IX/1** and 2, p. 880.
[24] R. W. Gallant: *Physical Properties of Hydrocarbons,* Gulf Publ. Comp., Houston 1968, 157–166.
[25] C. L. Yaws, *Chem. Eng. (N.Y.)* **83** (1976) no. 5, 107.
[26] J. E. Kilpatrick et al., *J. Res. Nat. Bur. Stand.* **42** (1949) 225.
[27] F. M. Fraser, E. J. Prosen, *J. Res. Nat. Bur. Stand.* **54** (1955) 143.
[28] T. Reis, *Chem. Process Eng. (London)* **53** (1972) no. 2, 68.
[29] C. A. Cociasu, A. Dumitrescu, *Rev. Chim. (Bucharest)* **23** (1972) 652.
[30] D. Forster, J. P. Sluka, A. Vavere, *CHEMTECH* **16** (1986) no. 12, 746.
[31] I.G. Farbenind., DE 741152, 1938 (W. Friedrichsen).
[32] *Hydrocarbon Process.* **50** (1971) no. 11, 167.
[33] F. Coussemant, M. Hellin, *Erdöl Kohle Erdgas Petrochem.* **15** (1962) no. 5, 274, 348. M. Hellin, M. Davidson, *Inf. Chim.* 1980 no. 206, 163, 181.
[34] W. Swodenk, W. Schwerdtel, P. Losacker, *Erdöl Kohle Erdgas Petrochem. Brennst. Chem.* **23** (1970) no. 10, 641.Z. Kriefalussy, *Chem. Ing. Tech.* **55** (1983) no.12, 1965.
[35] *Hydrocarbon Process.* **50** (1971) no. 11, 168.
[36] K. Naito, K. Ogino, *Pet. Petrochem. Int.* **13** (1973) no. 11, 44.
[37] Idemitsu Petrochem., JP 1 90 82, 1965.
[38] H. J. Peterson, J. O. Turner, *Hydrocarbon Process.* **53** (1974) no. 7, 121.
[39] *Hydrocarbon Process.* **50** (1971) no. 11, 171.
[40] D. W. Hall, F. L. Dormish, E. Hurley, Jr., *Ind. Eng. Chem. Prod. Res. Dev.* **9** (1970) no. 2, 234.
[41] Sumitomo Chem., DE-OS 2 004 355, 1970 (K. Takagi et al.).
[42] Sumitomo Chem., DE-OS 1 816 739, 1968 (Y. Watanabe et al.).
[43] Sumitomo Chem., DE-OS 1 941 949, 1969 (Y. Watanabe et al.).
[44] A. Mitsutani, *Am. Chem. Soc., Centennial Meeting,* New York 1976.
[45] A. Mitsutani, S. Kumano, *Chem. Econ. Eng. Rev.* **3** (1971) no. 3, 35.
[46] M. Ai, *J. Catal.* **106** (1987) 280.
[47] H. Pommer, A. Nürrenbach, *Pure Appl. Chem.* **43** (1975) 527.
[48] BASF, DE 1 275 049, 1967 (H. Müller, H. Overwien, H. Pommer). P. R. Stapp, *Ind. Eng. Chem. Prod. Res. Dev.* **15** (1976) no. 3 189.
[49] BASF, DE-OS 2 031 921, 1970 (H. Müller).
[50] Monsanto Co., EP 80 449, 1981 (D. Forster, G. E. Barker). D. A. Bolshakov et al., *Neftekhimiya* **23** (1983) no. 2, 183; *Chem. Abstr.* **98** (1983) 217 599 k.

[51] Goodyear Tire & Rubber Co., US 4 484 006, 1983 (H. R. Menapace).
[52] BASF, EP 58 927, 1982 (F. Merger, H. J. Förster). BASF, EP 92 097, 1983 (G. Dümbgen et al.).
[53] A. V. Irodov et al., *2H. Org. Khim.* **18** (1982) no. 7, 1401; *Chem. Abstr.* **97** (1982) 111 060 s.
[54] Erdölchemie GmbH, DE 2 163 396, 1971 (H. Fischer, G. Schnüchel).
[55] J. B. Moffat, *Rev. Chem. Intermed.* **8** (1987) no. 1, 1.
[56] Goodyear Tire & Rubber Co., US 4 524 233, 1984 (W.-L. Hsu et al.); US 4 587 372, 1985 (W.-L. Hsu); US 4 628 140, 1986 (L. G. Wideman).
[57] Monsanto Co., US 4 547 614, 1988 (A. Vavere).
[58] International Synthetic Rubber Co., GB-A 2 093 060, 1981 (D. E. Timm).
[59] Union Carbide, US 4 734 538, 1986 (G. L. O'Connor, St. W. Kaiser, J. M. McCain).
[60] Union Carbide, EP-A 272 662, 1987 (J. H. McCain, St. W. Kaiser, G. L. O'Connor). BASF, EP 162 385, 1984 (W. Hölderich et al.).W. Hölderich, *Pure Appl. Chem.* **58** (1986) no. 10, 1383.
[61] BASF, EP-A 219 042, 1985 (W. Hölderich, P. Hettinger, F. Merger).
[62] D. A. Bolshakov et al. *Prom-St. Sint. Kauch.* **1980** no. 8, 2; *Chem. Abstr.* **94** (1981) 84549 u.
[63] H. Ando et al., *Sekiyu Gakkaishi* 1969, 356, 362.
[64] C. A. Cocoiasu, *Rev. Chim. (Bucharest)* **23** (1972) 722.
[65] A. Fovorski, *Zh. Russ. Fiz.-Khim. O-va.* **32** (1902) 356, 652.
[66] Bayer, DE 246 241, 1910.
[67] *Hydrocarbon Process.* **54** (1975) no. 11, 154.
[68] A. Heath, *Chem. Eng. (N.Y.)* **80** (1973) no. 22, 448.
[69] R. Landau, G. S. Schaffel, A. C. Deprez, *Erdöl, Kohle, Erdgas, Petrochem.* 16 (1963) no. 7, 754.
[70] F. Andreas, K. Gröbe: *Propylenchemie*, Akademie-Verlag, Berlin 1969, Chap. 14.
[71] *Chem. Eng. News* **48** (1970) no. 10, 60.
[72] *Oil Gas J.* **69** (1971) no. 16, 17.
[73] R. L. Banks, R. B. Regier, *Ind. Eng. Chem. Prod. Res. Dev.* **10** (1971) no. 1, 46.
[74] R. L. Banks in B. E. Leach (ed.): *Appl. Industr. Catalysis* **2**, Academic Press, New York 1984, p. 218.
[75] S. Warwel, G. Pütz, *Chem. Ztg.* **112** (1988) no. 11, 15.
[76] K. L. Anderson, T. D. Brown, *Hydrocarbon Process.* **55** (1976) no. 8, 119.
[77] *Eur. Chem. News* **18** (1970) no. 452, 30; no. 453, 36.
[78] T. Reis, *Chem. Process Eng. (London)* **53** (1972) no. 3, 66.
[79] A. A. Digiacomo, J. B. Maerker, J. W. Schall, *Chem. Eng. Prog.* **57** (1961) no. 5, 35.
[80] R. G. Craig, J. M. Duffalo, *Chem. Eng. Prog.* **75** (1979) no. 2, 62.
[81] *Hydrocarbon Process.* **65** (1986) no. 4, 106.
[82] I. V. Garmonov, paper no. 1 presented at International Symposium on Isoprene Rubber, Moscow 1972.
[83] Y. A. Gorin, A. A. Wasilew, A. N. Makashina, *Khim. Promst. (Moskau)* 1958, no. 1, 1.
[84] R. L. Foster, D. K. Wunderlich, S. H. Patinkin, R. A. Sanford, *Pet. Refin.* **39** (1960) no. 11, 229.
[85] *Pet. Refin.* **40** (1961) no. 11, 257.
[86] *Hydrocarbon Process.* **52** (1973) no. 11, 140.
[87] R. A. Stowe, J. L. Mallory, I. J. Martin, J. V. Bishop, *Prepr. Div. Pet. Chem. Am. Chem. Soc.* **17** (1972) B 19; *Erdöl, Kohle, Erdgas, Petrochem. Brennst. Chem.* **27** (1974) 94.
[88] O. D. Sterligov, I. M. Arinich, T. V. Dudukina, *Izv. Akad. Nauk SSSR, Ser. Khim.* 1971, 2833.
[89] Petro-Tex, US 4 658 074, 1965 (L. Bajars).
[90] Phillips Petroleum, US 3 851 009, 1971 (R. S. Cichowski).
[91] A. Farkas, *Hydrocarbon Process.* **49** (1970) no. 7, 121.
[92] E. Nishikawa, T. Ueki, Y. Morita, *Bull. Jpn. Pet. Inst.* **12** (1970) 84.

[93] Petrotex, US 3 278 626, 1962 (L. Bajars).

[94] H. Pichler, H. Schulz, A. Zeinel Deen, *Erdöl, Kohle, Erdgas, Petrochem. Brennst. Chem.* **29** (1976) no. 9, 418.

[95] M. Sittig: *Diolefins Manufacture and Derivatives,* Noeyes Dev. Corp., Park Ridge, N.J. 1968, p. 2.

[96] S. M. Zulfugarova, D. B. Tagiev, *Neftekhimiya* **26** (1986) no. 2, 189; *Chem. Abstr.* **106** (1987) 101 699 x.

[97] K. E. Tokhadze et al., *Neftekhimiya* **23** (1983) no. 9, 511; *Chem. Abstr.* **99** (1983) 174 240 y.

[98] Halcon, DE-AS 1 568 764, 1966 (J. Kollar).

[99] D. L. Schultz, *Rubber World* **183** (1980) no. 10, 87.

[100] G. Zimmermann, D. Zschummel, S. Brändel, *Chem. Tech. (Leipzig)* **21** (1969) 97.

[101] M. J. Rhoad: *Proceedings,* 15th Annual Meeting, International Institute of Synthetic Rubber Producers, Inc., Kyoto 1974; *Rubber Ind. (London)* **9** (1975) no. 12, 68; *Rubber News* **14** (Jan. 23, 1978).

[102] Y. Wada, *Chem. Econ. Eng. Rev.* **6** (1974) no. 8, 36. H. Kuper, *Erdöl Kohle Erdgas Petrochem.* **35** (1982) no. 3, 119.

[103] J. W. Bennet, V. J. Guerico, *Chem. Eng. (London)* **1971**, 315. Esso, US 2 768 224, 1953 (L. B. Page, J. W. Rector).

[104] *Eur. Chem. News* **25** (1974) no. 642, 12, 27.

[105] Goodyear, DE-OS 2 444 232, 1974 (H. M. Lybarger).

[106] E. R. Gilliland: "Chemicals for Synthetic Rubbers" in: *The Science of Petroleum,* V/II, Oxford University Press, London 1953, p. 32.

[107] H. Kröper, H. M. Weitz, *Oil Gas J.* **65** (1967) no. 2, 98.

[108] U. Block, B. Hegner, D. Wolf, *Chem. Ing. Tech.* **47** (1975) 35.

[109] R. W. Swanson, J. A. Gerster, *J. Chem. Eng. Data* **7** (1962) 132.

[110] S. Takao, *Oil Gas J.* **77** (1979) no. 40, 81.

[111] T. Reis, *Chem. Process Eng. (London)* **53** (1972) no. 1, 34.

[112] B. A. Sarajew et al., *Khim. Promst. (Moskau)* 1975, 815.

[113] M. Enomoto, *Bull. Jpn. Pet. Inst.* **13** (1971) 169.

[114] G. H. Dale, W. C. MacCarthy, *World Pet. Congr. Proc. 7th* **4** (1967) 55.

[115] R. E. Lynn, J. C. Healy, *Chem. Eng. Prog.* **57** (1961) 46. H. Scherb, *Chem. Ing. Tech.* **44** (1972) no. 11, 704.

[116] Jap. Synthetic Rubber, DE 2 105 523, 1970 (T. Chikatsu et al.).

[117] M. J. Rhoad, *J. Inst. Rubber Ind.* **9** (1975) 68.

[118] S. J. Pawlow et al., *Khim. Promst. (Moskau)* 1971, no. 4, 256.

[119] *Isoprene, Chemical Economics Handbook,* SRI International, Menlo Park, Calif. 1988.

[120] J. Witte in *Houben-Weyl,* vol. **E 20/2,** 1987, 823.

[121] C. E. Morrell in G. S. Whitby: *Synthetic Rubber,* J. Wiley & Sons, New York 1954, p. 84.

[122] J. Pech, *österr. Chem. Ztg.* **68** (1967) 77.

[123] *Enjay, Storage and Handling of Liquefied Olefins and Diolefins,* company leaflet, New York 1962.

[124] M. L. Senyek in H. Mark et al. (eds.): *Encyclopedia of Polymer Science and Engineering,* vol. **8,** J. Wiley & Sons, New York 1987, p. 494.

[125] G. Hommel et al.: *Handbuch der gefährlichen Güter,* 3rd ed., Springer, Berlin – Heidelberg – New York 1987, Merkblatt 359.

[126] *Polyisoprene Elastomers, Chemical Economics Handbook,* SRI International, Menlo Park, California 1988.

[127] *Styrenic Thermoplastics, Elastomers, Chemical Economics Handbook*, SRI International, Menlo Park, California 1986.
[128] E. N. Kresee, R. H. Schatze, H. C. Wang in H. Mark et al. (eds.): *Encyclopedia of Polymer Science and Engineering*, vol. **8**, J. Wiley & Sons, New York 1987, p. 433.
[129] Exxon Chemical International Marketing Inc., *Polymers for Adhesives and Sealants*, company leaflet, Kraainem, Belgium 1989.
[130] *Chem. Week* **107** (1970) no. 21, 109; *Chem. Mark. Rep.*, 29th Apr. (1974) 7, 18.S. Wiseman, B. T. Bawden, *Chem. Eng. Progr.* 1977, no. 5, 60.
[131] *Chem. Age (London)* **111** (1975) no. 2399, 8. *Chem. Week* **121** (1977)no. 3, 17. *Jpn. Chem. Week* (1978) 6th June, 6. *Chem. Ind.* **34** (1982) no. 10, 252.
[132] J. Sibeud, *Perfum. Flavor.* **1** (June/July 1978) 7. Rhodia Inc., US 3 819 733, US 3 856 867, 1972 (H. Ramsden).
[133] H. Morikawa, S. Kitazome, *Ind. Eng. Chem. Prod. Res. Dev.* **18** (1988) no. 4, 254.
[134] A. J. Chalk, S. A. Magennis, *Ann. N.Y. Acad. Sci.* **333** (1980) 386.
[135] W. Keim, A. Behr, M. Röper, "Alkene and Alkyne Oligomerization, Cooligomerization and Telomerization Reactions," in G. Wilkinson et al. (eds.): *Comprehensive Organometallic Chemistry*, vol. **8**, Pergamon Press, Oxford – New York 1982, pp. 399, 445.
[136] A. Behr, *Aspects Homogeneous Catal.* **5** (1984) 48.
[137] W. Gaube, R. Berger, *Mitteilungsbl. Chem. Ges. D.D.R.* **34** (1987) no. 6, 122.
[138] W. Hoffmann, *Chem. Ztg.* **97** (1973) no. 1, 23.
[139] A. D. Josey, *J. Org. Chem.* **39** (1974) no. 2, 139.
[140] K. Takahashi, G. Hata, A. Miyake, *Bull. Chem. Soc. Jpn.* **46** (1973) no. 2, 600.
[141] K. Takabe et al., *Synthesis* 1977, no. 5, 307.
[142] Nissan Chem. Ind., DE-OS 2 542 798, 1974 (K. Takabe et al.).
[143] J. Barton et al., *Collect. Czech. Chem. Comm.* **48** (1983) no. 8, 2361.
[144] Takasago Perfumery Co., Jp-KK 81/5206, 1976 (S. Agutagawa).
[145] A. Mitsutani, *Hydrocarbon Process.* **58** (1979) no. 6, 56-C. *Chem. Eng.* **85** (1978) no. 2, 144.
[146] W. Keim, M. Röper, *J. Org. Chem.* **46** (1981) no. 18, 3702.
[147] W. Keim, M. Röper, M. Schieren, *J. Mol. Catal.* **20** (1983) 139.
[148] M. Röper, R. He, M. Schieren, *J. Mol. Catal.* **31** (1985) 335.
[149] K. Takabe, K. Katagiri, J. Tanaka, *Chem. Lett.*1977, no. 9, 1025.
[150] Nissan Chem. Ind., DE-OS 2 634 309, 1975 (J. Tanaka et al.).
[151] Nissan Chem. Ind., DE-OS 2 720 839, 1976 (A. Murata et al.).
[152] Takasago Perfumery Co., JA-AS 83/26 735, 1977 (H. Kumobayashi). *Chem. Abstr.* **90** (1979)23 327 jTakasago Perfumery Co. GB-A 2 088 355, 1980 (S. Akatugawa, T. Taketomi).
[153] K. Tani et al., *J. Chem. Soc. Chem. Commun.* 1982, no. 11, 600. K. Tani et al., *J. Am. Chem. Soc.* **106** (1984) no. 18, 5208. H. Takaya et al., *J. Am. Chem. Soc.* **109** (1987) no. 5, 1596.
[154] Takasago Perfumery Co., EP-A 0 170 470, 1984 (H. Kumobayashi et al.).
[155] G. W. Parshall, W. A. Nugent, *CHEMTECH* **18** (1988) no. 6, 376.
[156] R. Noyori, M. Kitamura, "Enantioselective Catalysis with Metal Complexes," in R. Scheffold (ed.): *Modern Synthetic Methods*, vol. **5**, Springer-Verlag, Berlin – Heidelberg – New York 1989, p. 115.
[157] K. Möbius, *Gummi, Fasern, Kunstst.* **42** (1989) no. 1, 6.
[158] M. Rosenzweig, *Chem. Eng. (N.Y.)* **79** (1972) no. 8, 52 F, 52 M.
[159] *Eur. Chem. News* **52** (1989) no. 1364, 7.
[160] *Hydrocarbon Process.* **67** (1988) no. 2, 17.
[161] S. Warwel, *Chem. Ztg.* **103** (1987) no. 5, 243.

[162] *Rubber Statistical Bulletin,* **43** (1988/89) no. 1, 45.
[163] *Worldwide Rubber Statistics,* Internat. Inst. of Synth. Rubber Producers, Inc., Houston, Texas 1989.
[164] H. W. Gerarde in G. D. Clayton, F. E. Clayton (eds.): *Patty's Toxicology,* 2nd rev. ed., vol. **II**, J. Wiley & Sons, New York – London 1963, p. 1205.
[165] Unpublished data, Bayer AG.
[166] R. Melnick: "Inhalation Toxicology Studies of Isoprene in F344 Rats and B6C3F1 Mice," *Abstract, Int. Symp. Toxicology, Carcinogenesis & Human Health Aspects of 1,3-Butadiene,* NIEHS, Research Triangle Park, N. C., April 1988.
[167] K. Mortelmans et al.: "Salmonella Mutagenicity Tests: II. Results From the Testing of 270 Chemicals," *Environ. Mutagen.* **8,** Supplement 7 (1986) 1–119.
[168] C. de Meester in I. Gut, M. Cikrt, G. L. Plaa (eds.): *Industrial and Environmental Xenobiotics,* Springer Verlag, Berlin 1981, pp. 195–203.
[169] A. Kushi et al.: "Mutagenicity of Gaseous Nitrogen Oxides and Olefins on Salmonella TA102 and TA104," *Mutat. Res.* **147** (1985) 263–264.
[170] P. G. Gervasi et al.: "Mutagenicity and Chemical Reactivity of Epoxide Intermediates of the Isoprene Metabolism and Other Structurally Related Compounds," *Mutat. Res.* **156** (1985) 77–82.
[171] R. R. Tice et al.: "Chloroprene and isoprene: cytogenetic studies in mice," *Mutagenesis* **3** (1988) 141–146.
[172] M. Del Monte et al.: "Isoprene Metabolism by Liver Microsomal Mono-oxygenases," *Xenobiotica* **15** (1985) 591–597.
[173] V. Longo et al.: "Hepatic Microsomal Metabolism of Isoprene in Various Rodents," *Toxicol. Lett.* **29** (1985) 33–37.
[174] H. Peter et al.: "Pharmacokinetics of Isoprene in Mice and Rats," *Toxicol. Lett.* **36** (1987) 9–14.
[175] A. R. Dahl et al.: "The Fate of Isoprene Inhaled by Rats: Comparison to Butadiene," *Toxicology Appl. Pharmacol.* **89** (1987) 237–248.

Ketenes

Raimund Miller, Lonza Inc., Fair Lawn, New Jersey 07410, United States
Claudio Abaecherli, Lonza AG, Visp, Switzerland
Adel Said, Lonza AG, Basel, Switzerland

1.	Introduction	3061	3.4.	Storage, Handling, and Transportation	3072
2.	Ketene	3062	3.5.	Uses	3073
2.1.	Physical Properties	3062	3.5.1.	First Generation Diketene Derivatives	3073
2.2.	Chemical Properties	3062	3.5.2.	Second Generation Diketene Derivatives	3076
2.3.	Production	3065	4.	Higher Ketenes	3076
2.4.	Uses	3066	4.1.	Properties	3076
3.	Diketene	3068	4.2.	Production and Uses	3077
3.1.	Physical Properties	3068	5.	Toxicology and Occupational Health	3078
3.2.	Chemical Properties	3069	6.	References	3079
3.3.	Production	3072			

1. Introduction

Ketenes are described by the general formula $R^1R^2C=C=O$. Of commercial and industrial importance are mainly the lowest member of this class, ketene itself, $H_2C=C=O$, and its dimer, diketene. The latter has attained a prominent role in industrial synthetic chemistry. Ketene and diketene lend themselves to a number of useful conversions because of their multifunctionality.

The parent substance, monomeric ketene, was first prepared in 1907 by Wilsmore by immersing a glowing platinum wire in liquid acetic anhydride. Staudinger subsequently described its preparation by the reaction of bromoacetyl bromide [598-21-0] with zinc. The first industrial use of ketene appeared in the 1920s, when larger quantities of acetic anhydride were required for the production of cellulose acetate.

2. Ketene

2.1. Physical Properties [1]–[8]

Ketene [463-51-4], $CH_2=C=O$, C_2H_2O, M_r 42.02, is a colorless, toxic gas with a characteristic odor; it is unstable at room temperature and atmospheric pressure. Ketene is very soluble in acetone and lower esters; it is only slightly soluble in hydrocarbons and halogenated hydrocarbons.

Some important physical properties of ketene are given below:

mp	−134.1 °C
bp at 101.3 kPa	−41 °C
Vapor pressure at −120 °C	0.53 kPa
−71 °C	19.5 kPa
20 °C	1.9985 MPa
30 °C	2.6355 MPa
40 °C	3.4137 MPa
50 °C	4.3506 MPa
90 °C	10.0253 MPa
Density	
gas	1.93 g/cm^3
liquid	0.7047 g/cm^3
Vapor density (theoretical, air = 1)	1.45
Free energy of formation, ΔG_f^0	−46.9 ± 1.6 kJ
Enthalpy of formation, ΔH_f^0	−47.7 kJ
Dipole moment	1.41 D

Crude ketene, as obtained from the cracking of acetic acid and used industrially, contains ca. 8 vol% acetic anhydride and 7 vol% inert gases (mainly carbon monoxide, carbon dioxide, ethylene, and methane). Ketene from the cracking of acetone also contains a stoichiometric amount of methane.

2.2. Chemical Properties [1]–[8]

Ketene is an extremely reactive and unstable compound: it dimerizes to diketene and polymerizes very quickly. The 4-methyleneoxetan-2-one structure of diketene was established in 1952 by X-ray diffraction [9]. Under the conditions for the industrial production of diketene, less than 5% of ketene dimerizes to 1,3-cyclobutanedione in its monoenolic form, which in turn is further acetylated. It can only be stored at liquid nitrogen temperature. In order to prevent polymerization, reactions with ketene are best conducted under reduced pressure. If this is not possible, the reaction vessel should be directly connected to the ketene source or compressor.

To avoid formation of acetic acid and acetic anhydride, reactions with ketene should be carried out in a dry atmosphere, but because it does not form peroxides (in contrast to the ketoketenes) ketene does not require an oxygen-free atmosphere.

The main reactive center in ketene is the carbonyl carbon atom, which is strongly electrophilic. Consequently, ketene reacts readily with nucleophiles. The electronic structure of ketene can be described by the following canonical forms:

$$H_2C=C=O \longleftrightarrow H_2C=C^+-\overset{..}{O}{}^- \longleftrightarrow H_2\overset{..}{C}{}^- -C^+=O$$

Ketene preferentially undergoes addition reactions with substances containing polar or easily polarized single or multiple bonds; addition usually takes place across the C=C double bond, the oxygen atom is seldom involved. In the absence of other reaction partners, ketene adds a second molecule of ketene to form diketene. The main types of reactions that ketene participates in are acetylation, cycloaddition, ketene insertion, and polymerization.

Acetylation. Compounds with a reactive hydrogen atom are acetylated by ketene:

$$H_2C=C=O + H-X \longrightarrow \underset{\underset{H\ \ X}{|\ \ \ |}}{H_2C-C=O}$$

Ketene acetylates substances containing hydroxyl groups, e.g., water is acetylated to acetic acid, aliphatic alcohols to alkyl acetates, phenols to aryl acetates, carboxylic acids to anhydrides, enols to enolacetates, hydrogen peroxide to diacetylperoxide, and hydroperoxides to peracetates. Acidic and basic catalysts strongly increase the reaction rate [10].

Sulfur analogs react with ketene similarly: hydrogen sulfide is acetylated to thioacetic anhydride via thioacetic acid and aliphatic thiols yield alkyl thioacetates. Hydrogen halides give the corresponding acetyl halides. The reaction with hydrogen sulfide needs catalysis, other sulfur compounds do not necessarily require catalysis.

Ketene reacts with ammonia, primary and secondary amines, amides, imides, hydrazines, hydrazones, ureas, and azomethines to give the corresponding monoacetates; sometimes the di- and triacetates are also formed. With strong bases the reaction can be quite violent. For weakly basic or substituted N–H groups, acids or amides are recommended as catalysts [11]. Ketene can selectively acetylate amino groups in the presence of hydroxyl groups. Examples are the acetylation of ammonia [1, p. 124] and the N-acetylation of ethanolamine in aqueous solution. Alcohols or even water can be used as solvents for such reactions.

Cycloadditions. Substances bearing a double bond (e.g., C=C, P=N, C=X, or N=X, where X = O, N, or S) can undergo [2+2] 1,2-cycloaddition with the ketene C=C bond. This synthetically useful reaction leads to versatile four-membered rings:

$$H_2C=C=O + X=Y \longrightarrow \underset{X-Y}{H_2C-C=O}$$

A catalyst is often necessary. Ketene itself does not react with nonpolar olefins (ketoketenes and halogenoketenes are more reactive), but it does react with 1,3-dienes and active olefins such as vinyl ethers, vinyl esters, enamines, alkoxyacetylenes, and allenes. Carbonyl compounds, especially those with electron-withdrawing groups, give 2-oxetanones (β-lactones). The dimerization of ketene to diketene falls into this category.

Optically active 2-oxetanones have been obtained with chiral tertiary amines as catalyst in cycloadditions with aldehydes and ketones [12]. Thietanones are obtained from thiocarbonyl compounds; nonenolizable imines (azomethines) give 2-azetidinones (β-lactams) [13]. 1,3- and 1,4-cycloadditions are also known, though rare; an example follows (yield ca. 16%):

For a review of cycloadditions with ketenes see [1], [8], [14].

Ketene Insertion. Substances with strongly polarized or polarizable single bonds add to the ketene C=C double bond in a reaction known as ketene insertion:

$$H_2C=C=O + X-Y \longrightarrow \underset{X}{\overset{O}{\|}}\!\!\diagup\!\!Y$$

Thus, active acyl chlorides with an electron-withdrawing substituent (e.g., $-CCl_3$ and $-CF_3$) give the corresponding 3-ketoacyl chlorides, phosgene inserts two moles of ketene to yield acetone dicarboxylic acid dichloride [15], and α-haloethers give 3-oxa-substituted propionyl halides. Other insertion reactions take place with chlorine (to give chloroacetyl chloride), bromine, thionylchloride, and with mineral acids.

Polymerization of ketene with various catalysts can give either polyesters (**1**) or polyketones (**2**); higher ketenes also give polyacetals (**3**) [16].

      1   2   3

2.3. Production

Ketenes are the internal anhydrides of the corresponding carboxylic acids, and as such can be made by removing a molecule of water from these acids, either directly or indirectly. Numerous methods have been used to convert a precursor to the corresponding ketene [8].

Ketene is produced by *pyrolysis of acetic acid* [64-19-7], which, although endothermic, is quite an efficient process. Decomposition products formed in the pyrolysis of acetic acid to ketene include methane, hydrogen, carbon monoxide, carbon dioxide, and ethylene. Higher acids can also be cracked to the corresponding ketenes, although the efficiency of conversion decreases as the chain length increases. The corresponding carboxylic anhydrides give better results [17], [18]. Long-chain alkyl ketenes are made industrially from the corresponding fatty acid chlorides. Ketene generators and ketene reaction apparatus for both the laboratory and production scale have been reviewed [19].

High-quality acetic acid is evaporated and the vapor passed continuously through a radiant coil under reduced pressure at 740–760 °C. Triethyl phosphate [78-40-0] catalyst is injected into the acetic acid vapor. Ammonia is added to the gas mixture leaving the furnace to prevent ketene and water from recombining and to neutralize the catalyst. The gas mixture is chilled to < 100 °C to remove water, unconverted acetic acid, and the acetic anhydride formed as a liquid [20], [21]. The gaseous ketene is absorbed in the appropriate reaction medium or dimerized in a liquid ring pump (liquid seal pump) if diketene is the desired reaction product. The medium consists of a liquid substance or a solution of a compound that is to be reacted with ketene (e.g., alcohols, amines, and thiols). Diketene itself is the preferred dimerization medium for ketene. The crude ketene gas is at a pressure of 20–30 kPa before entering the dimerization unit. It can, therefore, only be used for reactions with higher boiling reactants that will not vaporize at this pressure. In the case of more volatile reactants or when a higher reaction temperature is required, the crude ketene is compressed to ambient pressure. This is best done in a liquid ring pump using a liquid in which ketene is only sparingly soluble (e.g., decaline) [22].

A useful laboratory-scale synthesis of high-purity ketene has been described [23]. This method, which gives very pure ketene by *pyrolysis of diketene,* can also be employed on an industrial scale. Thus, a convenient and technically feasible process for producing pure ketene, uncontaminated by methane and other hydrocarbons and oxides, is available. Conversion of diketene to monomeric ketene is accomplished on an industrial scale by passing diketene vapor through a tube heated to 350–600 °C:

$$\text{diketene} \longrightarrow 2\ H_2C=C=O$$

The pyrolysis of dimeric ketenes, whether in the β-lactone or cyclobutane-1,3-dione form, is a general method for obtaining pure ketenes in high yield.

Ketene can also be generated on an industrial scale by *thermal decomposition of acetone or acetic anhydride*. Nevertheless, the former is only interesting as a raw material when the price is low. It is believed that acetone is still used in the Soviet Union for the production of ketene.

2.4. Uses

Ketene, the internal anhydride of acetic acid, is a powerful acetylating agent (see Section 2.2). It reacts generally at the C=C double bond with compounds containing an active hydrogen atom to give an acetyl derivative:

$$H_2C=C=O + H-A \longrightarrow \underset{H_3C}{\overset{A}{>}}=O$$

Production of Acetic Anhydride. Most of the ketene produced worldwide is used in the production of acetic anhydride.

$$H_2C=C=O + \underset{HO}{\overset{O}{\underset{\|}{C}}}CH_3 \longrightarrow CH_3\underset{\|}{\overset{O}{C}}O\underset{\|}{\overset{O}{C}}CH_3$$

In the early 1980s, the Tennessee Eastman Company developed a new process for the coal-based production of acetic anhydride by carbonylation of methyl acetate [79-20-9] [24]. This new process has economic advantages in times of high oil prices.

Production of Diketene. The second most important industrial use of ketene is the production of diketene by controlled dimerization (see Section 3.2).

$$2\ H_2C=C=O \longrightarrow \underset{H_2C}{\overset{O}{\square}}\hspace{-0.5em}-\hspace{-0.5em}O$$

The incorporation of the acetoacetate moiety into pigments, pharmaceuticals, and agricultural chemicals is an important area of synthetic chemistry.

The diketene – acetone adduct is a nonlachrymatory liquid, which can be safely handled at room temperature.

Production of Sorbic Acid. The third major use of ketene, albeit less important than the first two mentioned above, is in the production of sorbic acid [110-44-1] [25], [26]. Reaction of ketene with crotonaldehyde [123-73-9] gives, depending on the conditions, either the β-lactone **4** or the polyester **5**. Both products can be converted, thermally or by acidic catalysis, to sorbic acid:

$$\diagdown\diagup O + H_2C=C=O \longrightarrow$$
Crotonaldehyde

[structure 4] + [structure 5] → $\diagdown\diagup\diagdown\diagup$COOH
                                          Sorbic acid

The corresponding reaction of ketene with formaldehyde, once an important use of ketene in the production of acrylates via propiolactone, is no longer used on a commercial scale because the intermediate β-propiolactone [57-57-8] (2-oxetanone) is a suspected carcinogen.

$$H_2C=C=O + H_2C=O \xrightarrow{H^+} \text{[β-propiolactone]} \longrightarrow \diagup\text{COOH}$$

The most economical process for the production of acrylates is now based on the two-stage vapor-phase oxidation of propene to acrylic acid [27].

Production of Chloroacetyl Chloride. Chlorine adds to ketene to form chloroacetyl chloride [79-04-9] [28]:

$$H_2C=C=O + Cl_2 \longrightarrow Cl\diagdown\text{C(=O)}\diagup Cl$$

Chloroacetyl chloride has varied industrial uses, such as the production of herbicides of the alachlor type and local anesthetics of the lidocaine type.

4,4,4-Trihaloacetoacetates. In the presence of strong acid, such as boron trifluoride, appropriately substituted acyl chlorides add to ketene to form the corresponding acetoacetyl chlorides which can then be reacted with an alcohol to form the corresponding haloacetoacetates.

$$R^1\text{C(=O)}Cl + H_2C=C=O \longrightarrow R^1\text{C(=O)CH}_2\text{C(=O)}Cl \xrightarrow{R^2OH} R^1\text{C(=O)CH}_2\text{C(=O)}OR^2$$

Of industrial significance are ethyl 4,4,4-trifluoroacetoacetate [372-31-6] for the production of herbicides and antimalarial agents, and ethyl 4,4,4-trichloroacetoacetate [3702-98-5] for the production of pharmaceuticals.

Isopropenyl Acetate and Acetylacetone. Acetone, like other carbonyl compounds containing active hydrogen atoms, undergoes enol acetylation on reaction with ketene; a strong acid catalyst is required. The isopropenyl acetate [108-22-5] formed [29], [30] is the starting material for the production of acetylacetone [123-54-6], which is an intermediate for the sulfa drug sulfamethazine [57-68-1].

$H_2C=C=O + CH_3-CO-CH_3 \longrightarrow$ [structure showing CH$_3$ substituted β-lactone with CH$_3$]

Chiral Lactones. Chloral and 1,1,1-trichloroacetone react with ketene to form the corresponding 4-trichloromethyloxetanones. In the presence of an optically active catalyst, chiral lactones are obtained [12], [31], [32].

$H_2C=C=O + CCl_3-CO-R \xrightarrow{\text{Optically active catalyst}}$

[β-lactone with R and CCl$_3$] or [β-lactone with Cl$_3$C and R]

R = H, CH$_3$

3. Diketene

3.1. Physical Properties

Diketene [*674-82-8*], $C_4H_4O_2$, M_r 84.04, is a colorless liquid when pure; it possesses a pungent odor and is a strong lachrymator. The human odor perception threshold is 0.019 mg/m^3 [33].

Crude diketene obtained from the dimerization of ketene is often used without further purification; it is dark brown and contains up to 10% higher ketene polymers.

Some physical properties of diketene are listed below [1]–[7]:

mp	−7.5 °C
bp	
101.3 kPa	127 °C
13.3 kPa	69–71 °C
5.5 kPa	50.5 °C
3.1 kPa	38.5 °C
Vapor pressure (20 °C)	1.07 kPa
d_4^{20}	1.09471
d_4^{25}	1.088754
Vapor density (20 °C, theoretical, air = 1)	2.9
n_D^{20}	1.4379
Surface tension, mN/m	
20 °C	31.82
24 °C	30.37
25 °C	33.89
Critical pressure	5.47 MPa
Critical temperature	310 °C (583.16 K)
Viscosity	0.88 mPa · s
Flash point (open cup)	33 °C
Ignition temperature (air)	291 °C

Explosive limits in air, vol%
  lower   2 *
  upper   11.7
Dipole moment   3.53 D
Dielectric constant   16–17
Standard enthalpy of formation ΔG_f^0 −233.41 J

* A value of 0.2% has also been reported [34].

Diketene is miscible with acetic anhydride, with many ketones and esters (e.g., acetone, methyl acetate, ethyl acetate, butyl acetate) with diethyl ether, tetrahydrofuran, chlorinated hydrocarbons, acetonitrile, and toluene.

Furthermore, contrary to most literature reports, it is slightly soluble in aliphatic hydrocarbons such as hexane (ca. 12.5 wt% at 23 °C), octane, cyclohexane, and methylcyclohexane. It is very slightly soluble in water, in which it decomposes slowly.

3.2. Chemical Properties

Ketene itself forms primarily the unsymmetrical lactone dimer with less than 5% being converted to the symmetrical dimer, 1,3-cyclobutanedione [15506-53-3], which in turn is converted to the ketene trimer [38425-52-4] by enol acetylation [35].

1,3-Cyclobutane-dione ← 2 H₂C=C=O → β-Lactone

  ↘ H₂C=C=O

Ketene trimer

Diketene is a reactive, versatile compound, which cannot be stored indefinitely, even in the presence of stabilizing agents such as sulfates, borates, or sulfur [36]. On standing for several days or weeks, it turns yellow and then brown. Glass vessels are not recommended for storage because the basicity of glass promotes decomposition: stainless steel or aluminum is preferred.

Diketene reactions proceed mainly by ring opening and are highly exothermic. Traces of base or strong acid can lead to violent, almost explosive polymerization, and due care should be exercised when handling diketene.

Only the most important types of the large number of reactions of diketene are briefly discussed here. More detailed information can be found in [1]–[7].

Pyrolysis. Diketene is cleaved to two molecules of ketene by pyrolysis at 350–600 °C [20], and it is a convenient laboratory source of ketene (see Section 2.3). In a copper reactor at 550 °C and with admixture of nitrogen, decomposition can lead to allene and carbon dioxide [37].

Reactions at the Olefinic Double Bond (without Ring Opening) [38]–[40]. Hydrogen adds catalytically to the exocyclic olefinic double bond of diketene to form β-butyrolactone (4-methyl-2-oxetanone). Other reactions include radical addition (e.g., with halogenated compounds, secondary alcohols, alkylthiols, dialkyl phosphites) to give 4-substituted oxetanones. Diketene can also undergo addition of carbenes and nitrenes, and photochemical [2+2] and thermal [4+2] cycloadditions with olefins and other unsaturated compounds (Diels–Alder) to give spirocyclic compounds.

Polymerization of diketene under free radical conditions leads to a polymeric form with intact β-lactone rings. Ozonolysis to give the highly unstable malonic acid anhydride has been reported [41].

Ring-Opening Reactions. In most reactions, the strained β-lactone ring is opened, and diketene appears to react as acetylketene or one of its tautomeric forms:

Nucleophiles normally add to the acyl carbon–oxygen bond leading to acetoacetylation:

Nu = nucleophile

A variety of compounds with acidic hydrogen atoms react in this manner, e.g., alcohols, phenols, carboxylic acids, hydrogen halides, amines, carbon acids, amides, ureas, N-arylhydroxylamines, and sulfonamides. With the less reactive compounds, a catalyst is often required. When other functional groups are present (e.g., ureas, thioureas, hydrazines, hydroxylamines, amidines, urethanes, enamines, and amides), acetoacetylation can be followed by ring closure with either the keto group or the active methylene group of the acetoacetyl moiety. This leads to a variety of heterocyclic compounds [42], an example is:

Water hydrolyzes diketene only very slowly, leading to the unstable acetoacetic acid, which further decomposes to acetone and carbon dioxide [43].

Electrophiles react first with the exocyclic olefinic double bond of diketene. Subsequent nucleophilic attack at the lactone carbonyl group gives 4-substituted acetoacetic acid derivatives. Thus, chlorine gives the versatile building block 4-chloroacetoacetyl chloride [44]. Bromine reacts analogously.

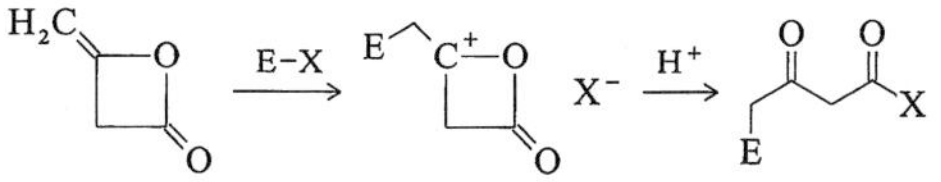

With *carbonyl compounds*, diketene gives 2,2-disubstituted 6-methyl-4H-1,3-dioxazin-4-ones (**6**):

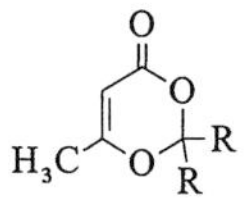

6

The most common adduct, diketene–acetone (**6**, R = CH$_3$), 2,2,6-trimethyl-4H-1,3-dioxin-4-one, is often used as a convenient diketene replacement, because it reacts similarly to diketene, but is safer to handle and to transport [45], [46].

Analogously, compounds with C = N groups, such as carbodiimides and imines, lead to 1,3-oxazines:

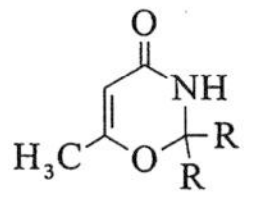

Diketene Polymers. Under various conditions, diketene dimerizes to give dehydroacetic acid [47]. This compound is frequently encountered as a byproduct in reactions of ketene and diketene.

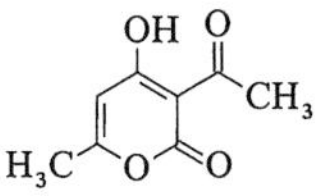

Dehydroacetic acid

With HgCl$_2$, ion-exchange resins, or alkoxides, a low molecular mass polyester of the form **1** (Section 3.1) is obtained. With BF$_3$ the formation of a polyketone of repeating –(CH$_2$–CO–) units **2** has been reported [48], [49]. With vinyl monomers, diketene affords copolymers with useful properties for adhesives, unsupported films, and molded objects [50].

3.3. Production

Ketene possesses a considerable tendency to polymerize. Under controlled conditions and in the appropriate dimerization medium, ketene can be dimerized to diketene, with yields up to 85%, based on acetic acid (see also Section 2.2). The remaining 15% is made up by higher ketene oligomers [35]. The preferred dimerization medium is diketene itself [51]. Crude diketene, containing 3–4% acetic anhydride, is a mobile, dark brown liquid due to 8–10% higher polymers. Diketene of >99.5% purity can be obtained by distillation [52]. Ultrapure diketene (99.99%) can be obtained by crystallization [53].

Diketene is now generally consumed at the site of production, because of its extreme reactivity and hazardous properties. In many cases, the 1:1 adduct of diketene and acetone, 2,2,6-trimethyl-4H-1,3-dioxin-4-one [*5394-63-8*] (**6**, R = CH$_3$) can be used in place of diketene [45].

3.4. Storage, Handling, and Transportation

Pure diketene is surprisingly stable at temperatures up to ambient in absence of acid or base catalysts. Liquid diketene is stable for several weeks at 0 °C, slow polymerization occurs above 15 °C. At elevated temperature, diketene undergoes slow decomposition and discoloration. Diketene can be stabilized with agents such as water [54], sulfur [55], alcohols, phenols, borate, and sulfate salts. In Japan, lactic acid is reportedly used. Diketene, stabilized with 1% anhydrous copper sulfate, retains its strength for up to one year at 25 °C.

Nevertheless, due to the high reactivity of diketene and transportation restrictions in certain countries, diketene is mainly used at the site of production. Hazard classifications for transportation follow:

IMDG code: class 3.3
RID/ADR: class 3, no. 31 c
CFR 49: 172.101 flammable liquid
UN number: 2521

In the United States and Japan, diketene is shipped in refrigerated tank trucks. Care must be taken to keep shipping time to a minimum and to ensure that the diketene is used upon arrival. Due to the potential hazard of shipping diketene, however, this practice is being discontinued in the United States.

Most reactions with diketene are strongly exothermic; it is, therefore, generally added in a small stream or in portions to other reactants. A build-up of unreacted diketene is thereby avoided. Diketene should not be stored in glass bottles due to the alkalinity of glass; stainless steel or aluminum containers should be used.

The total annual worldwide production of diketene is estimated at 80 000 – 100 000 t. The major producers are Lonza, Tennessee Eastman, Hoechst, Wacker, BASF, Nippon Gohsei, Daicel, and BP Chemicals.

3.5. Uses

Diketene is a powerful bactericidal agent [56] and can raise the octane rating of gasoline [57], but is not used for these purposes in the Western hemisphere.

The main reactions and uses of diketene have been comprehensively reviewed [7]. The most important industrial uses and some newer developments are highlighted in the following sections.

3.5.1. First Generation Diketene Derivatives

First and second generation diketene derivatives are depicted in Figure 1.

Acetoacetates. In the presence of tertiary amine catalysts, diketene reacts rapidly with lower alcohols. Industrially preferred catalysts include trimethylamine [75-50-3], triethylamine [121-44-8], and pyridine [110-86-1] [58]. Acidic catalysts such as sulfuric and sulfonic acids can also be used [59], [60]. The industrial acetoacetylation of alcohols is carried out continuously in a loop reactor at 50 – 80 °C without solvent. A comprehensive treatment of acetoacetate chemistry is given in [61]. Methyl and ethyl acetoacetates [141-97-9] are the most widely used esters, finding use in the pharmaceutical, agricultural, and allied industries.

2-Chloro- and 4-Chloroacetoacetates. The most specific chlorinating agent for acetoacetates in the 2-position is sulfurylchloride. Alternatively, diketene can be treated with hydrogen chloride at low temperature in chlorinated hydrocarbon solvents to afford acetoacetyl chloride [62], which is then chlorinated in the 2-position in excellent yield to give 2-chloroacetoacetyl chloride. The latter can be further treated with alcohols, aliphatic or aromatic amines to afford the corresponding 2-chloroacetoacetate derivatives [63].

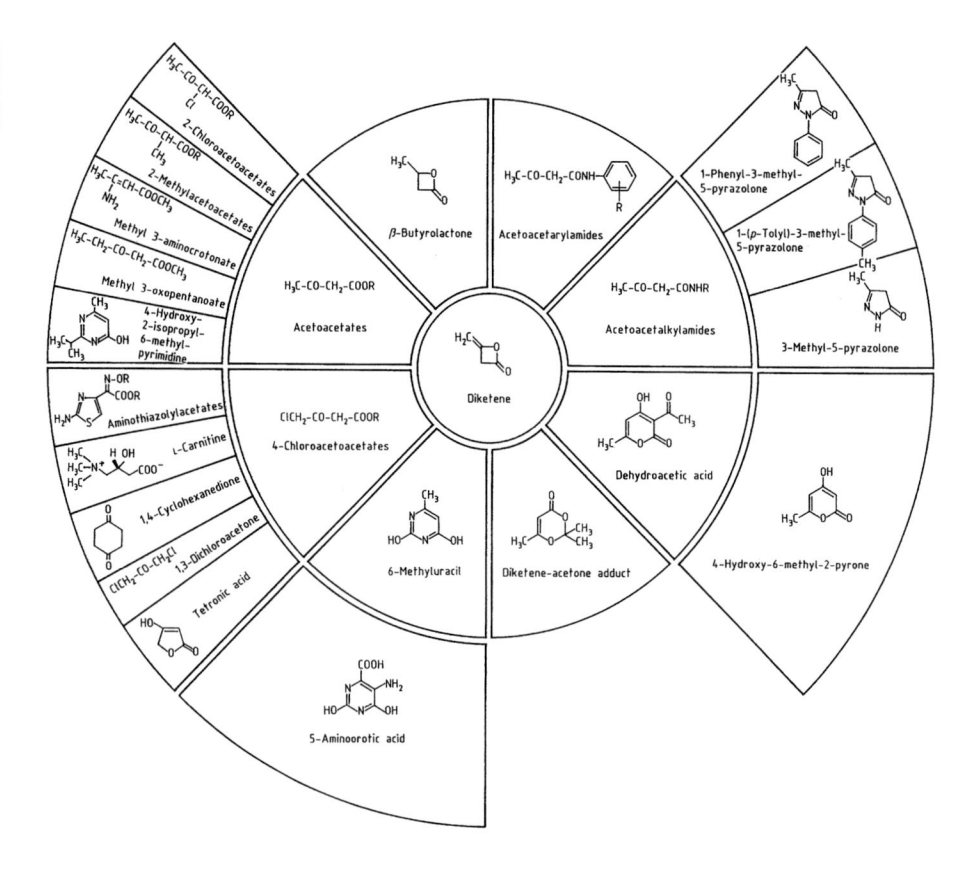

Figure 1. Diketene derivatives of the first generation (inner segments) and second generation (outer segments)

4-Chloroacetoacetates are conveniently prepared on an industrial scale by reaction of gaseous chlorine with diketene to form 4-chloroacetoacetyl chloride [64], which can react further with chlorine to afford 2,4-dichloroacetoacetyl chloride. Subsequent reaction with alcohols gives the corresponding 2,4-dichloroacetoacetates [65].

4-Chloroacetoacetates can be converted to the corresponding 4-alkoxyacetoacetates [66], useful starting materials for tetronic acid [67], [68]. Treatment of 2-chloro- and

2,4-dichloroacetoacetyl chlorides with water in the cold affords chloroacetone and 1,3-dichloroacetone, respectively [69].

6-Methyluracil [626-48-2] is produced by condensation of diketene with urea. It is used in the production of the analgesic mepirizole, the feed additive orotic acid, and 5-aminoorotic acid.

Diketene–acetone adduct [5394-63-8] can be considered as a stabilized form of diketene. It can be safely transported and may conveniently be used instead of diketene in many reactions. For example, it reacts with alcohols and amines to yield acetoacetates and acetoacetamides, respectively.

Diketene–acetone adduct generates unstable acetylketene together with acetone when heated >120 °C. Subsequent condensation with isocyanates, arylcyanates, or substituted cyanamides gives access to a wide range of 1,3-oxazine derivatives [70]. Furthermore, diketene–acetone adduct can also be used for the preparation of a variety of compounds that are not accessible from diketene [45].

Dehydroacetic Acid. Controlled dimerization of diketene in the presence of tertiary amine catalysts produces dehydroacetic acid [771-03-9], DHA, 3-acetyl-4-hydroxy-6-methyl-2-pyrone [71]. Many other bases are also useful dimerization catalysts, e.g., sodium acetate, potassium sorbate, and sodium alkoxides or sodium phenoxides.

On standing at room temperature, diketene is slowly converted to DHA, which is frequently observed as a byproduct in reactions of diketene. In most cases, DHA can be easily removed from the reaction mixture by bicarbonate extraction or by precipitation.

Dehydroacetic acid or its sodium salt is used as a food and cosmetic preservative, as a starting material for the veterinary drug clopidol [2971-90-6] (Coyden), and in the production of 4-hydroxy-6-methyl-2-pyrone.

Acetoacetalkylamides and Acetoacetarylamides. Aqueous ammonia and alkylamines react readily with diketene to form acetoacetamide and acetoacetalkylamides, respectively, whereas arylamines react more slowly to afford acetoacetanilides. Arylhydrazines yield 3-methyl-1-aryl-5-pyrazolones.

Acetoacetamide and acetoacetalkylamides are mainly used as aqueous solutions. The main uses for the alkylamides are in the production of insecticides such as monocrotophos, dicrotophos, phosphamidon, and oxamyl. N-(2-Hydroxyethyl)-3-oxobutanamide [24309-97-5] is used in the production of olaquindox, a growth stimulant for pigs.

Acetoacetarylamides are mainly used in the production of high quality organic pigments, particularly as azo coupling components, e.g., for a variety of yellow and orange pigments. The simplest compound of this series, acetoacetanilide [102-01-2], is also used in the production of the fungicides carboxin and methfuroxam, in addition to its use in the production of the pigment yellow 12.

β-Butyrolactone. Diketene undergoes catalytic hydrogenation to give *β*-butyrolactone, 4-methyl-2-oxethanone [*3068-88-0*] in ethyl acetate at 0 °C over a palladium catalyst [72]. *β*-Butyrolactone is used in the production of the analgesic bucetin.

3.5.2. Second Generation Diketene Derivatives

A number of second generation diketene derivatives (see Fig. 1) have already been mentioned in Section 3.5.1. A few are selectively described here.

3-Aminocrotonate esters, e.g., methyl 3-aminocrotonate [*14205-39-1*] (3-amino-2-butenoic acid, methyl ester) can be prepared by treatment of acetoacetates with aqueous ammonia [73]. They are mainly used in the production of a series of the dihydropyridine-type calcium antagonists.

2-Aminothiazol-4-yl Acetates. Diketene can be used in a number of ways for the preparation of substituted acetoacetamido side chains for *β*-lactam antibiotics, e.g., by conversion of the 4-chloroacetoacetamido side chain of 4-chloroacetoacetamidocephalosporanic acid with thiourea [74]. Alternatively, the aminothiazolylacetic acid side chains can be prepared separately for attachment to a variety of *β*-lactams [75] (e.g., semisynthetic cephalosporins of the third and fourth generation and of carbapenems [76]).

Acetoacetoxyethyl methacrylate [*21282-97-3*], AAEM, has a pendant acetoacetoxyethyl group which can modify the physical properties of acrylic polymers by decreasing solution viscosity and lowering the glass transition temperature. In addition, the acetoacetyl group contains an active methylene group and a carbonyl group, which can be used to modify or cross-link the polymer [77].

4. Higher Ketenes

4.1. Properties

Higher ketenes are divided into aldoketenes RCH=C=O and ketoketenes $R^1R^2C=C=O$ depending on whether one or two hydrogen atoms of ketene are substituted by an alkyl or aryl group. They are less important industrially than ketene or diketene.

Physical Properties. Methylketene is a gas at room temperature; most other aldo- or ketoketenes are yellow to reddish-brown liquids, some higher diarylketenes are

Table 1. Physical properties of higher ketenes

Name	CAS registry number	M_r	bp °C (kPa)	n_D^{20}
Methylketene	[6004-44-0]	56.06	−56 (101.3)	
Ethylketene	[20334-52-5]	70.09	36 (101.3)	
Phenylketene	[3496-32-0]	118.14		
Dimethylketene	[598-26-5]	70.09	34 (101.3) (mp 98 °C)	
Methylethylketene	[36854-53-2]	84.12	−26 (1.60)	
Diethylketene	[24264-08-2]	98.15	91 (101.3)	1.4126
Di-n-propylketene	[58844-38-5]	126.20	30 (1.47)	
Di-n-butylketene	[36638-69-4]	154.25	68 (1.40)	1.4308
Di-t-butylketene	[19824-34-1]	154.25	73 (6.0)	1.4370
Methylphenylketene	[3156-07-8]	132.16	74 (1.6) (mp −8 to −5 °C)	
Ethylphenylketene	[20452-67-9]	146.19	42 (0.04)	
Diphenylketene	[525-06-4]	194.23	119 (0.47)	

solids. Some physical properties of the most common aldo- and ketoketenes are given in Table 1.

Chemical Properties. The chemical behavior of the higher ketenes is essentially the same as that of ketene [1] although there are, however, some noticeable differences. Aldoketenes are extremely reactive, and it is very difficult to isolate the monomeric form. They are, therefore, generated and used in situ [1], [2].

Ketoketenes differ from ketene and aldoketenes in one important respect: they are oxidized by molecular oxygen even at temperatures well below 0 °C (− 80 °C has been reported for some aryl ketenes), forming hazardous peroxides that can explode at the slightest touch. Ketoketenes must, therefore, only be handled under an oxygen-free atmosphere. In all other reactions, including polymerization, ketoketenes are less reactive than aldoketenes or ketene itself: more forcing reaction conditions or a catalyst are often required. Ketoketenes, particularly diarylketenes, can be stored in oxygen- and catalyst-free solutions for longer periods, up to several months, due to their lower reactivity. Small amounts of aluminum trialkyl can be added to prevent peroxide formation [17].

4.2. Production and Uses

Various methods are available for the production of higher ketenes; the method used depends on the ketene [78].

Dialkylmalonic acids are useful starting materials for the preparation of dialkylketenes. When dimethylmalonic acid is heated with acetic anhydride, the intermediate cyclic acylal of dimethylketene and dimethylmalonic acid [595-46-0] can be isolated [79]. This compound gives dimethylketene on pyrolysis at 130 °C.

Diphenylketene is made by removal of hydrogen chloride from diphenylacetyl chloride with triethylamine [80]. This method is generally useful for the in situ preparation of ketenes, where the transient ketene can be captured by a ketenophile [81].

Long-chain monoalkylketenes (aldoketenes) and their lactone dimers, the alkylketene dimers, have been used for paper sizing for many years. They are prepared by removal of hydrogen chloride from long-chain fatty acid chlorides, e.g., stearic acid chloride. The intermediate highly reactive ketene dimerizes to the lactone dimer (*mp* ca. 50 °C) in the presence of acidic catalysts [82]:

$$2\ H_3C(CH_2)_{14}COCl$$
$$\downarrow$$
$$2\ H_3C(CH_2)_{13}HC=C=O$$
$$\downarrow$$

H₃C(CH₂)₁₃CH—(CH₂)₁₃CH₃ (β-lactone dimer)

Emulsions of these sizing agents form covalent bonds with the hydroxyl groups of cellulose below 100 °C, making fiber surfaces hydrophobic [83]–[85].

5. Toxicology and Occupational Health

Ketene is one of the most irritating gases to the respiratory tract, due to its great reactivity towards nucleophiles such as hydroxyl, amino, carboxyl, thiol, and phenolic groups, leading to acetylation of proteins. Its toxicity is considered to be of the same order of magnitude as that of phosgene [86]. Like the latter, it also causes latent damage to the respiratory tract (pulmonary edema) that may become acute only many hours after exposure.

Different, sometimes conflicting values, are found in the literature for inhalation toxicity [33], [86]–[88]. This could indicate a different sensitivity in different species, but is more likely to be due to the instability of ketene, which makes it difficult to maintain and measure a low concentration over a long time. The LCLo (mouse) is 23 ppm (30 min), 53 ppm (100 min) in the rat, rabbit, and guinea pig and 200 ppm (10 min) in the rat; other sources report an LC_{100} of 200 ppm (5 min) in the rat. No toxic effects were reported with a concentration of 1 ppm for 7 h/d during 55 days. No human casualties have been reported, although irritation of the respiratory tract and of the eye due to unknown concentrations of ketene among workers is known. After inhalation of even small quantities, a characteristic, long-lasting, unpleasant taste can be observed; smokers seem to be more sensitive to this phenomenon.

The LD_{50} of ketene (rat, oral) is 1300 mg/kg [88].

The TLV-TWA exposure limit for ketene is 0.5 ppm (0.9 mg/m^3), which is also the MAK value in the Federal Republic of Germany and in Switzerland. Regarding the maximal allowable instantaneous concentration, the TLV-STEL value is 1.5 ppm; the German MAK commission classifies ketene in category I with a limit twice the MAK (corresponding to 1 ppm) which is to be tolerated for no longer than 5 min, and no more than eight times a day [88]–[91]. No carcinogenic activity for ketene has been reported.

Diketene [86]–[88], [91], [92]. Although less acutely toxic than ketene, diketene is a strongly irritating agent and a powerful lachrimator. It is especially dangerous to the eye (in which it rapidly damages the corneal tissue) and to the respiratory tract; formation of pulmonary edema is possible for up to two days after exposure.

Symptoms of diketene exposure are a burning sensation in the eyes, nose, and throat, coughing, difficulty in respiration and ultimately loss of consciousness. A physician should be consulted immediately after inhalation of the vapor. Liquid diketene can be absorbed through the skin, where it causes itching and burning. No chronic toxic effects are known.

The acute oral toxicity (LD_{50}) is 560 mg/kg (rat) and 800 mg/kg (mouse). The reported LCLo is 20 000 ppm for an exposure of 1 h. The acute dermal toxicity LD_{50} in the rabbit has been given as 2.83 mL/kg and 6730 mg/kg.

No carcinogenic effects of diketene have been found on skin application in mice or subcutaneous injection and intragastric feeding in rats [93]–[95].

No permissible exposure limits have been set. Diketene has a pronounced "warning" effect due to its very low irritation threshold, and, when inhaled, it becomes unbearable before a lethal limit is reached.

Higher Ketenes. No toxicological data have been reported for higher ketenes, but it is prudent to consider them at least as hazardous and toxic as ketene itself.

6. References

General References

[1] *Houben-Weyl*, **VII/4**, 53–447.
[2] S. Patai (ed.): *The Chemistry of Ketenes, Allenes and Related Compounds (The Chemistry of Functional Groups Series)*, Wiley-Interscience, 1980.
[3] *Beilstein*, **H1** 724; **EI1** 376–377; **EII1** 779; **EIII1** 2942–2947; **EIV1** 3418–3420.
[4] *Beilstein*, **H7** 552; **EI7** 309; **EII7** 525; **EIII/IV17** 4297–4299; **EV17/9** 115–118.
[5] *Ullmann*, 4th ed., **14**, 181–189.

[6] *Kirk-Othmer*, **13**, 874–893.
[7] R. J. Clemens, *Chem. Rev.* **86** (1986) 241–318.
[8] B. Snider, *Chem. Rev.* **88** (1988) 793–811.

Specific References

[9] L. Katz, W. N. Lipscomb, *J. Org. Chem.* **17** (1952) 515.
[10] C. D. Hurd, A. S. Roe, *J. Am. Chem. Soc.* **61** (1939) 3355–3360. A. A. Ponomarev, Y. B. Isaev, *Z. Obsc. Khim.* **22** (1952) 652–654; *Chem. Abstr.* **47** (1953) 2695 a.
[11] R. E. Dunbar, G. C. White, *J. Org. Chem.* **23** (1958) 915–916.
[12] H. Wynberg, E. G. J. Staring, *J. Am. Chem. Soc.* **104** (1982) 166–168.
[13] J. C. Sheehan, E. J. Corey, Organic Reactions IX 395–399.
[14] H. Ulrich: *Cycloaddition Reactions of Heterocumulenes*, 9th ed., Academic Press, London–New York 1967.
[15] Lonza AG, CH 543 474, 1970 (F. Broussard).
[16] G. F. Pregaglia, M. Binaghi in N. M. B. Kaler (ed.): *Encyclopedia of Polymer Science and Technology*, "Ketene Polymers," 8th ed., Interscience, New York 1968, pp. 45–57.
[17] G. P. Pregaglia, M. Binaghi, *Macromol. Synth.* **3** (1968) 152–165.
[18] G. Sioli et al. *Chim. Ind.* (*Milan*) **53** (1971) 133–139.
[19] H. Stage, *Chem. Ztg.* **97** (1973) 67–73.
[20] Consortium für Elektrochem. Ind., DE 687 065, 1933 (J. Sixt).
[21] Consortium für Elektrochem. Ind., DE 734 349, 1934.
[22] Wacker, DE 1 079 623, 1963 (T. Altenschöpfer, H. Spes); DE 1 210 809, 1963; DE 1 203 248, 1964.
[23] S. Andreades, H. D. Carlson, *Org. Synth. Coll.* **5** (1973) 679–684.
[24] H. W. Coover, R. C. Hart, *Chem. Eng. Proc.* **78** (1982) 72–75; Halcon International, DE-OS 2 441 502, 1974 (C. Hewlett).
[25] Wacker, US 3 759 988, 1973 (G. Kunstle, H. Spes).
[26] Daicel, US 3 574 728, 1971 (I. Takasu, M. Higuchi, Y. Hijioka).
[27] *Kirk-Othmer*, 3rd ed., **1**, 337–338.
[28] Monsanto, US 3 758 569; US 3 758 571, 1973 (D. Bissing, V. Gash).
[29] H. Spes, *Chem. Ing. Tech.* **38** (1966) 955–962.
[30] B. H. Gwynn, E. F. Degering, *J. Am. Chem. Soc.* **64** (1942) 2216–2218. H. J. Hagemeyer, D. C. Hull, *Ind. Eng. Chem.* **41** (1949) 2920–2924. Wacker, DE 1 121 605, 1959 (T. Altenschoepfer, E. Enk, F. Knoerr, H. Spes). R. N. Lacey, *Adv. Org. Chem.* **2** (1960) 213–263. Hoffmann-La Roche, BE 621 067, 1962 (A. Sturzenegger).
[31] H. Wynberg, E. G. J. Staring, *PCT Int. Appl.* **WO 84/01 577** (1984); *NL Appl.* **82/4 070** (1982).
[32] Lonza AG, unpublished results.
[33] K. Verschueren: *Handbook of Environmental Data on Organic Chemicals*, Van Nostrand, New York 1977, pp. 258, 415.
[34] G. Hommel: *Handbuch der gefährlichen Güter*, 4th ed., Merkblatt 350, Springer Verlag, Berlin 1987.
[35] L. Tenud, M. Weilenmann, E. Dallwjk, *Helv. Chim. Acta* **60** (1977) 975–977.
[36] Lonza AG, CH 421 080, 1963 (H. Zima).
[37] Air Reduction Co. Inc., US 2 818 456, 1957 (R. T. Conley, T. F. Rutledge).
[38] T. Kato, T. Chiba, *Yuki Gosei Kagaku Kyokaishi* **39** (1981) 733–746; *Chem. Abstr.* **95** (1981) 168 405.

[39] T. Kato, *Acc. Chem. Res.* **7** (1974) 265.
[40] Wacker, US 2 763 664, 1956 (J. Sixt).
[41] C. L. Perrin, T. Arrhenius, *J. Am. Chem. Soc.* **100** (1978) 5249–5251.
[42] T. Kato, *Lect. Heterocycl. Chem.* **6** (1982) 105–119.
[43] J. M. Briody, D. P. N. Satchell, *J. Chem. Soc.* 1965, 3778–3785.
[44] Lonza AG, CH 642 611, 1979 (M. Gross); US 4 473 508, 1984 (M. Gross).
[45] R. J. Clemens: "The Chemistry of the Diketene Acetone Adduct," *Kodak Lab. Chem. Bull.* **55** (1984) no. 3.
[46] M. F. Carrol, A. R. Bader, *J. Am. Chem. Soc.* **75** (1953) 5400–5402.
[47] A. B. Steele, A. B. Boese, M. F. Dull, *J. Org. Chem.* **14** (1949) 460–469.
[48] K. Yoshida, Y. Yamashita, *Makromol. Chem.* **100** (1967) 177.
[49] W. R. Grace & Co., US 4 588 792, 1986 (Y. Okamoto, E. F. Hwang).
[50] Du Pont, US 2 585 537, 1952 (D. D. Coffman).
[51] Distillers Co. Ltd., DE 1 043 323, 1956 (R. N. Lacey). Consortium für Elektrochem. Ind., DE 700 218, 1937 (M. Mugdan, J. Sixt). Bayer, DE 832 440, 1949.
[52] Lonza AG, GB 852 865, 1960 (W. Moser, R. Perren). BASF, US 3 865 846, 1975 (G. Schulz, Y. Matthias, W. Kasper).
[53] Lonza AG, CH 423 754, 1967 (H. Keller).
[54] Celanese, US 3 759 955, 1973 (M. L. Jacobs, B. W. Higdon).
[55] Lonza AG, US 3 271 420, 1965 (H. Zima); GB 1 015 080, 1965 (H. Zima).
[56] US Army, US 3 733 413, 1973 (D. R. Spiner, R. Hoffmann). Allied, FR 1 579 873, 1969 (A. C. Pierce).
[57] Esso, GB 927 637, 1963.
[58] I.G. Farbenindustrie, GE 717 652, 1939 (F. J. Pohl, W. Schmidt); US 2 351 366, 1944 (F. Pohl, W. Schmidt).
[59] O. B. Boese Jr., *Ind. Eng. Chem.* **32** (1940) 16. Lonza AG, US 3 651 130, 1972 (O. Marti, W. Zimmerli, C. Zinstag, H. Keller); US 3 117 186, 1962; GB 903 613, 1962.
[60] Standard Oil, US 2 103 505, 1934 (P. J. Wiezerich, A. H. Gleason); US 2 228 452, 1941 (A. H. Gleason).
[61] *Kirk-Othmer*, 2nd ed., vol. **1**, 1963, pp. 153–159. Lonza AG, Technical Brochure, "Acetoacetates," 1987.
[62] C. D. Hurd, C. D. Kelso, *J. Am. Chem. Soc.* **62** (1940) 1548–1549.
[63] Lonza AG, EP 61 657, 1984 (L. Tenud, R. Miller, B. Jackson).
[64] Lonza AG, EP-A 28 709, 1981 (M. Gross); US 4 473 508, 1984 (M. Gross); Eastman Kodak, US 4 468 356, 1984 (D. E. van Sickle, G. C. Newland).
[65] Lonza AG, US 3 950 412, 1976 (K. J. Boosen).
[66] Lonza AG, EP 76 378, 1983 (C. Abächerli); EP 76 379, 1985 (E. Greth).
[67] Lonza AG, CH 649 294, 1981 (R. Miller, L. Tenud); CH 649 996, 1981 (R. Miller, L. Tenud).
[68] Lonza AG, CH 658 056, 1984 (T. Meul, L. Tenud, A. Huwiler).
[69] Carbide and Carbon Chemicals Co., US 2 209 683, 1941 (A. B. Boese, Jr.).
[70] G. Jäger, J. Wenzelburger, *Liebigs Ann. Chem.* 1976, 1689–1712.
[71] Distillers Co. Ltd., US 2 849 456, 1958 (S. J. Branch).
[72] Wacker, US 2 763 664, 1956 (J. Sixt).
[73] Armstrong Cork Co., US 4 046 803, 1977 (J. S. Heckles).
[74] Takeda Chem. Ind. Ltd., US 4 379 924, 1983 (S. Terao et al.); US 4 421 912, 1983 (I. Ninamida et al.).Fujisawa Pharm. Co. Ltd., US 4 254 260, 1976 (T. Takaya et al.).
[75] Lonza AG, US 4 391 979, 1983 (A. Huwiler, L. Tenud); EP-A 34 340, 1981.

[76] Lonza AG, Technical Brochure, "Aminothiazolylacetic acid derivatives – The new antibiotic side chains," 1986.
[77] Eastman Kodak Company, Technical Brochure, Publication No. N-319 (Oct. 1988).
[78] H. Eck, H. Spes, *Method. Chim.* **5** (1975) 493–510.
[79] H. Bestian, D. Günther, *Angew. Chem.* **75** (1963) 841–845.
[80] E. C. Taylor, A. McKillop, G. W. Hawks, *Org. Synth. Coll.* **6** (1988) 549–551.
[81] W. T. Brady, *Synthesis* 1971, 415–422.
[82] Deutsche Hydrierwerke, DE 885 834, 1937. *Chem. Eng. News* **33** (1955) 1018.
[83] C. A. Weisgerber, C. B. Hanford, *Tappi* **43** (1960) no. 12, 178 A.
[84] J. W. Dawis, W. H. Roberson, *Tappi* **39** (1956) no. 1, 21.
[85] D. H. Dumas, *TAPPI Papermakers Conf. [Proc.]* 1979, 67–72.
[86] H. A. Wooster, C. C. Luskbaugh, C. E. Redemann, *Ind. Hyg. Toxicol.* **29** (1947) 56–57.
[87] J. J. Gomer, *Zentralbl. Arbeitsmed. Arbeitsschutz* **1** (1951) 37–39, 58–63.
[88] Registry of Toxic Effects of Chemical Substances 1981–1982, 595 (Nr. OA7700000); 930 (Nr. RQ8225000), NIOSH Cincinnati, OH.
[89] DFG, *Maximale Arbeitsplatzkonzentration und biologische Arbeitsstofftoleranzwerte*, VCH Verlagsgesellschaft, Weinheim 1987.
[90] SUVA, *Zulässsige Werte am Arbeitsplatz*, Schweiz. Unfall Versicherungsanstalt, Luzern 1987.
[91] *Documentation of the Threshold Limit Values and Biological Exposure Index*, 5th ed., American Conference of Governmental Industrial Hygienists, Cincinnati, OH 1986.
[92] C. P. Carpenter, C. S. Weil, H. F. Smyth Jr., *Toxicol. Appl. Pharmacol.* **28** (1974) 313–319.
[93] B. L. Van Duuren et al., *J. Natl. Cancer Inst.* **31** (1963) 41–55.
[94] B. L. Van Duuren et al., *J. Natl. Cancer Inst.* **37** (1966) 825–838.
[95] B. L. Van Duuren et al., *J. Natl. Cancer Inst.* **39** (1967) 1213–1216, 1217–1228.

Ketones

Individual keywords: → *Acetone;* → *2-Butanone;* → *Cyclohexanol and Cyclohexanone.*

HARDO SIEGEL, BASF Aktiengesellschaft, Ludwigshafen, Federal Republic of Germany

MANFRED EGGERSDORFER, BASF Aktiengesellschaft, Ludwigshafen, Federal Republic of Germany

1.	Introduction	3084	5.	Unsaturated Ketones	3101
2.	Methyl Alkyl Ketones	3085	5.1.	3-Buten-2-one	3101
2.1.	3-Methyl-2-butanone	3085	5.2.	3-Methyl-3-buten-2-one	3103
2.2.	4-Methyl-2-pentanone	3087	5.3.	4-Methyl-4-penten-2-one	3104
2.3.	3-Methyl-2-pentanone	3089	5.4.	3,3,5-Trimethyl-2-cyclohexen-1-one	3105
2.4.	3,3-Dimethyl-2-butanone	3090	6.	Diketones	3106
2.5.	2-Heptanone	3091	6.1.	2,4-Pentanedione	3106
2.6.	5-Methyl-2-hexanone	3092	6.2.	2,3-Butanedione	3108
2.7.	2-Octanone	3093	6.3.	Higher α-Diketones	3109
2.8.	Higher Methyl Alkyl Ketones	3094	7.	Aromatic Ketones	3110
3.	Dialkyl Ketones	3095	7.1.	Methyl Phenyl Ketone	3110
3.1.	3-Pentanone	3095	7.2.	1-Phenyl-1-propanone	3111
3.2.	2,4-Dimethyl-3-pentanone	3096	7.3.	Diphenyl Ketone	3111
3.3.	2,6-Dimethyl-4-heptanone	3097	7.4.	4,4′-Diphenoxydiphenyl Ketone	3112
3.4.	3-Heptanone	3097	7.5.	1-Phenyl-2-propanone	3112
3.5.	5-Methyl-3-heptanone	3098	7.6.	2-Hydroxy-1,2-diphenylethanone	3113
4.	Cyclic Ketones	3099	8.	Toxicology and Occupational Health	3113
4.1.	Cyclopentanone	3099			
4.2.	Cycloheptanone	3100	9.	References	3115
4.3.	Higher Cyclic Ketones	3101			

1. Introduction

Ketones are organic compounds that contain a carbonyl group and two aliphatic or aromatic substituents.

$$R^1-\underset{\underset{O}{\|}}{C}-R^2$$

The substituents R^1 and R^2 may be saturated or unsaturated, linear or branched alkyl groups, or aromatic residues. They may be the same or different. Moreover, the alkyl or aryl substituents can also contain heteroatoms. More than one carbonyl group may be present in a particular molecule.

Chemical Properties. The alkyl and aryl groups of ketones act as electron donating groups toward the carbon atom of the carbonyl group. As a result, the reactivity of the carbonyl group is lower than that of, for example, aldehydes. This is also clear from a consideration of the limiting electronic structures with which ketones can be described:

$$R^1-\overset{\overset{:\ddot{O}:}{\|}}{C}-R^2 \rightleftharpoons R^1-\overset{\overset{:\ddot{O}:^-}{|}}{\underset{+}{C}}-R^2$$

In addition, the reactivity of the carbonyl group is influenced considerably by steric effects, e.g., branching at the α-carbon atom.

Industrially, the most important reactions of ketones are hydrogenation, reductive condensation, reductive amination, oxidation to peroxides, ketal formation, and condensation reactions. The acidic character of the α-hydrogen atoms flanking the carbonyl group enables numerous substitution and addition reactions to take place, e.g., the aldol reaction.

A typical property of ketones is the ability to enolize, this is particularly important for β-diketones:

$$R-\underset{\underset{O}{\|}}{C}-CH_2-R' \underset{+H^+}{\overset{-H^+}{\rightleftharpoons}} R-\underset{\underset{O}{\|}}{C}-\bar{C}H-R' \rightleftharpoons R-\underset{\underset{O^-}{|}}{C}=CH-R'$$

The carbon atom of the carbonyl group is sp^2 hybridized and so the substituents lie in the plane of the carbonyl group. Ketones are therefore prochiral, and enantioselective reactions at the carbonyl group can lead to chiral compounds.

Physical Properties. Aliphatic and cyclic ketones with ten carbon atoms or less are largely stable, colorless liquids that have a pleasant odor. They are miscible with organic solvents and are good solvents for paints, cellulose ethers, nitrocellulose, etc. Some are used as flavors and fragrances.

Production. As a result of the high stability of ketones, they can be synthesized in a multitude of ways, however only a few processes are practised industrially. Reactions such as condensation of methyl alkyl ketones with compounds containing acidic C–H groups or the ketonization of carboxylic acids or aldehydes are important. The functionalization of an existing carbon skeleton (e.g., by oxidation of alcohols, olefins, or alkanes) plays an important role.

Economic Aspects. Although solvent-poor or solvent-free systems have been developed and solvent recycling is becoming more widespread, ketones have retained their economic importance because they are increasingly replacing other solvents (e.g., chlorinated hydrocarbons) for aromatic compounds.

2. Methyl Alkyl Ketones

For acetone, see → Acetone. For 2-butanone (methyl ethyl ketone), see → 2-Butanone.

2.1. 3-Methyl-2-butanone

3-Methyl-2-butanone [563-80-4], methyl isopropyl ketone, $CH_3COCH(CH_3)_2$, $C_5H_{10}O$, M_r 86.13.

Properties. 3-Methyl-2-butanone is a colorless liquid with a characteristic ketone odor. Some physical properties are listed below:

mp	−92 °C
bp	94–95 °C
d_4^{20}	0.803
n_D^{20}	1.3890
Flash point	6 °C
Solubility in water (20 °C)	6.52 wt%

3-Methyl-2-butanone undergoes condensation reactions preferentially at the methyl group. The tertiary carbon atom is easily oxidized and the resulting hydroperoxide (*mp* 26–27 °C) can be produced by base-catalyzed air oxidation [1].

The use of condensation reactions of the methyl group and reactions of the carbonyl group for synthesis have been described.

Production. 3-Methyl-2-butanone is produced industrially by the following reactions:

1) Condensation of 2-butanone [78-93-3], formaldehyde, and hydrogen in a single step over palladium [2], or in two steps by condensation on a strongly acidic cation

exchanger, isolation of the 2-methyl-1-buten-3-one, followed by hydrogenation [3]:

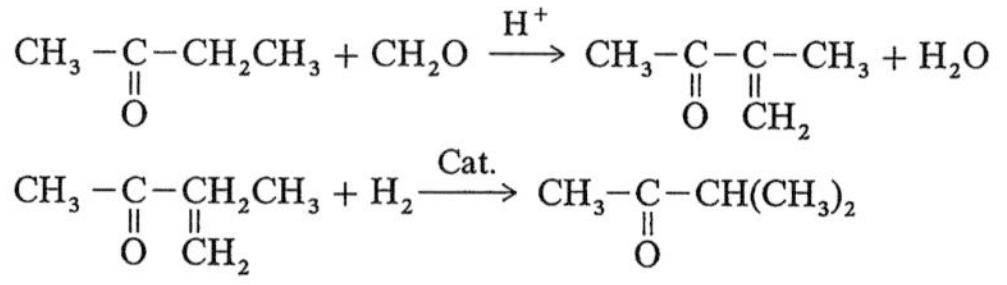

$$CH_3-\underset{O}{\underset{\|}{C}}-CH_2CH_3 + CH_2O \xrightarrow{H^+} CH_3-\underset{O}{\underset{\|}{C}}-\underset{CH_2}{\underset{\|}{C}}-CH_3 + H_2O$$

$$CH_3-\underset{O}{\underset{\|}{C}}-\underset{CH_2}{\underset{\|}{C}}CH_3 + H_2 \xrightarrow{Cat.} CH_3-\underset{O}{\underset{\|}{C}}-CH(CH_3)_2$$

2) By reaction of isobutyraldehyde with acetaldehyde in the gas phase over manganese oxide – aluminum oxide at 450 °C [4]:

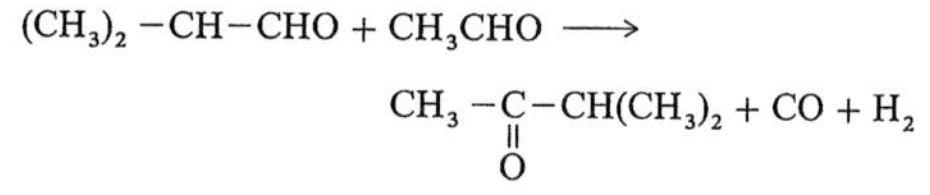

$$(CH_3)_2-CH-CHO + CH_3CHO \longrightarrow$$
$$CH_3-\underset{O}{\underset{\|}{C}}-CH(CH_3)_2 + CO + H_2$$

Isobutyric acid and acetic acid may be used in place of isobutyraldehyde and acetaldehyde [5].

3) By addition of water to isoprene at ca. 200 °C in the presence of phosphoric acid on silica [6]:

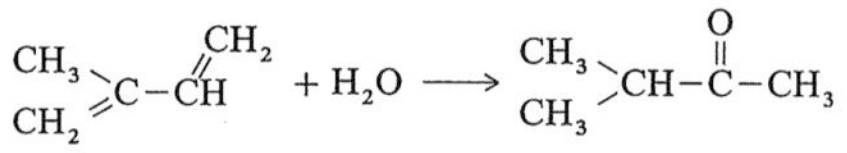

$$\underset{CH_2}{\overset{CH_3}{>}}C-\overset{CH_2}{\underset{\|}{CH}} + H_2O \longrightarrow \underset{CH_3}{\overset{CH_3}{>}}CH-\overset{O}{\underset{\|}{C}}-CH_3$$

4) By oxidation of 2-methyl-2-butene with oxygen [7] or hydroperoxides in the presence of molybdenum naphthenate at 150 °C [8]:

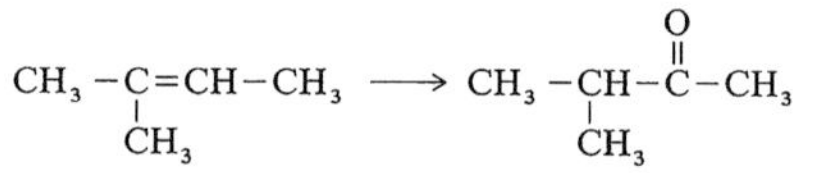

$$CH_3-\underset{CH_3}{\underset{|}{C}}=CH-CH_3 \longrightarrow CH_3-\underset{CH_3}{\underset{|}{CH}}-\overset{O}{\underset{\|}{C}}-CH_3$$

Storage and Transportation. 3-Methyl-2-butanone is available with a purity of >98 wt%. The following regulations apply to storage and transportation. Labelling according to the Gefahrstoffverordnung (GefStoffV, Federal Republic of Germany) [9]: F; R phrase: 11; S phrases: 9–16–33. Hazard codes for transportation are as follows [10]–[12]:

IMDG Code: class 3.2
RID/ADR: class 3, number 36
CFR 49: 172.102 flammable liquid
VbF: A 1
UN number: 2397
EEC number: 5788

Containers of V2A steel are used for transportation and storage.

Uses. 3-Methyl-2-butanone is employed predominantly as an intermediate for the production of pharmaceuticals, herbicides [13], and dye precursors [14]. It is also used in the synthesis of rubber auxiliaries and for the selective extraction of rare earth elements [15].

2.2. 4-Methyl-2-pentanone

4-Methyl-2-pentanone [108-10-1], methyl isobutyl ketone, MIBK, $CH_3COCH_2CH(CH_3)_2$, $C_6H_{12}O$, M_r 100.16.

Properties. 4-Methyl-2-pentanone is a colorless liquid with a less pronounced ketone-like odor. Some physical properties are given below:

mp	−80 °C
bp	116 °C
d_4^{20}	0.8004
n_D^{20}	1.3959
Enthalpy of evaporation	36.15 kJ/mol
Specific heat (20 °C)	1922 kJ/g
Enthalpy of combustion	3740 kJ/mol
Flash point (DIN 51755)	14 °C
Ignition temperature	475 °C
Ignition class (VED)	G1
Explosion class (VED)	1
Solubility at 25 °C	
4-Methyl-2-pentanone in water	1.7 wt %
Water in 4-methyl-2-pentanone	1.9 wt %

4-Methyl-2-pentanone is poorly soluble in water but is miscible with common organic solvents. It forms an azeotrope with water and with a large number of solvents. For example, the azeotrope with water (bp 87.9 °C) contains 75.7 wt% 4-methyl-2-pentanone, and that with n-butanol (bp 114.4 °C) 70 wt% 4-methyl-2-pentanone.

4-Methyl-2-pentanone undergoes autoxidation to form a peroxide:

$$CH_3-\underset{\underset{O}{\|}}{C}-CH_2-\underset{\underset{CH_3}{\diagdown}}{\overset{\overset{CH_3}{\diagup}}{CH}} \longrightarrow CH_3-\underset{\underset{O}{\|}}{C}-CH_2-\underset{\underset{OOH}{\diagdown}}{\overset{\overset{CH_3}{\diagup}}{C}}-CH_3$$

After evaporation of 4-methyl-2-pentanone – water mixtures in the presence of air, a dangerous, increasing concentration of peroxide in the aqueous phase has been reported [16]. The peroxide can also be obtained directly by oxidation with 50% hydrogen peroxide in the presence of acids [17].

In condensation reactions with carbonyl compounds such as acetone, the α-methyl group usually reacts. Only formaldehyde is able to undergo condensation reactions with the α-methylene group [18]. The usual industrial reactions such as hydrogenation and reductive amination may be carried out with the keto group.

Production. 4-Methyl-2-pentanone can be produced industrially from acetone or isopropyl alcohol according to the following four processes:

1) In three steps from acetone via diacetone alcohol and mesityl oxide, with subsequent hydrogenation.
2) In two steps from acetone via mesityl oxide.
3) In one step from acetone and hydrogen.
4) As a byproduct in the dehydrogenation of isopropyl alcohol to acetone.

In the *three-step process* acetone undergoes alkali-catalyzed condensation in the first step to form diacetone alcohol. Dehydration of the latter to mesityl oxide takes place in the liquid phase at 90–130 °C over an acid catalyst, such as phosphoric or sulfuric acid, with high selectivity [19]. Selective hydrogenation of the carbon–carbon double bond in mesityl oxide to give 4-methyl-2-pentanone may be carried out in both the liquid phase and the gas phase with various catalysts, preferably palladium [20]. The selectivity is excellent.

$$\begin{array}{c}CH_3\\ \end{array}\!\!\!\!\!\!>\!C=O \xrightarrow{OH^-} \begin{array}{c}CH_3\\ \end{array}\!\!\!\!\!\!>\!\!\underset{CH_3}{\overset{OH}{\underset{|}{C}}}\!-CH_2-\overset{O}{\overset{\|}{C}}-CH_3 \xrightarrow{H^+}$$

$$\begin{array}{c}CH_3\\ \end{array}\!\!\!\!\!\!>\!C=CH-\overset{O}{\overset{\|}{C}}-CH_3 \xrightarrow{H_2;\ Cat.} \begin{array}{c}CH_3\\ \end{array}\!\!\!\!\!\!>\!\!\underset{CH_3}{C}-CH_2-\overset{O}{\overset{\|}{C}}-CH_3$$

Dehydration and hydrogenation can also be combined into a single process step.

The *two-step process* is less important industrially. Acetone is converted in a single step to mesityl oxide, which is then hydrogenated to 4-methyl-2-pentanone in a second step. Catalysts such as copper chromite [21] or zirconium phosphate [22] are used for the condensation, and palladium on aluminum oxide for the hydrogenation. 4-Methyl-2-pentanol [*108-11-2*] is formed as a byproduct.

For the *single-step production* of 4-methyl-2-pentanone from acetone and hydrogen a combination of a cation exchanger with palladium is generally used as catalyst [23]. The acetone is fed with hydrogen over the palladium-charged catalyst at 130 °C and 0.5–5.0 MPa. A selectivity of > 95 % at a conversion of < 50 % is thereby attained. As a result of the increasing production of acetone by dehydrogenation of isopropyl alcohol, the production of 4-methyl-2-pentanone as a byproduct in this process is becoming more important [24] (see also → Acetone).

Storage and Transportation. 4-Methyl-2-pentanone is available with a purity of > 99 %. Typical specifications are given below:

Content	> 99 wt %
Water content (ASTM D 1364)	0.1 % max.
Hazen color no. (APHA, DIN 53 409)	10 max.
Nonvolatile residue (ASTM D 1353)	0.002 wt % max.
Boiling range (DIN 51 751)	114–117 °C
Acid content (as acetic acid)	0.002 wt % max.

Conditions that could lead to autoxidation (presence of air, light, heat, or heavy metals) or condensation (strongly acidic or basic media) should be avoided during storage. Containers of V2A steel are recommended for both storage and transportation.

Transportation is governed by the following regulations: Labelling according to GefStoffV [9]: F; R phrase: 11; S phrases: 9–16–23–33. Hazard codes for transportation are as follows [10]:

IMDG Code: class 3.2
RID/ADR: class 3, number 3 b
CFR 49: 172.101 flammable liquid
UN number: 1245

Uses. The major application of 4-methyl-2-pentanone is as a solvent for vinyl, epoxy, and acrylic resins, for natural resins, and for nitrocellulose. A further use is as a solvent for dyes in the printing industry. 4-Methyl-2-pentanone is also a versatile extracting agent, e.g., for the production of antibiotics, or the removal of paraffins from mineral oil for the production of lubricating oils. The importance of 4-methyl-2-pentanone as an intermediate for synthesis is relatively low. The most important product is 4-methyl-2-pentanol, obtained by hydrogenation of the ketone. 4-Methyl-2-pentanone peroxide has some importance as a polymerization initiator for ethylene [25] and for the hardening of unsaturated polyester resins [26].

Economic Aspects. In terms of production volume, 4-methyl-2-pentanone lies behind acetone and 2-butanone as the most important aliphatic ketone. In 1987 production in Europe was 70 000 t, in the United States 70 000 t, and in Japan 40 000 t. The total produced amounted to 180 000 t.

2.3. 3-Methyl-2-pentanone

3-Methyl-2-pentanone [565-61-7], methyl *sec*-butyl ketone, $CH_3COCH(CH_3)CH_2CH_3$, $C_6H_{12}O$, M_r 100.16.

Properties. 3-Methyl-2-pentanone is a colorless liquid with an odor reminiscent of peppermint. Some physical properties are listed below:

mp	−83 °C
bp	116 °C
d_4^{20}	0.8130
n_D^{20}	1.4012
Flash point	12 °C
Solubility in water (20 °C)	2.26 wt%

Production. 2-Butanone undergoes base-catalyzed aldolization with acetaldehyde to form 4-hydroxy-3-methylpentan-2-one [27]. Acid-catalyzed dehydration to 3-methyl-3-penten-2-one [28] is followed by hydrogenation over a palladium catalyst to give 3-methyl-2-pentanone.

$$CH_3-\underset{\underset{O}{\|}}{C}-CH_2-CH_3 + CH_3-CHO \longrightarrow CH_3-\underset{\underset{O}{\|}}{C}-\underset{\underset{HO}{|}}{CH}-CH_3$$
$$\underset{CH-CH_3}{|}$$

$$CH_3-\underset{\underset{O}{\|}}{C}-\underset{\underset{HO}{|}}{CH}-CH_3 \xrightarrow{H^+} CH_3-\underset{\underset{O}{\|}}{C}-\underset{\underset{HC-CH_3}{\|}}{C}-CH_3 + H_2O$$
$$\phantom{CH_3-C-CH-CH_3 \xrightarrow{H^+} CH_3}\underset{CH-CH_3}{|}$$

$$CH_3-\underset{\underset{O}{\|}}{C}-\underset{\underset{HC-CH_3}{\|}}{C}-CH_3 \xrightarrow[\text{Cat.}]{H_2} CH_3-\underset{\underset{O}{\|}}{C}-\underset{\underset{CH_2-CH_3}{|}}{CH}-CH_3$$

Uses. 3-Methyl-2-pentanone is used as a solvent. Its use as an intermediate for various syntheses has been reported, but the industrial importance is relatively low [29].

2.4. 3,3-Dimethyl-2-butanone

3,3-Dimethyl-2-butanone [75-97-8], *tert*-butyl methyl ketone, pinacolone, $CH_3COC(CH_3)_3$, $C_6H_{12}O$, M_r 100.16.

Properties. 3,3-Dimethyl-2-butanone is a colorless liquid with a light peppermint- or camphor-like odor.

mp	−52.5 °C
bp	106.2 °C
d_4^{20}	0.801
n_D^{20}	1.3964
Flash point	23 °C
Solubility in water (15 °C)	2.44 wt%

3,3-Dimethyl-2-butanone is stable toward autoxidation; under standard conditions it does not form hydroperoxides.

Only the α-methyl group can undergo condensation reactions. At the carbonyl group the usual reactions (hydrogenation, reductive amination) can take place.

Production. 3,3-Dimethyl-2-butanone can be produced by the following routes:

1) Hydrolysis of 4,4,5-trimethyl-1,3-dioxane, the product of the Prins reaction of isoprene with formaldehyde [30]:

$$\text{H}_3\text{C}\!\!\underset{\text{H}_2\text{C}}{\overset{}{\diagup}}\!\!\text{C}\!-\!\overset{\text{CH}_2}{\text{CH}} + 2\,\text{CH}_2\text{O} \xrightarrow{\text{H}^+} \text{H}_2\text{C}\underset{\text{O}\diagdown\;\diagup\text{O}}{\overset{\text{CH}_3\;\;\text{CH}_3}{\overset{|\;\;\;\;\diagup}{\text{CH}\!-\!\text{C}\!-\!\text{CH}_3}}}$$

$$\xrightarrow[\text{H}_2\text{O}]{\text{H}^+} \text{CH}_3\!-\!\underset{\overset{\|}{\text{O}}}{\text{C}}\!-\!\underset{\overset{|}{\text{CH}_3}}{\overset{\overset{\text{CH}_3}{|}}{\text{C}}}\!-\!\text{CH}_3 + \text{CH}_2\text{O}$$

2) The reductive dimerization of acetone under the influence of an acid catalyst with simultaneous rearrangement (the pinacol rearrangement) [31]:

$$2\,(\text{CH}_3)_2\text{CO} + \text{H}_2 \longrightarrow \begin{array}{c}(\text{CH}_3)_2\!-\!\text{C}\!-\!\text{OH}\\|\\(\text{CH}_3)_2\!-\!\text{C}\!-\!\text{OH}\end{array} \xrightarrow{\text{H}^+}$$

$$\text{CH}_3\!-\!\underset{\overset{\|}{\text{O}}}{\text{C}}\!-\!\underset{\overset{|}{\text{CH}_3}}{\overset{\overset{\text{CH}_3}{|}}{\text{C}}}\!-\!\text{CH}_3 + \text{H}_2\text{O}$$

3) Ketonization of pivalic acid with acetic acid or acetone over thorium, zirconium, or cerium catalysts [32]:

$$\text{CH}_3\!-\!\underset{\overset{|}{\text{CH}_3}}{\overset{\overset{\text{CH}_3}{|}}{\text{C}}}\!-\!\text{COOH} + \text{HOOCCH}_3 \xrightarrow{\text{Cat.}} \text{CH}_3\!-\!\underset{\overset{|}{\text{CH}_3}}{\overset{\overset{\text{CH}_3}{|}}{\text{C}}}\!-\!\overset{\overset{\text{O}}{\|}}{\text{C}}\!-\!\text{CH}_3$$

4) Oxidation of 2,3-dimethyl-2-butene with hydrogen peroxide [33]:

$$\underset{\text{CH}_3}{\overset{\text{CH}_3}{\diagdown}}\!\!\text{C}\!=\!\text{C}\!\!\underset{\text{CH}_3}{\overset{\text{CH}_3}{\diagup}} \xrightarrow[\text{H}_2\text{O},\,\text{H}^+]{\text{H}_2\text{O}_2} \text{CH}_3\!-\!\underset{\overset{|}{\text{CH}_3}}{\overset{\overset{\text{CH}_3}{|}}{\text{C}}}\!-\!\overset{\overset{\text{O}}{\|}}{\text{C}}\!-\!\text{CH}_3$$

Uses. 3,3-Dimethyl-2-butanone is produced in large amounts for fungicides, e.g., Triademefon (Bayer) [34]. An important intermediate for this type of fungicide is triazolylpinacolone [58905-32-1]. A further use is the synthesis of herbicides, e.g., Metribuzin (Bayer) [35].

2.5. 2-Heptanone

2-Heptanone [110-43-0], methyl pentyl ketone, methyl amyl ketone, $CH_3COCH_2(CH_2)_3CH_3$, $C_7H_{14}O$, M_r 114.18.

Properties. 2-Heptanone is a colorless liquid with a fruity odor.

mp	−35 °C
bp	149–150 °C
d_4^{20}	0.820
n_D^{20}	1.4085
Flash point	47 °C

Production. 2-Heptanone is produced industrially by reductive condensation of acetone with butyraldehyde in one or two steps [36]:

$$CH_3-\underset{\underset{O}{\|}}{C}-CH_3 + OHCCH_2CH_2CH_3 \xrightarrow{H_2, \text{ Cat.}}$$

$$CH_3-\underset{\underset{O}{\|}}{C}-CH_2CH_2CH_2CH_2CH_3 + H_2O$$

Storage and Transportation. Labelling according to GefStoffV (class II d) [9]: X_n; R phrases: 10–22; S phrase: 23. Hazard classifications are as follows:

IMDG Code: class 3.3
RID/ADR: class 3, number 31 c [10]
CFR 49: 172.101/102 flammable liquid
UN number: 1110

Uses. 2-Heptanone is used as a solvent and as a synthetic building block.

2.6. 5-Methyl-2-hexanone

5-Methyl-2-hexanone [*110-12-3*], methyl isoamyl ketone, $CH_3COCH_2CH_2CH(CH_3)_2$, $C_7H_{14}O$, M_r 114.18.

Properties. 5-Methyl-2-hexanone is a colorless liquid with an odor similar to that of amyl acetate.

mp	–41 °C
bp	145 °C
d_4^{20}	0.888
n_D^{20}	1.4062
Flash point	41 °C

Production. 5-Methyl-2-hexanone is produced by condensation of acetone with isobutyraldehyde [37]. This reaction may be carried out in one or two steps, in the liquid or in the gas phase [38]:

$$CH_3-\underset{\underset{O}{\|}}{C}-CH_3 + \underset{CH_3}{\overset{CH_3}{>}}CH-CHO \longrightarrow$$

$$CH_3-\underset{\underset{O}{\|}}{C}-CH_2-\underset{\underset{OH}{|}}{CH}-CH\underset{CH_3}{\overset{CH_3}{<}} \xrightarrow{-H_2O}$$

$$CH_3-\underset{\underset{O}{\|}}{C}-CH=CH-CH\underset{CH_3}{\overset{CH_3}{<}} \xrightarrow{H_2, \text{Cat.}}$$

$$CH_3-\underset{\underset{O}{\|}}{C}-CH_2-CH_2-CH\underset{CH_3}{\overset{CH_3}{<}}$$

Storage and Transportation. Labelling according to GefStoffV [9]: R phrase: 10; S phrase: 23. Hazard codes for transportation are as follows:

IMDG Code: class 3.3
RID/ADR: class 3, number 31 c
CFR 49: 172.102 flammable liquid
UN number: 2302

Uses. 5-Methyl-2-hexanone is used as a solvent for the production of high-solids coatings, because it possesses a high solubility for alkyd resins, acrylic resins, cellulose esters, cellulose nitrate, and vinylic copolymers.

2.7. 2-Octanone

2-Octanone [*111-13-7*], methyl hexyl ketone, $CH_3COCH_2(CH_2)_4CH_3$, $C_8H_{16}O$, M_r 128.21.

Properties. 2-Octanone is a colorless liquid with an odor reminiscent of apples and a taste similar to that of camphor.

mp	−16 °C
bp	173 °C
d_4^{20}	0.819
n_D^{20}	1.4150
Flash point	62 °C

Production. 2-Octanone is produced industrially by two methods:

Table 1. Physical properties of higher linear methyl alkyl ketones

Total number of carbon atoms	M_r	mp, °C	bp, °C (kPa)	Flash point, °C
C_9	142.2	−8 to −9	194 (101.3)	160
C_{10}	156.3	2−3	210 (101.3)	
C_{11}	170.3	11−12	223 (101.3)	192
C_{12}	184.3	20−21	144 (14.5)	
C_{13}	198.3	27−28	120 (6.5)	225
C_{16}	240.4	43	165 (16)	
C_{17}	254.4	45−46	156 (4)	248
C_{18}	268.5	52	251−252 (133)	

1) Reductive condensation of acetone with pentanal [39]:

$$CH_3-\underset{\underset{O}{\|}}{C}-CH_3 + OHCCH_2CH_2CH_2CH_3 \xrightarrow{H_2,\ Cat.}$$

$$CH_3-\underset{\underset{O}{\|}}{C}-CH_2(CH_2)_4CH_3 + H_2O$$

2) Oxidation of 1-octene with hydrogen peroxide in the presence of a catalyst containing rhenium oxide or a special palladium–copper catalyst in the presence of nitriles [40]:

$$CH_2=CH(CH_2)_5CH_3 \xrightarrow{H_2O_2,\ Cat.} CH_3-\underset{\underset{O}{\|}}{C}-CH_2(CH_2)_4CH_3$$

Uses. 2-Octanone is used as an intermediate and in fragrance compositions.

2.8. Higher Methyl Alkyl Ketones

Properties. The higher straight-chain methyl alkyl ketones (C_9–C_{18}) are colorless liquids or solids, with characteristic odors. The melting point of methyl alkyl ketones with 12 or more carbon atoms lies at room temperature or above. The most important physical properties are listed in Table 1.

Production. Higher methyl *n*-alkyl ketones can be prepared by the following processes:

1) Catalytic gas-phase ketonization of acetic acid with fatty acids [5]:

$$CH_3COOH + RCOOH \longrightarrow CH_3COR + CO_2 + H_2O$$

2) Catalytic gas-phase condensation of acetaldehyde with an *n*-aldehyde [41], [42]:

$$CH_3CHO + RCHO \longrightarrow CH_3COR + CO + H_2$$

3) Catalytic dehydrogenation of secondary alcohols [43]:

$$CH_3-\underset{OH}{\underset{|}{CH}}-CH_2R \xrightarrow{-H_2} CH_3-\underset{O}{\underset{\|}{C}}-CH_2R$$

4) Direct catalytic oxidation of α-olefins [40], [44]:

$$CH_2=CH-CH_2R \xrightarrow[Cat.]{[O_2]} CH_3-\underset{O}{\underset{\|}{C}}-CH_2R$$

5) Oxidation of α-olefins with organic hydroperoxides in the presence of palladium complex catalysts [45]:

$$CH_2=CH(CH_2)_9CH_3 + (CH_3)_3COOH \longrightarrow CH_3CO(CH_2)_9CH_3 + (CH_3)_3COH$$

Uses. Higher methyl alkyl ketones are used as nonvolatile solvents for resins and rubber formulations, and also as plasticizers. Condensation products with hexahydropyrimidines [46] or 1,4-phenylenediamine [47] are used in the production of antioxidants and vulcanization accelerators.

3. Dialkyl Ketones

3.1. 3-Pentanone

3-Pentanone [*96-22-0*], diethyl ketone, $CH_3CH_2COCH_2CH_3$, $C_5H_{10}O$, M_r 86.13.

Properties. 3-Pentanone is only slightly soluble in water, but miscible with organic solvents.

mp	−40 °C
bp	102 °C
d_4^{20}	0.814
n_D^{20}	1.3924
Flash point	12 °C
Ignition temperature	445 °C

Production. 3-Pentanone can be produced by catalytic ketonization of propionic acid over a thorium oxide or zirconium oxide catalyst at 350–380 °C [48]. Hydroformylation of ethylene in the presence of cobalt carbonyl complexes at 100 °C is a further route [49]:

$$2\,CH_2=CH_2 + CO + H_2 \xrightarrow{\text{Cat.}} CH_3CH_2-\underset{\underset{O}{\|}}{C}-CH_2CH_3$$

Storage and Transportation. The following regulations apply: Labelling according to GefStoffV [9]: F; R phrase: 11; S phrases: 9–16–33. Hazard classifications for transportation are as follows [10]–[12]:

IMDG Code: class 3.2
RID/ADR: class 3, number 3 b
CFR 49: 172.101 flammable liquid
VbF: A1
UN number: 1156
EEC number: 5789

Uses. 3-Pentanone is used as a solvent for paints and as a starting material for organic syntheses, e.g., for the production of trimethylphenol, a precursor of vitamin E. A further use is in fragrances.

3.2. 2,4-Dimethyl-3-pentanone

2,4-Dimethyl-3-pentanone [565-80-0], diisopropyl ketone, $(CH_3)_2CHCOCH(CH_3)_2$, $C_7H_{14}O$, M_r 114.19.

Properties. 2,4-Dimethyl-3-pentanone is a colorless liquid with a characteristic odor similar to that of camphor; it is insoluble in water but soluble in organic solvents.

mp	−69 °C
bp	125 °C
d_4^{20}	0.806
n_D^{20}	1.3990
Flash point	15 °C

Production. 2,4-Dimethyl-3-pentanone is preferably obtained by ketonization of isobutyric acid over a thorium oxide or zirconium oxide catalyst at 430 °C [50].

Storage and Transportation. Labelling according to GefStoffV [9]: F; R phrase: 11; S phrases: 16–23. Hazard classifications for transportation are as follows [10]–[12]:

RID/ADR: class 3, number 3 b
VbF: A1
EEC number: 5558

Uses. 2,4-Dimethyl-3-pentanone is used as a solvent and as an intermediate.

3.3. 2,6-Dimethyl-4-heptanone

2,6-Dimethyl-4-heptanone [*108-83-8*], diisobutyl ketone, $(CH_3)_2CHCH_2COCH_2CH(CH_3)_2$, $C_9H_{18}O$, M_r 142.24.

Properties. 2,6-Dimethyl-4-heptanone is a colorless liquid with a mild ketone-like odor.

mp	−49 °C
bp	169 °C
d_4^{20}	0.808
n_D^{20}	1.4114
Flash point	60 °C
Solubility (20 °C)	
2,6-Dimethyl-4-heptanone in water	0.06 wt %
Water in 2,6-dimethyl-4-heptanone	0.45 wt %

Production. This ketone can be prepared by the dehydrogenative condensation of isopropyl alcohol with 4-methyl-2-pentanone as byproduct [51], [52]. A mixture of acetone and isopropyl alcohol may be used instead of the latter alone [53]. 2,6-Dimethyl-4-heptanone can also be produced by the reductive condensation of acetone with 4-methyl-2-pentanone [54]:

$$(CH_3)_2CH-CH_2-CO-CH_3 + CH_3-CO-CH_3 \xrightarrow{H_2, Cat.} (CH_3)_2CH-CH_2-CO-CH_2-CH(CH_3)_2 + H_2O$$

Storage and Transportation. Labelling according to GefStoffV [9]: X_i; R phrases: 10 – 37; S phrase: 24. Hazard classifications for transportation are as follows:

IMDG Code: class 3.3
RID/ADR: class 3, number 31 c [10]
CFR 49: 172.101 flammable liquid
UN number: 1157

Uses. 2,6-Dimethyl-4-heptanone is used mainly as a solvent for paints, dyes, and adhesives, and as an extraction agent and solvent for recrystallization of organic compounds.

3.4. 3-Heptanone

3-Heptanone [*106-35-4*], butyl ethyl ketone, $CH_3CH_2COCH_2CH_2CH_2CH_3$, $C_7H_{14}O$, M_r 114.19.

Properties. 3-Heptanone is a colorless liquid with a green odor.

mp	−39 °C
bp	146 – 149 °C
d_4^{20}	0.818
n_D^{20}	1.4085
Flash point	41 °C

Production. 3-Heptanone is produced industrially by reductive condensation of propanal with 2-butanone [54]:

$$CH_3CH_2-\underset{\underset{O}{\|}}{C}-CH_3 + OHC-CH_2CH_3 \xrightarrow[\text{2) Cat., H}_2]{\text{1) KOH}}$$

$$CH_3CH_2-\underset{\underset{O}{\|}}{C}-CH_2CH_2CH_2CH_3 + H_2O$$

Storage and Transportation. Labelling according to GefStoffV [9]: X_n; R phrases: 10–22; S phrase: 23. Hazard codes for transportation are as follows:

IMDG Code: class 3.3
RID/ADR: class 3, number 31 c [10]
CFR 49: 172.101 flammable liquid
UN number: 1993 (nos)

Uses. 3-Heptanone is used as a perfume and fragrance, as a solvent for cellulose resins, nitrocellulose resins, and vinyl resins, and as a synthetic building block.

3.5. 5-Methyl-3-heptanone

5-Methyl-3-heptanone [*541-85-5*], amyl ethyl ketone, $CH_3CH_2COCH_2CH(CH_3)CH_2CH_3$, $C_8H_{16}O$, M_r 128.22.

Properties. 5-Methyl-3-heptanone is a colorless liquid with low solubility in water (0.3 wt%); water is poorly soluble in the ketone (0.9 wt%). 5-Methyl-3-heptanone is highly soluble in common organic solvents.

mp	<−70 °C
bp	157 – 162 °C
d_4^{20}	0.823
n_D^{20}	1.4142
Flash point	43 °C

Production. 5-Methyl-3-heptanone is obtained, together with 2-butanone, by the gas-phase dehydrogenation of *sec*-butyl alcohol. The industrial product generally contains ca. 4% of the isomeric 3,4-dimethyl-2-hexanone [*19550-10-8*].

$$CH_3-CH-CH_2CH_3 \xrightarrow{Cat.} CH_3-C-CH_2CH_3 + H_2$$
$$OH O$$

$$2\,CH_3-C-CH_2CH_3 \xrightarrow[+H_2]{-H_2O}$$
$$O$$

$$CH_3CH_2-C-CH_2CHCH_2CH_3$$
$$O CH_3$$

$$CH_3-C-CH-CHCH_2CH_3$$
$$O CH_3 CH_3$$

To increase the yield of 5-methyl-3-heptanone, the gas-phase hydrogenation of *sec*-butyl alcohol is carried out in the presence of noble metals on zinc oxide, thorium oxide, or titanium oxide supports [55]. Such catalysts promote the self-condensation of 2-butanone. 5-Methyl-3-heptanone is easily separated from 2-butanone by distillation because of the difference in boiling points.

5-Methyl-3-heptanone can also be obtained by base-catalyzed dimerization of 2-butanone, dehydration of the resulting 5-methyl-5-hydroxy-3-heptanone, and hydrogenation of the ensuing 5-methyl-4-hepten-3-one [56].

Storage and Transportation. Labelling according to GefStoffV [9]: X_i; R phrases: 10 – 36/37; S phrase: 23. Hazard codes for transportation are as follows:

IMDG Code: class 3.3
RID/ADR: class 3, number 31 c [10]
CFR 49: 172.102 flammable liquid
UN number: 2271

Uses. 5-Methyl-3-heptanone is used predominantly as a high-boiling solvent, e.g., for varnish resins.

4. Cyclic Ketones

For cyclohexanone, see → Cyclohexanol and Cyclohexanone.

4.1. Cyclopentanone

Cyclopentanone [*120-92-3*], C_5H_8O, M_r 84.12.

Properties. Cyclopentanone is a colorless liquid with a pleasant, slightly pepperminty odor. It is soluble in water and miscible with common organic solvents.

mp	−51 °C
bp	130–131 °C
d_4^{20}	0.951
n_D^{20}	1.4359
Flash point	30 °C

Production. Cyclopentanone is produced by decarboxylative cyclization of adipic acid [57]. This reaction may be carried out in the liquid phase at 350 °C over zeolites with adipic acid esters [58]. A further production route is oxidation of cyclopentene with oxygen in the presence of palladium(II) chloride or copper(II) chloride as catalyst [59].

Storage and Transportation. Labelling according to GefStoffV [9]: X_i; R phrases: 10–36/38; S phrase: 23. Hazard codes for transportation are as follows:

IMDG Code: class 3.3
RID/ADR: class 3, number 31 c [10]
CFR 49: 172.102 flammable liquid
UN number: 2245

Uses. Cyclopentanone is used in fragrances and as an intermediate in organic synthesis, e.g., for the synthesis of pharmaceuticals (such as cyclopentobarbital), jasmones, or cyclopentylamine. The latter is used for fungicides such as Pencycuron (Bayer). The use of cyclopentanone as a special solvent for polycarbonates has been reported [60].

4.2. Cycloheptanone

Cycloheptanone [*502-42-1*], $(CH_2)_6CO$, $C_7H_{12}O$, M_r 112.17.

Properties. Some physical properties are listed below:

mp	−21 °C
bp	179 °C
d_4^{20}	0.951
n_D^{20}	1.4611
Flash point	55 °C

Production. Cycloheptanone is produced by cyclization and decarboxylation of suberic acid (1,8-octanedioic acid) or suberic acid esters in the gas phase at 400–450 °C over alumina doped with zinc oxide or cerium oxide [61].

Table 2. Physical properties of higher cyclic ketones

	CAS registry number	Empirical formula	mp, °C	bp, °C (kPa)	d_4^{20}
Cyclooctanone	[502-49-8]	$C_8H_{14}O$	42	74 (1.6)	0.959
Cyclodecanone	[1502-06-3]	$C_{10}H_{18}O$	29	107 (1.7)	0.958
Cyclododecanone	[830-13-7]	$C_{12}H_{22}O$	61	125 (1.6)	0.906
Cyclotetradecanone	[3603-99-4]	$C_{14}H_{26}O$	53	155 (1.6)	

Uses. Cycloheptanone is a precursor for syntheses, in particular for pharmaceuticals (e.g., Bencyclane, a spasmolytic agent).

4.3. Higher Cyclic Ketones

Higher cyclic ketones include those compounds that are easily accessible from the products of olefin oligomerization over nickel catalysts (see also → Cyclododecanol, Cyclododecanone, and Laurolactam).

Properties. The higher cyclic ketones are colorless solids, each with a characteristic odor. Only cyclodecanone, used as a precursor for laurolactam, possesses significant importance; the others are used as fragrances or as intermediates. Physical properties are listed in Table 2.

5. Unsaturated Ketones

5.1. 3-Buten-2-one

3-Buten-2-one [78-94-4], methyl vinyl ketone, $CH_2=CHCOCH_3$, C_4H_6O, M_r 70.09.

Properties. 3-Buten-2-one is a colorless liquid with a penetrating odor. The compound is soluble in water and in organic solvents.

bp	81.4 °C
d_4^{20}	0.842
n_D^{20}	1.408
Flash point	−6 °C
Ignition temperature	370 °C

Pure 3-buten-2-one is stable only below 0 °C; it polymerizes rapidly at room temperature. It is stabilized for storage and transportation by the addition of hydroquinone and glacial acetic acid (0.1 wt % each).

A characteristic reaction of 3-buten-2-one is the addition of nucleophiles (e.g., amines [62] or alcohols) at the carbon–carbon double bond. Michael addition to 3-buten-2-one has been developed into a valuable preparative method in terpene and steroid synthesis; this specific reaction is known as Robinson annelation [63]. The dimerization to 2-acetyl-6-methyl-3,4-dihydro-2H-pyran is frequently observed as a side reaction in reactions with 3-buten-2-one.

Production. Two synthetic routes are industrially important [64]:

1) The Mannich reaction of diethylamine, acetone, and formaldehyde [65]:

$$CH_3COCH_3 + HCHO + (C_2H_5)_2NH \cdot HCl$$
$$\xrightarrow{100°C} CH_3COCH_2CH_2N(C_2H_5)_2 \cdot HCl$$
$$\xrightarrow{150-200°C} CH_3COCH=CH_2 + (C_2H_5)_2NH \cdot HCl$$

The diethylamine hydrochloride released by thermolysis of the Mannich base can be recycled for use in the reaction with acetone and formaldehyde.

2) Condensation of acetone with formaldehyde [66], [67] and subsequent elimination of water:

$$CH_3COCH_3 + HCHO \longrightarrow CH_3COCH_2CH_2OH$$
$$\xrightarrow[-H_2O]{H^+} CH_3COCH=CH_2$$

The water can be eliminated from 4-hydroxy-2-butanone in the liquid or gas phase. Singlestep processes are also known [68].

Storage and Transportation. 3-Buten-2-one can only be stored in stabilized form at room temperature. Hazard codes for transportation are as follows:

IMDG Code: class 3.2
RID/ADR: class 3, number 3 b
CFR 49: 172.101/102 flammable liquid
UN number: 1251

Uses. The Diels–Alder adduct with cyclopentadiene, methyl (5-norbornen-2-yl) ketone, is used as a precursor for pharmaceuticals such as Akineton (Knoll). Derivatives of α-vinyl lactic acid nitrile (from 3-buten-2-one and hydrocyanic acid) are used as fungicides, e.g., Ronilan (BASF).

The most important product based on 3-buten-2-one is 3-methyl-2-penten-4-in-1-ol, a building block in an industrial synthesis of vitamin A.

5.2. 3-Methyl-3-buten-2-one

3-Methyl-3-buten-2-one [*814-78-8*], methyl isopropenyl ketone, C_5H_8O, M_r 84.12.

$$CH_3-\underset{O}{\underset{\|}{C}}-\underset{CH_2}{\overset{\|}{C}}-CH_3$$

Properties. Methyl isopropenyl ketone is a water-clear, lachrymatory liquid with a pungent odor.

mp	−54 °C
bp	98 °C
d_4^{20}	0.855
n_D^{20}	1.4236
Flash point	5 °C
Solubility in water (20 °C)	5 wt%

3-Methyl-3-buten-2-one shows the typical properties of an α,β-unsaturated ketone, such as the Michael addition of alcohols and amines, and hetero Diels–Alder reactions. It possesses only limited stability; even at low temperature some cyclodimerization occurs [69].

Under the influence of radical initiators or light, methyl isopropenyl ketone polymerizes to poly(methyl isopropyl ketone) [70]. It is able to undergo copolymerization with suitable monomers such as styrene, butadiene, acrylonitrile, or methacrylic acid esters [71].

Production. 3-Methyl-3-buten-2-one is produced by acid-catalyzed condensation of 2-butanone with formaldehyde in the liquid phase [72]. A further possibility is the methylolization of 2-butanone under alkaline conditions followed by elimination of water:

$$CH_3-\underset{O}{\underset{\|}{C}}-CH_2CH_3 + CH_2O \longrightarrow CH_3-\underset{O}{\underset{\|}{C}}-\underset{CH_2}{\underset{\|}{C}}-CH_3 + H_2O$$

Storage and Transportation. Hazard codes for transportation are as follows:

IMDG Code: class 3.2
RID/ADR: class 3, number 3 b
CFR 49: 172.101 flammable liquid
UN number: 1246

Uses. 3-Methyl-3-buten-2-one is used predominantly as an intermediate in production of 3-methyl-2-butanone, together with a small amount for production of polymers.

5.3. 4-Methyl-4-penten-2-one

4-Methyl-4-penten-2-one [*141-79-7*], mesityl oxide, $(CH_3)_2C=CHCOCH_3$, $C_6H_{10}O$, M_r 98.15.

Properties. 4-Methyl-4-penten-2-one is a colorless, oily liquid with an odor reminiscent of peppermint.

mp	−53 °C
bp	129.5 °C
d_4^{20}	0.855
n_D^{20}	1.4458
Flash point	30 °C
Solubility in water (20 °C)	1.07 wt %

Mesityl oxide forms explosive peroxides on prolonged exposure to atmospheric oxygen. Mesityl oxide can be selectively hydrogenated at the olefinic double bond to give 4-methyl-2-pentanone.

Production. 4-Methyl-4-penten-2-one is the precursor of methyl isobutyl ketone (see Section 2.2) and can be produced from acetone in a single- or a two-step process (→ Acetone). Pure mesityl oxide is obtained by azeotropic distillation of the crude water-containing product. Subsequent purification by removal of the accompanying impurities, such as mesitylene and phorone [*504-20-1*], is effected by distillation [73].

Storage and Transportation. Labelling in accordance with GefStoffV [9]: X_n; R phrases: 10 – 20/21/22; S phrase: 25. Hazard classifications for transportation are as follows:

IMDG Code: class 3.1
RID/ADR: class 3, number 31 c [10]
CFR 49: 172.101 flammable liquid
UN number: 1229

Uses. 4-Methyl-4-penten-2-one is used to a small extent as an intermediate in synthesis. Most of it is converted to 4-methyl-2-pentanone by selective hydrogenation. Although 4-methyl-2-pentanone has excellent solvent properties for nitrocellulose, cellulose esters and ethers, copolymers of vinyl chloride – vinyl acetate and other synthetic resins, as well as natural and synthetic rubber, its use as a solvent is limited as a result of its toxicity and chemical instability. The fraction of 4-methyl-4-penten-2-

one permitted in solvent systems has been set at 5 vol% (Rule 66 of the Los Angeles, California, Air Pollution Control District).

Some of the numerous possible reactions that mesityl oxide may undergo have been used for the synthesis of products such as Budralazine, an antihypertensive agent [74].

5.4. 3,3,5-Trimethyl-2-cyclohexen-1-one

3,3,5-Trimethyl-2-cyclohexen-1-one [78-59-1], isophorone, α-isophorone, $C_9H_{14}O$, M_r 138.21.

α-Isophorone   β-Isophorone

Properties. Isophorone is a colorless liquid with a camphor-like odor. It usually contains 1–3% of the isomeric β-isophorone (3,5,5-trimethyl-3-cyclohexen-1-one). 3,3,5-Trimethyl-2-cyclohexen-1-one is miscible with aromatic and aliphatic hydrocarbons, aldehydes, ketones, alcohols, esters, and ethers.

mp	−8 °C
bp	215.5 °C
d_4^{20}	0.922
n_D^{20}	1.4775
Flash point	95 °C
Ignition temperature	470 °C
Solubility in water (20 °C)	1.2 wt%

Production. Isophorone is produced by condensation of acetone in the liquid phase at ca. 200 °C and 3.6 MPa in the presence of an aqueous potassium hydroxide solution (ca. 0.1%). The process steps condensation, separation of unreacted acetone, and hydrolysis of byproducts can be carried out in a single reactor [75]. Reaction in the gas phase at 350 °C over calcium–aluminum oxide has also been reported [76].

Storage and Transportation. The following regulations apply for storage and shipping: Labelling according to GefStoffV [9]: X_i (Appendix I.2.1); R phrase: 36/37/38; S phrase: 26; VbF: class A III [11]. Hazard classifications for transportation are as follows [10]:

RID/ADR: class 3, number 32 c.

Uses. As a cyclic unsaturated ketone, isophorone can be aromatized to 3,5-dimethylphenol or 2,3,5-trimethylphenol [77].

Hydrogenation leads, depending on conditions, to 3,3,5-trimethylcyclohexanone or 3,3,5-trimethylcyclohexanol, from which trimethyladipic acid is obtained by oxidation [78]. Trimethyladipic acid is used for the production of 2,2,4-trimethylhexamethylenediamine, which is a precursor for polyamides and polyurethanes.

Addition of hydrogen cyanide to the olefinic double bond [79], followed by amination of the keto group in the presence of hydrogen [80], gives 3,5,5-trimethyl-3-aminomethylcyclohexylamine [2855-13-2], isophorone diamine. Reaction of the latter with phosgene gives isophorone diisocyanate [4098-71-9]. Both compounds are used in large quantities for the production of polymers, mainly polyurethanes. The worldwide production capacity for isophorone currently stands at ca. 50 000 t/a.

6. Diketones

6.1. 2,4-Pentanedione

2,4-Pentanedione [123-54-6], acetylacetone, $CH_3COCH_2COCH_3$, $C_5H_8O_2$, M_r 100.11.

Properties. 2,4-Pentanedione is a colorless liquid with a mild ketone-like odor; it is completely miscible with common organic solvents.

mp	−23 °C
bp	140.4 °C
d_4^{20}	0.975
n_D^{20}	1.4510
Flash point	34.5 °C
Ignition temperature	350 °C
Solubility (water, 20 °C)	16 wt %

The chemical properties of 2,4-pentanedione are determined by the keto–enol tautomerism that exists in all states of aggregation:

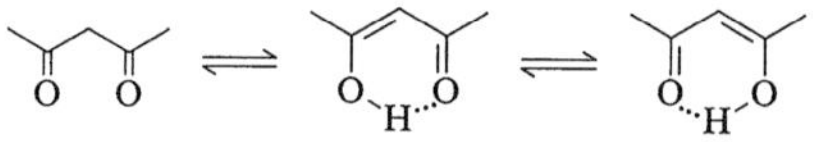

The equilibrium position is influenced by the state of the ketone (i.e., liquid, solid, gas), the temperature, and the solvent. For example, the liquid at 20 °C contains 18 % ketone and 82 % enol.

As a trifunctional compound with two carbonyl groups and an activated methylene group, 2,4-pentanedione is particularly suitable for the synthesis of heterocycles; these may be formed by reaction of both carbonyl groups or of one carbonyl group and the activated methylene group [81]. For example, reaction with hydrazine leads to 3,5-dimethyl pyrazoles (**1**); with hydroxylamine 3,5-dimethylisoxazole (**2**) is formed; with amidines, ureas, thioureas, or guanidines, 2,4-dimethylpyrimidines (**3**) are obtained;

with chloroacetaldehyde 2-methyl-3-acetyl furan (**4**) is formed. The best known reaction of 2,4-pentanedione is the formation of metal acetylacetonates (**5**) [82].

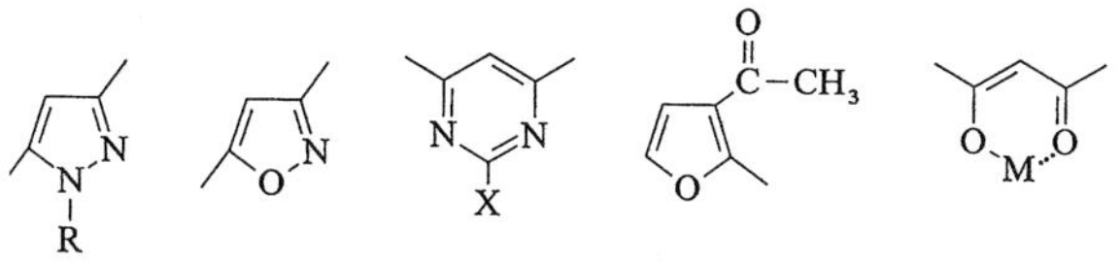

Production. 2,4-Pentanedione is produced by thermal or metal-catalyzed rearrangement of isopropenyl acetate [83] (obtained from acetone and ketene [84]):

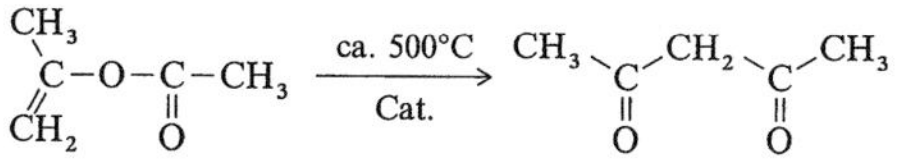

Isopropenyl acetate vapor is fed at atmospheric pressure through a V2A steel tube with an inner temperature of 520 °C. The hot reaction gases are quenched, condensed, and cooled to 20 °C, whereby the gaseous byproducts carbon monoxide, carbon dioxide, methane, and ketene are separated. The product is purified by fractional distillation.

Other industrially less important processes for the production of 2,4-pentanedione, include the Claisen ester condensation of ethyl acetate with acetone using sodium ethoxide as condensation agent [85] and the acetylation of acetoacetic acid esters with acetic anhydride in the presence of magnesium salts [86].

Storage and Transportation. 2,4-Pentanedione is sold with a purity of > 99 %. In addition to small amounts of acetone, 2,4-pentanedione contains 0.1 % water (max.) and 0.2 % acetic acid.

The preferred material for storage containers is enamel, because in the presence of atmospheric oxygen and water, 2,4-pentanedione can form acetylacetonates with most metals.

Labelling according to GefStoffV [9]: X_n (Appendix IIc); R phrases: 10–22; S phrases: 21–23–24/25. Hazard classifications for transportation are as follows:

IMDG Code: class 3.3
RID/ADR: class 3, number 31 c [10]
CFR 49: 172.102 flammable liquid
UN number: 2310

Containers of aluminum or V2A steel are suitable for transportation, although coloring of the product may occur.

Uses. 2,4-Pentanedione is used as an intermediate, in particular for the synthesis of heterocyclic substances for biologically active compounds and dyes, and for the production of metal acetylacetonates. Furthermore, 2,4-pentanedione is used as a solvent, and as an absorption and extraction agent, particularly for the separation and purification of metal ions. Another application is for the purification of metal-containing wastewater and for corrosion protection. 2,4-Pentanedione is also used as a component of catalyst systems for polymerization, copolymerization, oligo- and dimerization.

Metal Acetylacetonates. The industrial importance of metal acetylacetonates is due to their stability (e.g., in air) and their solubility in organic solvents, whereby homogenous catalytic or stoichiometric reactions of metal ions with organic compounds may be carried out.

Production. Metal acetylacetonates are usually produced by reaction of soluble metal salts with 2,4-pentanedione in aqueous or ethanolic solution. A base is required to neutralize and release the anion from the metal salt. Special production processes are used for some compounds.

Uses. Metal acetylacetonates are used predominantly for polymerization (e.g., olefins, styrene, methacrylates, acrylonitrile), for cross-linking (e.g., epoxy resins, olefins), or for the hardening of epoxy resins or unsaturated polyester resins.

Metal coatings formed with metal acetylacetonates (e.g., for solar cells [87]) have recently been described. A further application is their use as a stabilizer in poly(vinyl chloride) for minimizing changes in color [88].

Metal acetylacetonates can also catalyze reactions such as oxidation, isomerization, hydrogenation, dehydrogenation, esterification and transesterification [89]. A review of the uses of metal acetylacetonates has been published [90].

6.2. 2,3-Butanedione

2,3-Butanedione [*431-03-8*], diacetyl, $CH_3COCOCH_3$, $C_4H_6O_2$, M_r 86.09.

Properties. 2,3-Butanedione is a light, yellow–green liquid, which is readily soluble in all important organic solvents.

mp	−2.5 °C
bp	88 °C
d_4^{20}	0.9831
$n_D^{18.5}$	1.3933
Flash point	26 °C
Solubility in water (15 °C)	20 wt %

In the presence of alkali 2,3-butanedione readily undergoes self condensation to form compounds with a quinonoid structure.

Production. 2,3-Butanedione can be obtained by oxidation of 2-butanone over a copper catalyst at 300 °C in a yield of ca. 60% [91]. Production from glucose by fermentation has also been described [92], and also the acid-catalyzed condensation of 1-hydroxyacetone (obtained by dehydrogenation of 1,2-propanediol) with formaldehyde [93].

Dehydrogenation of 2,3-butanediol over a copper or silver catalyst in the presence of air is industrially important [94]. 3-Hydroxy-2-butanone (acetoin) is generally formed as a byproduct.

Storage and Transportation. Hazard codes for transportation are as follows:

IMDG Code: class 3.2
RID/ADR: class 3, number 3 b
CFR 49: 172.102 flammable liquid
UN number: 2346

Uses. 2,3-Butanedione is mainly important as a flavor component (buttery taste). It is used in low concentration in ice cream, baked goods, and margarine.

6.3. Higher α-Diketones

2,3-Pentanedione [600-14-6], $CH_3COCOCH_2CH_3$, $C_5H_8O_2$, M_r 100.12:

mp	−52 °C
bp	110 – 112 °C
Flash point	18 °C

2,3-Hexanedione [3848-24-6], $CH_3COCOCH_2CH_2CH_3$, $C_6H_{10}O_2$, M_r 114.5:

bp	129 – 130 °C
Flash point	28 °C
n_D^{20}	1.4137
d_4^{20}	0.934

Higher α-diketones are found in small amounts in various foods.

Production. Processes used include oxidation of the corresponding diols over a copper or silver catalyst [94], or selective oxidation of the methylene group in the corresponding methyl alkyl ketones with oxygen in the presence of catalysts such as nickel [95].

Uses. These compounds are used as aroma components in alcohol-free beverages and in baked goods.

7. Aromatic Ketones

7.1. Methyl Phenyl Ketone

Methyl phenyl ketone [98-86-2], acetophenone, $C_6H_5COCH_3$, C_8H_8O, M_r 120.15.

Properties. Methyl phenyl ketone is a colorless solid or liquid with a characteristic odor.

mp	20 °C
bp	202 °C
d_4^{20}	1.028
n_D^{20}	1.5325
Flash point	82 °C
Solubility in water (20 °C)	0.8 wt %

Methyl phenyl ketone is readily soluble in organic solvents. The usual condensation reactions can be carried out at the methyl group.

Production. Most methyl phenyl ketone originates from the Hock process for the production of phenol from isopropylbenzene (→ Phenol); it is isolated from the residue of this process. In addition, acetophenone can be obtained as a main product by selective decomposition of cumene hydroperoxide in the presence of copper catalysts at 100 °C [96]:

$$C_6H_5C(CH_3)_2OOH \xrightarrow[H_2O]{Cu(NO_3)_2} C_6H_5COCH_3 + CH_3OH$$

A second possibility is the oxidation of ethylbenzene with air or oxygen at 130 °C and 0.5 MPa. Catalysts used include cobalt salts or manganese salts of naphthenic or fatty acids. Conversion of ethylbenzene is limited to ca. 25 % to minimize the byproducts 1-phenylethanol and benzoic acid [97].

A third method is the Friedel–Crafts acetylation of benzene with acetic anhydride, but this is not of industrial importance.

Uses. Methyl phenyl ketone is mainly converted to resins (synthetic resins) by reaction with formaldehyde [98]. It is also used as a photoinitiator for special printing plates as well as for organic syntheses, particularly for pharmaceuticals. Certain derivatives, such as 1-phenylethanol (obtained by hydrogenation) and its acetate, are used as fragrances. Furthermore, methyl phenyl ketone is used for the synthesis of optical brighteners.

7.2. 1-Phenyl-1-propanone

1-Phenyl-1-propanone [*93-55-0*], ethyl phenyl ketone, propiophenone, $C_6H_5COCH_2CH_3$, $C_9H_{10}O$, M_r 134.12.

Properties. 1-Phenyl-1-propanone is a colorless liquid with a flowery odor, insoluble in water, readily soluble in organic solvents. Typical reactions can be carried out at the methylene group, the carbonyl group, and at the aromatic nucleus.

mp	18 °C
bp	218 °C
d_4^{20}	1.009
n_D^{20}	1.5258
Flash point	87 °C

Production. Propiophenone is produced by Friedel–Crafts acylation of benzene with propionic acid chloride in the presence of an equivalent amount of aluminum chloride [99]. Another industrial method is the catalytic ketonization of benzoic acid with propionic acid over a calcium acetate–aluminum oxide catalyst at 440–520 °C [100].

Uses. 1-Phenyl-1-propanone is used mainly as an intermediate for pharmaceuticals such as D-Propoxyphen [101], phenylpropanolamine, and Phenmetrazine [102].

7.3. Diphenyl Ketone

Diphenyl ketone [*119-61-9*], benzophenone, $C_6H_5COC_6H_5$, $C_{13}H_{10}O$, M_r 182.22.

Properties. The stable form of diphenyl ketone consists of colorless rhombic prismatic crystals. Several unstable forms with lower melting points also exist. Diphenyl ketone is insoluble in water and readily soluble in organic solvents.

mp	49–51 °C
bp	305 °C
d_4^{50}	1.0869
n_D^{45}	1.5975
Flash point	ca. 155 °C

Production. Benzophenone is usually produced by atmospheric oxidation of diphenylmethane in the presence of metal catalysts such as copper naphthenate [103], [104]. Other processes include Friedel–Crafts acylation of benzene with benzoyl chloride or of benzene with phosgene [105].

Storage. Benzophenone should be stored under cool, dry conditions. A temperature above 45 °C should be avoided (risk of caking).

Uses. Diphenyl ketone and substituted diphenyl ketones in particular are used mainly as photoinitiators for UV-curable printing inks and coatings. Diphenyl ketone is also used as an intermediate for pharmaceuticals and agricultural chemicals. It is employed in the perfume industry as a fixative and as a fragrance with a flowery note.

7.4. 4,4′-Diphenoxydiphenyl Ketone

4,4′-Diphenoxydiphenyl ketone [*14984-21-5*], 4,4′-diphenoxybenzophenone, $C_{25}H_{18}O_3$, M_r 366.42.

Properties. 4,4′-Diphenoxydiphenyl ketone, *mp* 148 °C, *bp* 270 °C, is a colorless solid, that is insoluble in water and soluble in organic solvents.

Production. 4,4′-Diphenoxydiphenyl ketone is produced from diphenyl ether and phosgene [106] in the presence of aluminum chloride as catalyst. It may also be prepared by the catalyzed atmospheric oxidation of 4,4′-diphenoxydiphenylmethane [107] or the reaction of diphenyl ether, tetrachloromethane, and aluminum chloride [108].

Uses. 4,4′-Diphenoxybenzophenone is used exclusively for the production of high-performance plastics, e.g., by Friedel–Crafts reaction with aromatic dicarboxylic acid chlorides, such as terephthalic acid dichloride [109]. The plastics display a series of industrially interesting properties, such as long-term heat resistance up to 270 °C, fire-extinguishing properties, and high chemical- and solvent resistance. The polymer is worked by injection molding or spinning.

7.5. 1-Phenyl-2-propanone

1-Phenyl-2-propanone [*103-79-7*], phenylacetone, benzyl methyl ketone, $C_6H_5CH_2COCH_3$, $C_9H_{10}O$, M_r 134.18.

Properties. 1-Phenyl-2-propanone is a colorless liquid with a pleasant odor; it is insoluble in water and soluble in organic solvents.

mp	−15 °C
bp (1.73 kPa)	100 °C
d_4^{20}	1.015
n_D^{20}	1.5158

Production. 1-Phenyl-2-propanone is produced by catalytic ketonization of phenylacetic acid with acetic acid over an alumina-supported cerium oxide catalyst at 400–500 °C [110]. It may also be produced by the rearrangement of phenylpropylene oxide on zeolites [111] or oxidation of 2-phenylpropanol.

Uses. 1-Phenyl-2-propanone is used as an intermediate for the synthesis of pharmaceuticals, e.g., Prenylamine, a coronary vasodilator (Hoechst) [112].

7.6. 2-Hydroxy-1,2-diphenylethanone

2-Hydroxy-1,2-diphenylethanone [579-44-2], benzoin, α-hydroxybenzyl phenyl ketone, $C_6H_5COCH(OH)C_6H_5$, $C_{14}H_{12}O_2$, M_r 212.25.

Properties. Benzoin is a colorless solid with an odor similar to that of bitter almonds and camphor, mp 133–134 °C, bp 194 °C (1.6 kPa). It occurs in two enantiomeric forms that can be separated by reacting the acid phthalates with (+)-quinidine in methanol to form diastereomic salts [113]:

S(+)-Benzoin [5928-67-6], $[\alpha]_D^{12}$ +120.5° (1.2 g/dL in acetone)
R(−)-Benzoin [5928-66-5], $[\alpha]_D^{16}$ −119.2° (2.8 g/dL in acetone)

Production. Benzoin is usually synthesized by reacting benzaldehyde in the presence of cyanide ions [114].

Uses. Benzoin is the starting compound for the synthesis of benzil [134-81-6]. Benzil derivatives are used as photoinitiators in polymerization reactions [115]. Benzoin and its alkyl ethers are also used for this purpose [116].

8. Toxicology and Occupational Health

The LD_{50} values of the ketones discussed in 2–7 are listed in Table 3.

Methyl Alkyl and Dialkyl Ketones. 3-Methyl-2-butanone is moderately irritating to the skin and eye [116], its TLV-TWA value is 200 ppm or 705 mg/m^3. The MAK value

Table 3. LD$_{50}$ values for ketones

Ketone	Animal	LD$_{50}$, mg/kg
Methyl Alkyl Ketones		
3-Methyl-2-butanone	mouse (oral)	2 572
	rabbit (dermal)	6 350
4-Methyl-2-pentanone	rat (oral)	2 080
	rabbit (inhalation)	8 000
2-Heptanone	mouse (oral)	730
	rat (oral)	1 670
	rabbit (dermal)	12 600
5-Methyl-2-hexanone	rat (oral)	4 760
2-Octanone	mouse (oral)	3 824
Dialkyl Ketones		
3-Pentanone	rat (oral)	2 140
2,6-Dimethyl-4-heptanone	rat (oral)	5 750
	rabbit (dermal)	16 000
Cyclic Ketones		
Cyclopentanone	mouse (subcutaneous)	2 600
Unsaturated Ketones		
3-Buten-2-one	rat (oral)	31
	rat (inhalation, 4 h)	7 *
4-Methyl-4-penten-2-one	mouse (oral)	710
	mouse (inhalation, 2 h)	10 *
3,3,5-Trimethyl-2-cyclohexen-1-one	rat (oral)	2 330
	rabbit (oral)	1 500
Diketones		
2,4-Pentanedione	rat (oral)	1 000
Aromatic Ketones		
Methyl phenyl ketone	rat (oral)	815
1-Phenyl-1-propanone	rat (oral)	4 490
Diphenylketone	mouse (oral)	>2 500
	rabbit (dermal)	3 500
4,4′-Diphenoxydiphenylketone	rat (oral)	>5 000
	rabbit (dermal)	>2 000
2-Hydroxy-1,2-diphenylethanone	rat (oral)	>6 400

* Values are given in milligrams per cubic meter.

for 4-methyl-2-pentanone is 100 ppm or 400 mg/m^3 [28]. The TLV-TWA value is 50 ppm or 205 mg/m^3. 2-Heptanone is classified as harmful to health. Inhalation of the vapor or aerosols containing the liquid should be avoided. Its TLV-TWA value is 50 ppm or 235 mg/m^3. 5-Methyl-2-hexanone has a TLV-TWA value of 50 ppm or 240 mg/m^3. The TLV-TWA value for 3-pentanone is 200 ppm or 705 mg/m^3. 2,6-Dimethyl-4-heptanone is registered according to GefStoffV as irritant; skin contact should therefore be avoided. The MAK value has been fixed at 290 mg/m^3 [28]; the TLV-TWA value is 25 ppm or 250 mg/m^3. 3-Heptanone has a TLV-TWA value of 50 ppm or 230 mg/m^3.

5-Methyl-3-heptanone is classified as an irritant. Inhalation of the vapor or aerosols containing the compound should be avoided. Its TLV-TWA value is 25 ppm or 130 mg/m^3.

Cyclic Ketones. Cyclopentanone is classified as an irritant (GefStoffV). Inhalation of the vapor or aerosols containing the liquid should be avoided.

Unsaturated Ketones. 3-Buten-2-one is a lachrymator and is highly toxic. Exposure of any kind should be avoided [16].

2-Methyl-2-penten-4-one (mesityl oxide) is regarded as harmful to health (class IId, "mindergiftig"), according to GefStoffV. It possesses a strong lachrymatory effect. Its MAK value is 100 mg/m^3 [28] and its TLV-TWA value is 15 ppm or 60 mg/m^3.

3,3,5-Trimethyl-2-cyclohexen-1-one (isophorone) is classified according to GefStoffV as an irritant (irritates the eyes, throat, and skin). Its MAK value is 28 mg/m^3 [28] and its TLV-TWA value is 5 ppm or 25 mg/m^3.

Diketones. 2,4-Pentanedione is classified as harmful to health (GefStoffV, "mindergiftig"). Contact with the skin and with the eye should be avoided.

Aromatic Ketones. Diphenyl ketone (benzophenone) has been employed for many years in the pharmaceutical and cosmetics industries and is regarded as essentially nontoxic. It is included in the FDA list of substances that may be added to foods. Benzoin is dangerous to health when inhaled or absorbed cutaneously.

9. References

[1] H. R. Gersmann, A. F. Bickel, *J. Chem. Soc. B* 1971, 2230.
[2] BASF, DE-OS 2 941 386, 1979 (G. Heilen, K. Halbritter, W. Gramlich).
[3] Rheinpreussen AG, DP 1 198 814, 1963; DP 1 233 848, 1964 (J. Wöllner, F. Engelhardt).
[4] Eastman Kodak, BP 1 194 058, 1967 (H. S. Young, J. W. Reynolds).
[5] Shell, EP 85996, 1982 (F. Wattimena). BASF, DE-OS 2 758 113, 1977 (H. Fröhlich et al.).
[6] Kuraray Co., JA 7 448 407, 1969.
[7] Shell, GB 2 167 406, 1984 (B. L. Feringa).
[8] Texaco, US 4 000 200, 1971 (J. K. Cox).
[9] *Gefahrstoffverordnung (GefStoffV)*, BGBl I, Ecomed Verlagsgesellschaft, Landsberg 1986, p. 224.
[10] *Gefahrenvermerke für Transport (GGVS)*, Verkehr-Verlag J. Fischer, Düsseldorf 1988.
[11] *VBF-Richtlinien*, Carl Heymanns Verlag, Köln 1980.
[12] *Arbeitsschutz Inland*, UVV 1, Berufsgenossenschaft der Chemischen Industrie, Jedermann Verlag, Heidelberg 1984.
[13] American Cyanamid, EP 205 879, 1986 (M. Los, D. W. Loug, G. P. Withers).
[14] A. E. Arbuzow, Yu P. Kitaev, *Zh. Obshch. Khim.* **27** (1957) 2328; *Chem. Abstr.* **52** (1958) 7281 h.
[15] W. G. Gruzensky, *US Bur. Mines Rep. Invest.* 1961, no. 5910; *Chem. Abstr.* **56** (1962) 8362 a.
[16] L. Bretherick, *Chem. Ind. (London)* 1972, no. 22, 863.
[17] *Chem. Abstr.* **77** (1972) 100 652 x.
[18] J. E. Dubois, R. Luft, *C. R. Hebd. Seances Acad. Sci.* **240** (1955) 1540.
[19] *Hydrocarbon Process.* **48** (1969) no. 11, 205.
[20] Scholven-Chemie, US 3 361 822, 1968; V. Macho, CS 131 903, 1969.

[21] Showa Denko, JP 45-41566, 1970.
[22] Tokuyama Soda, JP 45-28566, 1970.
[23] Rheinpreussen AG, DT 1 260 454, 1966 (J. Wöllner, W. Neier). VEBA-Chemie, DT 1 643 044, 1967 (J. Disteldorf, W. Flakus, W. Hübel).
[24] Showa Denko KK, JP 7 215 334, 1968.
[25] BASF, DE-OS 1 943 698, 1969 (H. Groper et al.).
[26] Japan Oil and Fats Co., JP 7023.414, 1966.
[27] J. E. Dubois, R. Luft, *C. R. Hebd. Seances Acad. Sci.* **238** (1954) 485.
[28] R. B. Wagner, *J. Am. Chem. Soc.* **71** (1949) 3214.
[29] Zoecon Corp., DE-OS 2 123 147, 1970 (J. B. Sidall). Schering, DE-OS 2 262 402, 1972 (L. Nuesslein, F. Arndt).
[30] Mobay, US 4 059 634, 1975 (D. N. Smith Jr.).
[31] Phillips Petroleum, US 2 596 212, 1952 (F. E. Gondon).Sumitomo, JP 57142-938, 1981.
[32] Daicel Chem. Ind., JP 7 197 237, 1981. Diamond Samrock Corp., DE-OS 2 737 511, 1977 (R. Cryberg, R. Bimber).
[33] Bayer, DE 3 211 306, 1982 (G. Rauleder, H. Waldmann).
[34] Bayer, DE 2 201 063, 1972 (W. Meisner et al.).
[35] Bayer, BE 697 083, 1976.
[36] BASF, DE 2 625 540, 1976 (A. Nissen et al.).
[37] R. Heilmann et al., *Bull. Soc. Chim. Fr.* 1957, 112.
[38] Labofina S.A., FR 2 094 808, 1970 (P. Camerman, J. Hanotien).
[39] BASF, DE 2 625 540, 1976 (A. Nissen et al.).
[40] Union Oil Comp., US 3 518 285, 1967 (D. M. Fenton, L. G. Wolgemuth); Catalytica Assoc., US 4 738 943, 1986 (J. Vasilevskis, P. L. Ridgway, E. R. Evitt).
[41] BASF, DE-OS 2 758 113, 1977 (H. Fröhlich et al.).
[42] Eastman Kodak, FR 1 533 651, 1966 (J. W. Reynolds et al.); FR 1 529 019, 1968 (J. W. Reynolds, H. S. Young).
[43] Wacker-Chemie, DE-OS 2 358 254, 1973 (L. O. Sommer, F. Knörr); Union Oil Co., US 3 836 553, 1970 (D. M. Fenton).
[44] Shell, GB 2 167 406, 1984 (B. L. Feringa).
[45] Burmah Oil Trading Ltd., DE-OS 2 403 411, 1973 (P. Field, D. A. Lock).
[46] Universal Oil Prod. Co., US 2 787 416, 1964 (H. A. Cyba).
[47] Pennwalt Corp., DE-OS 2 414 002, 1973 (I. G. Popoff, P. G. Haines).
[48] Daicel, JP 57197-237, 1981.BASF, DE-OS 2 111 722, 1971 (L. Schuster, L. Arnold).
[49] BASF, DP 2 445 193, 1974 (N. v. Kutepow).
[50] BASF, DP 2 758 113, 1977 (H. Fröhlich et al.).
[51] K. Kawamoto et al., *Bull. Chem. Soc. Jpn.* **41** (1968) 932.
[52] M. Araki, Y. Kotera, *Bull. Jpn. Pet. Inst.* **15** (1973) no. 1, 45.
[53] V. N. Ipatieff, V. Haensel, *J. Org. Chem.* **7** (1942) 195.
[54] BASF, DP 2 615 308, 1976 (A. Nissen et al.).
[55] H. Miyake et al., *Bull. Jpn. Pet. Inst.* **16** (1974) no. 1.
[56] W. Wayne, H. Adkins, *J. Am. Chem. Soc.* **62** (1940) 3401.
[57] *Bulletin of the Academy of Sciences of the USSR Div. of Chem. Science* (1968) 609.
[58] BASF, DE-OS 3 638 005, 1986 (H. Lermer, W. Hölderich, M. Schwarzmann).
[59] Agency of Ind. Sci., DE-OS 3 305 000, 1982 (K. Takehira et al.).
[60] *Plast. Eng.* **43** (1987) no. 10, 35.
[61] Hüls, DE-OS 2 637 787, 1986 (W. Kleine-Homann).

[62] S. I. Suminov, A. N. Kost, *Russ. Chem. Rev. (Engl. Transl.)* **38** (1969) no. 11, 884–899.
[63] B. P. Mundy, *J. Chem. Educ.* **50** (1973) 110–113.
[64] G. S. Mironov, M. I. Faberov, *Russ. Chem. Rev. (Engl. Transl.)* **33** (1964) no. 6, 311–319.
[65] Bayer, DT 877 606, 1953 (G. S. Mironov et al.); *Khim Promst. (Moscow)* **10** (1969) 727–729.
[66] Bayer, DT 222 551, 1909; Consortium F. Elektrochem. Ind., DT 730 117, 1943 (P. Halbig, A. Treibs); BASF, DT 1 269 121, 1968 (H. Pommer, H. Müller, H. Köhl).
[67] BASF, DT 1 277 235, 1968 (H. Müller et al.).
[68] Mitsubishi Chem., JP 4 828 211, 1971.
[69] J. Colonge, J. Dreux, *C. R. Hebd. Seances Acad. Sci.* **228** (1949) 583.
[70] C. S. Marvel et al., *J. Am. Chem. Soc.* **64** (1942) 93.
[71] Rheinpreussen AG, DT 1 004 807, 1955 (W. Grimme, F. Engelhardt).
[72] Rheinpreussen AG, DT 1 198 814, 1963; DT 1 233 848, 1964 (J. Wöllner, F. Engelhardt).
[73] Esso, GB 1 116 037, 1968 (L. J. Sirois, F. J. Hermann, M. E. Oldweiler).
[74] K. Keno et al., *Chem. Pharm. Bull.* **24** (1976) 1068.
[75] Bergwerksges. Hibernia, DT 1 165 018, 1960 (K. Schmitt, W. Baron, J. Disteldorf).
[76] Union Carbide Corp., EP 146927, 1983 (A. J. Papa, S. W. Kaiser).
[77] Shell, US 2 369 197, 1945 (De Loss E. Winkler, S. A. Ballard). Shell, US 2 628 985, 1952 (De Loss E. Winkler, H. de V. Finch).
[78] Bergwerksges. Hibernia, GB 915 510, 1963.
[79] Scholven-Chemie, DT 1 240 854, 1964 (K. Schmitt et al.).
[80] Hibernia-Chemie, DT 1 229 078, 1961 (K. Schmitt et al.).
[81] A. Heikel, *Chem. Zentralbl.* **1** (1936) 2076.
[82] *Houben Weyl*, Oxygen Compounds, part 2, vol. **6/2**, 1963.
[83] Wacker-Chemie, DT 1 001 249, 1955 (E. Enk, F. Büttner). Wacker-Chemie, DT 2 047 320, 1970 (G. Künstle et al.).
[84] Wacker-Chemie, DT 1 121 605, 1959 (E. Enk et al.).
[85] H. Adkins, J. L. Reiney, *Org. Synth. Coll.* **III** (1955) 16.
[86] UCC, US 2 369 250, 1943 (W. H. Reeder, G. A. Lescisin).
[87] Hitachi, DE 3 447 635, 1983 (H. Matsuyama et al.).
[88] *Jpn Chem. Week* 11/1984, 3.
[89] Merkblatt Wacker-Chemie, Acetonylaceton, München 1983.
[90] J. R. Harwood: *Industrial Applications of Organometallic Compounds*, Reinhold Publishing Corp., New York 1963.
[91] Mitsubishi Chemical Ind., JP 4 742 711, 1971.
[92] *Nature* **157** (1946) 336.
[93] DOW, US 2 799 707, 1956 (O. C. Dermer).
[94] BASF, DP 2 831 229, 1978 (W. Saner et al.).
[95] Mitsubishi Chem. Ind., JP 54 132 515, 1978.
[96] Rhone-Poulenc, BE 815 578, 1973.
[97] H. J. Sanders, H. F. Keag, H. S. McCulloug, *Ind. Eng. Chem.* **45** (1953) 2–14.
[98] Hüls, DP 3 324 287, 1983 (J. Dörffel).
[99] A. I. Vogel, *J. Chem. Soc.* 1948, 614.
[100] Union Carbide Corp., US 4 172 097, 1979 (C. A. Smith, L. F. Theiling).
[101] Eli Lilly, US 2 728 779, 1955 (A. Pohland).
[102] Boehringer Ingelheim, US 2 835 669, 1958 (O. Thoma).
[103] Universal Oil Products Co., US 3 642 906, 1969 (S. Kuhn).
[104] Nippon Shokubai Kagaku, JP 59 219 248, 1983.

[105] BASF, DT 403 507, 1920 (A. Mittasch, M. Müller-Cunradi).
[106] DOW Chemical, US 3 845 015, 1974 (R. J. Thomas).
[107] Nippon Shokubai Kagaku, JP 9 148 731, 1983.
[108] N. C. Deno, W. L. Evans, *J. Am. Chem. Soc.*, **79** (1957) 5804.
[109] Raychem, GB 1 471 171, 1974 (K. J. Dahl, V. Jansons).
[110] Diamond Shamrock Corp., BE 857 956, 1976 (R. Cryberg, R. Bimber).
[111] BASF, EP 162 387, 1984 (W. Hölderich et al.).
[112] *Optical Resolution Procedures for Chemical Compounds,* vol. 3, Optical Resolution Information Center, Manhattan College Riverdale, New York 10471, 1984.
[113] J. Solodar, *Tetrahedron Lett.* **2** (1971) 287–288. *Org. Synth. Coll.* **I** (1941) 94.
[114] M. B. Rubin, *Top. Curr. Chem.* **129** (1985) 1.
[115] H. J. Hagermann, *Prog. Org. Chem.* **13** (1985) 123.
[116] *Sigma-Aldrich,* Library of Chemical Safety Data, Edition I, 1984.
[117] *Registry of Toxic Effects of Chemical Substances 1985–86 Edition,* Vol. **1,** US: Department of Health and Human Services.

Lactic Acid

SURINDER P. CHAHAL, Croda Colloids Ltd., Widnes, United Kingdom

1.	Introduction	3119	5.	Quality Specifications	3127
2.	Properties	3120	6.	Analysis	3127
2.1.	Physical Properties	3120	7.	Storage and Transportation	3128
2.2.	Chemical Properties	3121	8.	Legal Aspects	3128
2.2.1.	Internal Esterification	3121	9.	Uses	3128
2.2.2.	Esterification	3122	10.	Derivatives	3129
2.2.3.	Oxidation	3122	10.1.	Salts	3129
2.2.4.	Reduction	3123	10.2.	Esters	3130
2.2.5.	Other Reactions	3123	10.3.	Lactamides	3130
3.	Production	3123	10.4.	Lactonitriles	3130
3.1.	Raw Materials	3124	11.	Economic Aspects	3131
3.2.	Lactic Acid Fermentation	3124	12.	Toxicology and Occupational Health	3131
3.3.	Lactic Acid Purification	3125	13.	References	3131
3.4.	Synthesis of Lactic Acid	3125			
3.5.	Construction Materials	3126			
4.	Environmental Protection	3127			

1. Introduction

Lactic acid, 2-hydroxypropionic acid, was discovered in 1780 by the Swedish chemist SCHEELE [1]; he isolated it from sour milk as an impure brownish syrup. The French chemists THENARD [2], [3], FOURCROY and VAUQUELIN [4], and BOUILLON-LAGRANGE [5] denied that SCHEELE's lactic acid was a single compound; they postulated that the acid was acetic acid combined with an animal material. Another Swedish chemist BERZELIUS repeated SCHEELE's experiments and concluded that lactic acid was a single compound [6]. He also found lactic acid in fresh milk [6], in ox meat [7], in blood and several other animal liquors. Lactic acid was rediscovered by the French chemist BRACONNOT [8] and, shortly after, the German chemist VOGEL [9] proved the identity of Scheele's acid. In 1839 FREMY [10] produced lactic acid by the fermentation of carbohydrates such as sucrose, lactose, mannitol, starch and dextrin. Industrial manufacture of lactic acid was established in 1881.

2. Properties

2.1. Physical Properties

Lactic acid, 2-hydroxypropionic acid [50-21-5], $CH_3CH(OH)COOH$, M_r 90.08, is the simplest hydroxycarboxylic acid with an asymmetrical carbon atom. It is found as a racemate (DL) and also in two optically active forms:

```
        COOH              COOH
         |                 |
    HO-C-H             H-C-OH
         |                 |
        CH₃               CH₃

   L(+) Lactic acid    D(−) Lactic acid
```

Pure and anhydrous racemic lactic acid is a white crystalline solid with a low melting point [11]. It is difficult to determine its melting point exactly because of the tremendous difficulty in producing anhydrous lactic acid; values range from 18 °C [11] to 33 °C [12]. Crystalline lactic acid, produced by careful distillation, undergoes internal esterification to produce lactoyllactic acid (see Section 2.2.1) [13]. If one of the optical isomers is present in excess in lactic acid, this isomer can be isolated by fractional crystallization of the freshly distilled acid from a mixture of equal parts diethyl ether and diisopropyl ether [14], [15]. The melting point of the pure optical isomers was found to be 52.7–52.8 °C. The boiling point of anhydrous lactic acid is approximately 125–140 °C at 27 kPa [16].

Because of the very hygroscopic nature of lactic acid, it is normally obtained as a concentrated solution, up to 90 wt%, which is colorless and virtually odorless. It is completely soluble in water, ethanol, diethyl ether and other organic solvents which are miscible with water; it is virtually insoluble in benzene and chloroform. The distribution of lactic acid between water and some organic solvents can be found in literature [17].

Synthetically produced lactic acid is optically inactive, i.e., a racemic mixture. Lactic acid produced by *fermentation* is optically active and specific production of optically active lactic acid, whether L(+) or D(−), can be determined by using an appropriate lactobacillus. The L(+) enantiomer is dextrorotatory and the D(−) enantiomer laevorotatory; however, the salts and esters of L(+) lactic acid are generally laevorotatory. L(+) Lactoyllactic acid is laevorotatory and an aqueous solution of L(+) lactic acid can therefore be laevorotatory by virtue of its content of lactoyllactic acid.

When measuring the angle of rotation of lactic acid, care must be taken to ensure that the acid is as free from lactoyllactic acid as possible. For a 20 wt% aqueous solution of L(+) lactic acid the optical rotation $[\alpha]_D^{25}$ was determined to be +2.53°, whereas an

Table 1. Physical data of aqueous lactic acid solutions at 25 °C

Concentration, wt%	Density, g/mL	Viscosity, mPa s	Refractive index	Conductivity, mS/cm
6.29	1.0115	1.042	1.3390	3.670
25.02	1.0570	1.725	1.3586	3.823
54.94	1.1302	4.68	1.3909	1.530
88.60	1.2006	36.9	1.4244	0.0567

Table 2. Thermodynamic data (at 25 °C)

	Crystalline L(+) lactic acid	Liquid racemic lactic acid	Reference
Heat of solution in water, ΔH, kJ/mol	+7.81		[25]
Heat of dilution in water, ΔH, kJ/mol		−4.19	[26]
Heat of fusion, ΔH, kJ/mol		+11.34	[25]
Heat of combustion, ΔH_C°, kJ/mol	−1344.3	−1363.7	[25], [26]
Heat of formation, ΔH_f°, kJ/mol	−694.3	−682.2	[25]
Free energy of formation, ΔG_f^0, kJ/mol	−523.1	−529.4	[25]

80 wt% solution gave a value of +5.10° [12]. The optical rotation of lactic acid also depends greatly on the solvent used [17].

The steam-volatility of lactic acid is very low; it increases as the concentration of lactic acid increases. Even at high lactic acid concentrations (>80 wt%), quantitative steam distillation at 100 °C is impracticable and uneconomical.

However, with the use of superheated steam (130 °C – 200 °C) quantitative entrainment of lactic acid can be obtained [18]–[21]. Superheated steam distillation is thus a method for purifying crude lactic acid [21].

Density [22], [23], viscosity [21], [23], refractive index [23] and conductivity [24] of aqueous lactic acid solutions of varying concentrations, are given in Table 2. The dissociation constant K of lactic acid is 1.38×10^{-4} at 25 °C which corresponds to a pK of 3.86. The pH of a 10 wt% aqueous solution of lactic acid is 1.75 [23]. Lactic acid – lactate buffers can be prepared in the pH range of ca. 2.75 – 4.75. Thermodynamic data of lactic acid are given in Table 2.

2.2. Chemical Properties

2.2.1. Internal Esterification

Lactic acid, being both an alcohol and an ester, can undergo intermolecular esterification to produce an ester; lactoyllactic acid

$$2\ CH_3CH(OH)COOH \longrightarrow CH_3CH(OH)COOCH(CH_3)COOH + H_2O$$

Lactoyllactic acid

Another bimolecular ester which can form if high temperature distillation of lactic acid is attempted is dilactide, a cyclic ester

$$\text{CH}_3\text{CH(OH)COOCH(CH}_3\text{)COOH} \longrightarrow \text{Dilactide} + \text{H}_2\text{O}$$

Lactoyllactic acid → Dilactide

On further dehydration of lactoyllactic acid, polylactic acids can be formed. These are linear polyesters of undefined chain length.

$$\text{HO-CH(CH}_3\text{)-CO-[O-CH(CH}_3\text{)-CO]}_n\text{-O-CH(CH}_3\text{)-COOH}$$

Polylactic acid

Commercially available aqueous lactic acid solutions contain varying amounts of lactoyllactic and polylactic acids. The quantities depend on the concentration and age of the solution. Solutions as weak as 6.5 wt% will contain $\approx$ 0.2 wt% lactoyllactic acid [23]; an 88 wt% lactic solution contains < 60% free lactic acid [23] and 100% lactic acid contains only 32% of the free monomeric acid [27].

2.2.2. Esterification

Synthesis of the lactates of lower alcohols occurs relatively easily by direct reaction with the corresponding alcohol. The yield can be increased by removing the water formed by azeotropic distillation. The amount of water produced during the reaction can be considerably reduced, and the yield of ester increased, if the alcohol is reacted with polylactic acid. Esters of higher alcohols can be produced efficiently by transesterification of methyl or ethyl lactates in the presence of catalysts (sulfuric acid, toluene sulfonic acid or aluminum isopropylate).

2.2.3. Oxidation

Oxidation of lactic acid yields different products according to the reaction conditions and oxidation methods. Treatment with chromic acid produces oxalic acid and carbon dioxide, oxidation with N-bromosuccinimide yields acetaldehyde and carbon dioxide. Hydrogen peroxide, in the presence of iron(II) sulfate, oxidizes lactic acid to pyruvic acid. Sulfuric acid gives rise to acetaldehyde, carbon monoxide, and water in the case of lactic acid, or acetaldehyde, carbon monoxide, and alcohol if a lactic ester is oxidized. The first step in all cases is probably the formation of the pyro-acid, which is very difficult to isolate as it is quickly further oxidized. Lactic acid reacts with bromal, produced from a haloform reaction, to give 4-methyl-2-(tribromomethyl)-5-dioxolone [28], [29].

2.2.4. Reduction

Hydrogen iodide reduces lactic acid to propionic acid [30]. Catalytic reduction with rhenium black catalyst yields propylene glycol (84%) and dilactide (16%) at 26 MPa and 150 °C [31].

2.2.5. Other Reactions

A quantitative yield of *2-chloropropionic acid* is obtained when equimolar amounts of lactic acid and thionyl chloride are heated together at 80 °C for a few hours. This reaction will produce 2-chloropropionyl chloride if an excess of thionyl chloride is used [32]. The action of diazomethane or dimethyl sulfate produces *methyl 2-methoxypropionate*.

Acylated lactic acids can be synthesized by the reaction of lactic acid with acylating agents such as the corresponding acid chlorides, or by hydrolysis of acylated lactic esters. More important are the acylated lactic esters which are prepared by the reaction of acid chlorides with lactic esters. The acylated lactic esters undergo pyrolysis at 500 °C by eliminating one molecule of the acylating acid to produce the corresponding derivative of acrylic acid

$$R-COO-CH(CH_3)-COOR' \longrightarrow CH_2=CH-COOR' + R-COOH$$

Ammonolysis of ethyl lactates, dilactide and short-chain length polylactic acids gives rise to *lactamide*. For example, lactamide is prepared by passing dry ammonia over heated dilactide [33] and polylactic acids can be treated similarly [34]. Alkyl lactates can be reacted with dry ammonia, under pressure and using ammonium chloride as a catalyst [35].

Many more different reactions of lactic acid are found in literature. Lactic acid is also of great biological importance and a tremendous amount of information is present in literature [17].

3. Production

Lactic acid is produced on an industrial scale either by *fermentation* or a *synthetic* method. In recent years the fermentation approach has become more successful because of the increasing market demand for lactic acid which is naturally produced.

3.1. Raw Materials

The raw materials used in *fermentation* consist predominantly of hexoses or compounds which can be easily split into hexoses, e.g., sugars, molasses, sugar beet juice, sulfite liquors, whey [36], as well as rice, wheat, and potato starches. Soluble proteins, phosphates and ammonium salts are also required as a source of bacterial nutrient. The nutrient requirements can be met by the addition of malt germs or yeast extracts. A pH controlling compound (caustic or lime solution or carbonate) is also required.

3.2. Lactic Acid Fermentation [36]–[39]

Two different types of lactic acid fermentation exist namely, *homolactic* (pure lactic) fermentation or *heterolactic* (mixed lactic) fermentation. *Homolactic fermentation* produces predominantly lactic acid (Pi = inorganic phosphate).

Glucose + 2 ADP + 2 Pi $\longrightarrow$ 2 Lactic acid + 2 ATP

Homolactic Lactobacteriaceae such as *Lactobacillus delbruckii, Lactobacillus bulgaricus* [36], *Lactobacillus Leichmannii, Lactobacillus casei,* and *Lactobacillus salivarius* can be used.

Heterolactic fermentation produces large amounts of other fermentation products, such as acetic acid, ethanol, formic acid and carbon dioxide, depending on the raw materials used. The heterolactic metabolism of glucose proceeds as follows:

Glucose + 2 ADP + 2 Pi $\longrightarrow$ Lactic acid + ethanol + CO_2 + 2 ATP

It can be accomplished by heterolactic Lactobacteriaceae, e.g., *Lactobacillus brevis, Lactobacillus buchneri, Lactobacillus bifidus* and also by various other bacteria, e.g., *Staphylococcus Bacillus,* Actinomyces, and Salmonella species as well as by the Rhizopus fungi species.

Homofermentative Lactobacteriaceae generally produce L(+) or racemic lactic acid, whereas there are two groups among the heterofermentative Lactobacteriaceae, those that produce only racemic lactic acid and those that produce mainly D(–) lactic acid. With both types of fermentation the L(+) : D(–) lactic acid ratio is influenced by the pH, and the age of the cultures in some bacteria (e.g., *Lactobacillus casei*).

Industrial-scale production of lactic acid requires large amounts of bacterial innoculum which has to be cultured. The fermentation takes place in large stainless steel or concrete vessels (50–100 m^3) and takes between 2–8 d to complete. During this time temperature, pH and progress of the reaction must be carefully monitored. The pH is particularly important and must be maintained between 5.5–6.5 using one of the neutralizing agents mentioned in Section 3.1. If levels of free lactic acid reach 1–2 wt% of total combined lactic acid then the bacteria will die.

As the fermentation is totally anaerobic, any air contamination and possible respiration processes must be suppressed. Many Lactobacteriaceae can be used at relatively high temperature, e.g., *Lactobacillus leichmannii* and *Lactobacillus delbruckii* strains at 50 °C; this stops any unwanted secondary fermentations produced by acetic, propionic and butyric acid bacteria. The lactic acid yields are between 85–95% based on fermentable sugars; secondary products such as acetic, propionic, and butyric acids, if produced at all, are found in small quantities, i.e., ca. 0.5%.

3.3. Lactic Acid Purification [40]

The fermented broth is generally heated to ca. 70 °C to kill the bacteria and then acidified with sulfuric acid to pH 1.8. The precipitated salts and biomass are removed by filtration and the resulting liquor is treated with activated charcoal to remove any dyes. The clarified lactic liquor is then ion exchanged and concentrated to 80%. Smell and taste can be further improved by oxidative treatment, e.g., with hydrogen peroxide. The lactic acid obtained at this stage is usually of consumable quality and is suitable for some food industries, for example, sugar confectionery and biscuit manufacturers. To produce pharmaceutical grade lactic acid, several additional purification steps are necessary. Good purification can be obtained by liquid–liquid extraction where lactic acid is extracted into an organic solvent or solvent mixture and then back extracted into water [41]–[48]. Good purification can also be achieved by (1) steam distillation [18]–[21], (2) esterification followed by distillation and hydrolysis [49]–[51], (3) by calcium salt formation and re-release of the acid [52], or (4) by reverse osmosis [53].

Fermentation of raw materials other than sugar–glucose solutions, e.g., whey and molasses can be carried out but is not an industrially used process. The lactic acid produced from these raw materials requires extensive purification operations and therefore can become economically unviable.

3.4. Synthesis of Lactic Acid

Since the 1960s lactic acid has also been produced in considerable amounts via synthetic routes. Several methods for the industrial production of synthetic lactic acid have been considered. The present industrial synthesis of lactic acid is known to be based on the reaction of acetaldehyde with hydrogen cyanide followed by the hydrolysis of the resultant lactonitrile:

$$CH_3-CHO + HCN \xrightarrow{HCl} CH_3-CH(OH)-CN$$
$$CH_3-CH(OH)-CN + 2H_2O \xrightarrow{HCl} CH_3-CH(OH)-COOH + NH_4Cl$$

The reaction of propene with liquid dinitrogen tetroxide [54]–[57], at 15–20 °C, gives rise to 1-nitropropan-2-ol which can be hydrolyzed with hydrochloric [58] or sulfuric acids [59] to lactic acid.

$$CH_3-CH=CH_2 \xrightarrow{N_2O_4} CH_3-\underset{OH}{CH}-CH_2-NO_2 \xrightarrow{HCl} CH_3-\underset{OH}{CH}-COOH$$

Another main product of the above reaction is nitratolactic acid ($H_3C-CH(ONO_2)-COOH$) which can undergo a saponification reaction to give lactic acid in 75–85% yields [60].

Another production method is based on a high pressure reaction of acetaldehyde with carbon monoxide and water in the presence of a nickel(II) iodide [61] or sulfuric acid [62] catalyst. Lactic acid can also be produced by chlorination and subsequent hydrolysis of propionic acid:

$$CH_3-CH_2-COOH \xrightarrow{Cl_2} CH_3-\underset{Cl}{CH}-COOH \xrightarrow{[OH]^-} CH_3-\underset{OH}{CH}-COOH$$

3.5. Construction Materials

Lactic acid is corrosive toward metallic materials, normally used in production and processing, especially at high temperatures. The precious metals as well as titanium, zirconium, tantalum, niobium, chromium, and molybdenum are resistant to corrosion. The corrosive properties of lactic acid also depend on its concentration; thus initial evaporation of lactic liquors can normally take place in 316 stainless steel evaporators, but final evaporation to 80% lactic acid normally takes place in titanium-lined evaporators.

As lactic acid fermentations are carried out at pH 5.5–6.5, it is not necessary to employ expensive, precious-metal-lined fermentors. Rubber-lined or even wooden fermentors can be used; however, in practice stainless steel fermentors are used for ease of sterilisation.

Table 3. Specifications of lactic acid

Quality	Pharmaceutical grade	Food Chemicals Codex	Typical edible grade
Assay, %	88.0	95 – 105 *	80
Chloride, %	0.008	0.2	0.02
Sulfate, %	0.02	0.25	0.05
Arsenic, ppm	4	3	0.2
Heavy metals, ppm	33	10	10
Iron, ppm	10	10	10
Ash, %	0.1	0.1	0.1
Calcium, %	0.02		

* Of specified concentration.

4. Environmental Protection

Lactic acid, a naturally occuring organic acid found in many different biological systems, is environmentally safe. It can be incorporated into food chains and is metabolized completely. Industrial fermentation of lactic acid produces byproducts and a waste effluent. The slightly acidic effluent is neutralized before emission into the environment; the byproduct is simply calcium sulfate mixed with a nitrogen-rich biomass and can be used in gardening and horticulture for breaking down clay soils and providing a nitrogenous nutrient for plants.

5. Quality Specifications

Pharmaceutical and edible grade lactic acids are considered to be of most importance. Although quality specifications for *pharmaceutical-grade* lactic acid are laid down in the national pharmacopeia in many countries, there is still no defined specification for *edible-grade* lactic acid. In the United States the specification for edible-grade lactic acid is laid down in the Food Chemicals Codex. Table 3 contains a comparison of specifications for pharmaceutical grade, Food Chemicals Codex, and typical edible lactic acid.

6. Analysis

Quantitative analysis of lactic acid can be carried out by titration against sodium hydroxide using phenolphthalein as the indicator. This method is only accurate for very weak lactic acid solutions. Stronger solutions contain lactoyllactic acid which must be previously hydrolyzed to lactic acid. This is done by boiling the lactic acid in a measured excess of sodium hydroxide, for a few minutes, and back titrating with hydrochloric acid using phenolphthalein as the indicator.

A modern technique for quantitative analysis, and for following the production of lactic acid by both fermentation and synthetic routes, is the use of HPLC systems. This method is accurate and can also quantitatively measure the presence of other entities, e.g., sugars and other organic acids.

Potentiometric and oxidimetric titration methods and photometric methods are known but they are of little importance compared to the methods above.

To determine the L(+) and D(−) lactic acid content of a solution, enzymatic methods are used. The enzymes and analyzers are commercially available together with lactic acid solutions of very high individual isomeric purity.

7. Storage and Transportation

Lactic acid is stored in glass, plastic or suitably lined tanks and is transported and supplied in well sealed plastic containers.

8. Legal Aspects

Lactic acid is permitted as a food acid by legislation in most countries. In Europe it has to conform to the specific criteria laid down in the EEC Council Directive 65/66/EEC and FCC.

9. Uses

Lactic acid and some of its derivatives (salts and esters) can be used in many applications. The food industry has become the most important outlet for lactic acid and lactates. About 25 000 t/a are used in almost every segment of the food industry. Lactic acid has a mild acid taste, does not overpower weaker aromatic flavors, and acts as a preservative. Its natural occurance means that it does not introduce a foreign element into the food and its salts are very soluble thus giving the possibility of partially replacing the acid in buffering systems.

Buffered lactic acid (lactic acid – sodium lactate mixture) is used in continuous production lines for sugar confectionery. In dairy products such as cottage cheese direct acidification with lactic acid is often preferred to fermentation with its associated risks of failure and infection. 1–2% lactic acid solutions are used as antimicrobial treatment for meat and poultry. It is a natural beer acid which improves microbial stability and flavor; it is preferred for the acidification of low-acid wines and ciders because it is not attacked by acid-degrading bacteria. Partial replacement of acetic acid by lactic acid in pickled products, such as gherkins, silverskin onions or in marinated fish, imparts a milder and more agreeable taste as well as improving microbial stability and flavor.

Other uses of lactic acid are in agriculture for silage manufacture; in animal feeds to promote correct fermentations in the gut; in the textile industry as a solubilizer, pH controlling agent – acidulant and as a biodegradable intermediate in chrome dying of wool [63]; in printing as an intermediate in duplicating ink manufacture and as a solvent in silk screen cleaning; in metal treatment as a biodegradable rust remover; in the manufacture of paints and resins. Polylactic acid is used as a biodegradable polymer for medical purposes.

10. Derivatives

Lactic acid is also used as an intermediate and raw material in the chemical industry for the preparation of derivatives such as lactic salts, esters, amides, and nitriles.

10.1. Salts

Lactic acid forms salts with most metals, ammonia, and a large number of organic bases. Several general methods have been applied to prepare lactates. Water-soluble lactates are concentrated and allowed to crystallize, whenever possible. However, some lactates are so soluble in water that they are hardly known in the crystalline state, e.g., sodium and potassium lactates. Sodium and potassium lactates are prepared by the reaction of lactic acid with the corresponding hydroxides. Lactates that can crystallize from water are generally prepared by boiling lactic acid with an excess of a slightly soluble oxide, hydroxide or carbonate of the metal. Many lactates can be prepared by the double decomposition of other lactates, e.g., the preparation of aluminum lactate from the reaction of calcium lactate and aluminum sulfate. Calcium and sodium lactates are by far the most important.

Calcium lactate [*814-80-2*], $(CH_3CH(OH)COO)_2Ca \cdot 5\,H_2O$, M_r 308.29, crystallizes from water as the pentahydrate. It is an intermediate during the fermentation of lactic acid. Anhydrous calcium lactate is produced if the pentahydrate is heated to 100 – 120 °C. The solubility of the lactate increases dramatically at elevated temperature.

Calcium lactate is used within the food industry as an egg extender and foaming agent because of its properties as a protein plasticizer (coagulant). In the medical profession it is used as a source of calcium.

Sodium lactate [72-17-3], $CH_3CH(OH)COONa$, M_r 112.06, is very hygroscopic and very soluble in water and alcohol. Because of this property, it is extremely difficult to isolate the crystalline form. It is used in many applications, for example, in pickling, as a preservative for ground meat, in cosmetics, and the textile industry.

In addition *aluminum, magnesium, strontium, copper, zinc, iron, zirconium* and *ammonium lactates* are also important, especially in the medical and cosmetic industry.

10.2. Esters

A great number of lactates (esters) are known. Particularly interesting are the methyl, ethyl, and *n*-butyl esters. These are used as solvents for varnishes, nitrocellulose, and polyvinyl compounds. The production of these compounds is discussed in Section 2.2.2.

Methyl lactate [547-64-8], $CH_3CH(OH)COOCH_3$, is a colorless liquid, miscible with water in any ratio; *bp* 145 °C (at 101.3 kPa), ϱ 1.0928 g/cm^3, n_D^{20} = 1.4141. It forms an azeotropic mixture with water at 99 – 99.5 °C; with a lactate content of 25 %.

Ethyl lactate [97-64-3], $CH_3CH(OH)COOCH_2CH_3$, is a colorless liquid, *bp* 154 °C (at 101.3 kPa), ϱ 1.030 g/cm^3, n_D^{20} 1.4124. It is miscible with water, ethanol, and diethyl ether in any ratio.

***n*-Butyl lactate**, $CH_3CH(OH)COOCH_2CH_2CH_2CH_3$, is a colorless liquid, *bp* 58 °C (at 2.0 kPa), ϱ 1.0324 g/cm^3, n_D^{20} 1.4125, its solubility in water is 4.36 wt% at 25 °C.

10.3. Lactamides

See Section 2.2.5 for method of preparation. Some lactamides such as *N*-cyclohexyl and *N*-(2-hydroxyethyl) lactamides, are of interest as plasticizers.

10.4. Lactonitriles

See Section 3.4 for method of preparation. The method, used for the production of synthetic lactic acid incorporates a lactonitrile as the intermediate, which is hydrolyzed to give lactic acid.

11. Economic Aspects

At present the world production of lactic acid is ca. 40 000 t/a; of which 65% is by fermentation and 35% by synthetic processes. All of Europe's production is via a *fermentation route*; the Dutch company CCA, together with its holdings in Spain and Brazil, produces 20 000–25 000 t/a. Croda, United Kingdom produces about 2 500 t/a, and in the People's Republic of China about the same amount is a produced.

Synthetic lactic acid is produced by Sterling in the United States (ca. 5 000–6 000 t/a) and Musashino in Japan (ca. 7 000 t/a).

12. Toxicology and Occupational Health

Lactic acid is a nontoxic naturally occuring edible acid used widely in the food industry as an acidulant. It is, however, acidic and will cause discomfort if it should come into contact with eyes or broken skin. Lactic acid may produce acetaldehyde on contact with strong oxidizing agents.

The problem of oral toxicity should not arise unless a large quantity is swallowed (LD_{50}: 3 730 mg/kg in rats). Lactic acid does not cause any visible necrosis of the skin tissue when tested according to the procedure outlined in the classification, packaging and labelling of dangerous substances 1984 [64]. It can cause eye irritation and must be washed out immediately.

13. References

[1] C. W. Scheele, Om Mjolk: *Kgl. Vetenskaps-Academiens nya Handlingar*, 1, Stockholm 1780, pp. 116–124.
[2] Thenard, *Bull. Sci. Soc. Philomatique* **3** (1805) no. 96, 283–284.
[3] Thenard, *Ann. Chim.* **59** (1806) 262–283.
[4] Fourcroy, Vauquelin, *Mem. Inst. Sci. lettres Arts. Sci. mathem. phys.* **6** (1806) 332–368.
[5] Bouillon-Lagrange, *Ann. Chim. (Paris)* **50** (1804) 272–296.
[6] J. Berzelius: *Econom. Ann. Kgl. Vetenskaps-Academien* (Stockholm) 7. Juli Månad; 84–100; Augusti Månad; 30–49 (1808).
[7] J. Berzelius: *Econom. Ann. Kgl. Vetenskaps-Academien* (Stockholm) 6. Maj Månad, 20–39 (1808).
[8] H. Braconnot, *Ann. Chim. (Paris)* **86** (1813) 84–100.
[9] J. Vogel, *Chem. Phys. (Schweigger)* **20** (1817) 425–429.
[10] E. Fremy, Transformation de la mannite, du sucre de lait, de rendus, **8** (1839) 960–961.
[11] F. Krafft, W. A. Dijes, *Ber. Deut. Chem. Ges.* **28** (1895) 2589–2597.

[12] L. B. Lockwood, D. E. Yoder, H. Zienty, *Ann. N. Y. Acad. Sci.* **119** (1965) 854.
[13] M. Brin, "Lactic acid – some definitions," *Ann. N. Y. Acad. Sci.* **119** (1965) 1084–1090.
[14] H. Borsook, H. M. Huffman, Y. P. Liv: "The preparations of crystalline lactic acid," *J. Biol. Chem.* **102** (1933) 449–460.
[15] H. A. Krebs, *Biochem. Prep.* **8** (1961) 75–79.
[16] D. Vorlander, R. Walter, *Z. Phys. Chem.* **118** (1925) 1–30.
[17] C. H. Holten: *Lactic Acid*, Verlag Chemie, Weinheim 1971.
[18] A. Partheil, *Z. Unters. Nahr. Genussm. Gebrauchsgegenstände* **5** (1902) 1053–1062.
[19] A. Partheil, W. Hubner, *Arch. Pharm. (Weinheim Ges.)* **241** (1903) 421–435.
[20] S. Suzukii, E. B. Hart, *J. Am. Chem. Soc.* **31** (1909) 1364–1367.
[21] A. Sepitka, *Prum. Potravin* **13** (1962) 385–388, 605–609; **14** (1963) 45–49, 82–85.
[22] R. A. Troupe, W. L. Aspy, P. R. Schrodt, *Ind. Eng. Chem.* **43** (1951) 1143–1146.
[23] W. Vidler, *Private Communication to Stichting Ilra*, 1959.
[24] A. W. Martin, H. V. Tartar, *J. Am. Chem. Soc.* **59** (1937) 2672–2675.
[25] G. Saville, H. A. Gundry, *Trans. Faraday Soc.* **55** (1959) 2036–2038.
[26] O. Meyerhof, *Biochem. Z.* **129** (1922) 594–604.
[27] S. Bezzi, L. Riccoboni, C. Sullam, *Reale arrad. d'Italia, Mem. cl Sci. Fis. Mat. nat.* **8** (1937) 181–200.
[28] E. Klimenko, *Ber. Deut. Chem. Ges.* **9** (1876) 967–968.
[29] E. Klimenko, *J. Prakt. Chem.* **13** (1876) no. 2, 98–102.
[30] E. Lautemann, *Justus Liebigs Ann. Chem.* **113** (1860) 217–220.
[31] H. S. Broadbent, G. C. Campbell, W. J. Bartley, J. H. Johnson, *J. Org. Chem.* **24** (1959) 1847–1854.
[32] G. Battaglino, *Ann. Chim. (Rome)* **56** (1966) 820–826.
[33] L. Henry, *Ber. Deut. Chem. Ges.* **7** (1874) 753–756.
[34] E. Jungfleisch, M. Godchot, *C. R. Seances Acad. Sci.* **140** (1905) 502–505.
[35] G. J. Wolf, J. G. Miller, A. R. Day, *J. Amer. Chem. Soc.* **78** (1956) 4372–4373.
[36] G. C. Cox, R. D. Macbean, *Aust. J. Dairy Technol.* **32** (1977) no. 1, 19–22.
[37] M. A. Cousin, E. H. Marth, *J. Food Prot.* **40** (1977) no. 6, 406–410.
[38] M. A. Cousin, E. H. Marth, *J. Food Prot.* **40** (1977) no. 7, 475–479.
[39] E. S. Mil'ko, O. V. Sperehip, I. L. Rabotnova, *Z. Allg. Mikrobiol.* **6** (1967) no. 4, 297–301.
[40] T. K. Bansal, M. S. Vishwanathan, T. K. Ghose, *Indian Chem. Eng.* **18** (1986) 24–28.
[41] Les Unives de Melle, FR 1494 968, 1967 (P. L. Rancon).
[42] Melle Besons S.A., DE 1 268 088, 1968 (C. Neuffer).
[43] W. P. Ratchford, US 2 539 472, 1951.
[44] V. B. Sevcenko, E. Renard, *J. Russ. Inorg. Chem.* **8** (1963) no. 2, 268–271.
[45] A. M. Baniel et al., US 4 275 234, 1981.
[46] Miles Laboratories Inc., EP 0 049 429, 1981 (A. M. Baniel).
[47] A. S. Kertes, C. J. King, *Biotechnol. Bioeng.* **28** (1986) 264–282.
[48] J. Chaudhuri, D. L. Pyle: *Separations for Biotechnology*, Harwood, London 1987, pp. 241–259.
[49] Bowmans Chem. Ltd., GB 907 322, 1962 (M. H. M. Arnold).
[50] A. Sepitka, M. Gartner, CS 104 398, 1962.
[51] A. Sepitka, *Prum. Potravin.* **14** (1963) 606; **17** (1966) 633.
[52] C. H. Boehringer, DE 744 755, 1943 (W. Stiller).
[53] B. R. Smith, R. D. Macbean, G. C. Cox, *Aust. J. Dairy Technol.* **32** (1977) no. 1, 23–26.
[54] B. F. Ustavscikov et al, *Dokl Chem. (Engl. Transl.)* **157** (1964) 685–688.
[55] B. F. Ustavscikov et al., *Probl. Org. Sint.* 1965, 18–23.

[56] B. F. Ustavscikov, V. A. Podgornova, M. I. Farberov, *Dokl. Chem. (Engl. Transl.)* **168** (1966) 637–639.
[57] B. F. Ustavscikov et al., *Pet. Chem. USSR (Engl. Transl.)* **6** (1967) no. 1, 20–29.
[58] V. V. Perekalin et al., *Dokl. Akad. Nauk SSSR* **116** (1966) 1129.
[59] F. G. Nasybullina, V. I. Burmistrov, *Zh. Prikl. Khim. (Leningrad)* **42** (1969) no. 6, 1436.
[60] Rhône-Poulenc, FR 92 116, 1968 (J. Boichard, B. Rossard, M. Gay, R. Janin); 90 556, 1968 (J. Boichard, B. Rossard, M. Gay, R. Janin).
[61] S. K. Bhattacharya, S. K. Palit, A. Das, *Ind. Eng. Chem. Prod. Res. Dev.* (1970) 992.
[62] Du Pont, US 2 265 945, 1942 (D. J. Loder).
[63] L. Benise, *J. Soc. Dyers Colour.* **94** (1978) 101–105.
[64] Information approved for the classification, packaging and labelling of dangerous substances for supply and conveyance by road. Her majesty's stationary office, London (1984).

Lactose and Derivatives

WILLEM A. ROELFSEMA, ccFriesland Cooperative Company, Leeuwarden, The Netherlands (Chap. 1, Section 2.2)

BEN F. M. KUSTER, Technische Universiteit Eindhoven, Eindhoven, The Netherlands (Chap. 1, Section 2.2)

MICHIEL C. HESLINGA, Solvay Pharmaceuticals, Weesp, The Netherlands (Section 2.1)

HENDRIK PLUIM, Solvay Pharmaceuticals, Weesp, The Netherlands (Section 2.1)

MARINUS VERHAGE, Solvay Pharmaceuticals, Weesp, The Netherlands (Section 2.1)

1.	**Lactose**	3135	2.	**Derivatives**	3142
1.1.	**Properties**	3136	2.1.	**Lactulose**	3142
1.1.1.	Physical Properties	3136	2.1.1.	Properties	3143
1.1.2.	Chemical Properties	3136	2.1.2.	Production	3143
1.2.	**Production**	3137	2.1.3.	Quality Specifications	3144
1.3.	**Quality Specifications**	3139	2.1.4.	Toxicology and Physiology	3144
			2.1.5.	Medical and Pharmaceutical Uses	3145
1.4.	**Uses**	3139	2.2.	**Miscellaneous Derivatives**	3146
1.5.	**Economic Aspects**	3140	2.2.1.	Lactobionic Acid	3146
1.6.	**Producers**	3141	2.2.2.	Hydrolyzed Lactose Syrup	3146
1.7.	**Physiological Properties**	3141	3.	**References**	3146

1. Lactose

Lactose, milk sugar, $C_{12}H_{22}O_{11}$, 4-O-β-D-galactopyranosyl-α-D-glucopyranose [63-42-3], M_r 342.30 (anhydrous); monohydrate [64044-51-5], M_r 360.31, is the principal carbohydrate constituent of milk of most mammals. Lactose concentration varies from 0 to 9 wt%, being zero in sea lions, 4.0% in goats, 4.8% in cows, and 6.7 wt% in humans [1].

α-Lactose

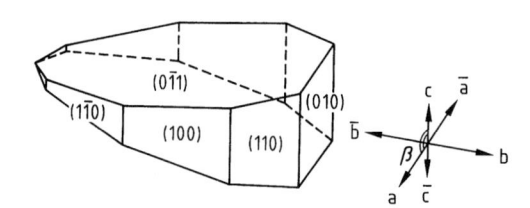

Figure 1. Crystal structure of α-lactose monohydrate

BARTOLETTI discovered lactose in milk in 1633, and in 1780 SCHEELE identified its sugar character. The structure was confirmed by synthesis in 1942 by HASKINS et al. [2].

1.1. Properties

Lactose exists in two anomeric forms: α- and β-lactose, differing only in the relative positions of the hydrogen and the hydroxyl group at the C^1-atom of the pyranosidic glucose-part. In aqueous solution lactose consists of 61.5% of β-pyranose and 38.5% of α-pyranose; the equilibrium between both forms is established by mutarotation ($K_\alpha + K_\beta = 1.74 \times 10^{-4}$ s^{-1} at 20 °C) [3]. Lactose usually crystallizes in the form of its α-lactose monohydrate. The anhydrous β-form can only be obtained at crystallization temperatures > 93 °C. Amorphous lactose, which is quite hygroscopic, is a mixture of α- and β-lactose; it can be obtained by rapid drying methods, e.g., spray drying.

1.1.1. Physical Properties

The physical properties are shown in Table 1 and Figures 1 and 2 [1], [4]–[9].

1.1.2. Chemical Properties

Lactose has the same principle molecular structure as most other disaccharides and differs only in its configuration. Thus, it generally reacts in the same way as other sugars (→Carbohydrates). Important points of attack are the reducing end of the glucose moiety, the various hydroxyl groups, and the glycosidic bond between the galactose and the glucose moiety [10].

Examples of reaction types are: (1) formation of anhydro derivatives under the influence of heat and radiation and, (2) formation of unsaturated heterocycles by heat in the presence of an acid [11], (3) oligomerization [12], (4) more or less extensive oxidation at various sites by chemical means or catalytically [11], (5) Maillard type reactions with amino compounds [13], (6) isomerization [14], (7) acetalation [15], (8) hydrogenation [16], (9) alkaline degradation [13], (10) complexation of metal cations [17], (11) glycosidation [11], (12) etherification [13], (13) halogenation [13], (14) esterification [14], and (15) hydrolysis by enzymatic or acid catalysis [18], [19].

Table 1. Physical properties of Lactose

Properties	α-Lactose monohydrate	β-Lactose
mp, °C	201–202 (decomp.)	253 (decomp.)
Specific rotation $[\alpha]_D^{20}$, 10^{-2} °·cm²/g	+89.4	+34.3
equilibrium value	+52.5	+52.5
Crystalline form	sphenoidal monoclinic 'tomahawk' (Fig. 1)	monoclinic 'uneven-sided diamond'
a, nm	0.7982	1.081
b, nm	2.1652	1.334
c, nm	0.4824	0.484
β, °	109.78	91.15
Density (293 K), g/cm³	1.54	1.59
Bulk density, g/cm³		
poured	0.5–0.9	
tapped	0.75–1.1	
Enthalpy of dissolution, J/g	−50.24	−9.62
Enthalpy of combustion, J/g	−16.1	−16.5
Specific heat (298 K), J K^{-1} g^{-1}	1.251	1.193
Entropy (298 K), J K^{-1} mol^{-1}	415	386
Entropy of formation (298 K), J K^{-1} mol^{-1}	−2453	−2248
Enthalpy of formation (298 K), kJ/mol	−2481	−2233
Gibbs free energy (298 K), kJ/mol	−1750	−1564
Solubility, g/100 g		
in H$_2$O	see Fig. 2	see Fig. 2
in DMSO	59 (20 °C)	30 (20 °C)
	73 (80 °C)	43 (80 °C)
pK_a at 3–5 °C	13.8	13.5
Diffusion constant in H$_2$O, m²/s	43×10^{-11}	

Properties of solutions	Concentration (g lactose per 100 grams water)				
	0	10	20	30	40
Density (20 °C), kg/m³	998	1043	1082	1124	1173
Viscosity, mPa·s					
(20 °C)	1.00	1.38	2.04	3.42	7.01
(60 °C)	0.47	0.70	0.90	1.29	2.19
Refractive index (25 °C)	1.3325	1.3484	1.3659		

1.2. Production

Generally cheese (sweet) whey or casein (acid) whey is used as raw material for production of lactose (see Fig. 3). Today, permeates from the ultrafiltration of milk or sweet whey are also coming into consideration. With all types of raw materials precautions have to be taken regarding the minerals present; calcium salts may give rise to deposits in evaporators.

The liquid whey (ca. 6.5% dry matter content) is clarified (a) and concentrated in falling-film evaporators (b) to 55–65% total solids. Upon cooling (c), the greater part of the lactose crystallizes as lactose monohydrate. The crystal slurry is transferred to a

Figure 2. Initial solubilities of α-lactose and β-lactose and final solubility at mutarotation equilibrium of α-lactose-hydrate vs. temperature

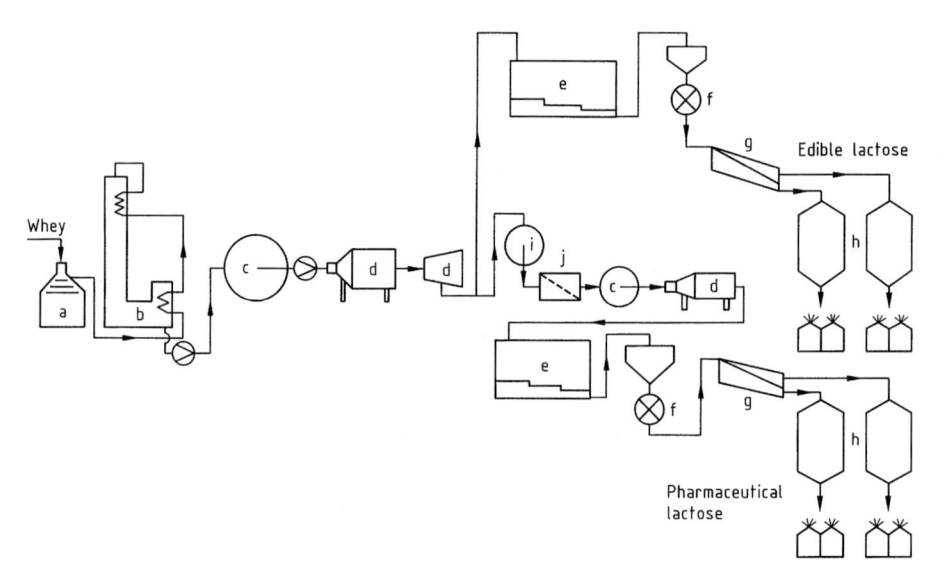

Figure 3. Production of lactose
a) Clarifier; b) Evaporator; c) Crystallization tank; d) Separator; e) Dryer; f) Mill; g) Sieve; h) Packing; i) Redissolving tank; j) Filter

Table 2. Specifications of various grades of lactose

Lactose grade	Crude	Edible	Pharmaceutical
Lactose, monohydrate, %	98	min. 99.0	min. 99.7
Free moisture, %	0.4	max. 0.3	max. 0.2
Protein (N content×6.38), %	1.0	max. 0.3	max. 0.1
Ash, %	0.5	max. 0.3	max. 0.1
Riboflavin, ppm	200	50	0

line of decanters and centrifuges and washed several times. Thus, after drying, milling and fractionation, yellowish lactose (edible grade) of different particle sizes is obtained.

After the washing stage the raw product can be redissolved (i), treated with activated carbon (j) and recrystallized to refined or pharmaceutical grades of lactose. From 1000 kg liquid whey 32.5 kg lactose and 32.5 kg delactosed wheypowder are obtained.

Environmental Protection. To avoid pollution the dryer should be provided with a cyclone to remove the powder from the exhaust air. Fluid cleaning streams must be purified by simple aerobic or anaerobic digestion before entering a public sewage treatment.

1.3. Quality Specifications

Lactose is produced in various grades of purity (see Table 2).

Chemical analyses of lactose are carried out by standard methods, complying with U.S.P./NF, EP and JP specifications. For details see the Pharmacopeias mentioned.

1.4. Uses

In the various uses of lactose the following functional properties are important [14], [18], [20], [21]:

1) free flowability,
2) flavor and taste enhancement,
3) nonhygroscopic, anti-caking, fat-coating properties,
4) adsorption and absorption capacities,
5) ease of solubility,
6) agglomerating and instantizing capacities,
7) slow fermentation plus acid regulation,
8) stability, long shelf life,
9) absence of odor,
10) good compressibility, and

11) moderate sweetness; the concentration-dependent sweetness of lactose is ca. 30% relative to sucrose

Food quality, i.e., *edible lactose* is used for:

1) infant and baby food,
2) frozen desserts,
3) ice cream,
4) skim milk, condensed milk,
5) dietetic foods,
6) coffee creamers,
7) chocolate and sugar confectionery,
8) bakery products, meat products,
9) canned fruit and vegetables,
10) dry soups, sauces and spices,
11) binder for liquid foods, and
12) fermentation processes

Pharmaceutical-grade lactose is used for: tableting (wet granulation, direct compressing, retableting), capsule filling, hypodermics, pan coating, and as raw material for lactulose and lactitol production.

1.5. Economic Aspects

Production of lactose is especially located in those areas which have a highly centralized cheese production such as Europe, Eastern and Central States of America and New Zealand. Production figures of 1988 are in 10^3 t/a:

World total	295
Europe	215
North America	65
Oceania	15

Areas of application can be divided into pharmaceutics 75 000 t/a (25%), baby and infant food 88 500 t/a (30%), human food 88 500 t/a (30%), fermentation and animal feed 43 000 t/a (15%).

1.6. Producers

The main producers of lactose are:

Meggle Milchindustrie GmbH & Co. KG, Wasserburg, Federal Republic of Germany
Milei GmbH, Stuttgart, Federal Republic of Germany
Laiteries E. Bridel SA, Retiers, France
Avonmore Creameries Ltd., Kilkenny, Ireland
Borculo Whey Products, Borculo, The Netherlands
ccFriesland Cooperative Company, Leeuwarden, The Netherlands
De Melkindustrie Veghel BV, Veghel, The Netherlands
DOMO Melkproductenbedrijven, Beilen, The Netherlands
Foremost Whey Products, Baraboo – Wisc., United States
Sheffield Products, Norwich – New York, United States
Lactose Company of New Zealand, Hawera, New Zealand

1.7. Physiological Properties [20]

The metabolizable energy of α-lactose monohydrate amounts to –15.7 kJ/g, that of β-lactose to –16.0 kJ/g.

Lactose is less cariogenic than other natural mono and disaccharides. This results from:

1) A slower rate of fermentation in the toothplaque,
2) A lower final acidity in the same environment, and
3) The lack of production of extra-cellular bacterial polysaccharides

Once lactose reaches the intestines it is enzymatically cleaved into glucose and galactose. The lactic acid formed by subsequent fermentation, and the resulting acid pH prevent the growth of undesirable microorganisms.

The intestinal absorption of calcium and other di and trivalent metal ions, such as magnesium, barium, zinc and iron is strongly improved by the presence of lactose. This can be explained by a decrease of the pH in the digestive tract, which increases the solubility and absorbability of the complexed metal ions.

Lactose has moderate laxative properties because its fermentation products increase the osmotic pressure in the large intestine, thus the loss of too much water is prevented by reabsorption.

In Asia, Africa, and South America large parts of the population have difficulties in absorbing lactose, whereas in Europe and North America, <8% of the population have such difficulties. This so-called lactose intolerance, i.e., clinical phenomena such as flatulence and diarrhoea, is normally caused by a strongly reduced lactase activity in the mucosa of the small intestine, giving rise to malabsorption of lactose. This does not imply, however, that a malabsorber is totally lactose intolerant [17].

Generally lactase activity in adults is ca. 10% of that of newborns.

2. Derivatives

2.1. Lactulose

Lactulose [4618-18-2], 4-O-β-D-galactopyranosyl-D-fructose, $C_{12}H_{22}O_{11}$, M_r 342.30, is a synthetic disaccharide in which galactose is linked to fructose. It was first synthesized in 1929 by isomerization of lactose [22].

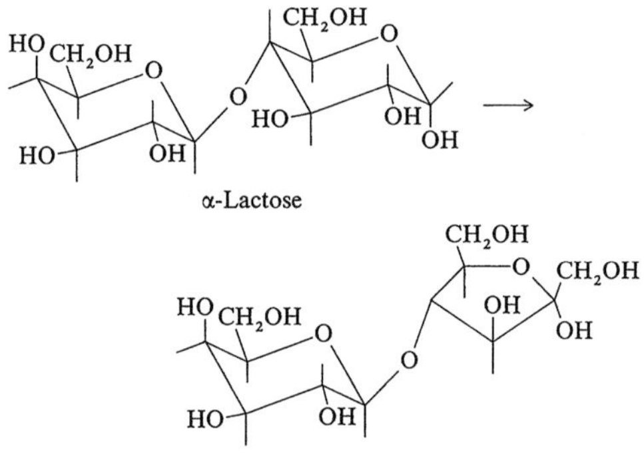

α-Lactose

Lactulose
(α-furanoid)

The first synthesis of lactulose was completely described by Montgomery and Hudson in 1930 [23]. They prepared lactulose from lactose in an alkaline solution of lime. Lactulose has not been detected in raw milk, but has been found in milk and milk products after heat treatment [24], [25].

Molecular Structure. The molecular structure of lactulose has been elucidated mainly by Hicks [26], [29].

The anomeric composition of lactose depends on the physical state. In solution the anomeric composition appears to be related to temperature [27] (Table 3). In the solid state two crystal modifications have been described; lactulose anhydrate [29] and lactulose trihydrate [28]. Both modifications differ in anomeric composition as shown in Table 4. The coexistence of different anomers in the same crystal structure is often observed in carbohydrate chemistry, because the ability to form hydrogen bonds of comparable energy facilitates cocrystallization.

Table 3. Variation in the percent distribution of anomeric forms of lactulose in water with temperature

	25 °C	58 °C	73 °C
ß-Pyranose	61.5 %	54.7 %	52.1 %
ß-Furanose	29.3 %	31.6 %	35.5 %
α-Furanose	7.6 %	11.2 %	12.4 %
Open ketoform	1.6 %	2.5 %	n.d.

Table 4. Percent distributions of anomeric forms in the crystal modifications of lactulose

	Anhydrate	Trihydrate
ß-Pyranose	16.0 %	
ß-Furanose	72.2 %	100 %
α-Furanose	11.8 %	

2.1.1. Properties

Lactulose anhydrate and trihydrate are white, odorless crystalline powders. Its most important properties are as follows:

mp anhydrate [30]	169 °C
mp trihydrate [31]	68 °C
Solubility in water (kg/kg) anhydrate [30]	3.2 (30 °C), 4.3 (60 °C), > 6.2 (90 °C)
Solubility in water (kg/kg) trihydrate [32]	1.3 (5 °C), 1.5 (10 °C), 2.1 (20 °C), 3.1 (30 °C)
Specific rotation $[\alpha]_D^{20}$ [30]	−51.4 °
Degree of sweetness [33]	0.48–0.62 (sucrose =1.0)
Concentration (% wt) related to refraction index [34]	0.97° Brix (sucrose index scale)

2.1.2. Production

Lactulose is available on the market both as a syrup and as a dry crystalline powder (anhydrate).

Lactulose can be prepared from an alkaline solution of lactose by the Lobry de Bruyn – Alberda van Ekenstein rearrangement (→ Carbohydrates). The first synthesis was performed with lime [23]. Since then many catalysts, such as alkali hydroxides [36], [37], magnesium oxide [38], alkali and alkaline-earth aluminates [39] or alkaline-earth sulfite [40], alkali borate [42], [43], alkali tetraborate plus ion-exchange resin [44], anion exchange resin [45], and amines [46] are reported to give good yields or to minimize secondary reactions.

On a industrial scale, lactulose syrup is usually produced by one of two ways:

1) Common lactulose syrup is prepared by dissolution of lactose in water and addition of sodium hydroxide solution while heating. Because of inevitable degradation reactions the conversion has to be limited. The excess of lactose is removed by crystallization. The solution is deionized using ion exchange resins. A subsequent concentration step leads to a second crop of unreacted lactose.

2) Alternative processes use coreagents such as borate [41] and aluminate [47]. These lactulose complexing agents enable the conversion of lactose to lactulose in a high yield by the prevention of degradation reactions [48]. So higher conversion and selectivity can be reached. Borate or aluminate is subsequently recycled.

The resulting syrup is the common marketed lactulose solution, containing 47–53 wt% of lactulose and some other sugars such as lactose, epilactose, galactose, fructose and tagatose. The amount of these related sugars depends on the process used and optional additional purification steps.

Two different processes for obtaining crystalline lactulose anhydrate are used:

1) Further purification of common lactulose syrup by crystallization from alcohols and/or water followed by drying [49].
2) Solidification of an already pure lactulose syrup by well controlled evaporation and stirring, optionally using seed crystals [50].

Two completely different approaches to the synthesis of lactulose, (1) an enzymatic coupling of galactose and fructose using β-galactosidase and (2) an Amadori rearrangement (→ Carbohydrates) of *p*-tolyl-*N*-lactosylamine followed by acid hydrolysis, have been reported [51], [52]. These processes have not found industrial application because of their low yield and intricacy.

Producers. The main manufacturers of lactulose are Solvay, (the Netherlands and Canada), Inalco (Italy), Leavosan (Austria), Morinaga (Japan) and Danipharm (Denmark) producing about 12 000 t/a for medical and pharmaceutical uses.

2.1.3. Quality Specifications

Lactulose, both the syrup and the crystalline anhydrate, is a pharmaceutical product and its specifications are laid down in, e.g., the U.S. Pharmacopeïa, European Pharmacopoeia, and Japanese Pharmacopoeia.

2.1.4. Toxicology and Physiology

Toxicology. Lactulose administration has not been reported to cause adverse side effects nor does it accumulate in the recipient. Therefore its intake is considered to be safe.

In oral administration, the acute toxicity is at the same level as found for sucrose (male rat 25 g/kg, female rat 28 g/kg) [53], [54].

Physiology. The significance of lactulose was recognized by PETUELY in 1957 when he discovered that lactulose plays an important role in infant nutrition as a "Bifidus Factor" to increase the prevalence of *Bifidobacterium* in the intestinal tract [55].

Most orally administered di-, oligo- and polysaccharides are split by intestinal enzymes into monosaccharides which are absorbed and enter the peripheral circulation. Lactulose, however is not hydrolyzed in the human small intestine and is consequently not absorbed [55]. Most of it is transported to the large intestine, where it is metabolized to low molecular mass organic acids by lactobacilli (*Bifidobacterium*).

Lactobacilli suppress harmful intestinal bacteria, assist in digestion and absorption, produce vitamins, and stimulate immune responses. Lactulose is effective in promoting the proliferation of lactobacilli. Moreover, lactulose does not raise blood glucose levels or influence insulin secretion [56], [57]. Since it is not hydrolyzed in the mouth, lactulose does not promote dental caries [58].

2.1.5. Medical and Pharmaceutical Uses

The main application of lactulose is in humans for the treatment and prevention of portal systemic encephalopathy (PSE) and chronic constipation [59]. Lactulose is an effective laxative because it is cleaved in the colon by lactobacilli to low molecular mass organic acids. This leads to a lowering of the pH of the colon contents and via an increase in osmotic pressure to an increase of the colon contents.

The efficacy of lactulose in the treatment of PSE probably also results from its effect on increasing the number of lactobacilli in the colon. As lactobacilli lack urease and other ammonia-generating enzymes, this might result in a decreased faecal urease activity, decreased urea hydrolysis, and therefore decreased ammonia formation.

Consistent with this hypothesis is the fact that when patients with PSE are treated with lactulose a reduction of the elevated arterial and venous blood ammonia levels occurs and lactulose notably improves the disease.

Other therapeutic uses suggested for lactulose include the treatment of salmonellosis [60].

The discovery of the important role of lactulose in the growth of lactobacilli has led to several areas of application in medicine (therapeutics, diagnostics), human nutrition (food additive), and even veterinary science and nutrition. These areas are rapidly increasing in importance.

2.2. Miscellaneous Derivatives

2.2.1. Lactobionic Acid

Lactobionic acid [96-82-2] is used as a carrier for antibiotics and prepared on a small scale by electrochemical oxidation [61].

Lactobionic acid

2.2.2. Hydrolyzed Lactose Syrup

Lactose is hydrolyzed by enzymatic processes and yields a syrup after concentrating. By this procedure solubility is increased to 60% total solids and sweetness becomes 0.70 relative to sucrose. Most hydrolysis procedures however are applied to deproteinized/demineralized permeates, (demineralized) whey and milkproducts [20].

3. References

[1] P. Linko in G. G. Birch, K. J. Parker (eds.): *Nutritive Sweeteners*, Applied Science Publishers, London–New Jersey 1982, Chap. 6.
[2] *Ullmann*, 4th ed. **24** 768–770.
[3] P. A. Morrissey in P. F. Fox (ed.): *Developments in Dairy Chemistry*, vol. **3**, Elsevier Applied Science Publishers, London–New York 1985, Chap. 1.
[4] T. A. Nickerson in B. H. Webb, A. H. Johnson, J. A. Alford (eds.): *Fundamentals of Dairy Chemistry*, 2nd ed., The Avi Publishing Company Inc., Westport 1974, Chap. 6.
[5] J. W. Noordik et al., *Z. Kristallogr.* **168** (1984) 59–65.
[6] R. A. Visser, doctoral thesis, Nijmegen, The Netherlands (1983).
[7] G. de Wit, A. P. G. Kieboom, H. van Bekkum, *Carbohydr. Res.* **127** (1984) 193–200.
[8] P. Walstra, R. Jenness: *Dairy Chemistry and Physics*, John Wiley & Sons, New York 1984, Chap. 3.
[9] T. J. Buma, G. A. Wiegers, *Neth. Milk Dairy J.* **21** (1967) 208–213.
[10] L. A. W. Thelwall in P. F. Fox (ed.): *Developments in Dairy Chemistry*, vol. **3**, Elsevier Applied Science Publishers, London–New York 1985, Chap. 2.
[11] J. R. Clamp, L. Hough, J. L. Hickson, R. L. Whistler, *Adv. Carbohydr. Chem.* **16** (1961) 159–206.
[12] B. W. Lew, US 3 974 138, 1976.
[13] L. A. W. Thelwall in C. K. Lee (ed.): *Developments in Food Carbohydrate*, vol. **2**, Applied Science Publishers London 1980, 275–326.
[14] J. G. Zadow, *J. Dairy Sci.* **67** (1984) 2654–2679.
[15] L. A. W. Thelwall, *J. Dairy Res.* **49** (1982) 713–724.

[16] J. A. van Velthuijzen, *J. Agric. Food Chem.* **27** (1979) 680–686.
[17] D. M. Paige, L. R. Davis in P. F. Fox (ed.): *Developments in Dairy Chemistry,* vol. **3,** Elsevier Applied Science Publishers, London–New York 1985, Chap. 4.
[18] P. G. Hobman, *J. Dairy Sci.* **67** (1984) 2630–2653.
[19] R. R. Mahoney in P. F. Fox (ed.): *Developments in Dairy Chemistry,* vol. **3,** Elsevier Applied Science Publishers, London–New York 1985, Chap. 3.
[20] B. F. Pritzwald-Stegmann, *J. Soc. Dairy Technol.* **39** (1986) 91–97.
[21] J. G. Zadow in *Milk, The Vital Force* (Proc. XXII International Dairy Congress, The Hague, 1986), D. Reidel Publishing Company, Dordrecht 1987, 737–748.
[22] E. M. Montgomery, C. S. Hudson, *Science* **69** (1929) 556.
[23] E. M. Montgomery, C. S. Hudson, *J. Am. Chem. Soc.* **52** (1930) 2101–2106.
[24] S. Adachi, *Nature* **181** (1958) 840–841.
[25] S. Adachi, S. Patton, *J. Dairy Sci.* **44** (1961) 1375–1393.
[26] P. E. Pfeffer, K. B. Hicks, W. L. Earl, *Carbohydr. Res.* **111** (1983) 181–194.
[27] P. E. Pfeffer et al., *Carbohydr. Res.* **102** (1982), 11–22.
[28] G. A. Jeffrey, *Carbohydr. Res.* **226** (1992) 29–42.
[29] G. A. Jeffrey, R. A. Wood, P. E. Pfeffer, K. B. Hicks, *J. Am. Chem. Soc.* **105** (1983) 2128–2130.
[30] B. J. Oosten, *Recl. Trav. Chim. Pays-Bas* **86** (1967) 675–676.
[31] G. A. Jeffrey et al., *Carbohydr. Res.* **226** (1992) 29–42.
[32] T. Mizota, *Carbohydr. Res.* **263** (1994) 163–166.
[33] F. W. Parrish et al., *J. Food Sci.* **44** (1979) 813–815.
[34] A. A. de Reijke, P. van Bemmel, H. W. Geluk, *J. Pharm. Sci.* **73** (1984) 1478–1479.
[35] E. M. Montgomery, *J. Am. Chem. Soc.* **52** (1930) 2101–2106.
[36] L. Gatzsche, H. Haenel, *Ernährungsforschung* **12** (1967) 641–647.
[37] Morinaga Milk Industry Co., US 3 816 174, 1974 (T. Nagasawa et al.).
[38] S. Galup, ES 397 810, 1974.
[39] Kraftco Corporation, US 3 546 206, 1970 (J. H. Guth, M. Prospect, L. Tumerman).
[40] Laevosan Gesellschaft, DE 2 002 385, 1970 (E. Nitsch, S. Mühlbock).
[41] Duphar International Research B. V. U.S. 5 071 530, (1991) (R. E. Krumbholz et al.).
[42] J. F. Mendicino, *J. Am. Chem. Soc.* **82** (1960) 4975–4979.
[43] Morinaga Milk Industry Co., JP 91/77, 1977 (T. Nagasawa et al.).
[44] R. Carubelli, *Carbohydr. Res.* **2** (1966) 480–485.
[45] H. Demaimay, C. Baron, *Lait* **575–576** (1978) 234–245.
[46] F. W. Parrish, US 3 514 327, 1970.
[47] Sirac SRL, US 4 957 564, 1990 (R. Carrobi et al.).
[48] R. van de Berg et al., *Carbohydr. Res.* **253** (1994) 1–12.
[49] F. Innocenti, EP 318 630, 1988.
[50] Duphar International Research B. V., EP 480 519, 1991 (H. Pluim et al.).
[51] M. Vaheri, V. Karppinen, *Acta Pharm. Technol.* **87** (1978) 75.
[52] S. Adachi, *Chem. Ind. (London)* (1957) 956.
[53] H. Okumura, T. Gomi, *Lab. Clin.* **7** (1973) 3517.
[54] E. M. Boyd, I. Godi, M. Abel, *Toxicol. Appl. Pharmacol.* **7** (1965) 609–618.
[55] F. Petuely, *Z. Kinderheilk.* **79** (1957) 174.
[56] H. Ruttloff et al., *Nahrung* **11** (1967) 39–47.
[57] K. Hoffman, D. A. A. Mossel, W. Konus, J. H. Kamer, *Klin. Wochschr.* **42** (1964) 126–130.
[58] K. K. Mäkinen, J. Rekola, *J. Dent. Res.* **54** (1975) 1244.
[59] U.S. Philips Corp., US 3 461 204, 1967 (J. Bircher).

[60] K. Hoffmann, *Dtsch. Med. Wochenschr.* **100** (1975) 1429–1431.
[61] S. K. Dutta, K. Sanat, US 4 137 397 (1979).

Lecithin

HIROYUKI TANNO, Central Research Laboratories, Ajinomoto Co. Inc., Kawasaki, Japan

1. Introduction 3149
2. Production 3151
3. Commercial Grades of Lecithin 3152
4. Physical Properties 3154
5. Chemical Properties 3155
6. Uses 3156
7. Quality Specifications and Analysis 3157
8. References 3157

1. Introduction

Lecithin was first isolated in 1850 by GOBLEY, who separated it from egg yolk; it was named lecithin after the greek word for egg yolk, *lekithos*.

Later, lecithin was identified as a mixture of many phosphatides. In the 1920s, a process which separates lecithin from crude soybean oil was developed, and today soybeans are still the main source for the mass production of lecithin. Generally, the term lecithin is used to describe soybean lecithin, which is composed mainly of soybean phosphatides and also contains soybean oil. The composition of soybean lecithin is given in Table 1. The uses of lecithin can be divided into foods, feeds, and general industrial uses. Lecithin is used widely in foods because it is both a natural emulsifier and a nutritious food in its own right (see Table 2) and since it gives rise to excellent flavors. In these applications, lecithin is mainly used as an emulsifier, and some treatment of lecithin (e.g., chemical modification and solvent extraction) is carried out. Recently, the physiological function of lecithin has been confirmed.

The term lecithin is also used in a chemical sense. In this case, lecithin refers to phosphatidylcholine (PC). The nomenclature of lecithin has been discussed by IUPAC and the International Union of Biochemistry (IUB). The chemical term for phosphatidylcholine was recommended to designate 1,2-diacylglycero-3-phosphorylcholine. The structures of the major phosphatides present in lecithin are shown in the following:

Table 1. Composition of soybean lecithin

Component	Content, wt%
Phosphatidylcholine	19–21
Phosphatidylethanolamine	8–20
Inositol phosphatides	20–21
Other phosphatides	5–11
Soybean oil	33–35
Sterols	2–5
Carbohydrates, free	5
Water	1

Table 2. Concentration of some minor components of soybean lecithin

Component	Concentration, µg/g
Tocopherol	1300
Biotin	0.42
Folic acid	0.60
Thiamin	0.12
Riboflavin	0.33
Pantothenic acid	5.59
Pyridoxine	0.29
Niacin	0.12

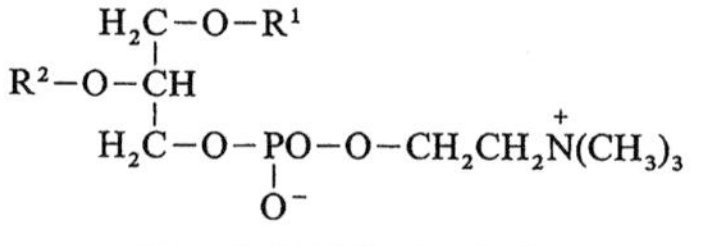

$$\begin{array}{l} H_2C-O-R^1 \\ R^2-O-CH \\ H_2C-O-PO-O-CH_2CH_2\overset{+}{N}(CH_3)_3 \\ O^- \end{array}$$

Phosphatidylcholine (PC)

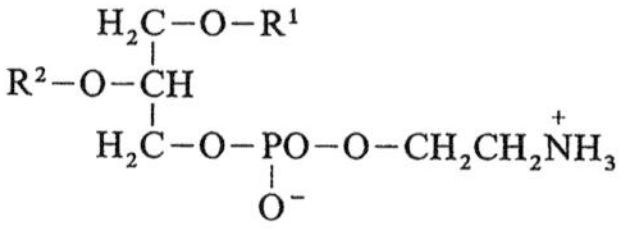

$$\begin{array}{l} H_2C-O-R^1 \\ R^2-O-CH \\ H_2C-O-PO-O-CH_2CH_2\overset{+}{N}H_3 \\ O^- \end{array}$$

Phosphatidylethanolamine (PE)

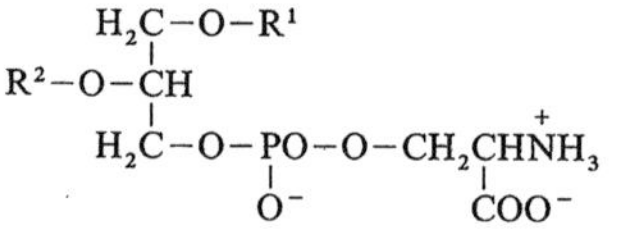

$$\begin{array}{l} H_2C-O-R^1 \\ R^2-O-CH \\ H_2C-O-PO-O-CH_2CH\overset{+}{N}H_3 \\ O^- COO^- \end{array}$$

Phosphatidylserine (PS)

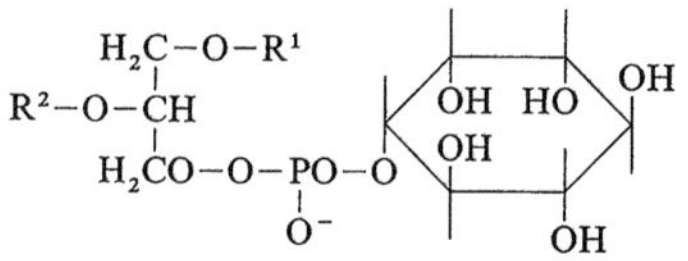

$$\begin{array}{l} H_2C-O-R^1 \\ R^2-O-CH \\ H_2CO-O-PO-O-\text{(inositol)} \\ O^- \end{array}$$

Phosphatidylinositol (PI)

where R^1 and R^2 represent fatty acid residues.

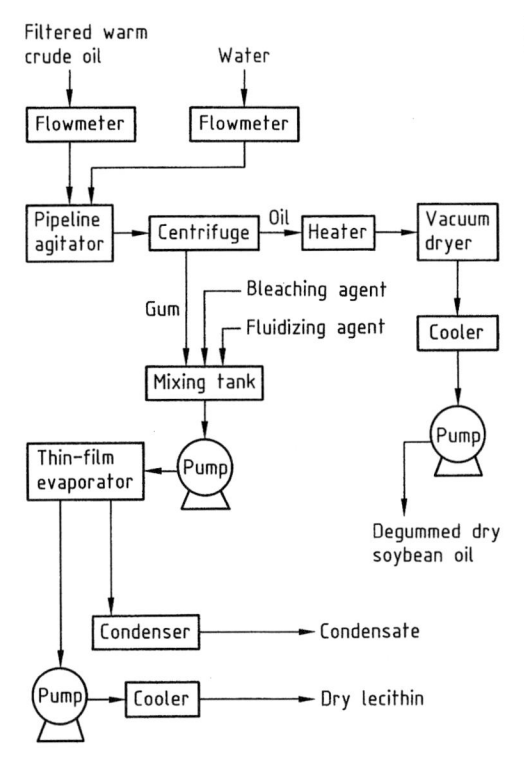

Figure 1. Flow sheet for the production of crude lecithin and degummed soybean oil

In this chapter, lecithin without special definition indicates soybean lecithin; the chemical term lecithin refers to phosphatidylcholine (PC).

2. Production

Modern production of lecithin is based mainly on plant seeds, especially soybeans. The quality of lecithin products is determined by the production process and the quality of the raw material soybeans. The flow sheet for a lecithin production process is shown in Figure 1.

Seed Preparation. Soybean oil from mature soybeans containing few burnt or fragmented beans gives the highest quality lecithins. Fragmented beans, or beans that have been stored for a long time, result in reduced lecithin yield. Processing of immature beans results in a high chlorophyll content so that the color of the lecithin is dark. This color is difficult to remove by bleaching.

Degumming. Crude oil which contains lecithin is obtained from raw material seeds by compression. The crude oil contains about 2–3 wt% of phosphatide. Water (1–2.5 vol.%) is added to the crude oil to hydrate the phosphatides. The method that is generally used in the United States involves adding 2 vol% water. The water–oil mixture is then heated at 70 °C for 30–60 min in a pipeline agitator that ensures thorough mixing. The oil-insoluble lecithin fraction (a wet gum known as lecithin hydrate) is separated by centrifugation. The gum is then transferred to a holding tank to allow addition of fluidizing and bleaching agents, if required.

Bleaching. To obtain a lighter colored product, hydrogen peroxide and benzoyl peroxide are used to bleach the lecithin. This is carried out either by using a 0.3–1.5 % hydrogen peroxide solution instead of water for the degumming process, or by addition of peroxide to the holding tank. The color of a sample is determined using the Gardner method.

Drying. The bleached gum contains 25–50% water. Since the high moisture content encourages the growth of microorganisms, it is necessary to dry the product. The method used is thin-film vacuum drying at a temperature of 80–105 °C and a pressure of 3.3–4 kPa. The drying is carried out within 1–2 min, because high temperatures promote the degeneration (darkening) of lecithin. After drying. it is necessary to cool the lecithin to 55–60 °C immediately, because of its instability at high temperature.

Molecular-Membrane Process. Recently, a molecular-membrane degumming process has been developed. The molecular mass of the triglycerides is ca. 900, while that of the phosphatides is ca. 800. However, in non-polar solvents such as hexane, the phosphatide molecules of lecithin aggregate to form micelles that have molecular masses of 20 000–30 000. In this way, lecithin can be separated from the triglycerides by a membrane.

3. Commercial Grades of Lecithin

The National Soybean Processors Association (NSPA) trading rules classify commercial lecithins into six grades according to color and viscosity. Fluid lecithin and plastic lecithin are classified by viscosity and each of these is further classified into three classes according to color: natural (unbleached), bleached, and double-bleached. Of these six, only the two unbleached grades are permitted as food additives. The specifications for the six types of lecithin are listed in Table 3.

The Food Chemical Codex (FCC) III classifies lecithin products as follows: 1) crude lecithin, 2) fluidized lecithin, 3) highly filtered lecithin, 4) compound lecithin, 5) chemically modified lecithin, 6) fractionated lecithin.

Crude Lecithin generally has high viscosity and is a brown fluid. The NSPA has standardized bleaching, the addition of minerals, and the moisture content.

Table 3. Commercial soybean lecithin specifications

Analysis	Grade		
	Natural	Bleached	Double bleached
Fluid lecithin			
Acetone insoluble, % (min)	62	62	62
Moisture, % (max)[a]	1	1	1
Benzene insoluble, % (max)	0.3	0.3	0.3
Acid value (max)	32	32	32
Color (max)[b]	10	7	4
Viscosity at 25 °C, Pa·s (max)	15	15	15
Plastic lecithin			
Acetone insoluble, % (min)	65	65	65
Moisture, % (max)[a]	1	1	1
Benzene insoluble, % (max)	0.3	0.3	0.3
Acid value (max)	30	30	30
Color (max)[b]	10	7	4
Penetration, mm (max)[c]	22	22	22

[a] Determined by toluene distillation.
[b] Measured on a 5% solution in mineral oil using the Gardner method.
[c] Using precision cone 73 525 and penetrometer 73 510.

Fluidized Lecithin is produced by addition of mineral salts, diluents, or fatty acids to crude lecithin. Fluidized lecithin can be handled more easily due to its lower viscosity. Although fluidized lecithins have a low content of acetone-insoluble material, they are graded as high class.

Highly Filtered Lecithin. High-grade lecithin products (e.g., for food) are made by removing the hexane-insoluble material by filtration.

Compound Lecithins. Compound lecithins are produced by blending lecithin with other components. These include oils, polysorbate, monoglyceride, emulsifiers, solvents, and plasticizers; they are generally blended with dry fluid lecithin.

Chemically Modified Lecithin. Lecithins are chemically modified by hydrogenation, hydroxylation, ethoxylation, halogenation, sulfonation, acylation, ozonization, and phosphorylation. Of the chemically modified lecithins, only hydroxylated lecithin is permitted as a food additive at present. Hydroxylation of lecithin increases the hydrophilicity and therefore also the tendency to form oil-in-water emulsions. Hydroxylated lecithins are produced by treating crude lecithin with water and hydrogen peroxide in lactic acid or acetic acid.

Fractionated Lecithin. The composition of crude lecithin, which contains many components, can be changed by fractionation with various solvents.
Acetone Fractionation. Most of the triglycerides and fatty acids can be separated from crude lecithin by acetone fractionation to give oil-free lecithin powders. Oil-free lecithin

costs more than ten times as much as regular lecithin. Because it is oil-free, acetone-fractionated lecithin has greater hydrophilicity and increased emulsifying activity.

Alcohol-Extracted Lecithin. Lecithin can be enriched by alcohol extraction. Phosphatidylcholine (PC) is concentrated in the alcohol-soluble fraction. This fraction has increased emulsifying activity for the formation of oil-in-water emulsions. The alcohol-insoluble fraction is rich in the hydrophobic phosphatidylinositol (PI) and therefore favors the formation of water-in-oil emulsions. Phosphatidylethanolamine (PE) is evenly divided between the alcohol soluble and insoluble fractions. As alcohol fractionation can change the phospholipid composition, its emulsifying activity is greater than that of acetone-fractionated oil-free lecithin.

Other Fractionation Methods. Lecithins containing large amounts of phosphatidylcholine can be obtained by adsorption and distribution chromatography. For example, products which contain up to 95% phosphatidylcholine are produced for pharmaceutical use by means of this method.

4. Physical Properties

There are many parameters which characterize the physical properties of lecithin. These are acetone-insoluble matter (AI), acid value (AV), moisture content, hexane-insoluble matter (HI), color, consistency, and clarity. Commercial lecithin is mainly composed of phospholipids (ca. 66%); the other major components are nonphospholipid compounds such as triglycerides. Since phospholipids are insoluble in acetone, and triglyceride is soluble in acetone, the amount of *acetone-insoluble material* in lecithin corresponds approximately to the phospholipid content.

The *acid value* is measured in miligrams of KOH per gram of lecithin and is determined by acid titration. The acid value is the phospholipid index number as defined in AOCS OM Ja-6-55.

The moisture content of lecithin products is typically less than 1.5%. The moisture content is determined using the Karl Fischer method.

The *hexane-insoluble matter* (HI) indicates the amount of residual fiber which originates from the production process. The HI measurement is the official Foods Chemicals Codex test method; the HI value must be suppressed to 0.1%. A high HI level indicates a low grade of lecithin. High-clarity lecithin can be produced by oil filtration to give HI levels of nearly zero. Although the major constituents of lecithin are colorless, the color of commercial lecithin is red-brown or dark-red. The colored components include carotinoid pigments and pigments formed by reaction of amino compounds with phosphatide sugars (Maillard reaction). Phosphatidylethanolamine also contributes to the color through formation of aldol products. The bleaching method is treatment with hydrogen peroxide or benzoyl peroxide. The color is measured by the Gardner method.

Table 4. Solubitity of lecithin and its components

	Solubility in		
	Water	Alcohol	Acetone
Soybean lecithin	insoluble	soluble	insoluble
Phosphatidylcholine	soluble	readily soluble	sparingly soluble
Phosphatidylethanolamine	readily soluble	soluble	insoluble
Phosphatidylinositol	readily soluble	insoluble	insoluble

Lecithin is classified as plastic or fluid according to its viscosity. The viscosity of fluid lecithin is decreased by addition of fatty acids. Commercial lecithin can be produced as a powder by treatment of the alcohol or acetone fractions. The acetone-insoluble grades consist of a mixture of isolated phospholipids which have either granular or powdered forms. The consistency of alcohol-fractionated lecithins depends on the phosphatidylcholine/phosphatidylethanolamine ratio, which in turn depends on the polarity of the alcohol used, the ratio of lecithin to alcohol, the temperature, and the extraction time.

Lecithin is soluble in oils and is hydrated in water, in which it disperses with formation of an emulsion. The solubility of crude lecithin and of the refined lecithin phospholipids is listed in Table 4.

5. Chemical Properties

The chemical properties of lecithin are largely determined by the presence of phosphatides and the fatty acid residues which they contain. Recently, the production of modified lecithin has increased because of the need for lecithins with specific properties, such as emulsifiers and wetting agents. The modification methods are either physical, enzymatic, or chemical.

The chemical structure of lecithin remains unchanged upon physical modification. Nevertheless, acetone or alcohol fractionation allows the ratio PC/PE and the amount of phospholipid to be changed, leading to changes in the properties and giving new functions which are lacked by crude lecithin.

Lecithin can be hydrolyzed enzymatically. On an industrial scale, the enzyme used is phospholipase A 2, which selectively hydrolyzes the fatty acid in the 2 position of the phospholipid and results in the lyso form of lecithin.

$$\begin{array}{c} CH_2-O-R_1 \\ | \\ R_2-O-CH \quad O \\ | \quad \parallel \\ CH_2-O-P-O-X \\ | \\ O^- \end{array} + H_2O \xrightarrow{50-70\,°C} \begin{array}{c} CH_2-O-R_1 \\ | \\ HO-CH \quad O \\ | \quad \parallel \\ CH_2-O-P-O-X \\ | \\ O^- \end{array} + HOR_2$$

Lysolecithin has increased solubility in water and greater emulsifying activity for the formation of oil-in-water emulsions.

Chemical hydrolysis by acids or alkalis is non-specific and affords products that are unsuitably colored.

The reaction of lecithin with acetic anhydride leads to acetylation of the amino group of phosphatidylethanolamine. Acetylated lecithin more readily forms oil-in-water emulsions. The treatment of lecithin with lactic acid and highly concentrated hydrogen peroxide introduces hydroxy groups into the unsaturated fatty acids of the phospholipids. Hydroxylated lecithin more readily forms dispersions with cold water.

6. Uses

Major food applications of lecithin are in chocolate, confectionery, margarine, and baked products. Non-food uses of lecithin include textiles, insecticides, paints, and many others. Lecithin functions as follows: 1) as an emulsifier, 2) as an antispattering agent, 3) crystallization control, 4) viscosity modifier, 5) antisticking agent, 6) wetting agent, 7) dispersing agent, 8) release agent. Usually, the amount used accounts for 0.1–1.0 wt% of the final product.

Important food applications of lecithin include baked goods, in which it improves the volume, disperses fats, and acts as an antioxidant, increasing shelf life; chocolate, in which it reduces viscosity and prevents crystallization; its addition to instant foods as a wetting, dispersing, emulsifying, and stabilizing agent; and margarine, in which it acts as an antispattering agent and dispersant.

Non-food uses of lecithin include lubricating oil, fuel oil, leather, textiles, dyes, rubber, soaps, and cosmetics. Lecithin is used in pharmaceuticals as a biological emulsifier and as a source of phosphatidylcholine; in paints as a pigment dispersant and for the stabilization of emulsions; in the production of leather to improve the absorption of fat liquor; and for mosquito control by reduction of surface tension.

Another use is as an additive to animal feeds and as a milk replacer for calves. Lecithin is said to improve the physical properties of the feed and to promote adsorption of some components. Furthermore, because lecithin contains choline and inositol, it promotes the metabolism of oils, fats, and cholesterol in the liver.

Storage. Long-term storage of lecithin at high temperature in the presence of air leads to oxidation of the unsaturated fatty acids, resulting in an off flavor and a black coloration. Therefore, it is necessary to exclude air and to maintain low temperature ($< 20\,°C$) when lecithin is stored for a long time.

Since lecithin contains nitrogen-containing compounds, such as ethanolamine, serine, and B vitamins, water suspensions of lecithin encourage the growth of microorganisms, and therefore require the addition of preservatives for long-term storage.

Table 5. Quality specifications for lecithin

Specification	Standard			
	FAO/WHO, 8431	FCC, 182 and 1400	EEC, E 332	Japan
Acid value	25–35	36	45	40
Benzene insoluble, %	0.3	0.3		0.3
Acetone insoluble, %	62	50	60	40
Peroxide value, meq/kg (max)	10		10	10
Heavy metals, ppm	40	40	40	20
Arsenic, ppm	3	3	3	4
Weight loss on drying, %	2	1.5	2	2

7. Quality Specifications and Analysis

Because lecithin compounds (including PC, PE, and PI) are too polar to be subjected to direct gas chromatographic analysis, liquid chromatographic methods are usually used for their analysis. For example, two-dimensional thin layer chromatography is the method most often used. Recently, high performance liquid chromatographic methods have become available.

Some quality specifications for lecithin are listed in Table 5.

8. References

[1] W. L. Erdarl et al., *J. Am. Oil Chem. Soc.* **50** (1973) 513.
[2] F. B. Jungalwala et al., *Biochem. J.* **145** (1975) 517.
[3] S. Chen et al., *J. Chromatogr.* **208** (1981) 339.
[4] F. L. Phillips et al., *J. Am. Oil Chem. Soc.* **58** (1981) 590.
[5] J. Holzl, *Biochem. Z.* **341** (1965) 168.
[6] B. W. Nichols, *Biochem. Biophys. Acta* **70** (1963) 417.
[7] S. Chen et al., *J. Chromatogr.* **276** (1983) 37.
[8] C. G. Duck-Chong et al., *Lipids* **18** (1983) 387.
[9] Van Nieuwenhuyzen, *J. Am. Oil Chem. Soc.* **58** (1981) 886.
[10] W. J. Strittmatter et al., *Nature (London)* **282** (1979) 857.
[11] A. R. Tall et al., *J. Med.* **299** (1978) 1232.
[12] R. J. Havel, *Ann. N. Y. Acad. Sci.* **348** (1980) 16.
[13] A. Kurode, *Food Chemical* **5** (1989) 73.
[14] L. R. Juneja et al., *J. Am. Oil Chem. Soc.* **66** (1989) 714.
[15] P. J. A. O'Doherty et al., *Lipids* **8** (1973) 249.
[16] P. Tso et al., *Gastroenterology* **80** (1981) 60.
[17] R. Hell et al., GB 1 215 868, 1970.

Maleic and Fumaric Acids

KURT LOHBECK, Hüls Aktiengesellschaft, Bottrop, Federal Republic of Germany (Chaps. 1–4)
HERBERT HAFERKORN, Hüls Aktiengesellschaft, Bottrop, Federal Republic of Germany (Chaps. 1–4)
WERNER FUHRMANN, Hüls Aktiengesellschaft, Bottrop, Federal Republic of Germany (Chaps.1–4)
NORBERT FEDTKE, Hüls Aktiengesellschaft, Marl, Federal Republic of Germany (Chap.5)

1.	Maleic Acid	3159	2.3.	Quality, Storage, and Transportation	3168
2.	Maleic Anhydride (MA)	3161			
2.1.	Properties	3161	2.4.	Uses	3169
2.2.	Production	3162			
2.2.1.	Oxidation of Benzene	3162	3.	Citraconic and Mesaconic Acids	3169
2.2.2.	Dehydration of Aqueous Maleic Acid Solutions	3164			
2.2.3.	Oxidation of C_4 Hydrocarbons	3165	4.	Fumaric Acid	3170
2.2.4.	Purification	3166	5.	Toxicology	3172
2.2.5.	Byproducts, Construction Materials, Effluent, and Discharge Air	3167	6.	References	3173

1. Maleic Acid

Maleic acid [*110-16-7*], $C_4H_4O_4$, *cis*-butenedioic acid, M_r 116.07, does not occur naturally. It was first produced by PELOUZE in 1834 by heating malic acid. Maleic acid became industrially important after the Barrett Co. in the United States successfully obtained the acid in 1919 by catalytic gas-phase oxidation of benzene. Maleic acid has been available commercially since 1928; maleic anhydride, since 1933.

R   COOH   HOOC   R
  \\//   \\//
  //\\   //\\
     COOH   COOH

R = H: Maleic acid   R = H: Fumaric acid
R = CH$_3$: Citraconic acid   R = CH$_3$: Mesaconic acid

Maleic acid itself is not very important economically. Small amounts are used for maleinate resins and, as maleic acid esters, for copolymers.

Physical Properties. The acid crystallizes in the form of monoclinic prisms. Other physical properties are listed below:

mp	130.5 °C
d_{20}^{20} (solid)	1.590

Heat of formation	−788.3 kJ/mol
Heat of combustion	−1358.9 kJ/mol
Solubility in 100 g of water	
at 25 °C	78.9 g
at 40 °C	112.3 g
at 60 °C	148.8 g
at 97.5 °C	392.6 g
Dissociation constants in water	
at 25 °C	K_1 1.14×10^{-2}
	K_2 5.95×10^{-7}

Chemical Properties. Maleic acid is very reactive at both its carboxyl groups and its double bond. Thus, on heating above 100 °C, water is eliminated with the formation of maleic anhydride. On further heating—preferably with the addition of catalysts—decarboxylation occurs with the formation of acrylic acid. Maleic acid also undergoes all normal reactions of *carboxylic acids*, such as esterification and amidation, but it does not form an acid chloride.

Examples of reactions involving the *double bond* are addition of water to form malic acid at elevated temperature and pressure; reaction with ozone to form glyoxylic acid; addition of halogens to form compounds such as dichlorosuccinic acid; and catalytic hydrogenation to succinic acid. Isomerization to fumaric acid occurs slowly already at temperatures as low as 100 °C; in the presence of a catalyst it proceeds almost quantitatively even at 60 °C.

Production. Maleic acid is generally obtained by heating maleic anhydride in a small amount of water. After the reaction mixture is cooled, the supernatant mother liquor is separated and the solid product is dried in vacuo. The mother liquor is diluted with water and returned to the hydration stage. It can, however, also be passed to the dehydration stage of maleic anhydride production (see Section 2.2.2), which avoids the problem of effluent purification. The apparatus must be made of corrosion-resistant steel.

Occasionally, maleic acid is also obtained directly from the wash water of maleic anhydride production; energy required for dehydration to the anhydride is thus saved. However, the wash water must first be purified with active charcoal; the charcoal must then be removed and disposed of. Furthermore, the purified solution must be concentrated in vacuo because fumaric acid is formed at elevated temperature and the chrome–nickel steel apparatus corrodes above 80 °C. Enriched wash water from *o*-xylene and naphthalene oxidation plants (phthalic anhydride production) is also used as a starting material for this process, as well as for the production of fumaric acid and maleic anhydride.

Quality Requirements. Maleic acid must meet stringent quality requirements. White crystals with a maximum fumaric acid content of 0.1–0.2 % are required, and the water content should be < 0.5 %.

2. Maleic Anhydride (MA)

Maleic anhydride [108-31-6], $C_4H_2O_3$, 2,5-furandione, M_r 98.06, has considerable industrial importance, in contrast to maleic acid.

Maleic anhydride

2.1. Properties

Physical Properties. Maleic anhydride crystallizes as orthorhombic needles. The most important physical properties may be summarized as follows:

mp	52.85 °C
bp (101.3 kPa)	202.0 °C
Heat of combustion	−1391.2 kJ/mol
Specific heat (liquid)	−1.67 kJ mol^{-1} K^{-1}
Heat of evaporation	54.8 kJ/mol
Heat of fusion	13.66 kJ/mol
Solubility in xylene (29.7 °C)	163.2 g/L

Vapor pressure of maleic anhydride at various temperatures:

t, °C	44.0	78.7	111.8	135.8	179.5	202.0
p, kPa	0.13	1.33	5.33	13.33	53.33	101.33

Chemical Properties. Maleic anhydride reacts readily with water to form maleic acid. At low temperatures, addition of alcohols leads to the formation of semiesters, whereas at elevated temperature in the presence of esterification catalysts, water is eliminated with the formation of diesters. Addition of ammonia or amines to maleic anhydride results in formation of the corresponding semiamides, which can be converted by elimination of water into cyclic imides.

The double bond in maleic anhydride is also extremely reactive. Addition of halogens leads to the formation of dihalogenated succinic anhydrides or mono- and dihalogenated maleic anhydrides, depending on reaction conditions. Hydrogenation yields succinic anhydride, 1,4-butanediol, tetrahydrofuran, or butyrolactone, depending on conditions.

Alkenylsuccinic anhydrides are formed by addition of olefins. Maleic anhydride undergoes Diels – Alder reactions with conjugated dienes. The anhydride is also suitable for homopolymerization and copolymerization. A summary of the reactions of maleic anhydride is given in [1], [2].

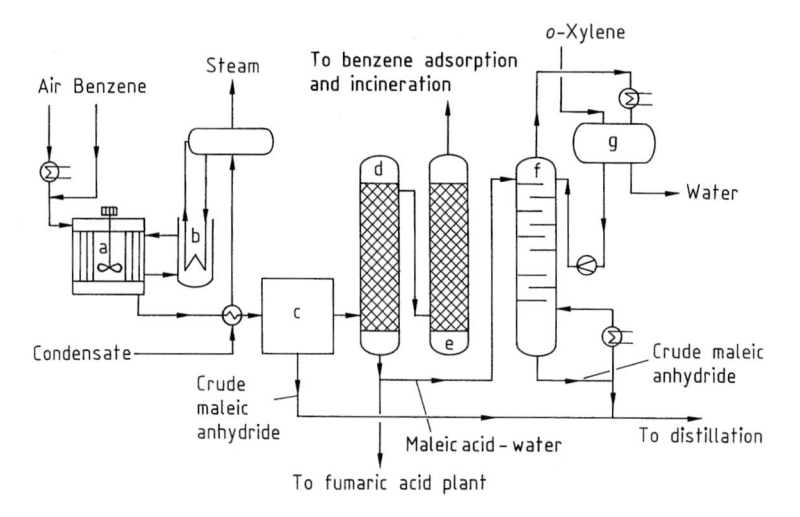

Figure 1. Plant for production of maleic anhydride by benzene oxidation
a) Reactor; b) Salt bath cooler; c) Partial condenser; d) Acid scrubber; e) Alkali scrubber; f) Dehydration column; g) Phase separator

2.2. Production

Maleic anhydride is produced industrially by catalytic oxidation of suitable hydrocarbons in the gas phase. Benzene used to be the predominant starting material, but in recent years oxidation of C_4 hydrocarbons has become increasingly important; the C_4-based capacities are continually increasing.

2.2.1. Oxidation of Benzene

To produce maleic anhydride (see Fig. 1), benzene is added to a preheated airstream so as to form a largely homogeneous mixture. If a large excess of air is used, the lower ignition limit of the mixture is not reached in the majority of methods.

Tubular reactors with a multiplicity of parallel vertical tubes of ca. 25-mm diameter are used for conventional fixed-bed methods. Reaction gas passes through the tubes over the catalyst, the process pressure being adjusted to an optimum value of 0.15 – 0.25 MPa. Supported catalysts are used, where an "active mass" of V_2O_5 and MoO_3 is applied on an inert carrier. The lifetime of the catalysts is up to four years and depends mainly on the operating temperature in the reactor; a uniform, interference-free operation; and the purity of starting materials. Both the desired main reaction

$C_6H_6 + 4\frac{1}{2} O_2 \rightarrow C_4H_2O_3 + 2 CO_2 + 2 H_2O$
$\Delta H = -1875$ kJ/mol

and the undesirable secondary reaction of combustion to carbon monoxide, carbon dioxide and water are strongly exothermic, and "hot spots" of 340–500 °C occur in the catalyst. A considerable amount of heat is produced during the reaction. Approximately 27 MJ of heat must be dissipated per tonne of reacted benzene, which corresponds to almost 10 t of saturated steam; in practice, ca. 80% of the steam can be utilized.

Heat is dissipated primarily by eutectic salt melts circulating around the reactor tubes. The hot salt melts are cooled with water in a secondary heat exchanger. Only part of the generated steam can be used in MA production. In many cases, excess steam—unless it can be utilized in other operations—is used to drive air compressors. The reaction can be optimized with regard to MA yield by means of the accurately adjustable (± 0.5 °C) salt bath temperature.

In a typical conversion of 100 mol benzene, 73 mol is oxidized to maleic anhydride and 23 mol is "burned." Unreacted benzene (4 mol) can largely be recovered by a subsequent adsorption step (e.g., with activated charcoal) and reused. Processes that involve recycling the major portion of reaction gas after MA separation are described in some patents; up to the present, they have not been used on an industrial scale [3], [4].

Separation of Crude Maleic Anhydride. After leaving the reactor, the gas mixture is initially cooled to 150–160 °C so that neither MA nor water condenses. The waste heat is utilized further. Different routes can then be used for further processing.

Partial Condensation. In liquid partial condensation the melting point of MA is the lower temperature limit. The reaction gas is cooled to ca. 55 °C in "partial condensers" [5] (e.g., bundles of vertical tubes 20–50 mm in diameter); MA condenses and can be removed as liquid from the collecting vessel. Discharge should occur as quickly as possible [6] because prolonged contact of the liquid product with the water-containing reaction gas leads to the formation of fairly large amounts of maleic acid. In this way, 40–60% of the MA contained in the reaction gas can be separated; the remainder is normally washed out with water as maleic acid and processed further to MA (see Section 2.2.2).

Separation by Water Scrubbing. In principle, all the MA contained in the reaction gas can also be recovered as maleic acid solution by water scrubbing. To obtain MA from this solution, the acid must be dehydrated. Because of the associated high energy requirement, this method is justified only to obtain MA from reaction gases with a high water content, such as occur in the C_4 oxidation.

For separation by washing with organic solvents, see Section 2.2.3.

2.2.2. Dehydration of Aqueous Maleic Acid Solutions

Maleic anhydride can be obtained by dehydration from the maleic acid solutions that are formed by washing the reaction gas with water (see Fig. 1). Above 150 °C even small amounts of alkali in the wash solution can decompose the resultant crude MA to decarboxylation products [7]; thus, alkali-free wash water must be used.

Temperatures above 100 °C are required for dehydration of maleic acid ($\Delta H = +34.88$ kJ/mol). A reaction rate sufficient for industrial processes is, however, achieved only above 130 °C. Isomerization of maleic acid to fumaric acid occurs concurrently ($\Delta H = -23.2$ kJ/mol).

Fumaric acid decomposes only when heated above 230 °C, without forming significant amounts of MA. The isomerization reaction must therefore be suppressed in the dehydration of maleic acid; this is achieved by having as short a residence time as possible at elevated temperature.

Entraining Agent Method. Aqueous maleic acid solution is added to the upper half of a distillation column (Fig. 1, f) whose bottom contains a boiling MA–entraining agent mixture. Aromatic hydrocarbons (e.g., xylene) are generally used as entrainers.

The entraining agent together with the water leaves the head of the column as an azeotropic mixture. Then it is separated as the organic phase in separator (g) and refluxed to the head of the column. The aqueous phase is returned to the reaction gas scrubber (d) of the oxidation plant [8].

The bottom product (MA, 10–40% entraining agent, 1–5% maleic acid, 1–3% fumaric acid) is distilled to yield pure MA (see Section 2.2.4). The process is performed continuously as well as batchwise.

Thermal Process Without Entraining Agent. The thin-layer evaporator technique has been used for some time for thermal dehydration of aqueous maleic acid solutions. By this technique the water can be separated as steam, and the residence time of maleic acid at 150–200 °C is kept very short. Conditions for isomerization thus exist for only a short time, and only a small amount of fumaric acid is formed (1–3%). Several thin-layer units are arranged in sequence [9], [10]. The MA separated as liquid in the partial condenser (see Fig. 1) is also fed to the second stage of this plant to dehydrate residual maleic acid (1–5%). The 99% MA obtained is distilled to yield pure product (see Section 2.2.4).

Fumaric acid and high-boiling residues are discharged from the bottom of the last thin-layer evaporator.

These continuously operating plants must be shut down relatively frequently for cleaning.

2.2.3. Oxidation of C₄ Hydrocarbons

Processes for the oxidation of C_4 hydrocarbons involve mainly the use of *n*-butane or *n*-butane–*n*-butene mixtures with a high paraffin content. The following types of process have been developed or are in the developmental stage:

fixed-bed process
fluidized-bed process
transport-bed process

In the United States the fixed-bed process with *n*-butane has been the only MA route used commercially since 1985; it is employed to a small but increasing extent in other parts of the world. With improved methods and sufficient raw material availability, C_4 oxidation has become increasingly important; the low raw material price has had a decisive influence on the economy of this process.

With C_4 oxidation lower conversion rates (80%) and selectivities (70%) contrast theoretically higher weight yields (+34 wt% compared to benzene oxidation) so that finally, weight yields for benzene and C_4 oxidation are comparable (ca. 95 wt% MA). Since the recovery of unreacted C_4 hydrocarbons is substantially more difficult than with benzene, reaction waste gases from C_4 oxidation must be passed to an incineration unit.

Fixed-Bed Process. In the fixed-bed process the catalyst is embedded in tubular reactors, similar to benzene oxidation. Since the publication of the first U.S. patents catalysts based on vanadium and phosphorus oxides (V–P–O catalysts) have mainly been used [11]–[15].

The concentration of hydrocarbons in the feed gas to the reactor is kept just below the explosion limit of 1.8 mol% [corresponding to 45 g/m³ (STP) air] in the commercial fixed-bed process. Conversion at the catalyst in the selective reaction proceeds as follows:

$C_4H_{10} + 3\frac{1}{2} O_2 \rightarrow C_4H_2O_3 + 4 H_2O$
$\Delta H = -1260$ kJ/mol

Because of the increased formation of water per formula conversion compared to benzene oxidation, with C_4 oxidation only a fairly small amount of MA can be directly liquefied from the reaction gas by partial condensation. The larger proportion (65–70%) is washed out as maleic acid in a subsequent water scrubber and must be dehydrated, with a high energy input.

In a modification of the water wash, MA contained in the reaction gas is alternatively absorbed by organic solvents [16]–[19]. More than 98% of the anhydride can be absorbed in this way, without noticeable amounts of maleic acid being formed. The solvent–anhydride mixture is then subjected to fractional distillation, in which MA is distilled from the high-boiling solvent and the latter is returned to the absorption column.

Fluidized-Bed Process. The fluidized-bed process has the advantage of a particularly uniform temperature profile (without "hot spots"). A uniform temperature profile is the necessary precondition for high reaction selectivity, it is, however, offset to some extent by intensive remixing of the products as a result of fluidization. A particular problem of this process is mechanical stress on the catalyst (V–P–O), its abrasion, and erosion at the heat-dissipating surfaces. On the other hand, the fluidized bed constitutes an extremely effective flame barrier, with the result that the process can operate at higher C_4 concentrations (i.e., within the explosion range) than the fixed-bed process. This results in process engineering advantages plus a considerably smaller investment for fluidized-bed technology [20].

The *ALMA process* (Alusuisse, Lummus) is currently the most technically advanced process; it combines a fluidized-bed reactor with a nonaqueous processing unit [18], [19].

In October 1988 a 10 000-t/a MA plant [21] combining fluidized-bed technology with an aqueous processing unit has come on stream at MTC Sohio, Japan.

Transport-Bed Process. The transport-bed process developed jointly by Monsanto and Du Pont is at the trial stage [22], [23]. The process involves two reactors: in the first, spent catalyst (V–P–O) removed from C_4 oxidation is regenerated by atmospheric oxygen. Regenerated (oxidized) catalyst is separated from the oxygen-containing gas; then it passes, with the addition of butane, to the second reactor. Here, catalytic conversion to MA takes place, the reaction being carried out with virtual exclusion of atmospheric oxygen and simply with the amount of oxygen transferred by the catalyst itself. A high selectivity of MA formation (75 mol%) is said to be achieved under the quasi-stoichiometric conversion conditions of this process, which also largely prevents remixing of reaction gases.

2.2.4. Purification

Batchwise Distillation. Batchwise distillation is the conventional method for processing crude MA. Directly separated MA, as well as mixtures from entraining agent dehydration, are added to the batch distillation column. The mixture is distilled under total reflux conditions until the remaining maleic acid is dehydrated. The small amounts of water eliminated are separated from the recirculating entraining agent (xylene). The entrainer is then distilled, and MA is recovered by distillation in vacuo. Batch stills of 50–150-m^3 capacity and distillation columns with 10–20 trays are used.

Continuous Distillation (see Fig. 2). For large-capacity MA plants, processing by continuous distillation is both possible and economical [24], [25]. The mixture of

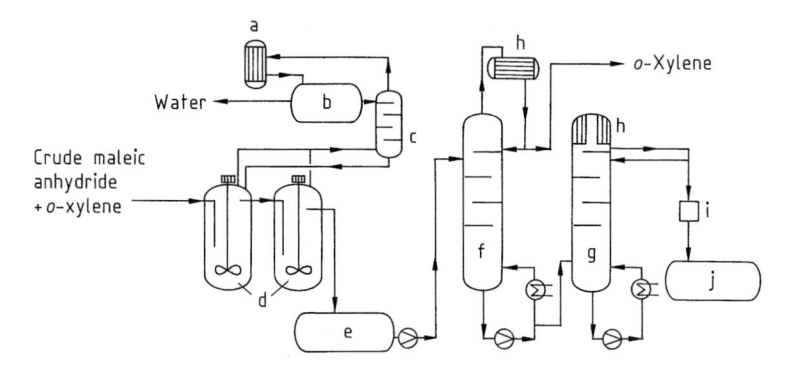

Figure 2. Continuously operating plant for production of pure maleic anhydride
a) Condenser; b) Separating vessel; c) Column; d) Postdehydration; e) Intermediate tank; f) o-Xylene column; g) Pure MA column; h) Cooler; i) Colorimeter; j) Pure MA vessel

directly separated and crude MA obtained by dehydration is continuously "postdehydrated". Residual xylene is separated in a column (f), and pure MA is finally distilled from the top of a second column (g).

2.2.5. Byproducts, Construction Materials, Effluent, and Discharge Air

Benzene Oxidation. *Byproducts.* Benzoquinone, formaldehyde, and formic acid occur in small amounts as byproducts of benzene oxidation.

Construction Materials. Directly condensed crude MA containing 1–5% maleic acid can be stored at ca. 60 °C in normal steel vessels. Special stainless steel is essential in the "wet" region of the plants (i.e., for scrubbers, vessels containing maleic acid solution, dehydration column, and apparatus used for postdehydration). The partial condensers should preferably also be made of special stainless steel. For reasons of quality, stainless steel is used for the pure MA distillation columns.

C_4 Oxidation. Because acetic, acrylic, and crotonic acids, etc., are formed as byproducts of C_4 oxidation, a suitable special stainless steel must be chosen with particular care for the relevant equipment.

Environmental Protection. The MA plants can be operated almost without any acid effluent if aqueous maleic acid solutions from the dehydration stage are recycled. Byproducts are contained in the distillation residue, which is normally combusted.

Environmental legislation differs very much in various countries. In the Federal Republic of Germany the TA Luft (Clean Air Regulations) was revised in 1986. The emission of benzene, which is classified as a carcinogen, must not exceed 5 mg/m^3 in waste gas from benzene oxidation [26]. This necessitates installation of a combustion unit downstream of the benzene adsorption stage, as in C_4 oxidation plants [27].

2.3. Quality, Storage, and Transportation

Quality Specifications. A typical specification for pure MA includes the following:

mp	52.5 °C (min.)
Acid number	1140 mg KOH/g (min.)
Purity	99.7 wt% (min.)
Melt color index	20 APHA (max.)
Ash content	10 ppm (max.)
Iron content	5 ppm (max.)

Storage of Pure MA. Pure MA in the molten state at 55 – 65 °C is stored in stainless steel or pure aluminum vessels. It can be stored for several weeks in an inert-gas atmosphere without any change in quality.

Commercial Forms and Transportation. *Liquid MA* is transported over large distances at 60 – 80 °C in well-insulated road tankers or tank containers. The vessels are provided with heating devices.

Solid MA is dispatched in the form of pastilles and briquettes weighing 0.5 – 20 g. Because of their high dust content, flakes are rarely produced. Multiply paper bags with welded polyethylene liners or bags of strong polyethylene film are normally used as packaging. Solid MA can be stored for several months in this packaging in a cool, dry place.

Maleic anhydride is included in the list of dangerous substances and preparations according to the EEC Directive 67/548/EEC and GefStoffV and must be specially packed and identified. However, on the basis of the most recent toxicological investigations, the identification according to GefStoffV [28] should be amended and amplified as follows [29]:

Danger symbol	C (corrosive)
R	22 – 34 – 42/43
S	22 – 28 water – 39

The hazard classification for the transport of MA are as follows:

IMDG Code (GGVSee)	Class 8
GGVS/GGVE, ADR/RID	Class 8, no. 31 c
UN no.	2215

2.4. Uses

Because of its properties as a dicarboxylic anhydride with a double bond, MA can be used for both polycondensation and polyaddition. Polyester and alkyd resins, lacquers, plasticizers, copolymers, and lubricants are the most important technical end products. Examples of industrially important copolymers are: MA – styrene (for engineering plastics) and MA – acrylic acid (used in the detergent industry). Polyester and alkyd resins, in particular, are used in the production of fiberglass reinforced plastics, in the construction and electrical industries, and in pipeline and marine construction.

Tetrahydrophthalic anhydride is formed by a Diels – Alder reaction with butadiene; hydrogenation of the anhydride yields hexahydrophthalic anhydride. Methylhexahydrophthalic anhydride is obtained in a similar manner from isoprene and MA. These products are used as anhydride curing agents for epoxy resins.

Smaller amounts of MA are used in the production of pesticides (captan, malathion) and growth inhibitors (maleic acid hydrazide). Surfactants are produced from maleic acid esters by reaction with sodium hydrogen sulfite. When added to drying oils, MA reduces the drying time and improves the coating quality of lacquers.

3. Citraconic and Mesaconic Acids

In 1881 ANSCHÜTZ obtained mesaconic [498-24-8], 2-methyl-*trans*-2-butenedioic acid and citraconic acid [498-23-7] (2-methyl-*cis*-2-butenedioic acid) for the first time by distilling citric acid. Citraconic acid forms an inner anhydride (citraconic anhydride [616-02-4], 3-methyl-2,5-furandione, *mp* 7 °C, *bp* (4 kPa) 109 °C, $d_4^{15.7}$ 1.2469), analogously to maleic acid. Citraconic anhydride is interesting particularly because of its low melting point (easier handling at lower temperature).

Many attempts have been made to develop industrial processes for obtaining *citraconic anhydride*, including its preparation from citric acid (1905), and gas-phase oxidation of alkyl-substituted olefins [30], methylbutanols [31], or mesityl oxide [32]. In the gas-phase oxidation of *o*-xylene, ca. 2 – 4 t of citraconic anhydride is formed per 1000 t of phthalic anhydride. Citraconic anhydride is present as citraconic acid after water washing. Three processes have been developed to isolate citraconic anhydride [33] – [35]. Its synthesis from succinic anhydride and formaldehyde has been proposed [36].

With regard to its industrial use, citraconic anhydride is frequently mentioned with MA in patents. Substitution of citraconic acid for MA, however, does not impart any special properties to the products. Citraconic anhydride is useful simply in producing liquid curing agents for epoxy resins: Diels – Alder synthesis with isoprene yields a nonhygroscopic curing agent that is liquid at room temperature [37].

4. Fumaric Acid

Fumaric acid [110-17-8], *trans*-butenedioic acid, M_r 116.07, occurs widely in nature. It received its name from WINCKLER, who isolated it from fumitory (*Fumaria officinalis*) in 1832. Subsequently, it was isolated many times under the names boletic acid, glaucinium acid, and lichen acid, and in 1834 it was described by PELOUZE who called it paramaleic acid. Fumaric acid is currently produced on an industrial scale, although it is less important than maleic anhydride.

Physical Properties. Fumaric acid crystallizes in the form of monoclinic prisms. Important physical properties are summarized below:

mp	286 – 287 °C (closed tube)
d^{25} (solid)	1.635
Heat of sublimation (92 °C)	136.1 kJ/mol
Dissociation constants in water	K_1 9.57 × 10^{-4}
	K_2 4.13 × 10^{-5}
Solubility in 100 g of water	
at 15.5 °C	0.428 g
at 100 °C	9.97 g

Chemical Properties. Fumaric acid sublimes without decomposition at 200 °C or slightly higher. Above 230 °C, however, it decomposes with the formation of maleic anhydride, water, and a considerable amount of residue. Addition of phosphorus pentoxide promotes the formation of maleic anhydride.

Its other chemical properties resemble those of maleic acid. Reactions proceed considerably more slowly, however, due partly to the lower water solubility of fumaric acid. In contrast to maleic acid, fumaric acid forms an acid chloride.

Production (see Fig. 3). Maleic acid or maleic anhydride, especially the maleic acid-containing wash water from the production of maleic anhydride or phthalic anhydride, serves as starting material for the manufacture of fumaric acid. The maleic acid concentration should be at least 30%. Maleic acid is converted almost quantitatively by thermal [38] or catalytic isomerization into the sparingly soluble fumaric acid, which is recovered by filtration. Various substances have been proposed as catalysts: mineral acids (e.g., hydrochloric acid) [39]; sulfur compounds such as thiocyanates, thiazoles, thiosemicarbazides, thioureas; or bromine compounds in combination with peroxides (e.g., persulfate) [40]. Thiourea is most commonly used in practice [41].

The maleic acid-containing wash water contains impurities that can affect quality and yield. This problem can be largely avoided (1) by thermal pretreatment of the wash water [42], (2) by adding urea [43] if thiourea is used as catalyst, and (3) by addition of sulfites or passaged of sulfur dioxide [44] and addition of mineral acids [45]. The crude fumaric acid obtained is purified by recrystallization from water, combined with purification by active charcoal. Losses during purification are ca. 10%.

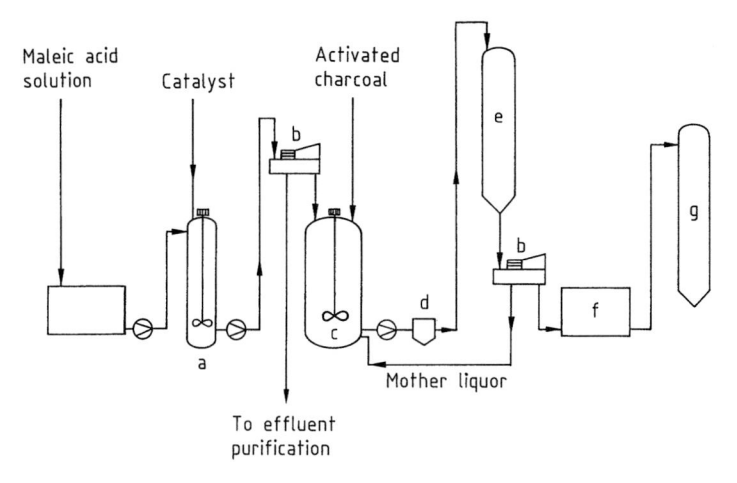

Figure 3. Plant for production of fumaric acid from aqueous maleic acid solutions
a) Isomerization vessel; b) Centrifuge; c) Dissolving tank; d) Filter; e) Crystallizer; f) Dryer; g) Silo

Fumaric acid production is generally carried out batchwise, although continuous processes are also known. If hydrogen chloride does not have to be added, an apparatus made of corrosion-resistant austenitic steel is satisfactory. Disposal of the mother liquor from the isomerization stage, which in addition to soluble impurities also contains the isomerization catalyst, is especially difficult. One possibility is to combust the effluent.

In the United States, fumaric acid is also produced by fermentation of monosaccharides or starch with selected fungal cultures (*Mucor, Aspergillus, Rhizopus*) [46].

Uses. Fumaric acid is often used instead of maleic acid to produce polyesters and copolymers. A nearly identical mixture of *trans*- and *cis*-dicarboxylic acids is formed by isomerization during the preparation of these products [47], regardless of whether maleic or fumaric acid is used as starting material. Fumaric acid is often preferred because of its innocuous nature. On the other hand, special effects are observed when fumaric acid is used, e.g., greater hardness of certain resins [48].

Because it is nontoxic, fumaric acid is used in the food industry, e.g., as an acidulant in baking powders and beverages, for which purpose particularly stringent requirements are placed on purity (U.S. Food Chemicals Codex). Fumaric acid (Fumaril) and iron(II) fumarate are used as special additives for animal feeds [49].

5. Toxicology

Maleic Anhydride. In aqueous solution, maleic anhydride slowly hydrolyzes to maleic acid; accordingly, the toxicological properties of the anhydride are in part determined by the toxicological activity profile of maleic acid. The *acute toxicity* is characterized by LD_{50} values of 481 mg (rat, oral), 465 mg (mice, oral), 2620 mg (rabbit, percutaneous), and 390 mg (guinea pig, oral) per kilogram of body weight [50]. The main acute toxic effects of this compound are its powerful local irritant and corrosive action on the skin, mucous membrane, and eyes. At a concentration of 1.5–2 ppm, nasal irritation was observed in workers after 1 min, and irritation of the eyes after 15–20 min [51]. Concentrations >2.5 ppm are extremely irritant. Asthmatic symptoms may occur after long-term exposure to maleic anhydride concentrations >1.2 ppm [52]. In addition to its acute irritant action, maleic anhydride is also characterized by a strongly sensitizing action, which has been observed experimentally in guinea pigs as well as in humans [52], [54].

No *systemic* toxic effects were observed in a 6-month inhalation exposure of rats, hamsters, and monkeys to up to 2.4 ppm of maleic anhydride [55]. Animal experimental studies did not reveal any evidence of a carcinogenic, teratogenic, or reproductive–toxicological action of maleic anhydride [55]–[57].

The currently valid MAK and TLV–TWA values were based on the acute irritant and sensitizing action of maleic anhydride and are 0.2 and 0.25 ppm, respectively [52], [53].

Maleic Acid. As an acid, the toxicological activity of maleic acid is characterized by marked irritant and corrosive action on the skin, mucous membrane, and eyes. A 20% aqueous solution produced a mild and reversible skin irritation in humans [58]. Lower concentrations (< 5%) are sufficient to produce serious ocular irritation [59]. Animal experiments confirm that the local irritant action is the most prominent toxicological feature of maleic acid. The *acute* oral and dermal toxicity is characterized by LD_{50} values of 708 mg (rat, oral), 2400 mg (mouse, oral), 1560 mg (rabbit, dermal), and >1000 mg (guinea-pig, dermal) per kilogram of body weight [60].

Kidney damage has been observed as a *systemic* toxic action of maleic acid. Morphological–functional changes in the kidneys were produced by intraperitoneal injection (100–350 mg/kg; rats, dogs) and also by 1-h inhalation exposure to ca. 720 mg/m^3 (rats) [61]. The damage to distal and proximal tubular resorption produced by maleic acid is reminiscent of the known (congenital or acquired) Fanconi syndrome in humans [62], [63]. This damage seems to be caused by the interaction of maleic acid with glutathione in tubular renal cells, leading to intolerable concentrations of free radicals and peroxides [64]. In addition to the resultant cellular damage, maleic acid appears to affect Na$^+$ and H$^+$ ion transport in the proximal tubes [65]. After chronic exposure of male rats (250, 500, 750 mg kg^{-1} d^{-1} for up to two years), an increased mortality as well as kidney damage and delayed growth was observed in all dosage groups [66]. Liver and testicular damage has also been found in the highest dosage group. No signs of *carcinogenic activity* of maleic acid were discovered with this long-term dosage, which however was not designed as a carcinogenicity study. No *genotoxic effects* were observed in the Ames test, with or without metabolizing fractions [67].

Fumaric Acid. Fumaric acid is a naturally occurring intermediate of the tricarboxylic acid cycle and is used as a food additive [68]. Accordingly the compound is found to be practically nontoxic in acute toxicity tests. An oral LD_{50} value of 10 g/kg of body weight was measured in rats, and in rabbits 20 g/kg of body weight applied dermally was not lethal [69]. In humans, no changes in the blood and urine parameters or in liver function were found after administration of 8 mg of fumaric acid per kilogram of body weight per day for one year [70]. Experimental investigations did not reveal any evidence of genotoxic [67], [71], carcinogenic [70], [72], or reproductive–toxicological [70] activity. The toxicological evaluation of fumaric acid is thus based primarily on the local irritant action on contact with the skin, mucous membrane, and eyes. In this respect, fumaric acid is classified as slightly irritant to the skin and moderately irritant to the eyes and mucous membranes.

6. References

[1] L. H. Flett, W. H. Gardner: *Maleic Anhydride Derivatives*, J. Wiley & Sons, New York 1952.
[2] B. C. Trivedi, B. M. Culbertson: *Maleic Anhydride*, Plenum Press, New York 1982.
[3] R. Klar, DE-AS 1069150, 1957.
[4] Chevron Res. Co., DE-OS 2658191, 1976 (K. R. Bakshi, D. M. Marquis, S. G. Paradis).
[5] United Eng. & Constr., US 2762449, 1953 (M. P. Sweeney).
[6] Halcon, DE-AS 1468586, 1962 (G. F. Vinci, M. Gans).
[7] C. E. Vogler et al., *J. Chem. Eng. Data* **8** (1963) no. 4, 620–623.
[8] Reichhold Chem. Inc., DE 833039, 1950 (C. H. B. Jarl).
[9] UCB, DE-OS 2003684, 1970 (E. Weyens).
[10] UCB S.A., *Hydrocarbon Processing* **64** (1985) no. II, 143.
[11] Princeton Chem. Res., US 3293268, 1966 (R. I. Bergman, N. W. Frisch).
[12] Chevron Res. Co., US 3864280, 1975 (R. A. Schneider).
[13] Monsanto Co., US 4456764, 1982 (J. T. Wrobleski).
[14] H. Bosch et al., *Appl. Catal.* **31** (1987) 323–337.
[15] G. Centi, J. R. Ebner et al., *Chem. Rev.* **88** (1988) 55–80.
[16] Ftalital S.p.A., US 4314946, 1982 (A. Neri, S. Sanchioni).
[17] F. Budi, A. Neri, G. Stefani, *Hydrocarbon Processing* **61** (1982) no. 1, 159–161.
[18] S. C. Arnold, L. Verde et al., *Hydrocarbon Processing* **64** (1985) no. 9, 123–126.
[19] *Hydrocarbon Processing* **64** (1985) no. 11, 142.
[20] G. Emig et al., *Catal. Today* **1** (1987) 477–498.
[21] *Eur. Chem. News* **50** (1988) Oct. 31, 8.
[22] Du Pont, EP-A 0189261 A1, 1986 (R. M. Contractor).
[23] *Eur. Chem. News* **50** (1988) Dec. 12, 36.

[24] VEBA-Chemie AG u. Metallges. AG, DE 2 061 336, 1970 (O. L. Garkisch, G. Ibing, G. Kammholz, H. Schirrmacher, K. Lohbeck).
[25] K. Lohbeck, *Erdöl Kohle Erdgas Petrochem. Suppl.* **1** (1976) 320–327.
[26] Technische Anleitung zur Reinhaltung der Luft, 27. 02. 1986, Ziff. 2.3 Krebserzeugende Stoffe.
[27] Ziff. 3.1.7 Organische Stoffe/Anhang E. [26]
[28] Verordnung über gefährliche Stoffe (GefStoffV), 26. 08. 1986, Anhang VI.
[29] Hüls AG, Tox. Bericht Nr. 1390, 1989, unpublished result.
[30] H. Pichler, DE-OS 2 206 713, 1972.
[31] Pfizer Inc., DE-OS 2 522 568, 1974 (R. G. Berg, B. E. Tate).
[32] Pfizer Inc., DE-OS 2 526 464, 1974 (R. G. Berg, D. S. Hetzel).
[33] Produits Chimiques, FR 1 489 787, 1966 (M. P. Juston).
[34] VEBA-Chemie, DE-AS 1 593 546, 1966 (G. Ibing, H. Haferkorn).
[35] BP Chemicals Int., GB 1 398 586, 1972 (L. E. Cooper).
[36] Pfizer Inc., DE-OS 2 352 468, 1972 (R. G. Berg, B. E. Tate).
[37] VEBA-Chemie, DE 1 593 548, 1966 (G. Ibing, K. Neubold, H. Haferkorn).
[38] California Res. Corp., US 2 816 922/3, 1954 (R. W. Stephenson).
[39] Allied Chem. Corp., US 2 393 352, 1942 (L. Winstrom).
[40] Halcon, US 3 262 972, 1963 (R. S. Barker).
[41] United States Rubber Comp., US 2 454 385/7, 1946 (L. H. Howland, W. F. Brucksch).
[42] Deutsche Texaco, DE 1 768 629, 1968 (P. Ackermann, H. Heinrichs).
[43] Allied Chem. & Dye Corp., US 2 483 576, 1947 (G. E. Neuman de Végvár).
[44] Montclair Res. Corp. + Ellis Foster Comp., US 2 843 628/9, 1953 (J. B. Rust).
[45] Halcon, US 3 389 173, 1966 (R. L. Russel, H. Olenberg).
[46] Chas. Pfizer & Co., Inc. + Merck & Co. Inc., US 2 327 191, 1939 (J. H. Kane, A. Finlay, P. F. Amann).
[47] S. S. Feuer et al., *Ind. Eng. Chem.* **46** (1954) 1643.
[48] R. C. Peter, *Ind. Chem.* **31** (1955) 564.
[49] H. Brune et al.: "Fumarsäure in der Tierernährung," *Landwirtschaftl. Schriftenreihe Boden Pflanze Tier*, Heft 8, Parey, Hamburg 1979.
[50] RTECS (Jan. 1988).
[51] ACGIH, TLVs, 4th ed. and Supplement (1980).
[52] D. Henschler, *MAK-Werte, Toxikologisch-arbeitsmedizinische Begründungen*, Verlag Chemie, Weinheim 1977.
[53] ACGIH *Treshold Limit Values and Biological Exposure Indices for 1989–1990*, Cincinnatti 1989.
[54] Hüls AG, unpublished results, Marl (1989).
[55] R. D. Short et al., *Fundam. Appl. Toxicol.* **10** (1988) 517.
[56] R. D. Short et al., *Fundam. Appl. Toxicol.* **7** (1986) 359.
[57] Chemical Industry Institute of Toxicology (1983), cited in R. D. Short et al., *Fundam. Appl. Toxicol.* **10** (1988) 517.
[58] M. B. Britz, H. I. Maibach, *Contact Dermatitis* **5** (1979) 375.
[59] L. Parmeggiani (ed.): *Encyclopedia of Occupational Health and Safety*, 3rd ed., International Labour Organisation, Geneve 1983.
[60] RTECS (Jan. 1988).
[61] R. R. Verani et al., *Lab. Invest.* **46** (1982) 79.
[62] E. D. Brewer et al., *Am. J. Physiol.* **245** (1983) F 339.
[63] E. C. Foulkes, *Toxicol. Appl. Pharmacol.* **71** (1983) 445.
[64] G. Gstraunthaler et al., *Biochem. Pharmacol.* **32** (1983) 2969.

[65] N. A. Rebouchas et al., *Pflugers Arch.* **401** (1984) 266.
[66] O. G. Fitzhugh, A. A. Nelson, *J. Am. Pharm. Assoc.* **36** (1947) 217.
[67] W. H. Rapson et al., *Bull. Environ. Contam. Toxicol.* **24** (1980) 590.
[68] Zusatzstoffverkehrs-VO, 7, Anlage 2, Liste 10, (20. Dezember 1977).
[69] E. H. Vernot et al., *Toxicol. Appl. Pharmacol.* **42** (1977) 417.
[70] S. Levey et al., *J. Am. Pharm. Assoc.* **35** (1946) 298.
[71] M. Ishidate Jr. et al., *Food Chem. Toxicol.* **22** (1984) 623.
[72] V. Saffioti, P. Shubik, *Natl. Cancer Inst. Monogr.* (1983) no. 10, 489.

Malonic Acid and Derivatives

Peter Pollak, Lonza AG, Basel, Switzerland

1.	Introduction 3177	7.	Quality Specifications and Analysis 3192
2.	Malonic Acid........... 3178		
3.	Malonates 3179	8.	Economic Aspects 3194
4.	Cyanoacetic Acid 3185	9.	Toxicology and Occupational
5.	Cyanoacetates 3187		Health................ 3194
6.	Malononitrile 3188	10.	References.............. 3195

1. Introduction

Compounds in the class of C_3 dicarboxylic acids are discussed in this article. The most important of these industrially are malonates, cyanoacetates, cyanoacetic acid, and malononitrile [1], [2]. The compounds are generally produced by reaction of a C_1 with a C_2 building block. Thus, malonic acid and cyanoacetic acid are produced from chloroacetic acid and hydrocyanic acid or carbon monoxide; malononitrile is produced from cyanogen chloride and acetonitrile.

The common feature of these compounds is the high reactivity of the central methylene group. Owing to the increasingly electron-withdrawing character of the substituents, the acidity of the hydrogen atoms in the 2-position increases in the order malonates ≪ cyanoacetates < malononitrile. Therefore, all these compounds undergo reactions typical of 1,3-dicarbonyl compounds. For example, they are easily alkylated or acylated, undergo aldol and Knoevenagel condensation, and take part in cycloaddition reactions to form pyrimidines and other nitrogen heterocycles.

The demand for malonic acid and its derivatives has increased considerably in the last few years because of the increasing importance of pyrimidines, pteridines, and purines in the life sciences [2]. The production capacity of malonic acid derivatives in the United States, Europe, and the Far East has increased accordingly.

The total annual production of malonic acid derivatives in the Western Hemisphere is ca. 20 000 t, about half of which is accounted for by diethyl malonate. However, only a fraction of the total production appears on the market, because most producers use these compounds themselves to make other products. The most important producers are Hüls (United States and Federal Republic of Germany), Lonza (Switzerland), Rhône-Poulenc (France), and Juzen (Japan).

2. Malonic Acid

Physical Properties. Malonic acid [*141-82-2*], propanedicarboxylic acid, HOOC–CH$_2$–COOH, M_r 104.06, mp 134–135 °C (decomp.), d_4^{25} 1.618, dissociation constants (water, 25 °C): K_1 1.67 × 10^{-3}, K_2 2.01 × 10^{-6}. The colorless, odorless, hygroscopic acid is dimorphic. Malonic acid sublimes in vacuum with some decomposition. Its solubility (20–22 °C) is 139 g in 100 g of water and 15 g in 100 g of pyridine. Malonic acid is only slightly soluble in alcohol or ether, and is insoluble in benzene.

Chemical Properties. Malonic acid is found in small amounts in sugar beet and green wheat, being formed by oxidative degradation of malic acid (Latin *malum*, apple). On heating the free acid above 130 °C or its aqueous solution above 70 °C, or upon irradiation with UV light, decomposition to acetic acid and carbon dioxide takes place. This facile decarboxylation is typical of β-keto acids; hence, malonic acid occupies a unique position among dicarboxylic acids. Other characteristic reactions are condensation and substitution reactions of the methylene group.

Malonic acid also behaves differently from homologous dicarboxylic acids in that it does not form a cyclic anhydride when heated with phosphorus pentoxide. Instead, linear carbon suboxide, C$_2$O$_3$ (M_r 68.03, mp –108 °C, bp 7 °C), is formed, a toxic gas that reacts violently with water to reform malonic acid. Substituted malonic acids also lose one mole of carbon monoxide at elevated temperature. Reaction with sulfuryl chloride gives mono- or dichloromalonic acid. Treatment with thionyl chloride or phosphorus pentachloride, in contrast, leads to the mono- or diacyl chloride. Reaction with bromine in chloroform yields mono- and dibromomalonic acids, whereas bromination in water at 100 °C leads to tribromoacetic acid and bromoform.

The methylene group condenses with aldehydes at low temperature to give arylidene or alkylidene malonic acids; at higher temperature, simultaneous decarboxylation occurs to yield cinnamic acid or homologues of acrylic acid. With benzaldehyde, benzylidenemalonic acid [*584-45-2*] or cinnamic acid is formed; with acetaldehyde, ethylidenemalonic acid, or, on heating, crotonic acid, is the product; furfural leads to furfurylacrylic acid, which can react further to give pimelic acid [3].

Production. The industrial production of malonic acid is less important than that of the dialkyl esters (malonates). Malonic acid is produced by alkaline saponification of cyanoacetic acid or by acid saponification of malonates.

Uses. Malonic acid is used instead of the cheaper malonates for the introduction of a –CH$_2$COOH group under mild conditions by Knoevenagel condensation and subse-

Table 1. Physical properties of dimethyl and diethyl malonate

Property	Dimethyl malonate	Diethyl malonate
M_r	132.12	160.17
mp, °C	−62	−50
bp, °C	181.4	199
d_4^{20}	1.1528	1.0551
n_D^{20}	1.4135	1.4134
Dipole moment, D	2.39	1.57

quent decarboxylation. For example, 3,4,5-trimethoxycinnamic acid, the key intermediate for the coronary vasodilators Cascoril and Vasodistal (Delalande), is prepared by Knoevenagel condensation of the corresponding substituted benzaldehyde with malonic acid, followed by decarboxylation.

Malonic Acid Derivatives. Meldrum's acid [2033-24-1], 2,2-dimethyl-1,3-dioxa-4,6-dione, M_r 144.12, mp 93–93 °C (decomp.), is produced from malonic acid and acetone in the presence of acetic anhydride. It is used commercially for the production of monoesters of malonic acid and certain β-diketo compounds.

Salts. Malonic acid forms acidic and neutral salts, as well as numerous double and complex salts. The sodium salt is soluble in water and reacts with barium chloride to give sparingly soluble, crystalline barium malonate; with calcium chloride, calcium malonate is formed. The silver salt, which also has low solubility, crystallizes as colorless needles. Platinum ammine malonate complexes are used in the treatment of cancer [4].

3. Malonates

Physical Properties. The most important esters are the dimethyl ester [108-59-8], $CH_3OOCCH_2COOCH_3$, and the diethyl ester [105-53-3], $C_2H_5OOCCH_2COOC_2H_5$ (see also Table 1). Both are colorless liquids that are sparingly soluble in water, soluble in benzene and chloroform, and miscible in all proportions with ether and alcohol. Diethyl malonate is traditionally the most important ester.

Chemical Properties. At room temperature, malonates exist almost completely in the keto form, both in the pure liquid and in alcohol solution; they are the least enolized of all the β-dicarbonyl compounds.

The most conspicuous property of malonates is the unusual reactivity of the methylene group. Treatment with sodium alkoxide gives the easily accessible sodium salt of the malonates, which can be converted to the corresponding alkyl- or acylacetic acid by reaction with alkyl or acyl halides, respectively, followed by saponification and decarboxylation.

$$H_2C\begin{matrix}COOR^1\\COOR^1\end{matrix} \xrightarrow[-ROH]{NaOR} Na-CH\begin{matrix}COOR^1\\COOR^1\end{matrix} \xrightarrow[-NaX]{+R^2X} R^2-CH\begin{matrix}COOR^1\\COOR^1\end{matrix}$$

$$R^2-CH\begin{matrix}COOR^1\\COOR^1\end{matrix} \xrightarrow[-2\,ROH]{+H_2O} R^2-CH\begin{matrix}COOR^1\\COOR^1\end{matrix} \xrightarrow{-CO_2} R^2-CH_2-COO$$

$R^1 = C_2H_5 \cdot CH_3$

R^2 = alkyl · acyl

X = halogen

This reaction, the malonic ester synthesis, is of great practical importance. Disubstituted carboxylic acids are also accessible by substitution of both methylene protons. Malonic ester synthesis can be used to prepare α-amino acids. As an alternative to the reaction of the sodium salts of the malonates with alkyl halides, alkyl malonic esters can also be prepared by piperidine-catalyzed Knoevenagel condensation with the corresponding aldehyde or ortho ester and subsequent hydrogenation. This synthetic variation is particularly suitable for the preparation of monosubstituted malonates. Malonates undergo Michael condensation with α,β-unsaturated ketones; for example, reaction of diethyl malonate with ethyl crotonate affords, after saponification and subsequent elimination of carbon dioxide, β-methylglutaric acid. Finally, malonates can be converted to pyrazolones and barbituric acids by reaction with hydrazines and urea, respectively.

Production and Uses. Two processes have become established for the industrial production of malonates.

The *hydrogen cyanide process* is based on hydrogen cyanide and chloroacetic acid; the intermediate cyanoacetic acid is simultaneously saponified and esterified in the presence of a large excess of mineral acid and alcohol.

$$Cl-CH_2COONa + NaCN \longrightarrow NC-CH_2COONa + NaCl$$

$$NC-CH_2COONa \xrightarrow[H_2SO_4,\,H_2O]{+ROH} CH_2\begin{matrix}COOR\\COOR\end{matrix}$$

A solution of sodium cyanide (ca. 25 %) in water is heated to 65 – 70 °C in a stainless steel reaction vessel. An aqueous solution of sodium chloroacetate is then added slowly with stirring; the temperature must not exceed 90 °C. Stirring is maintained at this temperature for an hour. Particular care must be taken to ensure that the hydrogen cyanide, which is formed continuously in small amounts, is trapped and neutralized. The solution of sodium cyanoacetate is concentrated by evaporation under vacuum and then transferred to a glass-lined reaction vessel for saponification and esterification.

The alcohol and mineral acid (weight ratio 1:2 to 1:3) are introduced in such a manner that the temperature does not rise above 60–80 °C. For each mole of ester, ca. 1.2 mol of alcohol is added.

Hydrochloric acid, which is formed as byproduct from unreacted chloroacetic acid, is fed into an absorption column. After the addition of acid and alcohol is complete, the mixture is heated at reflux for 6–8 h, whereby the intermediate malonic acid ester monoamide is hydrolyzed to a dialkyl malonate. The pure ester is obtained from the mixture of crude esters by extraction with benzene, toluene, or xylene. The organic phase is washed with dilute sodium hydroxide to remove small amounts of half ester. The diester is then separated from solvent by distillation at atmospheric pressure, and the malonic ester obtained by rectification under vacuum as a colorless liquid with a minimum assay of 99%. The aqueous phase contains considerable amounts of mineral acid and salts and must be treated before being fed to the waste treatment plant. The process is suitable for both the dimethyl and diethyl esters.

The yield based on sodium chloroacetate is 75–85%. Various low molecular mass hydrocarbons, some of them partially chlorinated, are formed as byproducts. Although a relatively simple plant is sufficient for the reaction itself, a sizeable investment is required for treatment of the wastewater and exhaust gas.

The *carbon monoxide process* involves the insertion of carbon monoxide into a chloroacetate ester to form an intermediate chloroformylacetate ester, which is then converted to dialkyl malonate by addition of alcohol:

$$\text{Cl-CH}_2\text{COOR} + \text{CO} \xrightarrow{\text{NaOR}} [\text{Cl-}\overset{\overset{\displaystyle O}{\|}}{\text{C}}\text{-CH}_2\text{COOR}] + \text{ROH}$$

$$\longrightarrow \text{CH}_2\begin{smallmatrix}\text{COOR}\\ \text{COOR}\end{smallmatrix} + \text{NaCl}$$

$$R = CH_3, C_2H_5, C_3H_7$$

According to the relevant patents [5], this process is carried out in a type of loop reactor. With a conversion of 90%, a selectivity of 95% is attained. The most important byproduct is ethyl acetate (see Fig. 1).

The reaction conditions are relatively mild: 20–80 °C, 0.12–1.0 MPa, pH 5–8. Dicobalt octacarbonyl is used as the catalyst.

Some other processes are based on readily available, relatively inexpensive raw materials and could be considered for industrial production:

1) Reaction of ketene with carbon monoxide and an alkyl nitrite in the presence of palladium(II) chloride [6]:

$$\text{CH}_2\text{=C=O} + \text{CO} \xrightarrow[\text{PdCl}_2]{2\,\text{RONO}} \text{H}_2\text{C}\begin{smallmatrix}\text{COOR}\\ \text{COOR}\end{smallmatrix}$$

This method is a development of the oxalic acid from carbon monoxide process of Ube Industries. A pilot plant has been built for malonic esters in Ube City, Japan.

Figure 1. Flow diagram for diethyl malonate production by the carbonylation process

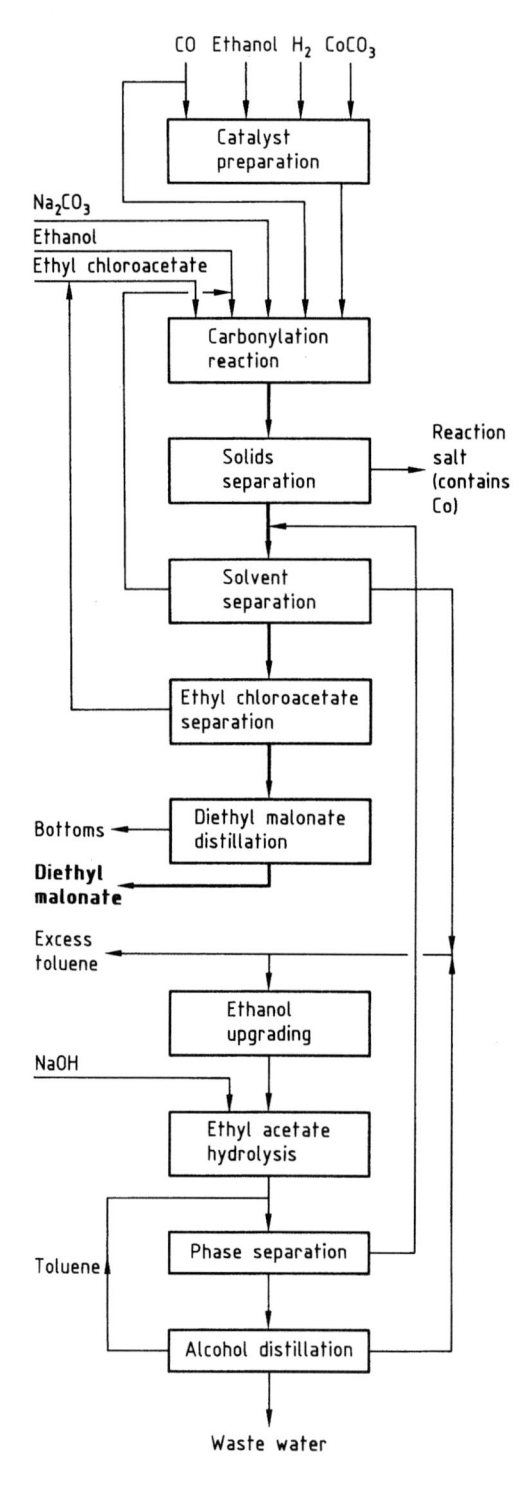

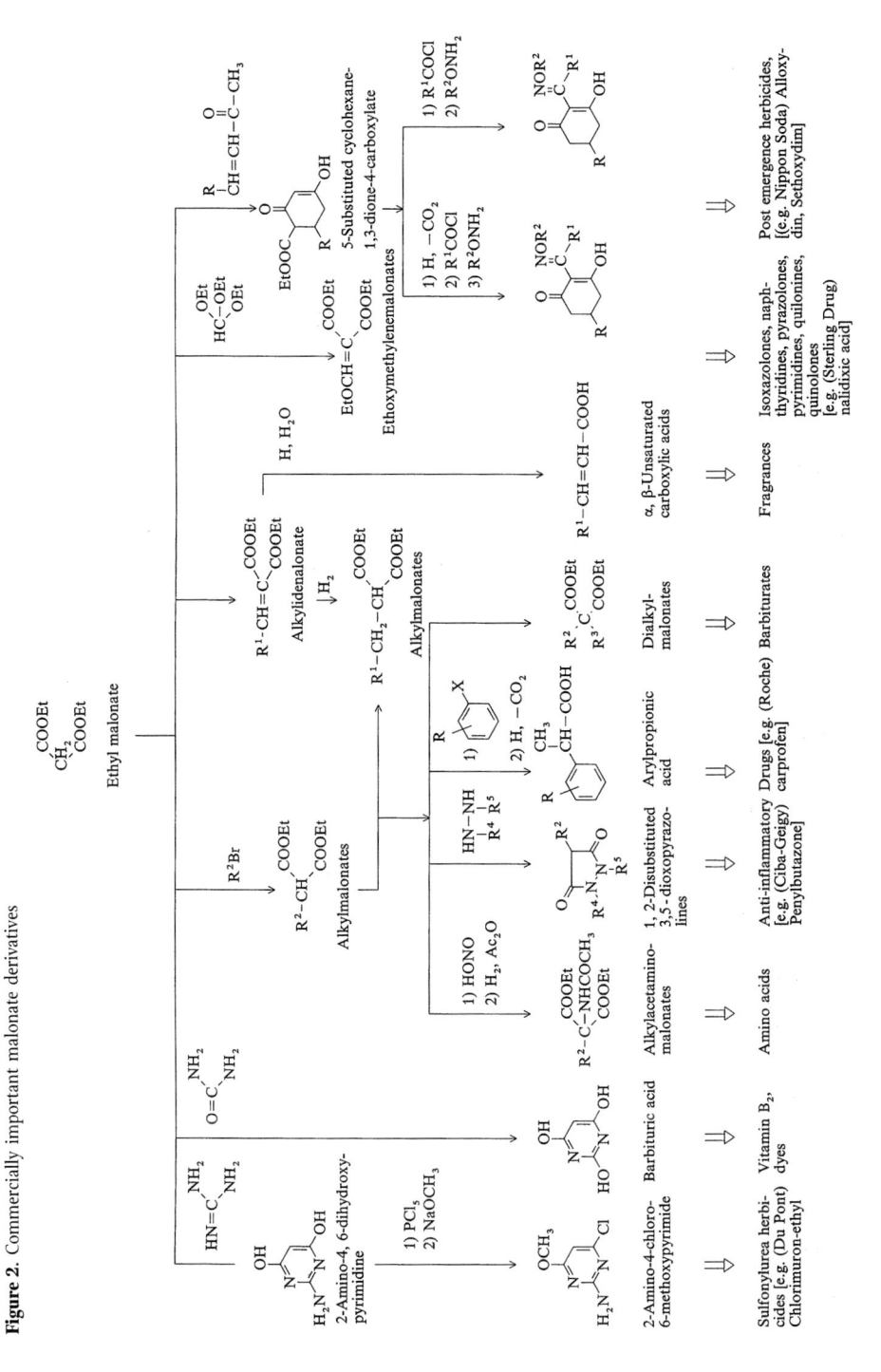

Figure 2. Commercially important malonate derivatives

Table 2. α-Substituted malonates $R^1R^2C(COOC_2H_5)_2$

R^1	R^2	Name	CAS registry number	Synthesis	Uses
CH_3	H	diethyl methylmalonate	[609-08-5]	malonic ester synthesis or Knoevenagel condensation	2-arylpropionic acids, e.g., pirprofen, carprofen
n-C_4H_9	H	diethyl n-butylmalonate	[133-08-4]	malonic ester synthesis or Knoevenagel condensation	phenylbutazone
$C_2H_5OCH=$		diethyl ethoxymethylene malonate	[87-13-8]	malonic ester + triethyl orthoformate	chloroquine, nalidixic acid
C_2H_5	C_2H_5	diethyl diethylmalonate	[77-25-8]	malonic ester synthesis or aldol condensation	barbital
C_6H_5	H	diethyl phenylmalonate	[83-13-6]	ethyl phenylacetate + carbon monoxide	phenobarbital, carbenicillin

2) Carboxylation of potassium acetate [7] or acetate esters in the presence of an alkali-metal phenoxide [8] or of 1,3-propanediol [9] under pressure

3) Electrocatalytic oxidation of propene [10]

These processes have not yet been used industrially.

Storage and Transportation. Malonates are stored in stainless steel tanks and shipped in road tankers, tank trucks, or 200-L plastic-coated steel drums.

Uses. The use of malonates is dominated by pharmaceuticals, crop protection, and fragrances. Commercially important products based on malonic esters are shown in Figure 2. A large part of the increased demand in the last ten years was accounted for by new herbicides, in particular the sulfonylureas (developed by Du Pont) and the postemergent herbicides based on cyclohexa-1,3-dione.

Malonate Derivatives. Of the malonates derived from higher alcohols, only diisopropyl malonate has any practical importance. Diisopropyl malonate [13195-64-7], M_r 188.23, bp 201–203 °C, is produced by Mitsubishi Chemical by means of the carbon monoxide process and is used for the production of the rice herbicide Fuji-One [11].

The most important malonates substituted at the methylene group are listed in Table 2. Particularly noteworthy is diethyl ethoxymethylene malonate, which is used for the synthesis of a number of drugs with a quinoline or naphthyridine structure [12].

Table 3. Physical properties of methyl and ethyl cyanoacetate

	Methyl cyanoacetate	Ethyl cyanoacetate
M_r	99.09	113.12
mp, °C	−23	−22.5
bp, °C	203	206
d_4^{20}	1.123	1.063
n_D^{20}	1.4181	1.4179

4. Cyanoacetic Acid

Cyanoacetic acid [372-09-8], $NCCH_2COOH$, M_r 85.06, mp 69–70 °C, bp 106–108 °C (partial decomp.), dissociation constant (water, 25 °C) 3.65×10^{-3}, forms colorless, hygroscopic crystals that undergo decarboxylation at ca. 165 °C to give acetonitrile. The compound dissolves readily in water, is moderately soluble in ethanol, and is poorly soluble in benzene. Because solid cyanoacetic acid and its solutions attack the skin, the acid should be handled carefully.

In addition to the usual reactions of carboxylic acids (see also → Carboxylic Acids, Aliphatic), the activated methylene group and the nitrile group are also available for reaction.

Production is based on sodium cyanoacetate solution as starting material (see Chap. 3), from which the free acid is obtained by acidification, extraction with an organic solvent, and evaporation of the solvent. In order to avoid the last two operations, most cyanoacetic acid derivatives are produced directly from the solution of the sodium salt. In addition to the classical method of production, the carboxylation of acetonitrile has been considered. For this route, acetonitrile must be deprotonated either by use of an alkali-metal phenolate [14], [15] or by electrochemical methods [16]. Because most cyanoacetic acid producers use the product captively, only a small fraction of the acid produced is traded.

The most important derivatives are the cyanoacetylureas. Cyanoacetylurea is the starting material for the sulfonamide Madribon (Roche) [17] and the diuretic amiloride (Merck, Sharp & Dohme) [18]; cyanoacetylmethylurea is used for the production of theophylline, and cyanoacetyl-N,N'-dimethylurea for caffeine.

N-Cyanoacetyl-N'-ethylurea is used for the production of the fungicide Curzate (Du Pont) [19]. Hydrogenation of cyanoacetic acid gives β-alanine. Finally, cyanoacetic acid is used as a chain extender: a 2-cyanoacrylic acid derivative is first formed by Michael addition to an aldehyde; subsequent decarboxylation gives a nitrile that contains two additional carbon atoms. Examples include the synthesis of 2-thiophenylacrylonitrile from thiophene-2-aldehyde, and 2-furfurylacrylonitrile from furfural [20]. Similar reactions are also employed in the synthesis of fragrances.

Salts. Sodium cyanoacetate, mp 176–178 °C, is sparingly soluble in ethanol. Copper cyanoacetate, produced from cyanoacetic acid and copper *tert*-butoxide in tetrahydro-

Malonic Acid and Derivatives

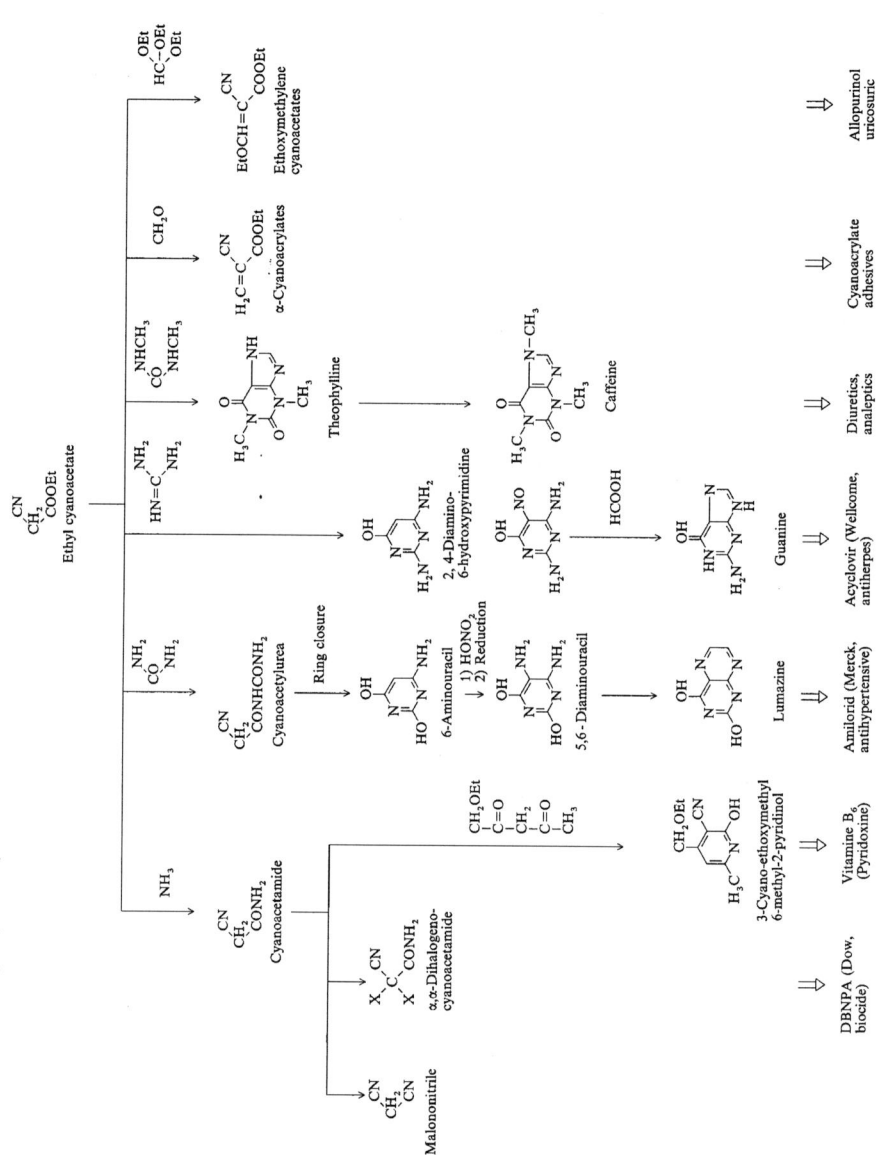

Figure 3. Commercially important cyanoacetate derivatives

furan (THF), decarboxylates on heating to form cyanomethylcopper, which is a cyanomethylating agent [21].

5. Cyanoacetates

Physical Properties. The most important derivatives are methyl cyanoacetate [*105-34-0*], NC–CH$_2$–COOCH$_3$, and ethyl cyanoacetate [*105-56-6*], NC–CH$_2$–COOC$_2$H$_5$, both of which are colorless liquids (see also Table 3).

Production. Cyanoacetates are produced industrially by the same route used for malonates, often in the same plant [22]. A process variation involves interchange of the cyanidation and esterification steps; i.e., a chloroacetate is produced from chloroacetic acid and methanol or ethanol, followed by reaction with sodium cyanide to give the corresponding cyanoacetate [23]. However, the following side reaction is difficult to suppress:

$$\text{ClCH}_2\text{COOR} + \text{NC–CH}_2\text{COOR} \longrightarrow \underset{\underset{\text{CH}_2\text{–COOR}}{|}}{\text{NC–CH–COOR}}$$

$$R = CH_3, C_2H_5$$

The crude esters are purified by distillation. Standard commercial quality today has a minimum assay of 99 %. Yields are ca. 80 %; considerable quantities of inorganic salts are also produced.

The world's largest cyanoacetate plant was brought onstream in 1987 in Mobile, Alabama, by Hüls America; annual production capacity is 4000 t.

Other synthetic routes, which are used in the laboratory but not on an industrial scale, include

1) cleavage of cyanoacetaldehyde dimethyl acetal, formed by reaction of acrylonitrile in the presence of palladium(II) chloride [12]; and
2) insertion of ketene into cyanogen chloride and alcoholysis of the intermediate cyanoacetyl chloride.

Cyanoacetate esters of higher primary alcohols are usually obtained by classical transesterification [23], whereas those of secondary alcohols are prepared in the presence of aluminum isopropoxide [24].

Uses. A selection of economically important applications of cyanoacetates is shown in Figure 3. The methyl ester can be used in place of the ethyl ester for most reactions.

6. Malononitrile

Physical Properties. Malononitrile [*109-77-3*], dicyanomethane, propanedinitrile, $CH_2(CN)_2$, M_r 66.07, *mp* 31 °C, *bp* (101.3 kPa) 218–219 °C, *bp* (1.47 kPa) 99 °C, d_4^{35} 1.0494, $n_D^{34.4}$ 1.41463, forms colorless crystals that are soluble in water, lower alcohols, diethyl ether, and acetonitrile, but insoluble in tetrachloromethane, petroleum ether, and xylene. After prolonged storage (several months) in air, the crystals become yellow to dark brown.

Chemical Properties. The chemistry of malononitrile is reviewed in [25], [26]. Malononitrile is a very reactive polar molecule, with a dipole moment of 3.56 D. The activated methylene group and the two nitrile groups, in which both the carbon and the nitrogen atoms can undergo addition reactions, allow many reactions to occur.

A characteristic reaction of malononitrile is formation of the dimer, 2-amino-1,1,3-tricyano-1-propene [27]; several trimers are also known [28], [29].

The ability of malononitrile to form metal salts by deprotonation of the methylene group is the basis for the synthesis of mono- and dialkyl derivatives by reaction with halogenoalkanes. Complexes and metal chelates are also known.

Reaction of carbon disulfide with two equivalents of malononitrile in aqueous alkali results in formation of 4,6-diamino-3,5-dicyano-2*H*-1-thiopyran-2-thione via the intermediate dimer [30]. Bromination of malononitrile in the presence of potassium bromide yields potassium bromide–tetrakis (dibromomalonitrile). Thermal decomposition of this complex affords tetracyanoethylene (TCNE), a highly electron deficient and strongly electrophilic reagent. TCNE undergoes two principle types of reactions: addition to its double bond—Diels–Alder additions of 1,3-dienes being particularly interesting—and replacement of the cyano group [31]. Reaction of the above-mentioned complex with potassium cyanide yields the potassium salt oftricyanomethane [32].

Hydrogen sulfide reacts with malononitrile to form cyanothioacetamide; further reaction leads to dithiomalonamide. Acylmalononitriles can be obtained by reaction with acid chlorides; orthoformates give ethoxymethylenemalononitrile. Mono- and diimidates, obtained by reaction of one or both nitrile groups of malononitrile with an alcohol under acid conditions, are important synthetic building blocks. The alkoxy group can be substituted by various nucleophiles. For instance, the mono and diimidates can be further reacted to give aminopyrimidines (cf. Minoxidil and Bensulfuronmethyl in Fig. 6). Depending on the protonating agent chosen, either 2-amino-4-chloro- or 2-chloro-4-aminopyrimidine derivatives are obtained [33].

In accordance with the high CH acidity, condensation of malononitrile with aldehydes (Knoevenagel condensation) is a generally applicable reaction. Only formaldehyde and acetaldehyde behave differently, leading to 2,2-dicyano-1,3-propanediol and 1,1,3,3-tetracyanopropane, respectively [34], [35].

Aromatic aldehydes with *o*-hydroxyl or *o*-amino groups condense with cyclization to give 2-imino-3-cyanocoumarins or 2-amino-3-cyanoquinolines, respectively [36]–[38].

The possibilities for reaction with ketones are similarly diverse. Condensation with acetone leads via the intermediate dimer to an aminocyclohexadiene [39]; under other conditions, 3,5-diaminophenol can be obtained [40]. With ninhydrin, malononitrile gives dicyanomethyleneindandione, which can also be obtained from tetracyanoethylene and indandione [41]. *o*-Quinones are equally suitable for condensation.

Nitrogen compounds also react with malononitrile. For example, benzenediazonium salts give mesoxalic acid hydrazones. The reaction with hydrazine is versatile: dimerization of the nitrile leads to an aminopyrazole, and alkylidene malonic acid dinitriles react with some elimination of hydrogen cyanide. Condensation with urea leads to 2-hydroxy-4,6-diaminopyrimidine.

Reaction of malononitrile with amidines (e.g., formamidine) does not lead as expected to 4,6-diaminopyrimidine, but to 4-amino-5-cyanopyrimidine [43]. A series of nitrogen heterocycles containing an amino group ortho to a cyano group can be synthesized analogously [44].

Nitroso compounds, as hetero carbonyl analogues, are also suitable for condensation reactions.

Addition reactions are also characteristic of malononitrile. Derivatives of malononitrile, malonates, and cyanoacetates all undergo Michael addition.

Production. Malononitrile was once obtained by elimination of water from cyanoacetamide in a batch process. It is now produced by a continuous high-temperature process [45] (Fig. 4) in which cyanogen chloride and acetonitrile are fed into a tube reactor above 700 °C.

Almost complete conversion is attained in one pass; the reaction products are malononitrile, excess acetonitrile, hydrochloric acid, and small amounts of maleic, succinic, and fumaric acids. The products leaving the reactor are immediately cooled to 40–80 °C, and gaseous hydrogen chloride is simultaneously separated, fed into a washer, and recovered as dilute hydrochloric acid. Excess acetonitrile is removed by a combination of vacuum distillation and thin-film evaporation. The recovered acetonitrile contains very little hydrogen chloride and can be recycled without risk of attendant corrosion.

Removal of maleic and fumaric acids from the crude malononitrile by fractional distillation is impractical because the boiling points differ only slightly. The impurities are therefore converted into high-boiling compounds in a conventional reactor by means of a Diels–Alder reaction with a 1,3-diene. Volatile and involatile byproducts are finally removed by two vacuum distillations and disposed of by combustion. The yield of malononitrile amounts to 66% based on cyanogen chloride or acetonitrile [45]. The first industrial plant to use this process was brought on stream by Lonza in Visp, Switzerland (see Fig. 5). The product has a minimum purity of 99%; the freezing point must be at least 30 °C.

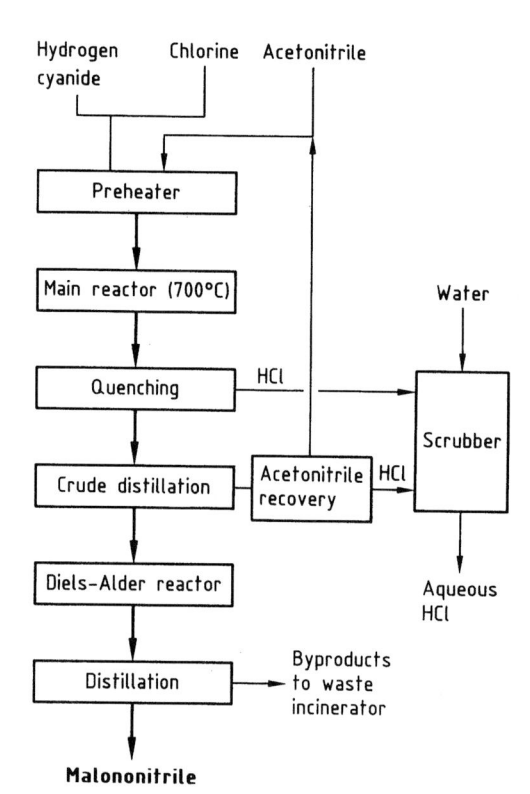

Figure 4. Process flow sheet for the production of malononitrile

Figure 5. Malononitrile plant (Lonza, Visp, Switzerland)

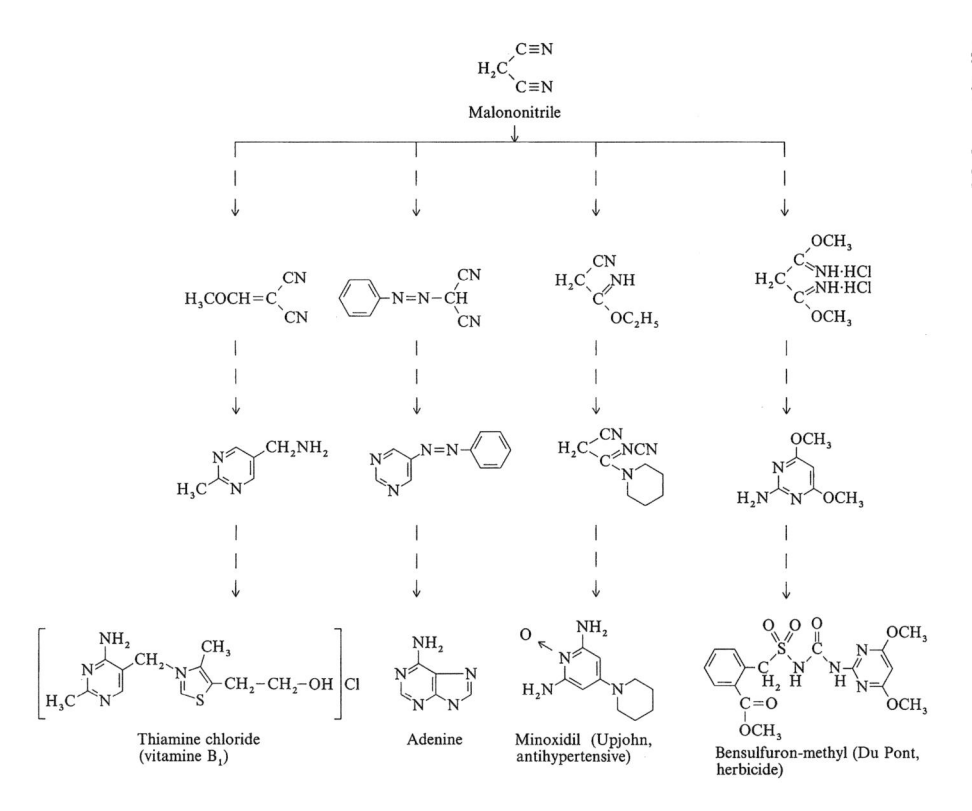

Figure 6. Commercially important malononitrile derivatives

For transportation, molten malononitrile is poured into 200-L plastic-lined drums. Care should be exercised when remelting the solidified product because spontaneous combustion can occur at elevated temperature, particularly above 100 °C. A water bath is recommended for this purpose.

Uses. Malononitrile is used predominantly as a building block for the production of pharmaceuticals and pesticides. The main steps in the production processes for the four important derivatives thiamine, adenine, minoxidil [13], and bensulfuron-methyl [104466-83-3] [14] are depicted in Figure 6.

Other important nitrogen heterocycles derived from malononitrile are the diuretic triamterene (Smith Kline) [47], the folic acid antagonists methotrexate [49] and aminopterin [50], and the gout remedy thiopurinol [52].

Furthermore, important methine dyes, in particular aminoaryl cyanines, are produced from malononitrile. In contrast to products derived from cyanoacetic acid as condensation component, these mainly yellow to yellow-green polyester dyes are characterized by high lightfastness. Notable members of this family of dyes, which all exhibit a dicyanovinyl functionality, include Resolin Brilliant Yellow 7GL, 10G, and

Table 4. Production of malonates and cyanoacetates

Company	Malonates		Cyanoacetates
	Annual capacity, 10^3 t	Process	Annual capacity, 10^3 t
United States			
Huels America			4000
Europe			
Hüls, Federal Republic of Germany	5000	CO	
Lonza, Switzerland	1000–1500	HCN	1000–1500
Rhône-Poulenc, France	2000	HCN	
Japan			
Mitsubishi	Unknown	CO	
Tateyama	500*	HCN	800*
Juzen	2000	HCN	2000
	2000	CO	

* Estimated.

10GN (Bayer); Palanil Yellow 7G (BASF); Terasil Brilliant Yellow 6G (Ciba-Geigy); Eastman Polyester Yellow 6G-LSW (Kodak); and Foiron Dark Blue RD-2RE (Sandoz).

A dicyanovinyl moiety is also present in tetracyanoquinodimethane [*1518-16-7*] (TCNQ), produced from malononitrile and 1,4-cyclohexadiene; TCNQ forms charge-transfer complexes with tetrathiafulvalene, which are being tested for use in the production of conducting films in photocopiers and telefax equipment, and for three-dimensional memories [60]. Condensation of *o*-chlorobenzaldehyde with malononitrile gives *o*-chlorobenzylidene malononitrile (CS gas), which is a well-known tear gas with a high safety factor [61].

7. Quality Specifications and Analysis

Assay (content) is the most important quality criterion for the compounds discussed in this article. The leading producers guarantee a minimum content of 99%.

The determination method used varies according to the functional groups present. For acids, potentiometric titration with sodium hydroxide is employed. With malonic acid, both equivalent points are measured, and the content is determined as follows:

$$\text{Malonic acid, \%} = \frac{(E_2 - E_1) \cdot M_r \cdot 10.406}{W}$$

where E_1 = volume of 1 M NaOH at the first equivalence point; E_2 = volume of 1 M NaOH at the second equivalence point; M = molarity of the NaOH solution; and W = weight of sample in grams.

Table 5. Product safety data

	CAS registry number	Main hazard identification	Flash point,* °C	LD$_{50}$ (oral, rat), mg/kg	Skin irritation (rabbit)	Transport classification
Malonic acid	[141-82-2]	corrosive		2 750	slight irritant	RID/ADR, IMDG Code, IATA/ICAO: not restricted
Dimethyl malonate	[595-46-0]	flammable	88	4 520	nonirritant	RID/ADR: 3; IMDG Code, IATA/ICAO: not restricted
Diethyl malonate	[105-53-3]	flammable	90	10 300	nonirritant	RID/ADR: 3; IMDG Code, IATA/ICAO: not restricted
Diisopropyl malonate	[13195-64-7]	flammable	95	>5 000	slight irritant	RID/ADR: 3; IMDG Code, IATA/ICAO: not restricted
Cyanoacetic acid	[372-09-8]	health		1 500	moderate irritant	RID/ADR: 8; IMDG Code: 8; IATA/ICAO: 8
Methyl cyanoacetate	[105-34-0]	health	110	3 062	slight irritant	RID/ADR: 6.1; IMDG Code: 6.1; IATA/ICAO: 6.1
Ethyl cyanoacetate	[105-56-6]	health	109	2 820	slight irritant	RID/ADR: 6.1; IMDG Code: 6.1; IATA/ICAO: 6.1
Malonodinitrile	[109-77-3]	health, explosive	>86	13.9	slight irritant	RID/ADR: 6.1; IMDG Code: 6.1; IATA/ICAO: 6.1

* Pensky–Martens, closed cup DIN 51 758.

In the assay of cyanoacetic acid, the hydrochloric acid present as an impurity is also included.

Gas chromatography is recommended for the quantitative analysis of malonates, cyanoacetates, and malononitrile.

In addition to assay, other important purity specifications include appearance, melting or boiling point, and water content.

8. Economic Aspects

The production capacity of malonates and cyanoacetates in the Western Hemisphere increased sharply during the last ten years and may now exceed 20 000 t annually (see Table 4). The United States is responsible for ca. 50% of the cyanoacetate capacity, although after Kay–Fries ceased production, diethyl malonate is no longer available for the free market. The figures given above do not include companies that use the products for internal consumption; these include Abbott (United States) for barbiturates and Hoffmann La Roche (Switzerland) for riboflavin. The stimulus for the increase in production capacity was the demand for new crop protection agents, e.g., cyclohexadione herbicides and Fuji-One (see also Fig. 2).

Considerable overcapacity exists today, which is intensified by imports from the East Bloc, particularly Rumania, and the People's Republic of China.

Apart from the malonates and cyanoacetates, cyanoacetic acid is the most important product in this group. The fraction of production used internally is particularly large. At the forefront are the leading producers of caffeine, Boehringer Ingelheim and Knoll (both Federal Republic of Germany). The main producer for the free market is Huels. Worldwide production of malonic acid does not amount to more than a few hundred tons per year.

9. Toxicology and Occupational Health

The most important safety data for the compounds discussed in this article are listed in Table 5. Values for toxicity (LD_{50}) and skin irritation stem from studies performed on commercially available products according to international standards. No MAK (or TLV) values have been set. Because of the low LD_{50} and its chemical reactivity (see also Chap. 6), particular care should be taken in handling malononitrile. Investigations of biological degradability are currently being conducted in the United States.

10. References

General References

Beilstein: Malonic acid II 566, II1 244, II2 516, II3 1597, II4 1874. Dimethyl malonate II 572, II$_1$ 247, II$_2$522, II$_3$ 1607, II$_4$ 1880. Diethyl malonate II 573, II$_1$ 247, II$_2$ 524, II$_3$ 1610, II$_4$ 1881. Cyanoacetic acid II 583, II$_1$ 253, II$_2$ 530, II$_3$ 1626, II$_4$ 1888. Methyl cyanoacetate II 584, II$_1$ 253, II$_2$ 530, II$_3$ 1628, II$_4$ 1889. Ethyl cyanoacetate II 585, II$_1$ 254, II$_2$ 531, II$_3$1628, II$_4$ 1889. Malononitrile II 589, II$_1$ 256, II$_2$ 535, II$_3$ 1634, II$_4$ 1892.

Specific References

[1] H. Henecka: Chemie der β-Dicarbonylverbindungen, Springer-Verlag, Berlin 1950.
[2] Company brochures: "Malonic and Cyanoacetic Esters" (1981), "Malononitrile" (1987) and "Pyrimidines, Pteridines and Purines" (1986) Lonza AG.
[3] P. D. Gardner, L. Rand, G. R. Haynes, *J. Am. Chem. Soc.* **78** (1956) 3425.
[4] Research Corp., DE-OS 2 329 485, 1973 (M. Cleare, J. Hoeschele, B. Rosenberg, L. Van Camp).
[5] Dynamit-Nobel, DE-OS 2 359 963, 1973; DE 2524 389, 1975 (E. Ch. Moustafa, U. Prange).
[6] Ube Industries, EP-A 6.611, 1980 (Kenji Nishimura et al.).
[7] Mitsui Toatsu, JP 49 056 911, 1974 (M. Kawamata, H. Tanabe).
[8] G. Bottacio, G. P. Chiusoli, M. G. Felicioli, *Gazz. Chim. Ital.* **103** (1973) 105. Mitsui Toatsu, JP 49 086-317, 1974 (M. Kawamata, T. Takahashi, et al.).
[9] Unilever, US 3 892 787, 1973 (E. N. Gutierrez, R. C. Reardon).
[10] H. Kinza, DE-OS 107 257, 1973; H. Kinza, *Z. Phys. Chem. (Leipzig)* **255** (1974) 517–537.
[11] Asahi Denka, JP-Kokai 78 112 807, 1977 (H. Yamamoto, Y. Shoji).
[12] A. Kleemann, J. Engel: Pharmazeutische Wirkstoffe, Erg. Bd., Thieme, Stuttgart–New York, 1060, 1123, 1156, 1234, 1236, 1240, 1258.
[13] A. Kleemann, Arzneimittel, 1972–1985, VCH Verlagsgesellschaft, Weinheim 1987, p. 320.
[14] E. I. Du Pont De Nemours And Company, EP 24 200, 1980 (J. J. Fuchs, E. Wat, Koon Wah).
[15] Mitsui Toatsu, JP 50 071 627, 1975 (M. Kawamata, et al.).
[16] Monsanto, US 3 945 896, 1974 (D. A. Tysee).
[17] The Merck Index, 9. Ed., Merck & Co., Rahway 1976, p. 1151.
[18] In [17] p. 55.
[19] J. M. Serres, G. A. Carraro, *Meded. Fac. Landbouwwet. Rijksuniv. Gent* 41/2, 645–650, 1976.
[20] J. M. Patterson, *Org. Synth.* **40** (1960) 46.
[21] T. Tsuda, T. Nakatsuka, T. Hirayama, T. Saegusa, *J. Chem. Soc., Chem. Commun.* 1974, 557–558.
[22] J. Mokry, *Chem. Zvesti* **8** (1954) no. 23, 63–75.
[23] Knapsack, DE 1 210 789, 1 272 914, 1964 (K. Sennewald, A. Hanser, W. Lork).
[24] A. Said, *Chimia* **28** (1974) 234.
[25] F. Freeman, *Chem. Rev.* **69** (1969) 591.
[26] A. J. Fatiadi, *Synthesis* **1978**, 163; 241.
[27] E. C. Taylor, K. S. Hartke, *J. Am. Chem. Soc.* **81** (1959) 2452.
[28] H. Junek, H. Sterk, *Z. Naturforsch.* **22** (1967) 732.
[29] J. D. Atkinson, M. C. Johnson, *J. Chem. Soc.,* C, 1969, 2181.
[30] T. Takeshima et al., *J. Org. Chem.* **35** (1970) 2434–2440.
[31] A. J. Fatiadi, *Synthesis* **1987**, 749–789; 959–978.

[32] FMC, US 4 059 433, 1976 (L. K. Gibbons).
[33] T. Hirayama et al., *Chem. Pharm. Bull.* **24** (1976) 507.
[34] Goodrich, US 2 541 350, 1951 (H. Gilbert).
[35] O. Diehls, B. Conn, *Ber. Dtsch. Chem. Ges.* **56** (1923) 2076.
[36] L. L. Woods, J. Sapp, *J. Org. Chem.* **30** (1965) 312.
[37] G. P. Schiemenz, *Chem. Ber.* **95** (1962) 483.
[38] H. Junek, *Monatsh. Chem.* **94** (1963) 890.
[39] J. K. Williams, *J. Org. Chem.* **28** (1963), 1054.
[40] *Arch. Inst. Pharmacodyn.* **179** (1969) 284; *Manuf. Chem.*, March 1970, p. 29.
[41] H. Junek, H. Sterk, *Tetrahedron Lett.* **40** (1968) 4309.
[42] H. Junek, H. Binder, H. Sterk, H. Hamböck, *Monatsh. Chem.* **99** (1968) 2195.
[43] D. J. Brown: *The Pyrimidines,* Interscience Publishers, New York 1962, p. 72.
[44] New Haven Chem. Corp., P.O.B. 3648, New Haven, Ct. 06525, catalogue.
[45] Lonza, CH 493 473, 1970 (U. Arni, A. Faucci); CH 493 474, 1970 (U. Arni, A. Stocker, K. Aufdereggen); CH 493 475, 1970 (A. Faucci); CH 502 992, 1971 (K. Aufdereggen, U. Arni, A. Faucci, A. Stocker); US 3 655 721, 1972 (U. Arni, A. Faucci, A. Stocker); US 3 683 003, 1972 (K. Aufdereggen, U. Arni, A. Faucci, A. Stocker). Degussa, DE 1 768 154, 1968 (T. Lüssling, F. Theissen, W. Wigert).
[46] H. J. Heinen, J. A. Eisele, B. J. Scheiner, US Dept. of Interior, Bureau of Mines Report of Investigation, Washington, December 1970.
[47] [17]p. 1233.
[48] Merck, DE-OS 2 443 076, 1974 (M. B. Miller, E. McManus).
[49] [17] p. 782.
[50] [17] p. 65.
[51] A/S GEA, Kopenhagen, DOS 2 442 080, 1974 (B. Alhede, St. Greve, N. Gelting).
[52] Dictionnaire Vidal, O.V.P. Paris 1971, p. 1520.
[53] Eastman Kodak, US 3 240 783, 1966 (J. M. Straley, D. J. Wallace, M. A. Weaver).
[54] Ciba, DE-OS 2 017 914 (1970).
[55] Ciba-Geigy, DE-OS 2 515 138, 1975 (N. B. Desai, V. Ramanathan).
[56] Bayer, DE-OS 2 445 583, 1974 (H. Beecken), DOS 2 446 759, 1974 (H. Beecken).
[57] American Color & Chem., US 3 948 938, 1976 (E. E. Renfrew).
[58] Bayer, BE 666 863, 1965 (K.-H. Schündehütte, K. Trautner).
[59] W. R. Hatchard, *J. Org. Chem.* **29** (1964) 660–668.
[60] J. H. Krieger, *Chem. Eng. News,* July 4 (1977) 14–17.
[61] T. S. Crockett: *Poice Chemical Agents Manual,* International Association of Chiefs of Police, Inc., Washington 1969.
[62] Gulf, US 3 781 446, 1973 (R. P. Cahoy, J. Sanjean); 3 825 663, 1974 (J. Sanjean, R. P. Cahoy); 3 890 367, 1975 (R. P. Cahoy).
[63] Cyanamid, US 3 981 717, 1976 (B. L. Walworth).
[64] Ciba-Geigy, DE-OS 2 535 769, 1975 (J. Drabek).
[65] Fisons, DE-OS 2 350 021, 1973 (P. L. Carter).
[66] Monsanto, US 3 691 227, 1972 (J. W. Baker, R. K. Howe).
[67] Chem. Eng. News, Jan 11, 1971, 14.Report A-172-2423-II Southern Research Institute, Birmingham, Ala. 35205, June 1970, unpublished.
[68] Sumitomo, DE 2 402 197, 1974 (I. Takagi, S. Kawanischi, Y. Nishizawa).
[69] FMC, US 4 032 321, 1977 (L. K. Gibbons).
[70] Lonza, personal communication, July 7, 1988.

Melamine and Guanamines

GEORGE M. CREWS, Melamine Chemicals, Inc., Donaldsonville, Louisiana 70346, United States, (Chaps. 1–9)

WILLI RIPPERGER, BASF Aktiengesellschaft, Ludwigshafen, Federal Republic of Germany (Chaps. 1–9)

DIETRICH BURKHARD KERSEBOHM, BASF Aktiengesellschaft, Ludwigshafen, Federal Republic of Germany (Chap. 10)

JOSEF SEEHOLZER, SKW Trostberg, Trostberg, Federal Republic of Germany (Chap. 11)

1.	Introduction	3197
2.	Physical Properties	3198
3.	Chemical Properties	3199
3.1.	Thermal Behavior	3199
3.2.	Hydrolysis	3201
3.3.	Salt Formation	3201
3.4.	Reaction with Aldehydes	3201
4.	Production	3202
4.1.	Low-Pressure Processes	3203
4.1.1.	BASF Process	3204
4.1.2.	Chemie Linz Process	3204
4.1.3.	Stamicarbon Process	3206
4.2.	High-Pressure Processes	3206
4.2.1.	Melamine Chemicals Process	3207
4.2.2.	Montedison (Ausind) Process	3208
4.2.3.	Nissan Process	3209
5.	Quality Specifications	3210
6.	Chemical Analysis	3210
7.	Storage and Transportation	3211
8.	Uses	3211
9.	Economic Aspects	3211
10.	Toxicology	3212
11.	Guanamines	3214
11.1.	Production	3215
11.2.	Uses	3216
11.3.	Toxicology	3217
12.	References	3218

1. Introduction

Melamine was first prepared and described in 1834 by LIEBIG, who obtained it from fusion of potassium thiocyanate with ammonium chloride. In 1885, A. W. VON HOFFMANN published its molecular structure. Melamine [108-78-1] (2,4,6-triamino-1,3,5-triazine), $C_3N_6H_6$, M_r 126.13, exists mainly in the amino form:

Not until 100 years later did melamine find industrial application in the production of melamine–formaldehyde resins. The first commercial plants came on stream in the late 1930s. Since that time melamine has become an increasingly important chemical commodity. In 1970, world capacity was estimated at 200 000 t. Current production is estimated to be 550 000 t/a. Most of the melamine produced is still used in the fabrication of melamine–formaldehyde resins.

Until about 1960, melamine was prepared exclusively from dicyandiamide [*461-58-5*]. This conversion was carried out in autoclaves at 10 MPa and 400 °C in the presence of ammonia, according to the equation

$$3\ H_2NC(NH)NHCN \longrightarrow 2\ C_3N_6H_6$$

In the early 1940s, MACKAY discovered that melamine could also be synthesized from urea [*57-13-6*] at 400 °C with or without catalyst [5]. Today, melamine is produced industrially almost exclusively from urea. Most processes using dicyandiamide as raw material were discontinued or replaced at the end of the 1960s.

2. Physical Properties

Melamine is manufactured and sold as fine, white, powdered crystals. The most important physical data for melamine are summarized in the following list [6], [7]:

mp	350 °C (subl.)
Density ϱ	1.573 g/cm^3
Dissociation constants	
$\quad K_{b1}$ (25 °C)	1.1×10^{-9}
$\quad K_{b2}$ (25 °C)	1.0×10^{-14}
Heat of formation ΔH_f^0 (25 °C)	-71.72 kJ/mol
Heat of combustion (25 °C)	-1967 kJ/mol
Heat of sublimation (25 °C)	-121 kJ/mol
Molar heat capacity (25 °C)	155 J K^{-1} mol^{-1}
Specific heat capacity c_p J kg^{-1} K^{-1}	
$\quad$ at 273–353 K	1470
$\quad$ at 300–450 K	1630
$\quad$ at 300–550 K	1720
Entropy (25 °C)	149 J K^{-1} mol^{-1}
Entropy of formation ΔS_f^0 (25 °C)	-835 J K^{-1} mol^{-1}
Free energy of formation G_f^0 (25 °C)	177 kJ/mol

Solubility (30 °C), g/100 mL, in
  Ethanol   0.06
  Acetone   0.03
  Dimethylformamide   0.01
  Ethyl cellosolve   1.12
  Water   0.5

The temperature dependence of the vapor pressure (in 10^5 Pa) in the range 417–615 K is described by the following equation [8]:

$\log p = 9.7334 - 6484.9/T$

Melamine solubility in water (in grams per 100 g of H_2O) over the range 20–100 °C coincides closely with the relationship [9]:

$\log L = 5.101 - 1642/T$

Reasonably reliable values can be obtained by means of this equation down to 0 °C.

Crystal Data. Melamine forms monoclinic crystals, space group $P2_1/a$ with $a = 1.0537$, $b = 0.7477$, $c = 0.7275$ nm, $\beta = 112°9'$, and $Z = 4$ [10].

3. Chemical Properties

The chemical properties of melamine are summarized in detail in [1] and [2]. The s-triazine ring is very stable and cleaves only under drastic conditions (e.g., heating above 600 °C or fusion with alkali compounds).

By X-ray diffraction studies, crystalline melamine has been shown to exist only in the symmetrical triamino structure; the same is true for the vapor phase and for both neutral and alkaline solutions. Although reactions are in some cases observed at the ring nitrogen atoms (the products being substituted isomelamines), the most important commercial reactions involve only the $-NH_2$ groups, which behave chemically as amido rather than amino functions.

3.1. Thermal Behavior

When melamine is heated above 300 °C in the absence of ammonia or at low ammonia partial pressure, deammoniation and condensation lead to compounds with higher molecular mass. Degradation starts with the release of ammonia and the formation of melem [*1502-47-2*] (2,5,8-triamino-1,3,4,6,7,9,9b-heptaazaphenalene):

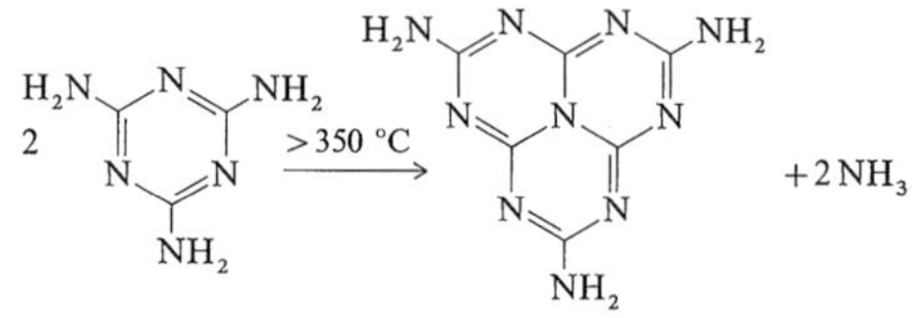

$$2 \; \text{Melamine (C}_3\text{H}_6\text{N}_6\text{)} \xrightarrow{>350\,°C} \text{Melem (C}_6\text{H}_6\text{N}_{10}\text{)} + 2\,\text{NH}_3$$

Further heating to ca. 600 °C yields more ammonia and melone [*32518-77-7*] [11]:

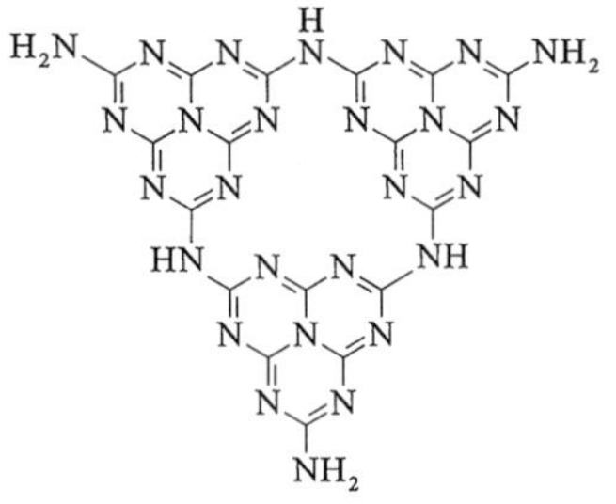

1

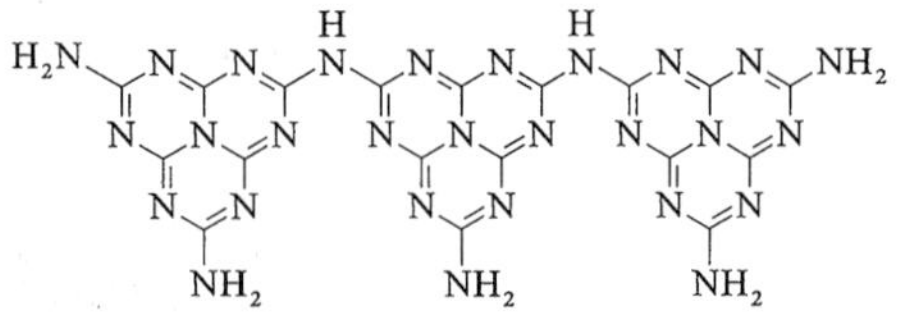

2

Various sources disagree on hydrogen analysis and other analytical data for melone. The material may represent a mixture of substances such as **1** and **2**.

Melam [(*N*-4,6-diamino-1,3,5-triazin-2-yl)-1,3,5-triazine-2,4,6-triamine], [*3576-88-3*] seems not to be an intermediate in the thermal degradation of melamine although the evidence is not entirely clear [11]–[13]. This substance can be prepared by heating melamine salts below 315 °C, and it arises as a byproduct of melamine synthesis.

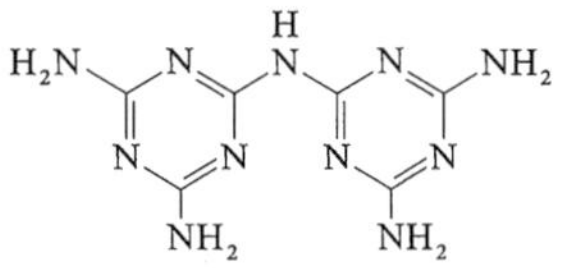

Melam ($C_6H_9N_{11}$)

The three deammoniation products—melam, melem and melone—are formed reversibly; addition of ammonia at high pressure and temperature regenerates melamine. Indeed, processes for melamine production invariably rely upon excess ammonia to suppress formation of these byproducts. If melamine is heated to 600 °C or higher it is partially cracked, leading to cyanamide along with other products. Hydrogen cyanide may also appear in the crack products, especially in the absence of oxygen [14].

3.2. Hydrolysis

Melamine is hydrolyzed by mineral acid or inorganic alkali. Hydrolysis proceeds stepwise, with loss of one, two, or all three amino groups:

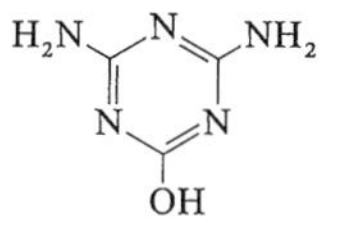

Ammeline 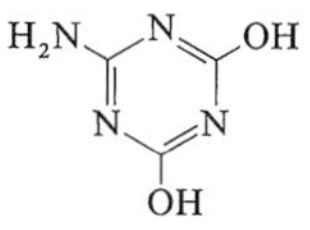   Ammelide

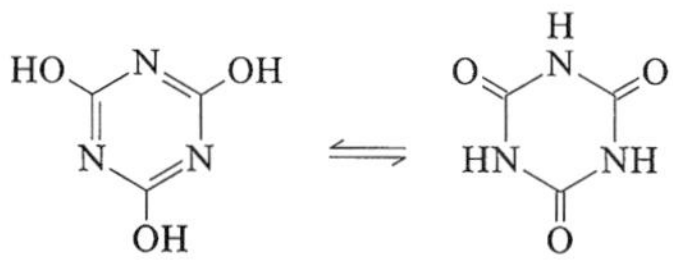

Cyanuric acid

The product spectrum varies with temperature, pH, and concentration; the end product is cyanuric acid [108-80-5]. Even small amounts of the oxotriazines (especially cyanuric acid and ammelide) markedly effect the condensation of melamine with formaldehyde by increasing the rate of condensation [15].

3.3. Salt Formation

Melamine is a weak base, forming well-defined salts with both organic and inorganic acids. The melamine ion is assigned the following structure:

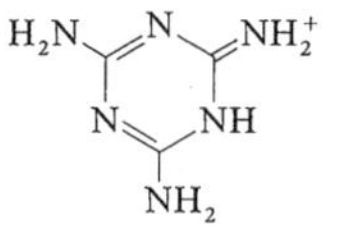

The water solubility of organic and inorganic salts of melamine is no higher than that of free melamine (see Table 1). Melamine cyanurate, melamine picrate, and melamine perchlorate are very insoluble in water, and are useful in the quantitative determination of melamine.

3.4. Reaction with Aldehydes

Melamine reacts with aliphatic and aromatic aldehydes to give a variety of products. Most important is the resinous material obtained from the reaction of formaldehyde with melamine:

Table 1. Industrially important melamine salts

Molecular formula	Solubility (20 °C), g per 100 g H_2O	M_r	mp, °C
$C_3N_6H_6 \cdot H_3PO_4$	0.43	224.12	223
$C_3N_6H_6 \cdot H_2SO_4$	0.19	350.33	380 *
$C_3N_6H_6 \cdot HNO_3$	0.68	189.14	298 **
$C_3N_6H_6 \cdot C_{17}H_{35}COOH$	0.17	410.42	154
$C_3N_6H_6 \cdot HCOOH$	1.56	172.16	250 (decomp.)

* Sublimes at 316 °C.
** Sublimes at 186 °C.

$$Me(NH_2)_3 + 6\ CH_2O \longrightarrow Me[N(CH_2OH)_2]_3$$

where Me represents that part of the melamine molecule that is not involved in the reaction.

All hydrogen atoms on the melamine molecule can be replaced by methylol groups, and products ranging from the monomethylol to the hexamethylol derivatives have been observed. The methylolmelamines are sparingly soluble in most solvents and are very unstable due to further condensation or resinification, e.g.,

$$MeNHCH_2OH + H_2N-Me \longrightarrow MeNHCH_2NHMe + H_2O$$
$$2\ MeNHCH_2OH \longrightarrow MeNHCH_2OCH_2NHMe + H_2O$$

Melamine – formaldehyde condensation products are characterized by good heat resistance and superior water-resisting properties. They are used, usually in combination with urea – formaldehyde resins, as glues in the woodworking industry, as impregnating resins for decorative laminates, and as a binder in molding materials containing a filler (e.g., cellulose or sawdust). Methylolmelamines can be etherified by heating with alcohol in the presence of an acid catalyst. Industrially most important are the products formed with methanol, *n*-butanol, and isobutanol. They are used as curing agents for surface coatings and as auxiliaries in the paper and textile industries [16].

4. Production

Melamine can be synthesized from urea at 390 – 410 °C:

$$6\ H_2N-CO-NH_2 \longrightarrow C_3N_3(NH_2)_3 + 6\ NH_3 + 3\ CO_2$$

The overall reaction is endothermic, requiring 649 kJ per mole of melamine starting with molten urea at 135 °C.

The processes themselves may be subdivided into two categories:

1) noncatalytic, high-pressure (≥ 8 MPa) processes, and
2) catalytic, low-pressure processes (ca. 1 MPa).

Each type includes three stages:

    synthesis
    melamine recovery and purification
    off-gas treatment

4.1. Low-Pressure Processes

Typical low-pressure processes utilize a fluidized catalyst bed at pressures from atmospheric to ca. 1 MPa and temperatures of 390–410 °C. The fluidizing gas is either pure ammonia or the ammonia–carbon dioxide mixture formed during the course of the reaction. Catalysts include alumina and materials of the silica–alumina type. Melamine leaves the reactor in gaseous form together with the fluidizing gas; it is separated from ammonia and carbon dioxide by quenching the gas stream either with water (followed by crystallization) or with cold reaction gas (desublimation).

In the catalytic processes the first reaction step is decomposition of urea to isocyanic acid and ammonia, after which the isocyanic acid is transformed into melamine:

$$6\,(NH_2)_2CO \longrightarrow 6\,HN=C=O + 6\,NH_3 \qquad \Delta H = 984 \text{ kJ/mol}$$
$$6\,HN=C=O \longrightarrow C_3N_3(NH_2)_3 + 3\,CO_2 \qquad \Delta H = -355 \text{ kJ/mol}$$
$$\overline{6\,(NH_2)_2CO \longrightarrow C_3N_3(NH_2)_3 + 6\,NH_3 + 3\,CO_2 \qquad \Delta H = 629 \text{ kJ/mol}}$$

The overall reaction mechanism is not yet fully understood, but isocyanic acid from the decomposition of urea is believed to be catalytically disproportionated into carbon dioxide and cyanamide or carbodiimide, which then trimerizes to melamine [17]–[19]:

$$6\,HOCN \longrightarrow 3\,[HNCNH] \rightleftarrows H_2NCN] + 3\,CO_2$$
$$\text{Carbodiimide} \quad \text{Cyanamide}$$
$$3\,HNCNH \longrightarrow C_3N_3(NH_2)_3$$

The melamine yield is ca. 90–95% based on urea. Byproducts include melam, melem, and melone, as well as oxotriazines such as ammeline, ammelide, and cyanuric acid. Ureidotriazine is also observed as a product of reaction between melamine and isocyanic acid.

Some byproducts are formed in the reactor during synthesis; others are not generated until the melamine recovery section, where deammoniation or hydrolysis occurs [19]–[28].

Worldwide, three low-pressure processes are in commercial operation: the BASF process, the Chemie Linz process, and the Stamicarbon process.

4.1.1. BASF Process

The BASF process (see Fig. 1) is a *one-stage*, low-pressure, catalytic vapor-phase process. Molten urea is fed to the fluidized catalytic bed reactor (a) at 395–400 °C and atmospheric pressure. Alumina is used as a catalyst, and fluidization is accomplished with an NH_3–CO_2 mixture (the process off-gas).

The reactor temperature is held at ca. 395 °C by molten salt circulated through internal heating coils (b). The fluidizing gas is also preheated to 400 °C. To secure an ammonia-rich atmosphere in the reaction zone, make-up ammonia is added to both the fluidizing gas and the urea nozzles.

Gas leaving the reactor is a mixture of gaseous melamine, traces of melem, and unreacted urea (in the form of its decomposition products isocyanic acid and ammonia), as well as ammonia and carbon dioxide (part newly formed, part fluidizing gas). In addition, the gas mixture contains entrained catalyst fines; coarser catalyst particles are retained by cyclone separators inside the reactor.

The gas mixture leaving the reactor is cooled in the gas cooler (d) to a temperature at which only the byproduct melem crystallizes. Precipitated melem, in the form of a fine powder, is removed together with the entrained catalyst fines in adjacent gas filters (e).

The filtered gas mixture enters the top of the crystallizer (f) where it is blended countercurrently with recycled off-gas (140 °C). The temperature in the crystallizer is thereby reduced to 190–200 °C, and more than 98% of the melamine crystallizes as fine crystals. Melamine is recovered from the gas in a cyclone (g), after which it is cooled and stored. It can be used without further treatment and has a minimum purity of 99.9%.

The nearly melamine-free gas stream from the cyclone is fed to the urea washing tower (i) where it is scrubbed with molten urea (135 °C), which provides both cooling and washing. Clean gas leaving the urea scrubber (after passing through droplet separators) is partially recycled to the reactor as fluidizing gas and partially recycled to the crystallizer as quenching gas. The surplus is fed to an off-gas treatment unit.

A single-stage reactor has the advantage of converting the corrosive intermediate isocyanic acid immediately to melamine; also, the heat of this exothermic reaction is used directly for the endothermic decomposition of urea, the first step in melamine synthesis.

4.1.2. Chemie Linz Process

The Chemie Linz process (see Fig. 2) is a *two-stage* process. In the first step, molten urea is decomposed in a fluidized sand-bed reactor (b) to ammonia and isocyanic acid at ca. 350 °C and 0.35 MPa. Ammonia is used as the fluidizing gas. Heat required for the decomposition is supplied to the reactor by hot molten salt circulated through internal heating coils. The gas stream is then fed to the fixed-bed catalytic reactor (c) where isocyanic acid is converted to melamine at ca. 450 °C and near-atmospheric pressure.

Melamine is recovered from the reaction gas by quenching with water and mother liquor from the centrifuges (h). The quencher (d) is specially designed to work quickly, thereby preventing significant hydrolysis of melamine to ammelide and ammeline. The melamine suspension from the quencher is

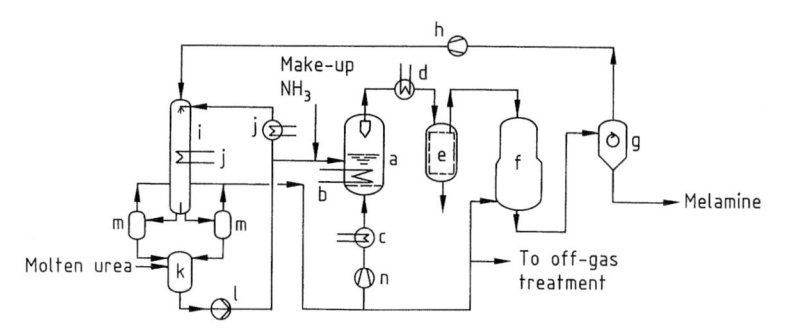

Figure 1. BASF process
a) Reactor; b) Heating coils; c) Fluidizing gas preheater; d) Gas cooler; e) Gas filter; f) Crystallizer; g) Cyclone; h) Blower; i) Urea washing tower; j) Heat exchanger; k) Urea tank; l) Pump; m) Droplet separator; n) Compressor

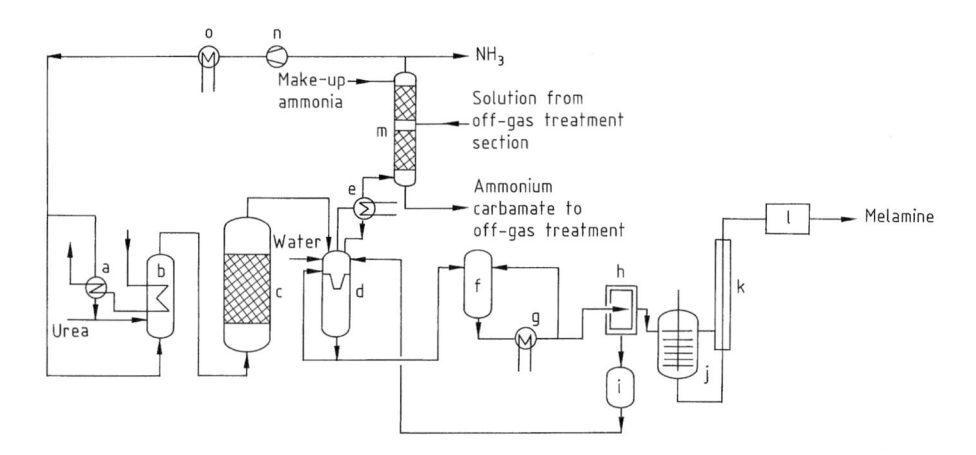

Figure 2. Chemie Linz process
a) Heat exchanger; b) Urea decomposer; c) Converter; d) Quencher; e) Heat exchanger; f) Suspension tank; g) Heat exchanger; h) Centrifuge; i) Mother-liquor vessel; j) Disk dryer; k) Elevator; l) Delumper; m) CO_2 absorption column; n) Compressor; o) Heat exchanger

cooled further to complete the melamine crystallization process. After being centrifuged, the crystals are dried, milled, and stored. A separate recrystallization step is not required.

Exhaust gas from the quencher is fed to an absorber (m) where carbon dioxide is removed as ammonium carbamate by washing with a lean carbamate solution from the off-gas treatment section. The wet ammonia gas is dried with make-up ammonia. Part of it is compressed and recycled to the urea decomposer, and part is exported. Remaining ammonia and carbon dioxide in the liquid effluent are then recovered in the off-gas treatment section.

4.1.3. Stamicarbon Process

Like the BASF process, the DSM Stamicarbon process (see Fig. 3) involves only a *single catalytic stage*. However, it differs from the former in that it is operated at 0.7 MPa, the fluidizing gas is pure ammonia, the catalyst is of the silica–alumina type, and melamine is recovered from the reactor outlet gas by water quench and recrystallization.

Urea melt is fed into the lower part of the reactor (b). The silica–alumina catalyst is fluidized by preheated (150 °C) ammonia, which enters the reactor at two points: at the bottom of the reactor to fluidize the catalyst bed, and at the urea nozzles to atomize the urea feed. The reaction is maintained at 400 °C by circulating molten salt through heating coils within the catalyst bed.

The melamine-containing reaction mixture from the reactor is quenched first in a quench cooler (f) and then in a scrubber (g) with recycled mother liquor from the crystallization section. The resulting melamine suspension is concentrated to ca. 35 wt % melamine in a hydrocyclone (h), after which it is fed to a desorption column (i) where part of the ammonia and carbon dioxide dissolved in the suspension is stripped off and returned to the scrubber. The preceding steps are all carried out at reaction pressure; for the following stages, the pressure is reduced.

The suspension leaving the bottom of the desorber is diluted with recycled and preheated mother liquor and water. Activated carbon and filter aids may also be added. The melamine dissolves completely, although separate dissolving vessels (n) are necessary to allow sufficient time for dissolution. The resulting solution is filtered using precoat-type filters (o). Crystallization of melamine is carried out in a vacuum crystallizer (p), and crystals are separated from the mother liquor by hydrocyclone (t) and centrifuge (u). The crystals are dried in a pneumatic dryer and then conveyed to product bins.

Surplus ammonia must be recovered as fluidizing gas from the wet ammonia–carbon dioxide mixture leaving the desorption column and the scrubber. The hot gas mixture is partly condensed by heat-exchange (k) with the mother liquor from melamine dissolution. The condensate and uncondensed gas are then passed at 0.7 MPa to an absorption column (x). Liquid make-up ammonia is fed to the top of this column to condense any carbon dioxide remaining in the ammonia gas. The ammonia is then compressed and recycled as fluidizing and urea-atomization gas for the reactor.

4.2. High-Pressure Processes

High-pressure melamine synthesis systems differ from low-pressure processes by producing melamine in the liquid instead of the vapor phase. They have the advantage of providing high-pressure off-gas more suitable for use in the urea synthesis facility. Liquid phase operation also lends itself to smaller reaction vessels, but the highly corrosive nature of the system dictates use of expensive, corrosion-resistant construction materials such as titanium.

High-pressure reactions occur without catalyst at >7 MPa and >370 °C. In general, molten urea is injected at high pressure into a molten melamine–urea mixture in the reactor, where it undergoes conversion to melamine. Sufficient residence time is provided in the reactor to ensure complete reaction, leading to melamine with a purity $>94\%$. Heat is supplied to the reactor either by electric heater elements or by a molten

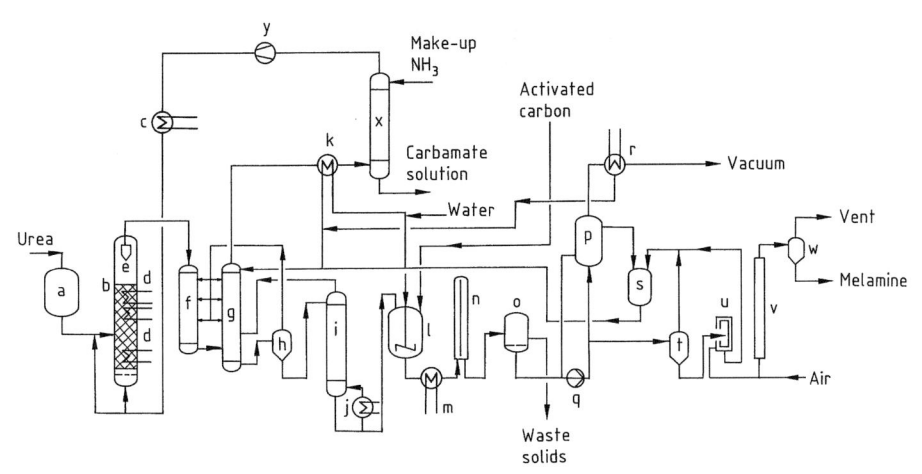

Figure 3. Stamicarbon process
a) Urea tank; b) Reactor; c) Preheater; d) Heating coils; e) Internal cyclone; f) Quench cooler; g) Scrubber; h) Hydrocyclone; i) Desorption column; j) Heat exchanger; k) Heat exchanger; l) Mixing vessel; m) Heat exchanger; n) Dissolving vessel; o) Precoat filter; p) Vacuum crystallizer; q) Pump; r) Heat exchanger; s) Mother-liquor vessel; t) Hydrocyclone; u) Centrifuge; v) Pneumatic dryer; w) Hydrocyclone; x) Absorption column; y) Compressor

salt heat-transfer system. Various types of off-gas separation and melamine purification follow.

High-pressure synthesis of melamine from urea proceeds via the intermediate cyanuric acid, which is subsequently converted to melamine under high pressure in an ammonia environment [29], [30]:

$$
\begin{array}{l}
3\,(NH_2)_2CO \longrightarrow 3\,HOCN + 3\,NH_3 \\
\quad \text{Urea} \qquad\qquad \text{Cyanic acid} \quad \text{Ammonia} \\
3\,HOCN \longrightarrow (NCOH)_3 \\
\quad \text{Cyanic acid} \quad \text{Cyanuric acid} \\
(NCOH)_3 + 3\,NH_3 \longrightarrow C_3N_3(NH_2)_3 + 3\,H_2O \\
\qquad\qquad\qquad\qquad\qquad \text{Melamine} \\
\underline{3\,(NH_2)_2CO + 3\,H_2O \longrightarrow 6\,NH_3 + 3\,CO_2} \\
6\,(NH_2)_2CO \longrightarrow C_3N_3(NH_2)_3 + 6\,NH_3 + 3\,CO_2
\end{array}
$$

The net reaction is the same as in the low-pressure process.

4.2.1. Melamine Chemicals Process

Melamine Chemicals uses a continuous high-pressure *single-stage* process that produces melamine with a purity of ca. 96–99.5 %. Molten urea is converted into melamine in a liquid-phase reactor. The off-gases (ammonia and carbon dioxide) are separated in a gas-separating vessel. Liquid melamine is then quenched in a cooling unit, where liquid ammonia is used to solidify the crystals.

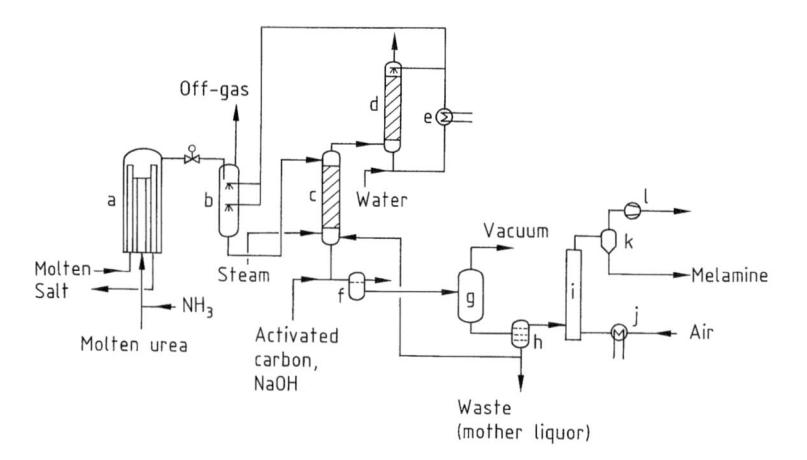

Figure 4. Montedison process
a) Reactor; b) Quencher; c) Stripper; d) Absorption column; e) Heat exchanger; f) Filter; g) Vacuum crystallizer; h) Filter; i) Pneumatic dryer; j) Heat exchanger; k) Cyclone; l) Blower

Process Description. Incoming urea is preheated by using it to scrub the reactor off-gas stream. This scrubber performs various functions, including (1) driving off any water that may be present in the urea feed, (2) preheating the molten urea, (3) removing melamine from the off-gases, and (4) recovering excess heat energy for subsequent use. The reactor is heated to ca. 370–425 °C with a heating-coil system and pressurized to about 11–15 MPa. Mixing is provided by heat convection and generation of gaseous reaction products.

Liquid melamine is separated from the off-gas in a gas separator, the product being collected at the bottom. The separator is held at about the same temperature and pressure as the reactor. The gaseous phase (ammonia and carbon dioxide, saturated with melamine vapor) is removed overhead to a urea scrubber. The melamine stream leaving this separator is then injected into the product cooling unit.

The product cooling unit employs liquid ammonia to both cool and solidify melamine. This is accomplished at a controlled temperature and pressure to minimize formation of such impurities as melam and melem. Product is removed from the pressurized cooling unit through a series of pressure-reducing hoppers. Depending on the required degree of purity, it may then be recrystallized [31].

The off-gas stream represents high-pressure (> 10 MPa) ammonia and carbon dioxide, which can be used directly as feed to the urea facility. Alternatively, this gas stream can be treated in a mono-ethanolamine (MEA) scrubber to recover ammonia and remove carbon dioxide.

4.2.2. Montedison (Ausind) Process

The Montedison process (see Fig. 4) operates at 370 °C and 7 MPa. The required temperature is maintained by a molten-salt heating system consisting of concentric bayonet-type tubes.

Molten urea at 150 °C is fed to the reactor (a) together with preheated ammonia. Average residence time of the mixture in the reactor is about 20 min. As the reaction mixture leaves the reactor, the pressure is lowered to 2.5 MPa, and the mixture is treated at 160 °C in a quencher (b) with an aqueous solution of ammonia and carbon dioxide to precipitate melamine. The water-saturated mixture of ammonia and carbon dioxide leaving the top of the quencher can be recycled to a plant for urea or fertilizer production.

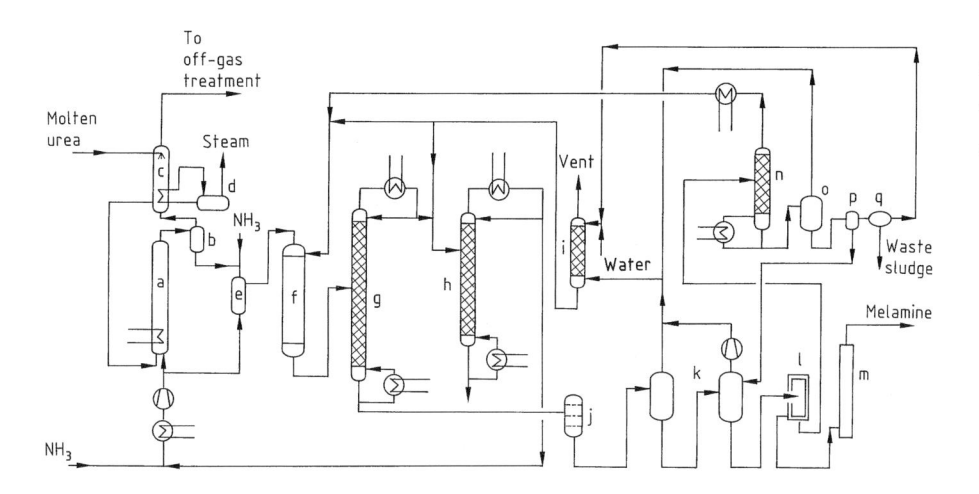

Figure 5. Nissan process
a) Reactor; b) Level tank; c) Off-gas washing tower; d) Steam drum; e) Cushion vessel; f) Quencher; g) NH_3-stripper; h) NH_3-distillation column; i) Absorber; j) Filter; k) Crystallizers; l) Centrifuge; m) Pneumatic dryer; n) Ammonia recovery tower; o) Crystallizer; p) Separator; q) Decanter

The aqueous melamine slurry remains in the quencher at 160 °C for some time to decompose unconverted urea and such byproducts as biuret and triuret to ammonia and carbon dioxide. It is then fed to a steam stripper (c), where any remaining ammonia and carbon dioxide are removed. Off-gas from the stripper is dissolved in water in an absorption column (d) and this solution is recycled to the quencher.

The ammonia- and carbon dioxide-free melamine slurry leaving the bottom of the stripper is diluted with mother liquor to dissolve melamine. Sodium hydroxide is also added, and the solution is then clarified with activated carbon (f). Melamine is crystallized from the clarified solution in a crystallizer (g) operated adiabatically under vacuum. Melamine crystals are separated from the mother liquor in a rotary filter (h), dried in a pneumatic conveyor–dryer (i), and stored.

4.2.3. Nissan Process

The Nissan melamine process (see Fig. 5) operates at 10 MPa and 400 °C. One characteristic feature is urea washing of the reactor off-gas. For this purpose, molten urea is also pressurized to 10 MPa and passed through a high-pressure washing tower (c) where it absorbs any melamine and unreacted urea present in off-gas leaving the reactor. The urea then flows into the reactor (a) by gravity. Ammonia is also fed to the reactor.

In a so-called level tank (b), effluent from the reactor is separated into gaseous and liquid phases. The gaseous phase passes through the previously described urea washing tower to an off-gas treatment facility. The liquid phase consists mainly of molten melamine. This melt is mixed with hot gaseous ammonia and fed to a "cushion" vessel (e) for aging (i.e., to allow byproducts to be reconverted to melamine).

After aging, the melamine melt is quenched (f) under pressure with aqueous ammonia, in which it dissolves. The resulting 20–30 wt% melamine solution is retained in the quencher at 180 °C until any remaining impurities have decomposed.

Most of the added ammonia is next removed from the solution in an ammonia stripper (g) (operated at 1.5 MPa) and the solution is filtered. Recovered ammonia is recycled. Crystallization takes places in two crystallizers (k) operated in series. Mother liquor and melamine crystals are separated in centrifuges (l), after which the crystals are dried and crushed before storage.

Further treatment of the mother liquor starts with an ammonia stripper (n), in which oxoaminotriazines precipitate. The slurry from this ammonia recovery tower is therefore alkalinized before being fed to a third crystallizer (o) operating at reduced temperature and pressure. Additional melamine crystallizes here, and after separation from the liquid (p) it is returned to the second crystallizer.

Lowering the pH of the mother liquor causes oxoaminotriazines to precipitate; these are removed by decantation (q). The clear mother liquor is used to absorb ammonia leaving the crystallizers and is subsequently recycled to the quencher together with ammonia released in the ammonia recovery tower.

5. Quality Specifications

Many consumers of melamine are satisfied with a purity of 99.9%. Some, however, specify additional measurable criteria such as: content of inorganic ash, moisture, and ammeline-related compounds (alkali solubles); particle-size distribution; pH; resin reaction time; and resin color. Average particle sizes ranging from 15 to 100 µm are available for various applications. The rate of dissolution of melamine in formaldehyde solutions depends on particle size, an important parameter that is generally also reported by producers.

6. Chemical Analysis

Melamine is difficult to characterize by traditional chemical methods. Purity specifications are usually based on differences obtained after subtracting determined impurity levels for moisture, ash, and alkali solubles. Instrumental analysis is possible using liquid chromatography [32] and spectroscopic methods, although melamine tends to form complexes with the pH-control buffers used in liquid chromatography, leading to variability in the UV absorption observed at different pHs. Melamine can be precipitated for quantitative determination as the perchlorate or picrate salt (see Section 3.3).

7. Storage and Transportation

Melamine is stable when stored under normal warehouse conditions. Although not particularly hygroscopic, powdered melamine must still be protected from wetting because, like most powders, it will pack and lump over extended storage periods.

In the VDI guideline 2263 melamine is classified as having a burning index of 2, i.e., during a fire it ignites quickly but the flame is rapidly extinguished. Flammability tests performed in accordance with the EEC guideline 84/449 A 10 showed that a glowing platinum wire (>1000 °C) was not able to produce continous burning of melamine.

Shipping considerations are typical of those for other nonhazardous powders. Melamine is available in standard-weight paper bags and semibulk bags. Bulk shipment is by truck and railroad car.

8. Uses

Most melamine is reacted with formaldehyde to produce resins for laminating and adhesive applications [15], [16], [33]. One of the major uses of melamine is in the upper sheet of laminated counter- and tabletops.

Another important use of melamine is as the amino cross-linker in heat-cured paint systems. In this case the methylated methylolmelamine is used, with varying molar ratios of melamine, formaldehyde, and alcohol for different paint system applications. High-solids paint systems for automotive applications also constitute a major market for melamine.

Other uses include the preparation of wet-strength resins for paper, water clarifying resins, ion-exchange resins, plastic molding compounds, adhesives, fire retardants in polyurethane foams, and intumescent paints. Important new applications are under development in the field of fire retardants for polymeric materials, especially polyurethane foams. Applications and uses of melamine differ widely among the main consumer countries or regions. Estimates are provided in Table 2.

9. Economic Aspects

About 18 melamine producers exist worldwide. Rated annual production capacity is ca. 550 000 t (data for 1990/1991; see Table 3). In the last decade the average annual growth rate for melamine consumption was ca. 2 %.

Table 2. Melamine applications, in percent, by region

Application	Europe	United States	Japan
Laminates	47	35	6
Glue, adhesives	25	4	62
Molding compounds	9	9	16
Coatings	8	39	12
Paper, textiles	11	5	3
Other		8	1
Total	100	100	100

Table 3. Melamine production capacity worldwide

Country	Company	Capacity, t/a
Fed. Rep. Germany	BASF	42 000
Austria	Chemie Linz	55 000
Netherlands	DSM	90 000
Italy	Ausind	28 000
France	Norsolor	15 000
Western Europe		*230 000*
Poland	Polimex Cekop	28 000
Rumania	Romchim	12 000
Soviet Union	Techmashimport	10 000
Eastern Europe		*50 000*
United States	American Melamine Ind.	50 000
	Melamine Chem.	47 000
America		*97 000*
Japan	Mitsubishi Petrochemical	32 000
	Mitsui Toatsu Chemical	38 000
	Nissan Chemical	42 000
Korea	Korea Fertilizer	16 000
Taiwan	Taiwan Fertilizer	10 000
Saudi Arabia	Safco	20 000
China	Sichuan Chemical Works	12 000
India	Gujarat State Fertilizer	5 000
Middle and Far East		*175 000*
Total		**552 000**

10. Toxicology

Acute Toxicity. From the standpoint of acute toxicity, melamine is not classified as a health risk. The oral LD_{50} for rats is >5000 mg per kilogram of body weight [34]. Melamine applied to the skin and eyes of rabbits is a non-irritant [34]. Skin sensitization could not be provoked by patch tests on humans [35] or guinea pigs [36].

Genetic Toxicity. Investigations into the potential genetic effects of melamine have included the following tests:

In vitro methods

1) Bacterial tests
   Ames test [37]
   Escherichia coli plate test [38]

2) Test with eukaryotes
   Saccharomyces cerevisiae (gene conversion) [38]
   Rat hepatocytes (DNA repair) [39]
   Mouse lymphoma test (point mutation) [40]
   CHO cells (chromosomal aberrations) [41]

In vivo methods

   Drosophila melanogaster (sex-related lethal test) [42]
   Mouse micronucleus test (oral administration of 1000 mg per kilogram of body weight) [43]

In none of these studies could melamine-induced mutagenicity or damage to genetic material be demonstrated.

Metabolism. Investigations into the metabolism and toxicokinetics of melamine showed that a single oral dose of 0.38 mg of ^{14}C-labeled melamine administered to a rat was eliminated unchanged in the urine to the extent of 90% [44]. The plasma half-life was found to be 2.7 h, with highest concentrations in the bladder and kidneys.

Chronic Toxicity and Carcinogenicity. The target organ system for melamine toxicity after prolonged administration to mice and rats is the urinary tract. Both species were administered, in some cases extremely high doses in the diet (750–30 000 ppm; ca. 62–2490 mg per kilogram of body weight per day) over periods of 14 d, 90 d, and 2 a. All studies led to changes in the kidney (inflammation, calcified concretions in the proximal tubuli) and bladder (inflammation, ulceration, epithelial hyperplasia, bladder stones). Daily dosages up to 15 000 ppm for mice and 5000 ppm for rats administered in the diet over 14 d produced no changes. Female rats tolerated doses of 9000 ppm/d for 13 weeks without symptoms. The incidence of bladder carcinoma increased only for male rats receiving melamine in the diet over a two-year period (dosage: 4500 ppm). Of eight rats displaying tumors, seven also had bladder stones, suggesting that chronic mechanical irritation of the mucous membranes of the bladder may have been responsible for the tumors. Dosages that do not result in bladder stones are not expected to be carcinogenic. Female rats in this study displayed neither bladder stones nor carcinoma [38], [45].

An initiation-promotion study on female mouse skin failed to reveal any tumor-inducing effect [46]. A single dermal application of 1 µmol of melamine followed by a

Table 4. Physical properties of the most important guanamines

	Acetoguanamine	Benzoguanamine	Caprinoguanamine
mp, °C	277	228	105–115
Solubility (at 20 °C), g/L, in			
Water	11.2	0.3	insoluble
Acetone	1.04	18.0	25.2
Benzene	0.07	0.3	11.2
Dimethylformamide	0.88	120.0	67.0

31-week treatment (twice a week) with 10 nmol of a promoter (12-O-tetradecanoylphorbol-13-acetate) resulted in no increased incidence of tumor-bearing animals.

Reproduction Toxicology. A study reported by THIERSCH on the reproduction toxicology of melamine was published in 1957. Intraperitoneal administration of a dose of 70 mg/kg to pregnant rats on days 4 and 5, 7 and 8, or 11 and 12 of pregnancy had no influence on the maternal or fetal development, nor was there any teratogenic effect [47].

11. Guanamines

In the course of heating guanidine acetate, NENCKI (1874) obtained a new compound to which he assigned the name "guanamine"; later he changed it to acetoguanamine after the discovery of other homologues. WEITH recognized in 1876 that this substance was in fact 2,4-diamino-6-methyl-1,3,5-triazine. The term guanamine has since been applied generally to 2,4-diamino-1,3,5-triazines substituted in the 6-position with alkyl, aryl, or alkaryl residues. Such substances are named on the basis of the carboxylic acid that contains one more carbon atom than is present in the substituent on the triazine ring:

$R = CH_3$: acetoguanamine
$R = C_3H_7$: butyroguanamine
$R = C_9H_{19}$: caprinoguanamine
$R = C_6H_5$: benzoguanamine

Compounds of this type that have major commercial significance are benzoguanamine [91-76-9], $C_9H_9N_5$, M_r 187.20, and acetoguanamine [542-02-9], $C_4H_7N_5$, M_r 125.14 [48].

The physical properties of aceto-, benzo-, and caprinoguanamine [5921-65-3], $C_{12}H_{23}N_5$, M_r 237.35, are summarized in Table 4. Figure 6 illustrates the most important reactions of aceto- and benzoguanamine [49]–[52]. For information regarding other guanamines and their reactions, see [53], [54].

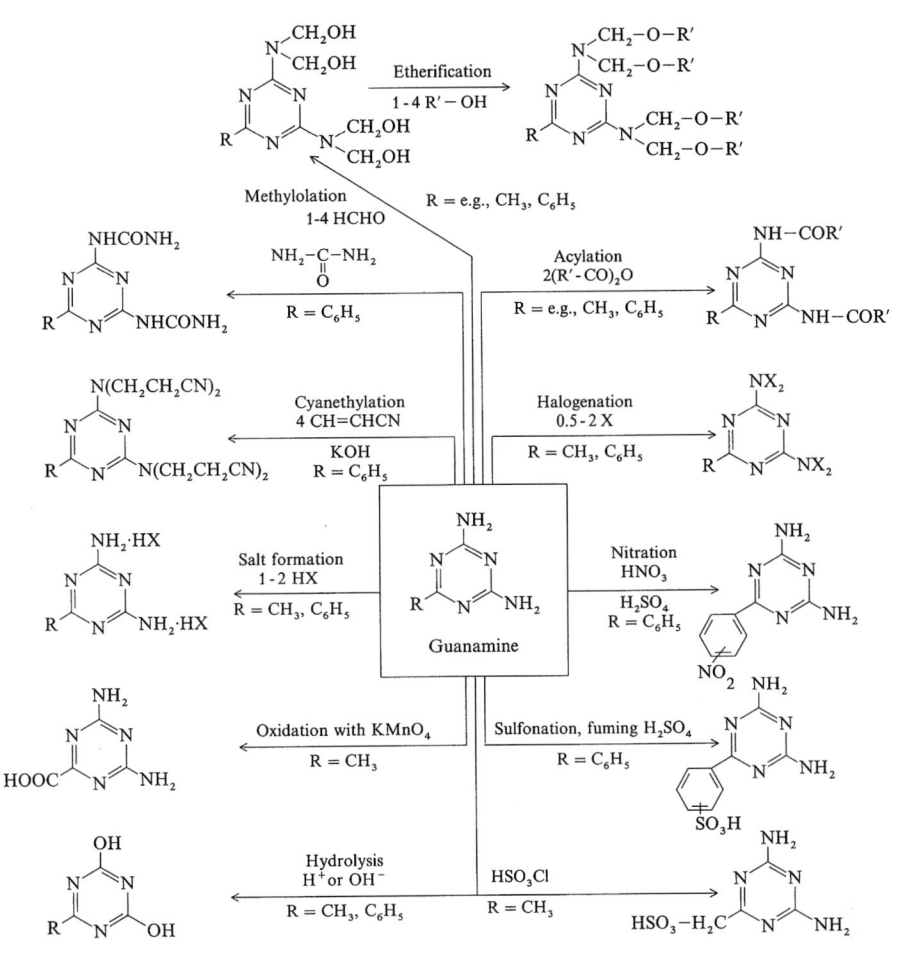

Figure 6. Reactions of guanamines

11.1. Production

Guanamines can be prepared by treatment of biguanides with esters [55], but the standard industrial method involves reaction of dicyandiamide with nitriles [56]:

$$\text{H}_2\text{N}-\text{C}=\text{N}-\text{CN} + \text{RCN} \xrightarrow{\text{Alkali}} \text{[triazine ring with R, NH}_2\text{, NH}_2\text{]}$$
$$\quad\quad\,|$$
$$\quad\,\text{NH}_2$$

Use of dinitriles leads to bisguanamines. Reaction occurs at 105–120 °C in the presence of alkaline catalysts (e.g., KOH) in polar solvents, usually alcohols, and high yields are obtained. The rate of reaction depends on the structure of the nitrile, the nature of the solvent, and the concentration of alkali. Aliphatic nitriles generally react more slowly than aromatic nitriles.

11.2. Uses

Acetoguanamine. Acetoguanamine is used as a condensation component in amino resins. Occasional use is made of the *pure* resins that result from reaction of acetoguanamine with formaldehyde. Such resins display a high degree of water tolerance and can be cured thermally. The rate of the *condensation* reaction is considerably less dependent on pH than in the case of melamine. Condensation may be carried out in weakly basic or weakly acidic media. Compared to melamine–formaldehyde resins, the *curing* of acetoguanamine resins occurs more slowly and at a more acidic pH.

Acetoguanamine is normally used to *modify* melamine–formaldehyde resins, conferring improved elasticity, higher gloss, and reduced risk of laminate shrinkage. A special advantage is the fact that these laminates are thermally post-formable. Even small amounts of acetoguanamine are sufficient to produce small bending radii without degradation of such important characteristics as surface hardness, stain receptivity, or thermal stability [57]. Acetoguanamine cyanurate, either alone or with melamine cyanurate, serves as an effective flame retardant in polyamides [58].

Benzoguanamine. Benzoguanamine is also used primarily in resins. Compared to acetoguanamine, benzoguanamine produces significantly more hydrophobic resins. It is a good elasticity- and flow-inducing agent for melamine–formaldehyde resins. In melamine molding compounds for tableware it inhibits staining by coffee or tea [59]. Melamine–polyester molding compounds containing benzoguanamine undergo very little shrinkage. The hydrophobic character of benzoguanamine–formaldehyde resins is useful in resins intended as chipboard glue because it enhances resistance to cold water. Co-condensation of benzoguanamine in phenolic resins improves the water resistance of laminates.

The most important application of benzoguanamine is in industrial paints. Benzoguanamine resins that have been etherified with C_4 alcohols are more compatible with hydrocarbons, oils, and various synthetic resins (e.g., alkyd, polyester, epoxy [60], and acrylate [61] resins) than are melamine ether resins. The addition of benzoguanamine to baking enamels based on melamine alkyd resins leads to increased gloss, elasticity, and stability toward alkali. Benzoguanamine is also used in the preparation of daylight fluorescent paints to increase their resistance to weathering [62]. A combination of finely dispersed dye in benzoguanamine resin can be used after curing as a filler or colorant in plastics [63].

Benzoguanamine ether resins are used to improve the flow characteristics of primers based on fluorocarbon and acrylic resins [64]. When added to curing agents for wood glues based on urea, benzoguanamine reduces the tendency of glued surfaces to emit formaldehyde [65].

Both benzoguanamine–formaldehyde resins and finely divided benzoguanamine are used as curing agents for epoxy resins [66]. Benzoguanamine is also an effective stabilizer for aqueous formaldehyde solutions, preventing the separation of formaldehyde polymers [67].

Tetramethoxymethylbenzoguanamine [4588-69-6] can be prepared by careful synthesis of tetrahydroxymethylbenzoguanamine followed by etherification with methanol [68]. This solid monomer is an effective cross-linking agent for powder coatings [69]. Liquid methylol–benzoguanamine ethers can be converted into solid cross-linking agents for polyester and acrylic ether resins by treatment with benzoguanamine (acetoguanamine) [70].

Nicotinoylaminotriazine derivatives, prepared from benzoguanamine and nicotinic anhydride, are pharmacologically active agents that accelerate the healing of abscesses. Similar properties are associated with benzoguanamine derivatives in which the phenyl group bears trifluoromethyl or methylsulfonyl substituents, as well as with 2,5-dichlorophenyldiamino-*s*-triazine [71].

Caprinoguanamine effectively inhibits the low-temperature precipitation of polymeric formaldehydes from concentrated formaldehyde solutions [72].

Other Guanamines. Guanamines with unsaturated groups in the substituent R can also be converted to polymers and copolymers [73]. Isophthalobisguanamine [5118-80-9] is utilized as a stabilizer for aqueous formaldehyde solutions [74]. The compound "CTU" (bis-3,9-cyanoethyl-2,4,8,10-tetraoxaspiro[5.5]-undecan)-guanamine [22535-90-6] (formula shown below) is employed (like benzoguanamine) primarily in paints [75].

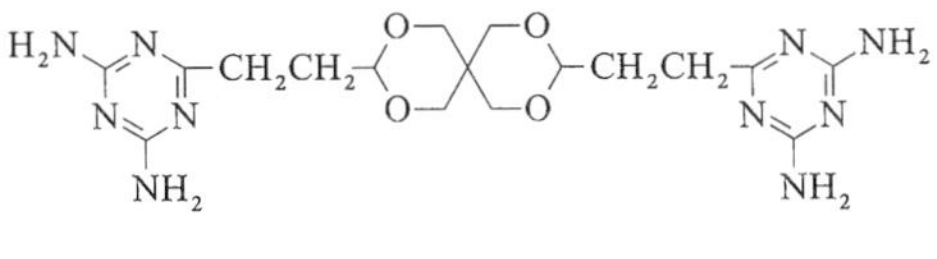

11.3. Toxicology

The acute oral LD_{50} for *acetoguanamine* in rats is 2740 mg/kg; for *benzoguanamine*, 1470 mg/kg. The corresponding value for *caprinoguanamine* is > 10 000 mg/kg, and its dermal LD_{50}(rabbit) is > 2800 mg/kg [76]. Caprinoguanamine is not an irritant with respect to either skin or mucous membranes [76]. An Ames test with benzoguanamine provided no indication of mutagenic activity [77], and long-term feeding experiments failed to reveal carcinogenic potential [78].

12. References

General References

[1] E. M. Smolin, L. Rapoport in A. Weissberger (ed.): *The Chemistry of Heterocyclic Compounds*, vol. **13**, Interscience, New York 1959,pp. 309–379.
[2] J. M. E. Quirke in A. R. Katritzky, C. W. Rees (eds.): *Comprehensive Heterocyclic Chemistry*, vol. **3** part 2 B, Pergamon Press, Oxford 1984, pp. 457–530.
[3] SRI International: *Melamine*, Report No 122, 1978.
[4] "The Manufacture of Non-Fertilizer Nitrogen Products," *Nitrogen* **139** (1982) 32–39.

Specific References

[5] Amer. Cyanamid, US 2 566 231, 1943 (J. H. Paden, J. S. Mackay).
[6] Melamine Chemicals, Donaldsonville, Louisiana, 70346 (USA): *Product Brochure*, 1980, pp. 4–5.
[7] C. C. Stephenson, D. J. Berets, *J. Am. Chem. Soc.* **74** (1952) 882–883.
[8] R. C. Hirt, J. E. Steger, G. L. Simard, *Polymer Sci.* **XLIII** (1960) 319–323.
[9] R. P. Chapmann, P. R. Averell, R. R. Harries, *Ind. Eng. Chem.* **35** (1943) 137.
[10] A. C. Larson, D. T. Cromer, *J. Chem. Phys.* **60** (1974) no. 1, 185–192.
[11] G. van der Plaats, H. Soons, R. Snellings, *Proc. Eur. Symp. Therm. Anal.* 2nd 1981, 215–218. L. Costa, G. Camino: "Thermal Behaviour of Melamine," *J. Therm. Anal.* **34** (1988) 423–429.
[12] H. May, *J. Appl. Chem.* **9** (1959) 340–344.
[13] A. I. Finkel'shtein, N. V. Spiridonova, *Russ. Chem. Rev. (Engl. Transl.)* **33** (1964) no. 7, 400–405.
[14] T. Morikawa: "Evolution of Hydrogen Cyanide During Combustion and Pyrolysis," *J. Combust. Toxicol.* **5** (1978) 315–330.
[15] R. I. Spasskaya, A. I. Finkel'shtein, E. N. Zil'bermann, Ts. N. Roginskaya, *Plastich. Massy*, (1980) no. 3, 7; *Int. Polym. Sci. Technol.* **7** (1980) no. 6, 69–71.
[16] W. Woebcken (ed.): *Kunststoff-Handbuch*, vol. 10, "Duroplaste," 2nd ed., Hanser Verlag, München 1988.
[17] M. Schwarzmann: "Make Melamine at Atmospheric Pressure," *Hydrocarbon Process.* **47** (1968) no. 10, 96–100.
[18] A. Schmidt: "über den Reaktionsmechanismus der Bildung von Melamin aus Isocyansäure," *Monatsh. Chem.* **99** (1968) 664–671.
[19] A. Schmidt: "Verfahrenstechnische Probleme bei der Herstellung von Melamin aus Harnstoff bei Atmosphärendruck", *Oesterr. Chem. Ztg.* **68** (1967) 175–179.
[20] K. Abe et al.: "Synthetic Production of Melamine by a High-Pressure Process," *Kagaku Kogaku* **40** (1976) no. 6, 298–302.
[21] sterreichische Stickstoffwerke, GB 1 054 502, 1967 (Stevens, Langner, Perry, Rollinson).
[22] M. Schwarzmann: "Make Melamine at Atmospheric Pressure," *Hydrocarbon Process.* **48** (1969) no. 9, 184–186.
[23] Allied Chemical, US 3 250 773, 1966 (I. Christoffel et al.).
[24] Montecatini, US 3 172 887, 1965 (E. Bondi).
[25] Amer. Cyanamid, US 2 615 019, 1952 (W. J. Klapproth, Jr.).
[26] *Kirk-Othmer*, 2nd ed., **2**, 366–367, **6**, 566.
[27] "Melamine OSW Process," *Br. Chem. Eng.* **14** (1969) no. 10, 1336.

[28] Nissan Chemical Industries, GB 1 032 507, 1966 (A. Murata et al.).
[29] Y. T. Nakajima: "Development of a High-Pressure Process for Melamine Manufacture from Urea," *Nikkakyo Geppo* **28** (1975) no. 7, 175–179.
[30] Office National Industriel de l'Azote, US 3 262 759, 1966 (J. L. Pomot et al.).
[31] Melamine Chemicals, US 4 565 867, 1984 (R. E. Thomas, D. E. Best).
[32] P. G. Stoks, A. W. Schwartz: "Determination of s-Triazine Derivatives at the nanogram level by Gas-Liquid Chromatography," *J. Chromatogr.* **168** (1979) 455–460.
[33] T. Götze, P. Dorries: "Neuentwicklungen auf dem Gebiet der technischen Melamin-Formaldehydharze,"*Plastverarbeiter* **33** (1982) 1118–1122.
[34] BASF Aktiengesellschaft, unpublished results.
[35] C. B. Shaffer: *Melamine: Acute and Chronic Toxicity,* Amer. Cyanamid, Central Medical Department, Wayne, N.J., June 15,1955.
[36] C. F. Reinhardt, M. R. Briselli in G. D. Clayton, F. E. Clayton (eds.): *Pattys Industrial Hygiene and Toxicology,* 3rd ed., John Wiley and Sons, New York 1981, p. 2770.
[37] S. Haworth et al., *Environ. Mutagen.* Suppl. 1 (1983) 3.
[38] ECETOC, Joint Assessement of Commodity Chemicals (JACC) no. 1, Melamine, CAS: 108-78-1, February (1983).
[39] J. Mirsalis et al., *Environ. Mutagen.* **5** (1983) 482.
[40] D. B. McGregor et al., *Environ. Mol. Mutagen.* **12** (1988) 85.
[41] S. M. Galloway et al., *Environ. Mol. Mutagen.* **10** (1987) 1.
[42] G. Röhrborn, *Vererbungs.* **93** (1962) 1.
[43] R. W. Mast et al., *Environ. Mutagen.* **4** (1982) 340.
[44] R. W. Mast et al., *Food Chem. Toxicol.* **21** (1983) 807.
[45] R. L. Melnick et al., *Toxicol. Appl. Pharmacol.* **72** (1984) 292.
[46] *IARC Monogr. Eval. Carcinog. Risk Chem. Man.* **39,** 11–18 June (1985) 333.
[47] J. B. Thiersch, *Proc. Soc. Exp. Biol. Med.* **94** (1957) 36.
[48] SKW Trostberg AG, Produktstudie *Guanamine,* 1979.
[49] Henkel, DE-OS 2 019 675, 1970 (J. Brüning, H.-W. Eckert, P. Krings).
[50] Henkel, DE-OS 1 964 793, 1969 (J. Brüning).
[51] BASF, DE-OS 2 018 719, 1970 (M. Schwarzmann).Henkel, DE-OS 1 445 907, 1964 (J. Schiefer).
[52] E. Riesz, C. Heutrel, *Rev. Gen. Caoutch. Plast.* **46** (1969) 237–241.
[53] Olin Mathieson Chem. Corp., DE-AS 1 245 385, 1963 (D. W. Kaiser, J. K. Zane).
[54] E. Smolin, L. Rapoport: *s-Triazines and Derivatives,* Interscience, New York 1959, pp. 222–258.
[55] Kaken Yakukako, DE-OS 2 121 694, 1971 (S. Murai, K. Yoshida, C. Tomioka).
[56] Olin Mathieson Chem. Corp., US 3 330 830, 1964 (D. W. Kaiser). Armour Pharmaceutical Comp., US 3 655 892, 1969 (C. D. Bossinger, T. Enkoji).Y. Ogata et al., *Tetrahedron* **20** (1964) 2755–2761, **22** (1966) 157–165. Y. Iwakura et al., *Bull. Chem. Soc. Jpn.* **38** (1965) 1820–1824.
[57] J. Seeholzer, *Kunststoffe* **69** (1979) 263–265.
[58] Toray, US 4 383 064, 1983 (H. Jida).
[59] American Cyanamid Company, US 3 367 917, 1968 (N. A. Granito). US 3 506 738, 1970 (A. J. Ross).
[60] General Mills Inc., US 3 271 350, 1963 (L. R. Vertnik).
[61] Rohm & Haas, US 3 082 184, 1959 (D. R. Falgiatore, A. M. Levantin).
[62] Lawter Chemicals, US 3 116 256, 1961 (G. F. D'Alelio, R. W. Voedisch).
[63] Nippon Shokubai Kagaku Kogyo, DE-AS 2 349 964, 1973 (T. Tsubakimoto, J. Fuzikawa, O. Toyonaka). DE-OS 2 634 415, 1976 (T. Tsubakimoto, J. Fuzikawa, O. Toyonaka).

[64] SCM Corporation, US 4 557 977, 1984 (T. J. Memmer, P. T. Abel).
[65] BASF, DE-OS 3 620 859, 1986 (G. Lehmann, H. Lehnert, M. Siegler, W. Pitteroff).
[66] Toyo Ink, JP-Kokai 62 218 454, 1962 (J. Katsumasa, T. Kojiro, K. Keiji). Matsushita Electric, JP-Kokai 61 246 227, 1961 (H. Hirohisa, F. Taro). F. Ueda, S. Aratani, K. Mimura, K. Kimura et al., *Arzneim. Forsch./Drug Res.* **34** (1984) 478–484.
[67] Süddeutsche Kalkstickstoff-Werke, DE-AS 1 205 072, 1962 (J. Seeholzer). Montecatini, DE-AS 1 205 071, 1962 (J. Dakli, N. Lupi, M. Morini).
[68] S. Sander, *J. Appl. Polym. Sci.* **13** (1969) 555.
[69] Hoechst, DE-OS 1 926 035, 1969 (H. Fischer, H.-J. Schubert).
[70] Ciba-Geigy, EP-A 0 011 049, 1979 (M. D. J. Rowland, K. Winterbottom).
[71] Nippon Shinyaku, US 4 103 009, 1977 (H. Murai, K. Ohata, Y. Aoyagi, F. Ueda, M. Kitano). DE-OS 2 711 746, 1977 (H. Murai, K. Ohata, Y. Aoyagi, F. Ueda).
[72] Süddeutsche Kalkstickstoff-Werke, DE 1 205 073, 1964 (P. Bornmann, H. Michaud).
[73] Cassella, GB 1 174 523, 1968 (H. v. Brachel, H. Kindler, H. Gattner). Rohm & Haas, US 3 554 684, 1969 (W. D. Emmous, J. G. Brodmyan).
[74] BASF, DE 2 365 180, 1973 (H. Diehm, C. Dudeck, G. Lehmann). DE-OS 2 358 856, 1973 (H. Diehm, H. Libowitzky, G. Matthias).
[75] Toka Shikiso and Ajinomoto, US 3 557 059, 1971 (H. Yoko, O. Tamiki, T. Kijchiro, M. Takao).
[76] Central Institute for Nutrition and Food Research TNO, unpublished results.
[77] Microtest Research Ltd., unpublished results.
[78] Bio-Research Consultants, Inc., Contract no. NIH-NCI-E68-1311, *Carcinogenicity of Chemicals Present in Man's Environment*, Final report (July 1973).